# SIGNALS, SYSTEMS & INFERENCE

## GLOBAL EDITION

ALAN V. OPPENHEIM & GEORGE C. VERGHESE
MASSACHUSETTS INSTITUTE OF TECHNOLOGY

# Pearson

Harlow, England • London • New York • Boston • San Francisco • Toronto • Sydney • Dubai • Singapore • Hong Kong
Tokyo • Seoul • Taipei • New Delhi • Cape Town • Sao Paulo • Mexico City • Madrid • Amsterdam • Munich • Paris • Milan

Vice President and Editorial Director, ECS: *Marcia J. Horton*
Acquisitions Editor: *Julie Bai*
Executive Marketing Manager: *Tim Galligan*
Marketing Assistant: *Jon Bryant*
Senior Managing Editor: *Scott Disanno*
Program Manager: *Joanne Manning*
Project Editor, Global Edition: *Radhika Raheja*
Acquisitions Editor, Global Edition: *Abhijit Baroi*
Global HE Director of Vendor Sourcing and Procurement: *Diane Hynes*
Director of Operations: *Nick Sklitsis*
Operations Specialist: *Maura Zaldivar-Garcia*
Cover Designer: *Lumina Datamatics*
Manager, Rights and Permissions: *Rachel Youdelman*
Associate Project Manager, Rights and Permissions: *Timothy Nicholls*
Senior Manufacturing Controller, Global Edition: *Kay Holman*
Media Production Manager, Global Edition: *Vikram Kumar*
Full Service Project Management: *Pavithra Jayapaul, Jouve India*

Pearson Education Limited

Edinburgh Gate
Harlow
Essex CM20 2JE
England

and Associated Companies throughout the world

Visit us on the World Wide Web at:
www.pearsonglobaleditions.com

© Pearson Education Limited 2017

British Library Cataloguing-in-Publication Data

A catalogue record for this book is available from the British Library

10 9 8 7 6 5 4 3 2 1

ISBN 10: 1292156201
ISBN 13: 9781292156200

Typeset by Jouve India
Printed and bound in Malaysia (CTP-VVP)

*We dedicate this book to*

Amar Bose, Bernard Gold, and Thomas Stockham

*&*

George Sr. and Mary Verghese, and Thomas Kailath

*These extraordinary people have had a profound impact on our lives and our careers*

# CONTENTS

# PREFACE

This book has grown out of an undergraduate course developed and taught by us in MIT's Department of Electrical Engineering and Computer Science. Our course is typically taken by third- and fourth-year undergraduate students from many engineering branches, as well as undergraduate and graduate students from applied science. There are two formal prerequisites for the course, and for this book: an introductory subject in time- and frequency-domain analysis of signals and systems, and an introductory subject in probability. These two subjects are typically taken by most engineering students early in their degree programs. The signals and systems subject almost invariably builds on an earlier course in differential equations, ideally with some basic linear algebra folded into it.

In many engineering departments, students with a strong interest in applied mathematics have then traditionally gone on to a more specialized undergraduate subject in control, signal processing, or communication. In addition to being specialized, such subjects often focus on deterministic signals and systems. Our aim instead was to build broadly on the prerequisite material, folding together signals, systems, and probability in ways that could make our course relevant and interesting to a wider range of students. The course could then serve both as a terminal undergraduate subject and as a sufficiently rigorous basis for more advanced undergraduate subjects or introductory graduate subjects in many engineering and applied science departments.

The course that gave rise to this book teaches students about signals and signal descriptions that are typically new to them, for example, random signals and their characterization through correlation functions and power spectral densities. It introduces them to new kinds of systems and system properties, such as state-space models, reachability and observability, optimum filters, and group delay. And it highlights model-based approaches to inference, particularly in the context of state estimation, signal estimation, and signal detection.

11

Although some parts of our course are well covered by existing textbooks, we did not find one that fit our needs across the range of topics. This led to lecture notes, which was the easier part, and then eventually this book. In the process, we continually experimented with and refined the content and order of presentation. Along the way we also at times included other material or excluded some that is now back in the book. Among the conclusions of these experiments was that we did not have time in a one-semester class to fold in even basic notions of information theory, despite its central importance to communication systems and, more generally, to inference.

As suggested in the Prologue to this book, signals, systems and probability have been and will continue to be usefully combined in studying fields such as signal processing, control, communication, financial engineering, biomedicine, and many others that involve dynamically varying processes operating in continuous or discrete time, and affected by disturbances, noise, or uncertainty. This premise forms the basis for the overall organization and content of our course and this text.

The book can be thought of as comprising four parts, outlined below. A more detailed overview of the individual chapters is captured in the table of contents. Chapters 1 and 2 present a brief review of the assumed prerequisites in signals and linear time-invariant (LTI) systems, though some portions of the material may be less familiar. A key intent in these chapters is to establish uniform notation and concepts on which to build in the chapters that follow. Chapter 3 discusses the application of some of this prerequisite material in the setting of digital communication by pulse amplitude modulation.

Chapters 4–6 are devoted to state-space models, concentrating on the single-input single-output LTI case. The development is largely built around the eigenmodes of such systems, under the simplifying assumption of distinct natural frequencies. This part of the book introduces the idea of model-based inference in the context of state observers for LTI systems, and examines associated feedback control strategies.

Chapters 7–9 provide a brief review of the assumed probability prerequisites, including estimation and hypothesis testing for static random variables. As with Chapters 1 and 2, we felt it important to set out our notation and perspectives on the concepts while making contact with what students might have encountered in their earlier probability subject. Again, some parts of this material, particularly on hypothesis testing, may be previously unfamiliar to some students.

In Chapters 10–13, we characterize wide-sense stationary random signals, and the outputs that result from LTI filtering of such signals. The associated properties and interpretations of correlation functions and power spectral densities are then used to study canonical signal estimation and signal detection problems. The focus in Chapter 12 is on linear minimum mean square error signal estimation, i.e., Wiener filtering. In Chapter 13, the emphasis is on signal detection for which optimum solutions involve matched filtering.

As is often said, the purpose of a course is to uncover rather than to cover a subject. In this spirit, each chapter includes a final section with some suggestions for further reading. Our intent in these brief sections is not to be exhaustive but rather to suggest the wealth of learning opened up by the material in this text. We have pointed exclusively to books rather than to papers in the research literature, and have in each case listed only a fraction of the books that could have been listed.

Each chapter contains a rich set of problems, which have been divided into Basic, Advanced, and Extension. Basic problems are likely to be easy for most students, while the Advanced problems may be more demanding. The Extension problems often involve material somewhat beyond what is developed in the chapter. Certain problems require simulation or computation using some appropriate computational package. Given the variety and ubiquity of such packages, we have intentionally not attempted to structure the computational exercises around any specific platform.

There is more material in this book than can be taught comfortably in a one-semester course. This allows the instructor or self-learner to choose different routes through the text, and over the years we have experimented with various paths. For a course that is more oriented towards communication or signal processing, Chapters 4, 5 and 6 (state-space models) can be omitted, or addressed only briefly. For a course with more of a control orientation, Chapter 3 (pulse amplitude modulation), Chapter 9 (hypothesis testing) and Chapter 13 (signal detection) can perhaps be considered optional.

A third version of the course, and the one that we currently teach, is outlined in a little more detail below. This version involves two weekly lectures over a semester of approximately thirteen weeks. The lectures are interleaved with an equal number of small-group recitation sections, devoted to more interactive discussion of specific problems that illustrate the lectures and help address the weekly homework. In addition, we staff optional small-group tutorials. Finally an optional evening "common room" that we run several times each week allows students in the class to congregate and interact with each other and with a member of the teaching staff while they work on their homework.

In our teaching in general, we like to emphasize that the homework is intended to provide an occasion for learning and engaging with the concepts and mechanics, rather than being an exam. We recommend that the end-of-chapter problems in this book be approached in the same spirit. In particular, we encourage students to work constructively together, sharing insights and approaches. Our grading of the problems is primarily for feedback to the students and to provide some accountability and motivation. The course does typically have a midterm quiz and a final exam, and many of the end-of-chapter problems in this text were first created as quiz or exam problems. There are also many possibilities for term projects that can grow out of the material in the class, if desired.

An introductory lecture in the same spirit as the Prologue to this text is followed by a brief review of the signals and systems material in Chapter 1. The focus in class is on what might be less familiar from the prerequisite subject, and students are tasked with reviewing the rest on their own, guided by appropriate homework problems. We then move directly to the state-space material in Chapters 4, 5 and 6. Even if students have had some prior exposure to state-space models, there is much that is likely to be new to them here, though they generally relate easily to the material. We have not held students responsible for the more detailed proofs, such as those on eigenvalue placement for LTI observers or state feedback, but do expect them to develop an understanding of the relevant results and how to apply them to small examples. An important lesson from the state-space observer framework is the role of a system model in going from measured signals to inferences about the system.

Our course then turns to probabilistic models and random signals. The probability review in Chapter 7 is mostly woven into lectures covering minimum mean square error (MMSE) and linear MMSE (LMMSE) estimation, which are dealt with in Chapter 8. In order to move more quickly to random signals rather than linger on review of material from the prerequisite probability course, we defer the study of hypothesis testing in Chapter 9 to the end of the course, using it as a lead-in to the signal detection material in Chapter 13. Part of the rationale is also that Chapters 9 and 13 are devoted to making inferences about discrete random quantities, namely the hypotheses, whereas Chapters 8 and 12 on (L)MMSE estimation deal with inferences about continuous random variables. We therefore move directly from Chapter 8 to Chapter 10, studying random signals, i.e., stochastic processes, focusing on the time-domain analysis of wide-sense stationary (WSS) processes, and LTI filtering of such processes.

The topic of power spectral density in Chapter 11 connects back to the development of transforms and energy spectral density in Chapter 1, and also provides the opportunity to refer to relevant sections of Chapter 2 on all-pass filters and spectral factorization. These topics are again important in Chapter 12, on LMMSE (or Wiener) filtering for WSS processes. In most offerings of the course, we omit the full causal Wiener filter development, instead only treating the case of prediction of future values of a process from past values of the same process.

The last part of the course refers strongly back to Chapter 3, using the context of digital communication via pulse amplitude modulation to motivate the hypothesis testing problem. The return to Chapter 3 can also involve reference to the material in Chapter 2 on channel distortions and group delay. The hypothesis testing paradigm is then treated as in Chapter 9. This serves as the foundation for the study of signal detection in the last chapter, Chapter 13.

The breadth of this book, and the different backgrounds we brought to the project, meant that we had much to learn from each other. We also learn each term from the very engaged students, teaching assistants and faculty colleagues who are involved in the course, as well as from the literature on the subjects treated here. This book will have amply met its objectives if it sparks and supports a similar voyage of discovery in its readers, as they construct their own individual re-synthesis of the themes of signals, systems and inference.

*Alan V. Oppenheim & George C. Verghese*
*Cambridge, Massachusetts*

# THE COVER

The choice of images for the front and back covers of both the North American Edition and this Global Edition originated in our desire to suggest some of the book's themes in a visually pleasing and striking way. Our explorations began with images of sundials, clocks, and astrolabes. The astrolabe (www.astrolabes.org), invented over two thousand years ago and used well into the 17th century, was an important instrument for astronomy and navigation. Our search for the front cover of this Global Edition eventually led to the photograph by Frans Lemmens (www.franslcmmens.com), taken inside the Eisinga Planetarium (www.planetarium-friesland.nl/en) in Franeker, Holland. This exquisite scale model of the solar system was meticulously built by the amateur astronomer Eise Eisinga in the ceiling of his living room, during the period 1774–1781, and is considered the oldest functioning planetarium.

The image of the dwarf planet Ceres on the back cover of this edition is derived from photographs taken by NASA's spacecraft Dawn (www.dawn.jpl.nasa.gov), which entered into orbit around Ceres in March 2015, after an eight-year journey from our planet. The mastery of signals, systems and inference that humankind has attained in the four centuries since the astrolabe faded from use is represented here: in the precisely controlled launch and trajectory of the Dawn spacecraft – which first included a rendezvous with the asteroid Vesta before moving on to Ceres – and in the subsequent recording, retrieval, and processing of data from it to yield such revealing and awe-inspiring images. But the image also evokes the boundless opportunities for new advances and horizons.

# ACKNOWLEDGMENTS

This text has its origins in an MIT subject that was first planned, designed and taught by us over twenty years ago. It has subsequently evolved to its current form through continual experimentation and with many variations of the material and presentation. The subject was conceived as part of the curriculum for a five-year Master of Engineering degree program that was being launched at that time in our Department of Electrical Engineering and Computer Science (EECS). We are grateful to Paul Penfield (as then department head), Jeffrey Shapiro (as associate head) and William Siebert for their part in defining the curricular structure that provided the opening for such a subject. Jeff Shapiro also worked with us on the initial definition of the content. Continued support of the curriculum, and of revisions to it, from subsequent department heads – John Guttag, Rafael Reif, Eric Grimson, Anantha Chandrakasan – and their administrations has been important, and we thank them for their support. More generally, we consider ourselves very fortunate to have had our academic careers develop in this highly collegial and vibrant department. MIT's culture of dedication to teaching and learning informed by research, and the Institute's recognition and celebration of excellence in teaching, have had a significant influence on us.

The staffing of the course, as taught in our department, includes a faculty member who gives two weekly lectures and has overall responsibility for running the course, as well as recitation instructors and teaching assistants who meet regularly with smaller groups of students. Numerous faculty colleagues in our department have collaborated with us over the years, as recitation instructors or as lecturers for the subject. Many students have served as able and enthusiastic teaching assistants. We have also benefited from the help of excellent administrative assistants. We take the opportunity in what follows to thank all these people for their multifaceted contributions to the development and running of the course, to the student experience in the course, and to this text.

In addition to each of us individually and jointly lecturing and overseeing the administration of the course many times, other colleagues who have served in that role are Bernard Lesieutre, Charles Rohrs, Jeffrey Shapiro, Gregory Wornell, and John Wyatt. In the process they have provided valuable feedback on the course content and course notes, as well as bringing new insights and developing new exam and homework problems.

Over the years, we have been privileged to work with a superbly talented and committed roster of faculty and senior graduate students serving as recitation instructors. The recitation instructors who have participated in the teaching of the subject are Jinane Abounadi, Elfar Adalsteinsson, Babak Ayazifar, Duane Boning, Petros Boufounos, John Buck, Mujdat Cetin, Jorge Goncalves, Julie Greenberg, Christoforos Hadjicostis, Peter Hagelstein, Thomas Heldt, Steven Isabelle, Franz Kaertner, James Kirtley, Amos Lapidoth, Bernard Lesieutre, Steve Massaquoi, Shay Maymon, Alexandre Megretski, Jose Moura, Asuman Ozdaglar, Michael Perrott, Rajeev Ram, Charles Rohrs, Melanie Rudoy, Jeffrey Shapiro, Ali Shoeb, William Siebert, Vladimir Stojanovic, Collin Stultz, Russell Tedrake, Mitchell Trott, Thomas Weiss, Alan Willsky, Gregory Wornell, John Wyatt, Laura Zager, and Lizhong Zheng. These colleagues have helped provide a rich experience for the students, and have made many contributions to the content of the course and this text.

Both we and the students in the class have been the beneficiaries of the dedication and energy of the stellar teaching assistants during this period: Irina Abarinov, Abubakar Abid, Anthony Accardi, Chalee Asavathiratham, Thomas Baran, Leighton Barnes, Soosan Beheshti, Ballard Blair, Petros Boufounos, Venkat Chandrasekaran, Jon Chu, Aaron Cohen, Roshni Cooper, Ujjaval Desai, Vijay Divi, Shihab Elborai, Baris Erkmen, Siddhartan Govindasamy, Hanhong Gao, James Geraci, Michael Girone, Carlos Gomez-Uribe, Christoforos Hadjicostis, Andrew Halberstadt, Nicholas Hardy, Everest Huang, Irena Hwang, Zahi Karam, Asif Khan, Alaa Kharbouch, Ashish Khisti, Lohith Kini, Alison Laferriere, Ryan Lang, Danial Lashkari, Adrian Lee, Karen Lee, Durodami Lisk, Karen Livescu, Lorenzo Lorilla, Zhipeng Li, Peter Mayer, Rebecca Mieloszyk, Jose Oscar Mur Miranda, Kirimania Murithi, Akshay Naheta, Kenny Ng, Tri Ngo, Paul Njoroge, Ehimwenma Nosakhare, Uzoma Orji, Tushar Parlikar, Pedro Pinto, Victor Preciado, Andrew Russell, Navid Sabbaghi, Maya Said, Peter Sallaway, Sridevi Sarma, Matthew Secor, Mariam Shanechi, Xiaomeng Shi, Andrew Singer, Lakshminarayan Srinivasan, Brian Stube, Eduardo Sverdlin-Lisker, Kazutaka Takahashi, Sayeed Tasnim, Afsin Ustundag, Kathleen Wage, Tianyu Wang, Keyuan Xu, HoKei Yee, and Laura Zager. Their inputs are reflected in myriad ways throughout this text.

Over the many years of offering this subject, we have been guided by the wisdom of our colleague Frederick Hennie in matters of instructional staffing. Agnes Chow's strategic yet detailed oversight of the EECS department's administrative and financial operations has allowed us and other faculty to focus on our teaching. Lisa Bella, as assistant to the department's education officers, attends almost single-handedly and with incredible responsiveness and good humor to the practical administrative aspects of supporting a hundred professors and over a hundred teaching assistants across the department's teaching enterprise each semester. For administrative assistance with our course in its many offerings, we would like to thank Alecia Batson, Margaret Beucler, Dimonika Bray, Susan Davco, Angela Glass, Vivian Mizuno, Sally Santiago, Darla Secor, Eric Strattman, and Diane Wheeler.

As the class subject has continued to evolve over the two-decade period, the accompanying course notes that ultimately led to this text have also grown and changed. The students in the class have been key participants in that process, through their questions, requests, challenges, suggestions, critiques, and encouragement. It is a continuing privilege to work with the gifted, engaged, thoughtful, and vocal students whom we have in our classrooms at MIT.

We have sometimes said, either ruefully or in jest, that the current text is the fourth edition of a book whose first three editions we never formally published. As any textbook author knows, however, the final phase of producing a polished text from what initially seem to be very good course notes is still a formidable task. Some of our teaching assistants and other students have more recently provided substantial help and feedback in advancing our lecture notes closer to a text. We would like to specifically acknowledge the efforts of Leighton Barnes and Ballard Blair, as well as Manishika Agaskar, Ganesh Ajjanagadde, Michael Mekonnen, Wan-Teh Chang and Guolong Su. For cheerfully, efficiently, and discerningly pulling together and keeping track of all the fragments and versions and edits as we advanced towards a text, we are enormously indebted to Laura von Bosau.

Our department leadership has consistently encouraged us to take the course notes beyond their role as a supplement to the classroom subject and into a published book, so that the material would be more widely and independently accessible. Anantha Chandrakasan's urging in recent years was a key catalyst in making this text happen. Also significant, and greatly appreciated, was the interest from several publishers. Tom Robbins saw the potential early on, and regularly offered helpful advice through the first decade of the course, during his time at Prentice Hall. Phil Meyler generously arranged for detailed feedback at a later stage. Our respect for the vision and integrity of vice president and editorial director Marcia Horton and executive editor Andrew Gilfillan at Pearson were major factors in our choice of publisher; their patience, commitment and confidence in the project meant a lot to us. Special thanks are due to the strong and accommodating editorial and production staff, particularly senior managing editor Scott Disanno at Pearson for his personal attention, and senior project manager Pavithra Jayapaul at Jouve for her outstanding and steady marshaling of the production of the North American edition through its countless details. For their post-publication support of the North American edition and their efforts in connection with the production of the present edition, we are grateful to Julie Bai, Joanne Manning, Michelle Bayman, Sandra Rodriguez and Radhika Raheja at Pearson.

In developing the cover design for the North American edition of the book, which has also informed the design for this edition, it was a pleasure to work closely with Krista Van Guilder, who was manager of media and design in MIT's interdisciplinary Research Laboratory of Electronics (RLE). RLE is the research home for both of us; the creative environment that it provides for research also impacts our teaching, including the development of this text. The forthright leadership of Yoel Fink, and Jeffrey Shapiro before him, and the exemplary competence and friendliness of the RLE headquarters staff, set the tone for RLE.

Getting to a bound book has naturally included weathering various challenges along the way. Not the least of these was reconciling our sometimes differing opinions, instincts, approaches, or styles on many minor and sometimes major issues. It helped that we started as friends, and as respectful colleagues. And the experience of working so closely and extensively together in coauthoring this text has, happily, deepened that respect and friendship.

In concluding, we express some individual and more personal thoughts and acknowledgments.

### Al Oppenheim

Much of the DNA in my contributions to this text derives, both literally and metaphorically, from my mother, as an extraordinary mentor and role model for me. It still astonishes me that as one of ten children in a poor immigrant family, whose parents arrived from Eastern Europe through Ellis Island, she managed to make her way through college and then medical school in the late 1920's. And then how, as a single parent, she very successfully raised three children while working full time in public health. An incredible and inspiring woman.

I landed at MIT, somewhat by accident, as a freshman in 1955, and shortly after wrote a letter home indicating that at the end of the first year I likely would leave for somewhere that was more fun. Clearly, before long MIT became fun and gratifying for me, and has been a wonderful place at which to have spent my entire academic life, first as a student and then as a faculty member. A tremendous expression of gratitude is due to MIT and more specifically to all of my teachers and mentors throughout this entire period at MIT. And, as indicated on the dedication page of this book, in particular and in very special ways to three mentors: Amar Bose, Ben Gold, and Tom Stockham, whose support and encouragement had a truly profound impact on me.

One of the most fortunate days of my life was the day I walked into the office of a then young assistant professor, Amar Bose, and subsequently signed on as his first teaching assistant. And he eventually signed on as my doctoral thesis advisor. What I learned from him about teaching, research, and life over the many decades of our relationship affected me in ways too numerous to describe. He set the highest standards in everything that he did, and his accomplishments as a teacher, an inventor, and an entrepreneur are legendary. Tom Stockham was another young assistant professor whom I met during my doctoral program. His excitement about and enthusiasm for my ideas gave me the courage to pursue them. During his years at MIT as a faculty member and then as research staff at MIT's Lincoln Laboratory, Tom was one of the pioneers of the then unknown field of digital signal processing. Through that and his later research at the University of Utah, Tom became widely acknowledged as the father of digital audio. Tom was an extraordinary teacher, researcher, practical engineer, and friend. I first met Bernard (Ben) Gold during my early days on the MIT faculty while he was a visiting faculty member in EECS. His work on speech compression was the context for his many pioneering contributions to digital signal processing. Ben's brilliance, creativity, and unassuming style were inspirational to me. He was as eager to learn from those around him as they were from him. Amar, Tom and Ben taught me so many things by example, including the importance of passion and extraordinary standards in every pursuit. Their influence on me is woven into the fabric of my life, my career, and this text. I miss them all, and their spirit remains deeply within me.

As any author knows, textbook writing is a long, difficult, but ultimately rewarding process. Throughout my career I've had the opportunity to write and edit a number of books, and in some cases through two or three editions. In that process, I've had the good fortune of collaborating with other wonderful co-authors in addition to George Verghese, specifically Ron Schafer and Alan Willsky. Such major collaborative projects can often strain relationships, but I'm delighted to say that in all cases, strong bonds and friendships have been the result.

I have often been asked whether I enjoy writing. My response typically has been that "writing is difficult and sometimes painful, but I enjoy *having* written." Projects of this magnitude inevitably require tolerance, patience, support and understanding from family and close friends. I've been incredibly fortunate to have had all of that throughout my career from my wife Phyllis, and from our children Justine and Jason, who have always been the source of tremendous joy. And I'm deeply appreciative of Nora Moran for her special friendship and encouragement (and chicken soup) during the completion of this book.

### *George Verghese*

My parents, George Sr. and Mary, grew up in small towns a mere fifteen miles apart in Kerala, India, but first met each other 2500 miles away in Addis Ababa, Ethiopia, where – young, confident, and adventurous – they had traveled in the early 1950's as teachers. Two further continents later, they continue to set a model for me, of lives lived gracefully. I have everything to thank them for, including the brothers they gave me.

Growing up with physics books to chew on at home surely played a part in landing me at the Indian Institute of Technology, Madras, for undergraduate studies. My favorite professors there, V.G.K. Murti (for network theory) and K. Radhakrishna Rao (for electronic circuits), treated their students with respect, and earned it back many times over with the clarity and integrity of their thinking and teaching, and with their friendly approachability. They are probably why becoming a professor began to seem an attractive proposition to me.

I was fortunate to be introduced to linear system theory by Chi-Tsong Chen at the State University of New York at Stony Brook, and still recall the excitement of my first course taught – and so elegantly – by the author of a textbook. A few months later I drove cross-country for a life-changing period at Stanford, to work under Thomas Kailath. It was an exceptional time to be there, particularly for the opportunity to learn from him as he completed his own text on linear systems, but also for the interactions with his other students, an amazing group. He undoubtedly thought forty years ago that he was only signing on to be my doctoral thesis advisor, but fifteen years later found himself a part of my family. I continue to learn from him on other fronts, and am still in awe of his acuity, bandwidth, energy, and generosity.

When I joined the faculty at MIT, I thought I would try it out for two years to see how I liked it. I've stayed for over 35. It has been a privilege to be affiliated with such an extraordinary institution, and with the people – students, faculty, and staff – who make it so. Working with Al Oppenheim has been a highlight.

My friends and extended family have helped me keep my labors on this text in necessary perspective, and I'm grateful to them for that. They will no doubt be relieved, the next time they ask, to hear that I'm not still working on the same book as the last time they checked. Throughout this, my dear wife Ann has been much more patient and understanding than I had any right to expect. And whenever she hit her limits, she hauled us off for a vacation that I invariably discovered I needed as much as she did. I could not have completed this project without her cheerful forbearance. Our daughters Deia and Amaya, now launched on trajectories of their own devising, keep us and each other smiling; they are our greatest blessings.

# Prologue

## SIGNALS, SYSTEMS AND INFERENCE

*Signals*, in the sense that we refer to them in this book, have been of interest at least since the time when human societies began to record and analyze numerical data, for example to track climate, commerce, population, disease, and the movements of celestial bodies. We are continually immersed in signals, registering them through our senses, measuring them through instruments, and analyzing, modifying, and interrelating them.

*Systems* and signals are intimately connected. In many contexts, it is important to understand the behavior of the underlying systems that generate the signals of interest. Furthermore, the challenges of collecting, interpreting, modeling, transforming, and utilizing signals motivate us to design and implement systems for these purposes, and to generate signals to control and manipulate systems.

*Inference*, as the term is used in this text, refers to combining prior knowledge and available measurements of signals to draw conclusions in the presence of uncertainty. The prior knowledge may take the form of partially specified models for the measured signals. Inference may be associated with the construction and refinement of such models. The implementation of algorithms for inference can also require designing systems to process the measured signals.

The application of concepts and methods involving signals, systems, and inference in combination is pervasive in science, engineering, medicine, and the social sciences. However, the mathematical, algorithmic, and computational underpinnings often evolve to become largely independent of the specific application. It is this common foundational material that is the focus of this text.

## A LITTLE HISTORY

An example of the sophistication attained centuries ago in signals, systems and inference is the astrolabe[1], the most popular astronomical instrument of the medieval world, used for navigation and time keeping in addition to charting the positions of celestial objects. Around 150 AD, Ptolemy of Alexandria described in detail the stereographic projection that forms the basis for the astrolabe; the trigonometric framework for this was developed even earlier, by Hipparchus of Rhodes around 180 BC. The instrument itself made its appearance around 400 AD, and was in widespread use well into the 1600s.

The interplay of signals, systems and inference is also nicely illustrated by Carl Friedrich Gauss's celebrated prediction[2] of the location of the asteroid Ceres, almost a full year after it had been lost to view. Ceres, whose image is on the back cover of this book, is now known to be the largest object in the asteroid belt, and – along with Pluto – is classified as a dwarf planet. The astronomer Giuseppe Piazzi in Palermo discovered the object on New Year's Day of 1801, but was only able to track its motion across the sky for a few degrees of arc before it faded six weeks later in the glare of the sun. There was at the time major interest in the possibility of this being a new planet that had been suspected to exist between Mars and Jupiter. The 24-year-old Gauss, using just three of Piazzi's observations, along with strategic combinations and simplifications of equations derived from Kepler's model of the trajectories of celestial objects, and with many days of hand calculation, was able to generate an estimate of the orbit of Ceres. The predictions made by other astronomers, who had typically assumed circular rather than elliptical orbits, failed to yield sightings of the asteroid. However, successful observations using Gauss's specifications were recorded in early December that year, and again on New Year's Eve. As Gauss put it, he had "restored the fugitive to observation." In later refinements of his method to account for all nineteen of Piazzi's observations rather than just three, and to apply to the motions of other celestial objects, Gauss also brought into play the method of least squares, which he had developed several years earlier. Chapter 8 of this text is devoted to the closely related topic of minimum mean square error estimation of random variables, while Chapter 12 extends this to estimation of random signals.

By 1805, and still motivated by the problem of interpolating measurements of asteroid orbits, Gauss had developed an efficient algorithm to compute the coefficients of finite trigonometric series[3]. He unfortunately never published his algorithm, though it was included in his posthumous collected works sixty years later. Variants of this algorithm were then independently rediscovered by others, as the problem of fitting harmonic series arose in diverse settings, for example to represent variations in barometric pressure or underground temperature, to calculate corrections to compasses on ships, or to model X-ray diffraction data from crystals. The most well known of these variants, commonly referred to collectively as the Fast Fourier Transform (FFT), was published by James Cooley and John Tukey[4] in 1965. Coming at a time when programmable electronic digital computers were beginning to enter routine use in science and engineering, the FFT soon found widespread application, and has had a profound impact.

Many of the foundational concepts and analytical tools discussed throughout this text for both deterministic and probabilistic systems, such as those reviewed in Chapters 1 and 7, have their origins in the work of mathematicians and scientists who lived around the time of Gauss, including Pierre-Simon Laplace and Jean-Baptiste Joseph Fourier, though later contributions also feature prominently, of course. Laplace today is most often associated with the transform that bears his name, but his place in probability theory is considerably more significant, for his 1812 treatise on the subject, and as the "discoverer" of the central limit theorem. Other parts of our text derive more directly from advances made in engineering and applied science since 1800.

The invention of the telegraph in the 1830s sparked a revolution[5] in communication, with subsequent major impact on theory and practice related to all of the topics in this book. It also led to advances in other areas such as transportation and weather prediction, in part because messages could now travel faster than horses, trains, and storms. Within a few years the dots and dashes of Morse code were being transmitted over electrical cables extended between and across continents. Telephony followed in the 1870s, wireless telegraphy and AM radio in the early 1900s, FM radio and television in the 1930s, and radar in the 1940s. Today we have satellite communication, wireless internet, and GPS navigation.

All these transformative technologies exploited and enhanced our ability to work with signals, systems and inference, and were significant catalysts for the creative development of electrical engineering in general. They presented the need to effectively generate electrical signals or electromagnetic waves, to characterize transmission media so that these signals could be propagated through them in predictable ways, to design any necessary filtering and amplification at various intermediate stages, and to develop appropriate signal processing circuits and systems for embedding information at the transmitter and extracting the intended information at the receiver. The modern study of signals and systems in engineering degree programs, with circuits as prime examples of systems, began to take root in the 1930s and '40s. Some of the notions that we describe in Chapter 2 arose primarily in the context of circuits and transmission lines for communication.

Occurring in parallel with advances in communication were developments relevant to the analysis and design of control systems. Among these were analog computation aimed at the simulation of differential equations that modeled various systems of interest. Though the concepts were described over fifty years earlier, the first practical mechanical implementation was the Differential Analyzer of Vannevar Bush and collaborators around 1930. More flexible and powerful electronic versions, namely analog computers using operational amplifiers, were widely used from the 1950s until they were supplanted by digital computers in the 1980's.

The design of self-regulating devices that utilize feedback dates back to at least around 250 BC, with the water clock of Ctesibius of Alexandria. One of the earliest and most important applications of feedback in the industrial age was James Watt's 1788 centrifugal governor for regulating the speed of steam engines, but it was only in 1868 that James Clerk Maxwell[6] showed how to analyze the dynamic stability of such governors. Feedback control began to be routinely incorporated in engineered systems from the beginning of the 20th century. Much of the associated mathematical theory that is in widespread application today – associated with people such as Harold Black, Harry Nyquist, and Hendrik Bode at Bell Labs in the 1920s and '30s – was actually developed in the context of designing stable and robust electronic amplifiers and oscillators for communication and signal processing. Other work on feedback control was motivated by servomechanism design for regulation in industrial manufacturing, chemical processes, power generation, transportation, and similar settings. Aleksandr Lyapunov's work in the 1890s on the stability of linear and nonlinear dynamic systems that were described in state-space form was not widely known till the 1960s, but is now an essential part of systems and control theory. These state-space models and methods, including the study of equilibrium, stability, measurement-driven simulations for state estimation, and feedback control, are treated in Chapters 4, 5, and 6.

Feedback mechanisms also play an essential role in living systems, as was explicitly described in 1865 by the physiologist Claude Bernard. As the mathematical study of communication and control developed in the early 20th century, Norbert Wiener and colleagues in such diverse fields as psychology, physiology, biology and the social sciences recognized the commonality and importance of feedback in these various disciplines. Their interactions in the 1940's eventually led to Wiener's definition and elaboration in 1948 of cybernetics as the study of control and communication in the animal and machine[7].

The treatment of signals, systems and inference in communication, control and signal processing inherently has to address distortion and errors introduced by non-ideal and poorly characterized components. Feedback is often introduced to overcome precisely such difficulties. A related issue, which inserts uncertainty in the behavior of the system, is that of random disturbances. These can corrupt the signal on a communication channel or at the receiver; can affect the performance of a feedback control system; and can affect the reliability of an inferred outcome. By showing how to model random disturbances in probabilistic terms, and characterizing them in the time and frequency domains, mathematical theory has made a significant impact on these applications. The work of Wiener[8] from the 1920's onward helped to set the foundations for engineering applications in these and related areas. A famous report of his on the extrapolation, interpolation and smoothing of time series[9] was a major advance in bringing the notions of Fourier analysis and stochastic processes into the setting of practical problems in signal processing and inference. Chapter 12, building on Chapters 8, 10 and 11, treats a class of filtering problems associated with Wiener's name, and shows how having a model for a random process provides a basis for filtering and prediction.

Claude Shannon went a step further in his revolutionary 1948 papers[10] that essentially gave birth to information theory. He modeled the communication source itself as a discrete random process, and introduced notions of information, entropy, channel capacity and coding that still form the frame of reference for the field. As noted in the Preface, a treatment of information theory is beyond the scope of this text. However, Shannon's work launched the era of digital communication, and the material we study in Chapter 3 on pulse amplitude modulation, including Nyquist's key contributions, is of considerable practical importance in digital communication. The task of signal detection in noise, addressed in Chapters 9 and 13, is also fundamental in this and many other applications.

As indicated at the beginning of this Prologue, another domain of investigation that has a long history and relationship to the material in this text is the study of time series, carried out not only in the natural sciences – astronomy and climatology, for example – and engineering but also in economics and elsewhere in the social sciences. A typical objective in time series analysis is to use measured noisy data to construct causal dynamic models, which can then be used to infer future values of these signals. There is particular interest in detecting and exploiting any trends or periodicities that might exist in the data. The considerations here are similar to those that motivated the work of Wiener and others, and the mathematical tools overlap, though the time-series literature tends to be more application driven and data centered. For example, the notion of a periodogram, which we encounter in Chapter 11, first appears in this literature, as a tool for detecting underlying periodicity in a random process[11].

The emergence over the past half-century of real-time digital computation capabilities has had major impact on the applications of signals, systems and inference, and has also given rise to new theoretical formulations. An important early example of how real-time computation can fundamentally change the approach to a central problem in signal processing and control is the Kalman filter, which generalized Wiener filtering in several respects and greatly extended its application. The seminal state-space formulation[12] introduced by Rudolf Kalman in 1960 for problems of signal filtering involves recursive least squares estimation of the state of a system whose output represents the signal of interest. The filter runs a computational algorithm in parallel with the operation of the system, with the results of the computation also available for incorporation into a feedback control law. The initial use of the Kalman filter was for navigation applications in the space program, but it is now much more widely applied. The treatment of state observers in Chapter 6 of this text makes connections with the Kalman filter, and the relation to the Wiener filter is outlined in Chapter 12.

## A GLANCE AHEAD

Among the most striking developments that the transition to the 21st century has brought to signals, systems and inference is vast distributed and networked computational power, including in small, inexpensive, and mobile packages. Advances in computing, communication, control, and signal processing have resulted in connection and action on scales that were imagined by only a few in the 1960s, at the dawn of the Internet, among them J. C. R. Licklider[13]. A transformational event on the path to making this vision a reality today for so much of humanity was Tim Berners-Lee's invention of the World Wide Web in 1989.

The close coupling of continuous- and discrete-time technologies is of growing importance. Digital signals, communication, and computation commonly mediate interactions among analog physical objects – in automotive systems, entertainment, robotics, human-computer interfaces, avionics, smart-grids, medical instrumentation, and elsewhere. It is also increasingly the case that a given engineered device or component is not easily classified as being intended specifically for communication or control or signal processing or something else; these aspects come together in different combinations at different times. The term "cyber-physical system"[14] is sometimes used to describe the combination of a networked interconnection of embedded computers and the distributed physical processes that they jointly monitor and control.

Our continuing exploration of the universe at both the smallest and largest scales relies in many ways on understanding how to work with signals, systems and inference. The invention of the microscope at the end of the 16th century had profound implications for the development of science at the cellular level and smaller. The invention of the telescope a few years later, at the beginning of the 17th century, similarly enlarged our view of the heavens, and had equally revolutionary consequences. The launch of the Hubble telescope in 1990 has led to our current ability to observe the cosmos at distances of hundreds of millions of light-years. The processing of images from the Hubble telescope incorporates sophisticated extensions of the basic concepts in this text. As one illustration, the techniques of deconvolution, an example of which is examined in Chapter 12, have played an important role in processing of Hubble telescope images, and most critically in initially helping to correct the distortions caused by spherical aberrations in the mirror until it was repaired. In 2003 and 2004, the Hubble telescope captured intriguing images of Ceres. And in March 2015, NASA's Dawn spacecraft, after a journey that lasted eight years, entered the orbit of Ceres, obtaining the most detailed and striking pictures yet of this dwarf planet, including the one that is incorporated into the back cover of this book. We imagine Gauss would be pleased.

Our intention in this book is to address foundational material for applications to signals, systems and inference across a broad set of domains in today's world. These applications are deeply embedded in so many of the systems that we see and use in our everyday lives, and yet are virtually invisible to, and taken for granted by, the casual observer or user. Automotive and entertainment systems, for example, are currently among the largest markets for specialized signal processing systems. Without question, this material will remain foundational for many years to come.

Speculations about the future are always subject to surprises. However, it is certain that new implementation platforms will continue to emerge from advances in such disciplines as quantum physics, materials science, photonics, and biology. And new mathematics will also emerge that will impact the study and application of signals, systems and inference. The novel directions that are opened up by these advances will undoubtedly still derive in part from concepts studied in this book, just as so much of what we use today is rooted in very specific ways on contributions from past centuries. The basic principles and concepts central to this text have a rich historical importance and an even richer future.

## NOTES

[1] J. E. Morrison, *The Astrolabe*, Janus 2007 (see also Morrison's rich website, astrolabes.org).

[2] D. Teets and K. Whitehead, "The discovery of Ceres: How Gauss became famous," *Mathematics Magazine*, vol. 72, no. 2, pp. 83–93, April 1999.

[3] M. T. Heideman, D. H. Johnson, and C. S. Burrus, "Gauss and the history of the Fast Fourier Transform," *IEEE Acoustics, Speech and Signal Processing Magazine*, pp. 14–21, October 1984.

[4] J. W. Cooley and J. W. Tukey, "An algorithm for the machine calculation of complex Fourier series," *Mathematics of Computation*, vol. 19, pp. 297–301, 1965.

[5] The launch of the modern information age, including the birth of telegraphy, is richly and vividly described in J. Gleick's *The Information: A History, A Theory, A Flood*, Vintage Books 2012.

[6] The same J. Clerk Maxwell whose equations launched wireless transmission. An informative description and assessment of his work on analyzing the stability of governors is given by O. Mayr, "Maxwell and the origins of cybernetics," *Isis*, vol. 62, no. 4, pp. 424–444, 1971.

[7] N. Wiener, *Cybernetics: or Control and Communication in the Animal and the Machine*, MIT Press 1948; 2nd edition 1961.

[8] An engaging account of Wiener's life and work can be found in F. Conway and J. Siegelman, *Dark Hero of the Information Age: In Search of Norbert Wiener, the Father of Cybernetics*, Basic Books 2005.

[9] N. Wiener, *Extrapolation, Interpolation, and Smoothing of Stationary Time Series, with Engineering Applications*, MIT Press 1949 (reprinted in 1964).

[10] See C. E. Shannon and W. Weaver, *The Mathematical Theory of Communication*, University of Illinois Press 1949 (reprinted in 1998).

[11] A. Schuster, "On the investigation of hidden periodicities with application to a supposed 26 day period of meteorological phenomena," *Terrestrial Magnetism*, vol. 3, no. 1, pp. 13–41, 1898. The title and venue of this paper are reflective of the sorts of interests that drove early studies in the time series literature.

[12] R. E. Kalman, "A new approach to linear filtering and prediction problems," *Transactions of the ASME-Journal of Basic Engineering*, vol. 82 (series D), pp. 35–45, 1960. Although our text does not include direct treatment of the Kalman filter, it does provide the foundation for reading and understanding this paper. The approachable and tutorial fashion in which the paper is written reflects the fact that it introduces an almost entirely new approach to signal filtering.

[13] J. C. R. Licklider, "Man-computer symbiosis," *IRE Transactions on Human Factors in Electronics*, vol. HFE-1, pp. 4–11, 1960.

[14] A lucid description of such systems and the challenges they present is given in the introduction to E. A. Lee and S. A. Seshia's *Introduction to Embedded Systems: A Cyber-Physical Systems Approach*, edition 1.5, LeeSeshia.org, 2014.

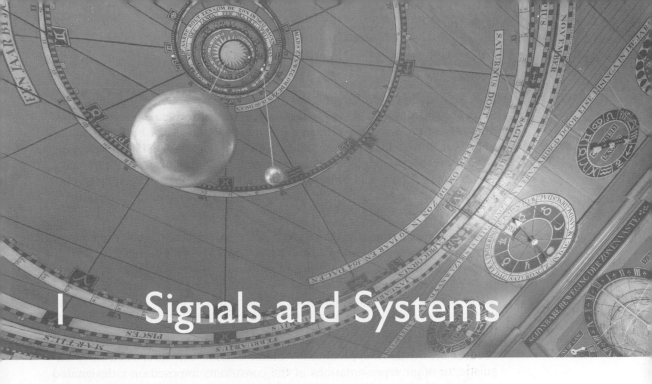

# 1 Signals and Systems

This text assumes a basic background in the representation of linear, time-invariant systems and the associated continuous-time and discrete-time signals, through convolution, Fourier analysis, Laplace transforms, and $z$-transforms. In this chapter, we briefly summarize and review this assumed background, in part to establish the notation that we will use throughout the text, and also as a convenient reference for the topics in later chapters.

## 1.1 SIGNALS, SYSTEMS, MODELS, AND PROPERTIES

Throughout this text we will be considering various classes of signals and systems, developing models for them, and studying their properties.

**Signals** are represented by real- or complex-valued functions of one or more independent variables. They may be one-dimensional, that is, functions of only one independent variable, or multidimensional. The independent variable may be continuous or discrete. For many of the one-dimensional signals, the independent variable is naturally associated with time although it may not correspond to "real time." When the independent variable is continuous, it is enclosed in curved parentheses, and when discrete in square parentheses to denote an integer variable. For example, $x(t)$ would correspond to a continuous-time (CT) signal and $x[n]$ to a discrete-time (DT) signal. The notations $x(\cdot)$ and $x[\cdot]$ will also be used to refer to the entire signal, suppressing the particular variable $t$ or $n$ used to denote time.

In the first six chapters, we focus entirely on deterministic signals. Starting with Chapter 7, we incorporate stochastic signals, that is, signals drawn from an ensemble of signals, any one of which can be the outcome of a given probabilistic process. To distinguish a signal ensemble representing a random process from a deterministic signal, we will typically use uppercase. For example, $X(t)$ would represent a CT random process whereas $x(t)$ would denote a specific signal in the ensemble. Similarly, $X[n]$ would correspond to a DT random process.

**Systems** are collections of software or hardware elements, components, or subsystems. A system can be viewed as mapping a set of input signals to a set of output or response signals. A more general view (which we don't incorporate in this text) is that a system is an entity imposing constraints on a designated set of signals without distinguishing specific ones as inputs or outputs. Any particular set of signals that satisfies the constraints is termed a behavior of the system.

**Models** are (usually approximate) mathematical, software, hardware, linguistic, or other representations of the constraints imposed on a designated set of signals by a system. A model is itself a system because it imposes constraints on the set of signals represented in the model, so we often use the words *system* and *model* interchangeably. However, it can sometimes be important to preserve the distinction between something truly physical and our representations of it mathematically or in a computer simulation.

The difference between representation as a mapping or in behavioral form can be illustrated by considering, for example, Ohm's law for a resistor. Expressed as $v(t) = R\,i(t)$, it suggests current $i(t)$ as an input signal and voltage $v(t)$ as the response, whereas expressed as

$$R\,i(t)/v(t) = 1 \tag{1.1}$$

it is more suggestive of a constraint relating these two signals. Similarly, the resistor-capacitor circuit in Figure 1.1 has constraints among the signals $v(t)$, $i_R(t)$, and $v_C(t)$ imposed by Kirchhoff's laws but does not identify which of the variables are input variables and which are output variables. More broadly, a behavioral representation comprises a listing of the constraints that the signals must satisfy. For example, if a particular system imposed a time-shift constraint between two signals without preference as to which would

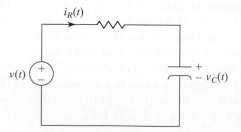

**Figure 1.1**   Resistor-capacitor circuit.

correspond to the input and which to the output, then a behavioral interpretation would be more appropriate. In this text, we will typically express systems as mappings from inputs to outputs.

The representation of a system or model as a mapping comprises the following: a set of input signals $\{x(\cdot)\}$, each of which can vary within some specified range of possibilities; similarly, a set of output signals $\{y(\cdot)\}$, each of which can vary; and a description of the mapping that uniquely defines the output signals as a function of the input signals.

One way of depicting a system as a mapping is shown in Figure 1.2 for the single-input, single-output CT case, with the interpretation that for each signal in the input set, $T\{\cdot\}$ specifies a mapping to a signal in the output set. Given the input $x(\cdot)$ and the mapping $T\{\cdot\}$, the output $y(\cdot)$ is unique. More commonly, the representation in Figure 1.3 is used to show the input and output signals at some arbitrary time $t$. With the notation in Figure 1.3, it is important to understand that the mapping $T\{\cdot\}$ is in general a mapping between sets of signals and not a memoryless mapping between a signal value $x(t)$ at a specific time instant to the signal value $y(t)$ at that same time instant. For example, if the system delays the input by $t_0$, then

$$y(t) = x(t - t_0) . \tag{1.2}$$

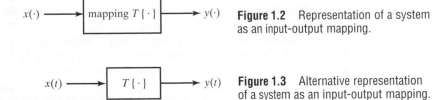

$x(\cdot) \longrightarrow$ mapping $T\{\cdot\}$ $\longrightarrow y(\cdot)$   **Figure 1.2**  Representation of a system as an input-output mapping.

$x(t) \longrightarrow T\{\cdot\} \longrightarrow y(t)$   **Figure 1.3**  Alternative representation of a system as an input-output mapping.

### 1.1.1 System Properties

For a system specified as a mapping, we use the following definitions of various properties, all of which we assume are familiar. They are stated here for the DT case but are easily modified for the CT case. We also assume a single-input, single-output system in our mathematical representation of the definitions that follow, for notational convenience.

- **Memoryless**: The output at any time instant does not depend on values of the input at any other time instant. The CT delay-by-$t_0$ system described in Eq. (1.2) is not memoryless. A simple example of a memoryless DT system is one for which

$$y[n] = x^2[n] \tag{1.3}$$

  for every $n$.

- **Linear**: The response to an arbitrary linear combination (or "superposition") of input signals is always the same linear combination of the individual responses to these signals.

- **Time-Invariant**: The response to any set of inputs translated arbitrarily in time is always the response to the original set, but translated by the same amount.

- **Linear and Time-Invariant (LTI)**: The system is both linear and time-invariant.

- **Causal**: The output at any instant does not depend on future inputs: for all $n_0$, $y[n_0]$ does not depend on $x[n]$ for $n > n_0$. Said another way, if $\widehat{x}[n], \widehat{y}[n]$ denotes another input-output pair of the system, with $\widehat{x}[n] = x[n]$ for $n \le n_0$ where $n_0$ is fixed but arbitrary, then it must be also true that $\widehat{y}[n] = y[n]$ for $n \le n_0$.

- **Bounded-Input, Bounded-Output (BIBO) Stable**: The output response to a bounded input is always bounded: $|x[n]| \le M_x < \infty$ for all $n$ implies that $|y[n]| \le M_y < \infty$ for all $n$.

---

**Example 1.1**    **System Properties**

As an example of these system properties, consider the system with input $x[n]$ and output $y[n]$ defined by the relationship

$$y[n] = x[4n + 1] \tag{1.4}$$

for all $n$. We would like to determine whether the system is memoryless, linear, time-invariant, causal, and/or BIBO stable.

*Memoryless*: A simple counterexample suffices to show that this system is not memoryless. Consider for example $y[n]$ at $n = 0$. From Eq. (1.4), $y[0] = x[1]$ and therefore depends on the value of the input at a time other than at $n = 0$. Consequently it is not memoryless.

*Linearity*: To check for linearity, we consider two arbitrary input signals, $x_A[n]$ and $x_B[n]$, and compare the output of their linear combination to the linear combination of their individual outputs. From Eq. (1.4), the response $y_A[n]$ to $x_A[n]$ and the response $y_B[n]$ to $x_B[n]$ are respectively (for all $n$):

$$y_A[n] = x_A[4n + 1] \tag{1.5}$$

and

$$y_B[n] = x_B[4n + 1] . \tag{1.6}$$

If with $x_C[n] = ax_A[n] + bx_B[n]$ for arbitrary $a$ and $b$ the output is $y_C[n] = ay_A[n] + by_B[n]$, then the system is linear. Applying Eq. (1.4) to $x_C[n]$ shows that this holds.

*Time Invariance*: To check for time invariance, we need to compare the output due to a time-shifted version of $x[n]$ to the time-shifted version of the output due to $x[n]$. The output $y[n]$ resulting from any specific input $x[n]$ is given in Eq. (1.4). The output $\widehat{y}[n]$ results from an input $\widehat{x}[n]$ that is a time-shifted (by $n_0$) version of the signal $x[n]$. Consequently

$$\widehat{y}[n] = \widehat{x}[4n + 1] = x[4n + 1 + n_0] . \tag{1.7}$$

If the system were time-invariant, then $\widehat{y}[n]$ would correspond to shifting $y[n]$ in Eq. (1.4) by $n_0$, resulting in replacing $n$ by $(n + n_0)$ in Eq. (1.4), which yields

$$y[n + n_0] = x[4n + 4n_0 + 1] . \tag{1.8}$$

Since the expressions on the right side of Eqs. (1.7) and (1.8) are not equal, the system is not time-invariant. To illustrate with a specific input, suppose that $x[n]$ is a unit impulse $\delta[n]$, which has the value 1 at $n = 0$ and the value 0 elsewhere. The output $y[n]$ of the system Eq. (1.4) would be $\delta[4n + 1]$, which is zero for all values of $n$, and $y[n + n_0]$ would likewise always be zero. However, if we consider $x[n + n_0] = \delta[n + n_0]$, the output will be $\delta[4n + 1 + n_0]$, which for $n_0 = 3$ will be 1 at $n = -1$ and zero otherwise.

*Causality*: Since the output at time $n = 0$ is the input value at $n = 1$, the system is not causal.

*BIBO Stability*: Since $|y[n]| = |x[4n + 1]|$ and the bound on $|x[n]|$ also bounds $|x[4n + 1]|$, the system is BIBO stable.

## 1.2 LINEAR, TIME-INVARIANT SYSTEMS

Linear, time-invariant (LTI) systems form the basis for engineering design in many contexts. This class of systems has the advantage of a rich and well-established theory for analysis and design. Furthermore, in many systems that are nonlinear, small deviations from some nominal steady operation are approximately governed by LTI models, so the tools of LTI system analysis and design can be applied incrementally around a nominal operating condition.

### 1.2.1 Impulse-Response Representation of LTI Systems

A very general way of representing an LTI mapping from an input signal to an output signal is through convolution of the input with the system impulse response. In CT the relationship is

$$y(t) = \int_{-\infty}^{\infty} x(v)h(t - v)\,dv = \int_{-\infty}^{\infty} x(t - \tau)h(\tau)\,d\tau \tag{1.9}$$

where $x(t)$ is the input, $y(t)$ is the output, and $h(t)$ is the unit impulse response of the system. In DT, the corresponding relationship is

$$y[n] = \sum_{k=-\infty}^{\infty} x[k]\,h[n - k] = \sum_{m=-\infty}^{\infty} x[n - m]\,h[m] \tag{1.10}$$

where $h[n]$ is the unit sample (or unit "impulse") response of the system.

The common shorthand notations for the convolution integral in Eq. (1.9) and the convolution sum in Eq. (1.10) are

$$y(t) = x(t) * h(t) \tag{1.11}$$

$$y[n] = x[n] * h[n] . \tag{1.12}$$

While these notations can be convenient, they can also easily lead to misinterpretation if not well understood. Alternative notations such as

$$y(t) = (x * h)(t) \tag{1.13}$$

have their advantages and disadvantages. We shall use the notations indicated in Eqs. (1.11) and (1.12) as shorthand for Eqs. (1.9) and (1.10), with the understanding that Eqs. (1.9) and (1.10) are the correct interpretations.

The characterization of LTI systems through convolution is obtained by representing the input signal as a superposition of weighted impulses. In the DT case, suppose we are given an LTI mapping whose impulse response is $h[n]$, that is, when its input is the unit sample or unit "impulse" function $\delta[n]$, its output is $h[n]$. A general input $x[n]$ can be assembled as a sum of scaled and shifted impulses, specifically:

$$x[n] = \sum_{k=-\infty}^{\infty} x[k]\,\delta[n-k]\,. \tag{1.14}$$

As a consequence of linearity and time invariance, the response $y[n]$ to this input is the sum of the similarly scaled and shifted impulse responses, and is therefore given by Eq. (1.10). What linearity and time invariance have allowed us to do is write the response to a general input in terms of the response to a special input. A similar derivation holds for the CT case.

It may seem that the preceding derivation indicates that all LTI mappings from an input signal to an output signal can be represented through a convolution sum. However, the use of infinite integrals or sums like those in Eqs. (1.9) and (1.10) actually involves some assumptions about the corresponding mapping. We make no attempt here to elaborate on these assumptions. Nevertheless, it is not hard to find "pathological" examples of LTI mappings—not significant for us in this text, or indeed in most engineering models—where the convolution relationship does not hold because these assumptions are violated.

It follows from Eqs. (1.9) and (1.10) that a necessary and sufficient condition for an LTI system to be BIBO stable is that the impulse response be absolutely integrable (CT) or absolutely summable (DT):

$$\text{BIBO stable (CT)} \iff \int_{-\infty}^{\infty} |h(t)|\,dt < \infty \tag{1.15}$$

$$\text{BIBO stable (DT)} \iff \sum_{n=-\infty}^{\infty} |h[n]| < \infty\,. \tag{1.16}$$

It also follows from Eqs. (1.9) and (1.10) that a necessary and sufficient condition for an LTI system to be causal is that the impulse response be zero for $t < 0$ (CT) or for $n < 0$ (DT).

## 1.2.2 Eigenfunction and Transform Representation of LTI Systems

Exponentials are eigenfunctions of LTI mappings, that is, when the input is an exponential for all time, which we refer to as an "everlasting" exponential, the output is simply a scaled version of the input. Therefore, computing the response to an everlasting exponential reduces to simply multiplying by the appropriate scale factor. Specifically, in the CT case, suppose

$$x(t) = e^{s_0 t} \tag{1.17}$$

for some possibly complex value $s_0$ (termed the complex frequency). Then from Eq. (1.9)

$$
\begin{aligned}
y(t) &= \int_{-\infty}^{\infty} h(\tau) x(t - \tau) \, d\tau \\
&= \int_{-\infty}^{\infty} h(\tau) e^{s_0(t - \tau)} \, d\tau \\
&= H(s_0) e^{s_0 t} ,
\end{aligned}
\tag{1.18}
$$

where

$$H(s) = \int_{-\infty}^{\infty} h(\tau) e^{-s\tau} \, d\tau , \tag{1.19}$$

provided the above integral has a finite value for $s = s_0$ (otherwise the response to the exponential is not well defined). Equation (1.18) demonstrates that $x(t)$ in the form of Eq. (1.17) is an eigenfunction with associated eigenvalue given by $H(s_0)$. Note that Eq. (1.19) is precisely the bilateral Laplace transform of the impulse response, or the transfer function of the system, and the set of values of $s$ in the complex plane for which the above integral takes a finite value constitutes the region of convergence (ROC) of the transform. We discuss the Laplace transform further in Section 1.4.

The fact that the everlasting exponential is an eigenfunction of an LTI system derives directly from the fact that time shifting an everlasting exponential produces the same result as scaling it by a constant factor. In contrast, the one-sided exponential $e^{s_0 t} u(t)$, where $u(t)$ denotes the unit step, is in general not an eigenfunction of an LTI mapping: time shifting a one-sided exponential does not produce the same result as scaling this exponential, as indicated in Example 1.2.

## Example 1.2    Eigenfunctions of LTI Systems

As demonstrated above, the everlasting complex exponential $e^{j\omega t}$ is an eigenfunction of any LTI system for which the integral in Eq. (1.19) converges at $s = j\omega$, while $e^{j\omega t} u(t)$ is not. Consider, as a simple example, a time delay:

$$y(t) = x(t - t_0) . \tag{1.20}$$

The output due to the input $e^{j\omega t} u(t)$ is

$$e^{-j\omega t_0} e^{j\omega t} u(t - t_0) .$$

This is not a simple scaling of the input, so $e^{j\omega t} u(t)$ is not in general an eigenfunction of LTI systems.

When $x(t) = e^{j\omega t}$, corresponding to having $s_0$ take the purely imaginary value $j\omega$ in Eq. (1.17), the input is bounded for all positive and negative time, and the corresponding output is of the form

$$y(t) = H(j\omega)e^{j\omega t} \tag{1.21}$$

provided that $H(s)$ in Eq. (1.19) converges for $s = j\omega$. Here $\omega$ is the (real-valued) frequency of the input. From Eq. (1.19), $H(j\omega)$ is given by

$$H(j\omega) = \int_{-\infty}^{\infty} h(t)e^{-j\omega t}\, dt \ . \tag{1.22}$$

The function $H(j\omega)$ in Eq. (1.22) is referred to as the system frequency response, and is also the continuous-time Fourier transform (CTFT) of the impulse response. The integral that defines the CTFT has a finite value for each $\omega$ (and can be shown to be a continuous function of $\omega$) if $h(t)$ is absolutely integrable, in other words if

$$\int_{-\infty}^{+\infty} |h(t)|\, dt < \infty \ . \tag{1.23}$$

This condition ensures that $s = j\omega$ is in the ROC of $H(s)$. Comparing Eq. (1.23) and Eq. (1.15), we note that this condition is equivalent to the system being BIBO stable. The CTFT can also be defined for certain classes of signals that are not absolutely integrable, as for $h(t) = (\sin t)/t$ whose CTFT is a rectangle in the frequency domain, but we defer examination of conditions for existence of the CTFT to Section 1.3.

Knowing the response to $e^{j\omega t}$ allows us to also determine the response to a general (real) sinusoidal input of the form

$$x(t) = A\cos(\omega t + \theta) = \frac{A}{2}\left[e^{j(\omega t+\theta)} + e^{-j(\omega t+\theta)}\right] . \tag{1.24}$$

Invoking superposition, and assuming $h(t)$ is real so $H(j\omega)$ is conjugate symmetric, some algebra shows that the corresponding output is

$$y(t) = \left|H(j\omega)\right| A\cos(\omega t + \theta + \angle H(j\omega)) \ . \tag{1.25}$$

Thus the output is again a sinusoid at the same frequency, but scaled in magnitude by the magnitude of the frequency response at the input frequency, and shifted in phase by the angle of the frequency response at the input frequency.

We can similarly examine the eigenfunction property in the DT case. A DT everlasting exponential is a geometric sequence or signal of the form

$$x[n] = z_0^n \tag{1.26}$$

for some possibly complex value $z_0$, termed the complex frequency. With this DT exponential input, the output of a convolution mapping follows by a simple computation that is analogous to what we showed above for the CT case. Specifically,

$$y[n] = h[n] * x[n] = H(z_0)z_0^n \ , \tag{1.27}$$

where

$$H(z) = \sum_{k=-\infty}^{\infty} h[k]z^{-k} \ , \tag{1.28}$$

provided the above sum has a finite value when $z = z_0$. Note that this sum is precisely the bilateral $z$-transform of the impulse response, and the set of values of $z$ in the complex plane for which the sum takes a finite value constitutes the ROC of the $z$-transform. As in the CT case, the one-sided exponential $z_0^n u[n]$ is not in general an eigenfunction. We discuss the $z$-transform further in Section 1.4.

Again, an important case is when $x[n] = (e^{j\Omega})^n = e^{j\Omega n}$, corresponding to $z_0$ in Eq. (1.26) having unit magnitude and taking the value $e^{j\Omega}$, where $\Omega$— the (real) "frequency"—denotes the angular position (in radians) around the unit circle in the $z$-plane. Such an $x[n]$ is bounded for all positive and negative time. Although we use a different symbol, $\Omega$, for frequency in the DT case, to distinguish it from the frequency $\omega$ in the CT case, it is not unusual in the literature to find $\omega$ used in both CT and DT cases for notational convenience. The corresponding output is

$$y[n] = H(e^{j\Omega})e^{j\Omega n} \tag{1.29}$$

provided that $e^{j\Omega}$ is in the ROC of $H(z)$. From Eq. (1.28), $H(e^{j\Omega})$ is given by

$$H(e^{j\Omega}) = \sum_{n=-\infty}^{\infty} h[n]e^{-j\Omega n} . \tag{1.30}$$

The function $H(e^{j\Omega})$ in Eq. (1.30) is the frequency response of the DT system, and is also the discrete-time Fourier transform (DTFT) of the impulse response. The sum that defines the DTFT has a finite value (and can be shown to be a continuous function of $\Omega$) if $h[n]$ is absolutely summable, in other words provided

$$\sum_{n=-\infty}^{\infty} |h[n]| < \infty . \tag{1.31}$$

This condition ensures that $e^{j\Omega}$ is in the ROC of $H(z)$. As in continuous time, this condition is equivalent to the system being BIBO stable. As with the CTFT, the DTFT can be defined for signals that are not absolutely summable; we will elaborate on this in Section 1.3.

Using Eq. (1.29), assuming $h[n]$ is real, and proceeding as in the CT case, it follows that the response to the sinusoidal input

$$x[n] = A\cos(\Omega n + \theta) \tag{1.32}$$

is

$$y[n] = \left| H(e^{j\Omega}) \right| A\cos(\Omega n + \theta + \angle H(e^{j\Omega})) . \tag{1.33}$$

Note from Eq. (1.30) that the frequency response for DT systems is always periodic, with period $2\pi$. The "low-frequency" response is found in the vicinity of $\Omega = 0$, corresponding to an input signal that is constant for all $n$. The "high-frequency" response is found in the vicinity of $\Omega = \pm\pi$, corresponding to an input signal $e^{\pm j\pi n} = (-1)^n$ that is the most rapidly varying DT signal possible.

When the input of an LTI system can be expressed as a linear combination of eigenfunctions, for instance (in the CT case)

$$x(t) = \sum_{\ell} a_\ell e^{j\omega_\ell t} , \tag{1.34}$$

then, by linearity, the output is the same linear combination of the responses to the individual exponentials. By the eigenfunction property of exponentials in LTI systems, the response to each exponential involves only scaling by the system's frequency response at the frequency of the exponential. Thus

$$y(t) = \sum_{\ell} a_\ell H(j\omega_\ell) e^{j\omega_\ell t} . \tag{1.35}$$

Similar expressions can be written for the DT case.

### 1.2.3 Fourier Transforms

A broad class of input signals can be represented as linear combinations of bounded exponentials through the Fourier transform. The synthesis/analysis formulas for the continuous-time Fourier transform (CTFT) are

$$x(t) = \frac{1}{2\pi} \int_{-\infty}^{\infty} X(j\omega) e^{j\omega t} d\omega \quad \text{(synthesis)} \tag{1.36}$$

$$X(j\omega) = \int_{-\infty}^{\infty} x(t) e^{-j\omega t} dt \quad \text{(analysis)}. \tag{1.37}$$

Note that Eq. (1.36) expresses $x(t)$ as a linear combination of exponentials, but this weighted combination involves a continuum of exponentials rather than a finite or countable number. If this signal $x(t)$ is the input to an LTI system with frequency response $H(j\omega)$, then by linearity and the eigenfunction property of exponentials the output is the same weighted combination of the responses to these exponentials, that is,

$$y(t) = \frac{1}{2\pi} \int_{-\infty}^{\infty} H(j\omega) X(j\omega) e^{j\omega t} d\omega . \tag{1.38}$$

By viewing this equation as a CTFT synthesis equation, it follows that the CTFT of $y(t)$ is

$$Y(j\omega) = H(j\omega) X(j\omega) . \tag{1.39}$$

The convolution relationship Eq. (1.9) in the time domain therefore becomes multiplication in the transform domain. Thus, to determine $Y(j\omega)$ at any particular frequency $\omega_0$, we only need to know the Fourier transform of the input at that single frequency, and the frequency response of the system at that frequency. This simple fact serves, in large measure, to explain why the frequency domain is virtually indispensable in the analysis of LTI systems.

The corresponding DTFT synthesis/analysis pair is defined by

$$x[n] = \frac{1}{2\pi} \int_{\langle 2\pi \rangle} X(e^{j\Omega}) e^{j\Omega n} d\Omega \quad \text{(synthesis)} \tag{1.40}$$

$$X(e^{j\Omega}) = \sum_{n=-\infty}^{\infty} x[n] e^{-j\Omega n} \quad \text{(analysis)} \tag{1.41}$$

where the notation $\langle 2\pi \rangle$ on the integral in the synthesis formula denotes integration over any contiguous interval of length $2\pi$. This is because the DTFT is always periodic in $\Omega$ with period $2\pi$, a simple consequence of the fact that $e^{j\Omega}$ is periodic with period $2\pi$. Note that Eq. (1.40) expresses $x[n]$ as a weighted combination of a continuum of exponentials.

As in the CT case, it is straightforward to show that if $x[n]$ is the input to an LTI mapping, then the output $y[n]$ has the DTFT

$$Y(e^{j\Omega}) = H(e^{j\Omega})X(e^{j\Omega}) \,. \tag{1.42}$$

# 1.3 DETERMINISTIC SIGNALS AND THEIR FOURIER TRANSFORMS

In this section, we review the DTFT of deterministic DT signals in more detail and highlight classes of signals that can be guaranteed to have well-defined DTFTs. We shall also devote some attention to the energy density spectrum of signals that have DTFTs. The section will bring out aspects of the DTFT that may not have been emphasized in your earlier signals and systems course. A similar development can be carried out for CTFTs.

## 1.3.1 Signal Classes and Their Fourier Transforms

The DTFT synthesis and analysis pair in Eqs. (1.40) and (1.41) hold for at least the three large classes of DT signals described below.

### Finite-Action Signals

Finite-action signals, which are also called absolutely summable signals or $\ell^1$ ("ell-one") signals, are defined by the condition

$$\sum_{k=-\infty}^{\infty} \left| x[k] \right| < \infty \,. \tag{1.43}$$

The sum on the left is often called the action of the signal. For these signals, the infinite sum that defines the DTFT is well behaved and the DTFT can be shown to be a continuous function for all $\Omega$. In particular, the values at $\Omega = +\pi$ and $\Omega = -\pi$ are well defined and equal to each other, which need not be the case when signals are not $\ell^1$.

### Finite-Energy Signals

Finite-energy signals, which are also referred to as square summable or $\ell^2$ ("ell-two") signals, are defined by the condition

$$\sum_{k=-\infty}^{\infty} \left| x[k] \right|^2 < \infty \,. \tag{1.44}$$

The sum on the left is called the energy of the signal.

In discrete time, an absolutely summable (i.e., $\ell^1$) signal is always square summable (i.e., $\ell^2$). However, the reverse is not true. For example, consider the signal $(\sin \Omega_c n)/\pi n$ for $0 < \Omega_c < \pi$, with the value at $n = 0$ taken to be $\Omega_c/\pi$, or consider the signal $(1/n)u[n-1]$, both of which are $\ell^2$ but not $\ell^1$. If $x[n]$ is such a signal, its DTFT $X(e^{j\Omega})$ can be thought of as the limit for $N \to \infty$ of the quantity

$$X_N(e^{j\Omega}) = \sum_{k=-N}^{N} x[k]e^{-j\Omega k} \tag{1.45}$$

and the resulting limit will typically have discontinuities at some values of $\Omega$. For instance, the transform of $(\sin \Omega_c n)/\pi n$ has discontinuities at $\Omega = \pm\Omega_c$.

### Signals of Slow Growth

Signals of slow growth are signals whose magnitude grows no faster than polynomially with the time index, for example, $x[n] = n$ for all $n$. In this case $X_N(e^{j\Omega})$ in Eq. (1.45) does not converge in the usual sense, but the DTFT still exists as a generalized (or singularity) function; for example, if $x[n] = 1$ for all $n$, then $X(e^{j\Omega}) = 2\pi\delta(\Omega)$ for $|\Omega| \le \pi$.

Within the class of signals of slow growth, those of most interest to us are bounded (or $\ell^\infty$) signals defined by

$$\left| x[k] \right| \le M < \infty \tag{1.46}$$

that is, signals whose amplitude has a fixed and finite bound for all time. Bounded everlasting exponentials of the form $e^{j\Omega_0 n}$, for instance, play a key role in Fourier transform theory. Such signals need not have finite energy, but will have finite average power over any time interval, where average power is defined as total energy over total time.

Similar classes of signals are defined in continuous time. Finite-action (or $L^1$) signals comprise those that are absolutely integrable, that is,

$$\int_{-\infty}^{\infty} \left| x(t) \right| dt < \infty . \tag{1.47}$$

Finite-energy (or $L^2$) signals comprise those that are square integrable, that is,

$$\int_{-\infty}^{\infty} \left| x(t) \right|^2 dt < \infty . \tag{1.48}$$

In continuous time, an absolutely integrable signal (i.e., $L^1$) may not be square integrable (i.e., $L^2$), as is the case, for example, with the signal

$$x(t) = \begin{cases} 1/\sqrt{t} & 0 < t \le 1 \\ 0 & \text{elsewhere.} \end{cases} \tag{1.49}$$

However, an $L^1$ signal that is bounded will also be $L^2$. As in discrete time, a CT signal that is $L^2$ is not necessarily $L^1$, as is the case, for example, with the signal

$$x(t) = \frac{\sin \omega_c t}{\pi t} . \tag{1.50}$$

In both continuous time and discrete time, there are many important Fourier transform pairs and Fourier transform properties developed and tabulated in basic texts on signals and systems. For convenience, we include here a brief table of DTFT pairs (Table 1.1) and one of CTFT pairs (Table 1.2). Other pairs are easily derived from these by applying various Fourier transform properties. Note that $\delta[\cdot]$ in the left column in Table 1.1 denotes unit sample functions, while $\delta(\cdot)$ in the right column are unit impulses. Also, the DTFTs in Table 1.1 repeat periodically outside the interval $-\pi < \Omega \leq \pi$.

In general, it is important and useful to be fluent in deriving and utilizing the main transform pairs and properties. In the following subsection we discuss Parseval's identity, a transform property that is of particular significance in our later discussion.

There are, of course, other classes of signals that are of interest to us in applications, for instance growing one-sided exponentials. To deal with such

**TABLE 1.1    BRIEF TABLE OF DTFT PAIRS**

| DT Signal $\longleftrightarrow$ DTFT for $-\pi < \Omega \leq \pi$ |
|---|
| $\delta[n] \longleftrightarrow 1$ |
| $\delta[n - n_0] \longleftrightarrow e^{-j\Omega n_0}$ |
| $1 \text{ (for all } n) \longleftrightarrow 2\pi\delta(\Omega)$ |
| $e^{j\Omega_0 n} \; (-\pi < \Omega_0 \leq \pi) \longleftrightarrow 2\pi\delta(\Omega - \Omega_0)$ |
| $a^n u[n], \; \lvert a \rvert < 1 \longleftrightarrow \dfrac{1}{1 - ae^{-j\Omega}}$ |
| $u[n] \longleftrightarrow \dfrac{1}{1 - e^{-j\Omega}} + \pi\delta(\Omega)$ |
| $\dfrac{\sin \Omega_c n}{\pi n} \longleftrightarrow \begin{cases} 1, & -\Omega_c < \Omega < \Omega_c \\ 0, & \text{otherwise} \end{cases}$ |
| $\left. \begin{array}{l} 1, \; -M \leq n \leq M \\ 0, \; \text{otherwise} \end{array} \right\} \longleftrightarrow \dfrac{\sin[\Omega(2M + 1)/2]}{\sin(\Omega/2)}$ |

**TABLE 1.2    BRIEF TABLE OF CTFT PAIRS**

| CT Signal $\longleftrightarrow$ CTFT |
|---|
| $\delta(t) \longleftrightarrow 1$ |
| $\delta(t - t_0) \longleftrightarrow e^{-j\omega t_0}$ |
| $1 \text{ (for all } t) \longleftrightarrow 2\pi\delta(\omega)$ |
| $e^{j\omega_0 t} \longleftrightarrow 2\pi\delta(\omega - \omega_0)$ |
| $e^{-at}u(t), \mathcal{R}e\{a\} > 0 \longleftrightarrow \dfrac{1}{a + j\omega}$ |
| $u(t) \longleftrightarrow \dfrac{1}{j\omega} + \pi\delta(\omega)$ |
| $\dfrac{\sin \omega_c t}{\pi t} \longleftrightarrow \begin{cases} 1, & -\omega_c < \omega < \omega_c \\ 0, & \text{otherwise} \end{cases}$ |
| $\left. \begin{array}{l} 1, \; -M \leq t \leq M \\ 0, \; \text{otherwise} \end{array} \right\} \longleftrightarrow \dfrac{\sin \omega M}{\omega/2}$ |

signals, we make use of $z$-transforms in discrete time and Laplace transforms in continuous time.

## 1.3.2 Parseval's Identity, Energy Spectral Density, and Deterministic Autocorrelation

An important property of the Fourier transform is Parseval's identity for $\ell^2$ signals. For discrete time, this identity takes the general form

$$\sum_{n=-\infty}^{\infty} x[n]y^*[n] = \frac{1}{2\pi}\int_{<2\pi>} X(e^{j\Omega})Y^*(e^{j\Omega})\,d\Omega \tag{1.51}$$

and for continuous time,

$$\int_{-\infty}^{\infty} x(t)y^*(t)\,dt = \frac{1}{2\pi}\int_{-\infty}^{\infty} X(j\omega)Y^*(j\omega)\,d\omega \tag{1.52}$$

where the superscript symbol * denotes the complex conjugate. Specializing to the case where $y[n] = x[n]$ or $y(t) = x(t)$, we obtain

$$\sum_{n=-\infty}^{\infty} |x[n]|^2 = \frac{1}{2\pi}\int_{<2\pi>} |X(e^{j\Omega})|^2\,d\Omega \tag{1.53}$$

$$\int_{-\infty}^{\infty} |x(t)|^2\,dt = \frac{1}{2\pi}\int_{-\infty}^{\infty} |X(j\omega)|^2\,d\omega . \tag{1.54}$$

Parseval's identity allows us to evaluate the energy of a signal by integrating the squared magnitude of its transform. What the identity tells us, in effect, is that the energy of a signal equals the energy of its transform (scaled by $1/2\pi$).

The right-hand sides of Eqs. (1.53) and (1.54) integrate the quantities $|X(e^{j\Omega})|^2$ and $|X(j\omega)|^2$. We denote these quantities by $\overline{S}_{xx}(e^{j\Omega})$ and $\overline{S}_{xx}(j\omega)$:

$$\overline{S}_{xx}(e^{j\Omega}) = |X(e^{j\Omega})|^2 \tag{1.55}$$

or

$$\overline{S}_{xx}(j\omega) = |X(j\omega)|^2 . \tag{1.56}$$

These are referred to as the energy spectral density (ESD) of the associated signal because they describe how the energy of the signal is distributed over frequency. To justify this interpretation more concretely, for discrete time, consider applying $x[n]$ to the input of an ideal bandpass filter of frequency response $H(e^{j\Omega})$ that has narrow passbands of unit gain and width $\Delta$ centered at $\pm\Omega_0$, as indicated in Figure 1.4. The energy of the output signal must then be the energy of $x[n]$ that is contained in the passbands of the filter. To calculate the energy of the output signal, note that this output $y[n]$ has the transform

$$Y(e^{j\Omega}) = H(e^{j\Omega})X(e^{j\Omega}) . \tag{1.57}$$

Consequently, by Parseval's identity, the output energy is

$$\sum_{n=-\infty}^{\infty} |y[n]|^2 = \frac{1}{2\pi}\int_{\langle 2\pi\rangle} |Y(e^{j\Omega})|^2\,d\Omega$$

$$= \frac{1}{2\pi}\int_{\langle 2\pi\rangle} |H(e^{j\Omega})|^2\,|X(e^{j\Omega})|^2\,d\Omega . \tag{1.58}$$

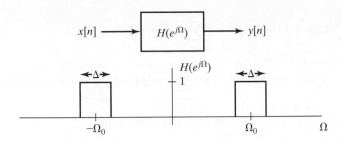

**Figure 1.4**    Ideal bandpass filter.

Since $|H(e^{j\Omega})|$ is unity in the passband and zero otherwise, Eq. (1.58) reduces to

$$\sum_{n=-\infty}^{\infty} |y[n]|^2 = \frac{1}{2\pi} \int_{\text{passband}} |X(e^{j\Omega})|^2 \, d\Omega$$

$$= \frac{1}{2\pi} \int_{\text{passband}} \overline{S}_{xx}(e^{j\Omega}) \, d\Omega \, . \tag{1.59}$$

Thus the energy of $x[n]$ in any frequency band is given by integrating $\overline{S}_{xx}(e^{j\Omega})$ over that band (and scaling by $1/2\pi$). In other words, the energy density of $x[n]$ as a function of $\Omega$ is $\overline{S}_{xx}(\Omega)/(2\pi)$ per radian. An exactly analogous discussion can be carried out for CT signals.

Since the ESD $\overline{S}_{xx}(e^{j\Omega})$ is a real function of $\Omega$, an alternate notation for it might be $\mathcal{E}_{xx}(\Omega)$. However, we use the notation $\overline{S}_{xx}(e^{j\Omega})$ in order to make explicit that it is the squared magnitude of $X(e^{j\Omega})$ and also the fact that the ESD for a DT signal is periodic with period $2\pi$.

The ESD also has an important interpretation in the time domain. In discrete time, for example, and assuming $x[n]$ is real, we obtain

$$\overline{S}_{xx}(e^{j\Omega}) = |X(e^{j\Omega})|^2 = X(e^{j\Omega})X(e^{-j\Omega}) \, . \tag{1.60}$$

Note that $X(e^{-j\Omega})$ is the transform of the time-reversed signal $\overleftarrow{x}[k] = x[-k]$. Thus, since multiplication of transforms in the frequency domain corresponds to convolution of signals in the time domain, we have

$$\overline{S}_{xx}(e^{j\Omega}) = |X(e^{j\Omega})|^2 \iff x[k] * \overleftarrow{x}[k] = \sum_{n=-\infty}^{\infty} x[n+k]x[n] = \overline{R}_{xx}[k] \, .$$

$$\tag{1.61}$$

The function $\overline{R}_{xx}[k]$ is referred to as the deterministic autocorrelation function of the signal $x[n]$, and we have just established that the transform of the deterministic autocorrelation function is the energy spectral density $\overline{S}_{xx}(e^{j\Omega})$. A basic Fourier transform property tells us that $\overline{R}_{xx}[0]$, which is the signal energy $\sum_{n=-\infty}^{\infty} x^2[n]$, is the area under the Fourier transform of $\overline{R}_{xx}[k]$, scaled by $1/(2\pi)$, namely the scaled area under $\overline{S}_{xx}(e^{j\Omega}) = |X(e^{j\Omega})|^2$; this, of course, corresponds directly to Eq. (1.53).

The deterministic autocorrelation function measures how alike a signal and its time-shifted version are in a total-squared-error sense. More specifically, in discrete time the total squared error between the signal and its time-shifted version is given by

$$\sum_{n=-\infty}^{\infty} (x[n+k] - x[n])^2 = \sum_{n=-\infty}^{\infty} x^2[n+k]$$

$$+ \sum_{n=-\infty}^{\infty} x^2[n] - 2 \sum_{n=-\infty}^{\infty} x[n+k]x[n]$$

$$= 2(\overline{R}_{xx}[0] - \overline{R}_{xx}[k]) . \tag{1.62}$$

Since the total squared error is always nonnegative, it follows that $\overline{R}_{xx}[k] \le \overline{R}_{xx}[0]$, and that the larger the deterministic autocorrelation $\overline{R}_{xx}[k]$ is, the closer the signal $x[n]$ and its time-shifted version $x[n+k]$ are.

Corresponding results hold in continuous time, and in particular

$$\overline{S}_{xx}(j\omega) = |X(j\omega)|^2 \iff x(\tau) * \overleftarrow{x}(\tau) = \int_{-\infty}^{\infty} x(t+\tau)x(t)dt = \overline{R}_{xx}(\tau)$$
$$\tag{1.63}$$

where $\overline{R}_{xx}(t)$ is the deterministic autocorrelation function of $x(t)$.

## 1.4 BILATERAL LAPLACE AND z-TRANSFORMS

Laplace and $z$-transforms can be thought of as extensions of Fourier transforms and are useful for a variety of reasons. They permit a transform treatment of certain classes of signals for which the Fourier transform does not converge. They also augment our understanding of Fourier transforms by moving us into the complex plane, where we can apply the theory of complex functions. We begin in Section 1.4.1 with a detailed review of the bilateral $z$- transform. In Section 1.4.2, we give a short review of the bilateral Laplace transform, paralleling the discussion in Section 1.4.1.

### 1.4.1 The Bilateral z-Transform

The bilateral $z$-transform is defined as

$$X(z) = \sum_{n=-\infty}^{\infty} x[n]z^{-n} . \tag{1.64}$$

Here $z$ is a complex variable, which we can also represent in polar form as

$$z = re^{j\Omega}, \quad r \ge 0, \quad -\pi < \Omega \le \pi \tag{1.65}$$

so

$$X(z) = \sum_{n=-\infty}^{\infty} x[n]r^{-n}e^{-j\Omega n} . \tag{1.66}$$

The DTFT corresponds to setting $r = 1$, in which case $z$ takes values on the unit circle. However, there are many useful signals for which the infinite sum does not converge (even in the sense of generalized functions) for $z$ confined to the unit circle. The term $z^{-n}$ in the definition of the z-transform introduces a factor $r^{-n}$ into the infinite sum, which permits the sum to converge (provided $r$ is appropriately restricted) for interesting classes of signals, many of which do not have DTFTs.

More specifically, note from Eq. (1.66) that $X(z)$ can be viewed as the DTFT of $x[n]r^{-n}$. If $r > 1$, then $r^{-n}$ decays geometrically for positive $n$ and grows geometrically for negative $n$. For $0 < r < 1$, the opposite happens. Consequently, there are many sequences for which $x[n]$ is not absolutely summable, but $x[n]r^{-n}$ is for some range of values of $r$.

For example, consider $x_1[n] = a^n u[n]$. If $|a| > 1$, this sequence does not have a DTFT. However, for any $a$, $x[n]r^{-n}$ is absolutely summable provided $r > |a|$. In particular, for example,

$$X_1(z) = 1 + az^{-1} + a^2 z^{-2} + \cdots \tag{1.67}$$

$$= \frac{1}{1 - az^{-1}}, \quad |z| = r > |a| . \tag{1.68}$$

As a second example, consider $x_2[n] = -a^n u[-n - 1]$. This signal does not have a DTFT if $|a| < 1$. However, provided $r < |a|$,

$$X_2(z) = -a^{-1}z - a^{-2}z^2 - \cdots \tag{1.69}$$

$$= \frac{-a^{-1}z}{1 - a^{-1}z}, \quad |z| = r < |a| \tag{1.70}$$

$$= \frac{1}{1 - az^{-1}}, \quad |z| = r < |a| . \tag{1.71}$$

The z-transforms of the two distinct signals $x_1[n]$ and $x_2[n]$ above get condensed to the same rational expressions, but for different regions of convergence. Hence the ROC is a critical part of the specification of the transform.

When $x[n]$ is a sum of left-sided and/or right-sided DT exponentials, with each term of the form illustrated in the examples above, then $X(z)$ will be rational in $z$ (or equivalently, in $z^{-1}$):

$$X(z) = \frac{Q(z)}{P(z)} \tag{1.72}$$

with $Q(z)$ and $P(z)$ being polynomials in $z$ or, equivalently, $z^{-1}$.

Rational z-transforms are typically depicted by a pole-zero plot in the z-plane, with the ROC appropriately indicated. This information uniquely specifies the signal, apart from a constant amplitude scaling. Note that there can be no poles in the ROC, since the transform is required to be finite in the ROC. z-transforms are often written as ratios of polynomials in $z^{-1}$. However, the pole-zero plot in the z-plane refers to the roots of the polynomials in $z$. Also note that if poles or zeros at $z = \infty$ are counted, then any ratio of polynomials always has exactly the same number of poles as zeros.

### Region of Convergence

To understand the complex-function properties of the $z$-transform, we split the infinite sum that defines it into nonnegative-time and negative-time portions. The nonnegative-time or one-sided $z$-transform is defined by

$$\sum_{n=0}^{\infty} x[n]z^{-n} \tag{1.73}$$

and is a power series in $z^{-1}$. The convergence of the finite sum $\sum_{n=0}^{N} x[n]z^{-n}$ as $N \to \infty$ is governed by the radius of convergence $R_1 \geq 0$ of the power series. The series converges (absolutely) for each $z$ such that $|z| > R_1$. The resulting function of $z$ is an analytic function in this region, that is, it has a well-defined derivative with respect to the complex variable $z$ at each point in this region, which is what gives the function its nice properties. The series diverges for $|z| < R_1$. The behavior of the sum on the circle $|z| = R_1$ requires closer examination and depends on the particular series; the series may converge (but may not converge absolutely) at all points, some points, or no points on this circle. The region $|z| > R_1$ is referred to as the ROC of the power series for the nonnegative-time part.

Next consider the negative-time part:

$$\sum_{n=-\infty}^{-1} x[n]z^{-n} = \sum_{m=1}^{\infty} x[-m]z^{m} \tag{1.74}$$

which is a power series in $z$, and has a radius of convergence $R_2$. The series converges (absolutely) for $|z| < R_2$, which constitutes its ROC; the series is an analytic function in this region. The series diverges for $|z| > R_2$. The behavior on the circle $|z| = R_2$ takes closer examination, and depends on the particular series; and the series may converge (but may not converge absolutely) at all points, some points, or no points on this circle. If $R_1 < R_2$, then the $z$-transform of $x[n]$ converges (absolutely) for $R_1 < |z| < R_2$; this annular region is its ROC. The transform is analytic in this region. The series that defines the transform diverges for $|z| < R_1$ and $|z| > R_2$. If $R_1 > R_2$, then the $z$-transform does not exist (for example, for $x[n] = 0.5^n u[-n-1] + 2^n u[n]$). If $R_1 = R_2$, then the transform may exist in a technical sense, but is not useful as a $z$-transform because it has no ROC. However, if $R_1 = R_2 = 1$, then we may still be able to compute and use a DTFT. For example, for $x[n] = 3$ for all $n$, or for $x[n] = (\sin \Omega_0 n)/(\pi n)$, the DTFT can be used by incorporating generalized functions such as impulses and step functions in the frequency domain.

### Relating the ROC to Signal Properties

For an absolutely summable sequence (such as the impulse response of a BIBO-stable system), that is, an $\ell^1$-signal, the unit circle must lie in the ROC or must be a boundary of the ROC. Conversely, we can conclude that a signal is $\ell^1$ if the ROC contains the unit circle because the transform converges absolutely in its ROC. If the unit circle constitutes a boundary of the ROC, then further analysis is generally needed to determine if the signal is $\ell^1$. Rational transforms always have a pole on the boundary of the ROC, as elaborated on below, so if the unit circle is on the boundary of the ROC of a rational transform, then there is a pole on the unit circle and the signal cannot be $\ell^1$.

For a right-sided signal, it is the case that $R_2 = \infty$, that is, the ROC extends everywhere in the complex plane outside the circle of radius $R_1$, up to (and perhaps including) $\infty$. The ROC includes $\infty$ if the signal is zero for negative time.

We can state a converse result if, for example, we know the signal comprises only sums of one-sided exponentials of the form obtained when inverse transforming a rational transform. In this case, if $R_2 = \infty$, then the signal must be right-sided; if the ROC includes $\infty$, then the signal must be causal, that is, zero for $n < 0$.

For a left-sided signal, $R_1 = 0$, that is, the ROC extends inward from the circle of radius $R_2$, up to (and perhaps including) zero. The ROC includes $z = 0$ if the signal is zero for positive time.

In the case of signals that are sums of one-sided exponentials, we have the converse: if $R_1 = 0$, then the signal must be left-sided; if the ROC includes $z = 0$, then the signal must be anticausal, that is, zero for $n > 0$.

As indicated earlier, the ROC cannot contain poles of the $z$-transform because poles are values of $z$ where the transform has infinite magnitude, while the ROC comprises values of $z$ where the transform converges. For signals with *rational* transforms, one can use the fact that such signals are sums of one-sided exponentials to show that the possible boundaries of the ROC are in fact precisely determined by the locations of the poles. Specifically:

(a) The outer bounding circle of the ROC in the rational case contains a pole and/or has radius $\infty$. If the outer bounding circle is at infinity, then (as we have already noted) the signal is right-sided, and is in fact causal if there is no pole at $\infty$.

(b) The inner bounding circle of the ROC in the rational case contains a pole and/or has radius 0. If the inner bounding circle reduces to the point 0, then (as we have already noted) the signal is left-sided, and is in fact anticausal if there is no pole at 0.

### *The Inverse z-Transform*

One method for inverting a rational $z$-transform is using a partial fraction expansion, then either directly recognizing the inverse transform of each term in the partial fraction representation or expanding the term in a power series that converges for $z$ in the specified ROC. For example, a term of the form

$$\frac{1}{1 - az^{-1}} \tag{1.75}$$

can be expanded in a power series in $az^{-1}$ if $|a| < |z|$ for $z$ in the ROC, and expanded in a power series in $a^{-1}z$ if $|a| > |z|$ for $z$ in the ROC. Carrying out this procedure for each term in a partial fraction expansion, we find that the signal $x[n]$ is a sum of left-sided and/or right-sided exponentials. For nonrational transforms, where there may not be a partial fraction expansion to simplify the process, it is still reasonable to attempt the inverse transformation by expansion into a power series consistent with the given ROC.

Although we will generally use partial fraction or power series methods to invert $z$-transforms, there is an explicit formula that is similar to that of the inverse DTFT, specifically,

$$x[n] = \frac{1}{2\pi} \int_{-\pi}^{\pi} X(z)z^n \, d\Omega \Big|_{z=\bar{r}e^{j\Omega}} \tag{1.76}$$

where the constant $\bar{r}$ is chosen to place $z$ in the ROC. This is not the most general inversion formula, but is sufficient for us, and shows that $x[n]$ is expressed as a weighted combination of DT exponentials.

As is the case for Fourier transforms, there are many useful $z$-transform pairs and properties developed and tabulated in basic texts on signals and systems. Appropriate use of transform pairs and properties is often the basis for obtaining the $z$-transform or the inverse $z$-transform of many other signals.

## 1.4.2 The Bilateral Laplace Transform

As with the $z$-transform, the Laplace transform is introduced in part to handle important classes of signals that do not have CTFTs, but it also enhances our understanding of the CTFT. The definition of the Laplace transform is

$$X(s) = \int_{-\infty}^{\infty} x(t) e^{-st} \, dt \tag{1.77}$$

where $s$ is a complex variable, $s = \sigma + j\omega$. The Laplace transform can thus be thought of as the CTFT of $x(t) e^{-\sigma t}$. With $\sigma$ appropriately chosen, the integral in Eq. (1.77) can exist even for signals that have no CTFT.

The development of the Laplace transform parallels closely that of the $z$-transform in the preceding section, but with $e^\sigma$ playing the role that $r$ did in Section 1.4.1. The interior of the set of values of $s$ for which the defining integral converges, as the limits on the integral approach $\pm\infty$, comprises the ROC for the transform $X(s)$. The ROC is now determined by the minimum and maximum allowable values of $\sigma$, say $\sigma_1$ and $\sigma_2$ respectively. We refer to $\sigma_1$, $\sigma_2$ as abscissas of convergence. The corresponding ROC is a vertical strip between $\sigma_1$ and $\sigma_2$ in the complex plane, $\sigma_1 < \text{Re}\{s\} < \sigma_2$. Equation (1.77) converges absolutely within the ROC; convergence at the left and right bounding vertical lines of the strip has to be separately examined. Furthermore, the transform is analytic (that is, differentiable as a complex function) throughout the ROC. The strip may extend to $\sigma_1 = -\infty$ on the left, and to $\sigma_2 = +\infty$ on the right. If the strip collapses to a line (so that the ROC vanishes), then the Laplace transform is not useful (except if the line happens to be the $j\omega$ axis, in which case a CTFT analysis may perhaps be recovered).

For example, consider $x_1(t) = e^{at}u(t)$; the integral in Eq. (1.77) evaluates to $X_1(s) = 1/(s-a)$ provided $\text{Re}\{s\} > a$. On the other hand, for $x_2(t) = -e^{at}u(-t)$, the integral in Eq. (1.77) evaluates to $X_2(s) = 1/(s-a)$ provided $\text{Re}\{s\} < a$. As with the $z$-transform, note that the expressions for the transforms above are identical; they are distinguished by their distinct regions of convergence.

The ROC may be associated with properties of the signal. For example, for absolutely integrable signals, also referred to as $L^1$ signals, the integrand in the definition of the Laplace transform is absolutely integrable on the $j\omega$ axis, so the $j\omega$ axis is in the ROC or on its boundary. In the other direction, if the $j\omega$ axis is strictly in the ROC, then the signal is $L^1$, because the integral converges absolutely in the ROC. Recall that a system has an $L^1$ impulse response if and only if the system is BIBO stable, so the result here is relevant to discussions of stability: if the $j\omega$ axis is strictly in the ROC of the system function, then the system is BIBO stable.

For right-sided signals, the ROC is some right half-plane (i.e., all $s$ such that $\text{Re}\{s\} > \sigma_1$). Thus the system function of a causal system will have an ROC that is some right half-plane. For left-sided signals, the ROC is some left half-plane. For signals with rational transforms, the ROC contains no poles, and the boundaries of the ROC will have poles. Since the location of the ROC of a transfer function relative to the imaginary axis relates to BIBO stability, and since the poles identify the boundaries of the ROC, the poles relate to stability. In particular, a system with a right-sided impulse response (e.g., a causal system) will be stable if and only if all its poles are finite and in the left half-plane, because this is precisely the condition that allows the ROC to contain the entire imaginary axis. Also note that a signal with a rational transform and no poles at infinity is causal if and only if it is right-sided.

A further property worth recalling is connected to the fact that exponentials are eigenfunctions of LTI systems. If we denote the Laplace transform of the impulse response $h(t)$ of an LTI system by $H(s)$, then $e^{s_0 t}$ at the input of the system yields $H(s_0)e^{s_0 t}$ at the output, provided $s_0$ is in the ROC of the transfer function.

# 1.5 DISCRETE-TIME PROCESSING OF CONTINUOUS-TIME SIGNALS

Many modern systems for applications such as communication, entertainment, navigation, and control are a combination of CT and DT subsystems, exploiting the inherent properties and advantages of each. In particular, the DT processing of CT signals is common in such applications, and we describe the essential ideas behind such processing here. As with the earlier sections, we assume that this discussion is primarily a review of familiar material, included here to establish notation and for convenient reference from later chapters in this text. In this section, and throughout this text, we will often relate the CTFT of a CT signal and the DTFT of a DT signal obtained from samples of the CT signal. We will use the subscripts $c$ and $d$ when necessary to help keep clear which signals are CT and which are DT.

## 1.5.1 Basic Structure for DT Processing of CT Signals

Figure 1.5 depicts the basic structure of this processing, which involves continuous-to-discrete (C/D) conversion to obtain a sequence of samples of the CT input signal; followed by DT filtering to produce a sequence of samples of the desired CT output; then discrete-to-continuous (D/C) conversion to reconstruct this desired CT output signal from the sequence of samples. We will often restrict ourselves to conditions such that the overall system in Figure 1.5 is equivalent to an LTI CT system. The necessary conditions for this typically include restricting the DT filtering to LTI processing by a system with frequency response $H_d(e^{j\Omega})$, and also requiring that the input $x_c(t)$ be appropriately bandlimited. To satisfy the latter requirement, it is typical to precede the structure in the figure by a filter whose purpose is to ensure that $x_c(t)$ is essentially bandlimited. While this filter is often referred to as an anti-aliasing filter, we can often allow some aliasing in the C/D conversion if the DT system removes the aliased components; the overall system can then still be a CT LTI system.

The ideal C/D converter in Figure 1.5 has as its output a sequence of samples of $x_c(t)$ with a specified sampling interval $T_1$, so that the DT signal is $x_d[n] = x_c(nT_1)$. Conceptually, therefore, the ideal C/D converter is straightforward. A practical analog-to-digital (A/D) converter also quantizes the signal to one of a finite set of output levels. However, in this text we do not consider the additional effects of quantization.

In the frequency domain, the CTFT of $x_c(t)$ and the DTFT of $x_d[n]$ can be shown to be related by

$$X_d(e^{j\Omega})\bigg|_{\Omega=\omega T_1} = \frac{1}{T_1} \sum_k X_c\left(j\omega - jk\frac{2\pi}{T_1}\right). \tag{1.78}$$

When $x_c(t)$ is sufficiently bandlimited so that

$$X_c(j\omega) = 0, \qquad |\omega| \geq \frac{\pi}{T_1} \tag{1.79}$$

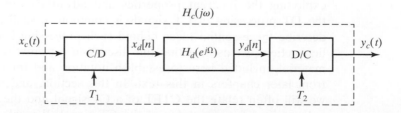

**Figure 1.5**   DT processing of CT signals.

that is, when the sampling is at or above the Nyquist rate, then Eq. (1.78) can be rewritten as

$$X_d(e^{j\Omega})\bigg|_{\Omega=\omega T_1} = \frac{1}{T_1}X_c(j\omega)\,, \qquad |\omega| < \pi/T_1 \qquad (1.80\text{a})$$

or equivalently

$$X_d(e^{j\Omega}) = \frac{1}{T_1}X_c\left(j\frac{\Omega}{T_1}\right)\,, \qquad |\Omega| < \pi\,. \qquad (1.80\text{b})$$

Note that $X_d(e^{j\Omega})$ is extended periodically outside the interval $|\Omega| < \pi$.

The ideal D/C converter in Figure 1.5 is defined through the interpolation relation

$$y_c(t) = \sum_n y_d[n]\frac{\sin\left(\pi\,(t - nT_2)/T_2\right)}{\pi(t - nT_2)/T_2}\,, \qquad (1.81)$$

which shows that $y_c(nT_2) = y_d[n]$. Since each term in the above sum is bandlimited to $|\omega| < \pi/T_2$, the CT signal $y_c(t)$ is also bandlimited to this frequency range, so this D/C converter is more completely referred to as the ideal bandlimited interpolating converter. The C/D converter in Figure 1.5, under the assumption Eq. (1.79), is similarly characterized by the fact that the CT signal $x_c(t)$ is the ideal bandlimited interpolation of the DT sequence $x_d[n]$.

Because $y_c(t)$ is bandlimited and $y_c(nT_2) = y_d[n]$, analogous relations to Eq. (1.80) hold between the DTFT of $y_d[n]$ and the CTFT of $y_c(t)$:

$$Y_d(e^{j\Omega})\bigg|_{\Omega=\omega T_2} = \frac{1}{T_2}Y_c(j\omega)\,, \qquad |\omega| < \pi/T_2 \qquad (1.82\text{a})$$

or equivalently

$$Y_d(e^{j\Omega}) = \frac{1}{T_2}Y_c\left(j\frac{\Omega}{T_2}\right)\,, \qquad |\Omega| < \pi\,. \qquad (1.82\text{b})$$

Figure 1.6 shows one conceptual representation of the ideal D/C converter. This figure interprets Eq. (1.81) to be the result of evenly spacing a sequence of impulses at intervals of $T_2$—the reconstruction interval—with impulse strengths given by the $y_d[n]$, then filtering the result by an ideal low-pass filter $L(j\omega)$ with gain $T_2$ in the passband $|\omega| < \pi/T_2$. This operation

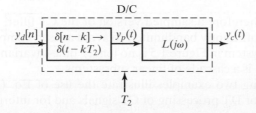

**Figure 1.6**  Conceptual representation of processes that yield ideal D/C conversion, interpolating a DT sequence into a bandlimited CT signal using reconstruction interval $T_2$.

produces the bandlimited CT signal $y_c(t)$ that interpolates the specified sequence values $y_d[n]$ at the instants $t = nT_2$, that is, $y_c(nT_2) = y_d[n]$.

## 1.5.2 DT Filtering and Overall CT Response

We now assume, unless stated otherwise, that $T_1 = T_2 = T$. If in Figure 1.5 the bandlimiting constraint of Eq. (1.79) is satisfied, and if we set $y_d[n] = x_d[n]$, then $y_c(t) = x_c(t)$. More generally, when the DT system in Figure 1.5 is an LTI DT filter with frequency response $H_d(e^{j\Omega})$, so

$$Y_d(e^{j\Omega}) = H_d(e^{j\Omega})X_d(e^{j\Omega}) , \qquad (1.83)$$

and provided any aliased components of $x_c(t)$ are eliminated by $H_d(e^{j\Omega})$, then assembling Eqs. (1.80), (1.82), and (1.83) yields

$$Y_c(j\omega) = H_d(e^{j\Omega})\Big|_{\Omega=\omega T} X_c(j\omega) , \quad |\omega| < \pi/T . \qquad (1.84)$$

The action of the overall system is thus equivalent to that of a CT filter whose frequency response is

$$H_c(j\omega) = H_d(e^{j\Omega})\Big|_{\Omega=\omega T} , \qquad |\omega| < \pi/T . \qquad (1.85)$$

In other words, under the bandlimiting and sampling rate constraints mentioned above, the overall system behaves as an LTI CT filter, and the response of this filter is related to that of the embedded DT filter through a simple frequency scaling. The sampling rate can be lower than the Nyquist rate, provided that the DT filter eliminates any aliased components.

If we wish to use the system in Figure 1.5 to implement a CT LTI filter with frequency response $H_c(j\omega)$, we choose $H_d(e^{j\Omega})$ according to Eq. (1.85), provided that $x_c(t)$ is appropriately bandlimited. If we define $H_c(j\omega) = 0$ for $|\omega| \geq \pi/T$, then Eq. (1.85) also corresponds to the following relation between the DT and CT impulse responses:

$$h_d[n] = T h_c(nT) . \qquad (1.86)$$

The DT filter is therefore a sampled version of the CT filter. When $x_c(t)$ and $H_d(e^{j\Omega})$ are not sufficiently bandlimited to avoid aliased components in $y_d[n]$, then the overall system in Figure 1.5 is no longer time-invariant. It is, however, still linear since it is a cascade of linear subsystems.

The following two examples illustrate the use of Eq. (1.85) as well as Figure 1.5, both for DT processing of CT signals and for interpretation of two important DT systems.

**Example 1.3**    **Digital Differentiator**

In this example we wish to implement a CT differentiator using a DT system in the configuration of Figure 1.5. We need to choose $H_d\left(e^{j\Omega}\right)$ so that $y_c(t) = \frac{dx_c(t)}{dt}$, assuming that $x_c(t)$ is bandlimited to $\pi/T$. The desired overall CT frequency response is therefore

$$H_c(j\omega) = \frac{Y_c(j\omega)}{X_c(j\omega)} = j\omega . \qquad (1.87)$$

Consequently, using Eq. (1.85) we choose $H_d(e^{j\Omega})$ such that

$$H_d(e^{j\Omega})\bigg|_{\Omega=\omega T} = j\omega , \qquad\qquad |\omega| < \frac{\pi}{T} \qquad (1.88a)$$

or equivalently

$$H_d(e^{j\Omega}) = j\Omega/T , \qquad\qquad |\Omega| < \pi . \qquad (1.88b)$$

A DT system with the frequency response in Eq. (1.88b) is commonly referred to as a digital differentiator. To understand the relation between the input $x_d[n]$ and output $y_d[n]$ of the digital differentiator, note that $y_c(t)$ — which is the bandlimited interpolation of $y_d[n]$ — is the derivative of $x_c(t)$, and $x_c(t)$ in turn is the bandlimited interpolation of $x_d[n]$. It follows that $y_d[n]$ can, in effect, be thought of as the result of sampling the derivative of the bandlimited interpolation of $x_d[n]$.

**Example 1.4**    **Half-Sample Delay**

In designing DT systems, a phase factor of the form $e^{-j\alpha\Omega}$, $|\Omega| < \pi$, is often included or required. When $\alpha$ is an integer, this has a straightforward interpretation: it corresponds simply to an integer shift of the time sequence by $\alpha$. When $\alpha$ is not an integer, the interpretation is not as immediate, since a DT sequence can only be directly shifted by integer amounts.

In this example we consider the case of $\alpha = \frac{1}{2}$, referred to as a half-sample delay. To provide an interpretation, we consider the implications of choosing the DT system in Figure 1.5 to have frequency response

$$H_d(e^{j\Omega}) = e^{-j\Omega/2} , \qquad\qquad |\Omega| < \pi . \qquad (1.89)$$

Whether or not $x_d[n]$ explicitly arose by sampling a CT signal, we can associate $x_d[n]$ with its bandlimited interpolation $x_c(t)$ for any specified sampling or reconstruction interval $T$. Similarly, we can associate $y_d[n]$ with its bandlimited interpolation $y_c(t)$ using the reconstruction interval $T$. With $H_d(e^{j\Omega})$ given by Eq. (1.89), the equivalent CT frequency response relating $y_c(t)$ to $x_c(t)$ is

$$H_c(j\omega) = e^{-j\omega T/2} \qquad (1.90)$$

representing a time delay of $T/2$, which is half the sample spacing; consequently, $y_c(t) = x_c(t - T/2)$. We therefore conclude that for a DT system with frequency response given by Eq. (1.89), the DT output $y_d[n]$ corresponds to samples of the half-sample delay of the bandlimited interpolation of the input sequence $x_d[n]$. Note that in this interpretation the choice for the value of $T$ is immaterial.

The preceding interpretation allows us to find the unit sample (or impulse) response of the half-sample delay system through a simple argument. If $x_d[n] = \delta[n]$, then $x_c(t)$ must be the bandlimited interpolation of this (with some $T$ that we could have specified to take any particular value), so

$$x_c(t) = \frac{\sin(\pi t/T)}{\pi t/T} \tag{1.91}$$

and therefore

$$y_c(t) = \frac{\sin\left(\pi(t - (T/2))/T\right)}{\pi(t - (T/2))/T} \tag{1.92}$$

which shows that the desired unit sample response is

$$y_d[n] = h_d[n] = \frac{\sin\left(\pi(n - (1/2))\right)}{\pi(n - (1/2))} \,. \tag{1.93}$$

This discussion of a half-sample delay also generalizes in a straightforward way to any integer or non-integer choice for the value of $\alpha$.

### 1.5.3 Nonideal D/C Converters

In Section 1.5.1 we defined the ideal D/C converter through the bandlimited interpolation formula Eq. (1.81), also illustrated in Figure 1.6, which corresponds to processing a train of impulses with strengths equal to the sequence values $y_d[n]$ through an ideal low-pass filter. A more general class of D/C converters, which includes the ideal converter as a particular case, creates a CT signal $y_c(t)$ from a DT signal $y_d[n]$ according to the following:

$$y_c(t) = \sum_{n=-\infty}^{\infty} y_d[n]\, p(t - nT) \tag{1.94}$$

where $p(t)$ is some selected basic pulse and $T$ is the reconstruction interval or pulse repetition interval. This too can be seen as the result of processing an impulse train of sequence values through a filter, but a filter that has impulse response $p(t)$ rather than that of the ideal low-pass filter. The CT signal $y_c(t)$ is thus constructed by adding together shifted and scaled versions of the basic pulse; the number $y_d[n]$ scales $p(t - nT)$, which is the basic pulse delayed by $nT$. Note that the ideal bandlimited interpolating converter of Eq. (1.81) is obtained by choosing

$$p(t) = \frac{\sin(\pi t/T)}{\pi t/T} \,. \tag{1.95}$$

In Chapter 3, we will discuss the interpretation of Eq. (1.94) as pulse-amplitude modulation (PAM) for communicating DT information over a CT channel.

The relationship in Eq. (1.94) can also be described quite simply in the frequency domain. Taking the CTFT of both sides, denoting the CTFT of $p(t)$

by $P(j\omega)$, and using the fact that delaying a signal by $t_0$ in the time domain corresponds to multiplication by $e^{-j\omega t_0}$ in the frequency domain, we get

$$Y_c(j\omega) = \left( \sum_{n=-\infty}^{\infty} y_d[n] \, e^{-jn\omega T} \right) P(j\omega)$$

$$= Y_d(e^{j\Omega}) \bigg|_{\Omega=\omega T} P(j\omega) . \qquad (1.96)$$

In the particular case where $p(t)$ is the sinc pulse in Eq. (1.95), with transform $P(j\omega)$ that has the constant value $T$ for $|\omega| < \pi/T$ and 0 outside this band, we recover the relation in Eq. (1.82).

In practice, the ideal frequency characteristic can only be approximated, with the accuracy of the approximation often related to cost of implementation. A commonly used simple approximation is the (centered) zero-order hold (ZOH), specified by the choice

$$p_z(t) = \begin{cases} 1 & \text{for } |t| < (T/2) \\ 0 & \text{elsewhere.} \end{cases} \qquad (1.97)$$

This D/C converter holds the value of the DT signal at time $n$, namely the value $y_d[n]$, for an interval of length $T$ centered at $nT$ in the CT domain, as illustrated in Figure 1.7. The centered ZOH is of course noncausal, but is easily replaced with the noncentered causal ZOH, for which the basic pulse is

$$p_{z'}(t) = \begin{cases} 1 & \text{for } 0 \le t < T \\ 0 & \text{elsewhere.} \end{cases} \qquad (1.98)$$

Such ZOH converters are commonly used.

Another common choice is a centered first-order hold (FOH), for which the basic pulse $p_f(t)$ is triangular as shown in Figure 1.8. Use of the FOH represents linear interpolation between the sequence values. Of course, the use of the ZOH and FOH will not be equivalent to exact bandlimited interpolation as required by the Nyquist sampling theorem. The transform of the centered ZOH pulse is

$$P_z(j\omega) = T \frac{\sin(\omega T/2)}{\omega T/2} \qquad (1.99)$$

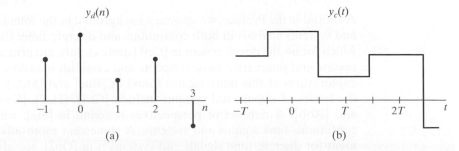

**Figure 1.7**  A centered zero-order hold (ZOH): (a) DT sequence; (b) the result of applying the centered ZOH to (a).

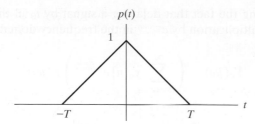

**Figure 1.8**   Basic pulse $p_f(t)$ for centered first-order hold (FOH).

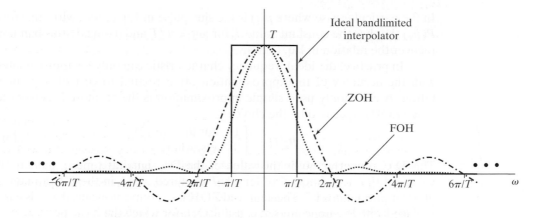

**Figure 1.9**   The Fourier transform amplitudes of the ideal bandlimited interpolator, the ZOH, and the FOH.

and that of the centered FOH pulse is

$$P_f(j\omega) = T\left(\frac{\sin(\omega T/2)}{\omega T/2}\right)^2 . \tag{1.100}$$

The Fourier transform amplitudes of the ideal bandlimited interpolator, the ZOH, and the FOH are shown in Figure 1.9.

## 1.6 FURTHER READING

As noted in the Preface, we assume a background in the foundations of signals and systems analysis in both continuous and discrete time. Chapters 1 and 2, which follow the development in [Op1] quite closely, are primarily intended to review and summarize basic concepts and establish notation. Computational explorations of this material are found in [Buc] and [McC]. Other texts on the basics of signals and systems include [Ch1], [Ha1], [Kwa], [La1], [Phi], and [Rob]. A rich set of perspectives is found in [Sie], which emphasizes continuous-time signals and systems. A somewhat more advanced development for discrete-time signals and systems is in [Op2], see also [Mit], [Ma1], [Pra] and [Pr1]. The geometric treatment in [Vet] exploits the view of signals as Hilbert-space vectors. Classic and fairly advanced books on signal analysis and Fourier transforms are [Bra], [Gui], [Pa1], [Pa2], [Pa3], all of which offer useful viewpoints. The treatment of Fourier theory in [Cha] is concise and illuminating.

## Problems

## Basic Problems

**1.1.** A simple physical model for the motion of a certain electric vehicle along a track is given by the following differential equation, with the position of the vehicle denoted by $y(t)$:

$$\frac{d^2 y(t)}{dt^2} = -\left(\frac{c_f}{m}\right)\frac{dy(t)}{dt} - \left(\frac{c_b}{m}\right)\frac{dy(t)}{dt} x_b(t) + x_a(t) ,$$

where $x_b(t)$ is the braking force applied to the wheels; $x_a(t)$ is the acceleration provided by the electric motor; $m$ is the mass of the car; and $c_f$ and $c_b$ are frictional constants for the vehicle and brakes, respectively. Assume that we have the constraint $x_b \geq 0$, but that $x_a$ can be positive or negative.

**(a)** Is the model linear? That is, do its nonzero solutions obey the superposition principle? Is the model time-invariant?

**(b)** How do your answers change if the braking force $x_b(t)$ is identically zero?

**1.2. (a)** Suppose the input signal to a stable LTI system with system function $H(s)$ is constant at some value $\alpha$ for all time $t$. What is the corresponding output at each $t$?

**(b)** Denote by $y(t)$ the output signal obtained from the system in (a) when the input to it is the signal $x(t) = t$ for all time. Now obtain two distinct expressions for the output corresponding to the input $t - \alpha$, where $\alpha$ is an arbitrary constant. *Hint*: Invoke the linearity and time invariance of the system, and use your result from (a). By choosing $\alpha$ appropriately, deduce that $y(t) = bt + y(0)$ for some constant $b$. Express $b$ in terms of $H(s)$.

**1.3.** Indicate whether the systems below satisfy the following system properties: linearity, time invariance, causality, and BIBO stability.

**(a)** A system with input $x(t)$ and output $y(t)$, with input-output relation

$$y(t) = x^4(t), \quad -\infty < t < \infty .$$

**(b)** A system with input $x[n]$ and output $y[n]$, and input-output relation

$$y[n] = \begin{cases} 0 & n \leq 0 \\ y[n-1] + x[n] & n > 0 . \end{cases}$$

**(c)** A system with input $x(t)$ and output $y(t)$, with input-output relation

$$y(t) = x(4t + 3) \quad -\infty < t < \infty .$$

**(d)** A system with input $x(t)$ and output $y(t)$, with input-output relation

$$y(t) = \int_{-\infty}^{\infty} x(\tau)\, d\tau \quad -\infty < t < +\infty .$$

**1.4.** We are given a certain LTI system with impulse response $h_0(t)$, and are told that when the input is $x_0(t)$, the output $y_0(t)$ is the waveform shown in Figure P1.4.

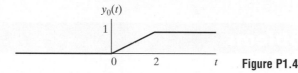

**Figure P1.4**

We are then given the following set of inputs $x(t)$ to LTI systems with the indicated impulse responses $h(t)$:

| | Input $x(t)$ | Impulse response $h(t)$ |
|---|---|---|
| (a) | $x(t) = 2x_0(t)$ | $h(t) = h_0(t)$ |
| (b) | $x(t) = x_0(t) - x_0(t-2)$ | $h(t) = h_0(t)$ |
| (c) | $x(t) = x_0(t-2)$ | $h(t) = h_0(t+1)$ |
| (d) | $x(t) = x_0(-t)$ | $h(t) = h_0(t)$ |
| (e) | $x(t) = x_0(-t)$ | $h(t) = h_0(-t)$ |
| (f) | $x(t) = \frac{dx_0(t)}{dt}$ | $h(t) = \frac{dh_0(t)}{dt}$ |

In each of these cases, determine whether or not we have enough information available to determine the output $y(t)$ uniquely. If it is possible to determine $y(t)$ uniquely, provide an analytical expression for it and a sketch of it. In those cases where you believe it is not possible to find $y(t)$ uniquely, see if you can prove that this is not possible.

**1.5.** **(a)** Consider an LTI system with input $x(t)$ and output $y(t)$ related through the equation

$$y(t) = \int_{-\infty}^{t} e^{-(t-\tau)}x(\tau - 2)\, d\tau \ .$$

What is the impulse response $h(t)$ for this system?
**(b)** Determine the response of this system when the input $x(t)$ is as shown in Figure P1.5-1.

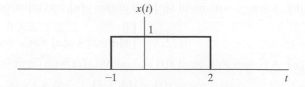

**Figure P1.5-1**

**(c)** Consider the interconnection of LTI systems shown in Figure P1.5-2. Here $h(t)$ is as in part (a). Determine the output $w(t)$ when the input $x(t)$ is the same as in part (b). Do this using the result of part (b), together with the properties of convolution; do *not* directly evaluate a convolution integral.

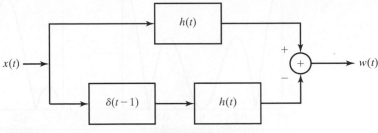

**Figure P1.5-2**

**1.6.** For each of the following pairs of signals, use convolution to find the response $y(t)$ of the LTI system with impulse response $h(t)$ to the input $x(t)$. Sketch your result.

**(a)** $x(t) = e^{-3t}u(t), h(t) = u(t-1)$.

**(b)** The signals are as shown in Figure P1.6.

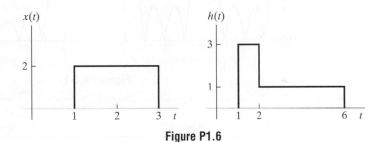

**Figure P1.6**

**1.7. (a)** Determine the Fourier transform of the sequence

$$r[n] = \begin{cases} 1, & 0 \le n \le M \\ 0, & \text{otherwise.} \end{cases}$$

Note that $M$ might not be an even number here. Sketch the magnitude of $R(e^{j\Omega})$.

**(b)** Consider the sequence

$$w[n] = \begin{cases} \dfrac{1}{2}\left(1 + \cos\dfrac{2\pi n}{M}\right), & 0 \le n \le M \\ 0, & \text{otherwise.} \end{cases}$$

Express the DTFT of $w[n]$ in terms of the DTFT of $r[n]$.

**1.8.** Figure P1.8-1 shows four DTFT magnitudes, numbered 1 through 4. Figure P1.8-2 shows the four associated DT signals, labeled A through D, but arranged in random order.

**(a)** Match the DTFTs with the signals. Explain why you choose each combination.

**(b)** The scales are missing from DTFT numbers 1 and 4. What should the scales or the maximum magnitude on the plot be and why?

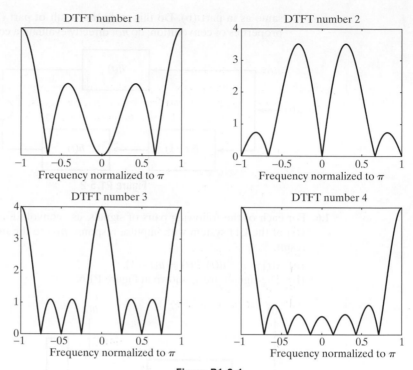

**Figure P1.8-1**

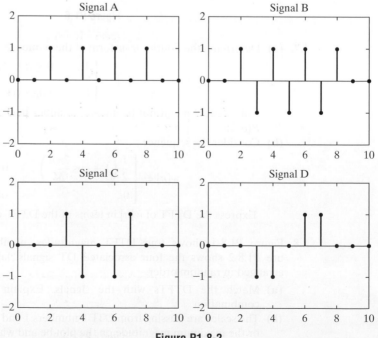

**Figure P1.8-2**

**1.9.** The Fourier transform of a DT signal $x[n]$ is

$$X(e^{j\Omega}) = e^{j2\Omega}\,(1 - e^{-j3\Omega})\,.$$

Completely specify $x[n]$.

**1.10.** For each of the signals in (a)–(g) below, determine which of the properties (1)–(6) listed here is satisfied by its Fourier transform. Compute as little as possible; instead invoke transform properties as necessary.

**(1)** $\Re e\{X(e^{j\Omega})\} = 0$.

**(2)** $\Im m\{X(e^{j\Omega})\} = 0$.

**(3)** There exists a real $\alpha$ such that $e^{j\alpha\Omega}X(e^{j\Omega})$ is real.

**(4)** $\int_{-\pi}^{\pi} X(e^{j\Omega})\,d\Omega = 0$.

**(5)** $X(e^{j\Omega})$ is periodic.

**(6)** $X(e^{j\Omega})\big|_{\Omega=0} = 0$.

**(a)** $x[n]$ in Figure P1.10-1.

**(b)** $x[n]$ in Figure P1.10-2.

**(c)** $x[n]$ in Figure P1.10-3.

**(d)** $x[n]$ in Figure P1.10-4.

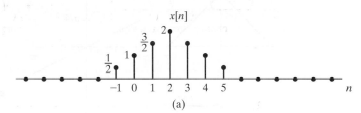

(a)

**Figure P1.10-1**

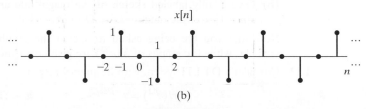

(b)

**Figure P1.10-2**

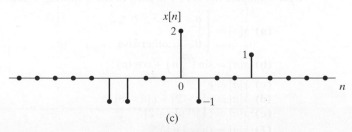

(c)

**Figure P1.10-3**

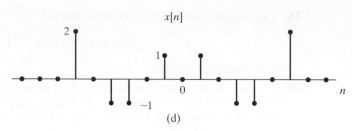

Figure P1.10-4

**(e)** $x[n] = \left(\frac{1}{2}\right)^n u[n]$.

**(f)** $x[n] = \delta[n-1] + \delta[n+2]$.

**(g)** $x[n] = \delta[n-1] + \delta[n+1]$.

**1.11.** Let $X(e^{j\Omega})$ be the Fourier transform of the DT signal $x[n]$. The magnitude and phase of $X(e^{j\Omega})$ are shown in Figure P1.11 for $|\Omega| < \pi$.

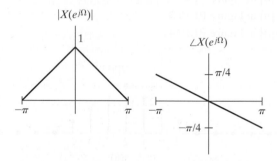

Figure P1.11

**(a)** Give a fully labeled sketch of the magnitude and phase of the DTFT of $x[-n]$ over the range $-2\pi < \Omega < 2\pi$.

**(b)** Give a fully labeled sketch of the magnitude and phase of the DTFT of $x[n-1]$ over the range $-2\pi < \Omega < 2\pi$.

Note that you are being asked to sketch the graphs over the range $-2\pi < \Omega < 2\pi$, while the graphs of $X(e^{j\Omega})$ are given over the range $-\pi < \Omega < \pi$.

**1.12.** Consider a DT LTI system with frequency response

$$H(e^{j\Omega}) = e^{-j3\Omega/2} , \qquad -\pi < \Omega < \pi .$$

Determine the output $y[n]$ if the input is $x[n] = \cos(\frac{4\pi}{3}n + \frac{\pi}{4})$.

**1.13.** Compute the DTFT of each of the following signals for $-\pi < \Omega \leq \pi$:

**(a)** $x[n] = \begin{cases} n, & -3 \leq n \leq 3 \\ 0 & \text{otherwise} \end{cases}$

**(b)** $x[n] = \sin\left(\frac{\pi}{2}n\right) + \cos(n)$

**(c)** $x[n] = e^{-2|n|}$

**(d)** $x[n] = u[n-2] - u[n-6]$

**(e)** $x[n] = (\frac{1}{3})^{|n|} u[n+2]$

**(f)** $x[n] = (n-1)(\frac{1}{3})^{|n|}$.

**1.14.** Compute the CTFT of each of the following signals:

   **(a)** $x(t) = e^{-2(t-1)}u(t-1)$;

   **(b)** $x(t) = e^{-|t|}$;

   **(c)** $x(t) = [e^{-\alpha t}\cos(\omega_0 t)]u(t)$.

**1.15.** Parseval's relation for CT signals states that

$$\int_{-\infty}^{+\infty} |x(t)|^2 \, dt = \frac{1}{2\pi} \int_{-\infty}^{+\infty} |X(j\omega)|^2 \, d\omega \; .$$

This says that the total energy of the signal can be obtained by integrating $|X(j\omega)|^2$ over all frequencies. Now consider a real-valued signal $x(t)$ processed by the ideal bandpass filter $H(j\omega)$ shown in Figure P1.15.

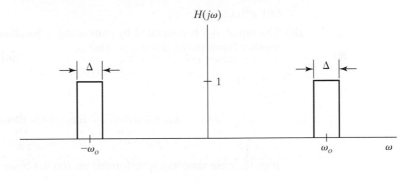

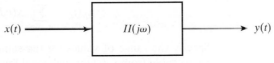

**Figure P1.15**

   **(a)** Express the energy in the output signal $y(t)$ as an integration over frequency of $|X(j\omega)|^2$.

   **(b)** For $\Delta$ sufficiently small so that $|X(j\omega)|$ is approximately constant over a frequency interval of width $\Delta$, show that the energy of the output $y(t)$ of the bandpass filter is approximately proportional to $\Delta|X(j\omega_0)|^2$.

On the basis of the foregoing result, $\Delta|X(j\omega_0)|^2$ is proportional to the energy of the signal in a bandwidth $\Delta$ around the frequency $\omega_0$. For this reason, $|X(j\omega)|^2$ is referred to as the energy spectral density of the signal $x(t)$.

   **(c)** Now consider a DT signal $x[n]$. Suppose we know that the DTFT $X(e^{j\Omega})$ of the deterministic signal $x[n]$ has a magnitude of 2 for $|\Omega| < 0.4\pi$, and unknown magnitude for $0.4\pi \leq |\Omega| \leq \pi$. This signal is applied to the input of an ideal low-pass filter whose frequency response $H(e^{j\Omega})$ is 3 in the interval $|\Omega| < 0.25\pi$ and is 0 for $0.25\pi \leq |\Omega| \leq \pi$. What is the energy $\sum y^2[n]$ of the output signal $y[n]$?

**1.16.** For each of the following, indicate whether the statement is true or false, and justify your conclusion for each.

**Statement:**

Exact reconstruction of a CT signal from its samples requires that the sampling frequency be greater than twice the highest frequency.

**Statement:**

Exact reconstruction of a CT signal from its samples is always possible if the sampling frequency is greater than twice the highest frequency.

**Statement:**

Exact reconstruction of a CT signal from its samples is always possible if the sampling frequency is greater than or equal to twice the highest frequency.

**1.17.** **(a)** Let $x(t)$ be a signal with Nyquist rate $\omega_c$. Determine the Nyquist rate for each of the following signals:

   (i)  $x(t) + 3x(t-4)$

   (ii) $x(t) * \frac{\sin(3000\pi t)}{\pi t}$

   (iii) $x^2(t)$.

**(b)** The signal $x(t)$ is generated by convolving a bandlimited signal $x_1(t)$ with another bandlimited signal $x_2(t)$, that is,

$$x(t) = x_1(t) * x_2(t) ,$$

where

$$X_1(j\omega) = 0 \qquad \text{for} \quad |\omega| > 1000\pi ,$$

$$X_2(j\omega) = 0 \qquad \text{for} \quad |\omega| > 2000\pi .$$

Impulse-train sampling is performed on $x(t)$ to obtain

$$x_p(t) = \sum_{n=-\infty}^{+\infty} x(nT)\delta(t - nT) .$$

Specify the range of values for the sampling period $T$ that ensures $x(t)$ is recoverable from $x_p(t)$ through ideal low-pass filtering.

**1.18.** Consider a general D/C converter of the form described in Section 1.5.3. Suppose

$$y_d[n] = \frac{\sin(\pi n/2)}{\pi n} .$$

Make fully labeled sketches of what $y_c(t)$ and $Y_c(j\omega)$ look like for:

**(a)** an ideal bandlimited interpolating D/C converter;

**(b)** a centered zero-order-hold D/C converter; and

**(c)** a linear interpolating D/C converter.

**1.19.** Consider the system shown in Figure P1.19-1 for filtering a CT signal using a DT filter. Here $x[n] = x_c(nT)$, $y[n] = y_c(nT)$, and $L_T(j\omega)$ is an ideal low-pass filter with cutoff frequency $\pi/T$ and gain $T$.

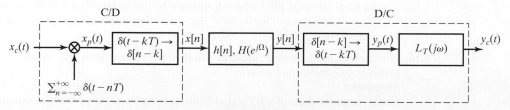

**Figure P1.19-1**

If $X_c(j\omega)$ and $H(e^{j\Omega})$ are as shown in Figure P1.19-2, and if $T$ is chosen to provide sampling of $x_c(t)$ at the Nyquist rate, sketch $X_p(j\omega)$, $X(e^{j\Omega})$, $Y(e^{j\Omega})$, $Y_p(j\omega)$, and $Y_c(j\omega)$.

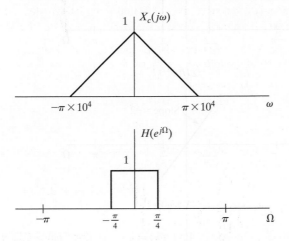

**Figure P1.19-2**

**1.20.** In the system shown in Figure P1.20-1, the magnitude and phase of $X_c(j\omega)$ and $H(e^{j\Omega})$ are shown in Figure P1.20-2. Draw neat sketches of the magnitude and phase of the following functions, with all relevant amplitudes and slopes labeled:

**(a)** $X(e^{j\Omega})$ as a function of $\Omega$ for $-2\pi < \Omega < 2\pi$;

**(b)** $Y(e^{j\Omega})$ as a function of $\Omega$ for $-2\pi < \Omega < 2\pi$;

**(c)** $Y_c(j\omega)$ as a function of $\omega$.

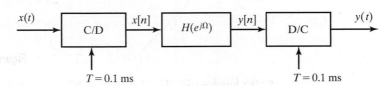

**Figure P1.20-1**

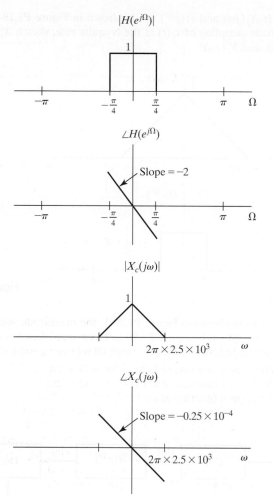

**Figure P1.20-2**

**1.21.** Figure P1.21-1 shows a CT filter that is implemented using an LTI DT filter with frequency response $H(e^{j\Omega})$.

**(a)** If the CTFT of $x_c(t)$, namely $X_c(j\omega)$, is as shown in Figure P1.21-2 and $\Omega_c = \frac{\pi}{5}$, sketch and label $X(e^{j\Omega})$, $Y(e^{j\Omega})$, and $Y_c(j\omega)$ for each of the following cases:

(i)   $\frac{1}{T_1} = \frac{1}{T_2} = 2 \times 10^4$

(ii)  $\frac{1}{T_1} = 4 \times 10^4$,    $\frac{1}{T_2} = 10^4$

(iii) $\frac{1}{T_1} = 10^4$,    $\frac{1}{T_2} = 3 \times 10^4$

**(b)** For $\frac{1}{T_1} = \frac{1}{T_2} = 6 \times 10^3$, and for input signals $x_c(t)$ whose spectra are band-limited to $|\omega| < 2\pi \times 5 \times 10^3$ (but otherwise unconstrained), what is the maximum choice of the cutoff frequency $\Omega_c$ of the filter $H(e^{j\Omega})$ for which the overall system is LTI? For this maximum choice of $\Omega_c$, specify $H_c(j\omega)$.

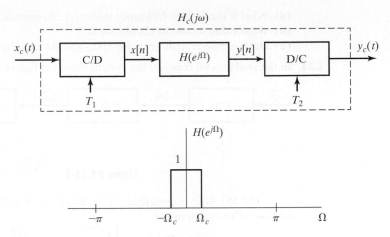

**Figure P1.21-1**

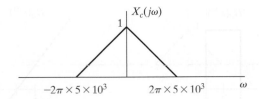

**Figure P1.21-2**

**1.22.** Consider our standard system for DT processing of CT signals, where the C/D converter samples the CT signal $x_c(t)$ with a sampling interval of $T$ seconds, while the ideal D/C converter at the output produces a bandlimited interpolation of the samples $y[n]$ using a reconstruction interval of $T$ seconds. Suppose the LTI DT system between these two converters is a notch filter, i.e., has a frequency response $H(e^{j\Omega})$ whose value is 0 at $\Omega = \pm\Omega_o$ (where $\Omega_o > 0$ is termed the notch frequency) and whose value is nonzero everywhere else in the interval $|\Omega| < \pi$. Suppose the input signal is of the form

$$x_c(t) = \cos(\omega_{in}t + \theta) \, .$$

Determine all values of $\omega_{in}$ for which the output $y_c(t)$ will be identically 0.

**1.23.** A bandlimited CT signal $x_c(t)$ is known to contain a 60-Hz component, which we want to remove using the system in Figure P1.23. Suppose $T = 2 \times 10^{-4}$ s and

$$H(e^{j\Omega}) = \frac{(1 - e^{-j(\Omega-\Omega_0)})(1 - e^{-j(\Omega+\Omega_0)})}{(1 - 0.5e^{-j(\Omega-\Omega_0)})(1 - 0.5e^{-j(\Omega+\Omega_0)})} \, .$$

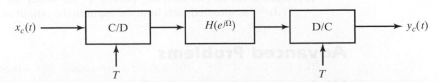

**Figure P1.23**

(a) What is the highest frequency that $x_c(t)$ can contain to avoid aliasing?

(b) What value should be chosen for $\Omega_0$?

(c) Draw a pole-zero diagram for $H(z)$ and sketch $|H(e^{j\Omega})|$.

1.24. In this problem we consider the system shown in Figure P1.24-1.

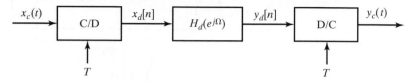

Figure P1.24-1

The DT filter is described on the interval $-\pi \leq \Omega \leq \pi$ by Figure P1.24-2 and the following equation:

$$H_d(e^{j\Omega}) = \begin{cases} e^{-j\Omega/3} & |\Omega| \leq \pi/3 \\ 0 & \pi/3 < |\Omega| \leq \pi \end{cases}$$

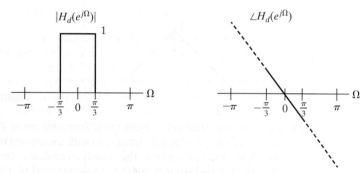

Figure P1.24-2

(a) Assume that we are sampling with period $T = \frac{1}{2} \times 10^{-6}$ sec. Suppose the input to the system is $x_c(t) = \cos(\omega_0 t)$. What is the output $y_c(t)$ for each of the following cases:

  (i) $\omega_0 = \frac{\pi}{2} \times 10^6$ rad/sec;

  (ii) $\omega_0 = \pi \times 10^6$ rad/sec;

  (iii) $\omega_0 = \frac{7}{2}\pi \times 10^6$ rad/sec.

Suppose now that the input $x_c(t)$ is bandlimited such that $X_c(j\omega) = 0$ for $|\omega| \geq 2\pi \times 10^6$ rad/sec.

(b) What is the largest sampling period, $T$, for which the output is identical to the input (except for a possible time shift $t_0$), i.e., $y_c(t) = x_c(t - t_0)$?

(c) What is the largest sampling period, $T$, for which the given system (with $H_d(e^{j\Omega})$ as specified) acts from the input to the output as an LTI CT system?

# Advanced Problems

1.25. Note: This problem requires utilizing an appropriate computational package. Consider the simple DT difference equation

$$x[n+1] = rx[n](1 - x[n]), \quad n \geq 0.$$

Assume $0 < x[0] < 1$ and $r \geq 0$.

**(a)** Is this system linear? Time-invariant?

**(b)** If $x[0] = 1/2$, for what values of $r$ is $x[n]$ bounded?

**(c)** Using an appropriate computational package, compute and plot $x[n]$ for different values of $r$, including $r = k/4, k = 1,2,\ldots,15$. Do the plots support your answer for (b)?

**1.26. (a)** The input $x(t)$ and output $y(t)$ of an LTI system satisfy the differential equation

$$4\frac{dy(t)}{dt} - 2y(t) = 3x(t) \,.$$

(i) Show that the input-output pairs

$$x(t) = \delta(t) \,, \qquad y(t) = \frac{3}{4}e^{0.5t}u(t)$$

and

$$x(t) = \delta(t) \,, \qquad y(t) = -\frac{3}{4}e^{0.5t}u(-t)$$

both satisfy the differential equation.

(ii) Specify *all* outputs $y(t)$ that satisfy the differential equation when $x(t) = \delta(t)$, and explicitly verify that the two special cases in (i) are included in your answer. *Hint:* Recall that the general solution to a linear differential equation is a particular solution plus a homogeneous solution. Also note that in general, a solution $y(t)$ does not need to have a Laplace transform, in fact, the only two solutions to this problem that do have Laplace transforms are the ones you found in part (i).

(iii) If the system is causal, what is its impulse response?

(iv) If the system is stable, what is its impulse response?

**(b)** The input and output of an LTI system satisfy the difference equation

$$y[n] - 3y[n - 1] = 2x[n] \,.$$

(i) Assuming the system is causal, determine its impulse response, i.e., its unit sample response. Is the system BIBO stable?

(ii) Assuming instead that the system is stable, determine its impulse response. Is the system causal?

**1.27.** Consider the causal cascade interconnection of three LTI systems in Figure P1.27-1.

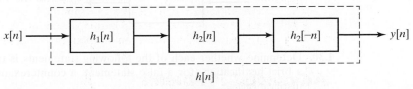

**Figure P1.27-1**

The impulse response $h_2[n]$ is given by $h_2[n] = u[n] - u[n - 2]$, and the overall impulse response $h[n]$ is as shown in Figure P1.27-2.

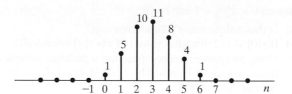

**Figure P1.27-2**

(a) Find the impulse response $h_1[n]$.

(b) Find the response of the overall system to the input

$$x[n] = \delta[n] - \delta[n - 2] \,.$$

**1.28.** If the response of a CT LTI system to the unit step $u(t)$ is $g(t)$, what is its response $y(t)$ to a general input $x(t)$? What condition on $g(t)$ is necessary and sufficient for the system to be BIBO stable?

**1.29.** For each of the DT systems $S_1$, $S_2$, $S_3$, and $S_4$ shown in Figure P1.29, the indicated input-output pair represents the results of one experiment with the corresponding system. Decide whether the output $y[n]$ and input $x[n]$ of this system definitely cannot, possibly could, or must satisfy a convolution relationship of the form $y[n] = h[n] * x[n]$ for some appropriate impulse response $h[n]$. Choose the statement that applies and explain your reasoning. In each case where your answer is "possibly could" or "must," determine an impulse response $h[n]$, frequency response $H(e^{j\Omega})$, or system function $H(z)$ that would account for the given input-output pair.

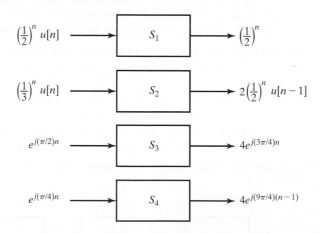

**Figure P1.29**

**1.30.** Determine whether each of the following statements is true or false, and give a brief justification (for a false statement, a counterexample will be adequate justification).

(i) If $e^{jt} + e^{j3t}$ is an eigenfunction of a CT system, then the system cannot be LTI.

(ii) If $h(t)$ is the nonzero impulse response of a stable LTI system, then $h(t)$ can be periodic.

(iii) There always exists a causal inverse of a causal and stable LTI system, although this inverse may not be stable.

**1.31.** Consider a real DT LTI system with frequency response

$$H(e^{j\Omega}) = \frac{1 + ae^{-j2\Omega} + 2e^{-j4\Omega}}{1 + be^{-j2\Omega}}.$$

Determine the constants $a$ and $b$, given the input/output pairs in Figure P1.31.

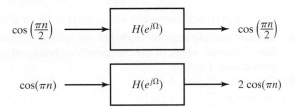

$$\cos\left(\frac{\pi n}{2}\right) \longrightarrow \boxed{H(e^{j\Omega})} \longrightarrow \cos\left(\frac{\pi n}{2}\right)$$

$$\cos(\pi n) \longrightarrow \boxed{H(e^{j\Omega})} \longrightarrow 2\cos(\pi n)$$

**Figure P1.31**

**1.32. (a)** A particular DT system maps its input signal $x[n]$ to the output signal $y[n]$. When the input is

$$x[n] = (-1)^n \quad \text{for all } n,$$

the output is

$$y[n] = 1 \quad \text{for all } n.$$

When the input is

$$x[n] = (-1)^{n+1} \quad \text{for all } n,$$

the output is again

$$y[n] = 1 \quad \text{for all } n.$$

(i) Could the system be linear? Explain.
(ii) Could the system be time-invariant? Explain.

**(b)** A particular DT system maps the input signal $x[n]$ to the output signal $y[n]$. When the input is

$$x[n] = (-1)^n \quad \text{for all } n,$$

the output is

$$y[n] = 1 \quad \text{for all } n.$$

When the input is

$$x[n] = (-1)^{n+1} \quad \text{for all } n,$$

the output is

$$y[n] = -1 \quad \text{for all } n.$$

Answer the same questions as in (i) and (ii) of (a) above.

**1.33.** A DT LTI system has frequency response

$$H(e^{j\Omega}) = \begin{cases} e^{-j\Omega 3} & |\Omega| < \dfrac{\pi}{5} \\[2mm] 0 & \dfrac{\pi}{5} \le |\Omega| \le \pi \end{cases}$$

The input to the system is the periodic unit sample "train" with period $N = 12$:

$$x[n] = \sum_{k=-\infty}^{\infty} \delta[n + 12k] .$$

Find the output $y[n]$ of the system.

**1.34.** Consider a DT LTI system with frequency response defined by

$$H_d(e^{j\Omega}) = |\Omega| \, e^{-j g(\Omega)} \quad \text{for } |\Omega| < \pi .$$

The function $g(\Omega)$ is known to be either $g(\Omega) = \Omega^2$ or $g(\Omega) = \Omega^3$.

(a) The unit sample response $h_d[n]$ of the system is known to be real. Use this fact to deduce which of the two possible choices of $g(\Omega)$ listed above is the correct one. Explain.

(b) Determine $\sum_{n=-\infty}^{\infty} h_d[n]$.

(c) Determine $\sum_{n=-\infty}^{\infty} h_d^2[n]$.

**1.35.** Consider an LTI system whose impulse response $h(t)$ is real, and whose associated system function is $H(s)$. Suppose throughout this problem that the input is $x(t) = e^{-3t} \cos t$, for all $t$, and that the corresponding output $y(t)$ is well defined for this input.

(a) If you were allowed to determine the value of $H(s)$ for only a single value of $s$, which value of $s$ would you pick in order to obtain an explicit time-domain expression for the output $y(t)$ corresponding to the above input? Now write down such an expression for $y(t)$ in terms of $H(s)$ evaluated at the $s$ that you selected.

(b) Suppose it is known that $y(0) = 0$ and $\dot{y}(0) = 1$. Reduce your expression for $y(t)$ in part (a) to the form $y(t) = e^{-3t}(A \cos t + B \sin t)$, and determine the constants $A$ and $B$.

**1.36.** Consider the causal LTI system described by the differential equation

$$\frac{d^2 y(t)}{dt^2} + 6\frac{dy(t)}{dt} + 9y(t) = \frac{d^2 x(t)}{dt^2} + 3\frac{dx(t)}{dt} + 2x(t) .$$

The inverse of this system is described by a differential equation. Find the differential equation describing the inverse. Also, find the impulse responses $h(t)$ and $g(t)$ of the original system and its causal inverse.

**1.37.** Suppose we want to design a DT LTI system which has the property that if the input is

$$x[n] = \left(\frac{1}{2}\right)^n u[n] - \frac{1}{4} \left(\frac{1}{2}\right)^{(n-1)} u[n-1] ,$$

then the output is

$$y[n] = \left(\frac{1}{3}\right)^n u[n] .$$

(a) Find the impulse response and frequency response of a DT LTI system that has the desired property.

(b) Find a difference equation relating the system input and output.

**1.38.** A DT LTI system has frequency response

$$H(e^{j\Omega}) = \frac{4 - 9e^{-j\Omega}}{4\cos\Omega + 2e^{-j\Omega} - 9} .$$

**(a)** The rational system function $H(z)$ corresponding to the frequency response above can be written as

$$H(z) = \frac{\alpha_1}{z - \frac{1}{2}} + \frac{\alpha_2}{z - 4}.$$

Find $\alpha_1$ and $\alpha_2$ and specify the ROC.

**(b)** Is the system causal?

**(c)** What is the system's impulse response $h[n]$?

**(d)** Suppose the input to the system is the signal $x[n] = 3^n$ for all $n$. What is the corresponding output, $y[n]$?

**(e)** Suppose the input to the system is the signal $x[n] = 3^n u[n]$, where $u[n]$ is the unit step function. As $n \to -\infty$, $y[n] \to \beta z_0^n$. What is $z_0$?

**1.39.** A CT LTI system with a rational system function $H(s)$ has the $s$-plane plot shown in Figure P1.39, where the ROC is $1 < Re\{s\} < 5$. The poles are at 1 and 5, and the zeros are at 0 and 6. We also know that $H(3) = 9$. We would like to find an inverse system $G(s)$ so that when $H(s)$ and $G(s)$ are cascaded, the overall output $y(t)$ equals the input $x(t)$.

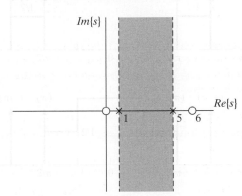

**Figure P1.39**

**(a)** Write the rational system function $H(s)$ for this system.

**(b)** Does a causal inverse exist? If so, write down its system function $G(s)$ and specify its ROC. If not, why not?

**(c)** Does a BIBO stable inverse exist? If so, write down its system function $G(s)$ and specify its ROC. If not, why not?

**(d)** Does a two-sided inverse exist? If so, write down its system function $G(s)$ and specify its ROC. If not, why not?

**1.40.** Suppose we are given a real and finite-energy (but otherwise arbitrary) DT signal $w[n]$, with associated DTFT $W(e^{j\Omega})$. We want to approximate $w[n]$ by another real, finite-energy DT signal $y[n]$ that is bandlimited to the frequency range $|\Omega| < \pi/4$; so $Y(e^{j\Omega})$ is zero for $|\Omega| \geq \pi/4$. Apart from this constraint on its bandwidth, we are free to choose $y[n]$ as needed to get the best approximation.

Suppose we measure the quality of approximation by the following sum-of-squared-errors criterion:

$$\mathcal{E} = \sum_{n=-\infty}^{\infty} (w[n] - y[n])^2.$$

Our problem is then to minimize $\mathcal{E}$ by appropriate choice of the bandlimited $y[n]$, given the signal $w[n]$. This problem leads you through to the solution.

**(a)** Express $\mathcal{E}$ in terms of a frequency-domain integral on the interval $|\Omega| < \pi$ that involves $W(e^{j\Omega}) - Y(e^{j\Omega})$.

**(b)** Write your integral from (a) as a sum of integrals, one over each of the ranges $-\pi/4 \leq \Omega \leq \pi/4$, $\pi/4 \leq \Omega < \pi$, and $-\pi \leq \Omega \leq -\pi/4$. Use this to deduce how $Y(e^{j\Omega})$ needs to be picked in order to minimize $\mathcal{E}$, and what the resulting minimum value of $\mathcal{E}$ is. (*Hint:* Resist the temptation in this case to expand out $|a - b|^2$, for complex $a$ and $b$, as $|a|^2 + |b|^2 - ab^* - a^*b$.)

**(c)** Using your result in (b), write down an explicit formula for the $y[n]$ that minimizes $\mathcal{E}$, expressing this $y[n]$ as a suitable integral involving $W(e^{j\Omega})$.

**1.41.** **(a)** What is the value at $n = 0$ of the signal $x[n]$ whose DTFT for $|\Omega| \leq \pi$ is shown in A in Figure P1.41-1?

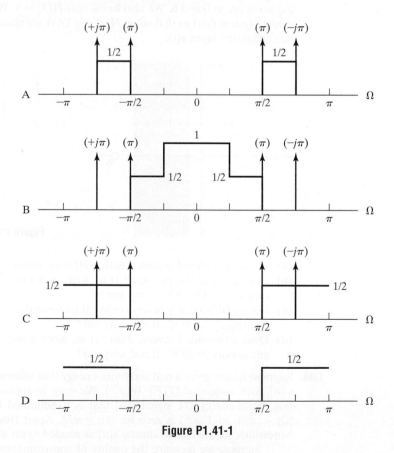

**Figure P1.41-1**

**(b)** List those DTFTs in A through H in Figure P1.41-1 and Figure P1.41-2 that correspond to a signal $x[n]$ for which $\sum x[n] = 0$. Explain. (The summation $\sum$ is over all $n$.)

**(c)** What is the value of $\sum (x[n])^2$ for the signal whose DTFT is shown in G in Figure P1.41-2? Explain. (The summation $\sum$ is over all $n$.)

**(d)** Determine and explain which of the DTFTs (A through H) in Figure P1.41-1 or Figure P1.41-2 is that of the signal

$$x[n] = \frac{\cos(3\pi n/4)\,\sin(\pi n/2)}{\pi n}.$$

**(e)** List those DTFTs in the set A through H in Figure P1.41-1 and Figure P1.41-2 that correspond to signals that are even in time.

**(f)** List those DTFTs in the set A through H in Figure P1.41-1 and Figure P1.41-2 that correspond to absolutely summable signals.

**(g)** List those DTFTs in the set A through H in Figure P1.41-1 and Figure P1.41-2 that correspond to a signal $x[n]$ for which $\sum x[n] \neq 0$, and for each such case compute the nonzero value of this sum.

**(h)** List those DTFTs in the set A through H in Figure P1.41-1 and Figure P1.41-2 that correspond to a signal $x[n]$ for which $\sum (-1)^n x[n] = 1$.

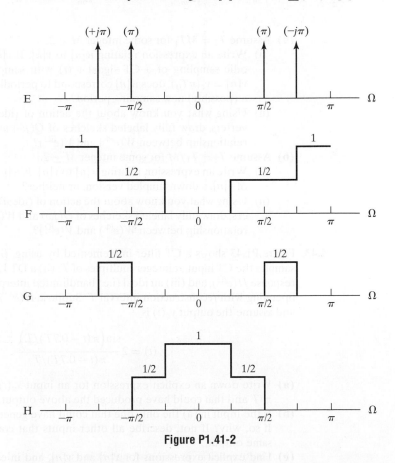

**Figure P1.41-2**

**1.42.** Consider the system shown in Figure P1.42-1, where the D/C converter is the standard ideal bandlimited interpolating converter. Suppose the DTFT of $v[n]$ is as shown in Figure P1.42-2.

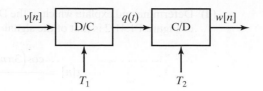

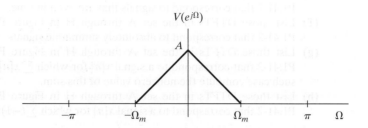

**Figure P1.42-1**

**Figure P1.42-2**

(a) Assume $T_2 = MT_1$ for some integer $M \geq 2$.
   (i) Write an expression relating $w[n]$ to $v[n]$. If $v[n]$ corresponds to periodic sampling of a CT signal $v_c(t)$ with sampling period $T_0$, i.e., if $v[n] = v_c(nT_0)$, does $w[n]$ correspond to periodic sampling of $v_c(t)$ with a (possibly) new sampling period?
   (ii) Using what you know about the action of (ideal) D/C and C/D converters, draw fully labeled sketches of $Q(j\omega)$ and $W(e^{j\Phi})$. What is the relationship between $W(e^{j\Phi})$ and $V(e^{j\Omega})$?
(b) Assume $T_2 = T_1/M$ for some integer $M \geq 2$.
   (i) Write an expression relating $w[n]$ to $v[n]$. Is $w[n]$ an upsampled version of $v[n]$, a downsampled version, or neither?
   (ii) Using what you know about the action of (ideal) D/C and C/D converters, draw fully labeled sketches of $Q(j\omega)$ and $W(e^{j\Phi})$. Again, what is the relationship between $W(e^{j\Phi})$ and $V(e^{j\Omega})$?

**1.43.** Figure P1.43 shows a CT filter implemented by using: (i) a C/D converter that samples the CT input at integer multiples of $T$; (ii) a DT LTI filter with frequency response $H(e^{j\Omega})$; and (iii) an ideal (i.e., bandlimited interpolating) D/C converter operating with reconstruction interval $T$. Suppose $H(e^{j\Omega}) = e^{-j0.4\Omega}$ for $|\Omega| < \pi$, and assume the output $y_c(t)$ is

$$y_c(t) = 2\frac{\sin\left(\pi(t - 0.7T)/T\right)}{\pi(t - 0.7T)/T}.$$

(a) Write down an explicit expression for an input $x_c(t)$ that is bandlimited to $\pi/T$ and that could have produced the above output.
(b) Is the input in (a) the only one that could have generated the given output? If so, why? If not, describe all other inputs that could have produced the same output.
(c) Find explicit expressions for $x[n]$ and $y[n]$, and in each case state whether these are the only possible $x[n]$ and $y[n]$ that could correspond to the given output $y_c(t)$.
(d) What is the unit sample response $h[n]$ of the DT system whose frequency response is the $H(e^{j\Omega})$ given above?

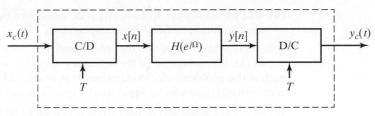

**Figure P1.43**

**1.44.** Figure P1.44-1 shows the standard configuration we have been using to discuss DT processing of CT signals by a DT LTI system. The C/D converter operates with sampling interval $T$, so that $x_d[n] = x(nT)$. The D/C converter is an ideal bandlimited interpolator operating with reconstruction (or interpolation) interval $T$, so that $y(t)$ is bandlimited to the frequency range $|\omega| < \frac{\pi}{T}$ and $y(nT) = y_d[n]$; and $H_d(e^{j\Omega})$ is the frequency response of the indicated DT LTI system. Assume for this problem that $T = 10^{-3}$ sec.

   Now consider two signals $b(t)$ and $c(t)$ whose plots for $0 < t < 6 \times 10^{-3}$ sec are shown in Figure P1.44-2, and which have the property that

$$c(nT) = b(nT)$$

for all $n$, not just those values of $n$ represented in the figure. These signals will be used as inputs to the system in Figure P1.44-1.

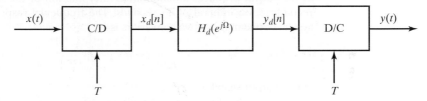

**Figure P1.44-1**

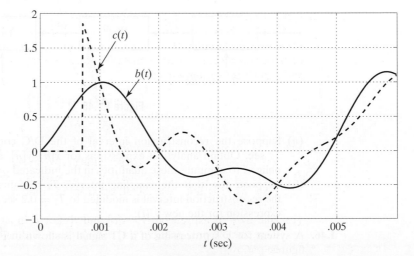

**Figure P1.44-2**

**(a)** One of the signals $b(t)$ and $c(t)$ is definitely not bandlimited. Which one is *not* bandlimited? Give a brief explanation.

Assume in what follows that the *other* signal, i.e., *not* the one you just identified in part (a), is bandlimited to the frequency range $|\omega| < \pi/T$. The two remaining parts of this problem refer to experiments in which $b(t)$ and $c(t)$ are used, but not necessarily in this order, as inputs to the system shown in Figure P1.44-2.

**(b)** When the input $x(t)$ to the system is chosen to be the bandlimited signal of the pair $b(t)$ and $c(t)$, the output $y(t)$ turns out to be $y(t) = x(t - 0.75T)$. Fully specify one possible choice of the frequency response $H_d(e^{j\Omega})$ that would yield this output for the specified input.

**(c)** If $h_d[n]$ denotes the unit sample response corresponding to the $H_d(e^{j\Omega})$ that you have picked, evaluate

$$h_d[0], \quad \sum_{n=-\infty}^{\infty} h_d[n], \quad \text{and} \quad \sum_{n=-\infty}^{\infty} (h_d[n])^2 .$$

**(d)** For the same system as in (b), determine what the output $y(t)$ of the system would be, in terms of $b(t)$ or $c(t)$, when the input $x(t)$ is the non-bandlimited signal of the pair $b(t)$ and $c(t)$.

**1.45.** In the system shown in Figure P1.45, the C/D converter samples the CT signal $x_c(t)$ with a sampling interval of $T_1 = 0.1$ seconds, while the ideal D/C converter at the output produces a bandlimited interpolation of the samples $y[n]$ using a reconstruction interval of $T_2$ seconds. The frequency response of the DT LTI system between these two converters is

$$H(e^{j\Omega}) = e^{-j\Omega/2}, \quad |\Omega| < \pi$$

and the input signal is

$$x_c(t) = \cos\left(22\pi t - \frac{\pi}{4}\right).$$

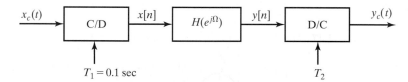

**Figure P1.45**

**(a)** Suppose the reconstruction interval of the D/C converter is $T_2 = 0.1 = T_1$ sec. Obtain analytical expressions for $x[n]$, $y[n]$, and $y_c(t)$. (Don't feel obliged to compute these quantities in the indicated order!).

**(b)** Suppose now that the sampling interval $T_1$ remains at 0.1 sec, but that the reconstruction interval is modified to $T_2 = 0.2$ sec. Obtain an analytical expression for the new $y_c(t)$.

**1.46.** A system for DT processing of a CT signal is shown in Figure P1.46-1. In this figure:
  (i) the C/D box is an ideal sampler whose output is $x_d[n] = x_c(nT_1)$;
  (ii) the output of the squarer is $z_d[n] = x_d^2[n]$;

(iii) $z_d[n]$ is filtered by a DT LTI filter whose frequency response is $H_d(e^{j\Omega})$; and

(iv) the D/C box is an ideal bandlimited interpolator whose output is

$$y_c(t) = \sum_{n=-\infty}^{\infty} y_d[n] \frac{\sin[\pi(t - nT_2)/T_2]}{\pi(t - nT_2)/T_2} .$$

(v) Formulas that may (or may not) be useful include:

$$\cos(2\theta) = \cos^2(\theta) - \sin^2(\theta)$$

$$\sin(2\theta) = 2\sin(\theta)\cos(\theta)$$

$$\cos^2(\theta) = \frac{1}{2} + \frac{1}{2}\cos(2\theta)$$

$$\sin^2(\theta) = \frac{1}{2} - \frac{1}{2}\cos(2\theta) .$$

Suppose that $x_c(t) = \cos(\pi t/3T)$, $T_1 = T_2/2 = T$, and $H_d(e^{j\Omega})$ is as shown in Figure P1.46-2 for $|\Omega| \leq \pi$. Determine $y_c(t)$.

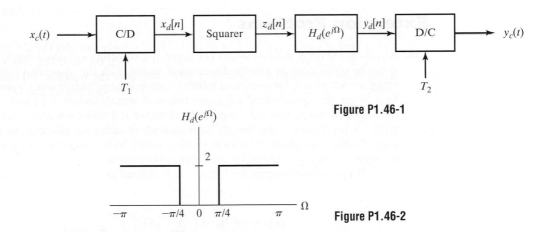

**Figure P1.46-1**

**Figure P1.46-2**

**1.47. (a)** Consider the time-invariant nonlinear deterministic system shown in Figure P1.47-1, with constant output $A$, whose value is determined by all time instances of the input signal $x[n]$. Consider the class of inputs of the form $x[n] = e^{j\Omega n}$, with $\Omega$ a real finite number. Varying the value of $\Omega$ at the input will change $A$, i.e., $A$ will be a function of $\Omega$. Specify whether $A$ will be periodic in $\Omega$, and if so, with what period. Explain.

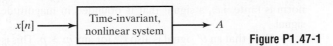

**Figure P1.47-1**

**(b)** Now consider the system shown in Figure P1.47-2. System 1 is a memoryless nonlinear system. System 2 determines the value of $A$ according to the relation

$$A = \sum_{n=0}^{100} |y[n]| .$$

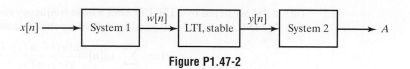

**Figure P1.47-2**

Again, the class of inputs being considered are of the form $x[n] = e^{j\Omega n}$, with $\Omega$ a real finite number. Varying the value of $\Omega$ at the input will change $A$; i.e., $A$ will be a function of $\Omega$.

(i)  If the LTI system is just the identity system and System 1 is defined by $w[n] = x^2[n]$, will $A$ be periodic in $\Omega$?

(ii)  More generally, if System 1 is a memoryless nonlinear system, and the LTI stable system is not necessarily the identity system, will $A$ be periodic in $\Omega$ and if so, with what period?

(c)  Explain in words why the frequency response $H(e^{j\Omega})$ of a DT LTI stable system is always periodic in frequency.

## Extension Problems

**1.48.**  A real-valued (or complex-valued) DT signal $x[\cdot]$, sometimes denoted simply by **x**, can be considered an infinite-dimensional vector, with the associated scalars being the set of real (or complex) numbers. Such a signal/vector can be scaled by $\alpha$ to get the signal/vector $\alpha x[\cdot]$, and two such signals/vectors $x_1[\cdot]$ and $x_2[\cdot]$ can be added component-wise to get a signal/vector $x[\cdot]$ whose $n$th component is $x_1[n] + x_2[n]$. To define the "length" or magnitude of such a signal/vector, we use a signal norm. This problem looks at a widely used family of signal norms, and at the norms of signals obtained by convolving other signals.

The $p$-norm of a signal for $1 \le p < \infty$ is defined as

$$\|\mathbf{x}\|_p = \|x[\cdot]\|_p = \left( \sum_{k=-\infty}^{\infty} \left| x[k] \right|^p \right)^{1/p} .$$

A signal whose $p$-norm is finite is said to be an $\ell^p$ signal.

Taking the limit $p \to \infty$ allows us to recognize the $\infty$-norm as

$$\|\mathbf{x}\|_\infty = \sup_k \left\{ \left| x[k] \right| \right\} ,$$

where "sup" denotes the supremum or least upper bound. A signal whose $\infty$-norm is finite, i.e., a signal that is bounded in magnitude, is said to be an $\ell^\infty$ signal.

Note that an $\ell^p$ signal is also $\ell^m$ for any $m > p$. This is because the values of an $\ell^p$ signal $x[n]$ for large $|n|$ must fall off in magnitude toward 0, and for $m > p$ this fall-off at large $|n|$ is even faster.

The $p$-norm for $1 \le p \le \infty$ satisfies the three properties required of a norm: (i) it is nonzero for all nonzero signals; (ii) it scales according to the relation

$$\|\alpha \mathbf{x}\|_p = |\alpha| . \|\mathbf{x}\|_p$$

when the signal is scaled by $\alpha$ (so we see, by taking $\alpha = 0$, that the norm of the zero signal is 0); and (iii) it satisfies the triangle inequality (also referred to as the Minkowski inequality),

$$\|\mathbf{x} + \mathbf{y}\|_p \leq \|\mathbf{x}\|_p + \|\mathbf{y}\|_p \ .$$

**(a)** For each of the following signals, determine (or numerically approximate) its $\ell^p$ norm for $p = 1, 2$, and $\infty$ in each case where this norm is finite:

(i) $x[n] = (-1)^n / n^2$ for $n > 0$, and 0 elsewhere;

(ii) the signal defined by

$$x[n] = \frac{\sin(\pi n/5)}{\pi n}$$

for $n \neq 0$, with $x[0]$ defined to be $1/5$; and

(iii) $x[n] = \big((0.2)^n - 1\big)u[n]$, where $u[n]$ is the unit step function (= 1 for $n \geq 0$, and 0 elsewhere).

It turns out that the convolution $\mathbf{h} * \mathbf{x}$ of two signals $\mathbf{h}$ and $\mathbf{x}$ satisfies the following inequality (called Young's inequality):

$$\|\mathbf{h} * \mathbf{x}\|_r \leq \|\mathbf{h}\|_p \, \|\mathbf{x}\|_q \ ,$$

where $1 \leq p \leq \infty$, and similarly for $q, r$, with

$$\frac{1}{r} = \frac{1}{p} + \frac{1}{q} - 1 \ .$$

Use this inequality to answer the following questions.

**(b)** Suppose $x[\cdot]$ is the input to an LTI system with a unit sample response of $h[\cdot]$, and an output signal denoted by $y[\cdot]$.

(i) If the input is bounded (i.e., $\ell^\infty$) and the unit sample response is absolutely summable (i.e., $\ell^1$), what can you say about the output signal? How does this change if the input is absolutely summable and the unit sample response is bounded?

(ii) If the input and the unit sample response are both square summable (i.e., $\ell^2$), what can you say about the output?

(iii) If the unit sample response is absolutely summable and the input is $\ell^s$ for some $1 \leq s \leq \infty$, what can you say about the output?

**1.49.** Suppose $x[n]$ is a known (real) signal of possibly infinite duration but finite (and nonzero) energy

$$\sum_{k=-\infty}^{\infty} x^2[k] = \mathcal{E}_x < \infty \ .$$

**(a)** With the deterministic autocorrelation defined as

$$\overline{R}_{xx}[m] = \sum_{n=-\infty}^{\infty} x[n]x[n-m] \ ,$$

the discussion around Eq. (1.62) shows that $\overline{R}_{xx}[0] \geq \overline{R}_{xx}[k]$ for all $k$. Modify the argument to show that in addition $\overline{R}_{xx}[0] \geq -\overline{R}_{xx}[k]$ for all $k$, and hence that

$$\overline{R}_{xx}[0] \geq \left| \overline{R}_{xx}[k] \right| \ .$$

(b) Is it possible to have equality in the preceding equation when $k$ takes some nonzero value $P$? (*Hint:* Show that if $\overline{R}_{xx}[0] = \left|\overline{R}_{xx}[P]\right|$ for some $P \neq 0$, then $x[n]$ would have to be periodic—with period $P$ or $2P$—which would contradict its having finite energy.)

Suppose now that we measure a signal $y[n]$ related to $x[\,\cdot\,]$ by

$$y[n] = x[n - L]\,,$$

where $L$ is a fixed but unknown lag. Since the signal $x[\,\cdot\,]$ is known, we can compute the deterministic cross-correlation function

$$\overline{R}_{yx}[m] = \sum_{n=-\infty}^{\infty} y[n]x[n - m]\,.$$

For instance, $x[n - L]$ may be the signal that a radar expects to have arrive back at its antenna in the noise-free case, after bouncing off a target, with $L$ being proportional to the distance of the target from the radar. We are assuming the radar knows the signal shape $x[k]$, and measures $y[n]$, so is in a position to compute the cross-correlation function of the nominal and received signals.

(c) Express $\overline{R}_{yx}[m]$ in terms of $\overline{R}_{xx}[m]$. Then find out for what value of $m$ the deterministic cross-correlation $\overline{R}_{yx}[m]$ takes its largest value, and determine what this largest value is (in terms of properties of the signal $x[\,\cdot\,]$). Explain how your results here could allow you to discover the unknown lag $L$ from $\overline{R}_{yx}[m]$ in the noise-free case.

Now suppose instead that the measured signal $y[n]$ is related to $x[\,\cdot\,]$ by

$$y[n] = x[n - L] + v[n]\,,$$

where $L$ is again a fixed but unknown lag, and $v[n]$ denotes a noise process whose value at each time $n$ is also not known, except for the fact that it is a zero-mean random variable of variance $\sigma_v^2 > 0$, and is chosen independently of the values of the noise process at other times.

(d) We can again compute the deterministic correlation function $\overline{R}_{yx}[m]$. Because of the noise, the value of this function at each $m$ will differ from the value computed in case (c) by a random amount, which we denote by $w[m]$. Determine the mean and standard deviation of $w[m]$.

(e) If in the noisy case (d) we apply the approach of (c) to guess at the right value of $L$, we might make an error because of the perturbations caused by the noise. Does your answer in (d) suggest that the task of determining $L$ can be performed more accurately when the ratio $\mathcal{E}_x/\sigma_v^2$ of signal energy to noise variance increases, or does increasing this ratio not help this task?

(f) Suppose $x[n]$ is allowed to be nonzero only at $D$ instants of time, and to take only the value 1 or $-1$ at each of these $D$ instants. This restriction causes the energy of the signal to be fixed at $\mathcal{E}_x = D$. It is often of interest (and the preceding parts of this problem should suggest why) to design such a signal $x[n]$ to have $\overline{R}_{xx}[0]$ be much larger than $\overline{R}_{xx}[m]$ for all $m \neq 0$. You might find it interesting in this connection to read about Barker codes, which have this feature. Compute and plot the deterministic autocorrelation and the energy spectral density of the Barker code of length 13 (which is the longest such code known), for which $x[n]$ takes the following values for $n = 0, 1, \cdots, 12$ respectively (and the value 0 everywhere else):

$$+1\,,\ +1\,,\ +1\,,\ +1\,,\ +1\,,\ -1\,,\ -1\,,\ +1\,,\ +1\,,\ -1\,,\ +1\,,\ -1\,,\ +1\,.$$

**1.50.** In this problem, you will compute and compare the DTFTs of several finite-length signals derived from the everlasting signal

$$x[n] = \cos(\pi n/4) \, .$$

If we know that $x[n]$ is a sinusoid, but do not know that its angular frequency is $\pi/4$, then one way to determine its frequency is to take the DTFT of a windowed segment of $x[n]$. However, the windowing leads to a spreading or blurring of the DTFT, thereby introducing some uncertainty into the frequency determination. Parts (a) and (b) below explore this issue. Parts (c) and (d) respectively deal with the spectra of systematically and randomly modulated (and then rectangularly windowed) versions of a sinusoid.

**(a)** Obtain an expression for the magnitude of the DTFT of the signal

$$y_1[n] = x[n]r[n] \, ,$$

where

$$r[n] = (u[n] - u[n-101])/101 \, ,$$

and $u[n]$ is the unit step. Compute and plot the magnitude of this DTFT in the following two ways:
  (i) by evaluating the expression you obtained; and
  (ii) by using an appropriate computational package that can compute samples of the DTFT of a finite sequence.

Verify that you get the same answer either way. In a sentence or two, summarize what this calculation tells you about how rectangular windowing affects the spectrum of a cosine.

**(b)** What happens if $x[n]$, instead of being a single cosine, is actually the sum of two cosines with closely spaced frequencies? Approximately how close together can the two frequencies get before the DTFT of the windowed signal fails to display a distinct peak for each of the cosines? *Hint:* You may find it helpful to first examine the DTFT of the rectangular window of length 101 samples used to obtain the segment $y_1[n]$ of $x[n]$.

**(c)** Now consider the signal

$$y_2[n] = x[n]t[n] \, ,$$

where

$$t[n] = \begin{cases} (n+1)/101^2, & 0 \le n < 100 \\ (201-n)/101^2, & 100 \le n \le 200 \\ 0, & \text{otherwise.} \end{cases}$$

Compute and plot the magnitude of the DTFT of $y_2[n]$. How does this plot relate to the DTFT in part (a)? In a sentence or two, summarize what this calculation tells you about how triangular windowing of a cosine compares with rectangular windowing. How does triangular windowing compare with rectangular windowing when the signal is the sum of two cosines with closely spaced frequencies?

**(d)** Now consider the signal

$$y_3[n] = y_1[n](-1)^n \, .$$

Compute and plot the magnitude of the DTFT of $y_3[n]$. How does this plot relate to the DTFT in part (a)? At what frequency is the signal energy concentrated now? Is this as expected?

**(e)** To begin developing a feel for random (or probabilistic or stochastic) signals, consider

$$y_4[n] = y_1[n]b[n] ,$$

where each value of $b[n]$ is independently set to either $-1$ or $+1$, each with probability equal to $1/2$, for $0 \leq n \leq 100$. You can generate $b[n]$ using any computational package that can generate random numbers. Compute and plot the magnitude of the DTFT of $y_4[n]$ using the same commands as in (a). Repeat for four different sets ("realizations") of the sequence $b[n]$. Do your DTFTs in this case look like they have any particular structure or do they look irregular? Is the energy of the signal concentrated in some part of the frequency spectrum, or is it spread out? Do you see any hint of the underlying cosine signal, $x[n]$, in the spectrum of $y_4[n]$? Do your results for the four different realizations appear to have anything in common?

If a friend received (or an enemy intercepted!) your signal $y_4[n]$, and happened to know the particular sequence $b[n]$ that was used to generate $y_4[n]$, can you see a way she could determine the signal multiplying $b[n]$ (which happens to be $y_1[n]$ in our example)? What do you think are the prospects of recovering $y_1[n]$ from $y_4[n]$ if one didn't know $b[n]$?

**(f)** The ideas touched on in part (e) underlie code-division multiple access (CDMA) schemes for wireless communication. To explore this more directly, suppose

$$y_5[n] = 7\,b_1[n] + 3\,b_2[n] + 4\,b_3[n] ,$$

where the *codes* $b_1[n]$, $b_2[n]$, and $b_3[n]$ are each obtained independently in the same fashion as $b[n]$ in part (e). In a CDMA communication context, $y_5[n]$ would be the signal transmitted from the base station to mobile units in its area. Mobile unit 1 only knows the code $b_1[n]$ and, after receiving $y_5[n]$, wants to determine the scale factor (in this case 7) multiplying $b_1[n]$. Similarly, mobile unit 2 knows only the code $b_2[n]$ and wants to determine the scale factor (in this case 3) multiplying $b_2[n]$. And similarly again for mobile unit 3. In actual systems, these constants represent the information transmitted to the respective mobile units over a time interval comparable with the duration of the code $b_i[n]$; the constants are actually varied on a time scale that is slow compared to the code duration.

Independently generate, in the same way that you generated $b[n]$ in (e) above, codes $b_1[n]$, $b_2[n]$, and $b_3[n]$. Also use these to construct $y_5[n]$ according to the above equation. Now compare the sums

$$\sum_n b_1[n]b_2[n] , \quad \sum_n b_2[n]b_3[n] , \quad \sum_n b_3[n]b_1[n]$$

with the values of

$$\sum_n b_1^2[n] , \quad \sum_n b_2^2[n] , \quad \sum_n b_3^2[n] .$$

Does this calculation suggest how mobile unit $i$ can estimate the constant that multiplies $b_i[n]$, for $i = 1, 2, 3$, using just knowledge of $b_i[n]$ and $y_5[n]$? Implement whatever scheme you come up with, and see how well you can estimate the various scale factors.

**1.51.** In the system shown in Figure P1.51, the C/D converter samples the CT signal $x_c(t)$ with a sampling interval of $T_1 = 0.1$ sec, while the D/C converter at the output produces a bandlimited interpolation of the samples $y[n]$ using a reconstruction interval of $T_2$ seconds. The frequency response of the DT LTI system between these two converters is

$$H(e^{j\Omega}) = \frac{j\Omega}{T_1}, \quad |\Omega| < \pi$$

and the input signal is

$$x_c(t) = \cos\left(25\pi t - \frac{\pi}{4}\right).$$

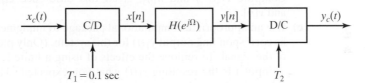

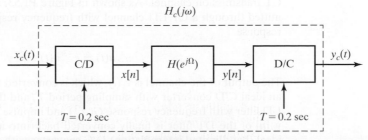

**Figure P1.51**

**(a)** Suppose the reconstruction interval of the D/C converter is $T_2 = 0.1$ seconds. We would like to determine whether $y_c(t)$ is the derivative of the given $x_c(t)$.

    (i) Obtain analytical expressions for $x[n]$, $y[n]$, and $y_c(t)$, and provide fully labeled sketches of each for the range of $0 \le n \le 6$ and $0 \le t \le 6$, respectively.

    (ii) Is $y_c(t)$ equal to the derivative of $x_c(t)$?

    (iii) Is the overall system linear? Explain.

**(b)** Suppose now that the sampling interval $T_1$ remains at 0.1 seconds, but that the reconstruction interval is modified to $T_2 = 0.2$ sec. Obtain an analytical expression for the new $y_c(t)$, and provide a sketch.

**1.52.** Figure P1.52 shows the block diagram of a system for DT processing of CT signals. The overall CT system is linear and time-invariant, with frequency response $H_c(j\omega)$ for inputs $x_c(t)$ that are appropriately bandlimited. In the following parts, we will use the particular input $x_1(t) = \sin(9t)$ to illustrate properties of this system. We would like to design $H(e^{j\Omega})$ such that the overall system, $H_c(j\omega)$, is a CT differentiator, in which $y_c(t) = dx_c(t)/dt$.

**Figure P1.52**

(a) First, find an expression for $dx_1(t)/dt$.

(b) A common approximation of the derivative of a signal with respect to time, for small values of $T$, is as follows:

$$y_c(t) = \frac{dx_c(t)}{dt} \approx \frac{x_c(t+T) - x_c(t-T)}{2T}.$$

In order to implement this, we can use a DT system that imposes the following input-output relation on $x[n]$ and $y[n]$:

$$y[n] = \frac{x[n+1] - x[n-1]}{2T}.$$

Compute $H(e^{j\Omega})$ and $h[n]$, given this difference equation relating $x[n]$ and $y[n]$.

(c) Use an appropriate computational package to implement and plot $x_1(t)$ and its corresponding output $y_1(t)$ for this system. (Only plot the middle section of the signals to remove the effects of using a finite length sample of $x_1(t)$ as input.) Is the resulting $y_1(t)$ what you expected? How is the amplitude related to that of the expression you found in (a)?

(d) Another possible way of computing the derivative of $x_c(t)$ is to use a DT system with frequency response $H(e^{j\Omega}) = j\Omega/T$ for $|\Omega| < \pi$. In this case, as we know from Example 1.3, the output $y_c(t)$ will be equal to the derivative of an appropriately bandlimited input $x_c(t)$. Show that the impulse response of this DT system, $h[n]$, is

$$h[n] = \begin{cases} \dfrac{(-1)^n}{nT}, & n \neq 0 \\ 0, & n = 0. \end{cases}$$

*Hint:* Remember that if $x_c(t) = T\sin(\pi t/T)/\pi t$, then $x[n] = \delta[n]$. Find the response of the CT system to this input and sample to get $y[n]$.

(e) Now, using only the values of $h[n]$ for $-40 \leq n \leq 40$, compute and plot an approximation to $y_1(t)$ again, using an appropriate computational package. How close is this $y_1(t)$ to the one you calculated in part (a)? (Just check a few values.)

(f) Plot both approximations. Which one is better? Is this what you expected? You may also want to plot the actual derivative you found in part (a) along with the two approximations.

**1.53.** This problem examines the use of a DT filter to compensate for the effects of a CT transmission channel. As shown in Figure P1.53, a CT signal $x_c(t)$ is transmitted through a CT LTI channel with frequency response $H(j\omega)$ and impulse response

$$h(t) = e^{-3t}u(t).$$

The output of the channel is $v_c(t)$, which is converted to a DT signal $v_d[n]$ using an ideal C/D converter with sampling period $T$, and then processed using a DT LTI filter with frequency response $G(e^{j\Omega})$ and impulse response $g[n]$. The output $y_d[n]$ of the DT filter is finally converted back into a CT signal $y_c(t)$ using an ideal (bandlimited interpolation) D/C converter with reconstruction interval $T$. We wish to choose the compensator so as to obtain $y_c(t) = x_c(t)$ for appropriately bandlimited inputs $x_c(t)$.

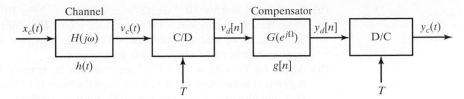

**Figure P1.53**

(a) What is $H(j\omega)$?

(b) What is the highest frequency that $x_c(t)$ can contain, i.e., to what frequency must $X_c(j\omega)$ be bandlimited, if aliasing is to be avoided in the C/D conversion? Assume from now on that $x_c(t)$ satisfies this condition.

(c) Determine $G(e^{j\Omega})$ such that $y_c(t) = x_c(t)$.

(d) For $G(e^{j\Omega})$ as in (c), determine:
   (i) $\sum_{n=-\infty}^{+\infty} g^2[n]$; and
   (ii) $g[n]$.

**1.54.** Consider the system shown in Figure P1.54-1. The anti-aliasing filter is a CT filter with the frequency response $L(j\omega)$ shown in Figure P1.54-2.

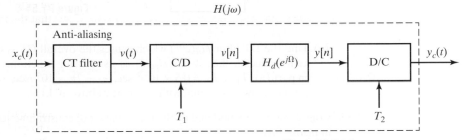

**Figure P1.54-1**

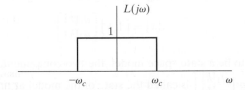

**Figure P1.54-2**

For parts (a), (b), and (c), $H_d(e^{j\Omega})$ is as shown in Figure P1.54-3, where $0 < \Omega_c < \pi$.

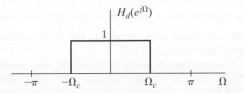

**Figure P1.54-3**

**(a)** Let $T_1 = T_2 = 0.5 \times 10^{-4}$ sec, $\Omega_c = \pi/4$, and $\omega_c = 2\pi \times 10^4$ /sec. Is the overall system that processes $x_c(t)$ to produce $y_c(t)$ LTI or not? If not, explain why not. If so, find and plot the CT transfer function $H(j\omega)$. Give the frequency in hertz at which $H(j\omega)$ goes to zero.

**(b)** Again, let $T_1 = T_2 = 0.5 \times 10^{-4}$ sec, but now $\Omega_c$ is variable. Let $\omega_{c,\max}$ be the largest value of $\omega_c$ for which the overall system is LTI for all inputs $x_c(t)$. Find and plot $\omega_{c,\max}$ as a function of $\Omega_c$ for $0 < \Omega_c < \pi$.

**(c)** Is the overall system linear under the conditions in part (a) if $T_2$ is reduced to $0.25 \times 10^{-4}$ sec? Is it time-invariant? Explain why or why not in each case.

For the remainder of the problem, suppose $x_c(t)$ is an audio signal $r(t)$ recorded in an environment with an echo that is a delayed version of $r(t)$:

$$x_c(t) = r(t) + \alpha r(t - T_o), \quad T_o > 0, \ 0 < \alpha < 1 .$$

**(d)** Find the CT transfer function $H_{ec}(s)$ for an echo canceller that removes the echo as shown in Figure P1.54-4.

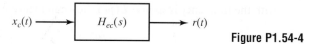

**Figure P1.54-4**

**(e)** Find the DT filter $H_d(e^{j\Omega})$ which causes the overall CT system to act as an echo canceller, i.e., it has $H_{ec}(s)$ as its transfer function. Assume that, as in part (a), $T_1 = T_2 = 0.5 \times 10^{-4}$ sec, $\omega_c = 2\pi \times 10^4$ /sec. Also assume that $x_c(t)$ is a low-pass signal with no energy above 10 kHz.

**1.55.** Consider a system modeled by the following set of constraint equations, arranged in matrix form:

$$\begin{bmatrix} q_1[k+1] \\ q_2[k+1] \end{bmatrix} = \begin{bmatrix} \frac{1}{3} & 1 \\ 0 & \frac{1}{2} \end{bmatrix} \begin{bmatrix} q_1[k] \\ q_2[k] \end{bmatrix} + \begin{bmatrix} 1 \\ 1 \end{bmatrix} x[k]$$

$$y[k] = \begin{bmatrix} 1 & 1 \end{bmatrix} \begin{bmatrix} q_1[k] \\ q_2[k] \end{bmatrix}$$

This is said to be a state-space model; the two-component, or two-dimensional, signal $\mathbf{q}[k] \equiv \begin{bmatrix} q_1[k] \\ q_2[k] \end{bmatrix}$ is called the state of the model at time $k$, while $x[k]$ is its input and $y[k]$ is its output. The noteworthy features of a state-space model are that (i) the state at the next time instant, $k + 1$, is given as a function of the state and input at the present time $k$; and (ii) the output at the present time is given as a function of the state and input at the present time (although in the above example, the input at time $k$ happens to not affect the output at time $k$).

**(a)** Is the system linear with respect to the four signals $q_1, q_2, x$, and $y$? Is it time-invariant? Memoryless?

**(b)** Suppose the system is known to be causal. Find its output response $y[n]$ for $n \geq 0$, if its input is the unit sample $x[n] = \delta[n]$ and its initial condition is $\mathbf{q}[0] = \begin{bmatrix} 2 \\ 0 \end{bmatrix}$. This can be done in the time domain, or by using $z$-transforms.

(c) Suppose that the first equation in (a) is replaced by

$$\begin{bmatrix} q_1[k+1] \\ q_2[k+1] \end{bmatrix} \begin{bmatrix} \frac{1}{3} & 2\cos\left(\frac{\pi}{3}k\right) \\ 0 & \frac{1}{2} \end{bmatrix} \begin{bmatrix} q_1[k] \\ q_2[k] \end{bmatrix} \begin{bmatrix} 1 \\ 1 \end{bmatrix} x[k]$$

while the second equation is kept the same. Is the system linear? Memoryless? Time-invariant?

(d) Show that there is a natural sense in which the model in (c) is periodically varying, and determine its period. Does the periodicity mean that any signals $q_1$, $q_2$, $x$, and $y$ which simultaneously satisfy the model constraints have to be periodic?

# 2 Amplitude, Phase, and Group Delay

As you have seen in your prior studies of signals and systems, and as emphasized in Chapter 1, transforms play a central role in characterizing and representing signals and linear, time-invariant (LTI) systems in both continuous and discrete time. In this chapter, we discuss some specific aspects of transform representations that may not be as familiar but will play an important role in later chapters. One aspect is the interpretation of Fourier transform phase through the concept of group delay. A second involves the conditions under which the Fourier transform phase is uniquely specified by the magnitude. We also discuss one particular approach to recovering the phase from the magnitude, which is referred to as spectral factorization.

## 2.1 FOURIER TRANSFORM MAGNITUDE AND PHASE

The Fourier transform of a signal or the frequency response of an LTI system is in general a complex-valued function. A magnitude-phase representation of a Fourier transform $H(j\omega)$ takes the form

$$H(j\omega) = |H(j\omega)| \, e^{j\angle H(j\omega)} . \tag{2.1}$$

In Eq. (2.1), $|H(j\omega)|$ denotes the (nonnegative) magnitude and $\angle H(j\omega)$ denotes the (real-valued) phase. For example, if $H(j\omega)$ is the sinc function, $\sin(\omega)/\omega$, then $|H(j\omega)|$ is the absolute value of this function, while $\angle H(j\omega)$ is 0 in frequency ranges where the sinc function is positive and $\pm\pi$ in frequency ranges where the sinc function is negative. An alternative representation is an amplitude-phase representation

$$H(j\omega) = A(\omega)\, e^{j\angle_A H(j\omega)} \tag{2.2}$$

in which $A(\omega) = \pm|H(j\omega)|$ is real but can be positive for some frequencies and negative for others. Correspondingly, $\angle_A H(j\omega) = \angle H(j\omega)$ when $A(\omega) = +|H(j\omega)|$, and $\angle_A H(j\omega) = \angle H(j\omega) \pm \pi$ when $A(\omega) = -|H(j\omega)|$. This representation is often preferred when its use can eliminate discontinuities of $\pi$ radians in the phase as $A(\omega)$ changes sign. In the case of the sinc function above, for instance, we can choose $A(\omega) = \sin(\omega)/\omega$ and $\angle_A H(j\omega) = 0$. A similar discussion applies also in discrete time.

In either a magnitude-phase representation or an amplitude-phase representation, the phase is ambiguous, as any integer multiple of $2\pi$ can be added at any frequency without changing $H(j\omega)$ in Eq. (2.1) or Eq. (2.2). A typical phase computation resolves this ambiguity by generating the phase modulo $2\pi$, that is, as the phase passes up through $+\pi$ it "wraps around" to $-\pi$ (or down through $-\pi$ it wraps around to $+\pi$). In Section 2.2, we will find it convenient to resolve this ambiguity by choosing the phase to be a continuous function of frequency. This is referred to as the unwrapped phase, since the discontinuities at $\pm\pi$ are unwrapped to obtain a continuous phase curve. The unwrapped phase is obtained from $\angle H(j\omega)$ by adding steps of height equal to $\pm\pi$ or $\pm 2\pi$ wherever needed, in order to produce a continuous function of $\omega$. The steps of height $\pm\pi$ are added at points where $H(j\omega)$ passes through 0, to absorb sign changes as needed; the steps of height $\pm 2\pi$ are added wherever else is needed, invoking the fact that such steps make no difference to $H(j\omega)$, as is evident from Eq. (2.1). We shall proceed as though $\angle H(j\omega)$ is indeed continuous and differentiable at the points of interest, understanding that continuity can indeed be obtained by adding in the appropriate steps of height $\pm\pi$ or $\pm 2\pi$.

Typically, our intuition for the time-domain effects of the frequency response magnitude or amplitude of an LTI filter on a signal is rather well developed. For example, if the frequency response magnitude is small at high frequencies, then we expect the output signal to vary slowly and without sharp discontinuities even when the input might have these features. On the other hand, an input signal whose low frequencies are attenuated relative to the high frequencies will tend to vary rapidly and without slowly varying trends.

Visualizing the effect on a signal of the phase of the frequency response of a system is more subtle, but equally important. We begin the discussion by first considering several specific examples that are helpful in treating the more general case. Throughout this discussion, we will consider the system to be an all-pass system with unity gain, that is, the amplitude of the frequency response $A(j\omega) = 1$ (continuous time) or $A(e^{j\Omega}) = 1$ (discrete time), so that we can focus entirely on the effect of the phase. The unwrapped phase associated with the frequency response will be denoted as $\angle_A H(j\omega)$ for continuous time and $\angle_A H(e^{j\Omega})$ for discrete time.

## Example 2.1    Linear Phase

Consider an all-pass system with the frequency response

$$H(j\omega) = e^{-j\alpha\omega} . \tag{2.3}$$

In an amplitude-phase representation, $A(j\omega) = 1$ and $\angle_A H(j\omega) = -\alpha\omega$. The unwrapped phase for this example is linear with respect to $\omega$, with slope of $-\alpha$. For input $x(t)$ with Fourier transform $X(j\omega)$, the Fourier transform of the output is $Y(j\omega) = X(j\omega)e^{-j\alpha\omega}$ and correspondingly the output $y(t)$ is $x(t - \alpha)$. In other words, linear phase with a slope of $-\alpha$ in the system frequency response corresponds to a time delay of $\alpha$ (or a time advance of $|\alpha|$ if $\alpha$ is negative).

For a discrete-time (DT) system with

$$H(e^{j\Omega}) = e^{-j\alpha\Omega} , \qquad |\Omega| < \pi \tag{2.4}$$

the phase is again linear with slope $-\alpha$. When $\alpha$ is an integer, the time-domain interpretation of the effect on an input sequence $x[n]$ is again straightforward and is a simple delay (for $\alpha$ positive) or advance (for $\alpha$ negative) of $|\alpha|$. When $\alpha$ is not an integer, the effect is still commonly referred to as "a delay of $\alpha$," but the interpretation is more subtle. If we think of $x[n]$ as being the result of sampling a bandlimited, continuous-time (CT) signal $x(t)$ with sampling period $T$, the output $y[n]$ will be the result of sampling the signal $y(t) = x(t - \alpha T)$ with sampling period $T$. In fact, we saw this result in Example 1.4 of Chapter 1 for the specific case of a half-sample delay, that is, $\alpha = \frac{1}{2}$.

## Example 2.2    Constant Phase Shift

As a second example, we again consider an all-pass system with frequency-response amplitude $A(j\omega) = 1$ and unwrapped phase

$$\angle_A H(j\omega) = \begin{cases} -\phi_0 & \text{for } \omega > 0 \\ +\phi_0 & \text{for } \omega < 0 \end{cases} \tag{2.5}$$

as indicated in Figure 2.1.

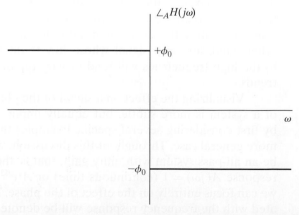

**Figure 2.1**    Phase plot of all-pass system with constant phase shift $\phi_0$.

Note that the phase will be an odd function of $\omega$ because we assume that the system impulse response is real valued. In this example, we consider $x(t)$ to be of the form

$$x(t) = s(t)\cos(\omega_0 t + \theta)\,, \quad \omega_0 > 0\,, \tag{2.6}$$

that is, an amplitude-modulated signal at a positive carrier frequency of $\omega_0$. Consequently, $X(j\omega)$ can be expressed as

$$X(j\omega) = \frac{1}{2}S(j\omega - j\omega_0)e^{j\theta} + \frac{1}{2}S(j\omega + j\omega_0)e^{-j\theta} \tag{2.7}$$

where $S(j\omega)$ denotes the Fourier transform of $s(t)$.

We also assume that $S(j\omega)$ is bandlimited to $|\omega| < \Delta$, with $\Delta$ sufficiently small so that the term $S(j\omega - j\omega_0)e^{j\theta}$ is zero for $\omega < 0$ and the term $S(j\omega + j\omega_0)e^{-j\theta}$ is zero for $\omega > 0$, that is, we assume that $(\omega_0 - \Delta) > 0$. Thus $x(t)$ is characterized by a slowly varying modulation of its carrier. The associated spectrum of $x(t)$ is depicted in Figure 2.2.

With these assumptions on $x(t)$, it is relatively straightforward to determine the output $y(t)$. Specifically, the system frequency response $H(j\omega)$ is

$$H(j\omega) = \begin{cases} e^{-j\phi_0} & \omega > 0 \\ e^{+j\phi_0} & \omega < 0\,. \end{cases} \tag{2.8}$$

Since the term $S(j\omega - j\omega_0)e^{j\theta}$ in Eq. (2.7) is nonzero only for $\omega > 0$, it is simply multiplied by $e^{-j\phi_0}$, and similarly the term $S(j\omega + j\omega_0)e^{-j\theta}$ is multiplied only by $e^{+j\phi_0}$. Consequently, the output Fourier transform $Y(j\omega)$ is given by

$$\begin{aligned} Y(j\omega) &= X(j\omega)H(j\omega) \\ &= \frac{1}{2}S(j\omega - j\omega_0)e^{+j\theta}e^{-j\phi_0} + \frac{1}{2}S(j\omega + j\omega_0)e^{-j\theta}e^{+j\phi_0}\,, \end{aligned} \tag{2.9}$$

which we recognize as a simple phase shift by $\phi_0$ of the carrier in Eq. (2.6), that is, replacing $\theta$ in Eq. (2.7) by $\theta - \phi_0$. Consequently,

$$y(t) = s(t)\cos(\omega_0 t + \theta - \phi_0)\,. \tag{2.10}$$

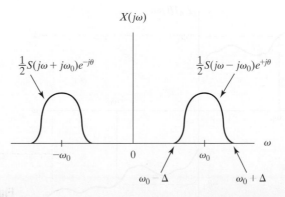

**Figure 2.2**   Spectrum of $x(t)$ with $s(t)$ a narrowband signal.

This change in phase of the carrier can also be expressed in terms of a time delay for the carrier by rewriting Eq. (2.10) as

$$y(t) = s(t)\cos\left[\omega_0\left(t - \frac{\phi_0}{\omega_0}\right) + \theta\right]$$

$$= s(t)\cos\left[\omega_0\left(t - \tau_p\right) + \theta\right] \tag{2.11}$$

where $\tau_p$, the negative of the ratio of the phase at $\omega_0$, i.e., $(-\phi_0)$, to the frequency $\omega_0$, is referred to as the phase delay of the system at frequency $\omega_0$:

$$\tau_p = -\frac{\angle H(j\omega_0)}{\omega_0} = \frac{\phi_0}{\omega_0}. \tag{2.12}$$

## 2.2 GROUP DELAY AND THE EFFECT OF NONLINEAR PHASE

In Example 2.1, we saw that a phase characteristic which is linear with frequency corresponds in the time domain to a time shift. In this section, we consider the effect of a nonlinear phase characteristic. We again assume that the system is an all-pass system with frequency response

$$H(j\omega) = A(j\omega)\,e^{j\angle_A[H(j\omega)]} \tag{2.13}$$

with $A(j\omega) = 1$. A general nonlinear phase characteristic that is an odd function of $\omega$ and is unwrapped for $|\omega| > 0$ is depicted in Figure 2.3. In Section 2.2.1, we first consider the case of narrowband signals. In Section 2.2.2, we extend that result to broadband signals.

### 2.2.1 Narrowband Input Signals

As we did in Example 2.2, we again assume that $x(t)$ is narrowband of the form in Eq. (2.6) and with the Fourier transform depicted in Figure 2.2.

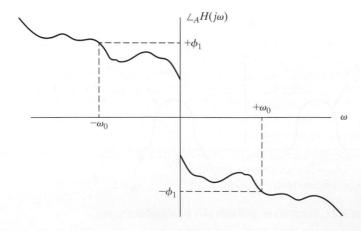

**Figure 2.3** Nonlinear unwrapped phase characteristic.

We next assume that $\Delta$ in Figure 2.2 is sufficiently small so that in the vicinity of $\pm\omega_0$ $\angle_A H(j\omega)$ can be approximated sufficiently well by the zeroth and first-order terms of a Taylor series expansion, that is,

$$\angle_A H(j\omega) \approx \angle_A H(j\omega_0) + (\omega - \omega_0)\left[\frac{d}{d\omega}\angle_A H(j\omega)\right]_{\omega=\omega_0} . \tag{2.14}$$

Defining $\tau_g(\omega)$ as

$$\tau_g(\omega) = -\frac{d}{d\omega}\angle_A H(j\omega) \tag{2.15}$$

our approximation to $\angle_A H(j\omega)$ in a small region around $\omega = \omega_0$ is expressed as

$$\angle_A H(j\omega) \approx \angle_A H(j\omega_0) - (\omega - \omega_0)\tau_g(\omega_0) . \tag{2.16}$$

Similarly, in a small region around $\omega = -\omega_0$ we make the approximation

$$\angle_A H(j\omega) \approx \angle_A H(-j\omega_0) - (\omega + \omega_0)\tau_g(-\omega_0) . \tag{2.17}$$

As we will see shortly, the quantity $\tau_g(\omega)$ plays a key role in our interpretation of the effect of a nonlinear phase characteristic on a signal.

With the Taylor series approximations in Eqs. (2.16) and (2.17), and for input signals with frequency content for which the approximation is valid, we can replace Figure 2.3 with Figure 2.4. Note that in Figure 2.4

$$-\phi_1 = \angle_A H(j\omega_0) \tag{2.18a}$$

and

$$-\phi_0 = \angle_A H(j\omega_0) + \omega_0\tau_g(\omega_0) . \tag{2.18b}$$

For LTI systems in cascade, the frequency responses multiply, and correspondingly their amplitudes multiply and their phases add. Consequently, we can represent the all-pass frequency response $H(j\omega)$ in Figure 2.4 as the cascade of two all-pass systems, $H_I(j\omega)$ and $H_{II}(j\omega)$, with unwrapped phase as depicted in Figure 2.5.

We recognize $H_I(j\omega)$ as corresponding to Example 2.2. Consequently, with $x(t)$ narrowband of the form in Eq. (2.6), we have

$$x(t) = s(t)\cos(\omega_0 t + \theta)$$

$$x_I(t) = s(t)\cos\left[\omega_0\left(t - \frac{\phi_0}{\omega_0}\right) + \theta\right] . \tag{2.19}$$

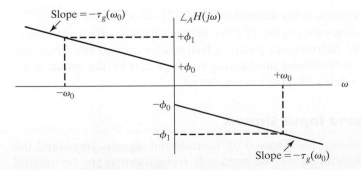

**Figure 2.4**  Taylor series approximation of nonlinear phase in the vicinity of $\pm\omega_0$.

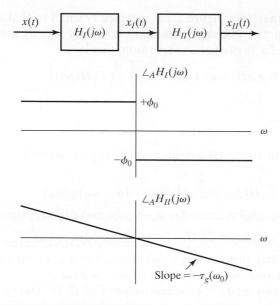

**Figure 2.5** An all-pass system with frequency response $H(j\omega)$ represented as the cascade of two all-pass systems with frequency responses $H_I(j\omega)$ and $H_{II}(j\omega)$.

Next we recognize $H_{II}(j\omega)$ as corresponding to Example 2.1 with $\alpha = \tau_g(\omega_0)$. Consequently,

$$x_{II}(t) = x_I\left(t - \tau_g(\omega_0)\right) \tag{2.20}$$

or equivalently

$$x_{II}(t) = s(t - \tau_g(\omega_0))\cos\left[\omega_0\left(t - \frac{\phi_0 + \omega_0\tau_g(\omega_0)}{\omega_0}\right) + \theta\right]. \tag{2.21}$$

From Eqs. (2.18a) and (2.18b), we see that

$$\phi_0 + \omega_0\tau_g(\omega_0) = \phi_1 = -\angle_A H(j\omega_0). \tag{2.22}$$

Equation (2.21) can therefore be rewritten as

$$x_{II}(t) = s(t - \tau_g(\omega_0))\cos\left[\omega_0\left(t - \frac{\phi_1}{\omega_0}\right) + \theta\right] \tag{2.23a}$$

or

$$x_{II}(t) = s(t - \tau_g(\omega_0))\cos\left[\omega_0\left(t - \tau_p(\omega_0)\right) + \theta\right] \tag{2.23b}$$

where $\tau_p(\omega_0)$ is the phase delay defined in Eq. (2.12), i.e., $\tau_p(\omega_0) = -\frac{\angle_A H(j\omega_0)}{\omega_0}$.

In summary, according to Eq. (2.23b), the time-domain effect of the nonlinear phase for the narrowband group of frequencies around the frequency $\omega_0$ is to delay the narrowband modulating envelope $s(t)$ by the group delay $\tau_g(\omega_0)$, and apply a delay of $\tau_p(\omega_0)$ to the carrier.

### 2.2.2 Broadband Input Signals

Thus far our discussion has focused on narrowband signals. To extend the discussion to broadband signals, we need only recognize that any broadband

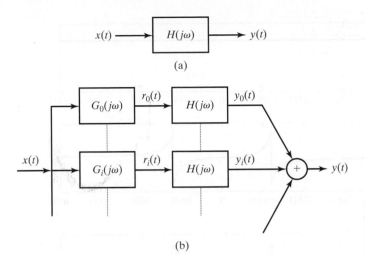

**Figure 2.6**    (a) A general LTI system with frequency response $H(j\omega)$. (b) An equivalent system in which the input is divided into frequency groups.

signal can be viewed as a superposition of narrowband signals. This representation can in fact be developed formally by recognizing that the system in Figure 2.6(a) is equivalent to the parallel combination in Figure 2.6(b) as long as

$$\sum_i G_i(j\omega) = 1 .$$    (2.24)

By choosing the filters $G_i(j\omega)$ to satisfy Eq. (2.24) and to be narrowband around center frequencies $\omega_i$, each of the intermediate signals $r_i(t)$ is a narrowband signal. Consequently, the time-domain effect of the phase of $H(j\omega)$ is to apply the group delay and phase delay to each of the narrowband components (i.e., frequency groups) $r_i(t)$. If the group delay is different at the different center (i.e., carrier) frequencies $\omega_i$, then according to Eq. (2.23b) in the time domain different frequency groups will arrive at the output at different times and with carrier phase related to the phase delay.

**Example 2.3    Illustration of the Effect of Nonlinear Phase**

As a first example, consider $H(j\omega)$ in Figure 2.6 to be the CT all-pass system with frequency-response magnitude, phase, and group delay as shown in Figure 2.7. The corresponding impulse response has an impulse at $t = 0$ followed by the response shown in Figure 2.8.

   If the phase of $H(j\omega)$ were linear with frequency, the impulse response would simply be a delayed impulse, that is, all the narrowband components would be delayed by the same amount and correspondingly would add up to a delayed impulse. However, as we see in Figure 2.7, the group delay is not constant since the phase is nonlinear. In particular, frequencies around 1200 Hz are delayed significantly more than around other frequencies. Correspondingly, in Figure 2.8 we see that specific frequency group dominant in the impulse response until around 2.5 ms, and fading beyond.

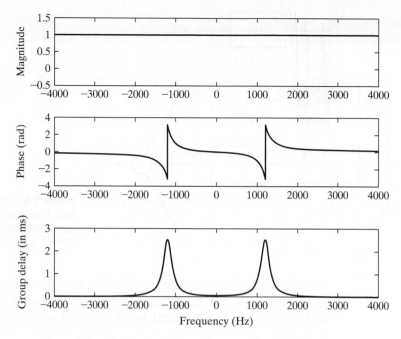

**Figure 2.7**     Magnitude, phase, and group delay of an all-pass filter.

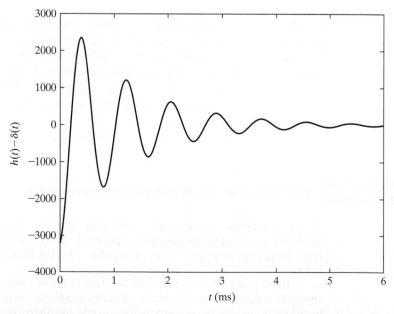

**Figure 2.8**     Non-impulsive component of the impulse response for the filter in Figure 2.7.

**Example 2.4** **A Second Example of the Effect of Nonlinear Phase**

A second example is shown in Figure 2.9, in which $H(j\omega)$ is again an all-pass system with nonlinear phase and consequently nonconstant group delay. With this example, we would expect to see different delays in the frequency groups around 50 Hz, 150 Hz, and 300 Hz. Specifically, the group delay in Figure 2.9(c) suggests that the frequency group at 300 Hz will be delayed to about 0.02 seconds, that the group at 150 Hz will be delayed further, and that the tail of the impulse response will consist primarily of the frequency group at 50 Hz, fading by 0.1 sec.

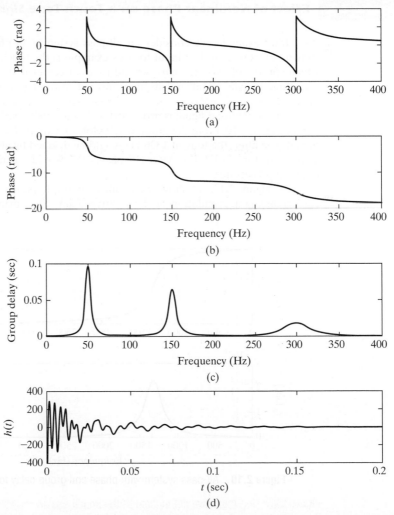

**Figure 2.9**   Phase, group delay, and impulse response for an all-pass system: (a) principal phase; (b) unwrapped phase; (c) group delay; (d) impulse response. (From A. V. Oppenheim and A. S. Willsky, *Signals and Systems*, Prentice Hall, 1997, Figure 6.5.)

In both of these examples, the input is highly concentrated in time (i.e., an impulse) and the response is dispersed in time because of the nonconstant group delay, which results from the nonlinear phase. In general, the effect of nonlinear phase is referred to as dispersion. In communication systems and many other application contexts, even when a channel has a relatively constant frequency-response magnitude characteristic, nonlinear phase can result in significant distortion and other negative consequences because of the resulting time dispersion. For this reason, it is often essential to incorporate phase equalization to compensate for nonconstant group delay.

---

**Example 2.5**     **Effect of Nonlinear Phase on a Touch-Tone Signal**

As a third example, we consider an all-pass system with phase and group delay as shown in Figure 2.10. The input for this example is the touch-tone phone digit "five," which consists of two very narrowband tones at center frequencies 770 and 1336 Hz. The time-domain signal and its two narrowband component signals are shown in Figure 2.11.

The touch-tone signal is processed with multiple passes through the all-pass system of Figure 2.10. From the group delay plot, we expect that, in a single pass through the all-pass filter, the tone at 1336 Hz would be delayed by about 2.5 milliseconds relative to the tone at 770 Hz. After 200 passes, this would accumulate to a relative delay of about 0.5 seconds.

In Figure 2.12, we show the result of multiple passes through a filter and the corresponding accumulation of the relative delays, in a manner consistent with our expectations.

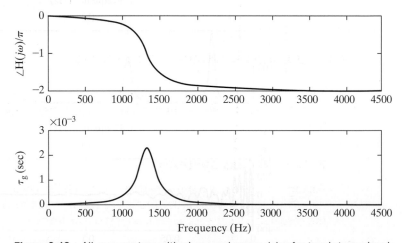

**Figure 2.10**   All-pass system with phase and group delay for touch-tone signal.

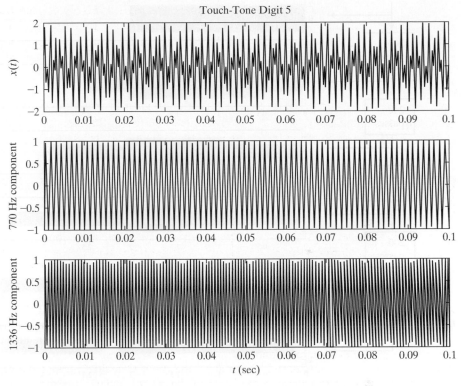

**Figure 2.11**   Touch-tone signal with its two narrowband component signals.

## 2.3 ALL-PASS AND MINIMUM-PHASE SYSTEMS

Two particularly interesting and useful classes of stable LTI systems are all-pass systems and minimum-phase systems. We define and discuss them in this section.

### 2.3.1 All-Pass Systems

An all-pass system is a stable system for which the magnitude of the frequency response is a constant, independent of frequency. The frequency response in the case of a CT all-pass system is thus of the form

$$H_{ap}(j\omega) = A e^{j\angle H_{ap}(j\omega)} \, , \tag{2.25}$$

where $A$ is a constant, not varying with $\omega$. Assuming the associated transfer function $H(s)$ is rational in $s$, it will have the corresponding form

$$H_{ap}(s) = A \prod_{k=1}^{M} \frac{s + a_k^*}{s - a_k} \, . \tag{2.26}$$

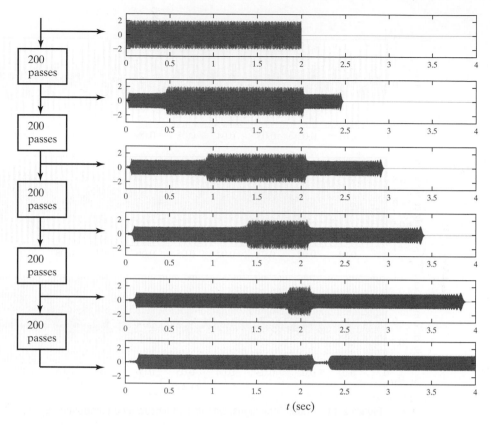

**Figure 2.12**   Effect of passing the touch-tone signal of Figure 2.11 multiple times through an all-pass filter.

For each pole at $s = +a_k$, there is a zero at the mirror image across the imaginary axis, namely at $s = -a_k^*$; and if $a_k$ is complex and the system impulse response is real-valued, every complex pole and zero will occur in a conjugate pair, so there will also be a pole at $s = +a_k^*$ and a zero at $s = -a_k$. It is straightforward to verify that each of the $M$ factors in Eq. (2.26) has unit magnitude for $s = j\omega$. An example of a pole-zero diagram (in the $s$-plane) for a causal CT all-pass system is shown in Figure 2.13.

For a DT all-pass system, the frequency response is of the form

$$H_{ap}(e^{j\Omega}) = A e^{j\angle H_{ap}(e^{j\Omega})} . \qquad (2.27)$$

If the associated transfer function $H(z)$ is rational in $z$, it will have the form

$$H_{ap}(z) = A \prod_{k=1}^{M} \frac{z^{-1} - b_k^*}{1 - b_k z^{-1}} . \qquad (2.28)$$

The poles and zeros in this case occur at conjugate reciprocal locations: for each pole at $z = b_k$, there is a zero at $z = 1/b_k^*$. A zero at $z = 0$ (and associated pole at $\infty$) is obtained by setting $b_k = \infty$ in the corresponding factor above,

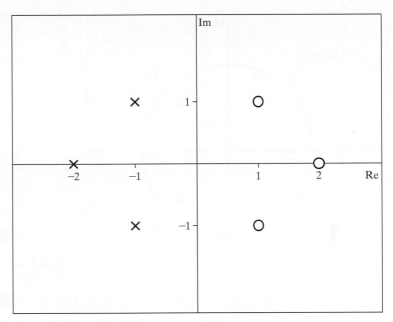

**Figure 2.13**    Typical pole-zero plot for a CT all-pass system.

after first dividing both the numerator and denominator by $b_k$; this results in the corresponding factor in Eq. (2.28) being just $z$. Again, if the impulse response is real-valued, then every complex pole and zero will occur in a conjugate pair, so there will be a pole at $z = b_k^*$ and a zero at $z = 1/b_k$. It is again straightforward to verify that each of the $M$ factors in Eq. (2.28) has unit magnitude for $z = e^{j\Omega}$. An example of a pole-zero diagram (in the $z$-plane) for a causal DT all-pass system is shown in Figure 2.14.

The phase of a CT all-pass system will be the sum of the phases associated with each of the $M$ factors in Eq. (2.26). Assuming the system is causal (in addition to being stable), then for each of these factors $\text{Re}\{a_k\} < 0$. With some algebra it can be shown that each factor of the form $\frac{s+a_k^*}{s-a_k}$ now has positive group delay at all frequencies, a property that we will make reference to shortly. Similarly, assuming causality (in addition to stability) for the DT all-pass system in Eq. (2.28), each factor of the form $\frac{z^{-1}-b_k^*}{1-b_k z^{-1}}$ with $|b_k| < 1$ contributes positive group delay at all frequencies (or zero group delay in the special case of $b_k = 0$). Thus, in both continuous and discrete time, the frequency response of a causal all-pass system has constant magnitude and positive group delay at all frequencies.

## 2.3.2 Minimum-Phase Systems

In classical network theory, control systems, and signal processing, a CT LTI system with a rational transfer function is defined as minimum phase if it is stable, causal, and has all its finite zeros strictly within the left-half plane. The

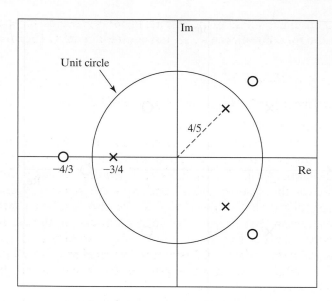

**Figure 2.14**   Typical pole-zero plot for a DT all-pass system.

notion arises in the context of deducing the transfer function to within a sign factor, from its associated magnitude on the imaginary axis, that is, from the frequency response magnitude. We shall see in the next section, Example 2.7, how this can be accomplished algebraically for a minimum-phase CT system, using what we have learned about all-pass system in Section 2.3.1.

For a DT LTI system with a rational transfer function, one can similarly deduce the transfer function from its magnitude on the unit circle if the system is stable, causal, has only finite zeros (i.e., has no zeros at infinity), and all these zeros are strictly inside the unit circle. This is equivalent to requiring that the system be stable, causal, and have a stable and causal inverse, which we shall use in this text as the definition of minimum phase for a DT LTI system. Minimum-phase DT models will be particularly important for our discussion in Chapter 12 of causal Wiener filtering for random signals.

### Group Delay of Minimum-Phase Systems

The use of the term *minimum phase* is historical, and the property should perhaps more appropriately be termed "minimum group delay." This interpretation utilizes the fact, which we establish below, that any causal and stable CT system with a rational transfer function $H_{cs}(s)$ and no finite zeros on the imaginary axis, can be represented as the cascade of a minimum-phase system and a causal all-pass system:

$$H_{cs}(s) = H_{min}(s)H_{ap}(s) \, . \tag{2.29}$$

Similarly, in the DT case, provided the transfer function $H_{cs}(z)$ has no zeros on the unit circle, it can be written as

$$H_{cs}(z) = H_{min}(z)H_{ap}(z) \, . \tag{2.30}$$

The frequency-response magnitude of the all-pass factor is constant, independent of frequency, and for convenience we set this constant to unity. Then from Eq. (2.29)

$$|H_{cs}(j\omega)| = |H_{min}(j\omega)| \quad \text{and} \tag{2.31a}$$

$$\text{grpdelay}[H_{cs}(j\omega)] = \text{grpdelay}[H_{min}(j\omega)] + \text{grpdelay}[H_{ap}(j\omega)] . \tag{2.31b}$$

Similar equations hold in the DT case.

We will see in the next section that the minimum-phase term in Eq. (2.29) can be uniquely determined from the magnitude of $H_{cs}(j\omega)$. Consequently all causal, stable CT systems with rational transfer functions and no zeros on the imaginary axis, and with the same frequency-response magnitude, differ only in the choice of the causal all-pass factor in Eq. (2.29). However, we have shown previously that causal all-pass factors must contribute positive group delay. Therefore we conclude from Eq. (2.31b) that within this class of CT systems, the one with no all-pass factors in Eq. (2.29) will have the minimum group delay. The corresponding result is established similarly in the DT case.

To illustrate the factorization of Eq. (2.29), we consider a simple example.

---

**Example 2.6    Causal, Stable System as Cascade of Minimum-Phase and All-Pass**

Consider the causal, stable system with transfer function

$$H_{cs}(s) = \frac{(s-1)(s+2)}{(s+3)(s+4)} . \tag{2.32}$$

Since it has a zero in the right half-plane, specifically at $s = 1$, it is not minimum phase. However, consider the cascade of $H_{cs}(s)$ with an identity factor $\frac{(s+1)}{(s+1)}$ to express $H_{cs}(s)$ as

$$H_{cs}(s) = \frac{(s-1)(s+2)}{(s+3)(s+4)} \cdot \frac{(s+1)}{(s+1)} \tag{2.33}$$

or equivalently

$$H_{cs}(s) = \frac{(s+1)(s+2)}{(s+3)(s+4)} \cdot \frac{(s-1)}{(s+1)} . \tag{2.34}$$

Equation (2.34) corresponds to a cascade of a minimum-phase factor $H_{min}(s)$ and a causal all-pass factor $H_{ap}(s)$ where

$$H_{min}(s) = \frac{(s+1)(s+2)}{(s+3)(s+4)} \tag{2.35a}$$

and

$$H_{ap}(s) = \frac{(s-1)}{(s+1)} . \tag{2.35b}$$

In effect, the factor $\frac{(s+1)}{(s+1)}$ in Eq. (2.33) reflects the zero at $s = 1$ to the location $s = -1$ and then also cancels it with a pole at the same location. This approach generalizes easily for multiple zeros in the right half-plane and also to DT systems.

Generalizing from Example 2.6, consider a causal, stable transfer function $H_{cs}(s)$ expressed in the form

$$H_{cs}(s) = A \frac{\prod_{k=1}^{M_1}(s - l_k) \prod_{i=1}^{M_2}(s - r_i)}{\prod_{n=1}^{N}(s - d_n)}, \tag{2.36}$$

where $d_n$ represents the poles of the system, $l_k$ represents the zeros in the left half-plane, and $r_i$ represents the zeros in the right half-plane. Since $H_{cs}(s)$ is stable and causal, all of the poles are in the left half-plane and would be associated with the factor $H_{min}(s)$ in Eq. (2.29), as would be all of the zeros $l_k$. We next represent the right half-plane zeros as

$$\prod_{i=1}^{M_2}(s - r_i) = \prod_{i=1}^{M_2}(s + r_i) \prod_{i=1}^{M_2} \frac{(s - r_i)}{(s + r_i)}. \tag{2.37}$$

Since $\text{Re}\{r_i\}$ is positive, the first factor on the right side in Eq. (2.37) represents the left half-plane zeros. The second factor corresponds to all-pass terms with left half-plane poles, and with zeros at mirror-image locations to the poles. Thus, combining Eqs. (2.36) and (2.37), $H_{cs}(s)$ has been decomposed according to Eq. (2.29), where

$$H_{min}(s) = A \frac{\prod_{k=1}^{M_1}(s - l_k) \prod_{i=1}^{M_2}(s + r_i)}{\prod_{n=1}^{N}(s - d_n)} \tag{2.38a}$$

$$H_{ap}(s) = \prod_{i=1}^{M_2} \frac{(s - r_i)}{(s + r_i)}. \tag{2.38b}$$

The corresponding result in Eq. (2.30) for discrete time follows in a very similar manner.

## 2.4 SPECTRAL FACTORIZATION

The approach used for the minimum-phase/all-pass decomposition developed above is useful in a variety of contexts. One that is of particular interest to us in later chapters arises when we are given or have measured the magnitude of the frequency response of a stable system with a rational transfer function $H(s)$ (and real-valued impulse response), and our objective is to recover $H(s)$ from this information. A similar task may be posed in the DT case, but we focus on the CT version here. We are thus given

$$|H(j\omega)|^2 = H(j\omega)H^*(j\omega) \tag{2.39}$$

or, since $H^*(j\omega) = H(-j\omega)$,

$$|H(j\omega)|^2 = H(j\omega)H(-j\omega). \tag{2.40}$$

Now $H(j\omega)$ is $H(s)$ for $s = j\omega$, and therefore

$$|H(j\omega)|^2 = H(s)H(-s)\Big|_{s=j\omega}. \tag{2.41}$$

For any numerator or denominator factor $(s - a)$ in $H(s)$, there will be a corresponding factor $(-s - a)$ in $H(s)H(-s)$. Thus $H(s)H(-s)$ will consist of factors in the numerator or denominator of the form $(s - a)(-s - a) = -s^2 + a^2$, and will therefore be a rational function of $s^2$. Consequently $|H(j\omega)|^2$ will be a rational function of $\omega^2$. Thus, if we are given or can express $|H(j\omega)|^2$ as a rational function of $\omega^2$, we can obtain the product $H(s)H(-s)$ by making the substitution $\omega^2 = -s^2$.

The product $H(s)H(-s)$ will always have its zeros in pairs that are mirrored across the imaginary axis of the $s$-plane, and similarly for its poles. For any pole or zero of $H(s)H(-s)$ at the real value $a$, there will be another at the mirror image $-a$, while for any pole or zero at the complex value $q$, there will be others at $q^*$, $-q$, and $-q^*$, forming a complex conjugate pair $(q, q^*)$ and its mirror image $(-q^*, -q)$. We then need to assign one of each mirrored real pole and zero and one of each mirrored conjugate pair of poles and zeros to $H(s)$, and the mirror image to $H(-s)$.

If we assume (or know) that $H(s)$ is causal, in addition to being stable, then we would assign the left half-plane poles of each pair to $H(s)$. With no further knowledge or assumption, we have no guidance on the assignment of the zeros other than the requirement of assigning one of each mirror image pair to $H(s)$ and the other to $H(-s)$. If we further know or assume that the system has all its zeros in the left half-plane, then the left half-plane zeros from each mirrored pair are assigned to $H(s)$, and the right half-plane zeros to $H(-s)$. This process of factoring $H(s)H(-s)$ to obtain $H(s)$ is referred to as spectral factorization.

## Example 2.7    Spectral Factorization

Consider a frequency-response magnitude that has been measured or approximated as

$$|H(j\omega)|^2 = \frac{\omega^2 + 1}{\omega^4 + 13\omega^2 + 36} = \frac{\omega^2 + 1}{(\omega^2 + 4)(\omega^2 + 9)}. \tag{2.42}$$

Making the substitution $\omega^2 = -s^2$, we obtain

$$H(s)H(-s) = \frac{-s^2 + 1}{(-s^2 + 4)(-s^2 + 9)} \tag{2.43}$$

which we further factor as

$$H(s)H(-s) = \frac{(s + 1)(-s + 1)}{(s + 2)(-s + 2)(s + 3)(-s + 3)}. \tag{2.44}$$

It now remains to associate appropriate factors with $H(s)$ and $H(-s)$. Assuming the system is causal in addition to being stable, the two left half-plane poles at $s = -2$ and $s = -3$ must be associated with $H(s)$. With no further assumptions, either one of the numerator factors can be associated with $H(s)$ and the other with $H(-s)$. However, if we know or assume that $H(s)$ is minimum phase, then we would assign the left half-plane zero to $H(s)$, resulting in the choice

$$H(s) = \frac{(s+1)}{(s+2)(s+3)} . \tag{2.45}$$

In the DT case, a similar development leads to an expression for $H(z)H(1/z)$ from knowledge of $|H(e^{j\Omega})|^2$. The zeros of $H(z)H(1/z)$ occur in conjugate reciprocal pairs, and similarly for the poles. We again have to split such conjugate reciprocal pairs, assigning one of each to $H(z)$, the other to $H(1/z)$, based on whatever additional knowledge we have. For instance, if $H(z)$ is known to be causal in addition to being stable, then all the poles of $H(z)H(1/z)$ that are inside the unit circle are assigned to $H(z)$; and if $H(z)$ is known to be minimum phase as well, then all the zeros of $H(z)H(1/z)$ that are in the unit circle are assigned to $H(z)$, along with as many additional zeros at the origin as needed to ensure a causal inverse.

## 2.5 FURTHER READING

Chapter 2 has continued the review of the basic concepts of signals and systems, and the references listed at the end of Chapter 1 are generally useful for this chapter as well. For further reading specifically about group delay and its effects on signal transmission and audio, see [Op1], [Op2], [La1], and [Zad]. Minimum-phase systems and spectral factorization are treated in [Gui], [He1], [Moo], [Op2], [Pa3], [Pa4], and [Th1], for example, sometimes in the setting of the power spectral densities arising in Chapters 11, 12, and 13.

# Problems

## Basic Problems

**2.1.** The output $y(t)$ of a causal LTI system is related to the input $x(t)$ by the differential equation

$$\frac{dy(t)}{dt} + 2y(t) = x(t) .$$

**(a)** Determine the frequency response

$$H(j\omega) = \frac{Y(j\omega)}{X(j\omega)}$$

of the system and plot the magnitude and phase as a function of $\omega$.
**(b)** For the frequency response $H(j\omega)$ of part (a), determine and plot the group delay $\tau_g(\omega)$.

**2.2.** A CT system with frequency response $H_1(j\omega)$ has associated group delay $\tau_{g1}(\omega)$, and another system with frequency response $H_2(j\omega)$ has group delay $\tau_{g2}(\omega)$. What is the group delay associated with the cascade of the two systems, whose frequency response is $H_1(j\omega)H_2(j\omega)$? Explain your reasoning.

**2.3.** Consider the signal

$$s(t) = \sum_{n=-\infty}^{\infty} a_n p(t - nT) \cos(\omega_1 t) + \sum_{n=-\infty}^{\infty} b_n p(t - nT) \cos(\omega_2 t) ,$$

where $f_1 = \omega_1/2\pi = 1$ kHz and $f_2 = \omega_2/2\pi = 3$ kHz, and

$$p(t) = \text{sinc}(0.5 \cdot 10^3 t) ,$$

where $t$ is in seconds.

Suppose this signal is transmitted over the channel whose frequency response, $H(j\omega)$, is shown in Figure P2.3.

**(a)** Sketch the group delay as a function of $\omega$ for this channel.

**(b)** Give an expression for the channel output $y(t)$.

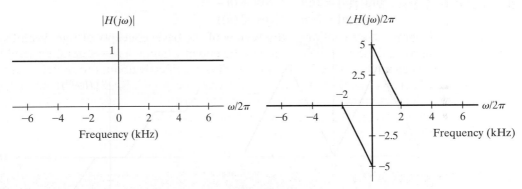

**Figure P2.3**

**2.4.** For each of the following statements, specify whether it is true or false:

**(a)** For the system with frequency response $H(j\omega) = 3e^{-j3\omega}$, the phase delay and the group delay are equal.

**(b)** Time-shifted versions of eigenfunctions of an LTI system (both CT and DT) are always eigenfunctions of the LTI system.

**(c)** Linear combinations of eigenfunctions of an LTI system (both CT and DT) are always eigenfunctions of the LTI system.

For the following three questions consider the DT LTI system described by the unit sample response and corresponding frequency response

$$h[n] = \left(\frac{2}{3}\right)^n u[n] - \delta[n] , \qquad H(e^{j\Omega}) = \frac{\frac{2}{3}e^{-j\Omega}}{1 - \frac{2}{3}e^{-j\Omega}} .$$

**(d)** The system is causal.

**(e)** The system is bounded-input, bounded-output (BIBO) stable.

**(f)** The response $y[n]$ of the system to the input $x[n] = (-1)^n$ (for all $n$) takes
the form

$$y[n] = K_1(-1)^n + K_2 \left(\frac{2}{3}\right)^{n-1} u[n-1],$$

where $K_1 \neq 0$ and $K_2 \neq 0$.

**2.5.** For each of the following multiple-choice questions, select the correct answer
and provide a short explanation. The frequency response of a system is plotted
in Figure P2.5 ($-\pi < \Omega < \pi$).

**(a)** For the input $x[n] = \cos(\frac{\pi}{3}n)$, the output is given by

  (i) $y[n] = 2x[n-2]$
  (ii) $y[n] = 2x[n-1]$
  (iii) $y[n] = 2x[n]$
  (iv) $y[n] = 2x[n+1]$.

**(b)** For the input $x[n] = s[n]\cos(\frac{2\pi}{3}n)$, where $s[n]$ is a very low frequency
bandlimited signal (compared to $2\pi/3$), the output may be approximated by

  (i) $y[n] = 2s[n+1]\cos(\frac{2\pi}{3}(n))$
  (ii) $y[n] = 2s[n]\cos(\frac{2\pi}{3}(n-1))$
  (iii) $y[n] = 2s[n-1]\cos(\frac{2\pi}{3}(n-1))$
  (iv) $y[n] = 2s[n-1]\cos(\frac{2\pi}{3}(n))$.

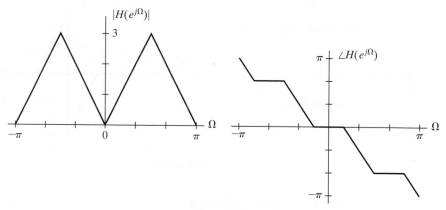

**Figure P2.5**

**2.6.** Consider a DT LTI system whose unit sample response $h[n]$ is $\delta[n] - \delta[n-1]$.

**(a)** The system's frequency response $H(e^{j\Omega})$ can be written in the form

$$H(e^{j\Omega}) = 2e^{j\Theta(\Omega)} \sin(\Omega/2).$$

Determine $\Theta(\Omega)$ for $-\pi \leq \Omega \leq \pi$, and also the phase delay and group delay
of the system in this frequency range.

**(b)** If the input to the system is

$$x[n] = p[n] \cos\left(\frac{\pi n}{3}\right),$$

where $p[n]$ is a slowly varying (narrowband) envelope, the approximate
form of the output signal is

$$y[n] = q[n] \cos\left(\frac{\pi(n-n_0)}{3}\right).$$

Determine $n_0$, and describe in words how $q[n]$ is related to $p[n]$.

**2.7.** The first plot in Figure P2.7-1 shows a signal $x[n]$ that is the sum of three narrowband pulses which do not overlap in time. Its transform magnitude $|X(e^{j\Omega})|$ is shown in the second plot. The frequency-response magnitude and group delay for filter A, a DT LTI system, are shown in the third and fourth plots, respectively.

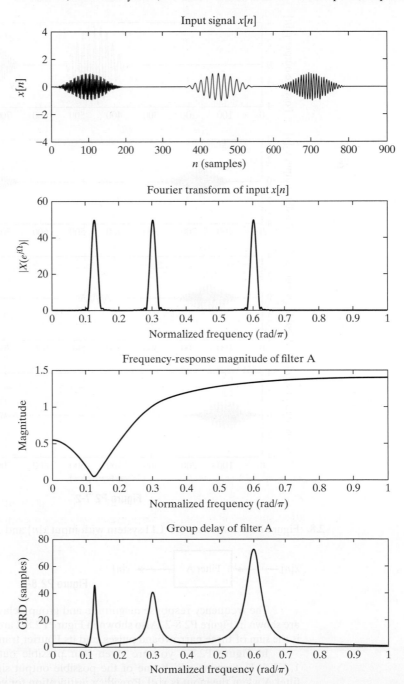

**Figure P2.7-1**

The plots in Figure P2.7-2 show four possible output signals $y_i[n]$: $i = 1, 2, 3, 4$.

Determine which of the possible output signals is the output of filter A when the input is $x[n]$. Clearly state your reasoning.

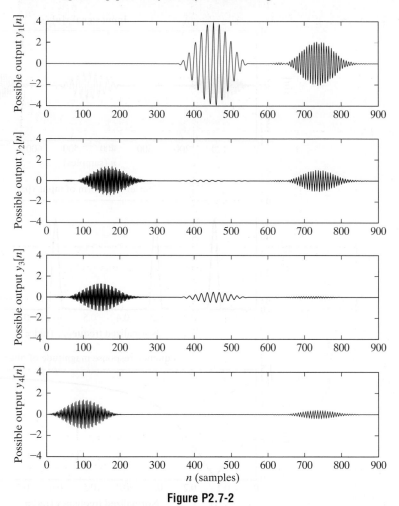

**Figure P2.7-2**

**2.8.** Figure P2.8-1 shows a DT LTI system with input $x[n]$ and output $y[n]$.

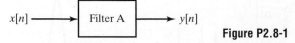

$x[n] \longrightarrow$ Filter A $\longrightarrow y[n]$

**Figure P2.8-1**

The frequency-response magnitude and group delay functions for filter A are shown in Figure P2.8-2. Also shown in Figure P2.8-2 are the signal $x[n]$, which is the sum of three narrowband pulses, and its Fourier transform magnitude.

In Figure P2.8-3 you are given four possible output signals $y_i[n]$: $i = 1, 2, 3, 4$. Determine which one of the possible output signals is the output of filter A when the input is $x[n]$. Provide a justification for your choice.

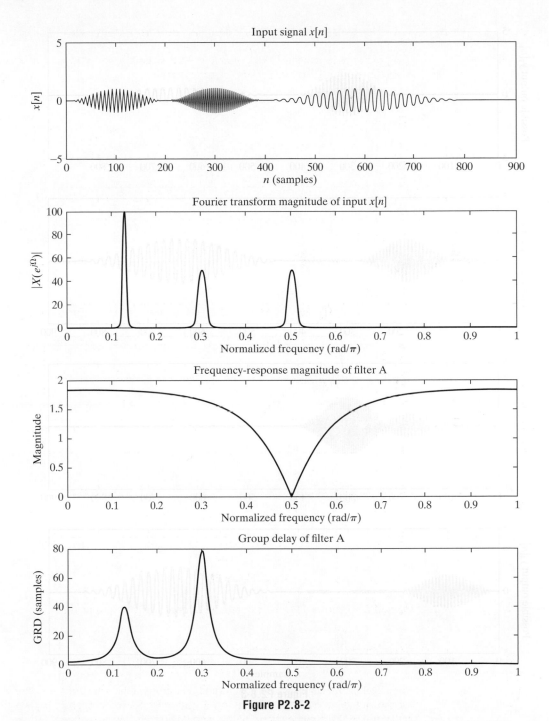

**Figure P2.8-2**

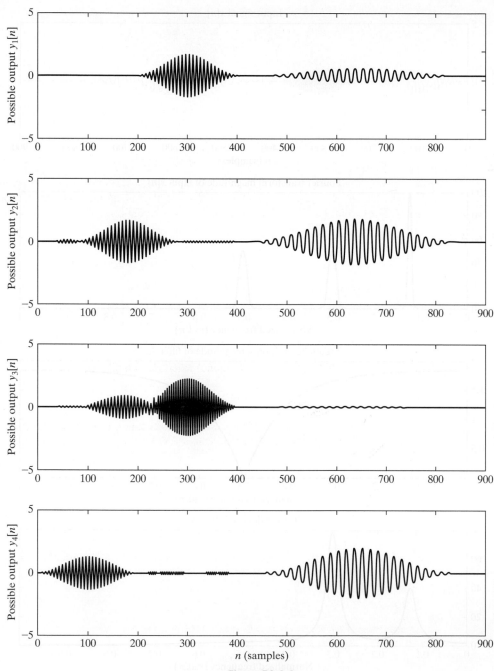

**Figure P2.8-3**

**2.9.** Does the system represented by

$$G(z) = \frac{z^{-1}}{1 - 0.7z^{-1}}, \quad |z| > 0.7$$

have a causal and stable inverse?

   If yes, specify the transfer function $G_I(z)$ (including its region of convergence) that represents the causal, stable inverse.

   If no, give the transfer function (including its region of convergence) of an all-pass system $G_{AP}(z)$ and the transfer function (including its region of convergence) of a minimum phase system $G_{MP}(z)$ such that

$$G(z) = G_{MP}(z)G_{AP}(z) .$$

**2.10.** Write each of the following stable system functions as the product of an all-pass system function and a minimum-phase system function. Note that part (a) pertains to a CT system, while part (b) pertains to a DT system.

**(a)**

$$G(s) = \frac{s - 2}{s + 1} = A(s)M(s) ,$$

where $A(s)$ is all-pass and $M(s)$ is minimum phase.

**(b)**

$$H(z) = 1 - 6z = B(z)N(z) ,$$

where $B(z)$ is all-pass and $N(z)$ is minimum phase.

**2.11.** The squared magnitude of the frequency response of a filter is

$$|H(j\omega)|^2 = \frac{\omega^2 + 1}{\omega^2 + 100} .$$

Determine $H(j\omega)$, given that it is stable and causal, and has a stable and causal inverse. Assume that $H(j\omega)$ is positive at $\omega = 0$.

# Advanced Problems

**2.12.** Suppose we apply the modulated signal

$$x(t) = m(t) \cos(\omega_0 t)$$

to the input of the LTI communication channel shown in Figure P2.12-1 with frequency response $H(j\omega)$, where the modulating signal is

$$m(t) = \frac{\sin(\pi t/T)}{\pi t/T} .$$

Assume $(1/T) = 75$ kHz and $(\omega_0/2\pi) = 1300$ kHz.

Channel

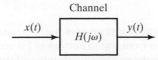

**Figure P2.12-1**

(a) Make a neat and fully labeled sketch of $X(j\omega)$.

(b) Find a time-domain expression for the output $y(t)$ of the channel if the channel frequency response is

$$H(j\omega) = e^{-j\omega(4\times10^{-6})}.$$

(c) Find an approximate (but reasonably accurate) time-domain expression for the output $y(t)$ of the channel if the channel characteristics are actually as shown in Figure P2.12-2 rather than as specified in (b). Also state what features of the signal and/or channel make your approximation reasonable.

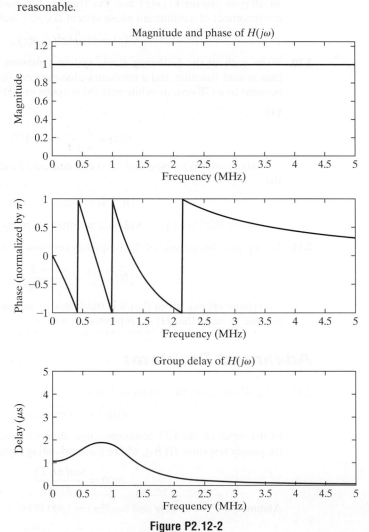

Figure P2.12-2

2.13. The impulse responses $h_1(t)$ through $h_4(t)$ of four different all-pass systems are shown from top to bottom in Figure P2.13-1. In Figure P2.13-2 are the associated group delay plots, randomly ordered, and labeled A through D from top to bottom. For each impulse response in Figure P2.13-1 specify the associated group delay plot from Figure P2.13-2, and explain your reasoning.

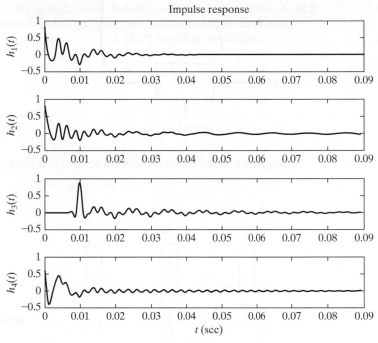

**Figure P2.13-1**

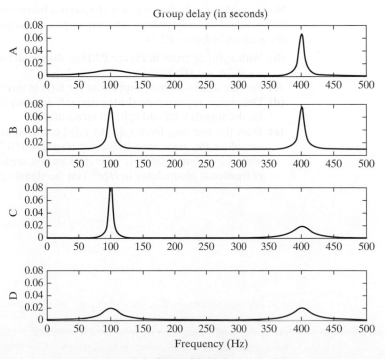

**Figure P2.13-2**

**2.14.** Consider the ideal (bandlimited interpolating) discrete-to-continuous (D/C) converter with input $s_d[n]$, output $s_c(t)$, and reconstruction interval $T = 0.5 \times 10^{-3}$ sec, shown in Figure P2.14-1.

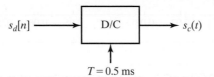

$T = 0.5$ ms                                        **Figure P2.14-1**

Suppose that the Fourier transform of $s_d[n]$ is as shown in Figure P2.14-2.

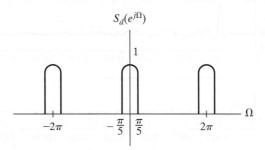

**Figure P2.14-2**

**(a)** Express $S_c(j\omega)$ in terms of $S_d(e^{j\Omega})$ and give a fully labeled sketch of $S_c(j\omega)$ in the interval $|\omega| < 2\pi \times 2000$ rad/sec.

Now consider a scenario, in which $H_c(j\omega)$ is a frequency response whose magnitude is equal to 1 everywhere, and whose phase and group delay characteristics are as given in Figure P2.14-3.

**(b)** With $x_d[n]$ as given in Figure P2.14-4, draw a detailed sketch of $X_c(j\omega)$ for $|\omega| < 2\pi \times 2,000$ rad/sec.

**(c)** Provide a time-domain expression for $x_c(t)$ in terms of $s_c(t)$.

**(d)** Determine approximate (but reasonably accurate) time-domain expressions for the signals $y_c(t)$ and $y_d[n]$ in terms of $s_c(t)$.

**(e)** Does the mapping from $x_d[n]$ to $y_d[n]$ correspond to an LTI system? If so, how does the associated frequency response $H(e^{j\Omega})$ relate to $H_c(j\omega)$, and what is its group delay at $\Omega = 4\pi/5$? Can you see how to interpret the effect of fractional group delay in $H(e^{j\Omega})$ on the signal $x_d[n]$?

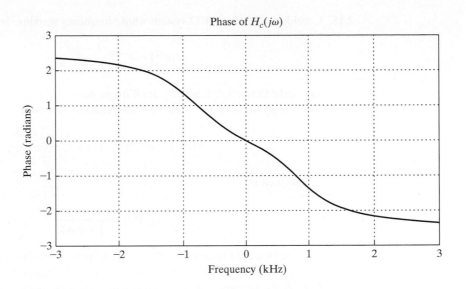

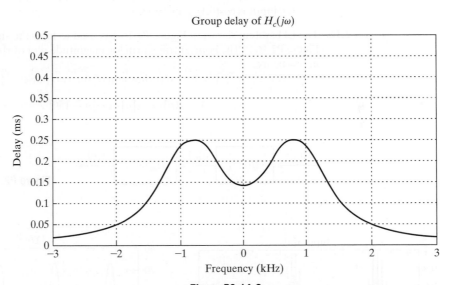

**Figure P2.14-3**

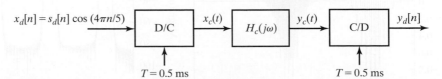

**Figure P2.14-4**

**2.15.** Consider a DT causal LTI system whose frequency response is

$$H(e^{j\Omega}) = e^{-j4\Omega}\frac{1 - \frac{1}{2}e^{j\Omega}}{1 - \frac{1}{2}e^{-j\Omega}} \ .$$

**(a)** Show that $|H(e^{j\Omega})|$ is unity at all frequencies.

**(b)** The group delay for a system with frequency response

$$F(e^{j\Omega}) = e^{j3\Omega}H(e^{j\Omega}) = e^{-j\Omega}\frac{1 - \frac{1}{2}e^{j\Omega}}{1 - \frac{1}{2}e^{-j\Omega}}$$

is given by

$$\tau_{g,F}(\Omega) = \frac{\frac{3}{4}}{\frac{5}{4} - \cos\Omega} \ .$$

Use this fact to determine the group delay $\tau_{g,H}(\Omega)$ associated with the given system $H(e^{j\Omega})$. Plot the resulting expression for $\tau_{g,H}(\Omega)$.

**(c)** Obtain an approximate expression for the output of the system $H(e^{j\Omega})$ when the input is $\cos(0.1n)\cos(\pi n/3)$ .

**2.16.** The LTI system shown in Figure P2.16-1 is used to filter the signal $x[n]$ shown in Figure P2.16-2. The Fourier transform log magnitude plot of $x[n]$ is also displayed in this figure.

The input $x[n]$ is the following sum of the two pulses $x_1[n]$ and $x_2[n]$:

$$x[n] = x_1[n] + x_2[n - 150] \ .$$

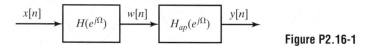

**Figure P2.16-1**

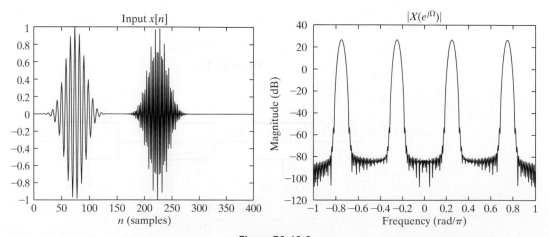

**Figure P2.16-2**

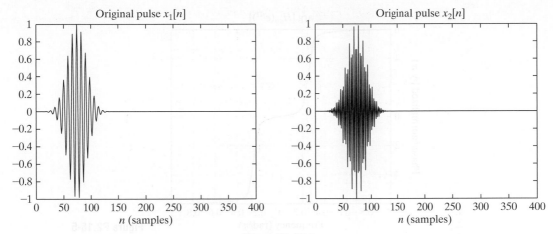

**Figure P2.16-3**

The two pulses, $x_1[n]$ and $x_2[n]$, are shown in Figure P2.16-3. The first filter in Figure P2.16-1 has an impulse response $h[n]$ and frequency response $H(e^{j\Omega})$ that are purely real. The log magnitude, $20\log_{10}|H(e^{j\Omega})|$, for this filter is shown in Figure P2.16-4. The causal and stable filter $H_{ap}(e^{j\Omega})$ in the system is an all-pass filter, i.e., $|H_{ap}(e^{j\Omega})| = 1$. The phase of the all-pass filter is shown in Figure P2.16-5.

Assume that all sequences are zero outside of the intervals shown in the figures. Also, all of the time signals are DT sequences, with plots showing consecutive points connected by straight lines.

**(a)** Calculate the group delay of the filter with frequency response $H(e^{j\Omega})$, and specify (by a careful sketch, or in some other appropriate way) $w[n]$ and $|W(e^{j\Omega})|$, the magnitude of the Fourier transform of $w[n]$, for $|\Omega| < \pi$.

**(b)** Make a labeled sketch of what you would expect the output $y[n]$ to look like.

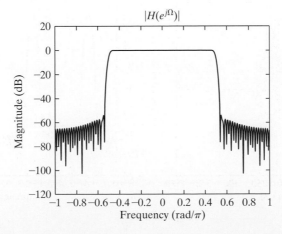

**Figure P2.16-4**

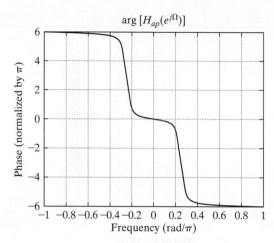

**Figure P2.16-5**

**2.17.** The first plot in Figure P2.17-1 shows a signal $x[n]$ that is the sum of three narrow-band pulses which do not significantly overlap in time. Its transform magnitude $|X(e^{j\Omega})|$ is shown in the second plot. The group delay and frequency-response magnitude functions of filter A, a DT LTI system, are shown in the third and fourth plots, respectively.

   The remaining plots in Figures P2.17-2 and P2.17-3 show eight possible output signals $y_i[n]$: $i = 1, 2, \ldots, 8$. Determine which of the possible output signals is the output of filter A when the input is $x[n]$. Clearly state your reasoning.

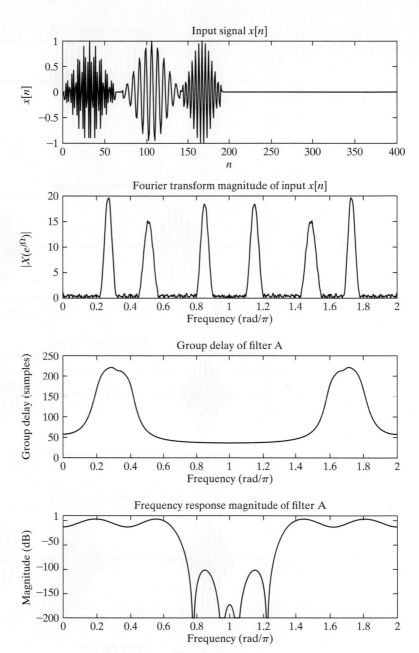

**Figure P2.17-1**

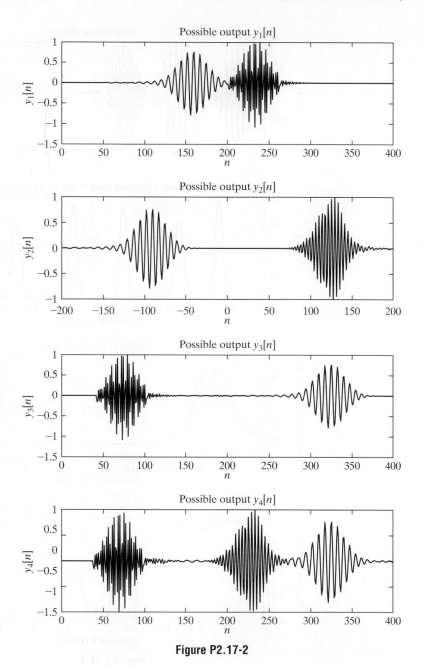

**Figure P2.17-2**

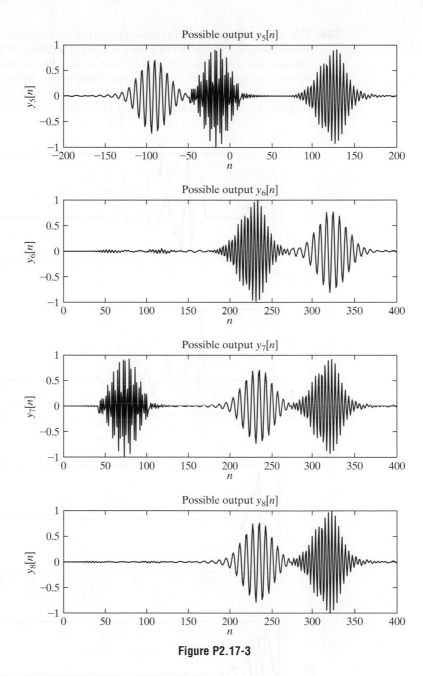

**Figure P2.17-3**

**2.18.** The impulse responses $h_1[n]$ through $h_4[n]$ of four different all-pass systems are shown from top to bottom in Figure P2.18-1 (the values of $h_i[n]$ for consecutive values of $n$ are connected by straight lines in these figures to improve readability). Shown in Figure P2.18-2 are the associated group delay plots, but randomly ordered, and labeled A through D from top to bottom. Specify which impulse response goes with which group delay plot, being sure to explain your reasoning.

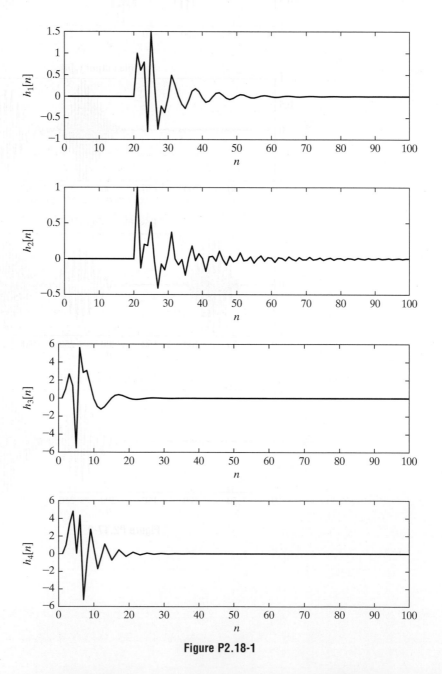

**Figure P2.18-1**

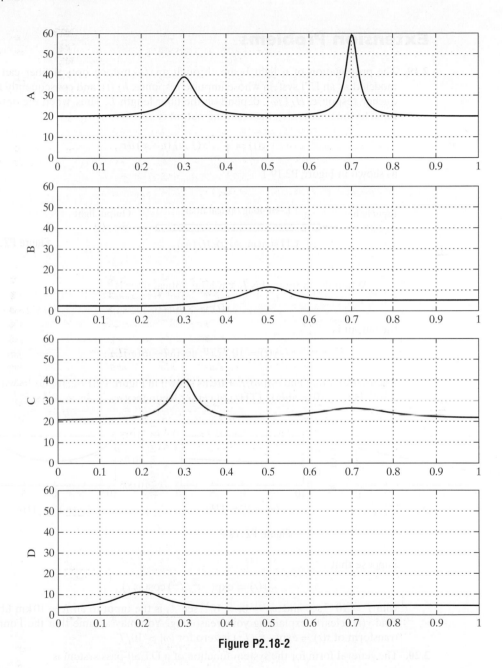

**Figure P2.18-2**

## Extension Problems

**2.19.** The propagation of a light beam through an $L$-km-long optical fiber can be modeled as an LTI system whose impulse response, $h_L(t)$, and consequently frequency response, $H_L(j\omega)$, depend on the fiber length $L$. Thus, when the optical input to the fiber is $x(t)$, the resulting output is

$$y(t) = \int_{-\infty}^{\infty} x(\tau) h_L(t - \tau) \, d\tau \,,$$

as shown in Figure P2.19-1.

Input light
$x(t)$ → [ L-km-long optical fiber ] → Output light
$y(t)$

LTI system: $h_L(t)$, $H_L(j\omega)$

**Figure P2.19-1**

When the input to this $L$-km-long optical fiber is

$$x(t) = \cos(\omega t) \,,$$

the output is

$$y(t) = 10^{-\alpha(\omega)L} \cos(\omega t - \beta(\omega)L) \,,$$

where $\alpha(\omega)$, $\beta(\omega)$, and $d\beta(\omega)/d\omega$ are as shown in Figure P2.19-2, for $\omega/2\pi$ within $\pm 10^{13}$ Hz $= \pm 10$ terahertz (THz) of center frequency $\omega_o = 4\pi \times 10^{14}$ sec$^{-1}$.

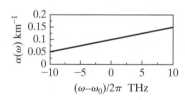
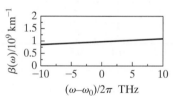
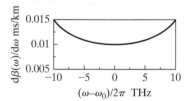

**Figure P2.19-2**

Suppose that

$$x(t) = \exp(-t^2/T^2) \cos(\omega_o t) \,,$$

with $T = 10^{-6}$ sec and $\omega_o = 4\pi \times 10^{14}$ sec$^{-1}$, is the input to an $L = 10$ km fiber. Find $y(t)$, clearly explaining your reasoning. You may assume that the Fourier transform of $s(t) = \exp(-t^2/T^2)$ is zero for $|\omega| > 10/T$.

**2.20.** The general form for the system function of a DT all-pass system is

$$H_{AP}(z) = A \prod_{k=1}^{M} \frac{z^{-1} - a_k^*}{1 - a_k z^{-1}}$$

and for a CT all-pass is

$$H_{AP}(s) = B \prod_{k=1}^{M} \frac{s + b_k^*}{s - b_k} \,.$$

While we phrase this problem in terms of the CT all-pass, the conclusions are identical for discrete time.

**(a)** For $M = 1$ and $b_k$ real in the given equation, show the pole-zero plot for $H_{AP}(s)$.

**(b)** For $M = 1$, $b_k$ possibly complex, and with $H_{AP}(s)$ corresponding to a causal, stable system, show that the group delay associated with $H_{AP}(s)$ will always be greater than zero for all frequencies. You can do this analytically by explicitly differentiating the expression for the phase, but it is also simple (and more intuitive) to argue it from geometrical consideration of the pole-zero plot.

**(c)** By referring to (b), construct an argument to conclude that a stable, causal, CT all-pass system always has positive group delay (though of course the specific value of the group delay will be different at different frequencies).

# 3 Pulse Amplitude Modulation

In Chapter 1 we discussed the discrete-time (DT) processing of continuous-time (CT) signals, and in that context reviewed and discussed D/C conversion for reconstructing a CT signal from a DT sequence. The basic structure embodied in Eq. (1.94) involves, in effect, generating a train of equally spaced CT pulses, most typically the impulse response of an ideal low-pass filter as in Eq. (1.81), with amplitudes corresponding to the values of the DT sequence. This constitutes pulse amplitude modulation (PAM).

Another context in which it is useful and important to generate a CT signal from a DT sequence is in communication systems, in which DT data—perhaps digital or quantized data—is to be transmitted over a channel as a CT signal, after which the DT sequence is recovered at the receiver. For example, consider transmitting a binary sequence of 1s and 0s from one computer to another over a telephone line or cable, or from a digital cell phone to a base station over a high-frequency electromagnetic channel. These instances correspond to having analog channels that require the transmission of a CT signal compatible with the bandwidth and other constraints of the channel. Such requirements impact the choice of the CT waveform that the discrete sequence is modulated onto.

The translation of a DT signal to a CT signal appropriate for transmission, and the translation back to a DT signal at the receiver, are both accomplished by modems (modulators/demodulators), many of which are based on the use of PAM. In addition to the fact that PAM plays an essential role in signal processing and in communications systems, it offers an opportunity to further analytically understand the relationship between the CT and DT domains.

## 3.1 BASEBAND PULSE AMPLITUDE MODULATION

The structure of the modulator and demodulator for PAM is shown in Figure 3.1. The signal $x(t)$ in Figure 3.1(a) is referred to as a pulse amplitude modulated signal.

There are strong similarities between PAM in the context of D/C conversion and in the context of communication systems, as is evident in comparing Figures 1.5 and 3.1. For example, $y_c(t)$ as specified in Figure 1.5 and Eq. (1.81) has the same form as $x(t)$ in Figure 3.1(a) but with $p(t)$ in the context of ideal D/C conversion specifically given by

$$p(t) = \frac{\sin(\pi t/T)}{\pi t/T}. \qquad (3.1)$$

The DT sequence $x_d[n]$ in Figure 1.5 is obtained from the CT signal to be processed in the same way that $\hat{a}[n]$ is obtained from the PAM signal in Figure 3.1(b). Consequently, similar basic analysis applies, although the contexts are very different.

### 3.1.1 The Transmitted Signal

As indicated above, the idea in PAM for communication over a CT channel is to transmit a sequence of CT pulses of some prespecified shape $p(t)$, with the sequence of pulse amplitudes carrying the information. The resulting PAM signal is

$$x(t) = \sum_n a[n]\,p(t - nT). \qquad (3.2)$$

This baseband signal at the transmitter is then usually further modulated onto a high-frequency sinusoidal or complex exponential carrier to form a bandpass signal before actual transmission, although we ignore this aspect for

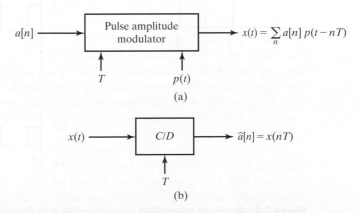

**Figure 3.1**    (a) Pulse amplitude modulation (PAM) of a DT sequence $a[n]$. (b) Recovery of a discrete sequence from a PAM signal when $p(t)$ has the form of Eq. (3.1).

now. The sequence values $a[n]$ are the pulse amplitudes, and $T$ is the pulse repetition interval or the intersymbol spacing, so $1/T$ is the symbol rate (or "baud" rate). The pulse $p(t)$ may be confined in length to $T$ so that the terms in Eq. (3.2) do not overlap, or it may extend over several intervals, as we will see in several examples shortly. The DT signal $a[n]$ may comprise samples of a bandlimited CT signal (taken at the Nyquist rate or higher, and perhaps quantized to a specified set of levels); or 1 and 0 for on/off or "unipolar" signaling; or 1 and $-1$ for antipodal or "polar" signaling. Each of these possibilities is illustrated in Figure 3.2, with the basic pulse being a simple rectangular pulse. These pulses would require substantial channel bandwidth (of the order of $1/\Delta$ at least) in order to be transmitted without significant distortion, so it is often necessary to find alternative choices that use less bandwidth to accommodate the constraints of the channel. Such considerations are important in designing appropriate pulse shapes, and we will elaborate on them shortly.

If $p(t)$ is chosen such that $p(0) = 1$ and $p(nT) = 0$ for $n \neq 0$, then it is possible to recover the amplitudes $a[n]$ from the PAM waveform $x(t)$ just by sampling $x(t)$ at times $nT$, since $x(nT) = a[n]$ in this case. However, our interest is in recovering the amplitudes from the signal at the receiver, rather than directly from the transmitted signal, so we need to consider how the communication channel affects $x(t)$. Our objective will be to recover the DT signal

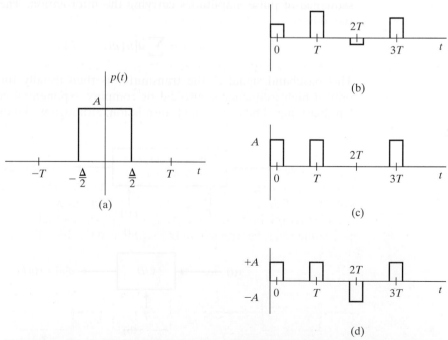

**Figure 3.2**  Baseband transmitted signal (a) pulse $p(t)$; (b) $x(t)$ when $a[n]$ are samples of a CT signal; (c) $x(t)$ for $a[n]$ from on/off signaling; (d) antipodal signaling.

using a C/D system as in Fig. 3.1(b), after first compensating for distortion and noise in the channel.

### 3.1.2 The Received Signal

When a PAM signal is transmitted through a channel, the characteristics of the channel affect our ability to accurately recover the pulse amplitudes $a[n]$ from the received signal $r(t)$. We might model $r(t)$ as

$$r(t) = h(t) * x(t) + v(t) \qquad (3.3)$$

corresponding to the channel being modeled as linear and time-invariant (LTI) with impulse response $h(t)$, and channel noise being represented through the additive noise signal $v(t)$. We would still like to recover the pulse amplitudes $a[n]$ from samples of $r(t)$—or from samples of an appropriately filtered version of $r(t)$—with the samples taken at intervals of $T$.

The overall model is shown in Figure 3.3, with $f(t)$ representing the impulse response of an LTI filter at the receiver. This receiver filter will play a key role in filtering out the part of the noise that lies outside the frequency bands in which the signal information is concentrated. In Chapters 12 and 13, we address in detail strategies to account for and filter the noise. In this chapter, we focus only on the noise-free case. Even in the absence of noise, it may be desirable to utilize a filter before sampling to compensate for channel distortion characteristics such as group delay and dispersion.

### 3.1.3 Frequency-Domain Characterizations

Assuming that the noise $v(t)$ is zero and that the combined frequency response of the channel and receiver filter is unity over the frequency range where $P(j\omega)$, the Fourier transform of $p(t)$, is significant, then $b(t)$ and $x(t)$ will be equal. However, this does not guarantee that $\hat{a}[n]$ and $a[n]$ will be equal because without additional conditions on $p(t)$, even in the absence of noise, periodic sampling of $b(t)$ will not recover $a[n]$. To explore this, we directly examine $x(t)$, $r(t)$, $b(t)$, and their respective Fourier transforms $X(j\omega)$, $R(j\omega)$, and $B(j\omega)$. The understanding that results is a prerequisite for designing $P(j\omega)$ and picking the intersymbol time $T$ for a given channel, and also allows us to determine the influence of the DT signal $a[n]$ on the CT signals $x(t)$ and $r(t)$.

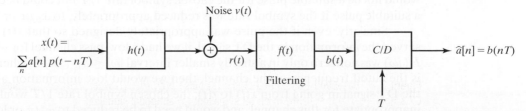

**Figure 3.3**   Transmitter, channel, and receiver model for a PAM system.

To compute $X(j\omega)$, we take the transform of both sides of Eq. (3.2):

$$X(j\omega) = \left( \sum_n a[n]\, e^{-j\omega nT} \right) P(j\omega)$$

$$= A(e^{j\Omega})|_{\Omega=\omega T}\, P(j\omega) \tag{3.4}$$

where $A(e^{j\Omega})$ denotes the DTFT of the sequence $a[n]$. The quantity $A(e^{j\Omega})|_{\Omega=\omega T}$ that appears in the above expression is a uniform rescaling of the frequency axis of the DTFT; in particular, the point $\Omega = \pi$ in the DTFT is mapped to the point $\omega = \pi/T$ in the expression $A(e^{j\Omega})|_{\Omega=\omega T}$.

The expression in Eq. (3.4) describes $X(j\omega)$, assuming the DTFT of the sequence $a[n]$ is well defined. For example, if $a[n] = 1$ for all $n$, corresponding to periodic repetition of the basic pulse waveform $p(t)$, then $A(e^{j\Omega}) = 2\pi\delta(\Omega)$ for $|\Omega| \leq \pi$, and repeats with period $2\pi$ outside this range. Hence $X(j\omega)$ comprises a train of impulses spaced apart by $2\pi/T$; the strength of each impulse is $2\pi/T$ times the value of $P(j\omega)$ at the location of the impulse since from the scaling property of impulses

$$\delta(\Omega)|_{\Omega=\omega T} = \delta(\omega T) = \frac{1}{T}\delta(\omega) \quad \text{for positive } T. \tag{3.5}$$

In the absence of noise, the received signal $r(t)$ and the signal $b(t)$ that results from filtering at the receiver are both easily characterized in the frequency domain:

$$R(j\omega) = H(j\omega)X(j\omega)\,, \qquad B(j\omega) = F(j\omega)H(j\omega)X(j\omega)\,. \tag{3.6}$$

Some important constraints emerge from Eqs. (3.4) and (3.6). Note first that for a general DT signal $a[n]$, necessary information about the signal will be distributed in its DTFT $A(e^{j\Omega})$ at frequencies $\Omega$ throughout the interval $|\Omega| \leq \pi$; knowing $A(e^{j\Omega})$ only in a smaller range $|\Omega| \leq \Omega_a < \pi$ will in general be insufficient to allow reconstruction of the DT signal. Setting $\Omega = \omega T$ as specified in Eq. (3.4), we see that $A(e^{j\omega T})$ will contain necessary information about the DT signal at frequencies $\omega$ that extend throughout the interval $|\omega| \leq \pi/T$. Thus, if $P(j\omega) \neq 0$ for $|\omega| \leq \pi/T$, then $X(j\omega)$ preserves the information in the DT signal; and if $H(j\omega)P(j\omega) \neq 0$ for $|\omega| \leq \pi/T$, then $R(j\omega)$ preserves the information in the DT signal; and if $F(j\omega)H(j\omega)P(j\omega) \neq 0$ for $|\omega| \leq \pi/T$, then $B(j\omega)$ preserves the information in the DT signal.

The above constraints have some design implications. A pulse for which $P(j\omega)$ is nonzero only in a strictly smaller interval $|\omega| \leq \omega_p < \pi/T$ would cause loss of information in going from the DT signal to the PAM signal $x(t)$, and would not be a suitable pulse for the chosen symbol rate $1/T$ but could become a suitable pulse if the symbol rate was reduced appropriately, to $\omega_p/\pi$ or less.

Similarly, even if the pulse was appropriately designed so that $x(t)$ preserved the information in the DT signal, if we had a low-pass channel for which $H(j\omega)$ was nonzero only in a strictly smaller interval $|\omega| \leq \omega_c < \pi/T$, where $\omega_c$ is the cutoff frequency of the channel, then we would lose information about the DT signal in going from $x(t)$ to $r(t)$; the chosen symbol rate $1/T$ would be inappropriate for this channel, and would need to be reduced to $\omega_c/\pi$ or lower in order to preserve the information in the DT signal.

## Example 3.1    Determination of Symbol Rate in a PAM System

In the PAM system of Figure 3.3, assume that the noise $v(t) = 0$ and that $d(t)$ represents the combination of $h(t)$ and $f(t)$, that is,

$$d(t) = h(t) * f(t) \tag{3.7}$$

so that

$$b(t) = x(t) * d(t) . \tag{3.8}$$

Also, assume that the DTFT of the DT sequence $a[n]$ is bandlimited and specifically is nonzero only for $|\Omega| \leq \pi/2$ in the interval $|\Omega| \leq \pi$ and as usual, periodically repeats at integer multiples of $2\pi$, as illustrated in Figure 3.4.

Suppose that the channel and receiver filter combined constitute an ideal low-pass filter with a cutoff frequency of 1 kHz and a constant gain of $C$ in the passband, that is,

$$D(j\omega) = \begin{cases} C & |\omega| < 2\pi \times 10^3 \\ 0 & |\omega| > 2\pi \times 10^3 . \end{cases} \tag{3.9}$$

We want to determine the maximum symbol rate $1/T$ and constraints on the modulating pulse $p(t)$ so that no information is lost about $a[n]$ in the process of modulation, transmission through the channel, and filtering at the receiver, that is, so that $a[n]$ can be recovered from $b(t)$.

Applying Eq. (3.4) to Figure 3.4, we illustrate separately in Figure 3.5 the two terms to be multiplied in Eq. (3.4) to produce $X(j\omega)$.

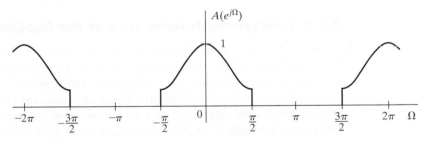

**Figure 3.4**   DTFT of the sequence $a[n]$ at the input of the PAM system in Figure 3.3.

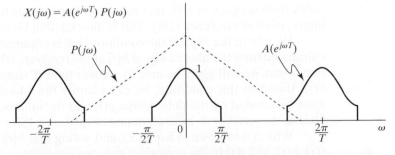

**Figure 3.5**   $P(j\omega)$ superimposed on $A(e^{j\omega T})$. The product represents $X(j\omega)$ in Eq. (3.4).

From Figure 3.5, it should be clear that, for this example, as long as $P(j\omega) \neq 0$ for $|\omega| \leq \frac{\pi}{2T}$, no information is lost due to the pulse modulation. This is different from the condition stated prior to this example because we assumed in Figure 3.4 that the spectrum of the signal $a[n]$ does not cover the full frequency range $|\omega| < \pi$.

Next, $b(t)$ is the result of applying the low-pass filter $d(t)$ to $x(t)$. Specifically,

$$B(j\omega) = \begin{cases} C\, X(j\omega) & |\omega| < 2\pi \cdot 10^3 \\ 0 & \text{otherwise.} \end{cases} \tag{3.10}$$

Assuming that $P(j\omega)$ is nonzero for $|\omega| \le \frac{\pi}{2T}$, all of the frequency information about $a[n]$ is contained in the frequency range $|\omega| \le \frac{\pi}{2T}$. Consequently, as long as the low-pass filter cutoff frequency is larger than this, all of the information is retained. If not, then some information is lost. Specifically, then, we require that

$$2\pi \cdot 10^3 > \frac{\pi}{2T} \tag{3.11}$$

or

$$\frac{1}{T} < 4 \cdot 10^3 . \tag{3.12}$$

At a symbol rate lower than $4 \cdot 10^3$, $x(t)$ will be recoverable from $b(t)$. However, since $P(j\omega)$ and $D(j\omega)$ may have imposed some frequency shaping, recovery of $a[n]$ from $b(t)$ may not be straightforward. Specifically, it may require additional frequency shaping to equalize for the frequency shaping imposed by $P(j\omega)$ and by $D(j\omega)$. In the next section, we consider conditions under which $a[n]$ can be recovered from $b(t)$ through the simple use of a C/D system as indicated in Figure 3.3.

### 3.1.4 Intersymbol Interference at the Receiver

In the absence of channel impairments, signal values can be recovered from the transmitted pulse trains shown in Figure 3.2 by resampling at time instants that are integer multiples of $T$. However, these pulses, while nicely time-localized, have infinite bandwidth. Since any realistic channel will have a limited bandwidth, one effect of a communication channel on a PAM waveform is to "delocalize," or disperse, the energy of each pulse through low-pass filtering and/or nonconstant group delay. As a consequence, pulses that may not have overlapped (or that overlapped only benignly) at the transmitter may overlap at the receiver in a way that impedes the recovery of the pulse amplitudes from samples of $r(t)$, that is, in a way that leads to error referred to as intersymbol interference (ISI). This is illustrated in Figure 3.6.

We now make explicit the condition that is required in order for ISI to be eliminated from the filtered signal $b(t)$ at the receiver. When this no-ISI condition is met, we will again be able to recover the DT signal simply by sampling $b(t)$. Based on this condition, we can identify the additional constraints that must be satisfied by the pulse shape $p(t)$ and the impulse response $f(t)$ of the filter at the receiver so as to eliminate or minimize ISI.

With $x(t)$ as given in Eq. (3.2), and noting that $b(t)$ is the convolution of $f(t)$, $h(t)$, and $x(t)$ in the noise-free case, we can write

$$b(t) = \sum_n a[n]\, g(t - nT) , \tag{3.13}$$

where $g(t)$ is the convolution of $f(t)$, $h(t)$, and $p(t)$.

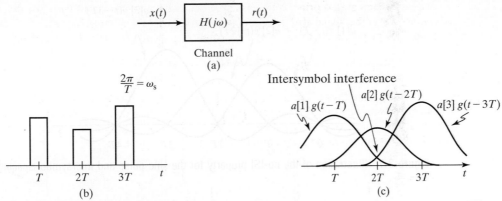

**Figure 3.6**    Illustration of intersymbol interference (ISI). (a) Representation of the channel as an LTI system; (b) channel input PAM signal $x(t)$; (c) channel output with ISI.

We assume that $g(t)$ is continuous (i.e., has no discontinuity) at the sampling times $nT$. Our requirement for signal recovery with no ISI is then that

$$g(0) = c, \quad \text{and} \quad g(nT) = 0 \text{ for nonzero integers } n, \tag{3.14}$$

where $c$ is some nonzero constant. If this condition is satisfied, then it follows from Eq. (3.13) that $b(nT) = ca[n]$, and consequently the DT signal is exactly recovered to within the known scale factor $c$.

---

**Example 3.2    Maximum Symbol Rate for No ISI with a Sinc Pulse**

In this example, we consider $g(t)$ in Eq. (3.13) as

$$g(t) = \frac{\sin \omega_c t}{\omega_c t}, \tag{3.15}$$

with corresponding CTFT $G(j\omega)$ given by

$$G(j\omega) = \frac{\pi}{\omega_c} \quad \text{for} \quad |\omega| < \omega_c$$

$$= 0 \quad \text{otherwise.} \tag{3.16}$$

Choosing the intersymbol spacing to be $T = \dfrac{\pi}{\omega_c}$, we can avoid ISI in the received samples, since $g(t) = 1$ at $t = 0$ and is zero at other integer multiples of $T$, as illustrated in Figure 3.7. We are thereby able to transmit at a symbol rate $1/T$ that is twice the cutoff frequency of $g(t)$, or $1/T = (2\omega_c)/(2\pi)$. From the earlier comments, in the discussion following Eq. (3.4) on constraints involving the symbol rate and the channel cutoff frequency, we cannot expect to do better, that is, transmit at a higher rate, in general.

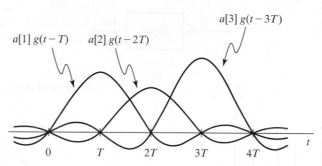

**Figure 3.7** Illustration of the no-ISI property for the sinc pulse and intersymbol spacing of $T = \frac{\pi}{\omega_c}$.

In the next section, we translate the no-ISI time-domain condition in Eq. (3.14) to a frequency-domain condition that is useful in designing the pulse $p(t)$ and the receiver filter $f(t)$ for a given channel. The frequency-domain interpretation of the no-ISI condition of Eq. (3.14) was originally explored by Nyquist in 1924 and extended by him in 1928 to a statement of the sampling theorem. This theorem then waited almost 20 years to be brought to prominence by Gabor and Shannon.

## 3.2 NYQUIST PULSES

Consider the function $\widehat{g}(t)$ obtained by sampling $g(t)$ with a periodic impulse train:

$$\widehat{g}(t) = g(t) \sum_{n=-\infty}^{+\infty} \delta(t - nT) . \tag{3.17}$$

Our requirements on $g(t)$ in Eq. (3.14) imply that $\widehat{g}(t) = c\,\delta(t)$, an impulse of non-zero strength $c$, whose transform is $\widehat{G}(j\omega) = c$. Taking transforms of both sides of Eq. (3.17), and utilizing the fact that multiplication in the time domain corresponds to convolution in the frequency domain, we obtain

$$\widehat{G}(j\omega) = c = \frac{1}{T} \sum_{m=-\infty}^{+\infty} G\left(j\omega - jm\frac{2\pi}{T}\right) . \tag{3.18}$$

The expression on the right-hand side of Eq. (3.18) represents a replication of $G(j\omega)$ (scaled by $1/T$) at every integer multiple of $2\pi/T$ along the frequency axis. The Nyquist requirement is thus that $G(j\omega)$ and all its replications, spaced $2\pi/T$ apart, add up to a nonzero constant. Some illustrations of $G(j\omega) = F(j\omega)H(j\omega)P(j\omega)$ that satisfy this condition are given in Example 3.3.

The particular case of the sinc function of Eqs. (3.15) and (3.16) in Example 3.2 certainly satisfies the Nyquist condition of Eq. (3.18) when

$T = \pi/\omega_c$. For example, for an ideal low-pass channel $H(j\omega)$ with band-width $\omega_c$ or greater, with the carrier pulse chosen to be the sinc pulse of Eq. (3.15), and the receiver filter chosen as $F(j\omega) = 1$ for $|\omega| \leq \omega_c$ so that there was no additional filtering, there would be no intersymbol interference at the sampler. However, there are several problems with the sinc pulse. First, the signal extends indefinitely in time in both directions. Second, the sinc has a very slow roll-off in time (as $1/t$). The slow roll-off in time is coupled with the sharp cutoff of the transform of the sinc in the frequency domain. This is a familiar manifestation of time-frequency duality: quick transition in one domain means slow transition in the other.

It is highly desirable in practice to have pulses that taper off more quickly in time than a sinc pulse. One reason is that, given the inevitable inaccuracies in sampling times due to timing jitter, there will be some unavoidable ISI, and this ISI will propagate for unacceptably long times if the underlying pulse shape decays too slowly. Also, a faster roll-off allows better approximation of an infinitely long two-sided signal by a finite-length signal, which can then be made causal by appropriate delay, as would be required for a causal implementation. The penalty for more rapid pulse roll-off in time is that the transition in the frequency domain will tend to be more gradual, necessitating a larger bandwidth for a given symbol rate or a reduced symbol rate for a given bandwidth.

**Example 3.3    Illustration of Alternative Nyquist Pulses**

The two choices in Figure 3.8 have smoother transitions in frequency than the case in Example 3.2, and correspond to pulses that fall off in time as $1/t^2$. It is evident that both can be made to satisfy the Nyquist condition by appropriate choice of $T$.

Specifically, to satisfy the Nyquist condition requires that

$$\frac{1}{T} \sum_{m=-\infty}^{+\infty} G\left(j\omega - jm\frac{2\pi}{T}\right) = c. \tag{3.19}$$

In Figure 3.9 we illustrate $\widehat{G}(j\omega)$ for the choices of $G(j\omega)$ in Figure 3.8, such that the Nyquist condition is satisfied.

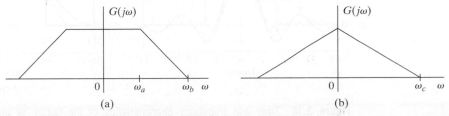

**Figure 3.8**    Two possible choices for the Fourier transform of pulses that decay in time as $1/t^2$ and satisfy the Nyquist zero-ISI condition for appropriate choice of $T$.

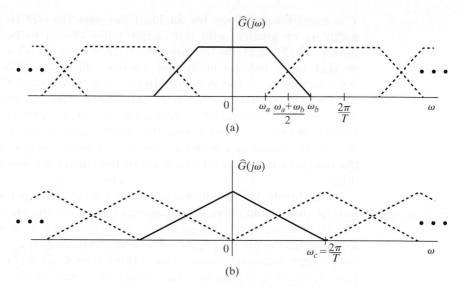

(a)

(b)

**Figure 3.9**  Equation (3.19) applied to the pulses represented in Figure 3.8.

For the choice in Figure 3.8(a), the Nyquist condition is satisfied if $\omega_a = \left(\frac{2\pi}{T} - \omega_b\right)$ or, equivalently, $\frac{\pi}{T} = \left(\frac{\omega_a + \omega_b}{2}\right)$. For the choice in Figure 3.8(b), the Nyquist condition is satisfied if $\frac{2\pi}{T} = \omega_c$.

Still smoother transitions can be obtained when $G(j\omega)$ is chosen from the family of frequency-domain characteristics associated with $g(t)$ specified by

$$g(t) = \frac{\sin \frac{\pi}{T}t}{\frac{\pi}{T}t} \frac{\cos \beta \frac{\pi}{T}t}{1 - (2\beta t/T)^2}, \tag{3.20}$$

where $\beta$ is termed the roll-off parameter.

The case $\beta = 1$ is a raised cosine pulse, in the frequency domain, with $g(t)$ that falls off as $1/t^3$ for large $t$, as in Figure 3.10(a).

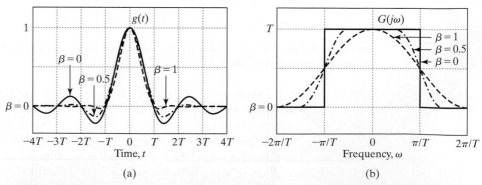

(a)                                                      (b)

**Figure 3.10**  Time and frequency characteristics of the family of pulses in Eq. (3.20). (a) Time waveform; (b) Fourier transform.

Once $G(j\omega)$ is specified, knowledge of the channel characteristic $H(j\omega)$ allows us to determine the corresponding pulse transform $P(j\omega)$, if we fix $F(j\omega) = 1$. In the presence of channel noise that corrupts the received signal $r(t)$, it is often preferable to do only part of the pulse shaping at the transmitter, with the rest done at the receiver prior to sampling. For instance, if the channel has no distortion in the passband (i.e., if $H(j\omega) = 1$ in the passband) and if the noise intensity is uniform in this passband, then the optimal choice of pulse is $P(j\omega) = \sqrt{G(j\omega)}$, assuming that $G(j\omega)$ is purely real and nonnegative, and this is also the optimal choice of receiver filter $F(j\omega)$. We will say more about this when we discuss matched filtering in Chapter 13.

## 3.3 PASSBAND PULSE AMPLITUDE MODULATION

The previous discussion centered on the design of baseband pulses. For transmission over phone lines, wireless links, satellites, cable, or most other physical channels, the baseband signal needs to be modulated onto a higher frequency carrier, that is, converted to a passband signal. This also opens opportunities for augmentation of PAM. Over the years, there has been considerable evolution of digital modem standards, speeds, and robustness, based on the principles of pulse amplitude modulation. Variation and augmentation include frequency-shift keying (FSK), phase-shift keying (PSK), and quadrature amplitude modulation (QAM), each of which we describe in more detail below. The resulting increase in symbol rate and bit rates over the years reflects improvements in signal processing, better modulation schemes, the use of better conditioned channels, and more elaborate coding with correspondingly sophisticated decoding.

For baseband PAM, the transmitted signal takes the form of Eq. (3.2), that is,

$$x(t) = \sum_{n} a[n]\, p(t - nT)\,, \qquad (3.21)$$

where $p(t)$ is a low-pass pulse. When this is amplitude-modulated onto a sinusoidal carrier, the transmitted signal takes the form

$$s(t) = \sum_{n} a[n]\, p(t - nT) \cos(\omega_c t + \theta_c)\,, \qquad (3.22)$$

where $\omega_c$ and $\theta_c$ are the carrier frequency and phase.

In the simplest form of Eq. (3.22), specifically with $\omega_c$ and $\theta_c$ fixed, Eq. (3.22) corresponds to using amplitude modulation to shift the frequency content from baseband to a band centered at the sinusoidal modulation frequency $\omega_c$. However, since two additional parameters, $\omega_c$, and $\theta_c$, have been introduced, there are additional possibilities for embedding data in $s(t)$. Specifically, in addition to changing the amplitude in each symbol interval, we can consider changing the modulation frequency and/or the phase in each symbol interval. These alternatives lead to frequency-shift keying (FSK) and phase-shift keying (PSK).

### 3.3.1 Frequency-Shift Keying (FSK)

With frequency-shift keying, Eq. (3.22) takes the form

$$s(t) = \sum_n a[n] p(t - nT) \cos((\omega_0 + \Delta_n)t + \theta_c) , \qquad (3.23)$$

where $\omega_0$ is the nominal carrier frequency and $\Delta_n$ is the shift in the carrier frequency in symbol interval $n$. In principle, both $a[n]$ and $\Delta_n$ can incorporate data, although it is typically the case that in FSK the amplitude $a[n]$ does not change.

### 3.3.2 Phase-Shift Keying (PSK)

In phase-shift keying, Eq. (3.22) takes the form

$$s(t) = \sum_n a[n] p(t - nT) \cos(\omega_c t + \theta_n) . \qquad (3.24)$$

In each symbol interval, information can now be incorporated in both the pulse amplitude $a[n]$ and the carrier phase $\theta_n$. In what is typically referred to as PSK, information is only incorporated in the phase, that is, as with FSK the amplitude $a[n]$ does not change.

For example, choosing

$$\theta_n = \frac{2\pi b_n}{M} \quad ; \quad 0 \leq b_n \leq M - 1 , \qquad (3.25)$$

where $b_n$ is an integer, one of $M$ symbols can be encoded in the phase in each symbol interval. For $M = 2$, $\theta_n = 0$ or $\pi$, resulting in what is commonly referred to as binary PSK (BPSK). With $M = 4$, $\theta_n$ takes on one of the four values $0$, $\frac{\pi}{2}$, $\pi$, or $\frac{3\pi}{2}$.

To interpret PSK somewhat differently, and as a prelude to expanding the discussion to a further generalization (quadrature amplitude modulation, or QAM), it is convenient to express Eq. (3.24) in an alternate form. Using the identity

$$\cos(\omega_c t + \theta_n) = \cos(\theta_n) \cos(\omega_c t) - \sin(\theta_n) \sin(\omega_c t) \qquad (3.26)$$

we can write

$$s(t) = I(t) \cos(\omega_c t) - Q(t) \sin(\omega_c t) , \qquad (3.27)$$

with

$$I(t) = \sum_n a_i[n] p(t - nT) \qquad (3.28)$$

$$Q(t) = \sum_n a_q[n] p(t - nT) \qquad (3.29)$$

and

$$a_i[n] = a[n] \cos(\theta_n) \qquad (3.30)$$

$$a_q[n] = a[n] \sin(\theta_n) . \qquad (3.31)$$

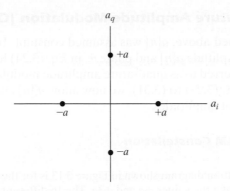

**Figure 3.11**    *I-Q* constellation for $\theta_n = 0, \frac{\pi}{2}, \pi, \frac{3\pi}{2}$.

Equation (3.27) is referred to as the quadrature form of Eq. (3.24), with $I(t)$ and $Q(t)$ referred to as the in-phase and quadrature components. For BPSK, $a_i[n] = \pm a$ and $a_q[n] = 0$.

For PSK with $\theta_n$ in the form of Eq. (3.25) and $M = 4$, we noted that $\theta_n$ can take on any of the four values $0, \frac{\pi}{2}, \pi$, or $\frac{3\pi}{2}$. In the notation of Eqs. (3.30) and (3.31), $a_i[n]$ will then be either $+a, -a$, or zero and $a_q[n]$ will be either $+a, -a$, or zero. However, clearly the choice of $M = 4$ can only encode four symbols, not nine, that is, the various possibilities for $a_i[n]$ and $a_q[n]$ are not independent. Specifically, for $M = 4$, if $a_i[n] = +a$, then $a_q[n]$ must be zero since $a_i[n] = +a$ implies that $\theta_n = 0$. A convenient way of looking at this is through what is referred to as an *I-Q* constellation, as shown in Figure 3.11. Each point in the constellation represents a different symbol that can be encoded. With the constellation of Figure 3.11, one of four symbols can be encoded in each symbol interval.

An alternative form of four-phase PSK is to choose

$$\theta_n = \frac{2\pi b_n}{4} + \frac{\pi}{4} \quad ; \quad 0 \le b_n \le 3 \tag{3.32}$$

in which case $a_i[n] = \pm \frac{a}{\sqrt{2}}$ and $a_q[n] = \pm \frac{a}{\sqrt{2}}$, resulting in the constellation in Figure 3.12. In this case, the amplitude modulation of $I(t)$ and $Q(t)$ as defined in Eqs. (3.28) and (3.29) can be done independently. Modulation with this constellation is commonly referred to as QPSK (quadrature phase-shift keying).

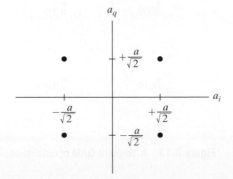

**Figure 3.12**    *I-Q* constellation for quadrature phase-shift keying (QPSK).

### 3.3.3 Quadrature Amplitude Modulation (QAM)

In PSK as described above, $a[n]$ was assumed constant. Incorporating encoding in both the amplitude $a[n]$ and phase $\theta_n$ in Eq. (3.24) leads to a richer form of modulation referred to as quadrature amplitude modulation (QAM). In the expressions in Eqs. (3.27) to (3.31), we now allow $a_i[n]$ and $a_q[n]$ to be chosen from a larger set of amplitudes.

| Example 3.4 | A 16-Point QAM Constellation |
|---|---|

The QAM constellation diagram shown in Figure 3.13 is for the case in which each set of amplitudes can take the values $\pm a$ and $\pm 3a$. The 16 different combinations that are available in this case can be used to code 4 bits, as shown in the figure. In one current voice channel modem standard of this form, the carrier frequency is 1800 Hz, and the symbol frequency or baud rate $(1/T)$ is 2400 Hz. With 4 bits per symbol, this results in 9600 bits/second. One baseband pulse shape $p(t)$ that may be used is the square root of the cosine-transition pulse in Eq. (3.20), say with $\beta = 0.3$. This pulse contains frequencies as high as $1.3 \times 1200 = 1560$ Hz. After modulation of the 1800 Hz carrier, the signal occupies the band from 240 Hz to 3360 Hz, which is in the passband of the voice telephone channel.

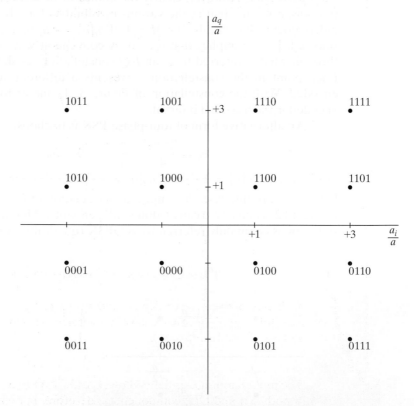

**Figure 3.13**   A 16-point QAM constellation.

Faster modems use more elaborate QAM-based schemes. One voice channel standard involves 128 QAM, which could in principle convey 7 bits per symbol, but at the price of greater sensitivity to noise because the constellation points are more tightly clustered for a given signal power. However, QAM in combination with so-called trellis-coded modulation (TCM) incorporates some redundancy by introducing dependencies among the modulating amplitudes for successive bits, leading to greater noise immunity and an effective rate of 6 bits per symbol. The symbol rate is still 2400 Hz, so the transmission is at $6 \times 2400 = 14{,}400$ bits/second. Yet another standard involves 1024 QAM, which could convey 10 bits per symbol, although with more noise sensitivity. The combination with TCM introduces redundancy for error control, and the resulting bit rate is 28,800 bits/second, corresponding to 9 effective bits per symbol at a symbol frequency of 3200 Hz attained by better exploitation of the channel bandwidth.

**Example 3.5    An 8-Point QAM Constellation**

In this example, we consider the 8-point QAM constellation shown in Figure 3.14 with the eight points lying on a circle of radius $a$. In the expression in Eq. (3.24), $\theta_n$ can be one of the eight values

$$0, \frac{\pi}{4}, \frac{\pi}{2}, \frac{3\pi}{4}, \pi, \frac{5\pi}{4}, \frac{6\pi}{4}, \frac{7\pi}{4}. \tag{3.33}$$

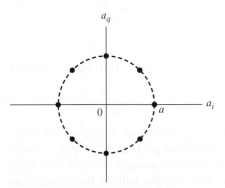

**Figure 3.14**    An 8-point QAM constellation.

With each of these eight values, $a_i$ and $a_q$ in Eqs. (3.30) and (3.31) are related such that $a_i^2[n] + a_q^2[n] = a^2$. For example, as indicated in Figure 3.14, with $\theta_n = \frac{3\pi}{4}$, we have $a_i[n] = -\frac{a}{\sqrt{2}}$, and $a_q[n] = +\frac{a}{\sqrt{2}}$.

The modulated signals defined by Eqs. (3.27) to (3.31) can carry encoded data in both $I(t)$ and $Q(t)$ components. Therefore, in demodulation we must be able to extract these separately. This can be done through quadrature demodulation, as shown in Figure 3.15.

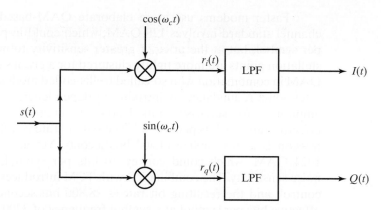

**Figure 3.15**   Demodulation scheme for a quadrature modulated PAM signal.

In both the modulation and demodulation, it is assumed that the band-width of $p(t)$ is low compared with the carrier frequency $\omega_c$ so that the bandwidths of $I(t)$ and $Q(t)$ are less than $\omega_c$. The input signal $r_i(t)$ in Figure 3.15 is

$$r_i(t) = I(t)\cos^2(\omega_c t) - Q(t)\sin(\omega_c t)\cos(\omega_c t) \tag{3.34}$$

$$= \frac{1}{2}I(t) + \frac{1}{2}I(t)\cos(2\omega_c t) - \frac{1}{2}Q(t)\sin(2\omega_c t) . \tag{3.35}$$

Similarly,

$$r_q(t) = I(t)\cos(\omega_c t)\sin(\omega_c t) - Q(t)\sin^2(\omega_c t) \tag{3.36}$$

$$= \frac{1}{2}I(t)\sin(2\omega_c t) - \frac{1}{2}Q(t) + \frac{1}{2}Q(t)\cos(2\omega_c t) . \tag{3.37}$$

Choosing the cutoff frequency of the low-pass filters to be greater than the bandwidth of $p(t)$, and therefore also greater than the bandwidth of $I(t)$ and $Q(t)$, but low enough to eliminate the components in $r_i(t)$ and $r_q(t)$ around $2\omega_c$, the outputs will be the quadrature signals $I(t)$ and $Q(t)$ within a scale factor that depends on the gains of the low-pass filters.

## 3.4 FURTHER READING

Texts that deal with digital communication – for example [An1], [An2], [Bar], [Gib], [Ha2], [La2], [Pr2] and [Zie] – include treatments of pulse amplitude modulation and the extension to quadrature amplitude modulation. They also discuss the Nyquist zero-ISI condition and its role in pulse shaping, which surfaces again in the context of signal detection in Chapter 13.

## Problems

### Basic Problems

**3.1.** Suppose a PAM signal $x(t)$ is

$$x(t) = \sum_n a[n]p(t - nT_0)$$

where $p(t)$ is the sinc pulse

$$p(t) = \frac{\sin(\pi t/T_1)}{\pi t/T_1} .$$

The signal $x(t)$ is sent through a channel with an ideal low-pass frequency response given by $H(j\omega) = 1$ for $|\omega| < \omega_c$, and $H(j\omega) = 0$ otherwise. Let the symbol repetition interval $T_0$ be chosen as the *smallest* one that will yield zero ISI at the output of the channel. Determine $T_0$ in terms of $T_1$ and $\omega_c$ for the following two cases:

(i) when $\omega_c > (\pi/T_1)$;
(ii) when $\omega_c < (\pi/T_1)$.

**3.2.** A signal $p(t)$ has transform $P(j\omega)$ given by

$$P(j\omega) = 1 - \frac{|\omega|}{2\omega_m} \qquad \text{for } |\omega| < \omega_m ,$$

and 0 outside this frequency range (leading to a "raised roof" shape). What is the smallest $T$ for which you can guarantee $p(nT) = 0$ for all integer $n \neq 0$? And what is $p(0)$?

**3.3.** Consider a PAM system as indicated in Figure P3.3-1, with

$$r(t) = \sum_{k=-\infty}^{\infty} a[k]g(t - kT)$$

where $g(t) = p(t) * h(t)$.

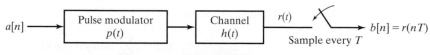

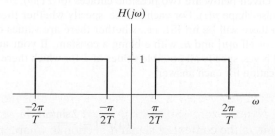

**Figure P3.3-1**

The channel frequency response is shown in Figure P3.3-2. Give one choice for $P(j\omega)$, the Fourier transform of the transmitted pulse $p(t)$, so that $b[n] = a[n]$ for all $n$. Clearly show your reasoning.

**Figure P3.3-2**

**3.4.** Consider a PAM system in which we wish to transmit symbols quickly while avoiding intersymbol interference (ISI) when sampling at the output of the channel. The usual block diagram describing this process is shown in Figure P3.4-1, and the frequency response $H(j\omega)$ of the particular channel of interest in this problem is shown in Figure P3.4-2.

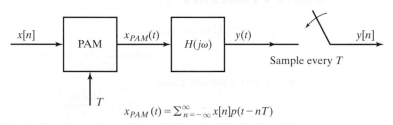

$$x_{PAM}(t) = \sum_{n=-\infty}^{\infty} x[n]p(t-nT)$$

**Figure P3.4-1**

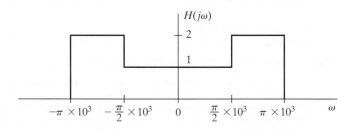

**Figure P3.4-2**

If the pulse shape is

$$p(t) = \frac{\sin(2\pi \times 10^3 t)}{\pi t}$$

what is the fastest symbol rate, $1/T$, for which there is no ISI, i.e., for which $y[n] = cx[n]$ for some constant $c$? Also determine the value of $c$.

**3.5.** A PAM transmitter sends

$$s(t) = \sum_{n=-\infty}^{\infty} a[n]p(t-nT)$$

to convey a message sequence $a[n]$. At the receiver, a sequence $b[n]$ is obtained by sampling $s(t)$ at integer multiples of $T$, i.e., the receiver's output sequence is $b[n] = s(nT)$.

Given below are two possible choices for $P(j\omega)$, the Fourier transform of the pulse shape $p(t)$. For each choice, specify whether there are values of $T$ for which there will be no ISI, i.e., whether there are values of $T$ for which $b[n] = ca[n]$ for all $a[n]$ and $n$, with $c$ being a constant. If your answer for a particular $P(j\omega)$ is yes, specify all possible values of $T$ for which there is no ISI. Give a brief justification for each answer.

**(a)**

$$P_1(j\omega) = \frac{2\sin(\omega)}{\omega}$$

**(b)**

$$P_2(j\omega) = \begin{cases} e^{-j\omega/2} & \text{for } |\omega| \leq \pi \\ 0 & \text{otherwise.} \end{cases}$$

**3.6.** Consider the system in Figure P3.6 for sending data at a rate of one value every $T$ seconds.

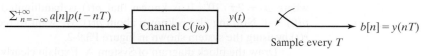

$$\sum_{n=-\infty}^{+\infty} a[n]p(t-nT) \longrightarrow \boxed{\text{Channel } C(j\omega)} \xrightarrow{y(t)} \times \quad \longrightarrow b[n] = y(nT)$$

Sample every $T$

**Figure P3.6**

We are interested in designing the pulse shape $p(t)$ so that for an arbitrary data sequence $a[n]$ there is no ISI, i.e., so that $b[n] = a[n]$. The channel is an ideal low-pass filter:

$$C(j\omega) = \begin{cases} 1 & |\omega| \leq B \\ 0 & |\omega| > B . \end{cases}$$

**(a)** Construct a reasonable but brief argument to show that it is not possible to find a pulse $p(t)$ that satisfies the no-ISI requirement when $B = \frac{\pi}{2T}$.
**(b)** When $B = \frac{\pi}{T}$ specify a pulse shape $p(t)$ that satisfies the no-ISI requirement. Is the $p(t)$ you found unique or are there other pulse shapes that also meet the no-ISI requirement for the channel? Explain your answer.

**3.7.** Consider a binary PAM communication system in which antipodal signaling is used with the pulse

$$p(t) = \begin{cases} 1 - |t|/T & |t| < T \\ 0 & \text{otherwise} \end{cases}$$

and transmitted at a rate of $f_b$ pulses per second.

**(a)** Determine the maximum value of $f_b$ such that the $N$ pulses in the following signal do not overlap:

$$y(t) = \sum_{k=1}^{N} p(t - k/f_b) .$$

**(b)** What is the maximum value of $f_b$ such that samples taken at the bit rate and located at the pulse peaks are not affected by ISI, i.e.,

$$y(n/f_b) = \sum_{k=1}^{N} p((n-k)/f_b) = p(0) ?$$

**3.8.** In Figure 3.13, we showed a 16-point QAM constellation. Suppose that the binary numbers to be transmitted in the first three symbol times are

Symbol time $n = 0$: 1100
Symbol time $n = 1$: 0011
Symbol time $n = 2$: 0110

**(a)** Determine $a_i[n]$ and $a_q[n]$ $(n = 0, 1, 2)$ for $I(t)$ and $Q(t)$ in the quadrature-form representation of $s(t)$.

**(b)** Determine $a[n]$ and $\theta_n$ $(n = 0, 1, 2)$ in the PSK representation of $s(t)$.

**3.9.** Consider the QAM communication system with the constellation shown in Figure P3.9-1, where each symbol represents 2 bits as indicated.
Suppose the received signal is

$$r(t) = a_i \cdot p(t) \cos(\omega_0 t) + a_q \cdot p(t) \sin(\omega_0 t) ,$$

where $\omega_0 = 2\pi \cdot 10^6$. It is known that $p(t)$ is bandlimited to $2\pi \cdot 50$ and that $p(0) = 1$. Nothing else is known about $p(t)$. We want to recover $a_i$ and $a_q$ from $r(t)$ by using the system shown in Figure P3.9-2.
Draw the block diagram of System A. Explain clearly why it works.

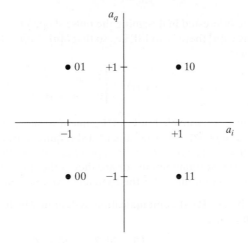

**Figure P3.9-1**

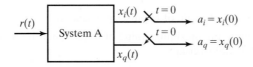

**Figure P3.9-2**

**3.10.** The signal that is transmitted in QAM has the form

$$x(t) = x_C(t) \cos \omega_0 t + x_S(t) \sin \omega_0 t ,$$

where $x_C(t)$ and $x_S(t)$ have bandwidths less than the carrier frequency $\omega_0$. Describe what sort of signal processing you would do to recover $x_C(t)$ and $x_S(t)$ separately.

## Advanced Problems

**3.11.** Suppose the PAM signal

$$x(t) = \sum_n a[n] p(t - nT_0)$$

with

$$p(t) = \left[\frac{\sin(\pi t/T_1)}{\pi t/T_1}\right]^2$$

is sent through a channel with the ideal low-pass frequency response $H(j\omega) = 1$ for $|\omega| < \omega_c$, and $H(j\omega) = 0$ otherwise. Let the symbol repetition interval $T_0$ be chosen as the smallest one that will yield zero ISI at the receiving end.

**(a)** Determine $T_0$ in terms of $T_1$ and $\omega_c$ for the following two cases:
  (i) when $\omega_c > (2\pi/T_1)$; and
  (ii) when $\omega_c < (2\pi/T_1)$.
**(b)** For both the cases in (a), determine what the output of the channel will be at time $t = 0$ if the input is simply $x(t) = p(t)$, corresponding to the input DT sequence being $a[n] = \delta[n]$.
**(c)** How do your answers to part (a) change if the channel frequency response is actually $e^{-j\omega D} H(j\omega)$, where $D$ is a fixed positive constant and $H(j\omega)$ is as defined earlier? And at what time would you have to sample the channel output to get the same value as you did in (b)?

**3.12.** Consider two candidate pulses to be used in a PAM communication system for binary data. The time waveform of the first pulse, $p_1(t)$, is shown in Figure P3.12, and the spectrum of the second pulse is

$$P_2(j\omega) = \frac{2\sin(5 \times 10^{-5}\omega)}{\omega}.$$

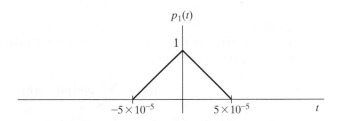

**Figure P3.12**

Suppose that antipodal signaling is used. That is, a 1 is represented by transmitting the pulse $p_i(t)$ and a 0 is represented by transmitting $-p_i(t)$, the negative of the pulse. Also assume that the channel is sufficiently wideband that it does not distort either of the candidate pulse shapes.

**(a)** The pulses are transmitted at a rate of $10^4$ pulses per second, referred to as the signaling rate, symbol rate, or baud rate. For each pulse, specify whether there will be ISI at the receiver. You can answer this by thinking in the time domain, but also try to check that your answer is consistent with the frequency-domain statement given in the Nyquist theorem on zero-ISI pulse shapes.
**(b)** Repeat part (a) for a signaling rate of $2 \times 10^4$ pulses per second.
**(c)** Assume a signaling rate of $2 \times 10^4$ pulses per second. For each of the two pulses, provide a labeled sketch of the signal at the receiving end of the channel when the symbol sequence "1 1 0 1 0" is sent.

**3.13.** Suppose $p(t)$ is a triangular pulse of the form

$$p(t) = 1 - \frac{|t|}{T} \quad \text{for} \quad |t| < T,$$

and $p(t) = 0$ for $|t| \geq T$. Note that $p(0) = 1$ and $p(nT) = 0$ for $n \neq 0$, and also that the Fourier transform of $p(t)$ is

$$P(j\omega) = T \frac{\sin^2\left(\frac{\omega T}{2}\right)}{\left(\frac{\omega T}{2}\right)^2} .$$

Using these facts, or in some other way, evaluate the following infinite sum (i.e., find its value as a function of $\omega$):

$$\sum_{k=-\infty}^{\infty} \frac{\sin^2\left(\frac{\omega T}{2} - k\pi\right)}{\left(\frac{\omega T}{2} - k\pi\right)^2} .$$

**3.14.** A PAM communication system is shown in Figure P3.14-1.

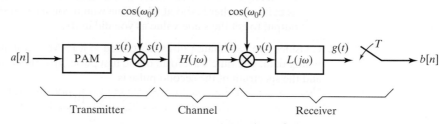

**Figure P3.14-1**

In this figure:

- the transmitter input is $\{a[n]: -\infty < n < \infty\}$, the data sequence to be sent;
- the output of the PAM system is

$$x(t) = \sum_{n=-\infty}^{\infty} a[n] p(t - nT) ,$$

where $p(t)$ is the pulse shape whose Fourier transform, $P(j\omega)$, is shown in Figure P3.14-2;
- the transmitter output is $s(t) = x(t) \cos(\omega_0 t)$, where $\omega_0 \gg 2\pi/T$;
- the channel is the passband filter $H(j\omega)$, shown in Figure P3.14-2, whose output is $r(t) = s(t) * h(t)$;
- the receiver applies the low-pass filter $L(j\omega)$, shown in Figure P3.14-2, to $y(t) = r(t) \cos(\omega_0 t)$ to obtain the baseband waveform $g(t)$; and
- the receiver output is the sequence $\{b[n]: -\infty < n < \infty\}$, where $b[n] = g(nT)$.

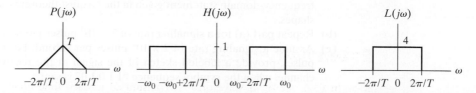

**Figure P3.14-2**

**(a)** For this part only, assume that $a[n] = \delta[n]$, i.e., $a_0 = 1$ and $a[n] = 0$ for $n \neq 0$. Make a labeled sketch of $R(j\omega)$, the Fourier transform of $r(t)$.

**(b)** Will the receiver's output suffer from ISI? Justify your answer in detail.

**3.15.** Figure P3.15-1 shows a PAM system.

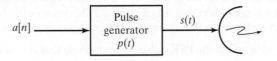

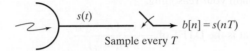

**Figure P3.15-1**

**(a)** If $p(t) = \delta(t)$, express $S(j\omega)$ in terms of $A(e^{j\Omega})$.

For the remainder of this problem, assume:

$$p(t) = \frac{\sin(\pi \cdot 10^4 t)}{\pi \cdot 10^4 t} \qquad P(j\omega) = \begin{cases} 10^{-4} & |\omega| < \pi \cdot 10^4 \\ 0 & |\omega| > \pi \cdot 10^4 \end{cases}$$

Also assume that there is no distortion or noise introduced by the channel and that the decoded sequence $b[n]$ is obtained by sampling the received signal $s(t)$ at integer multiples of $T$.

**(b)** Suppose that $T = 10^{-4}$ and that $a[n]$ is an arbitrary input sequence. Determine whether or not $b[n] = a[n]$ for all $n$.

**(c)** Now assume that $A(e^{j\Omega})$ is as sketched in Figure P3.15-2 and $T = \frac{1}{2} \times 10^{-4}$.
  (i) Sketch $S(j\omega)$.
  (ii) Determine whether or not $b[n] = a[n]$ for all $n$ in this case.

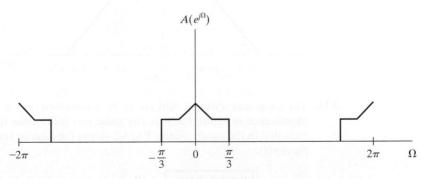

**Figure P3.15-2**

# Extension Problems

**3.16.** In PSK one of $M$ waveforms is transmitted in each time block, where the $M$ waveforms are defined by

$$s_m(t) = A\cos\left(\omega_0 t + \frac{2\pi}{M}(m-1)\right) \quad m = 1, 2, \ldots, M \quad 0 \leq t \leq T.$$

PSK can also be described as a form of QAM, i.e., the above equation can also be expressed in the form

$$s_m(t) = a_c \cos \omega_0 t + a_s \sin \omega_0 .$$

**(a)** Draw the constellation diagram for PSK with $M = 16$.

**(b)** Consider a QAM system with a 16-point constellation where $a_c$ and $a_s$ can each take on values $\pm 1, \pm 3$ and a PSK system with $M = 16$. The amplitude $A$ is chosen so that the maximum energy over one time interval in the QAM pulse and in the PSK pulse is the same, so that $A = \max \left\{ \sqrt{a_c^2 + a_s^2} \right\}$. Which of the two systems do you think would have better noise immunity? Clearly explain your reasoning.

**3.17.** **(a)** What is the DTFT of $\dfrac{\sin \left( \pi \left( n - \frac{1}{4} \right) \right)}{\pi \left( n - \frac{1}{4} \right)}$?

Explicitly check by verifying that the inverse transform of your answer gives the above time function. Also determine the value of the expression

$$\sum_n \cos \left( \frac{\pi n}{3} \right) \frac{\sin(\pi(n - \frac{1}{4}))}{\pi(n - \frac{1}{4})} ,$$

explaining your reasoning.

**(b)** Consider the signal $x(t) = \sum_n v[n] w(t - nT)$, where $v[n]$ is an arbitrary DT sequence and $w(t)$ is a CT signal whose Fourier transform (CTFT) is as shown in Figure P3.17. As $t \to \infty$, the magnitude of $w(t)$ falls off as $t^{-k}$ for some integer $k$. Determine the value of $k$, and explain your reasoning. Also specify what value of $T$ will result in the relationship $x(nT) = Cv[n]$, for some constant $C$. Explain your reasoning.

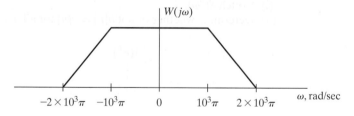

Figure P3.17

**3.18.** The sequences $a[n]$ and $b[n]$ are to be transmitted over a single PAM channel as indicated in Figure P3.18-1. The pulse $p(t)$ is the same for both sequences as indicated in the figure. Figure P3.18-2 shows the Fourier transform $P(j\omega)$ of the pulse $p(t)$.

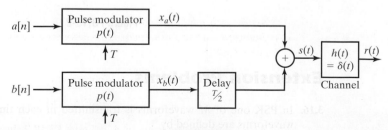

**Figure P3.18-1**

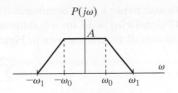

Figure P3.18-2

The signals $x_a(t)$, $x_b(t)$, $s(t)$, and $r(t)$ are

$$x_a(t) = \sum_{n=-\infty}^{\infty} a[n]p(t - nT)$$

$$x_b(t) = \sum_{n=-\infty}^{\infty} b[n]p(t - nT)$$

$$s(t) = r(t) = x_a(t) + x_b\left(t - \frac{T}{2}\right).$$

Figure P3.18-3 shows the block diagram of the demodulator.

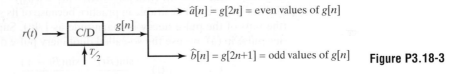

Figure P3.18-3

The sequence $g[n]$ is

$$g[n] = r\left(n\frac{T}{2}\right).$$

**(a)** Determine a condition on $p(t)$ in the time domain so that

$$\hat{a}[n] = a[n]$$
$$\hat{b}[n] = b[n].$$

Justify your answer clearly and succinctly.

**(b)** Determine a condition on $\omega_0$ and $\omega_1$ consistent with your answer in part (a), again so that

$$\hat{a}[n] = a[n]$$
$$\hat{b}[n] = b[n].$$

Also determine the value of $A$ in Figure P3.18-2. Justify your answers clearly and succinctly.

**3.19.** We wish to use a PAM communication system, with pulse shape $p(t)$, to send the DT signal $a[n] = \pm 1$ for $n \geq 0$ through an ideal channel bandlimited to $|\omega| \leq B$. The overall system is shown in Figure P3.19. In this figure,

$$s(t) = \sum_{n=0}^{+\infty} a[n] p(t - nT) ,$$

where $T$ is the intersymbol time (which is to be chosen), and the frequency response of the channel is

$$H(j\omega) = \begin{cases} 1, & |\omega| \leq B \\ 0, & |\omega| > B . \end{cases}$$

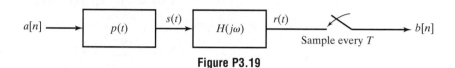

**Figure P3.19**

**(a)** If we use the sinc pulse

$$p(t) = \frac{\sin(Bt)}{Bt} ,$$

what is the smallest possible intersymbol time $T$ (expressed in terms of $B$) for which there is no ISI (i.e., for which $b[n] = a[n]$)?

**(b)** The sinc pulse is not desirable in practice because of its slow decay over time (the tails of the pulse decay in magnitude as $1/|t|$). Suppose instead of the sinc pulse in (a), we use the so-called duobinary pulse defined by

$$p(t) = \frac{\sin(Bt)}{Bt} + \frac{\sin(Bt - \pi)}{Bt - \pi} .$$

Determine $|P(j\omega)|$. What bandwidth does the duobinary pulse occupy?

**(c)** Show that, as $|t|$ tends toward $\infty$, the tails of the duobinary pulse $p(t)$ in (b) decay in magnitude as $1/|t|^2$.

**(d)** For the duobinary pulse $p(t)$ in (b), what is the smallest possible intersymbol time $T$ (expressed in terms of $B$) for which there is no ISI? How does the corresponding symbol rate $(1/T)$ compare with the one obtained with the sinc pulse in (a)?

The result of (d) shows that we have to pay a penalty in symbol rate if we want no ISI with the duobinary pulse (although we do obtain the benefit of pulse tails that fall off as $1/|t|^2$, and hence a better behaved pulse for practical implementation). However, it turns out that ISI is not bad if it only occurs in a predictable way and affects only some adjacent samples, because its effects can then be undone quite simply within the receiver. By allowing some limited ISI of this form, we can increase the rate at which we send symbols in our PAM system, as the next two parts of this problem show.

**(e)** With the duobinary pulse in (b), suppose we transmit symbols at the rate you determined in (a) (i.e., using the value of $T$ you determined in part (a)). How is the received sequence $b[n]$ for $n \geq 0$ related to $a[n]$ in this case? Explain how to determine the transmitted sequence from the received sequence.

**(f)** Although you find in (e) that the $b[n]$ are simply obtained from the $a[n]$, and vice versa, starting from $n = 0$, what should be clear is that if there is an error, then all subsequent estimates are corrupted. To overcome this defect, what we do is, rather than transmitting the desired sequence $a[n]$, we transmit a *precoded* version of it, namely the sequence $c[n]$ defined by $c[n] = a[n]c[n-1]$, with $c[-1] = 1$ by definition. Now show that if $b[n] = \pm 2$ then we can infer that $a[n] = 1$, and if $b[n] = 0$ we can infer that $a[n] = -1$. The receiver decision for each $n$ is no longer coupled to the decisions at other times, so there is no error propagation in decoding.

**3.20.** *Note: This problem relies on a basic understanding of random variables and random processes discussed in Chapters 7 and 10.* In a particular DT communication system similar to the CT PAM system discussed in this chapter, the essential problem at the receiver involves a received signal $r[n]$ of the form

$$r[n] = Ap[n] + v[n] ,$$

where $A$ is a random variable that is chosen by the transmitter; the receiver only knows the mean $\mu_A$ and variance $\sigma_A^2$ of $A$. Assume $A$ is uncorrelated with the wide-sense stationary (WSS) white-noise process $v[\cdot]$, which represents noise on the communication channel, whose intensity is $\sigma_v^2$, i.e., with power spectral density $S_{vv}(e^{j\Omega}) = \sigma_v^2$ for all $\Omega$. Also consider $p[\cdot]$ to be a known signal of finite energy

$$\mathcal{E} = \sum_n p^2[n] .$$

The received signal in the absence of noise thus has a known shape $p[n]$ but random amplitude $A$. The receiver filters the received signal $r[n]$ using a stable LTI filter with unit sample response $f[n]$, producing a signal $b[n] = f[n] * r[n]$ at its output, where $*$ denotes convolution. We are particularly interested in the random variable $B = b[0]$ obtained by sampling the output of the filter at time $0$:

$$B = b[0] = \sum_{n=-\infty}^{\infty} f[n]r[-n] = A\left(\sum f[n]p[-n]\right) + \left(\sum f[n]v[-n]\right) = \alpha A + V ,$$

where we have introduced the symbols

$$\alpha = \sum f[n]p[-n] , \qquad V = \sum f[n]v[-n] ,$$

to simplify notation. Note that $\alpha$ is a deterministic constant but $V$ is a random variable. It will also be helpful in what follows to denote the energy of $f[\cdot]$ simply by

$$\mathcal{F} = \sum_n f^2[n] .$$

The stability of the filter guarantees that $\mathcal{F}$ is finite. This problem looks at using a measurement of $B$ to estimate $A$, and choosing the filter $f[n]$ to make this estimation as accurate as possible.

**(a)** Determine the mean and variance of $V$, and the cross-covariance $\sigma_{AV}$ of $A$ and $V$. All your answers can be written in terms of $\sigma_v$ and $\mathcal{F}$.

**(b)** Determine the mean and variance of $B$, the cross-covariance $\sigma_{AB}$ of $A$ and $B$, and their correlation coefficient $\rho_{AB}$, all expressed in terms of the problem parameters and the simplified notation introduced above.

**(c)** Describe the linear minimum mean square error (LMMSE) estimator of $A$ that uses a measurement of $B$, i.e., find $\gamma$ and $\mu$ in

$$\widehat{A} = \gamma B + \mu ,$$

so as to minimize $E[(A - \widehat{A})^2]$. Again, your answers should be expressed in terms of the problem parameters and the simplified notation introduced above.

**(d)** The minimum mean square error (MMSE) associated with the estimator in (c) can be written as

$$\sigma_A^2(1 - \rho_{AB}^2) \,.$$

Express this in terms of the problem parameters and the simplified notation above, and note that only $\alpha$ and $\mathcal{F}$ in your expression are affected by how we choose the filter $f[n]$. Use your expression to show that the MMSE is minimized if $\alpha^2/\mathcal{F}$ is made as large as possible.

**(e)** The Cauchy–Schwarz inequality can be used to show that

$$\frac{\alpha^2}{\mathcal{F}} \le \mathcal{E} \,,$$

with equality if and only if

$$f[n] = c\,p[-n] \,,$$

for any nonzero constant $c$, which we can take to be 1 without loss of generality here. Hence the MMSE is minimized if the filter impulse response at the receiver is *matched* to the shape of the signal from the transmitter. Note that with $f[n] = p[-n]$, we get $\alpha = \mathcal{E} = \mathcal{F}$. Using this, rewrite your expressions for the MMSE and for the constants $\gamma$ and $\mu$ in the LMMSE estimator in (c), in terms of $\mu_A$, $\sigma_A$, $\sigma_v$, and $\mathcal{E}$. As a check on your answers, explicitly verify that your expressions behave reasonably as the parameters take various extreme values. Pick at least three sets of extreme cases to check.

**3.21.** *Note: This problem relies on a basic understanding of random processes as discussed in Chapter 10.* Figure P3.21-1 depicts a PAM system in which the transmitted sequence $a[n]$ is a zero-mean wide-sense stationary (WSS) Gaussian random sequence with autocorrelation function

$$R_{aa}[m] = \left(\frac{1}{2}\right)^{|m|} \,.$$

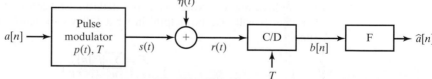

**Figure P3.21-1**

The channel introduces additive noise $\eta(t)$. The received signal $r(t)$ is sampled to obtain $b[n]$; $b[n]$ is then processed with a memoryless affine system F whose output $\widehat{a}[n]$ is an estimate of $a[n]$. The associated relationships are:

- $s(t) = \sum_{n=-\infty}^{\infty} a[n]p(t - nT)$;
- $r(t) = \sum_{n=-\infty}^{\infty} a[n]p(t - nT) + \eta(t)$;
- $b[n] = r(nT)$;
- $\eta(t)$ is zero-mean WSS noise with autocorrelation function $R_{\eta\eta}(\tau) = Ne^{-|\tau|}$ and is independent of $a[n]$; and
- $\widehat{a}[n] = k_0 + k_1 b[n]$.

**(a)** If $p(t)$ is as shown in Figure P3.21-2, determine whether there is ISI present in $r(t)$.

**(b)** For this part assume that $p(t)$ is chosen so that there is no ISI in $r(t)$, and that $p(0) = 1$. The output $\widehat{a}[n]$ of the system F has the form $\widehat{a}[n] = k_0 + k_1 b[n]$. Determine $k_0$ and $k_1$ to minimize the mean square error $\varepsilon$, given as:

$$\varepsilon = E\left[(a[n] - \widehat{a}[n])^2\right] .$$

**(c)** For this part assume that $p(t)$ is chosen so that there is no ISI in $r(t)$, and that $p(0) = 1$. You are at the transmitter, therefore you know what $a[n]$ is, and you are trying to estimate what $b[n]$ will be at the receiver. With $\widehat{b}[n]$ denoting the estimate of $b[n]$, at the transmitter determine the estimate $\widehat{b}[n]$ that will minimize the mean square error $\varepsilon_T$ defined as

$$\varepsilon_T = E\left[(b[n] - \widehat{b}[n])^2\right] .$$

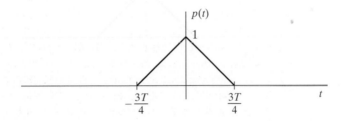

**Figure P3.21-2**

**3.22.** *Note: This problem relies on a basic understanding of random processes as discussed in Chapter 10.* In Figure P3.22-1, we show a PAM system in which the transmitted sequence $a[n]$ is continuous in amplitude.

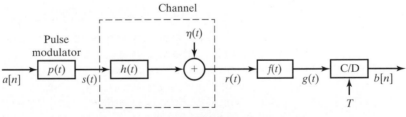

**Figure P3.22-1**

The channel is modeled as an LTI system with impulse response $h(t)$ and additive output noise $\eta(t)$. The received signal $r(t)$ is processed with an LTI filter $f(t)$ and then sampled to obtain $b[n]$. The associated relationships are

- $s(t) = \sum_{n=-\infty}^{\infty} a[n]p(t - nT)$;
- $p_c(t)$ is defined as $p(t) * h(t)$;
- $r(t) = \sum_{n=-\infty}^{\infty} a[n]p_c(t - nT) + \eta(t)$;
- $g(t) = f(t) * r(t)$;
- $b[n] = g(nT)$; and
- $\eta(t)$ is a zero-mean wide-sense stationary (WSS) random process with autocorrelation function $R_{\eta\eta}(\tau) = N\delta(\tau)$.

**(a)** For this part only
  - $\eta(t) = 0$ (i.e., $N = 0$);
  - $H(j\omega) = e^{-j\omega/2}$;
  - $f(t) = \delta(t)$; and
  - $p(t)$ is as shown in Figure P3.22-2.

Determine the fastest symbol rate $(1/T)$ so that there is no ISI in $g(t)$, i.e., so that $b[n] = ca[n]$ where $c$ is a constant. Also, determine the value of $c$.

For the remainder of this problem assume that there is no ISI in $r(t)$ or $g(t)$.

**(b)** Determine the mean and variance of $b[0]$ in terms of $p_c(t), f(t), N$, and $a[n]$.

**(c)** Determine $f(t)$ in terms of $p_c(t)$ so that $E\{b[0]\} = a[0]$ and the variance of $b[0]$ is minimized.

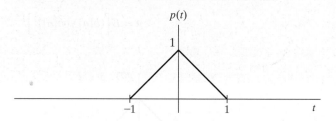

**Figure P3.22-2**

# 4 State-Space Models

The discussion of system descriptions up to this point has emphasized and used models that represent the transformation of input signals into output signals. In the case of linear and time-invariant (LTI) models, we have focused on their impulse response, frequency response, and transfer function. Such input-output models do not directly consider the internal behavior of the systems they represent.

Internal behavior can be important for a variety of reasons. For instance, in examining issues of stability, a system model can be stable from an input-output perspective, yet internal variables may display unstable behavior. This chapter begins a discussion of system models that display the internal dynamical behavior of the system as well as the input-output characteristics. The discussion is illustrated by numerous examples. The study of such models and their applications continues through Chapters 5 and 6 as well.

## 4.1 SYSTEM MEMORY

In this chapter we introduce an important model description—the state-space model—that highlights the internal behavior of a system and is especially suited to representing causal systems, particularly for real-time applications such as control. These models arise in both continuous-time (CT) and discrete-time (DT) forms. In general they can be nonlinear and time-varying, although we will focus on the LTI case.

A state-space model for a causal system answers a question asked about such systems in many settings. We pose the question for the causal DT case, though it can also be asked for causal CT systems: given the input value $x[n]$ at some arbitrary time $n$, how much needs to be known about past values of the input, that is, about $x[k]$ for $k < n$, in order to determine the present output $y[n]$? As the system is causal, having all past values $x[k]$, in addition to $x[n]$, will suffice, but the issue is whether all past $x[k]$ are actually needed.

The above question addresses the issue of memory in the system, and is worthwhile for a variety of reasons. For example, the answer conveys an idea of the complexity, or number of degrees of freedom, associated with the dynamic behavior of the system. The more we need to know about past inputs in order to determine the present output, the richer the variety of possible output behaviors, and the more ways one can be surprised in the absence of knowledge of the past. We will only consider systems with a finite number of degrees of freedom, or with finite-dimensional memory; these are often referred to as lumped systems.

One application in which the above question arises is in implementing a computer algorithm that acts causally on a data stream. Thinking of the algorithm as a system, the answer to the question indicates how much memory will be needed to run the algorithm. In a control application, the answer to the memory question above suggests the required level of complexity for the controller of a given system. The controller has to remember enough about the past to determine the effects of present control actions on the response of the system.

With a state-space description, everything about the past that is relevant to the present and future is summarized in the present values of a finite set of state variables. These values together specify the present state of the system. We are interested in the case of real-valued state variables. The number of state variables, also referred to as the order of the state-space description, indicates the number of degrees of freedom, or the dimension of the memory, associated with the system or model.

## 4.2 ILLUSTRATIVE EXAMPLES

As a prelude to developing the general form of a state-space model, this section presents in some detail a few CT and DT examples. In addition to illustrating the process of building a state-space model, these examples will suggest how state-space descriptions arise in a variety of contexts. This section may alternatively be read after the more general presentation of state-space models in Section 4.3. Several further examples appear later in the chapter.

To begin, we examine a mechanical system that, despite its simplicity, is rich enough to bring out typical features of a CT state-space model, and serves as a prototype for a variety of other systems.

## Example 4.1    *Inverted Pendulum*

Consider the inverted pendulum shown in Figure 4.1. The pendulum is rigid, with mass $m$, and can rotate about the pivot at its base, moving in the plane orthogonal to the pivot axis. The distance from the pivot to the center of mass is $\ell$, and the pendulum's moment of inertia about the pivot is $\mathcal{I}$. These parameters are all assumed constant.

The line connecting the pivot to the center of mass is at an angle $\theta(t)$ at time $t$, measured clockwise from the vertical. An external torque is applied to the pendulum around the axis of the pivot. We treat this torque as the input to our system, and denote it by $x(t)$, taken as positive when it acts counterclockwise.

Suppose the system output variable of interest, $y(t)$, is just the pendulum angle, so that $y(t) = \theta(t)$. In a typical control application, one might want to manipulate $x(t)$—in response to measurements that are fed back to the controller—so as to maintain $y(t)$ near the value 0, thus balancing the inverted pendulum vertically.

The external torque is opposed by the torque due to the acceleration $g$ of gravity acting on the mass, which produces a clockwise torque of value $mg\ell \sin(\theta(t))$. Finally, assume a frictional torque that opposes the motion in proportion to the magnitude of the angular velocity. This torque is thus given by $-\beta\dot{\theta}(t)$, where $\dot{\theta}(t) = d\theta(t)/dt$ and $\beta$ is some nonnegative constant.

Although the inverted pendulum is a simple system in many respects, it captures some essential features of systems that arise in diverse balancing applications, for instance, supporting the body on a human ankle or a mass on a robot joint or wheel axle. There are also control applications in which the pendulum is intended to move in the vicinity of its normal hanging position rather than the inverted position, that is, with $\theta(t) \approx \pi$. One might alternatively want the pendulum to rotate through full circles around the pivot. All of these motions are described by the equations below.

**A Conventional Model**    The rotational form of Newton's law says the rate of change of angular momentum equals the net torque. We can accordingly write

$$\frac{d}{dt}\left(\mathcal{I}\frac{d\theta(t)}{dt}\right) = mg\ell \sin(\theta(t)) - \beta\frac{d\theta(t)}{dt} - x(t) . \tag{4.1}$$

Since $\mathcal{I}$ is constant, the preceding expression can be rewritten in a form that is closer to what is typically encountered in an earlier differential equations course:

$$\mathcal{I}\frac{d^2 y(t)}{dt^2} + \beta\frac{dy(t)}{dt} - mg\ell \sin(y(t)) = -x(t) , \tag{4.2}$$

which is a single second-order nonlinear differential equation relating the output $y(t)$ to the input $x(t)$.

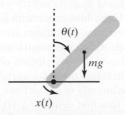

**Figure 4.1**    Inverted pendulum.

**State Variables**   To get at the notion of state variables, we examine what constitutes the memory of the system at some arbitrary time $t_0$. Assume the parameters $\mathcal{I}$, $m$, $\ell$, and $\beta$ are all known, as is the external input $x(t)$ for $t \geq t_0$. The question is, what more needs to be known about the system at $t_0$ in order to solve for the behavior of the system for $t > t_0$.

Solving Eq. (4.1) for $\theta(t)$ in the interval $t > t_0$ ultimately requires integrating the equation twice, which in turn requires knowledge of the initial position and velocity, $\theta(t_0)$ and $\dot{\theta}(t_0)$ respectively. Another way to recognize the special role of these two variables is by considering the energy of the pendulum at the starting time. The energy is the result of past inputs to the system, and is reflected in the ensuing motion of the system. The potential energy at $t = t_0$ is determined by $\theta(t_0)$ and the kinetic energy by $\dot{\theta}(t_0)$, so these variables are key to understanding the behavior of the system for $t > t_0$.

**State-Space Model**   The above discussion suggests that two natural memory variables of the system at any time $t$ are $q_1(t) = \theta(t)$ and $q_2(t) = \dot{\theta}(t)$. Taking these as candidate state variables, a corresponding state-space description is found by trying to express the rates of change of these variables at time $t$ entirely in terms of the values of these variables and of the input at the same time $t$. For this simple example, a pair of equations of the desired form can be obtained quite directly. Invoking the definitions of $q_1(t)$ and $q_2(t)$, as well as Eq. (4.1), and still assuming $\mathcal{I}$ is constant, we obtain

$$\frac{dq_1(t)}{dt} = q_2(t)\,, \tag{4.3}$$

$$\frac{dq_2(t)}{dt} = \frac{1}{\mathcal{I}}\Big(mg\ell \sin(q_1(t)) - \beta q_2(t) - x(t)\Big)\,. \tag{4.4}$$

This description comprises a pair of coupled first-order differential equations, driven by the input $x(t)$. These are referred to as the state evolution equations. The corresponding output equation expresses the output $y(t)$ entirely in terms of the values of the state variables and of the input at the same time $t$; in this case, the output equation is simply

$$y(t) = q_1(t)\,. \tag{4.5}$$

The combination of the state evolution equations and the output equation constitutes a state-space description of the system. The fact that such a description of the system is possible in terms of the candidate state variables $\theta(t)$ and $\dot{\theta}(t)$ confirms these as state variables—the "candidate" label can now be dropped.

Not only does the ordinary differential equation description in Eq. (4.1) or equivalently in Eq. (4.2) suggest what is needed to obtain the state-space model, but the converse is also true: the differential equation in Eq. (4.1), or equivalently in Eq. (4.2), can be obtained from Eqs. (4.3), (4.4), and (4.5).

**Some Variations**   The choice of state variables above is not unique. For instance, the quantities defined by $q_1(t) = \theta(t) + \dot{\theta}(t)$ and $q_2(t) = \theta(t) - \dot{\theta}(t)$ could have functioned equally well. Equations expressing $\dot{q}_1(t)$, $\dot{q}_2(t)$, and $y(t)$ as functions of $q_1(t)$, $q_2(t)$, and $x(t)$ under these new definitions are easily obtained, and yield a different but entirely equivalent state-space representation.

The state-space description obtained above is nonlinear but time-invariant. It is nonlinear because the state variables and input, namely $q_1(t)$, $q_2(t)$, and $x(t)$, are combined nonlinearly in at least one of the functions defining $\dot{q}_1(t)$, $\dot{q}_2(t)$, and $y(t)$—in this case, the function defining $\dot{q}_2(t)$. The description is time-invariant because all the functions defining $\dot{q}_1(t)$, $\dot{q}_2(t)$, and $y(t)$ are time-invariant, that is, they combine their arguments $q_1(t)$, $q_2(t)$, and $x(t)$ according to a prescription that does not depend on time.

For small enough deviations from the fully inverted position, $q_1(t) = \theta(t)$ is small, so $\sin(q_1(t)) \approx q_1(t)$. With this approximation, Eq. (4.4) is replaced by

$$\frac{dq_2(t)}{dt} = \frac{1}{\mathcal{I}}\Big(mg\ell q_1(t) - \beta q_2(t) - x(t)\Big). \tag{4.6}$$

The function defining $\dot{q}_2(t)$ is now an LTI function of its arguments $q_1(t)$, $q_2(t)$, and $x(t)$, so the resulting state-space model is now also LTI.

For linear models, matrix notation allows a compact representation of the state evolution equations and the output equation. We will use bold lowercase letters for vectors and bold uppercase for matrices. Defining the state vector and its derivative by

$$\mathbf{q}(t) = \begin{bmatrix} q_1(t) \\ q_2(t) \end{bmatrix}, \qquad \dot{\mathbf{q}}(t) = \frac{d\mathbf{q}(t)}{dt} = \begin{bmatrix} \dot{q}_1(t) \\ \dot{q}_2(t) \end{bmatrix}, \tag{4.7}$$

the linear model becomes

$$\dot{\mathbf{q}}(t) = \begin{bmatrix} \dot{q}_1(t) \\ \dot{q}_2(t) \end{bmatrix} = \begin{bmatrix} 0 & 1 \\ mg\ell/\mathcal{I} & -\beta/\mathcal{I} \end{bmatrix} \begin{bmatrix} q_1(t) \\ q_2(t) \end{bmatrix} + \begin{bmatrix} 0 \\ -1/\mathcal{I} \end{bmatrix} x(t)$$

$$= \mathbf{A}\mathbf{q}(t) + \mathbf{b}x(t), \tag{4.8}$$

where the definitions of the matrix $\mathbf{A}$ and vector $\mathbf{b}$ should be clear by comparison with the preceding equality. The corresponding output equation can be written as

$$y(t) = \begin{bmatrix} 1 & 0 \end{bmatrix} \begin{bmatrix} q_1(t) \\ q_2(t) \end{bmatrix} = \mathbf{c}^T \mathbf{q}(t), \tag{4.9}$$

with $\mathbf{c}^T$ denoting the transpose of a column vector, that is, a row vector. The time invariance of the system is reflected in the fact that the coefficient matrices $\mathbf{A}$, $\mathbf{b}$, and $\mathbf{c}^T$ are constant rather than time-varying.

The ideas in the above example can be generalized to much more elaborate settings. In general, a natural choice of state variables for a mechanical system is the set of position and velocity variables associated with each component mass. For example, in the case of $N$ point masses in three-dimensional space that are interconnected with each other and to rigid supports by massless springs, the natural choice of state variables would be the associated $3N$ position variables and $3N$ velocity variables. If these masses were confined to move in a plane, we would instead have $2N$ position variables and $2N$ velocity variables.

The next example suggests how state-space models arise in describing electrical circuits.

## Example 4.2    Electrical Circuit

Consider the resistor-inductor-capacitor (RLC) circuit shown in Figure 4.2. All the component voltages and currents are labeled in the figure.

We begin by listing the characteristics of the various components, which we assume are linear and time-invariant. The defining equations for the inductor,

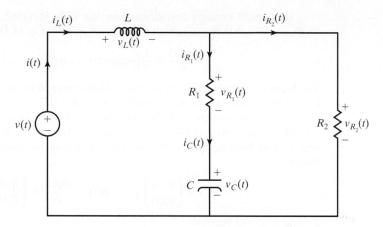

**Figure 4.2**   RLC circuit.

capacitor, and the two resistors take the form, in each case, of an LTI constraint relating the voltage across the element and the current through it. Specifically, we have

$$v_L(t) = L\frac{di_L(t)}{dt}$$

$$i_C(t) = C\frac{dv_C(t)}{dt}$$

$$v_{R_1}(t) = R_1 i_{R_1}(t)$$

$$v_{R_2}(t) = R_2 i_{R_2}(t) \ . \tag{4.10}$$

The voltage source is defined by the condition that its voltage is a specified or arbitrary $v(t)$, regardless of the current $i(t)$ that is drawn from it.

The next step is to describe the constraints on these variables that arise from interconnecting the components. The interconnection constraints for an electrical circuit are imposed by Kirchhoff's voltage law (KVL) and Kirchhoff's current law (KCL). Both KVL and KCL produce additional LTI constraints relating the variables associated with the circuit. Here, KVL and KCL yield the following equations:

$$v(t) = v_L(t) + v_{R_2}(t)$$

$$v_{R_2}(t) = v_{R_1}(t) + v_C(t)$$

$$i(t) = i_L(t)$$

$$i_L(t) = i_{R_1}(t) + i_{R_2}(t)$$

$$i_{R_1}(t) = i_C(t) \ . \tag{4.11}$$

Other such KVL and KCL equations can be written for this circuit, but turn out to be consequences of the equations above, rather than new constraints.

Equations (4.10) and (4.11) together represent the individual components in the circuit and their mutual connections. Any set of signals that simultaneously satisfies all these constraint equations constitutes a valid solution—or behavior—of the circuit. Since all the constraints are LTI, it follows that weighted linear combinations or superpositions of behaviors are themselves behaviors of the circuit, and time-shifted behaviors are again behaviors of the circuit, so the circuit itself is LTI.

**Input, Output, and State Variables**    Let us take the source voltage $v(t)$ as the input to the circuit, and also denote this by $x(t)$, our standard symbol for an input. Any of the circuit voltages or currents can be chosen as the output. Choose $v_{R_2}(t)$, for instance, and denote it by $y(t)$, our standard symbol for an output.

As in the preceding example, a good choice of state variables is established by determining what constitutes the memory of the system at any time. Apart from the parameters $L$, $C$, $R_1$, $R_2$, and the external input $x(t)$ for $t \geq t_0$, we ask what needs to be known about the system at a starting time $t_0$ in order to solve for the behavior of the system for $t > t_0$.

The existence of the derivatives in the defining expressions in Eq. (4.10) for the inductor and capacitor suggests that at least $i_L(t_0)$ and $v_C(t_0)$ are needed, or quantities equivalent to these. Note that, similarly to what was observed in the previous example, these variables are also associated with energy storage in the system, in this case the energy stored in the inductor and capacitor respectively. We accordingly identify the two natural memory variables of the system at any time $t$ as $q_1(t) = i_L(t)$ and $q_2(t) = v_C(t)$, and these are our candidate state variables.

**State-Space Model**    We now develop a state-space description for the RLC circuit of Figure 4.2 by trying to express the rates of change of the candidate state variables at time $t$ entirely in terms of the values of these variables and of the input at the same time $t$. This is done by reducing the full set of relations in Eqs. (4.10) and (4.11), eliminating all variables other than the input, output, candidate state variables, and derivatives of the candidate state variables.

This process for the present example is not as transparent as in Example 4.1, and some attention is required in order to carry out the elimination efficiently. A good strategy—and one that generalizes to more complicated circuits—is to express the inductor voltage $v_L(t)$ and capacitor current $i_C(t)$ as functions of just the allowed variables, namely $i_L(t)$, $v_C(t)$, and $x(t) = v(t)$. Once this is accomplished, we make the substitutions

$$v_L(t) = L\frac{di_L(t)}{dt} \qquad \text{and} \qquad i_C(t) = C\frac{dv_C(t)}{dt} , \tag{4.12}$$

then rearrange the resulting equations to get the desired expressions for the rates of change of the candidate state variables. Following this procedure, and introducing the definition

$$\alpha = \frac{R_2}{R_1 + R_2} \tag{4.13}$$

for notational convenience, we obtain the desired state evolution equations. These are written below in matrix form, exploiting the fact that these state evolution equations turn out to be linear:

$$\begin{bmatrix} di_L(t)/dt \\ dv_C(t)/dt \end{bmatrix} = \begin{bmatrix} -\alpha R_1/L & -\alpha/L \\ \alpha/C & -1/(R_1+R_2)C \end{bmatrix} \begin{bmatrix} i_L(t) \\ v_C(t) \end{bmatrix} + \begin{bmatrix} 1/L \\ 0 \end{bmatrix} x(t) . \tag{4.14}$$

This is of the form

$$\dot{\mathbf{q}}(t) = \mathbf{A}\mathbf{q}(t) + \mathbf{b}x(t) , \tag{4.15}$$

where

$$\mathbf{q}(t) = \begin{bmatrix} q_1(t) \\ q_2(t) \end{bmatrix} = \begin{bmatrix} i_L(t) \\ v_C(t) \end{bmatrix} \tag{4.16}$$

and the definitions of the coefficient matrices $\mathbf{A}$ and $\mathbf{b}$ are determined by comparison with Eq. (4.14). The fact that these matrices are constant establishes that the description is LTI. The key feature here is that the model expresses the rates of change of the state variables at any time $t$ as constant linear functions of their values and that of the input at the same time instant $t$.

As we will see in the next chapter, the state evolution equations in Eq. (4.14) can be used to solve for the state variables $i_L(t)$ and $v_C(t)$ for $t > t_0$, given the input $x(t) = v(t)$ for $t \geq t_0$ and the initial conditions on the state variables at time $t_0$. Furthermore, knowledge of $i_L(t)$, $v_C(t)$, and $v(t)$ suffices to reconstruct all the other voltages and currents in the circuit at time $t$. Having picked the output of interest to be $v_{R_2}(t) = y(t)$, we can write (again in matrix notation)

$$y(t) = v_{R_2}(t) = \begin{bmatrix} \alpha R_1 & \alpha \end{bmatrix} \begin{bmatrix} i_L(t) \\ v_C(t) \end{bmatrix} = \mathbf{c}^T \mathbf{q}(t) . \tag{4.17}$$

**Input-Output Behavior**    Transforming Eqs. (4.10) and (4.11) using the bilateral Laplace transform, and noting that differentiation in the time domain maps to multiplication by $s$ in the transform domain, we can solve for the transfer function $H(s)$ of the system from $x(t)$ to $y(t)$. Alternatively, we can obtain the same transfer function from Laplace transformation of the state-space description in Eqs. (4.14) and (4.17). The next chapter presents an explicit formula for this transfer function in terms of the coefficient matrices $\mathbf{A}$, $\mathbf{b}$, and $\mathbf{c}^T$.

For our RLC example, this transfer function $H(s)$ from input to output is

$$H(s) = \frac{Y(s)}{X(s)} = \frac{\alpha\left(\frac{R_1}{L}s + \frac{1}{LC}\right)}{s^2 + \alpha\left(\frac{1}{R_2 C} + \frac{R_1}{L}\right)s + \alpha\frac{1}{LC}} . \tag{4.18}$$

The corresponding input-output second-order LTI differential equation is

$$\frac{d^2 y(t)}{dt^2} + \alpha\left(\frac{1}{R_2 C} + \frac{R_1}{L}\right)\frac{dy(t)}{dt} + \alpha\left(\frac{1}{LC}\right)y(t) = \alpha\left(\frac{R_1}{L}\right)\frac{dx(t)}{dt} + \alpha\left(\frac{1}{LC}\right)x(t) . \tag{4.19}$$

The procedure for obtaining a state-space description that is illustrated in Example 4.2 can be used even if some of the circuit components are non-linear. It can then often be helpful to choose inductor flux rather than current as a state variable, and similarly to choose capacitor charge rather than voltage as a state variable. It is generally the case, just as in the Example 4.2, that the natural state variables in an electrical circuit are the inductor currents or fluxes, and the capacitor voltages or charges. The exceptions occur in degenerate situations, for example where a closed path in the circuit involves only capacitors and voltage sources. In the latter instance, KVL applied to this path shows that the capacitor voltages are not all independent.

State-space models arise naturally in many problems that involve tracking subgroups of some population of objects as they interact in time. For instance, in chemical reaction kinetics the interest is in determining the expected molecule numbers or concentrations of the various interacting chemical constituents as the reaction progresses in continuous time. Another instance involves modeling, in either continuous time or discrete time, the spread of a fashion, opinion, idea, or disease through a human population, or of a software virus through a computer network. The following example develops one such DT model and begins to explore its behavior. Some later examples extend the analysis further.

---

**Example 4.3**  **Viral Propagation**

The DT model presented here captures some essential aspects of viral propagation in a variety of settings. The model is one of a large family of such models, both deterministic and stochastic, that have been widely studied. Though much of the terminology derives from modeling the spread of disease by viruses, the paradigm of viral propagation has been applied to understanding how, for example, malicious software, advertisements, gossip, or cultural memes spread in a population or network.

The deterministic model here tracks three component subpopulations from the $n$th DT epoch to the $(n+1)$th. Suppose the total population of size $P$ is divided into the following subgroups, or "compartments," at integer time $n$:

- $s[n] \geq 0$ is the number of susceptibles, currently virus-free but vulnerable to acquiring the virus;
- $i[n] \geq 0$ is the number of infectives, carrying the virus and therefore capable of passing it to the susceptibles by the next epoch; and
- $r[n] \geq 0$ is the number of recovered, no longer carrying the virus and no longer susceptible, because of acquired immunity.

The model below assumes these variables are real-valued rather than integer-valued, which results in substantial simplification of the model, and may be a satisfactory approximation when $P$ is very large.

We assume the birth rate in these three subgroups has the same value $\beta$; this is the (deterministic) fractional increase in the population per unit time due to birth. Suppose the death rate is also $\beta$, so the total size of the population remains constant at $P$. Assume $0 \leq \beta < 1$.

Let the rate at which susceptibles become infected be proportional to the concentration of infectives in the general population, hence a rate of the form $\gamma(i[n]/P)$ for some $0 < \gamma \leq 1$. The rate at which infectives move to the recovered compartment is denoted by $\rho$, with $0 < \rho \leq 1$. We take newborns to be susceptible, even if born to infective or recovered members of the population. Suppose also that newborns are provided immunity at a rate $0 \leq v[n] \leq 1$, for instance by vaccination, moving them directly from the susceptible compartment to the recovered compartment. We consider $v[n]$ to be the control input, and denote it by the alternative symbol $x[n]$.

With the above notation and assumptions, we arrive quite directly at the very simple (and undoubtedly simplistic) model below, for the change in each subpopulation over one time step:

$$s[n+1] - s[n] = -\gamma(i[n]/P)s[n] + \beta(i[n] + r[n]) - \beta P x[n]$$

$$i[n+1] - i[n] = \gamma(i[n]/P)s[n] - \rho i[n] - \beta i[n]$$

$$r[n+1] - r[n] = \rho i[n] - \beta r[n] + \beta P x[n] . \tag{4.20}$$

A model of this type is commonly referred to as an SIR model, as it comprises susceptible, infective, and recovered populations. We shall assume that the initial conditions, parameters, and control inputs are chosen so as to maintain all subpopulations at nonnegative values throughout the interval of interest. The actual mechanisms of viral spread are of course much more intricate and complicated than captured in this elementary model, and also involve substantial randomness and uncertainty.

If some fraction $\phi$ of the infectives gets counted at each time epoch, then the aggregate number of infectives reported can be taken as our output $y[n]$, so

$$y[n] = \phi i[n] \,. \tag{4.21}$$

Notice that the expressions in Eq. (4.20) have a very similar form to the CT state evolution equations we arrived at in the earlier two examples. For the DT case, take the rate of change of a variable at time $n$ to be the increment over one time step forward from $n$. Then Eq. (4.20) expresses the rates of change of the indicated variables at time $n$ as functions of these same variables and the input at time $n$. It therefore makes sense to think of $s[n]$, $i[n]$, and $r[n]$ as state variables, whose values at time $n$ constitute the state of the system at time $n$.

The model here is time-invariant because the three expressions that define the rates of change all involve combining the state variables and input at time $n$ according to prescriptions that do not depend on $n$. The consequence of this feature is that any set of $s[\cdot]$, $i[\cdot]$, and $r[\cdot]$ signals that simultaneously satisfy the model equations will also satisfy the model equations if they are all shifted arbitrarily by the same time offset. However, the model is not linear; it is nonlinear because the first two expressions involve a nonlinear combination of $s[n]$ and $i[n]$, namely their product. The expression in Eq. (4.21) writes the output at time $n$ as a function of the state variables and input at time $n$—though it happens in this case that only $i[n]$ is needed.

It is conventional in the DT case to rearrange the state evolution equations into a form that expresses the state at time $n+1$ as a function of the state variables and input at time $n$. Thus Eq. (4.20) would be rewritten as

$$s[n+1] = s[n] - \gamma(i[n]/P)s[n] + \beta(i[n] + r[n]) - \beta Px[n]$$
$$i[n+1] = i[n] + \gamma(i[n]/P)s[n] - \rho i[n] - \beta i[n]$$
$$r[n+1] = r[n] + \rho i[n] - \beta r[n] + \beta Px[n] \,. \tag{4.22}$$

In this form, the equations give a simple prescription for obtaining the state at time $n+1$ from the state and input at time $n$. Summing the three equations also makes clear that for this example

$$s[n+1] + i[n+1] + r[n+1] = s[n] + i[n] + r[n] = P \,. \tag{4.23}$$

Thus, knowing any two of the subgroup populations suffices to determine the third, if $P$ is known. Examining the individual relations in Eqs. (4.20) or (4.22), and noting that the term $i[n] + r[n]$ in the first equation of each set could equivalently have been written as $P - s[n]$, we see that the first two relations in fact only involve the susceptible and infective populations, in addition to the input, and therefore comprise a state evolution description of lower order, namely

$$s[n+1] = s[n] - \gamma(i[n]/P)s[n] + \beta(P - s[n]) - \beta Px[n]$$
$$i[n+1] = i[n] + \gamma(i[n]/P)s[n] - \rho i[n] - \beta i[n] \,. \tag{4.24}$$

Figure 4.3 shows a few state-variable trajectories produced by stepping the model in Eq. (4.24) forward from a particular $s[0]$, fixed at 8000 out of a population $(P)$ of 10,000, using different initial values $i[0]$. Note that in each case the number of

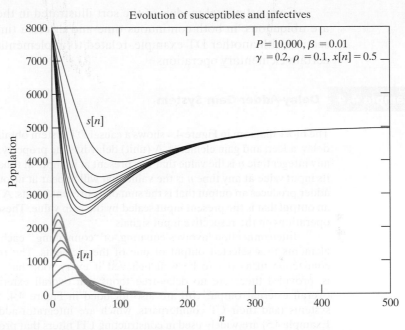

**Figure 4.3**  Response of SIR model for a particular choice of parameter values and a variety of initial conditions.

infectives, $i[n]$, initially increases from its value at the starting time $n - 0$, before eventually decaying. This initial increase would correspond to "going viral" in the case of a rumor, advertisement, or fashion that spreads through a social network, or to an epidemic in the case of disease propagation. The second equation in Eq. (4.24) shows that $i[n+1] > i[n]$ precisely when

$$\frac{s[n]}{P} > \frac{\rho + \beta}{\gamma} = \frac{1}{R_0} \; . \tag{4.25}$$

Here

$$R_0 = \frac{\gamma}{\beta + \rho} \tag{4.26}$$

is a parameter that typically arises in viral propagation models, and is termed the basic reproductive ratio (referring to "reproduction" of infectives, not to population growth). Thus $i[n]$ increases at the next time step whenever the fraction of susceptibles in the population, $s[n]/P$, exceeds the threshold $1/R_0$. As $s[n]/P$ cannot exceed 1, there can be no epidemic if $R_0 \leq 1$. The greater the amount by which $R_0$ exceeds 1, the fewer the number of susceptibles required in order for an epidemic to occur.

Figure 4.3 also shows that the system in this case, with the immunization rate fixed at $x[n] = 0.5$, reaches a steady state in which there are no infectives. This is termed an infective-free steady state. In Examples 4.8, 4.10, and 5.5, we explore further characteristics of the model in Eq. (4.24). In particular, it will turn out that it is possible—for instance by dropping the immunization rate to $x[n] = 0.2$ while keeping the other parameters as in Figure 4.3—for the attained steady state to have a nonzero number of infectives. This is termed an endemic steady state.

Compartmental models of the sort illustrated in the preceding example are ubiquitous, in both continuous time and discrete time. We conclude this section with another DT example, related to implementation of a filter using certain elementary operations.

## Example 4.4    Delay-Adder-Gain System

The block diagram in Figure 4.4 shows a causal DT system obtained by interconnecting delay, adder, and gain elements. A (unit) delay has the property that its output value at any integer time $n$ is the value that was present at its input at time $n - 1$; or equivalently, its input value at any time $n$ is the value that will appear at its output at time $n + 1$. An adder produces an output that is the sum of its present inputs. A gain element produces an output that is the present input scaled by the gain value. These all correspond to LTI operations on the respective input signals.

Interconnection involves equating, or "connecting," each input of these various elements to a selected output of one of the elements. The result of such an interconnection turns out to be well behaved if every loop has some delay in it, that is, provided there are no delay-free loops. An overall external input $x[n]$ and an overall external output $y[n]$ are also included in Figure 4.4. Such delay-adder-gain systems (and their CT counterparts, which are integrator-adder-gain systems, as in Example 4.5) are widely used in constructing LTI filters that produce a signal $y[\cdot]$ from a signal $x[\cdot]$.

The memory of this system is embodied in the delay elements, so it is natural to consider the outputs of these elements as candidate state variables. Accordingly, we label the outputs of the memory elements in this example as $q_1[n]$ and $q_2[n]$ at time $n$. For the specific block diagram in Figure 4.4, the detailed component and interconnection equations relating the indicated signals are

$$q_1[n + 1] = q_2[n]$$

$$q_2[n + 1] = p[n]$$

$$p[n] = x[n] - 0.5q_1[n] + 1.5q_2[n]$$

$$y[n] = q_2[n] + p[n] \,. \tag{4.27}$$

The response of the system for $n \geq n_0$ is completely determined by the external input $x[n]$ for times $n \geq n_0$ and the values $q_1[n_0]$ and $q_2[n_0]$ that are stored at the

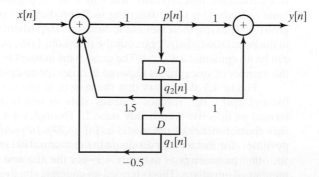

**Figure 4.4**    Delay-adder-gain block diagram.

outputs of the delay elements at time $n_0$. The delay elements capture the state of the system at each time step, that is, they summarize all the past history that is relevant to how the present and future inputs to the system determine the present and future response of the system.

The relationships in Eq. (4.27) need to be condensed in order to express the values of the candidate state variables at time $n+1$ in terms of the values of these variables at time $n$ and the value of the external input at the same time instant $n$. This corresponds to expressing the inputs to all the delay elements at time $n$ in terms of all the delay outputs at time $n$ as well as the external input at this same time. The result for this example is captured in the following matrix equation:

$$\mathbf{q}[n+1] = \begin{bmatrix} q_1[n+1] \\ q_2[n+1] \end{bmatrix} = \begin{bmatrix} 0 & 1 \\ -0.5 & 1.5 \end{bmatrix} \begin{bmatrix} q_1[n] \\ q_2[n] \end{bmatrix} + \begin{bmatrix} 0 \\ 1 \end{bmatrix} x[n]$$

$$= \mathbf{A}\mathbf{q}[n] + \mathbf{b}x[n] . \tag{4.28}$$

Similarly, the output at time $n$ can be written in terms of the values of the candidate state variables at time $n$ and the value of the external input at the same time instant $n$:

$$y[n] = \begin{bmatrix} -0.5 & 2.5 \end{bmatrix} \begin{bmatrix} q_1[n] \\ q_2[n] \end{bmatrix} + x[n] = \mathbf{c}^T\mathbf{q}[n] + \mathrm{d}x[n] . \tag{4.29}$$

Notice that in this example, unlike in the previous examples, the output $y[n]$ at any time $n$ depends not only on the state variables at time $n$ but also on the input at that time $n$.

Equations (4.28) and (4.29) establish that $q_1[n]$ and $q_2[n]$ are indeed valid state variables. Specifically, the equations explicitly show that if one is given the values $q_1[n_0]$ and $q_2[n_0]$ of the state variables at some initial time $n_0$, and also the input trajectory from $n_0$ onward, that is, $x[n]$ for times $n \geq n_0$, then we can compute the values of the state variables and the output for times $n \geq n_0$. All that is needed is to iteratively apply Eq. (4.28) to find $q_1[n_0+1]$ and $q_2[n_0+1]$, then $q_1[n_0+2]$ and $q_2[n_0+2]$, and so on for increasing time arguments, and to use Eq. (4.29) at each time to find the output.

Transforming the relationships in Eq. (4.27) using the bilateral $z$-transform, and noting that time-advancing a signal by one step maps to multiplication by $z$ in the transform domain, we can solve for the transfer function $H(z)$ of the system from $x[\cdot]$ to $y[\cdot]$. Alternatively, the same transfer function can be obtained from $z$-transformation of the state-space description; the next chapter presents an explicit formula for this transfer function in terms of the coefficient matrices $\mathbf{A}$, $\mathbf{b}$, $\mathbf{c}^T$, and d. Either way, the resulting transfer function for our example is

$$H(z) = \frac{Y(z)}{X(z)} = \frac{1 + z^{-1}}{1 - \frac{3}{2}z^{-1} + \frac{1}{2}z^{-2}} , \tag{4.30}$$

which corresponds to the following input-output difference equation:

$$y[n] - \frac{3}{2}y[n-1] + \frac{1}{2}y[n-2] = x[n] + x[n-1] . \tag{4.31}$$

The development of CT state-space models for integrator-adder-gain systems follows a completely parallel route. Integrators replace the delay elements. Their outputs at time $t$ constitute a natural set of state variables for the system; their values at any starting time $t_0$ establish the initial conditions for integration over the interval $t \geq t_0$. The state evolution equations result from expressing the inputs to all the integrators at time $t$ in terms of all the integrator outputs at time $t$ as well as the external input at this same time.

## 4.3 STATE-SPACE MODELS

As illustrated in the examples of the preceding section, it is often natural and convenient, when studying or modeling physical systems, to focus not just on the input and output signals but rather to describe the interaction and time evolution of several key variables or signals that are associated with the various component processes internal to the system. Assembling the descriptions of these components and their interconnections leads to a description that is richer than an input-output description. In particular, the examples in Section 4.2 describe system behavior in terms of the time evolution of a set of state variables that completely capture at any time the past history of the system as it affects the present and future response. We turn now to a more formal definition of state-space models in the DT and CT cases, followed by a discussion of the two defining characteristics of such models.

### 4.3.1  DT State-Space Models

A state-space model is built around a set of state variables; we mostly limit our discussion to real-valued state variables. The number of state variables in a model or system is referred to as its order. We shall only deal with state-space models of finite order, which are also referred to as lumped models.

For an $L$th-order model in the DT case, we generically denote the values of the $L$ real state variables at time $n$ by $q_1[n], q_2[n], \cdots, q_L[n]$. It is convenient to gather these variables into a state vector

$$\mathbf{q}[n] = \begin{bmatrix} q_1[n] \\ q_2[n] \\ \vdots \\ q_L[n] \end{bmatrix}. \tag{4.32}$$

The value of this vector constitutes the state of the model or system at time $n$.

**DT LTI State-Space Model**    A DT LTI state-space model with single or scalar input $x[n]$ and single output $y[n]$ takes the following form, written in compact matrix notation

$$\mathbf{q}[n+1] = \mathbf{A}\mathbf{q}[n] + \mathbf{b}x[n]\,, \tag{4.33}$$

$$y[n] = \mathbf{c}^T\mathbf{q}[n] + \mathrm{d}x[n]\,. \tag{4.34}$$

In Eqs. (4.33) and (4.34), $\mathbf{A}$ is an $L \times L$ matrix, $\mathbf{b}$ is an $L \times 1$ matrix or column vector, and $\mathbf{c}^T$ is a $1 \times L$ matrix or row vector, with the superscript $^T$ denoting transposition of the column vector $\mathbf{c}$ into the desired row vector. The quantity d is a $1 \times 1$ matrix, or a scalar. The entries of all these matrices in the case of an LTI model are numbers, constants, or parameters, so they do not vary with $n$.

The next value of each state variable and the present value of the output are all expressed as LTI functions of the present state and present input. We refer to Eq. (4.33) as the state evolution equation, and to Eq. (4.34) as the output equation. The model obtained for the delay-adder-gain system in Example 4.4 in the previous section has precisely the above form.

The system in Eqs. (4.33) and (4.34) is termed LTI because of its structure: the next state and current output are LTI functions of the current state and current input. However, this structure also gives rise to a corresponding behavioral sense in which the system is LTI. A particular set of input, state, and output signals—$x[\cdot]$, $\mathbf{q}[\cdot]$, and $y[\cdot]$, respectively—that together satisfy the above state evolution equation and output equation is referred to as a behavior of the DT LTI system. It follows from the linear structure of the above equations that scaling all the signals in a behavior by the same scalar constant again yields a behavior of this system. Also, summing two behaviors again yields a behavior. More generally, a weighted linear combination of behaviors again yields a behavior, so the behaviors of the system have the superposition property. Similarly, it follows from the time invariance of the defining equations that an arbitrary time shift of a behavior—shifting the input, state, and output signals in time by the same amount—again yields a behavior. Thus, the LTI structure of the equations is mirrored by the LTI properties of its solutions or behaviors.

**Delay-Adder-Gain Realization**   A delay-adder-gain system of the form encountered in Example 4.4 can be used to simulate, or "realize," any $L$th-order, DT LTI model of the type given in Eqs. (4.33) and (4.34). Key to this is the fact that adders and gains suffice to implement the additions and multiplications associated with the various matrix multiplications in the LTI state-space description.

To set up the simulation, we begin with $L$ delay elements, and label their outputs at time $n$ as $q_j[n]$ for $j = 1, 2, \cdots, L$; the corresponding inputs are then $q_j[n + 1]$. The $i$th row of Eq. (4.33) shows what LTI combination of these $q_j[n]$ and $x[n]$ is required to compute $q_i[n + 1]$, for each $i = 1, 2, \cdots, L$. Similarly, Eq. (4.34) shows what LTI combination of the variables is required to compute $y[n]$. Each of these LTI combinations can now be implemented using gains and adders.

The implementation produced by the preceding prescription is not unique: there are multiple ways to implement the linear combinations, depending, for example, on whether there is special structure in the matrices, or on how computation of the various terms in the linear combination is grouped and sequenced. In the case of the system in Example 4.4, for example, starting with the model in Eqs. (4.28) and (4.29) and following the procedure outlined in this paragraph will almost certainly lead to a different realization than the one in Figure 4.4.

**Generalizations**   Although our focus in the DT case will be on the above LTI, single-input, single-output, state-space model, there are various natural generalizations of this description that we mention for completeness. A multi-input DT LTI state-space model replaces the single term $\mathbf{b}x[n]$ in Eq. (4.33) by a sum of terms, $\mathbf{b}_1 x_1[n] + \cdots + \mathbf{b}_M x_M[n]$, where $M$ is the number of inputs. This corresponds to replacing the scalar input $x[n]$ by an $M$-component vector $\mathbf{x}[n]$ of inputs, with a corresponding change of $\mathbf{b}$ to a matrix $\mathbf{B}$ of dimension $L \times M$. Similarly, for a multi-output DT LTI state-space model, the single output quantity in Eq. (4.34) is replaced by a collection of such output equations, one for each of the $P$ outputs. Equivalently, the scalar output $y[n]$ is replaced by a $P$-component vector $\mathbf{y}[n]$ of outputs, with a corresponding change of $\mathbf{c}^T$ and d to matrices $\mathbf{C}^T$ and $\mathbf{D}$ of dimensions $P \times L$ and $P \times M$ respectively.

A linear but time-varying DT state-space model takes the same form as in Eqs. (4.33) and (4.34), except that some or all of the matrix entries are time-varying. A linear but periodically varying model is a special case of this, with matrix entries that all vary periodically with a common period.

All of the above generalizations can also be simulated or realized by delay-adder-gain systems, except that the gains will need to be time-varying for the case of time-varying systems. For the nonlinear systems described below, more elaborate simulations are needed, involving nonlinear elements or combinations.

A nonlinear, time-invariant, single input, single output model expresses $\mathbf{q}[n+1]$ and $y[n]$ as nonlinear but time-invariant functions of $\mathbf{q}[n]$ and $x[n]$, rather than as the LTI functions embodied by the matrix expressions on the right-hand sides of Eqs. (4.33) and (4.34). Our full and reduced models for viral propagation in Example 4.3 were of this type. A third-order nonlinear time invariant state-space model, for instance, comprises state evolution equations of the form

$$q_1[n+1] = f_1\Big(q_1[n], q_2[n], q_3[n], x[n]\Big)$$

$$q_2[n+1] = f_2\Big(q_1[n], q_2[n], q_3[n], x[n]\Big)$$

$$q_3[n+1] = f_3\Big(q_1[n], q_2[n], q_3[n], x[n]\Big) \qquad (4.35)$$

and an output equation of the form

$$y[n] = g\Big(q_1[n], q_2[n], q_3[n], x[n]\Big), \qquad (4.36)$$

where the state evolution functions $f_1(\cdot)$, $f_2(\cdot)$, $f_3(\cdot)$ and the output function $g(\cdot)$ are all time-invariant nonlinear functions of the three state variables $q_1[n]$, $q_2[n]$, $q_3[n]$, and the input $x[n]$. Time invariance here means that the functions combine their arguments in the same way, regardless of the time index $n$. In vector notation,

$$\mathbf{q}[n+1] = \mathbf{f}\Big(\mathbf{q}[n], x[n]\Big), \qquad y[n] = g\Big(\mathbf{q}[n], x[n]\Big), \qquad (4.37)$$

where for the third-order case

$$\mathbf{f}(\cdot) = \begin{bmatrix} f_1(\cdot) \\ f_2(\cdot) \\ f_3(\cdot) \end{bmatrix}. \qquad (4.38)$$

The notation for an $L$th-order description follows the same pattern.

Finally, a nonlinear, time-varying model expresses $\mathbf{q}[n+1]$ and $y[n]$ as nonlinear, time-varying functions of $\mathbf{q}[n]$ and $x[n]$. In other words, the manner in which the state evolution and output functions combine their arguments can vary with $n$. For this case, we would write

$$\mathbf{q}[n+1] = \mathbf{f}\Big(\mathbf{q}[n], x[n], n\Big), \qquad y[n] = g\Big(\mathbf{q}[n], x[n], n\Big). \qquad (4.39)$$

Nonlinear, periodically varying models can also be defined as a particular case in which the time variations are periodic with a common period.

### 4.3.2 CT State-Space Models

Continuous-time state-space descriptions take a very similar form to the DT case. The state variables for an $L$th-order system may be denoted as $q_i(t)$, $i = 1, 2, \ldots, L$, and the state vector as

$$\mathbf{q}(t) = \begin{bmatrix} q_1(t) \\ q_2(t) \\ \vdots \\ q_L(t) \end{bmatrix}. \qquad (4.40)$$

In the DT case the state evolution equation expresses the state vector at the next time step in terms of the current state vector and input values. In the CT case the state evolution equation expresses the rates of change or derivatives of each of the state variables as functions of the present state and inputs.

**CT LTI State-Space Model**    The general $L$th-order CT LTI state-space representation takes the form

$$\frac{d\mathbf{q}(t)}{dt} = \dot{\mathbf{q}}(t) = \mathbf{A}\mathbf{q}(t) + \mathbf{b}x(t), \qquad (4.41)$$

$$y(t) = \mathbf{c}^T\mathbf{q}(t) + \mathrm{d}x(t), \qquad (4.42)$$

where $d\mathbf{q}(t)/dt = \dot{\mathbf{q}}(t)$ denotes the vector whose entries are the derivatives of the corresponding entries of $\mathbf{q}(t)$. The entries of all these matrices are numbers or constants or parameters that do not vary with $t$. Thus, the rate of change of each state variable and the present value of the output are all expressed as LTI functions of the present state and present input. As in the DT LTI case, the LTI structure of the above system is mirrored by the LTI properties of its solutions or behaviors, a fact that will become explicit in Chapter 5. The models in Eqs. (4.8) and (4.9) of Example 4.1 and Eqs. (4.14) and (4.17) of Example 4.2 are precisely of the above form.

**Integrator-Adder-Gain Realization**    Any CT LTI state-space model of the form in Eqs. (4.41) and (4.42) can be simulated or realized using an integrator-adder-gain system. The approach is entirely analogous to the DT LTI case that was described earlier. We begin with $L$ integrators, labeling their outputs as $q_j(t)$ for $j = 1, 2, \cdots, L$. The inputs of these integrators are then the derivatives $\dot{q}_j(t)$. The $i$th row of Eq. (4.41) now determines what LTI combination of the $q_j(t)$ and $x(t)$ is required to synthesize $\dot{q}_i(t)$, for each $i = 1, 2, \cdots, L$.

We similarly use Eq. (4.42) to determine what LTI combination of these variables is required to compute $y(t)$. Finally, each of these LTI combinations is implemented using gains and adders. We illustrate this procedure with a specific example below.

**Generalizations**  The basic CT LTI state-space model can be generalized to multi-input and multi-output models, to nonlinear time-invariant models, and to linear and nonlinear time-varying or periodically varying models. These generalizations can be described just as in the case of DT systems, by appropriately relaxing the restrictions on the form of the right-hand sides of Eqs. (4.41) and (4.42). The model for the inverted pendulum in Eqs. (4.3), (4.4), and (4.5) in Example 4.1 was nonlinear and time-invariant, of the form

$$\dot{\mathbf{q}}(t) = \mathbf{f}\big(\mathbf{q}(t), x(t)\big), \qquad y(t) = g\big(\mathbf{q}(t), x(t)\big). \tag{4.43}$$

A general nonlinear and time-varying CT state-space model with a single input and single output takes the form

$$\dot{\mathbf{q}}(t) = \mathbf{f}\big(\mathbf{q}(t), x(t), t\big), \qquad y(t) = g\big(\mathbf{q}(t), x(t), t\big). \tag{4.44}$$

## Example 4.5    Simulation of Inverted Pendulum for Small Angles

For sufficiently small angular deviations from the fully inverted position for the inverted pendulum considered in Example 4.1, the original nonlinear state-space model simplifies to the LTI state-space model described by Eqs. (4.8) and (4.9). This LTI model is repeated here for convenience, but with the numerical values of a specific pendulum inserted:

$$\dot{\mathbf{q}}(t) = \begin{bmatrix} \dot{q}_1(t) \\ \dot{q}_2(t) \end{bmatrix} = \begin{bmatrix} 0 & 1 \\ 8 & -2 \end{bmatrix} \begin{bmatrix} q_1(t) \\ q_2(t) \end{bmatrix} + \begin{bmatrix} 0 \\ -1 \end{bmatrix} x(t)$$

$$= \mathbf{A}\mathbf{q}(t) + \mathbf{b}x(t) \tag{4.45}$$

and

$$y(t) = \begin{bmatrix} 1 & 0 \end{bmatrix} \begin{bmatrix} q_1(t) \\ q_2(t) \end{bmatrix} = \mathbf{c}^T \mathbf{q}(t). \tag{4.46}$$

To simulate this second-order LTI system using integrators, adders, and gains, we begin with two integrators and denote their outputs at time $t$ by $q_1(t)$ and $q_2(t)$. The inputs to these integrators are then $\dot{q}_1(t)$ and $\dot{q}_2(t)$, respectively, at time $t$. The right-hand sides of the two expressions in Eq. (4.45) now show how to synthesize $\dot{q}_1(t)$ and $\dot{q}_2(t)$ from particular weighted linear combinations of $q_1(t)$, $q_2(t)$, and $x(t)$. We use gain elements to obtain the appropriate weights, then adders to produce the required weighted linear combinations of $q_1(t)$, $q_2(t)$, and $x(t)$. By feeding these weighted linear combinations to the inputs of the respective integrators, $\dot{q}_1(t)$ and $\dot{q}_2(t)$ are set equal to these expressions. The output $y(t) = q_1(t)$ is directly read from the output of the first integrator. The block diagram in Figure 4.5 shows the resulting simulation.

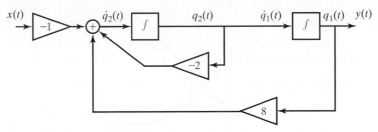

**Figure 4.5**  Integrator-adder-gain simulation of inverted pendulum for small angular deviations from vertical.

### 4.3.3 Defining Properties of State-Space Models

The two defining characteristics of state-space models are the following:

- **State Evolution Property** The state at any initial time, along with the inputs over any interval from that initial time onward, determine the state trajectory, that is, the state as a function of time, over that entire interval. Everything about the past that is relevant to the future state is embodied in the present state.

- **Instantaneous Output Property** The outputs at any instant can be written in terms of the state and inputs at that same instant.

The state evolution property is what makes state-space models particularly well suited to describing causal systems. In the DT LTI case, the validity of this state evolution property is evident from Eq. (4.33), which allows $\mathbf{q}[n]$ to be updated iteratively, moving from time $n$ to time $n + 1$ using only knowledge of the present state and input. The same argument can also be applied to the general DT state evolution expression in Eq. (4.39).

The state evolution property in the general CT case is more subtle to establish, and actually requires that the function $\mathbf{f}(\mathbf{q}(t), x(t), t)$ defining the rate of change of the state vector satisfy certain mild technical conditions. These conditions are satisfied by all the models of interest to us in this text, so we shall not discuss the conditions further. Instead, we describe how the availability of a CT state-space model enables a simple numerical approximation of the state trajectory at a discrete set of times spaced an interval $\Delta$ apart. This numerical algorithm is referred to as the forward-Euler method.

The algorithm begins by using the state and input information at the initial time $t_0$ to determine the initial rate of change of the state, namely $\mathbf{f}(\mathbf{q}(t_0), x(t_0), t_0)$. As illustrated in Figure 4.6, this initial rate of change is tangent to the state trajectory at $t_0$. The approximation to the actual trajectory is obtained by stepping forward a time increment $\Delta$ along this tangent—the forward-Euler step—to arrive at the estimate

$$\mathbf{q}(t_0 + \Delta) \approx \mathbf{q}(t_0) + \mathbf{f}(\mathbf{q}(t_0), x(t_0), t_0)\Delta . \qquad (4.47)$$

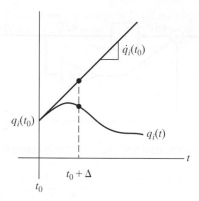

**Figure 4.6**   Using the CT state evolution equations to obtain the state trajectories over an interval.

This is equivalent to using a first-order Taylor series approximation to the trajectory, or using a forward-difference approximation to $\dot{\mathbf{q}}(t_0)$.

With the estimate of $\mathbf{q}(t_0 + \Delta)$ now available, and knowing the input $x(t_0 + \Delta)$ at time $t_0 + \Delta$, the same procedure can be repeated at this next time instant, thereby getting an approximation to $\mathbf{q}(t_0 + 2\Delta)$. This iteration can be continued over the entire interval of interest. Under the technical conditions alluded to above, the algorithm accumulates an error of order $\Delta^2$ at each time step, and takes $T/\Delta$ time steps in an interval of length $T$, thereby accumulating an error of order $T\Delta$ by the end of the interval. This error can be made arbitrarily small by choosing a sufficiently small $\Delta$.

The forward-Euler algorithm suffices to suggest how a CT state-space description gives rise to the state evolution property. For actual numerical computation, more sophisticated numerical routines would be used, based for example on higher-order Taylor series approximations, and using variable-length time steps for better error control. The CT LTI case is, however, much simpler than the general case. We shall demonstrate the state evolution property for this class of state-space models in detail in the Chapter 5, when we show how to explicitly solve for their behavior.

The instantaneous output property is evident in the LTI case from the output expressions in Eqs. (4.34) and (4.42). It also holds for the various generalizations of basic single-input, single-output LTI models that we listed earlier, most broadly for the output relations in Eqs. (4.39) and (4.44).

The state evolution and instantaneous output properties are the defining characteristics of a state-space model. In setting up a state-space model, we introduce the additional vector of state variables $\mathbf{q}[n]$ or $\mathbf{q}(t)$ to supplement the input variables $x[n]$ or $x(t)$ and output variables $y[n]$ or $y(t)$. This supplementation is done precisely in order to obtain a description that satisfies these properties.

Often there are natural choices of state variables suggested directly by the particular context or application. As already noted, and illustrated by the preceding examples in both DT and CT cases, state variables are related to the "memory" of the system. In many physical situations involving CT models, the state variables are associated with energy storage because this is what is carried over from the past to the future.

One can always choose any alternative set of state variables that together contain exactly the same information as a given set. There are also situations in which there is no particularly natural or compelling choice of state variables, but in which it is still possible to define supplementary variables that enable a valid state-space description to be obtained.

Our discussion of the two key properties above—and particularly of the role of the state vector in separating past and future—suggests that state-space models are particularly suited to describing causal systems. In fact, state-space models are almost never used to describe noncausal systems. We shall always assume here, when dealing with state-space models, that they represent causal systems. Although causality is not a central issue in analyzing many aspects of communication or signal processing systems, particularly in non-real-time contexts, it is generally central to control design and operation for dynamic systems, and this is where state-space descriptions find their greatest value and use.

## 4.4 STATE-SPACE MODELS FROM LTI INPUT-OUTPUT MODELS

State-space representations can be very naturally and directly generated during the modeling process in a variety of settings, as the examples in Section 4.2 demonstrated. Other—and perhaps more familiar—descriptions can then be derived from them, for instance input-output descriptions.

It is also possible to proceed in the reverse direction, constructing state-space descriptions from transfer functions, unit sample or impulse responses, or input-output difference or differential equations, for instance. This is often worthwhile as a prelude to simulation, filter implementation, in control design, or simply in order to understand the initial description from another point of view. The state variables associated with the resulting state-space descriptions do not necessarily have interesting or physically meaningful interpretations, but still capture the memory of the system.

The following two examples illustrate this reverse process, of synthesizing state-space descriptions from input-output descriptions, for the important case of DT LTI systems. Analogous examples can be constructed for the CT LTI case. The first example below also makes the point that state-space models of varying orders can share the same input-output description, a fact that we will understand better following the structural analysis of LTI systems developed in the next chapter. That structural analysis actually ends up also relating quite closely to the second example in this section.

---

**Example 4.6**    *State-Space Models from an Input-Output Difference Equation*

Consider the LTI input-output difference equation

$$y[n] + a_1 y[n-1] + a_2 y[n-2] = b_1 x[n-1] + b_2 x[n-2] . \tag{4.48}$$

Building on the idea of state variables as memory variables, consider using the following array of "past" variables as a candidate state vector:

$$\mathbf{q}[n] = \begin{bmatrix} y[n-1] \\ y[n-2] \\ x[n-1] \\ x[n-2] \end{bmatrix} . \tag{4.49}$$

To obtain the corresponding state-space model, $\mathbf{q}[n+1]$ has to be related to $\mathbf{q}[n]$ and $x[n]$. Given that the initial difference equation is linear and time-invariant, we might anticipate obtaining an LTI state-space description in the matrix form shown in Eqs. (4.33) and (4.34). Using those equations as the template, consider what entries are required in the matrices $\mathbf{A}$, $\mathbf{b}$, $\mathbf{c}^T$, and d to satisfy the equations for the above choice of $\mathbf{q}[n]$, also taking account of the relationship embodied in the given difference equation. The resulting fourth-order state-space model takes the form

$$\mathbf{q}[n+1] = \begin{bmatrix} y[n] \\ y[n-1] \\ x[n] \\ x[n-1] \end{bmatrix} = \begin{bmatrix} -a_1 & -a_2 & b_1 & b_2 \\ 1 & 0 & 0 & 0 \\ 0 & 0 & 0 & 0 \\ 0 & 0 & 1 & 0 \end{bmatrix} \begin{bmatrix} y[n-1] \\ y[n-2] \\ x[n-1] \\ x[n-2] \end{bmatrix} + \begin{bmatrix} 0 \\ 0 \\ 1 \\ 0 \end{bmatrix} x[n]$$

$$= \mathbf{A}\mathbf{q}[n] + \mathbf{b}x[n] ,$$

$$y[n] = \begin{bmatrix} -a_1 & -a_2 & b_1 & b_2 \end{bmatrix} \begin{bmatrix} y[n-1] \\ y[n-2] \\ x[n-1] \\ x[n-2] \end{bmatrix}$$

$$= \mathbf{c}^T \mathbf{q}[n] . \tag{4.50}$$

If we are somewhat more careful about our choice of state variables, it is possible to get more economical models. For a third-order model, suppose we pick as state vector

$$\mathbf{q}[n] = \begin{bmatrix} y[n] \\ y[n-1] \\ x[n-1] \end{bmatrix} . \tag{4.51}$$

The corresponding third-order state-space model takes the form

$$\mathbf{q}[n+1] = \begin{bmatrix} y[n+1] \\ y[n] \\ x[n] \end{bmatrix} = \begin{bmatrix} -a_1 & -a_2 & b_2 \\ 1 & 0 & 0 \\ 0 & 0 & 0 \end{bmatrix} \begin{bmatrix} y[n] \\ y[n-1] \\ x[n-1] \end{bmatrix} + \begin{bmatrix} b_1 \\ 0 \\ 1 \end{bmatrix} x[n] ,$$

$$y[n] = \begin{bmatrix} 1 & 0 & 0 \end{bmatrix} \begin{bmatrix} y[n] \\ y[n-1] \\ x[n-1] \end{bmatrix} . \tag{4.52}$$

A more subtle choice of state variables can yield a second-order state-space model. For instance, picking

$$\mathbf{q}[n] = \begin{bmatrix} y[n] \\ -a_2 y[n-1] + b_2 x[n-1] \end{bmatrix} , \tag{4.53}$$

the corresponding second-order state-space model takes the form

$$\begin{bmatrix} y[n+1] \\ -a_2y[n]+b_2x[n] \end{bmatrix} = \begin{bmatrix} -a_1 & 1 \\ -a_2 & 0 \end{bmatrix}\begin{bmatrix} y[n] \\ -a_2y[n-1]+b_2x[n-1] \end{bmatrix} + \begin{bmatrix} b_1 \\ b_2 \end{bmatrix}x[n]$$

$$y[n] = \begin{bmatrix} 1 & 0 \end{bmatrix}\begin{bmatrix} y[n] \\ -a_2y[n-1]+b_2x[n-1] \end{bmatrix}. \tag{4.54}$$

It turns out to be impossible in general to get a state-space description of order lower than 2 in this case, if we want our state-space model to display all the behavior that the original difference equation is able to. This should not be surprising, in view of the fact that Eq. (4.48) is a second-order difference equation, which we know requires two initial conditions in order to solve forward in time.

Notice how, in each of the above cases, the information contained in Eq. (4.48) — the original difference equation — has been incorporated into the state-space model.

For an LTI system, the most fundamental description of input-output behavior is provided by the system's impulse response. In the case of a causal LTI system, the impulse response is zero for negative times. The following example suggests, for the DT case, what additional constraints on the unit sample response or impulse response $h[n]$ will ensure that it is realizable as the impulse response of a causal DT LTI state-space system. The class of causal DT LTI state-space systems considered in the next chapter turns out to have impulse responses of precisely the form shown in this example. The example illuminates the relation between exponential components of the impulse response and state variables of the underlying realization.

## Example 4.7    *State-Space Model from a Unit Sample Response*

Consider the impulse response $h[n]$ of a causal DT LTI system. Causality requires that $h[n] = 0$ for $n < 0$. The output $y[n]$ can be related to past and present inputs $x[k]$, $k \le n$, through the convolution sum

$$y[n] = \sum_{k=-\infty}^{n} h[n-k]x[k] \tag{4.55}$$

$$= \left( \sum_{k=-\infty}^{n-1} h[n-k]x[k] \right) + h[0]x[n]. \tag{4.56}$$

The first term in Eq. (4.56), namely

$$q[n] = \sum_{k=-\infty}^{n-1} h[n-k]x[k], \tag{4.57}$$

represents the effect of the past on the present at time $n$, and would therefore seem to have some relation to the notion of a state variable. Updating $q[n]$ to the next time step, we obtain

$$q[n+1] = \sum_{k=-\infty}^{n} h[n+1-k]x[k]. \tag{4.58}$$

In general, if the impulse response has no special form, the successive values of $q[n]$ have to be recomputed from Eq. (4.57) for each $n$. When we move from $n$ to $n+1$, none of the past inputs $x[k]$ for $k \leq n$ can be discarded because the expression for $q[n+1]$ involves a different linear combination of the present and past $x[\cdot]$ than was used to compute $q[n]$. Since all past inputs have to be retained, the memory of the system is infinite.

However, consider the class of systems for which $h[n]$ has the exponential form

$$h[n] = \beta \lambda^{n-1} u[n-1] + \mathrm{d}\,\delta[n]\,, \tag{4.59}$$

where $\beta$, $\lambda$, and d are constants. This $h[n]$ is shown in Figure 4.7. The corresponding transfer function is

$$H(z) = \frac{\beta}{z - \lambda} + \mathrm{d} \tag{4.60}$$

(with region of convergence $|z| > |\lambda|$ ). What is important about this impulse response is that a time-shifted version of it is simply related to a scaled version of it, because of its DT-exponential form. For this case,

$$q[n] = \beta \sum_{k=-\infty}^{n-1} \lambda^{n-k-1} x[k] \tag{4.61}$$

and

$$q[n+1] = \beta \sum_{k=-\infty}^{n} \lambda^{n-k} x[k] \tag{4.62}$$

$$= \lambda \left( \beta \sum_{k=-\infty}^{n-1} \lambda^{n-k-1} x[k] \right) + \beta x[n]$$

$$= \lambda q[n] + \beta x[n]\,. \tag{4.63}$$

Gathering Eqs. (4.56) and (4.61) with (4.63) results in a pair of equations that together constitute a state-space description for this system:

$$q[n+1] = \lambda q[n] + \beta x[n] \tag{4.64}$$

$$y[n] = q[n] + \mathrm{d}x[n]\,. \tag{4.65}$$

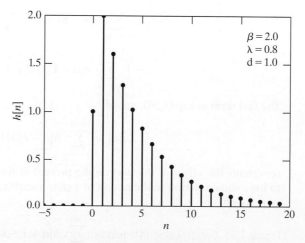

**Figure 4.7**   DT-exponential unit sample response.

Consider next a similar but higher-order system with impulse response

$$h[n] = (\beta_1 \lambda_1^{n-1} + \beta_2 \lambda_2^{n-1} + \cdots + \beta_L \lambda_L^{n-1})u[n-1] + d\,\delta[n] \quad (4.66)$$

with the $\beta_i$, $\lambda_i$, and d being constants. The corresponding transfer function is

$$H(z) = \left( \sum_{i=1}^{L} \frac{\beta_i}{z - \lambda_i} \right) + d\,. \quad (4.67)$$

If the transfer function $H(z)$ of a causal DT LTI system is a rational function of $z$ with distinct (i.e., nonrepeated) poles, then it can be written in this form using a partial fraction expansion, with appropriate choices of the $\beta_i$, $\lambda_i$, $L$, and d. Note that although we only treat rational transfer functions $H(z)$ whose numerator and denominator polynomials have real coefficients, the poles of $H(z)$ may include some complex $\lambda_i$ (and associated complex $\beta_i$), but in each such case its complex conjugate $\lambda_i^*$ will also be a pole, with associated weighting factor $\beta_i^*$, and the sum

$$\beta_i(\lambda_i)^n + \beta_i^*(\lambda_i^*)^n \quad (4.68)$$

will be real.

The block diagram in Figure 4.8 shows that the LTI system specified by Eqs. (4.66) or (4.67) can be obtained through the parallel interconnection of subsystems with transfer functions corresponding to the simpler case of Eq. (4.60), or equivalently paralleled subsystems with impulse responses corresponding to Eq. (4.59). Motivated by this structure and the treatment of the first-order example, we define a state variable for each of the $L$ subsystems:

$$q_i[n] = \beta_i \sum_{k=-\infty}^{n-1} \lambda_i^{n-k-1} x[k], \quad i = 1, 2, \ldots, L\,. \quad (4.69)$$

With these definitions, state evolution equations for the subsystems take the same form as in Eq. (4.63):

$$q_i[n+1] = \lambda_i q_i[n] + \beta_i x[n], \quad i = 1, 2, \ldots, L\,. \quad (4.70)$$

Also, combining Eqs. (4.57), (4.65), and (4.66) with the definitions in Eq. (4.69), the output equation becomes

$$y[n] = q_1[n] + q_2[n] + \cdots + q_L[n] + d\,x[n]\,. \quad (4.71)$$

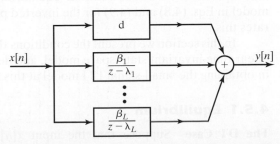

**Figure 4.8**  Decomposition of the rational transfer function of a causal DT LTI system, for the case of distinct poles.

Equations (4.70) and (4.71) together comprise an $L$th-order state-space description of the given system. This state-space description can be written in the standard
matrix form in Eqs. (4.33) and (4.34), with

$$
\mathbf{A} = \begin{bmatrix} \lambda_1 & 0 & 0 & \cdots & 0 & 0 \\ 0 & \lambda_2 & 0 & \cdots & 0 & 0 \\ \vdots & \vdots & \vdots & \ddots & \vdots & \vdots \\ 0 & 0 & 0 & \cdots & 0 & \lambda_L \end{bmatrix}, \qquad \mathbf{b} = \begin{bmatrix} \beta_1 \\ \beta_2 \\ \vdots \\ \beta_L \end{bmatrix} \tag{4.72}
$$

$$
\mathbf{c}^T = \begin{bmatrix} 1 & 1 & \cdot & \cdots & \cdot & 1 \end{bmatrix}. \tag{4.73}
$$

The diagonal form of $\mathbf{A}$ in Eq. (4.72) reflects the fact that the state evolution
equations in this example are decoupled, with each state variable being updated independently according to Eq. (4.70). We will see later how a general description of
the form given in Eqs. (4.33) and (4.34), with a distinct-eigenvalue condition that
we shall impose, can actually be transformed to a completely equivalent description in which the new $\mathbf{A}$ matrix is diagonal, as in Eq. (4.72). When there are
complex eigenvalues, this diagonal state-space representation will of course have
complex entries.

## 4.5 EQUILIBRIA AND LINEARIZATION OF NONLINEAR STATE-SPACE MODELS

Chapters 5 and 6 will focus on LTI state-space models. One justification for this
focus is that LTI systems have rich structure and behavior that are amenable to
detailed study. This allows the development of powerful analytical approaches
and computational tools for dealing with them. As a consequence, it is
common for various modules in engineered systems—for instance, electrical
circuits as in Example 4.2 or DT filters as in Example 4.4—to be designed
within an LTI framework.

A second reason for our focus is that an LTI model arises naturally
as an approximate description of the local or "small-signal" behavior of a
nonlinear time-invariant model, for small deviations of its state variables
and inputs from a set of constant equilibrium values. The LTI state-space
model in Eqs. (4.8) and (4.9) for the inverted pendulum in Example 4.1 illustrates this.

In this section we present the conditions that define equilibrium in a nonlinear time-invariant state-space model, and describe the role of linearization
in obtaining the small-signal LTI model at this equilibrium.

### 4.5.1 Equilibrium

**The DT Case**   Suppose that the input $x[n]$ in the state-space model of
Eq. (4.37) is kept constant at the value $\bar{x}$ for all $n$. The corresponding state
equilibrium is a state value $\bar{\mathbf{q}}$ with the property that if $\mathbf{q}[n] = \bar{\mathbf{q}}$ with $x[n] = \bar{x}$,

then $\mathbf{q}[n+1] = \overline{\mathbf{q}}$. Equivalently, the point $\overline{\mathbf{q}}$ in the state space is an equilibrium (or equilibrium point) if, with $x[n] = \overline{x}$ for all $n$ and with the system initialized at $\overline{\mathbf{q}}$, the system subsequently remains fixed at $\overline{\mathbf{q}}$. The equilibrium is thus a steady state of the system. From Eq. (4.37), this is equivalent in the DT case to requiring that

$$\overline{\mathbf{q}} = \mathbf{f}(\overline{\mathbf{q}}, \overline{x}) . \tag{4.74}$$

The corresponding equilibrium output is

$$\overline{y} = g(\overline{\mathbf{q}}, \overline{x}) . \tag{4.75}$$

In defining an equilibrium, no consideration is given to what the system behavior is in the vicinity of the equilibrium point, that is, to how the system will behave if initialized close to—rather than exactly at—the point $\overline{\mathbf{q}}$. This issue of the stability of an equilibrium or steady state is considered when we discuss local behavior, and in particular local stability, around the equilibrium.

For specificity, consider a DT second-order nonlinear time-invariant state-space system, with state evolution equations of the form

$$q_1[n+1] = f_1\Big(q_1[n], q_2[n], x[n]\Big)$$

$$q_2[n+1] = f_2\Big(q_1[n], q_2[n], x[n]\Big) , \tag{4.76}$$

and with the output $y[n]$ defined by the equation

$$y[n] = g\Big(q_1[n], q_2[n], x[n]\Big) . \tag{4.77}$$

The state evolution functions $f_1(\cdot)$, $f_2(\cdot)$ and the output function $g(\cdot)$ at time $n$ are all time-invariant nonlinear functions of the two state variables $q_1[n]$ and $q_2[n]$ and of the input $x[n]$. These equations are the second-order version of the general expressions in Eq. (4.37).

In this second-order case, and given $\overline{x}$, we would find the equilibrium by solving the following system of two simultaneous nonlinear equations in two unknowns:

$$\overline{q}_1 = f_1(\overline{q}_1, \overline{q}_2, \overline{x})$$

$$\overline{q}_2 = f_2(\overline{q}_1, \overline{q}_2, \overline{x}) . \tag{4.78}$$

There is no guarantee in general that an equilibrium exists for the specified constant input $\overline{x}$, and there is no guarantee of a unique equilibrium when an equilibrium does exist.

---

**Example 4.8    Equilibrium in the Viral Propagation Model**

The state evolution equations of the reduced second-order model for viral propagation that we considered in Example 4.3 took the form

$$s[n+1] = s[n] - \gamma(i[n]/P)s[n] + \beta(P - s[n]) - \beta Px[n]$$

$$i[n+1] = i[n] + \gamma(i[n]/P)s[n] - \rho i[n] - \beta i[n] , \tag{4.79}$$

where $s[n]$ denoted the number of susceptibles at time $n$, $i[n]$ the number of infectives at that time, and $x[n]$ the rate at which newborns are immunized. The parameter $P$ denoted the total population size, including the third subpopulation, namely that of the recovered group, $r[n]$. The birth rate and death rate were both $\beta$; the recovery rate was $\rho$; and $\gamma$ denoted the coupling coefficient that determined how effectively the infectives caused the susceptibles to contract the virus. We also introduced there the basic reproductive ratio $R_0 = \gamma/(\beta + \rho)$.

Suppose the immunization rate is held constant at $x[n] \equiv \bar{x}$, where $0 \le \bar{x} \le 1$. The corresponding pair of equilibrium state variables, namely $\bar{s}$ susceptibles and $\bar{i}$ infectives, must then satisfy

$$\bar{s} = \bar{s} - \gamma(\bar{i}/P)\bar{s} + \beta(P - \bar{s}) - \beta P\bar{x}$$
$$\bar{i} = \bar{i} + \gamma(\bar{i}/P)\bar{s} - \rho\bar{i} - \beta\bar{i} . \tag{4.80}$$

Some straightforward computation shows that the above equations have two possible solutions, corresponding to two possible equilibria, which we label IFE and EE for "infective-free equilibrium" and "endemic equilibrium," respectively:

$$\text{IFE:} \quad \bar{s} = P(1 - \bar{x}) \text{ and } \bar{i} = 0; \quad \text{or} \tag{4.81}$$

$$\text{EE:} \quad \bar{s} = P/R_0 \text{ and } \bar{i} = (\beta P/\gamma)[R_0(1 - \bar{x}) - 1] . \tag{4.82}$$

In the IFE there are no infectives. This is the equilibrium seen in steady state in Figure 4.3, and it does indeed occur at $\bar{s} = P(1 - \bar{x}) = 10{,}000 \times 0.5 = 5000$.

In the EE, on the other hand, the number of infectives is positive, under the condition $R_0 > 1/(1 - \bar{x})$. If $R_0 = 1/(1 - \bar{x})$ then there are no infectives in equilibrium, and in fact under this condition the IFE and EE equilibria coalesce into a single equilibrium. If $R_0 < 1/(1 - \bar{x})$, as with the system simulated in Figure 4.3, then the only possible equilibrium with nonnegative populations is the IFE.

**The CT Case** We can apply the same idea to compute equilibria of CT nonlinear time-invariant state-space systems. Define the equilibrium $\bar{\mathbf{q}}$ for the system in Eq. (4.43) as a state value that the system does not move from when initialized there, and when the input is fixed at $x(t) = \bar{x}$. In the CT case, what this requires is that the rate of change of the state, namely $\dot{\mathbf{q}}(t)$, be zero at the equilibrium, which yields the condition
$$\mathbf{0} = \mathbf{f}(\bar{\mathbf{q}}, \bar{x}) . \tag{4.83}$$

(In general we shall use $\mathbf{0}$ for any vectors or matrices whose entries are all 0, with the correct dimensions being apparent from the context.)

Again consider the concrete case of a second-order system:
$$\dot{q}_1(t) = f_1\Big(q_1(t), q_2(t), x(t)\Big)$$
$$\dot{q}_2(t) = f_2\Big(q_1(t), q_2(t), x(t)\Big) , \tag{4.84}$$

with

$$y(t) = g\Big(q_1(t), q_2(t), x(t)\Big) . \tag{4.85}$$

For this second-order case, this condition takes the form
$$0 = f_1(\bar{q}_1, \bar{q}_2, \bar{x})$$
$$0 = f_2(\bar{q}_1, \bar{q}_2, \bar{x}), \tag{4.86}$$

which is again a set of two simultaneous nonlinear equations in two unknowns, with possibly no solution for a specified $\bar{x}$, or one solution, or many.

---

**Example 4.9**    **Equilibrium in the Inverted Pendulum**

We saw in Example 4.1 that the state evolution equations for the inverted pendulum are

$$\frac{dq_1(t)}{dt} = q_2(t) , \tag{4.87}$$

$$\frac{dq_2(t)}{dt} = \frac{1}{\mathcal{I}}\Big(mg\ell \sin(q_1(t)) - \beta q_2(t) - x(t)\Big) . \tag{4.88}$$

If the input torque $x(t)$ is held constant at the value $\bar{x}$, and the corresponding equilibrium values of the state variables are $\bar{q}_1$ and $\bar{q}_2$, then setting the rates of change of the state variables to 0 at the equilibrium yields

$$0 = \bar{q}_2 , \tag{4.89}$$

$$0 = \frac{1}{\mathcal{I}}\Big(mg\ell \sin(\bar{q}_1) - \beta\bar{q}_2 - \bar{x}\Big) . \tag{4.90}$$

The equilibrium velocity $\bar{q}_2$ is therefore 0, and the equilibrium position $\bar{q}_1$ satisfies

$$\sin(\bar{q}_1) = \bar{x}/(mg\ell) . \tag{4.91}$$

Since the maximum attainable magnitude of the function $\sin(\bar{q}_1)$ is 1, Eq. (4.91) has solutions $\bar{q}_1$ if and only if $|\bar{x}| \leq mg\ell$. For $|\bar{x}| > mg\ell$, there can be no equilibrium, and in fact the pendulum exhibits continuous rotations.

If $\bar{x} = mg\ell$, then Eq. (4.91) has the unique solution $\bar{q}_1 = \pi/2$ in the range $[-\pi, \pi]$, corresponding to the pendulum being stationary in the horizontal position, with its mass providing the maximum possible torque to counterbalance the externally imposed torque $\bar{x}$. If $\bar{x} = -mg\ell$, the equilibrium position is at $\bar{q}_1 = -\pi/2$, which is again horizontal.

Otherwise, for $mg\ell > \bar{x} > 0$, there are two solutions: the first is at the angle $\bar{q}_1 = \arcsin(\bar{x}/(mg\ell))$ in the range $[0, \pi/2]$, and the second at $\pi$ minus this angle. These two equilibria correspond to the pendulum being stationary at some angle above a horizontal line through the pivot, or stationary at the same angle below the horizontal line; in both cases, the torque due to the mass of the pendulum balances the externally imposed torque $\bar{x}$. For $0 > \bar{x} > mg\ell$, there is a symmetric pair of equilibria at the negatives of these angles. If the applied external torque is 0, that is, $\bar{x} = 0$, then the possible equilibrium positions are $\bar{q}_1 = 0$ and $\bar{q}_1 = \pi$, corresponding respectively to the pendulum balancing straight up, or hanging straight down.

---

## 4.5.2 Linearization

We now examine system behavior in the vicinity of an equilibrium point of a time-invariant nonlinear state-space model. For concreteness, consider once more the second-order DT nonlinear system in Eq. (4.76). However, the development below generalizes directly to $L$th-order DT systems, and also to the CT case, which is described separately below. Suppose that instead of

$x[n]$ being fixed at the constant value $\bar{x}$ associated with an equilibrium, $x[n]$ deviates from this by a value $\tilde{x}[n]$, so

$$\tilde{x}[n] = x[n] - \bar{x} . \tag{4.92}$$

Suppose the state variables are correspondingly perturbed from their respective equilibrium values by amounts denoted by

$$\tilde{q}_1[n] = q_1[n] - \bar{q}_1 , \quad \tilde{q}_2[n] = q_2[n] - \bar{q}_2 , \tag{4.93}$$

and the output is perturbed by

$$\tilde{y}[n] = y[n] - \bar{y} . \tag{4.94}$$

Our objective is to find a model that exactly or closely describes the behavior of these various perturbations from an equilibrium point. This is possible if the perturbations or deviations from equilibrium are small, because that allows truncated Taylor series to provide good approximations to the various nonlinear functions. Linearization corresponds to truncating the Taylor series to first order (i.e., to terms that are linear in the deviations), and produces an LTI state-space model in the setting considered here. This LTI model is referred to as the linearized, or small-signal, model at the equilibrium.

To linearize the original DT second-order nonlinear model in Eq. (4.76), rewrite the variables appearing in that model in terms of the perturbations, using the quantities defined in Eqs. (4.92) and (4.93), and then expand in Taylor series to first order around the equilibrium values:

$$\bar{q}_i + \tilde{q}_i[n+1] = f_i\big(\bar{q}_1 + \tilde{q}_1[n] , \ \bar{q}_2 + \tilde{q}_2[n] , \ \bar{x} + \tilde{x}[n]\big) \qquad \text{for } i = 1, 2$$

$$\approx f_i(\bar{q}_1, \bar{q}_2, \bar{x}) + \frac{\partial f_i}{\partial q_1}\tilde{q}_1[n] + \frac{\partial f_i}{\partial q_2}\tilde{q}_2[n] + \frac{\partial f_i}{\partial x}\tilde{x}[n] . \tag{4.95}$$

All the partial derivatives above are evaluated at the equilibrium values, and are therefore constants, not dependent on the time index $n$. Also note that the partial derivatives above are with respect to the continuously variable state and input arguments; there are no "derivatives" taken with respect to $n$, the discretely varying time index.

The definition of the equilibrium values in Eq. (4.78) shows that the term $\bar{q}_i$ on the left of the above set of expressions exactly equals the term $f_i(\bar{q}_1, \bar{q}_2, \bar{x})$ on the right, so what remains is the approximate relation

$$\tilde{q}_i[n+1] \approx \frac{\partial f_i}{\partial q_1}\tilde{q}_1[n] + \frac{\partial f_i}{\partial q_2}\tilde{q}_2[n] + \frac{\partial f_i}{\partial x}\tilde{x}[n] \tag{4.96}$$

for $i = 1, 2$. Replacing the approximate equality sign ($\approx$) with the equality sign ($=$) in this set of expressions produces the linearized model at the equilibrium point. This linearized model may be written in matrix form as

$$\underbrace{\begin{bmatrix} \tilde{q}_1[n+1] \\ \tilde{q}_2[n+1] \end{bmatrix}}_{\tilde{\mathbf{q}}[n+1]} = \underbrace{\begin{bmatrix} \partial f_1/\partial q_1 & \partial f_1/\partial q_2 \\ \partial f_2/\partial q_1 & \partial f_2/\partial q_2 \end{bmatrix}}_{\mathbf{A}} \underbrace{\begin{bmatrix} \tilde{q}_1[n] \\ \tilde{q}_2[n] \end{bmatrix}}_{\tilde{\mathbf{q}}[n]} + \underbrace{\begin{bmatrix} \partial f_1/\partial x \\ \partial f_2/\partial x \end{bmatrix}}_{\mathbf{b}} \tilde{x}[n] . \tag{4.97}$$

We have therefore arrived at a standard DT LTI state-space description of the state evolution of the linearized model, with state and input variables that are the respective deviations from equilibrium of the underlying nonlinear model. The corresponding output equation is derived similarly, and takes the form

$$\tilde{y}[n] = \underbrace{\left[\begin{array}{cc} \partial g/\partial q_1 & \partial g/\partial q_2 \end{array}\right]}_{\mathbf{c}^T} \tilde{\mathbf{q}}[n] + \underbrace{[\partial g/\partial x]}_{\mathrm{d}} \tilde{x}[n] \,. \tag{4.98}$$

The matrix of partial derivatives denoted by $\mathbf{A}$ in Eq. (4.97) is also called a Jacobian matrix, and denoted in matrix notation by

$$\mathbf{A} = \left[\frac{\partial \mathbf{f}}{\partial \mathbf{q}}\right]_{\bar{\mathbf{q}}, \bar{x}} \,. \tag{4.99}$$

The entry in its $i$th row and $j$th column is the partial derivative $\partial f_i(\cdot)/\partial q_j$, evaluated at the equilibrium values of the state and input variables. Similarly,

$$\mathbf{b} = \left[\frac{\partial \mathbf{f}}{\partial x}\right]_{\bar{\mathbf{q}}, \bar{x}} \,, \quad \mathbf{c}^T = \left[\frac{\partial g}{\partial \mathbf{q}}\right]_{\bar{\mathbf{q}}, \bar{x}} \,, \quad \mathrm{d} = \left[\frac{\partial g}{\partial x}\right]_{\bar{\mathbf{q}}, \bar{x}} \,. \tag{4.100}$$

---

**Example 4.10    Linearized Model for Viral Propagation**

We return to the state evolution equations of the second-order model for viral propagation considered in Example 4.3, namely

$$s[n+1] = s[n] - \gamma(i[n]/P)s[n] + \beta(P - s[n]) - \beta P x[n]$$

$$i[n+1] = i[n] + \gamma(i[n]/P)s[n] - \rho i[n] - \beta i[n] \,. \tag{4.101}$$

The two sets of state equilibrium values $\bar{s}$ and $\bar{\imath}$ for this system were determined in Example 4.8. Computing the appropriate Jacobians, the linearized model is

$$\begin{bmatrix} \tilde{s}[n+1] \\ \tilde{\imath}[n+1] \end{bmatrix} = \begin{bmatrix} 1 - (\gamma\bar{\imath}/P) - \beta & -\gamma\bar{s}/P \\ \gamma\bar{\imath}/P & 1 + (\gamma\bar{s}/P) - \rho - \beta \end{bmatrix} \begin{bmatrix} \tilde{s}[n] \\ \tilde{\imath}[n] \end{bmatrix} + \begin{bmatrix} -\beta P \\ 0 \end{bmatrix} \tilde{x}[n] \,. \tag{4.102}$$

The output for the original system was defined in Eq. (4.21) and is already linear, so the linearized output equation is simply

$$\tilde{y}[n] = \phi\tilde{\imath}[n] = \begin{bmatrix} 0 & \phi \end{bmatrix} \begin{bmatrix} \tilde{s}[n] \\ \tilde{\imath}[n] \end{bmatrix} \,. \tag{4.103}$$

The state evolution matrix in Eq. (4.102) can be rewritten in terms of the problem parameters by substituting in the expressions for the equilibrium values $\bar{s}$ and $\bar{\imath}$ that were obtained in Example 4.8. In the case of the infective-free equilibrium (IFE), where $\bar{\imath} = 0$, the state evolution matrix evaluates to

$$\mathbf{A}_{IFE} = \begin{bmatrix} 1 - \beta & -\gamma(1 - \bar{x}) \\ 0 & 1 + \gamma[(1 - \bar{x}) - (1/R_0)] \end{bmatrix} \,. \tag{4.104}$$

For the endemic equilibrium (EE), where $\bar{\imath} > 0$, and which can only exist for the case $R_0 > 1/(1 - \bar{x})$, the state evolution matrix evaluates to

$$\mathbf{A}_{EE} = \begin{bmatrix} 1 - \beta R_0(1 - \bar{x}) & -\gamma/R_0 \\ \beta[R_0(1 - \bar{x}) - 1] & 1 \end{bmatrix} \,. \tag{4.105}$$

The derivation of linearized state-space models in continuous time follows exactly the same route, except that CT equilibrium is specified by the condition in Eq. (4.83) rather than Eq. (4.74). The result is a model with a state evolution equation of the form

$$\frac{d}{dt}\widetilde{\mathbf{q}}(t) = \left[\frac{\partial \mathbf{f}}{\partial \mathbf{q}}\right]_{\overline{\mathbf{q}},\overline{x}} \widetilde{\mathbf{q}}(t) + \left[\frac{\partial \mathbf{f}}{\partial x}\right]_{\overline{\mathbf{q}},\overline{x}} \widetilde{x}(t) \qquad (4.106)$$

and output equation

$$\widetilde{y}(t) = \left[\frac{\partial g}{\partial \mathbf{q}}\right]_{\overline{\mathbf{q}},\overline{x}} \widetilde{\mathbf{q}}(t) + \left[\frac{\partial g}{\partial x}\right]_{\overline{\mathbf{q}},\overline{x}} \widetilde{x}(t) . \qquad (4.107)$$

**Example 4.11    Linearized Model for the Inverted Pendulum**

In this example, we return to the second-order state evolution equations for the inverted pendulum considered in Example 4.1:

$$\frac{dq_1(t)}{dt} = q_2(t) , \qquad (4.108)$$

$$\frac{dq_2(t)}{dt} = \frac{1}{\mathcal{I}}\Big(mg\ell \sin(q_1(t)) - \beta q_2(t) - x(t)\Big) . \qquad (4.109)$$

The equilibria associated with this system were computed in Example 4.9. Evaluating the relevant Jacobians, the linearized model is

$$\frac{d}{dt}\begin{bmatrix} \widetilde{q}_1(t) \\ \widetilde{q}_2(t) \end{bmatrix} = \begin{bmatrix} 0 & 1 \\ (mg\ell/\mathcal{I})\cos(\overline{q}_1) & -\beta/\mathcal{I} \end{bmatrix}\begin{bmatrix} \widetilde{q}_1(t) \\ \widetilde{q}_2(t) \end{bmatrix} + \begin{bmatrix} 0 \\ -1/\mathcal{I} \end{bmatrix}\widetilde{x}(t) . \qquad (4.110)$$

Note that for the case of linearization around the vertical equilibrium, with $\overline{q}_1 = 0$, we recover the linearized model obtained in Eq. (4.8) of Example 4.1.

The output in the original nonlinear model was defined by the LTI expression $y(t) = \theta(t) = q_1(t)$, which is already linear in the state variable, so the output of the small-signal model is simply

$$\widetilde{y}(t) = \begin{bmatrix} 1 & 0 \end{bmatrix}\begin{bmatrix} \widetilde{q}_1(t) \\ \widetilde{q}_2(t) \end{bmatrix} . \qquad (4.111)$$

Chapter 5 is devoted to more detailed analysis of the solution structure of both DT and CT LTI systems. We will examine the internal behavior of such a system, its stability, and its coupling to the input and output of the system.

## 4.6 FURTHER READING

An excellent introduction to state-space models is [Lue], which illustrates the value of such models for the study of dynamic behavior, stability and control in various interesting applications. The very accessible [Str] focuses on complex behavior (including limit cycles, bifurcations and chaos) in low-order nonlinear state-space models, and includes examples from diverse application domains. A more computationally oriented text that covers a similar range of nonlinear behavior in a variety of applications is [Lyn]. The text [Clo] shows how to develop state-space and other models for mechanical, electrical, electromechanical, thermal and fluid systems, the focus being on LTI models and their analysis by transform methods. Specialized books in areas such as circuits [Chu], epidemic modeling [Dal] or population modeling [Cas] also include material on state-space modeling and analysis in their respective domains. The texts cited here constitute useful further reading for Chapters 5 and 6 as well.

# Problems

## Basic Problems

**4.1.** A mass on a straight frictionless track is attached to one end of the track by a nonlinear spring, and pulled on by an external force $x(t)$ from the other end. Its motion is described by the equation

$$\frac{d^2 y(t)}{dt^2} = -y^3(t) + x(t) ,$$

where $y(t)$ is the distance from the end where the mass is attached.

   **(a)** Taking $x(t)$ as the input and $y(t)$ as the output, choose appropriate state variables and write down a (nonlinear) state-space description for this system.
   **(b)** Determine the equilibrium values of the state variables in your model if the input is $x(t) = 8$ for all $t$.
   **(c)** Obtain a linearized state-space model for the system to describe small deviations of the state variables and output from their equilibrium values in (b), in response to perturbations in the input and initial values of the state variables.

**4.2.** Suppose that instead of the differential equation in Problem 4.1, the system was described by

$$\frac{d^2 y(t)}{dt^2} = -y^3(t) + \frac{dx(t)}{dt} + x(t) .$$

   **(a)** Will the choice of state variables you made in Problem 4.1 still work?
   **(b)** Repeat parts (a)–(c) of Problem 4.1 for this new differential equation model.

**4.3.** Suppose the motion of a small object is modeled by the equation

$$\frac{d^2 p(t)}{dt^2} + \phi \frac{dp(t)}{dt} + p^3(t) - \mu^2 p(t) = x(t) ,$$

where $p(t)$ denotes the position of the object (with respect to some reference point), $x(t)$ is the force applied to it, and $\phi$ (like $\mu^2$) is a positive constant.

(a) Write down a state-space model for the system, using an appropriate choice of state variables, taking $x(t)$ as the input and the velocity $dp(t)/dt$ as the output $y(t)$. Is your model nonlinear? Is it time-invariant?

(b) There are three possible equilibria when $x(t) = 0$ for all $t$. For each of these three equilibria, specify the corresponding values of your state variables and output.

(c) Of the three equilibria you found in (b), only one of them has the corresponding equilibrium position of 0. For this particular equilibrium, write down the linearized state-space model that describes small deviations $\tilde{q}(t)$ and $\tilde{y}(t)$ of the state variables and output from their respective equilibrium values, in response to small deviations $\tilde{x}(t)$ of $x(t)$ from its equilibrium value of 0. The entries of the matrices in your model should be expressed in terms of the parameters $\phi$ and $\mu$.

(d) Similarly determine the linearized models at the other two equilibria.

**4.4.** (a) Obtain a second-order CT state-space description for the circuit shown in Figure P4.4, choosing appropriate state variables. The resistor, inductor, and capacitor are all linear and time-invariant components, $x(t)$ is a current source, and $y(t)$ is the resistor current. Write your description in the standard form

$$\frac{d\mathbf{q}(t)}{dt} = \mathbf{A}\mathbf{q}(t) + \mathbf{b}x(t)$$

$$y(t) = \mathbf{c}^T\mathbf{q}(t) + \mathrm{d}x(t) \ .$$

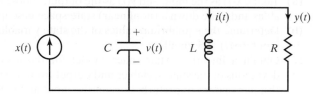

**Figure P4.4**

(b) Find the differential equation relating $x(t)$ and $y(t)$. The direct (and intended) approach is to eliminate variables appropriately, using your state equations. However, also confirm your answer by computing the transfer function from $x(t)$ to $y(t)$ using impedance methods (if you are familiar with these), and then seeing what that transfer function implies about the input-output differential equation.

**4.5.** Obtain a second-order CT state-space description for the circuit shown in Figure P4.5, using appropriate state variables, and writing it in the standard single-input single-output form

$$\frac{d\mathbf{q}(t)}{dt} = \mathbf{A}\mathbf{q}(t) + \mathbf{b}x(t)$$

$$y(t) = \mathbf{c}^T\mathbf{q}(t) + \mathrm{d}x(t) \ .$$

The resistor, inductor, and capacitor are all linear and time-invariant components, $x(t)$ is the source voltage, and $y(t)$ is the resistor voltage.

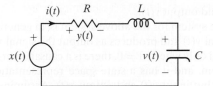

**Figure P4.5**

**4.6.** Figure P4.6 is a block diagram for a second-order CT LTI system. The boxes in
this figure are integrators and the triangles are amplifiers or gains. The output
of an integrator is the integral of its input. Alternatively, its input can be seen
as the derivative of its output. For example, if the input to the top integrator is
$\dot{q}_1(t)$, then its output is $q_1(t)$. The output of a gain element is simply the input
scaled by the gain. Specify the values of the nine gains so that the system in the
figure has the same state-space description as the one you have calculated for the
circuit in Problem 4.5.

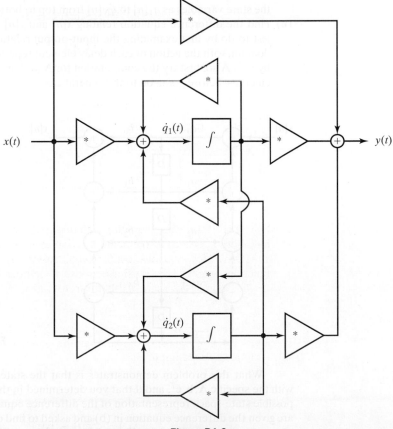

**Figure P4.6**

**4.7.** Consider the causal single-input, single-output $L$th-order CT LTI state-space system

$$\dot{\mathbf{q}}(t) = \mathbf{A}\mathbf{q}(t) + \mathbf{b}x(t) , \quad y(t) = \mathbf{c}^T\mathbf{q}(t) + \mathrm{d}x(t) ,$$

with input $x(t)$ and output $y(t)$.

An inverse system for the above system is, in general terms, one that takes as input the signal $y(t)$ and produces as output the signal $x(t)$.

When (and only when) $\mathrm{d} \neq 0$, there is a causal inverse system for the above state-space system, and it has a state-space representation involving the same state vector $\mathbf{q}(t)$ but input $y(t)$ and output $x(t)$. Assuming $\mathrm{d} \neq 0$, determine this state-space representation, that is, express the quantities $\mathbf{A}_{\text{in}}, \mathbf{b}_{\text{in}}, \mathbf{c}_{\text{in}}^T$, and $\mathrm{d}_{\text{in}}$ in the state-space representation below in terms of $\mathbf{A}, \mathbf{b}, \mathbf{c}^T$, and d:

$$\dot{\mathbf{q}}(t) = \mathbf{A}_{\text{in}}\mathbf{q}(t) + \mathbf{b}_{\text{in}}y(t) , \quad x(t) = \mathbf{c}_{\text{in}}^T\mathbf{q}(t) + \mathrm{d}_{\text{in}}y(t) .$$

**4.8. (a)** Obtain a state-space description of the form

$$\mathbf{q}[n + 1] = \mathbf{A}\mathbf{q}[n] + \mathbf{b}x[n]$$

$$y[n] = \mathbf{c}^T\mathbf{q}[n] + \mathrm{d}x[n]$$

for the system shown in Figure P4.8. Choose as state variables the outputs of the delay elements, which have been shown as boxes marked with $D$. Label the state variables as $q_1[n]$ to $q_N[n]$ from top to bottom.

**(b)** Find the difference equation relating $x[n]$ and $y[n]$. You may find this easiest to do by first examining the input-output relationship in the transform domain, with the action of each delay element represented by multiplication by $z^{-1}$. Also, first try the computation for $N = 2$ or 3 to see the pattern that emerges, before going on to the general case.

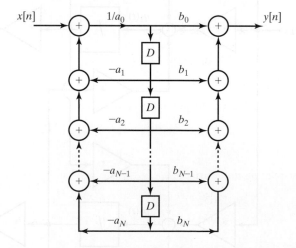

**Figure P4.8**

What this problem demonstrates is that the state-space equation in (a), with the specific $\mathbf{A}, \mathbf{b}, \mathbf{c}^T$, and d that you determined in that part, constitutes one possible state-space representation of the difference equation in (b). Thus, if you are given the difference equation in (b) and asked to find one possible state-space representation or realization of it, you are now in a position to respond.

**4.9. (a)** Find the possible equilibrium states (that is, find the possible equilibrium values $\bar{q}_1$ and $\bar{q}_2$ of the state variables) for the following DT second-order system, when the input $x[n]$ is held constant at a nonzero value $\bar{x}$:

$$q_1[n+1] = \left(q_1[n]\right)^2 x[n]$$

$$q_2[n+1] = q_1[n] .$$

**(b)** If you did things correctly in (a), you should have found that there are two possible equilibrium states, one being at the origin of the state space. For this part, we are interested in the *other* equilibrium state, that is, the nonzero one.

Obtain a linearized state-space model that governs small perturbations from the nonzero equilibrium, using the notation $\tilde{q}_i[n] = q_i[n] - \bar{q}_i$ for $i = 1, 2$, and similarly $\tilde{x}[n] = x[n] - \bar{x}$.

**4.10.** The following pair of DT difference equations has been used as a model for certain types of population growth:

$$p[n+1] = q[n]$$

$$q[n+1] = \mu\, q[n] \left(1 - p[n]\right) , \quad \mu > 0 .$$

This model is in state-space form.

**(a)** Is this system linear? Time-invariant?

**(b)** Find the equilibrium points of the model, i.e., values $\bar{p}$ and $\bar{q}$ of $p[n]$ and $q[n]$ respectively such that $p[n+1] = p[n] = \bar{p}$ and $q[n+1] = q[n] = \bar{q}$.

**(c)** Assume $p[0] = q[0] = \frac{1}{2}$. Compute and plot the state vector $(p[n], q[n])$ for $0 \le n \le 500$ for $\mu = 1.8,\ 1.9,\ 2.0,\ 2.1,$ and $2.2$ (five separate plots). Describe the behavior of the state trajectories as $n$ increases, for each of these values of $\mu$.

## Advanced Problems

**4.11.** Consider the LTI difference equation

$$y[k] + a_{n-1}y[k-1] + \cdots + a_0 y[k-n] = b_{n-1}x[k-1] + \ldots + b_0 x[k-n] .$$

Show that this can be put in the state-space form

$$\mathbf{q}[k+1] = \mathbf{A}\mathbf{q}[k] + \mathbf{b}x[k]$$

$$y[k] = \mathbf{c}^T\mathbf{q}[k] ,$$

by defining state variables

$$q_1[k] = -a_0 y[k-1] + b_0 x[k-1]$$

$$q_2[k] = -a_1 y[k-1] + b_1 x[k-1] + q_1[k-1]$$

$$\vdots$$

$$q_{n-1}[k] = -a_{n-2}y[k-1] + b_{n-2}x[k-1] + q_{n-2}[k-1]$$

$$q_n[k] = y[k] .$$

Determine the matrix $\mathbf{A}$ and vectors $\mathbf{b}$ and $\mathbf{c}^T$.

**4.12.** **(a)** Consider the LTI differential equation

$$w^{(n)}(t) + a_{n-1}w^{(n-1)}(t) + \cdots + a_0 w(t) = x(t) .$$

Show that if we define $q_i(t) = w^{(i-1)}(t)$, then we can obtain an $n$th-order state-space description of the form

$$\dot{\mathbf{q}}(t) = \mathbf{A}\mathbf{q}(t) + \mathbf{b}x(t) ,$$

where the $i$th entry of the vector $\mathbf{q}(t)$ is $q_i(t)$, and the $i$th entry of $\dot{\mathbf{q}}(t)$ is $\dot{q}_i(t)$. Determine the matrix $\mathbf{A}$ and vector $\mathbf{b}$.

**(b)** Define

$$y(t) = b_{n-1}w^{(n-1)}(t) + \cdots + b_0 w(t) .$$

Use the fact that the differential equation given in (a) is LTI to show that

$$y^{(n)}(t) + a_{n-1}y^{(n-1)}(t) + \cdots + a_0 y(t) = b_{n-1}x^{(n-1)}(t) + \cdots + b_0 x(t) .$$

Also show that

$$y(t) = \mathbf{c}^T \mathbf{q}(t) ,$$

where $\mathbf{c}^T$ is a row vector that you should determine.

The preceding development establishes that the state evolution and output equations we have defined, together with the specific $\mathbf{A}, \mathbf{b}$, and $\mathbf{c}^T$ determined in (a) and (b), constitute one possible state-space representation of the $n$th-order differential equation given here in (b).

**4.13.** A variety of important biochemical reactions involve four chemical species, with respective concentrations $q_i(t) \geq 0$ for $i = 1, 2, 3, 4$, whose interactions are approximately—according to the law of mass action—governed by the following state-space model with input $x(t)$ and output $y(t)$:

$$\frac{dq_1(t)}{dt} = -k_f q_1(t)q_2(t) + k_r q_3(t) + x(t)$$

$$\frac{dq_2(t)}{dt} = -k_f q_1(t)q_2(t) + (k_r + k_c)q_3(t)$$

$$\frac{dq_3(t)}{dt} = k_f q_1(t)q_2(t) - (k_r + k_c)q_3(t)$$

$$\frac{dq_4(t)}{dt} = k_c q_3(t)$$

$$y(t) = q_4(t) .$$

The most well-known example involves a substrate (species 1, typically denoted by S) to which an enzyme (species 2, or E) binds, forming a complex (species 3, or ES), which is then converted into a product (species 4, or P) and the enzyme. The net effect is the conversion of substrate to product, with the enzyme serving as a catalyst but not being consumed or produced. The quantities $k_f$, $k_r$, and $k_c$ are reaction rates, and are constant if the temperature is held constant. The input $x(t) \geq 0$ in this case describes the rate of addition of substrate.

**(a)** Assuming temperature is held constant, is this model time varying or time-invariant, and is it linear or nonlinear?

**(b)** Show that $q_2(t) + q_3(t)$ stays constant at its initial value $q_2(0) + q_3(0)$. In the specific example above, this sum is the total enzyme concentration, so the result here shows that the enzyme is not consumed or produced—it merely catalyzes the reaction.

Assume for the rest of this problem that $x(t)$ is fixed at some value $\bar{x} \geq 0$.

   (c) With $\bar{x} = 0$, what are the equilibrium values of the state variables, if $q_2(0) + q_3(0) = E_0 > 0$?

   (d) Show that no equilibrium of the full system is possible with $\bar{x} > 0$, but that the first 3 state variables can have constant values in this case. Express these three constant values in terms of $\bar{x}$ and the rate constants, and determine the rate of increase of $y(t)$ (which is the product P in the particular example above).

   (e) Numerically explore the behavior of the system for various choices of the rate parameters, input value $\bar{x}$, and initial conditions.

**4.14.** Consider a causal CT system with input $x(t)$ and output $y(t)$, obtained by cascading two CT causal subsystems as shown in Figure P4.14.

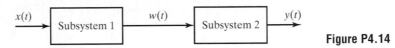

**Figure P4.14**

The intermediate variable $w(t)$ is related to the system input $x(t)$ by the differential equation

$$\frac{dw(t)}{dt} + w(t) = \frac{dx(t)}{dt} + \gamma x(t) ,$$

where $\gamma$ is a real parameter. The system output $y(t)$ is related to the intermediate variable $w(t)$ by the differential equation

$$\frac{dw(t)}{dt} - \epsilon w(t) = \frac{dy(t)}{dt} ,$$

where $\epsilon$ is another real parameter.

   (a) Three vectors have been respectively suggested by your friends Esmeralda, Faustus, and Gayatri as candidates for a state vector $\mathbf{q}(t)$ in a state-space model of this system, that is, a model of the form

$$\dot{\mathbf{q}}(t) = \mathbf{f}\big(\mathbf{q}(t), x(t)\big) , \qquad y(t) = g\big(\mathbf{q}(t), x(t)\big) .$$

   The three candidates are

$$\mathbf{q}_E(t) = \begin{bmatrix} w(t) \\ y(t) \end{bmatrix} , \quad \mathbf{q}_F(t) = \begin{bmatrix} w(t) - x(t) \\ y(t) - w(t) \end{bmatrix} , \quad \mathbf{q}_G(t) = \begin{bmatrix} x(t) - y(t) \\ y(t) - w(t) + 1 \end{bmatrix} .$$

   Which of these candidates can actually serve as a system state vector? (At least one of them will work fine, but you should check whether more than one might.) Give a brief explanation for your choice or choices of valid state vectors.

   (b) Only one state vector candidate from (a) yields an LTI state-space model for this system, in the standard form

$$\frac{d\mathbf{q}(t)}{dt} = \mathbf{A}\mathbf{q}(t) + \mathbf{b}x(t), \quad y(t) = \mathbf{c}^T\mathbf{q}(t) + dx(t) .$$

   Determine which one of the three state vector candidates it is, and fully specify the corresponding entries of $\mathbf{A}, \mathbf{b}, \mathbf{c}^T$, and $d$ (possibly as functions of the parameters $\gamma$ and $\epsilon$).

# Extension Problems

**4.15. (a)** Verify by simulation the claim at the end of Example 4.3 on viral propagation, regarding what happens in steady state when the immunization rate drops to $x[n] = 0.2$, keeping all other parameters the same. Provide plots similar to those in Figure 4.3

**(b)** Check that your simulation in (a) actually settles to the endemic equilibrium computed in Example 4.8.

**(c)** For the parameter values you used in (a), compute the matrix $\mathbf{A}_{EE}$ given in Eq. (4.105), which governs small deviations from the endemic equilibrium. Also compute the eigenvalues and associated eigenvectors for this matrix.

**4.16.** Consider the DT LTI second-order model

$$\underbrace{\begin{bmatrix} q_1[n+1] \\ q_2[n+1] \end{bmatrix}}_{\mathbf{q}[n+1]} = \underbrace{\begin{bmatrix} -a_1 & 1 \\ -a_2 & 0 \end{bmatrix}}_{\mathbf{A}} \underbrace{\begin{bmatrix} q_1[n] \\ q_2[n] \end{bmatrix}}_{\mathbf{q}[n]} + \underbrace{\begin{bmatrix} b_1 \\ b_2 \end{bmatrix}}_{\mathbf{b}} x[n],$$

$$y[n] = \underbrace{\begin{bmatrix} 1 & 0 \end{bmatrix}}_{\mathbf{c}^T} \mathbf{q}[n] .$$

**(a)** By working with the relationships embodied in the state-space model, find a second-order difference equation relating the input and output, $x[n]$ and $y[n]$.

**(b)** Let $Q_1(z)$, $Q_2(z)$, $X(z)$, and $Y(z)$ denote the unilateral or one-sided $z$-transforms of $q_1[n]$, $q_2[n]$, $x[n]$, and $y[n]$, respectively, where the unilateral $z$-transform of a signal $v[n]$ is defined as

$$V(z) = \sum_{n=0}^{+\infty} v[n] z^{-n} .$$

Verify that the unilateral $z$-transform of the time-shifted signal $v[n+1]$ is

$$z V(z) - z v[0] .$$

**(c)** Apply the result in (b) to the given state-space model and verify that

$$\mathbf{Q}(z) = \begin{bmatrix} Q_1(z) \\ Q_2(z) \end{bmatrix} = (z\mathbf{I} - \mathbf{A})^{-1} \mathbf{b} X(z) + z(z\mathbf{I} - \mathbf{A})^{-1} \mathbf{q}[0] ,$$

and

$$Y(z) = \mathbf{c}^T \mathbf{Q}(z) .$$

Then find an expression in terms of $\mathbf{A}$, $\mathbf{b}$, and $\mathbf{c}^T$ for $H(z)$, the transfer function from $X(z)$ to $Y(z)$ when $\mathbf{q}[0] = 0$.

**(d)** Now evaluate your expression for $H(z)$ in (c) in terms of $a_1$, $a_2$, $b_1$, and $b_2$, and verify that it is consistent with the difference equation you obtained in (a).

**4.17.** The following simple DT state-space model attempts to describe the season-to-season variations in the population densities of two interacting animal species. One of them (the predator) feeds on the other (the prey). The classical example is that of the Canadian lynx and the snowshoe hare. Denoting the respective population densities in season $n$ by $\ell[n]$ and $r[n]$ respectively, the model is

$$\ell[n+1] = (1-d)\ell[n] + \mu\ell[n]r[n],$$

$$r[n+1] = (1+b)r[n] - \nu\ell[n]r[n] - \beta x[n].$$

Here $0 < d < 1$ represents the rate at which the lynx population would decay in the absence of hares, and $b > 0$ represents the rate at which the hare population would grow in the absence of lynx. The coupling coefficients $\mu > 0$ and $\nu > 0$ respectively determine the advantage to the lynx population and the jeopardy to the hare population of interactions between the two species, with these interactions assumed to occur in proportion to the product of the population densities. The coefficient $\beta > 0$ determines the effectiveness of external control actions, represented by the input $x[n]$, that are aimed at limiting the hare population.

**(a)** Compute the two equilibrium points of this model, when $x[n]$ is fixed at the value $\bar{x}$. You can think of an equilibrium point as an initial state $\ell[0] = \bar{\ell}$ and $r[0] = \bar{r}$ from which the system does not move for $n > 0$.

**(b)** For each equilibrium point $\bar{\ell}$ and $\bar{r}$, compute the associated linearized (and LTI, DT state-space) model that approximately describes small deviations $\tilde{\ell}[n] = \ell[n] - \bar{\ell}$ and $\tilde{r}[n] = r[n] - \bar{r}$ of the species population densities away from this equilibrium. The linearized model retains only first-order deviations from equilibrium, under the assumption that higher-order terms are negligible, so a product term of the form $\tilde{\ell}[n]\,\tilde{r}[n]$, e.g., would be neglected.

Write your linearized model in the form

$$\begin{bmatrix} \tilde{\ell}[n+1] \\ \tilde{r}[n+1] \end{bmatrix} = \mathbf{A} \begin{bmatrix} \tilde{\ell}[n] \\ \tilde{r}[n] \end{bmatrix},$$

where $\mathbf{A}$ is a $2 \times 2$ matrix. You will, in general, have a different $\mathbf{A}$ associated with each equilibrium point.

**(c)** Now start with some initial values $\ell[0] > 0$ and $r[0] > 0$ in the close vicinity of the nontrivial equilibrium analyzed in (b), then simulate and plot the resulting trajectories of $\ell[n]$ and $r[n]$ over 700 time steps using the original nonlinear model. Begin by fixing $b = d = 0.05$, $\mu = \nu = 0.001$, and $\beta = 0$ for your simulations. Then explore how changes in these parameter values affect the results.

# 5 LTI State-Space Models

Chapter 4 introduced state-space models for causal dynamical systems. The state evolution property of these models allows numerical solution to advance the state forward in time from a given initial state, provided the input is known from that initial time onward. This will generate the entire state trajectory associated with the particular initial state and input. For nonlinear models, we cannot in general make broad qualitative conclusions, such as whether the state trajectory will always be bounded to some finite region of the state space when the input magnitude is bounded. However, for linear, time-invariant (LTI) models, the structure and behavior can be analyzed in great detail and generality, both in discrete time and continuous time. This chapter is devoted to the analysis of LTI state-space models. The insights and tools developed from the study of LTI systems provide a powerful basis for the design of various modules and components of engineering systems, and are also relevant to the behavior of nonlinear time-invariant systems for small perturbations around an equilibrium.

## 5.1 CONTINUOUS-TIME AND DISCRETE-TIME LTI MODELS

Throughout this chapter, we restrict ourselves to single-input, single-output $L$th-order continuous-time (CT) LTI state-space models of the form

$$\frac{d\mathbf{q}(t)}{dt} = \dot{\mathbf{q}}(t) = \mathbf{A}\mathbf{q}(t) + \mathbf{b}x(t) \tag{5.1}$$

$$y(t) = \mathbf{c}^T \mathbf{q}(t) + dx(t) , \tag{5.2}$$

or discrete-time (DT) LTI state-space models of the form

$$\mathbf{q}[n+1] = \mathbf{A}\mathbf{q}[n] + \mathbf{b}x[n] \tag{5.3}$$

$$y[n] = \mathbf{c}^T \mathbf{q}[n] + dx[n] . \tag{5.4}$$

Equation (5.1) represents CT LTI system dynamics as a set of coupled, first-order, linear, constant-coefficient differential equations for the $L$ variables in $\mathbf{q}(t)$, driven by the input $x(t)$. The symbol $\dot{\mathbf{q}}(t)$ denotes the component-wise derivative of $\mathbf{q}(t)$, with its $i$th component given by $\dot{q}_i(t)$. Equation (5.3) provides a similar representation of DT LTI system dynamics using coupled first-order difference equations.

**Solving LTI Models**    Determining the general solution of an LTI state-space model follows the process used in solving linear constant-coefficient differential or difference equations in one variable. The general solution is written as the sum of a homogeneous solution, that is, a solution for the unforced or zero-input case, and a particular solution of the system. In the case of state-space models, the homogeneous or zero-input response (ZIR) of most interest is the response to the specified nonzero initial state at some starting time, with the input identically zero at and after this starting time. The ZIR is treated in Section 5.2. Section 5.3 describes the full solution as the sum of the ZIR and a specific particular solution, the zero-state response (ZSR) due to the nonzero input alone, when the initial state is zero. Understanding the ZIR and the full ZIR-plus-ZSR solution from the starting time onward gives insight into system stability, and into how the internal behavior relates to the input-output characteristics of the system.

The CT and DT LTI state-space models in Eqs. (5.1)–(5.4) are completely specified by the associated matrix $\mathbf{A}$, vectors $\mathbf{b}$ and $\mathbf{c}^T$, and scalar d; their respective states are embodied in the vectors $\mathbf{q}(t)$ or $\mathbf{q}[n]$. The analysis of these models thus involves vector and matrix notation and operations, and some related linear algebra. The following subsection briefly summarizes the notation and facts we need in order to get started.

**Vector and Matrix Operations**    We assume familiarity with the idea of a vector space, where the associated operations of vector addition and scalar multiplication again yield vectors in the space. A column or row array of $L$ real numbers specifies a point or vector in an $L$-dimensional real Euclidean space, commonly represented in two and three dimensions as an arrow from the origin to the specified point. Associated with this graphical picture is the computation of the sum of two vectors using the "parallelogram rule," which simply corresponds to component-wise addition of the entries of the associated arrays. Similarly, scalar multiplication just scales the array, that is, multiplies each component of the array by the given scalar.

It will also be necessary, at least at intermediate stages of computations involving complex eigenvalues, to consider complex Euclidean space. The vectors in this case are column or row arrays of $L$ complex numbers, and the scalars are also complex numbers. However, in our setting, no special attention is needed when making this extension from real to complex Euclidean spaces; the algebraic computations proceed as in the real case, with the obvious modifications. Furthermore, our final answers in each instance can be written and represented graphically in terms of real Euclidean vectors, because the complex eigenvalues in the problems of interest to us arise in complex conjugate pairs. We shall therefore generally rely on the context to make clear whether we are working with real or complex vectors and scalars.

The span of a set of vectors $\mathbf{v}_1, \ldots, \mathbf{v}_k$ denotes all vectors obtained as weighted sums, or linear combinations, of the form $\alpha_1 \mathbf{v}_1 + \cdots + \alpha_k \mathbf{v}_k$ for arbitrary scalars $\alpha_1, \ldots, \alpha_k$, with $\alpha_i \mathbf{v}_i$ here denoting scalar multiplication of a vector, and $+$ denoting vector addition. The vectors in the span themselves form a vector space, so we more commonly refer to the vectors $\alpha_1 \mathbf{v}_1 + \cdots + \alpha_k \mathbf{v}_k$ as comprising the space spanned by the vectors $\mathbf{v}_1, \ldots, \mathbf{v}_k$.

A set of vectors is termed independent if no vector in the set can be expressed as a weighted sum of the others. The largest number of vectors that can form an independent set in a vector space constitutes the dimension of the space. Such a set of vectors is called a maximal independent set. In a two-dimensional plane, any two vectors will be independent if and only if one is not a scalar multiple of the other; any three vectors in the plane will be dependent, that is, not independent. The space spanned by two independent vectors in a three-dimensional space is itself a two-dimensional space—a plane through the origin—embedded in the larger space, and is an example of a subspace, or a subset of a vector space that itself forms a vector space. Our use of vector space methods will be restricted to the case of finite-dimensional spaces, though we have actually encountered important infinite-dimensional vector spaces in earlier chapters, most notably the space of $\ell^2$ DT signals.

The vectors in a maximal independent set within a vector space form a basis for the space, that is, every other vector in the space can be written as a weighted sum of the vectors in the basis set, and these weights are uniquely determined. Though we omit the quite straightforward proof of the preceding statement, it is an important result, and one that will be invoked repeatedly in the following development.

We will in addition rely on an elementary knowledge of matrix operations. A matrix is usually viewed as a rectangular array of scalars, but it is often more helpful to see it as an array of column vectors arranged side by side. Thus, the $L \times k$ matrix $\mathbf{P}$ may be viewed as

$$\mathbf{P} = \begin{bmatrix} \mathbf{p}_1 & \mathbf{p}_2 & \cdots & \mathbf{p}_k \end{bmatrix}, \tag{5.5}$$

where the $i$th column $\mathbf{p}_i$ is a column vector in an $L$-dimensional space. With this picture, the matrix-vector product $\mathbf{Pw}$ can be interpreted as forming a weighted linear combination of the columns of $\mathbf{P}$, with weights that are given by the corresponding entries of the vector $\mathbf{w}$:

$$\mathbf{Pw} = \begin{bmatrix} \mathbf{p}_1 & \mathbf{p}_2 & \cdots & \mathbf{p}_k \end{bmatrix} \begin{bmatrix} w_1 \\ w_2 \\ \vdots \\ w_k \end{bmatrix} = \sum_{1}^{k} \mathbf{p}_i w_i . \tag{5.6}$$

We shall invoke this interpretation frequently. The set of vectors of the form $\mathbf{Pw}$, as $\mathbf{w}$ varies over all possibilities, is a vector space that is referred to as the range of the matrix $\mathbf{P}$; the dimension of the range is called the rank of the matrix.

There are also situations in which it is helpful to view a matrix as an array of row vectors arranged one on top of the other, or simply as an operator that maps vectors or matrices onto other vectors or matrices. We shall also assume familiarity with the computation and use of the determinant and inverse of a square matrix, at least for $2 \times 2$ and $3 \times 3$ matrices. Some examples later in this chapter illustrate these computations.

Finally, we will use the fact that the independence of $L$ column vectors (or row vectors) in an $L$-dimensional space can be tested by forming the $L \times L$ matrix whose columns (or rows) are these vectors, and then computing the determinant of this matrix. The vectors are independent if and only if the determinant is nonzero.

## 5.2 ZERO-INPUT RESPONSE AND MODAL REPRESENTATION

### 5.2.1 Undriven CT Systems

We begin with the case of a CT LTI state-space model. For notational convenience, we consider the time at which the analysis starts—the initial time—to be $t = 0$. There is no loss of generality in doing so when dealing with time-invariant models. If the initial time is actually some other time $t_0$, then we only need to account for the fact that the elapsed time since the starting time is $t - t_0$ rather than $t$.

Consider the response of the undriven system corresponding to Eq. (5.1), that is, the response with $x(t) = 0$ for $t \geq 0$, but with some nonzero initial condition $\mathbf{q}(0)$. This is the ZIR of the system, and is a solution of the undriven (or unforced, or homogeneous) system

$$\dot{\mathbf{q}}(t) = \mathbf{Aq}(t) . \tag{5.7}$$

**CT Exponential Solutions and Modes**    It is natural when analyzing an undriven LTI system to look for a solution in exponential form. This is because undriven LTI systems must have solutions that are invariant to shifting and scaling, and exponentials have the unique property that shifting them in time is equivalent to scaling them in amplitude. We accordingly look for a nonzero solution of the form

$$\mathbf{q}(t) = \mathbf{v}e^{\lambda t} , \quad \mathbf{v} \neq \mathbf{0} , \tag{5.8}$$

where each state variable is a scalar multiple of the same exponential $e^{\lambda t}$, with these scalar multiples assembled into the vector $\mathbf{v}$. The boldface $\mathbf{0}$ in Eq. (5.8) denotes an $L$-component column vector whose entries are all 0. Writing $\mathbf{v} \neq \mathbf{0}$ signifies that at least one component of $\mathbf{v}$ is nonzero.

Substituting Eq. (5.8) into the undriven model in Eq. (5.7) results in the relation

$$\lambda \mathbf{v}e^{\lambda t} = \mathbf{A}\mathbf{v}e^{\lambda t} , \tag{5.9}$$

from which it follows that the vector $\mathbf{v}$ and scalar $\lambda$ must satisfy

$$\lambda \mathbf{v} = \mathbf{A}\mathbf{v} , \quad \text{or equivalently,} \tag{5.10}$$

$$(\lambda \mathbf{I} - \mathbf{A})\mathbf{v} = \mathbf{0} , \quad \mathbf{v} \neq \mathbf{0} . \tag{5.11}$$

Here $\mathbf{I}$ denotes the identity matrix, in this case of dimension $L \times L$. The identity matrix has 1 in all its diagonal positions (i.e., positions where the row index equals the column index), and 0 everywhere off-diagonal, so $\mathbf{I}\mathbf{v} = \mathbf{v}$ for all vectors $\mathbf{v}$. This matrix is introduced in Eq. (5.11) because the $L \times L$ matrix $\mathbf{A}$ can only be subtracted from an $L \times L$ matrix, not from the scalar $\lambda$.

Equation (5.11) can be interpreted as requiring some nonzero weighted linear combination of the columns of the matrix $(\lambda \mathbf{I} - \mathbf{A})$ to add up to the zero vector $\mathbf{0}$, where the weights are the entries of the vector $\mathbf{v}$. In other words, $\lambda$ must be such that the columns of the matrix $(\lambda \mathbf{I} - \mathbf{A})$ are dependent. As noted earlier, a simple analytical test for dependence of the columns of a square matrix is that the determinant of the matrix be 0:

$$\det(\lambda \mathbf{I} - \mathbf{A}) = 0 . \tag{5.12}$$

For an $L$th-order system, the above determinant is always a monic polynomial in $\lambda$ of degree $L$, called the characteristic polynomial of the system or of the matrix $\mathbf{A}$:

$$\det(\lambda \mathbf{I} - \mathbf{A}) = a(\lambda) = \lambda^L + a_{L-1}\lambda^{L-1} + \cdots + a_0 . \tag{5.13}$$

(The label *monic* denotes the fact that the coefficient of the highest-degree term is 1.) It follows that the assumed exponential-form solution $\mathbf{v}e^{\lambda t}$ in Eq. (5.8) is a nonzero solution of the undriven system if and only if $\lambda$ is one of the $L$ roots $\{\lambda_i\}_{i=1}^{L}$ of the characteristic polynomial. These roots are referred to as characteristic roots or natural frequencies of the system, and as eigenvalues of the matrix $\mathbf{A}$.

The vector $\mathbf{v}$ in Eq. (5.8) is correspondingly a nonzero solution $\mathbf{v}_i$ of the system of equations

$$(\lambda_i \mathbf{I} - \mathbf{A})\mathbf{v}_i = \mathbf{0} , \quad \mathbf{v}_i \neq \mathbf{0} , \tag{5.14}$$

and is termed the characteristic vector or eigenvector associated with $\lambda_i$. Note from this equation that multiplying any eigenvector by a nonzero scalar again yields an eigenvector for the same eigenvalue. Consequently eigenvectors are only defined up to a nonzero scaling, and any convenient nonzero scaling or normalization can be used. Thus, it is the one-dimensional space (or line through the origin) spanned by the eigenvector that is characteristic of the system, not the actual eigenvector itself.

In summary, the undriven system has a solution of the exponential form $\mathbf{v}e^{\lambda t}$ if and only if $\lambda$ equals some characteristic value or eigenvalue of $\mathbf{A}$, and the nonzero vector $\mathbf{v}$ is an associated characteristic vector or eigenvector.

We shall only be dealing with state-space models for which all the signals and the coefficient matrices $\mathbf{A}$, $\mathbf{b}$, $\mathbf{c}^T$, and d are real-valued (though we may subsequently transform these models into the diagonal form seen in the previous chapter, which may then have complex entries, but occurring in very structured ways). The coefficients $\{a_i\}$ defining the characteristic polynomial $a(\lambda)$ in Eq. (5.13) are therefore real, and thus the complex roots of this polynomial occur in conjugate pairs. Also, if $\mathbf{v}_i$ is an eigenvector associated with a complex eigenvalue $\lambda_i$, then $\mathbf{v}_i^*$—the vector whose entries are the complex conjugates of the corresponding entries of $\mathbf{v}_i$—is an eigenvector associated with $\lambda_i^*$, the complex conjugate of $\lambda_i$. This can be seen by taking the complex conjugate of both sides of the equality in Eq. (5.10) for the case where $\lambda = \lambda_i$, $\mathbf{v} = \mathbf{v}_i$, and $\mathbf{A}$ is real, to get $\lambda_i^* \mathbf{v}_i^* = \mathbf{A}\mathbf{v}_i^*$.

| **Example 5.1** | **Eigenvalues and Eigenvectors for the Linearized Pendulum Model** |
|---|---|

For the inverted pendulum introduced in Examples 4.1, 4.5, 4.9, and 4.11 in Chapter 4, consider the undriven LTI model obtained for small deviations from the fully inverted and stationary equilibrium position, and assume specific numerical values of the parameters that result in the model

$$\dot{\mathbf{q}}(t) = \begin{bmatrix} 0 & 1 \\ 8 & -2 \end{bmatrix} \mathbf{q}(t) , \tag{5.15}$$

as in Example 4.5. Recall that the first state variable denoted the clockwise angular deviation from the vertical, while the second state variable denoted the clockwise angular velocity. The linearized model approximately describes the time evolution of these quantities, as long as they both remain close to their equilibrium values of 0.

To determine the ZIR, we require the eigenvalues and eigenvectors of the matrix

$$\mathbf{A} = \begin{bmatrix} 0 & 1 \\ 8 & -2 \end{bmatrix} , \tag{5.16}$$

so we begin by computing its characteristic polynomial:

$$\det(\lambda \mathbf{I} - \mathbf{A}) = \det \begin{bmatrix} \lambda & -1 \\ -8 & \lambda + 2 \end{bmatrix} = \lambda(\lambda + 2) - (-8)(-1)$$

$$= \lambda^2 + 2\lambda - 8 = (\lambda - 2)(\lambda + 4) . \tag{5.17}$$

In computing the determinant, we have used the fact that the determinant of an arbitrary $2 \times 2$ matrix is given by

$$\det \begin{bmatrix} a & b \\ c & d \end{bmatrix} = ad - bc . \tag{5.18}$$

The roots of the characteristic polynomial are $\lambda_1 = 2$ and $\lambda_2 = -4$. To find $\mathbf{v}_1$, we look for a nonzero solution to the equation

$$(\lambda_1 \mathbf{I} - \mathbf{A})\mathbf{v}_1 = \begin{bmatrix} 2 & -1 \\ -8 & 4 \end{bmatrix} \begin{bmatrix} v_{11} \\ v_{21} \end{bmatrix} = \begin{bmatrix} 0 \\ 0 \end{bmatrix} = \mathbf{0} . \tag{5.19}$$

Since eigenvectors are only defined to within a nonzero scale factor, we can take $v_{11} = 1$, as long as it is not 0, and then solve for $v_{21}$. If the assumption of nonzero $v_{11}$ is invalid, then we will discover that there is no choice of $v_{21}$ that satisfies the above pair of equations, in which case we would set $v_{11} = 0$ and then solve for $v_{21}$. In the present example, setting $v_{11} = 1$ in Eq. (5.19) shows that choosing $v_{21} = 2$ will satisfy both equations. Hence

$$\mathbf{v}_1 = \begin{bmatrix} 1 \\ 2 \end{bmatrix} . \tag{5.20}$$

Higher-order systems require proceeding more systematically to solve $L - 1$ linear equations for the remaining $L - 1$ components of the eigenvector.

To solve for the eigenvector $\mathbf{v}_2$ associated with $\lambda_2 = -4$, the relevant equation is

$$(\lambda_2 \mathbf{I} - \mathbf{A})\mathbf{v}_2 = \begin{bmatrix} -4 & -1 \\ -8 & -2 \end{bmatrix} \begin{bmatrix} v_{12} \\ v_{22} \end{bmatrix} = \begin{bmatrix} 0 \\ 0 \end{bmatrix} = \mathbf{0} . \tag{5.21}$$

Again choosing $v_{12} = 1$, it follows that $v_{22} = -4$, so

$$\mathbf{v}_2 = \begin{bmatrix} 1 \\ -4 \end{bmatrix} . \tag{5.22}$$

Suppose instead that the equilibrium of interest had been the normal hanging position rather than the inverted position. Then the linearized model would have had an entry of $-8$ instead of 8 in the matrix $\mathbf{A}$ of Eq. (5.16), resulting in the matrix

$$\mathbf{A}' = \begin{bmatrix} 0 & 1 \\ -8 & -2 \end{bmatrix} . \tag{5.23}$$

The eigenvalues of this matrix are then the roots of

$$\lambda^2 + 2\lambda + 8 = (\lambda + 1)^2 + 7 , \tag{5.24}$$

namely $\lambda_1 = -1 + j\sqrt{7}$ and $\lambda_2 = \lambda_1^* = -1 - j\sqrt{7}$. Carrying out the eigenvector computation as before, with the appropriate modifications, we find

$$\mathbf{v}_1 = \begin{bmatrix} 1 \\ -1 + j\sqrt{7} \end{bmatrix} = \begin{bmatrix} 1 \\ -1 \end{bmatrix} + j \begin{bmatrix} 0 \\ \sqrt{7} \end{bmatrix} = \mathbf{u} + j\mathbf{w} , \quad \mathbf{v}_2 = \mathbf{v}_1^* = \mathbf{u} - j\mathbf{w} , \tag{5.25}$$

where $\mathbf{u}$ and $\mathbf{w}$ are the real and imaginary parts of the vector $\mathbf{v}_1$.

When computing the eigenvalues of $3 \times 3$ matrices, it helps to know the following expression for the determinant of a $3 \times 3$ matrix:

$$\det \begin{bmatrix} a & b & c \\ d & e & f \\ g & h & i \end{bmatrix} = (aei + bfg + cdh) - (gec + dbi + ahf) . \tag{5.26}$$

A general and very useful result is that the eigenvalues of an upper triangular matrix—one that has all its nonzero entries on the diagonal and above—are simply the elements along the diagonal of the matrix. The same holds for a lower triangular matrix, and also for a diagonal matrix, that is, one whose nonzero entries are on the diagonal. These results follow from the fact that if $\mathbf{A}$ is such a matrix, then the determinant of $\lambda\mathbf{I} - \mathbf{A}$ is simply the product of the elements along the diagonal.

A nonzero solution of the single-exponential form in Eq. (5.8) with $\lambda = \lambda_i$ and $\mathbf{v} = \mathbf{v}_i$ is referred to as the $i$th mode of the undriven system in Eq. (5.7). The associated $\lambda_i$—which we have already termed a characteristic frequency or natural frequency—is thus often called the $i$th mode frequency or modal frequency, and $\mathbf{v}_i$ is termed the $i$th mode shape or modal shape because its various entries indicate the proportion in which the different state variables are excited in the mode.

Note that if the solution is precisely the $i$th mode,

$$\mathbf{q}(t) = \mathbf{v}_i e^{\lambda_i t} , \tag{5.27}$$

then the corresponding initial condition is $\mathbf{q}(0) = \mathbf{v}_i$. It follows from the state evolution property—applied to this special case of zero input—that for the initial condition $\mathbf{q}(0) = \mathbf{v}_i$, only the $i$th mode will be excited. Thus, to excite only a single mode, the initial condition must be the associated eigenvector, or equivalently, must lie in the one-dimensional eigenspace defined by the eigenvector.

It can also be shown that eigenvectors associated with distinct eigenvalues are linearly independent, that is, none of them can be written as a weighted linear combination of the remaining ones. For simplicity, we shall restrict ourselves throughout to the case in which all $L$ eigenvalues of $\mathbf{A}$ are distinct. This will guarantee that $\mathbf{v}_1, \mathbf{v}_2, \ldots, \mathbf{v}_L$ form an independent set, and hence a basis for the $L$-dimensional state space. (In some cases in which $\mathbf{A}$ has repeated eigenvalues, it is possible to find a full set of $L$ independent eigenvectors, but this is not generally true.) We shall repeatedly use the fact that any vector in an $L$-dimensional space, such as our state vector $\mathbf{q}(t)$ at any time $t$, can be written as a unique linear combination of any $L$ independent vectors in that space, such as our $L$ eigenvectors.

**Modal Representation of the CT ZIR**    Because the undriven system in Eq. (5.7) is linear, a weighted linear combination of modal solutions of the form in Eq. (5.27), one for each eigenvalue $\lambda_i$, will also satisfy the equation. Consequently a more general solution for the zero-input response is

$$\mathbf{q}(t) = \sum_{i=1}^{L} \alpha_i \mathbf{v}_i e^{\lambda_i t} . \tag{5.28}$$

To verify that this expression—for arbitrary weights $\alpha_i$—serves as a solution of the undriven system in Eq. (5.7), simply substitute the expression into the undriven equation, and invoke the eigenvalue/eigenvector relation $\lambda_i \mathbf{v}_i = \mathbf{A}\mathbf{v}_i$:

$$\frac{d}{dt}\left(\sum_{i=1}^{L} \alpha_i \mathbf{v}_i e^{\lambda_i t}\right) = \sum_{i=1}^{L} \alpha_i \mathbf{v}_i \lambda_i e^{\lambda_i t} = \mathbf{A}\left(\sum_{i=1}^{L} \alpha_i \mathbf{v}_i e^{\lambda_i t}\right) . \tag{5.29}$$

The initial condition corresponding to the expression for $\mathbf{q}(t)$ in Eq. (5.28) is

$$\mathbf{q}(0) = \sum_{i=1}^{L} \alpha_i \mathbf{v}_i . \tag{5.30}$$

Since the $L$ eigenvectors $\mathbf{v}_i$ are independent under our assumption of distinct eigenvalues, the right side of Eq. (5.30) can be made equal to any desired $\mathbf{q}(0)$ by proper choice of the coefficients $\alpha_i$, and these coefficients are unique. Hence specifying the initial condition of the undriven system in Eq. (5.7) specifies the $\alpha_i$ via Eq. (5.30). With these coefficients thereby determined, the expression in Eq. (5.28) becomes an explicit general solution of the ZIR for the undriven system in Eq. (5.7), under our assumption of distinct eigenvalues. The expression on the right side of Eq. (5.28) is referred to as the modal representation or decomposition of the ZIR.

---

**Example 5.2     ZIR of the Linearized Inverted Pendulum Model**

In Example 5.1 we computed the eigenvalues and eigenvectors associated with the $\mathbf{A}$ matrix that governed small perturbations of a pendulum from its inverted equilibrium. The two modes of the undriven LTI model were of the form $\mathbf{v}_1 e^{2t}$ and $\mathbf{v}_2 e^{-4t}$, hence respectively one that grows exponentially in magnitude and another that decays exponentially in magnitude. The particular combination needed to construct the ZIR depends on the initial condition $\mathbf{q}(0)$.

Suppose the initial position is given by $q_1(0) = 1.1$ (in the applicable units), and the initial velocity by $q_2(0) = -4$. To determine how much the two modes are excited by this initial condition, we solve Eq. (5.30) for $\alpha_1$ and $\alpha_2$, after making the appropriate numerical substitutions:

$$\mathbf{q}(0) = \begin{bmatrix} 1.1 \\ -4 \end{bmatrix} = \alpha_1 \mathbf{v}_1 + \alpha_2 \mathbf{v}_2 = \alpha_1 \begin{bmatrix} 1 \\ 2 \end{bmatrix} + \alpha_2 \begin{bmatrix} 1 \\ -4 \end{bmatrix} . \tag{5.31}$$

Solving these two simultaneous equations for the two unknowns yields $\alpha_1 = \frac{1}{15}$ and $\alpha_2 = \frac{31}{30}$. The ZIR initiated by this initial condition $\mathbf{q}(0)$ is therefore

$$\begin{aligned} \mathbf{q}(t) &= \alpha_1 \mathbf{v}_1 e^{\lambda_1 t} + \alpha_2 \mathbf{v}_2 e^{\lambda_2 t} \\ &= \frac{1}{15} \begin{bmatrix} 1 \\ 2 \end{bmatrix} e^{2t} + \frac{31}{30} \begin{bmatrix} 1 \\ -4 \end{bmatrix} e^{-4t} . \end{aligned} \tag{5.32}$$

In the LTI model, this yields a solution in which position and velocity both eventually grow exponentially in magnitude. However, when the position or velocity becomes sufficiently large, and thereby sufficiently far from its equilibrium value, the behavior of the linearized LTI model is no longer a good approximation of the small-signal behavior of the nonlinear inverted pendulum model.

It is illuminating to interpret the modal solution graphically. Figure 5.1 shows the state space—also called the phase plane in the case of a second-order system—with coordinates $q_1$ and $q_2$. The eigenvectors $\mathbf{v}_1$ and $\mathbf{v}_2$ are indicated as arrows from

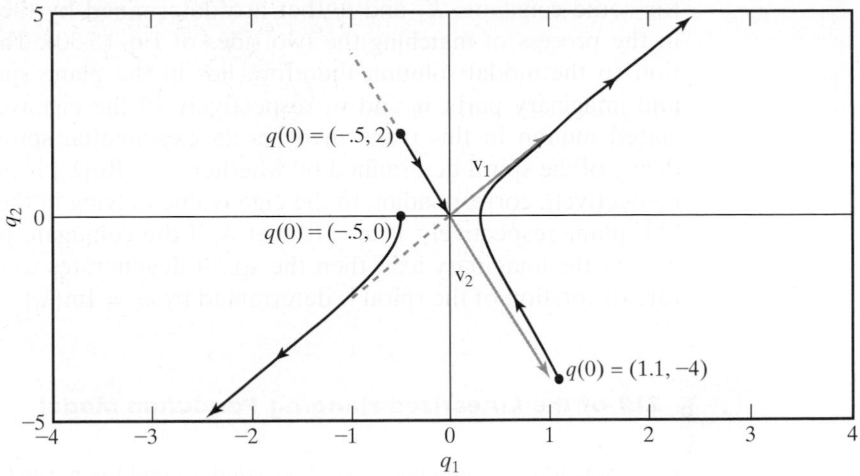

**Figure 5.1** Modal decomposition of the ZIR in the phase plane, for three different initial conditions of the linearized inverted pendulum model in Example 5.2.

the origin to the appropriate points. The solution described above corresponds to the rightmost trajectory in the figure. The initial condition has a large component along $\mathbf{v}_1$ and a small component along $\mathbf{v}_1$; these components can be constructed graphically using the parallelogram-rule decomposition. However, the component along $\mathbf{v}_2$ decays with time as $e^{-4t}$, while the component along $\mathbf{v}_1$ increases with time as $e^{2t}$. By around $t = 1$, corresponding to four time constants of the associated exponential, the component along $\mathbf{v}_2$ is negligible, so the motion is essentially an exponential outward motion along $\mathbf{v}_1$.

To excite only the stable mode of the model requires an initial condition of the form $\mathbf{q}(0) = \alpha_2 \mathbf{v}_2$ for some nonzero weighting factor $\alpha_2$, so that the subsequent motion of the system is $\alpha_2 \mathbf{v}_2 e^{-4t}$. Figure 5.1 shows a trajectory corresponding to one such initial condition, with $\alpha_2 = -0.5$, hence an initial angular position of $q_1(0) = -0.5$ and an initial angular velocity of $q_2(0) = 2$. The motion of the system is an exponential decay of both the angular position and the angular velocity to 0, with a decay factor of the form $e^{-4t}$. The pendulum thus converges exponentially to rest in the inverted position.

The third trajectory shown in Figure 5.1 corresponds to the pendulum starting from rest, so $q_2(0) = 0$, with an initial position of $q_1(0) = -0.5$. Decomposing this initial condition into its components along the eigenvectors, one component decays in magnitude as $e^{-4t}$ and the other grows as $e^{2t}$. Again, after about $t = 1$, the motion is essentially an exponential outward motion, this time along $-\mathbf{v}_1$.

**Complex Mode Pairs in CT**   The eigenvalues in Example 5.2 are both real. The contribution to the modal decomposition from a conjugate pair of complex eigenvalues $\lambda_i$ and $\lambda_i^*$ will be a real term of the form

$$\alpha_i \mathbf{v}_i e^{\lambda_i t} + \alpha_i^* \mathbf{v}_i^* e^{\lambda_i^* t} \,. \tag{5.33}$$

Writing $\lambda_i = \sigma_i + j\omega_i$, so $\sigma_i$ is the real part and $\omega_i$ the imaginary part of the eigenvalue, and similarly writing $\mathbf{v}_i = \mathbf{u}_i + j\mathbf{w}_i$, some algebra shows that

$$\alpha_i \mathbf{v}_i e^{\lambda_i t} + \alpha_i^* \mathbf{v}_i^* e^{\lambda_i^* t} = K_i e^{\sigma_i t} [\mathbf{u}_i \cos(\omega_i t + \theta_i) - \mathbf{w}_i \sin(\omega_i t + \theta_i)] \tag{5.34}$$

for some constants $K_i$ and $\theta_i$ that are determined by the initial conditions in the process of matching the two sides of Eq. (5.30). The above contribution to the modal solution therefore lies in the plane spanned by the real and imaginary parts, $\mathbf{u}_i$ and $\mathbf{w}_i$ respectively, of the eigenvector $\mathbf{v}_i$. The associated motion in this plane involves an exponential spiral, with growth or decay of the spiral determined by whether $\sigma_i = \mathrm{Re}\{\lambda_i\}$ is positive or negative respectively, corresponding to the eigenvalue $\lambda_i$ lying in the open right or left half-plane respectively. If $\sigma_i = 0$, that is, if the conjugate pair of eigenvalues lies on the imaginary axis, then the spiral degenerates to a closed loop. The rate of rotation of the spiral is determined by $\omega_i = \mathrm{Im}\{\lambda_i\}$.

**Example 5.3**    **ZIR of the Linearized Hanging Pendulum Model**

Returning to the pendulum model linearized around the normal hanging position, as presented in Example 5.1, the eigenvalues were found to be

$$\lambda_1 = -1 + j\sqrt{7} = \lambda_2^*$$

and the corresponding eigenvectors were given by

$$\mathbf{v}_1 = \begin{bmatrix} 1 \\ -1 \end{bmatrix} + j \begin{bmatrix} 0 \\ \sqrt{7} \end{bmatrix} = \mathbf{u} + j\mathbf{w} = \mathbf{v}_2^* .$$

Directly applying the result in Eq. (5.34), the ZIR of the system in this case is

$$\begin{bmatrix} q_1(t) \\ q_2(t) \end{bmatrix} = Ke^{-t}\left( \begin{bmatrix} 1 \\ -1 \end{bmatrix} \cos(\sqrt{7}t + \theta) - \begin{bmatrix} 0 \\ \sqrt{7} \end{bmatrix} \sin(\sqrt{7}t + \theta) \right) . \qquad (5.35)$$

This corresponds to a contracting clockwise spiral in the phase plane, as seen in Figure 5.2. Physically, the trajectory corresponds to the pendulum undergoing damped oscillations whose amplitude decays exponentially.

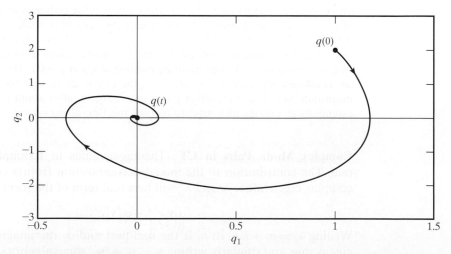

**Figure 5.2**  Phase-plane plot of ZIR for pendulum around normal hanging position in Example 5.3.

The constants $K$ and $\theta$ are determined by the initial condition, on solving for them in the following pair of equations:

$$\begin{bmatrix} q_1(0) \\ q_2(0) \end{bmatrix} = K\left(\begin{bmatrix} 1 \\ -1 \end{bmatrix}\cos(\theta) - \begin{bmatrix} 0 \\ \sqrt{7} \end{bmatrix}\sin(\theta)\right). \tag{5.36}$$

### 5.2.2 Undriven DT Systems

A similar development can be carried out in the DT case for the ZIR of the system in Eq. (5.3), namely the solution of the undriven system

$$\mathbf{q}[n+1] = \mathbf{A}\mathbf{q}[n] . \tag{5.37}$$

One difference from the CT case is that the ZIR trajectories of the DT system in state space comprise a sequence of discrete points rather than a continuous trajectory. Another difference is that Eq. (5.37) is easily stepped forward from time 0 to obtain the explicit solution

$$\mathbf{q}[n] = \mathbf{A}^n\mathbf{q}[0] . \tag{5.38}$$

However, this expression does not convey much insight into the qualitative behavior of the state trajectory over time unless we understand how the entries of $\mathbf{A}$ relate to the entries of $\mathbf{A}^n$.

---

**Example 5.4**    **The Entries of $\mathbf{A}^n$ as $n$ Increases**

The relation between the entries of a matrix $\mathbf{A}$ and the entries of its $n$th power $\mathbf{A}^n$ is not direct. The eigenvalues play a central role in this. Consider the following examples:

$$\mathbf{A}_1 = \begin{bmatrix} 0.6 & 0.6 \\ 0.6 & 0.6 \end{bmatrix}, \qquad \mathbf{A}_2 = \begin{bmatrix} 101 & 100 \\ -101 & -100 \end{bmatrix},$$

$$\mathbf{A}_3 = \begin{bmatrix} 100.5 & 100 \\ -100.5 & -100 \end{bmatrix}, \qquad \mathbf{A}_4 = \begin{bmatrix} 0.6 & 100 \\ 0 & 0.5 \end{bmatrix}. \tag{5.39}$$

All four matrices have enough special structure that one can write simple expressions for their $n$th powers. The matrix $\mathbf{A}_1$ has all its entries less than 1 in magnitude, yet its $n$th power for $n \geq 1$ is

$$\mathbf{A}_1^n = (1.2)^n\begin{bmatrix} 0.5 & 0.5 \\ 0.5 & 0.5 \end{bmatrix}, \tag{5.40}$$

whose entries grow exponentially with $n$. This behavior is due to the eigenvalues of $\mathbf{A}_1$ being 1.2 and 0. On the other hand, even though $\mathbf{A}_2$, $\mathbf{A}_3$, and $\mathbf{A}_4$ have large entries in them, for $n \geq 1$ their respective $n$th powers are

$$\mathbf{A}_2^n = \mathbf{A}_2$$

$$\mathbf{A}_3^n = (0.5)^n\begin{bmatrix} 201 & 200 \\ -201 & -200 \end{bmatrix}$$

$$\mathbf{A}_4^n = \begin{bmatrix} 0.6^n & 1000(0.6^n - 0.5^n) \\ 0 & 0.5^n \end{bmatrix} . \tag{5.41}$$

Thus $\mathbf{A}_2^n$, whose eigenvalues are at 0 and 1, remains nonzero but bounded; $\mathbf{A}_3^n$, whose eigenvalues are at 0.5 and 0, has entries going to 0 as $n \to \infty$; and $\mathbf{A}_4^n$, whose eigenvalues are at 0.6 and 0.5, again has all entries going to 0 as $n \to \infty$.

The conclusion from the above observations is that the individual entries of a matrix do not directly reveal much about how powers of the matrix behave. The eigenvalues of a matrix are what determine whether—and how—the entries of the matrix power grow, decay, or remain bounded. When all the eigenvalues have magnitude less than 1, the entries all decay to 0; if any eigenvalues have magnitude greater than 1, the entries grow exponentially; and if some eigenvalues have magnitude 1, with the rest of smaller magnitude, then the entries of the matrix power stay bounded but do not all decay to 0.

**Modal Decomposition of the DT ZIR**   Equation (5.38) shows that the behavior of $\mathbf{A}^n$ determines the ZIR of a DT LTI system. To further elucidate this ZIR, again assume an exponential solution, which in the DT case takes the form

$$\mathbf{q}[n] = \mathbf{v}\lambda^n , \quad \mathbf{v} \neq \mathbf{0} . \tag{5.42}$$

Substituting this into the undriven system of Eq. (5.37) produces exactly the same eigenvalue/eigenvector condition as in the CT case, namely

$$\lambda\mathbf{v} = \mathbf{A}\mathbf{v} , \quad \text{or equivalently,}$$

$$(\lambda\mathbf{I} - \mathbf{A})\mathbf{v} = \mathbf{0} , \quad \mathbf{v} \neq \mathbf{0} . \tag{5.43}$$

Assuming that $\mathbf{A}$ has $L$ distinct eigenvalues $\lambda_1, \cdots, \lambda_L$ guarantees that the associated eigenvectors $\mathbf{v}_1, \cdots, \mathbf{v}_L$ are independent. The modal decomposition of the general ZIR solution in the DT case then takes the form

$$\mathbf{q}[n] = \sum_{i=1}^{L} \alpha_i \mathbf{v}_i \lambda_i^n , \tag{5.44}$$

where the coefficients $\alpha_i$ are determined by the initial condition, through the equation

$$\mathbf{q}[0] = \sum_{i=1}^{L} \alpha_i \mathbf{v}_i . \tag{5.45}$$

**Complex Mode Pairs in DT**   If the real matrix $\mathbf{A}$ has a complex eigenvalue

$$\lambda_i = \rho_i e^{j\Omega_i} , \quad \rho_i > 0 , \quad 0 < \Omega_i < \pi , \tag{5.46}$$

written here in a form that is better suited to the DT case, then the complex conjugate of this eigenvalue, namely $\lambda_i^* = \rho_i e^{-j\Omega_i}$, also appears among the eigenvalues of $\mathbf{A}$. Furthermore, if $\mathbf{v}_i = \mathbf{u}_i + j\mathbf{w}_i$ is the eigenvector associated with $\lambda_i$, then its complex conjugate is the eigenvector associated with the eigenvalue $\lambda_i^*$. The contribution of this complex conjugate pair of modes to the ZIR is a real term of the form

$$\alpha_i \mathbf{v}_i \lambda_i^n + \alpha_i^* \mathbf{v}_i^* (\lambda_i^*)^n . \tag{5.47}$$

With some algebra, the real expression in Eq. (5.47) can be reduced to the form

$$\alpha_i \mathbf{v}_i \lambda_i^n + \alpha_i^* \mathbf{v}_i^* (\lambda_i^*)^n = K_i \rho_i^n [\mathbf{u}_i \cos(\Omega_i n + \theta_i) - \mathbf{w}_i \sin(\Omega_i n + \theta_i)] . \qquad (5.48)$$

The constants $K_i$ and $\theta_i$ are determined by the initial conditions in the process of matching the two sides of Eq. (5.45). This pair of modes thus has an oscillatory contribution to the ZIR.

The case of an eigenvalue $\lambda_i$ that has angle $\Omega_i = \pm\pi$ merits special attention. This corresponds to $\lambda_i$ being the negative real number $-\rho_i$, rather than one of a complex pair. Its contribution to the ZIR is nevertheless oscillatory, due to the alternating sign of $(-\rho_i)^n$. In fact, this sign alternation represents the fastest oscillation a DT system can have.

### 5.2.3 Asymptotic Stability of LTI Systems

An LTI state-space system is termed asymptotically stable or internally stable if its ZIR decays to zero for all initial conditions. The system is sometimes called marginally stable if it is not asymptotically stable but has a ZIR that remains bounded for all time with all initial conditions. The system is called unstable if it is not asymptotically stable; a marginally stable system is thus also considered unstable. The stability of an LTI system is directly related to the behavior of its modes, and more specifically to the values of the characteristic or natural frequencies $\lambda_i$, which are the roots of the characteristic polynomial.

**CT Systems**    For the CT case, we can write $\lambda_i = \sigma_i + j\omega_i$, so $\sigma_i$ and $\omega_i$ denote the real and imaginary parts of the natural frequency. The complex signal

$$e^{\lambda_i t} = e^{\sigma_i t} e^{j\omega_i t} \qquad (5.49)$$

decays exponentially in magnitude to 0 if and only if $\sigma_i < 0$, that is, if and only if $\text{Re}\{\lambda_i\} < 0$. The modal decomposition of the ZIR in Eq. (5.28) now shows that the condition $\text{Re}\{\lambda_i\} < 0$ for all $1 \leq i \leq L$ is necessary and sufficient for asymptotic stability of a CT LTI system. Thus all natural frequencies have to be in the open left half of the complex plane, that is, strictly in the left half-plane.

An associated observation follows from considering the relative magnitudes of the various terms in the modal decomposition of the ZIR. Note that for sufficiently large values of time $t$, the dominant terms in the modal decomposition will be those involving the eigenvalues of maximum real part $\sigma_{max}$, known as the dominant eigenvalues, as represented in Figure 5.3; the other modal terms will be much smaller relative to these. The quantity $\sigma_{max}$ is referred to as the spectral abscissa of the matrix $\mathbf{A}$. The modal decomposition in Eq. (5.28) now shows that the state trajectory $\mathbf{q}(t)$ for sufficiently large $t$ will be essentially confined to the space spanned by the eigenvectors associated with the dominant eigenvalues. The system is marginally stable if $\sigma_{max} = 0$.

If the LTI system is the result of linearizing around an equilibrium point of an underlying nonlinear time-invariant system, then asymptotic stability of the linearization indicates that the nonlinear system is locally asymptotically

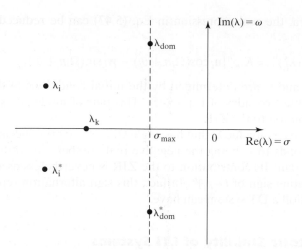

**Figure 5.3**  Dominant modes for an asymptotically stable system.

stable in a sense that can be made precise but which we do not elaborate on here. On the other hand, if the linearization has at least one mode that grows exponentially—if $\sigma_{max} > 0$—then the nonlinear system is locally unstable. In the marginal case, where the linearization is not asymptotically stable but has no growing modes—when $\sigma_{max} = 0$—the linearization does not contain enough information to make a conclusion about the local behavior of the non-linear system; higher-order terms, beyond those retained in the linearization, are needed to establish what the local behavior is.

The linearized models of the pendulum examined in Examples 5.1 and 5.2 illustrate some of these results. The linearized model around the inverted equilibrium had characteristic frequencies at 2 and −4, and was therefore not asymptotically stable. The graphical analysis in Example 5.2 and Figure 5.1 also showed that the dominant term in the ZIR was associated with the eigenvalue at 2, and that the state trajectory for sufficiently large $t$ was essentially along the eigenvector associated with this dominant eigenvalue. In contrast, the linearized model around the normal hanging equilibrium had characteristic frequencies of $-1 \pm \sqrt{7}$, and therefore was asymptotically stable. This pair of complex eigenvalues together constituted the dominant set in this case.

**DT Systems**  For the DT case, we write $\lambda_i = \rho_i e^{j\Omega_i}$ with $\rho_i \geq 0$, so $\rho_i$ is the magnitude $|\lambda_i|$ and $\Omega_i$ is the angle $\angle \lambda_i$ of the natural frequency. Then the complex signal

$$\lambda_i^n = \rho_i^n e^{jn\Omega_i} \tag{5.50}$$

decays exponentially in magnitude to 0 if and only if $\rho_i = |\lambda_i| < 1$. For the special case of an eigenvalue at 0, the associated mode drops to 0 in a single step. The modal decomposition of the ZIR in Eq. (5.44) now shows that the condition $|\lambda_i| < 1$ for all $1 \leq i \leq L$ is necessary and sufficient for asymptotic stability of a DT LTI system. Thus all natural frequencies have to be in the open unit circle of the complex plane, that is, strictly within the unit circle.

The notion of dominant modes carries over to the DT setting as well. In the DT case, the eigenvalues of maximum magnitude $\rho_{max}$ are the dominant ones. The quantity $\rho_{max}$ is called the spectral radius of the matrix $\mathbf{A}$. For $n$ sufficiently large, the trajectory $\mathbf{q}[n]$ will be essentially confined to the space spanned by the eigenvectors associated with the dominant eigenvalues. The system is marginally stable if $\rho_{max} = 1$.

As in the CT case, if the DT LTI system is the result of linearizing around an equilibrium point of an underlying nonlinear time-invariant system, then asymptotic stability of the linearization indicates that the nonlinear system is locally asymptotically stable. If the linearization has at least one mode that grows exponentially—if $\rho_{max} > 1$—then the nonlinear system is locally unstable. In the marginal case, where the linearization is not asymptotically stable but has no growing modes—when $\rho_{max} = 1$—the linearization does not contain enough information to make a conclusion about the local behavior of the nonlinear system; higher-order terms, beyond those retained in the linearization, are needed to establish what the local behavior is.

The modal decompositions in Eqs. (5.28) and (5.44) served to validate the claims regarding conditions for asymptotic stability in CT and DT systems respectively, but these modal decompositions were obtained under the assumption of distinct eigenvalues. Nevertheless, it can be shown that the conditions for asymptotic stability in the general case are identical to those above.

---

**Example 5.5**    **Asymptotic Stability of Linearized Viral Propagation Model**

For the SIR viral propagation model introduced in Example 4.3 of Chapter 4, we determined in Example 4.8 that the model had two equilibrium points, the infective-free equilibrium (IFE) and the endemic equilibrium (EE). Subsequently, in Example 4.10, we derived the linearized state-space description at each of these equilibrium points.

Now consider the parameter values and input associated with the simulations that yielded the plots in Figure 4.3:

$$\beta = 0.01, \quad \gamma = 0.2, \quad \rho = 0.1, \quad P = 10,000, \quad x[n] = \bar{x} = 0.5 \ .$$

Here $\beta$ denotes the common birth and death rates, $\gamma$ specifies the infection rate, $\rho$ is the recovery rate, $P$ is the total population size, and $x[n]$ is the immunization rate.

With these parameters, $R_0 = \gamma/(\beta + \rho) = 1.818 < 1/(1 - \bar{x}) = 2$, and hence there is no EE, as noted in Example 4.8. The small-signal behavior around the IFE is governed by the state evolution matrix

$$\mathbf{A}_{IFE} = \begin{bmatrix} 1 - \beta & -\gamma(1 - \bar{x}) \\ 0 & 1 + \gamma[(1 - \bar{x}) - (1/R_0)] \end{bmatrix} = \begin{bmatrix} 0.99 & -0.1 \\ 0 & 0.99 \end{bmatrix} . \tag{5.51}$$

The two eigenvalues of this matrix are at 0.99, hence of magnitude $< 1$, so the linearized model is asymptotically stable. We therefore expect that trajectories initiated sufficiently close to this equilibrium in the nonlinear system will actually converge to the IFE, and this is seen in Figure 4.3. Since $0.99^{300} = 0.049$, we expect the excursions to be under 5% of their initial values by the time $n = 300$, and this is consistent with that figure. The specific features of the transient behavior prior to convergence—in particular, the single excursion in a direction away from the equilibrium values before settling exponentially (or geometrically) back to these values—are consistent with the repeated eigenvalues at 0.99. We omit further analysis of this repeated eigenvalue case (but see the discussion of $\mathbf{A}_2$ in Example 5.13).

For a case that settles to the EE, we use the same parameters as above, but with the lower immunization rate $x[n] = \bar{x} = 0.2$, so $R_0 = 1.818 > 1/(1 - \bar{x}) = 1.25$, thus allowing a positive equilibrium value of infectives, $\bar{\imath} > 0$. A simulation of the system for this case, and for several choices of $i[0]$, is shown in Figure 5.4. The state variables settle in steady state to the equilibrium values given for the EE in Eq. (4.82):

$$\bar{s} = P/R_0 = 5500$$

and

$$\bar{\imath} = (\beta P/\gamma)[R_0(1 - \bar{x}) - 1] \approx 227 \ .$$

The small-signal behavior around the EE is governed by the state evolution matrix

$$\mathbf{A}_{EE} = \begin{bmatrix} 1 - \beta R_0(1 - \bar{x}) & -\gamma/R_0 \\ \beta[R_0(1 - \bar{x}) - 1] & 1 \end{bmatrix} = \begin{bmatrix} 0.9855 & -0.11 \\ 0.0045 & 1 \end{bmatrix} . \tag{5.52}$$

The eigenvalues of this matrix are a complex pair at $0.993e^{\pm j0.021}$. We thus expect to see an oscillatory settling to the equilibrium value, with a period of $2\pi/0.021 \approx 296$ time steps. Since $0.993^{296} = 0.125$, the amplitude at the end of one period of oscillation is expected to decay to around 12% of its value at the beginning of the period. These expectations are borne out by the plots in Figure 5.4.

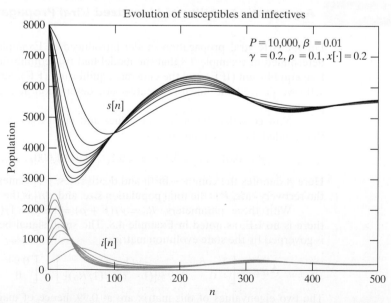

**Figure 5.4**　Response of SIR model for the same choice of parameter values as in Example 4.3, but with the immunization rate dropped to $x[n] = 0.2$. The attained steady state now corresponds to an endemic equilibrium.

# 5.3 GENERAL RESPONSE IN MODAL COORDINATES

In this section we develop the solution of the full driven system in both the CT and DT cases, again obtaining a modal representation.

## 5.3.1 Driven CT Systems

Consider the CT LTI state-space description in Eqs. (5.1) and (5.2), with the input $x(t)$ now being some arbitrary function of time rather than identically zero. Motivated by the fact that the ZIR was simply described in terms of modes, that is, in terms of the behavior of components of the state vector along each of the $L$ independent eigenvectors of the system, we now pursue the same approach for the driven case.

The state vector $\mathbf{q}(t)$ at any time $t$ is a vector in the $L$ dimensional state space, and can therefore be uniquely written as a weighted linear combination of the $L$ independent eigenvectors $\{\mathbf{v}_i\}$, with weights denoted by $\{r_i(t)\}$. Accordingly,

$$\mathbf{q}(t) = \sum_{i=1}^{L} \mathbf{v}_i r_i(t) = \mathbf{V}\mathbf{r}(t) , \qquad (5.53)$$

where the second equality expresses the summation as a matrix-vector multiplication. The $i$th entry of the vector $\mathbf{r}(t)$ is $r_i(t)$, and the $i$th column of the $L \times L$ matrix $\mathbf{V}$ is the $i$th eigenvector, $\mathbf{v}_i$:

$$\mathbf{V} = \begin{bmatrix} \mathbf{v}_1 & \mathbf{v}_2 & \cdots & \mathbf{v}_L \end{bmatrix} . \qquad (5.54)$$

The matrix $\mathbf{V}$ is termed the modal matrix. The quantities $r_i(t)$ are the modal coordinates at time $t$.

In the special case of the ZIR, the modal expansion in Eq. (5.28) shows that the functions $r_i(t)$ are given by

$$r_i(t) = \alpha_i e^{\lambda_i t} , \qquad (5.55)$$

where the values $\{\alpha_i\}$ are determined by the initial condition $\mathbf{q}(0)$, using Eq. (5.30). For the driven case considered here, one should anticipate a more complicated expression for $r_i(t)$ that involves the input trajectory $x(\cdot)$.

We proceed by substituting the expression for $\mathbf{q}(t)$ from Eq. (5.53) into the state evolution description in Eq. (5.1), to determine what constraints the modal coordinate functions $\{r_i(t)\}$ have to satisfy in order for the state evolution equation to be satisfied. An important preliminary step, however, is to also express the input vector $\mathbf{b}x(t)$ that appears in Eq. (5.1) in terms of its components along the eigenvectors. Since $\mathbf{b}$ is an $L$-vector in the same $L$-dimensional space as $\mathbf{q}(t)$, we can express it as

$$\mathbf{b} = \sum_{i=1}^{L} \mathbf{v}_i \beta_i = \mathbf{V}\boldsymbol{\beta} \qquad (5.56)$$

for some uniquely determined weights $\beta_i$, where $\boldsymbol{\beta}$ is a vector whose $i$th component is $\beta_i$. With this notation, substitution of Eq. (5.53) into the state evolution equation, just as was done in the case of the undriven system to obtain Eq. (5.29), yields

$$\sum_{i=1}^{L} \mathbf{v}_i \dot{r}_i(t) = \mathbf{A}\left(\sum_{i=1}^{L} \mathbf{v}_i r_i(t)\right) + \left(\sum_{i=1}^{L} \mathbf{v}_i \beta_i\right) x(t)$$

$$= \sum_{i=1}^{L} \mathbf{v}_i \Big(\lambda_i r_i(t) + \beta_i x(t)\Big) , \tag{5.57}$$

where we have invoked the fact that $\mathbf{A}\mathbf{v}_i = \lambda_i \mathbf{v}_i$ to obtain the second equality.

The left and right sides of Eq. (5.57) express a vector in two different ways as a weighted linear combination of the eigenvectors $\{\mathbf{v}_i\}$. However, since the eigenvectors form an independent set, the weights required to express the vector are unique. It follows that

$$\dot{r}_i(t) = \lambda_i r_i(t) + \beta_i x(t) , \quad i = 1, 2, \ldots, L . \tag{5.58}$$

Each of these $L$ equations is a first-order LTI differential equation in a single variable. It can be solved explicitly, and independently of the others, to yield the corresponding modal coordinate as a function of time. As shown below, to solve the $i$th equation from time 0 onward over any interval of time requires only the initial value $r_i(0)$ and the input $x(\cdot)$ over that interval. Thus, though we began with a system of $L$ coupled first-order equations, we can now use Eqs. (5.53) and (5.58) to write the solution in terms of $L$ decoupled scalar first-order equations, each of which can be solved directly.

The initial conditions $\{r_i(0)\}$ that are needed to solve the scalar LTI differential equations in Eq. (5.58) are obtained from the initial state $\mathbf{q}(0)$, since Eq. (5.53) shows that

$$\mathbf{q}(0) = \sum_{i=1}^{L} \mathbf{v}_i r_i(0) . \tag{5.59}$$

Thus the weights $r_i(0)$ are those required to express the initial state of the system in terms of the eigenvectors. Comparing with Eq. (5.30) shows that these are precisely the quantities $\alpha_i$ introduced in developing the modal representation of the ZIR, so

$$r_i(0) = \alpha_i , \quad i = 1, 2, \ldots, L . \tag{5.60}$$

The explicit solution of the $i$th scalar first-order LTI differential equation in Eq. (5.58) for $t \geq 0$ can now be found by any of the standard approaches for solving a forced LTI differential equation in one variable. For instance, the total solution is the sum of a solution of the homogeneous or undriven system, and any particular solution of the forced or driven system; the amplitude of the homogeneous component is chosen such that the total solution matches the specified initial condition. The solution can also be obtained by Laplace transform methods, as shown later, in Section 5.4. For now, the solution of Eq. (5.58) is most usefully written as

$$r_i(t) = \underbrace{e^{\lambda_i t} r_i(0)}_{\text{ZIR}} + \underbrace{\int_0^t e^{\lambda_i(t-\tau)} \beta_i x(\tau) \, d\tau}_{\text{ZSR}}, \quad t \geq 0, \quad 1 \leq i \leq L. \tag{5.61}$$

The braces above show the separate contributions to the total solution made by (i) the response due to the initial state alone, namely the zero-input response or ZIR; and (ii) the response due to the system input alone, namely the zero-state response or ZSR. The ZIR is a solution of the undriven system, and the ZSR is a particular solution, namely the one corresponding to an initial condition of 0. The correctness of the solution in Eq. (5.61) can be established by directly verifying that it satisfies Eq. (5.58) and has the right initial condition $r_i(0)$ at time $t = 0$.

One route to the full solution of the LTI state evolution description in Eq. (5.1) is now apparent: determine the initial values $r_i(0)$ of the modal coordinates from the given $\mathbf{q}(0)$ using Eq. (5.59), then use these values and the given $x(t)$ to determine the time functions $r_i(t)$ using Eq. (5.61), and finally use these time functions to determine the state trajectory $\mathbf{q}(t)$ via the general modal decomposition in Eq. (5.53).

To solve for the corresponding output $y(t)$, note from Eq. (5.2) that

$$
\begin{aligned}
y(t) &= \mathbf{c}^T \mathbf{q}(t) + dx(t) = \mathbf{c}^T \left( \sum_{i=1}^{L} \mathbf{v}_i r_i(t) \right) + dx(t) \\
&= \sum_{i=1}^{L} (\mathbf{c}^T \mathbf{v}_i) r_i(t) + dx(t) \\
&= \sum_{i=1}^{L} \xi_i r_i(t) + dx(t) , \tag{5.62}
\end{aligned}
$$

where

$$\xi_i = \mathbf{c}^T \mathbf{v}_i . \tag{5.63}$$

The expression in Eq. (5.61) also serves to show that for an asymptotically stable system, where all $\lambda_i$ have negative real parts, a bounded input always results in a bounded state trajectory for arbitrary initial conditions. The magnitude of the first term on the right of Eq. (5.61) is bounded by $|r_i(0)|$ for an asymptotically stable system. If $|x(t)| \leq M < \infty$ for all $t$, then the magnitude of the second term on the right of Eq. (5.61) can also be bounded as follows:

$$\left| \int_0^t e^{\lambda_i(t-\tau)} \beta_i x(\tau) \, d\tau \right| \leq |\beta_i| M \int_0^t |e^{\lambda_i(t-\tau)}| \, d\tau , \tag{5.64}$$

and the integral on the right in this expression is also bounded for an asymptotically stable system. This establishes the claim that asymptotic stability is sufficient to ensure bounded-input, bounded-state stability, and thus also bounded-input, bounded-output (BIBO) stability.

### 5.3.2 Driven DT Systems

A closely parallel development can be made for the DT LTI state-space system in Eqs. (5.3) and (5.4), with CT exponentials replaced by DT exponentials. We write the state vector $\mathbf{q}[n]$ in terms of modal coordinates as follows:

$$\mathbf{q}[n] = \sum_{i=1}^{L} \mathbf{v}_i r_i[n] = \mathbf{V}\mathbf{r}[n] , \tag{5.65}$$

where $\mathbf{V}$ is again the modal matrix, whose columns are the eigenvectors of the matrix $\mathbf{A}$. The given initial condition $\mathbf{q}[0]$ serves to determine the initial values $r_i[0]$ of the modal variables via

$$\mathbf{q}[0] = \sum_{i=1}^{L} \mathbf{v}_i r_i[0] . \tag{5.66}$$

Substituting the modal representation of Eq. (5.65) into the state evolution equations shows that the modal coordinates $r_i[n]$ are governed by decoupled first-order DT LTI state-space descriptions of the form

$$r_i[n+1] = \lambda_i r_i[n] + \beta_i x[n] , \quad i = 1, 2, \ldots, L \tag{5.67}$$

where the $\beta_i$ are defined exactly as before, via the relation in Eq. (5.56).

The solution of each such equation can be determined by simply stepping forward iteratively. The general expression for the solution is thereby determined to be

$$r_i[n] = \underbrace{\lambda_i^n r_i[0]}_{\text{ZIR}} + \underbrace{\sum_{k=0}^{n-1} \lambda_i^{n-k-1} \beta_i x[k]}_{\text{ZSR}} , \quad n \geq 1, \ 1 \leq i \leq L . \tag{5.68}$$

The corresponding output can be expressed in terms of the modal variables as

$$y[n] = \sum_{i=1}^{L} \xi_i r_i[n] + \mathrm{d}x[n] , \tag{5.69}$$

where the $\xi_i$ are defined exactly as before through Eq. (5.63). Again, asymptotic stability of the system is sufficient to ensure that the system is bounded-input, bounded-state stable, and hence BIBO stable.

The following example brings together CT and DT systems in a manner that is relevant for DT control of CT systems.

### Example 5.6     Sampled-Data Model

Suppose the input $x(t)$ to the CT LTI state-space model in Eqs. (5.1) and (5.2) is produced by a computer that updates the value of the input every $T$ seconds, starting at time $t = 0$, and holds $x(t)$ constant at that value until the next update. Thus

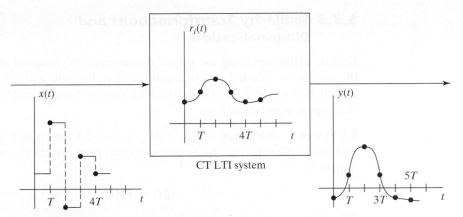

**Figure 5.5** Relation between underlying CT system and the DT sampled-data representation in Example 5.6.

$$x(t) = x[n] \quad \text{for} \quad nT \le t < nT + T,\tag{5.70}$$

where $x[n]$ is the sequence of input values determined by the computer. We show in this example that $\mathbf{q}(nT + T)$ and $y(nT)$ can be written as LTI functions of $\mathbf{q}(nT)$ and $x[n]$, thus providing a DT LTI state-space model for the sampled state vector $\mathbf{q}(nT) = \mathbf{q}[n]$. The situation is shown schematically in Figure 5.5.

The key to establishing this result is to relate $r_i(nT + T) = r_i[n + 1]$ to $r_i(nT) = r_i[n]$ and $x[n]$ for each $i$ from 1 to $L$. For this, rewrite the solution in Eq. (5.61) to apply to a starting time of $nT$ and an ending time of $nT + T$, obtaining

$$r_i(nT + T) = r_i[n + 1] = e^{\lambda_i(nT+T-nT)}r_i(nT) + \int_{nT}^{nT+T} e^{\lambda_i(nT+T-\tau)}\beta_i x(\tau)\, d\tau$$

$$= e^{\lambda_i T}r_i[n] + e^{\lambda_i T}\Big(\int_0^T e^{-\lambda_i \sigma}\, d\sigma\Big)\beta_i x[n].\tag{5.71}$$

The second equality results from a change of variables to $\sigma = \tau - nT$, and moving the quantities $e^{\lambda_i T}$ and $\beta_i x[n]$ outside the integral sign, as they do not depend on $\sigma$. If $\lambda_i = 0$, this simplifies to the DT LTI scalar state evolution equation

$$r_i[n + 1] = r_i[n] + (T\beta_i)x[n],\tag{5.72}$$

and if $\lambda_i \ne 0$, then to the DT LTI scalar state evolution equation

$$r_i[n + 1] = (e^{\lambda_i T})r_i[n] + \Big(\frac{e^{\lambda_i T} - 1}{\lambda_i}\beta_i\Big)x[n].\tag{5.73}$$

Having the sampled values of the modal coordinates $r_i(t)$ at integer multiples of $T$ then yields the sampled values of the state vector $\mathbf{q}(t)$ at integer multiples of $T$, and thereby also the sampled values of the output $y(t)$. With the matrix notation that we develop shortly, these various scalar expressions can be assembled into a simple matrix form that can be recognized as a DT LTI state-space model for the sampled state and output. Example 5.7 below presents the details.

### 5.3.3 Similarity Transformations and Diagonalization

Before further applying the modal decomposition obtained above, we revisit the arguments that led to this decomposition, but now using matrix algebra to describe what was done. This provides a different and valuable perspective on state-space representations.

**CT Systems**   We return to Eq. (5.53), repeated below, which related the original state vector $\mathbf{q}(t)$ to the vector of modal coordinates $\mathbf{r}(t)$ using the modal matrix $\mathbf{V}$:

$$\mathbf{q}(t) = \mathbf{V}\mathbf{r}(t) . \tag{5.74}$$

The entries of $\mathbf{r}(t)$ represent the components of a point (i.e., a state) in the state space when the eigenvectors are used as the basis for the state space. In these modal coordinates, the eigenvector $\mathbf{v}_i$ constitutes the basis vector for the $i$th modal coordinate, as represented earlier in the phase plane diagram of Figure 5.1.

Under our assumption of distinct eigenvalues, the eigenvectors $\mathbf{v}_i$ are independent, which guarantees that $\mathbf{V}$ is invertible, so Eq. (5.74) implies that

$$\mathbf{r}(t) = \mathbf{V}^{-1}\mathbf{q}(t) . \tag{5.75}$$

Equations (5.74) and (5.75) establish that the modal coordinate vector $\mathbf{r}(t)$ is completely equivalent to the original state vector $\mathbf{q}(t)$; each can be recovered from the other. The equivalence, however, does not immediately make clear whether $\mathbf{r}(t)$ itself satisfies an LTI state-space equation. We establish this next, but do so more generally, for the case of a linear transformation to an arbitrary set of new coordinates, not specifically to modal coordinates.

Consider a coordinate transformation in the state space to a new set of coordinates that are the components of a vector $\mathbf{z}(t)$, related to the original state vector $\mathbf{q}(t)$ by

$$\mathbf{q}(t) = \mathbf{M}\mathbf{z}(t) , \tag{5.76}$$

where the constant matrix $\mathbf{M}$ is chosen to be invertible. By considering the case where $\mathbf{z}(t)$ has 1 in its $i$th position and 0 everywhere else, that is, where $\mathbf{z}(t)$ is the $i$th unit vector in the new coordinates, we see that the $i$th column of the matrix $\mathbf{M}$ is the representation of the $i$th unit vector of the new $\mathbf{z}$ coordinates in terms of the old $\mathbf{q}$ coordinates. Substituting the preceding equation in the original state-space description in Eqs. (5.1) and (5.2), and then solving for $\dot{\mathbf{z}}(t)$, we obtain

$$\mathbf{M}\dot{\mathbf{z}}(t) = \mathbf{A}\mathbf{M}\mathbf{z}(t) + \mathbf{b}x(t) , \quad \text{or equivalently,}$$

$$\dot{\mathbf{z}}(t) = (\mathbf{M}^{-1}\mathbf{A}\mathbf{M})\mathbf{z}(t) + (\mathbf{M}^{-1}\mathbf{b})x(t) , \quad \text{and} \tag{5.77}$$

$$y(t) = (\mathbf{c}^T\mathbf{M})\mathbf{z}(t) + \mathrm{d}x(t) . \tag{5.78}$$

Equations (5.77) and (5.78) are still in the form of the state evolution description and the output description of an LTI state-space model, but with state vector $\mathbf{z}(t)$, and with modified coefficient matrices. This model is entirely equivalent to the original one, since Eq. (5.76) permits $\mathbf{q}(t)$ to be obtained from $\mathbf{z}(t)$, and the invertibility of $\mathbf{M}$ permits $\mathbf{z}(t)$ to be obtained from $\mathbf{q}(t)$.

The invertible transformation above is termed a similarity transformation, and the LTI state-space description that results from applying it is said to be similarity equivalent—or similar—to the original description. Since a similarity transformation is simply a transformation to a new coordinate system in the state space, it should not be surprising that the essential dynamical properties of the transformed description in Eqs. (5.77) and (5.78) are unchanged from those of the original one in Eqs. (5.1) and (5.2). For instance, since the ZIR of the transformed description is related to the ZIR of the original through the constant matrix transformation in Eq. (5.76), the two ZIR expressions will display the same modal frequencies. This tells us that the eigenvalues of $\mathbf{M}^{-1}\mathbf{AM}$ are identical to those of $\mathbf{A}$; only the eigenvectors change, with $\mathbf{v}_i$ transforming to $\mathbf{M}^{-1}\mathbf{v}_i$. These statements are easy to verify algebraically, as

$$\lambda\mathbf{v} = \mathbf{A}\mathbf{v} \quad \text{if and only if} \quad \lambda(\mathbf{M}^{-1}\mathbf{v}) = (\mathbf{M}^{-1}\mathbf{AM})(\mathbf{M}^{-1}\mathbf{v}) . \tag{5.79}$$

Note also that the input $x(t)$ and output $y(t)$ are unaffected by this state transformation. For a given input, and assuming an initial state $\mathbf{z}(0)$ in the transformed description that is related to $\mathbf{q}(0)$ via Eq. (5.76), one obtains the same output as would have resulted from the original description in Eqs. (5.1) and (5.2). Thus, the input-output relationships of the system are unaffected by a similarity transformation.

**DT Systems**   Similarity transformations can be defined in exactly the same way for the DT case in Eqs. (5.3) and (5.4), writing

$$\mathbf{q}[n] = \mathbf{Mz}[n] \tag{5.80}$$

for an invertible matrix $\mathbf{M}$, and deducing that

$$\mathbf{z}[n+1] = (\mathbf{M}^{-1}\mathbf{AM})\mathbf{z}[n] + (\mathbf{M}^{-1}\mathbf{b})x[n] , \quad \text{and} \tag{5.81}$$

$$y[n] = (\mathbf{c}^T\mathbf{M})\mathbf{z}[n] + \mathrm{d}x[n] . \tag{5.82}$$

**Similarity Transformation to Modal Coordinates**   The transformation $\mathbf{q}(t) = \mathbf{Vr}(t)$ in Eq. (5.53) can now be recognized as a similarity transformation, using the modal matrix $\mathbf{V}$ as the transformation matrix. What is interesting and useful about this modal transformation is that it is a diagonalizing transformation, in the sense that the state evolution matrix $\mathbf{A}$ transforms to a diagonal matrix $\mathbf{\Lambda}$:

$$\mathbf{V}^{-1}\mathbf{AV} = \text{diagonal } \{\lambda_1, \cdots, \lambda_L\} = \begin{bmatrix} \lambda_1 & 0 & \cdots & 0 \\ 0 & \lambda_2 & \cdots & 0 \\ \vdots & \vdots & \ddots & \vdots \\ 0 & 0 & \cdots & \lambda_L \end{bmatrix} = \mathbf{\Lambda} . \tag{5.83}$$

An easy way to prove this is by verifying that $\mathbf{V}\boldsymbol{\Lambda} = \mathbf{A}\mathbf{V}$:

$$
\begin{bmatrix} \mathbf{v}_1 & \mathbf{v}_2 & \cdots & \mathbf{v}_L \end{bmatrix}
\begin{bmatrix}
\lambda_1 & 0 & \cdots & 0 \\
0 & \lambda_2 & \cdots & 0 \\
\vdots & \vdots & \ddots & \vdots \\
0 & 0 & \cdots & \lambda_L
\end{bmatrix}
= \mathbf{A} \begin{bmatrix} \mathbf{v}_1 & \mathbf{v}_2 & \cdots & \mathbf{v}_L \end{bmatrix} , \quad (5.84)
$$

which is simply the equality in Eq. (5.10), written for $i = 1, \cdots, L$ and stacked up as the columns of the matrices on the right and left. Premultiplying both sides of this equality by $\mathbf{V}^{-1}$ then yields the relation in Eq. (5.83).

Postmultiplying both sides of the preceding equality by $\mathbf{V}^{-1}$ shows

$$
\mathbf{A} = \mathbf{V}\boldsymbol{\Lambda}\mathbf{V}^{-1} , \quad (5.85)
$$

from which it follows that, for integer $n > 0$,

$$
\begin{aligned}
\mathbf{A}^n &= (\mathbf{V}\boldsymbol{\Lambda}\mathbf{V}^{-1})(\mathbf{V}\boldsymbol{\Lambda}\mathbf{V}^{-1})\cdots(\mathbf{V}\boldsymbol{\Lambda}\mathbf{V}^{-1}) \\
&= \mathbf{V}\boldsymbol{\Lambda}^n\mathbf{V}^{-1} , \quad (5.86)
\end{aligned}
$$

where there are $n$ terms being multiplied on the right of the first equality, and where the cancellation of adjacent $\mathbf{V}$ and $\mathbf{V}^{-1}$ in this extended product yields the second equality. Note that because $\boldsymbol{\Lambda}$ is a diagonal matrix, $\boldsymbol{\Lambda}^n$ is also diagonal, with $\lambda_i^n$ in the $i$th diagonal position. The relation in Eq. (5.86) can be used to verify the expressions given in Example 5.4 for the powers of the matrices $\mathbf{A}_1, \mathbf{A}_2, \mathbf{A}_3$, and $\mathbf{A}_4$.

**Diagonalized CT Systems**    The system description that results from using the modal similarity transformation is determined by substituting $\mathbf{V}$ for $\mathbf{M}$ in Eqs. (5.77) and (5.78). The result is

$$
\begin{aligned}
\dot{\mathbf{r}}(t) &= (\mathbf{V}^{-1}\mathbf{A}\mathbf{V})\mathbf{r}(t) + (\mathbf{V}^{-1}\mathbf{b})x(t) \\
&= \boldsymbol{\Lambda}\mathbf{r}(t) + \boldsymbol{\beta}x(t) , \quad (5.87)
\end{aligned}
$$

$$
\begin{aligned}
y(t) &= (\mathbf{c}^T\mathbf{V})\mathbf{r}(t) + \mathrm{d}x(t) \\
&= \boldsymbol{\xi}^T\mathbf{r}(t) + \mathrm{d}x(t) , \quad (5.88)
\end{aligned}
$$

where the column vector $\boldsymbol{\beta}$ and row vector $\boldsymbol{\xi}^T$ are defined via

$$
\boldsymbol{\beta} = \mathbf{V}^{-1}\mathbf{b} = \begin{bmatrix} \beta_1 \\ \beta_2 \\ \vdots \\ \beta_L \end{bmatrix} , \quad \boldsymbol{\xi}^T = \mathbf{c}^T\mathbf{V} = \begin{bmatrix} \xi_1 & \xi_2 & \cdots & \xi_L \end{bmatrix} . \quad (5.89)
$$

Rewriting the first of these definitions as $\mathbf{b} = \mathbf{V}\boldsymbol{\beta} = \sum \mathbf{v}_i\beta_i$ shows that the scalars $\beta_i$ are precisely those defined earlier in Eq. (5.56). The second of these definitions shows that $\xi_i = \mathbf{c}^T\mathbf{v}_i$, which is precisely the definition given earlier in Eq. (5.63). Taking account of these definitions and of the fact that $\boldsymbol{\Lambda}$ in Eq. (5.87) is diagonal shows that the individual rows of Eq. (5.87) yield the decoupled set of first-order LTI differential equations obtained earlier in Eq. (5.58), while the output equation in (5.88) is exactly Eq. (5.62). The earlier decoupled

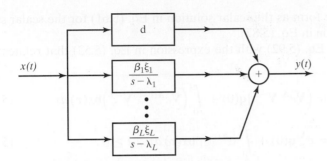

**Figure 5.6**   Decoupled structure of CT LTI system in modal coordinates.

description has thus been recovered using matrix operations on the original description. The decoupling is a consequence of the fact that in these modal coordinates the matrix governing state evolution, namely $\Lambda$, is diagonal.

Figure 5.6 depicts the decoupled description. This is essentially the CT version of Figure 4.8, which displayed a natural state-space realization for a system whose unit sample response was a sum of DT exponentials. What we have now established for the CT case, but the DT case is exactly parallel to this, is that any LTI state-space system with distinct natural frequencies can be thought of in terms of such a decoupled or diagonalized representation. Example 5.13 at the end of this chapter illustrates how the picture can change when there are repeated natural frequencies.

**The CT Matrix Exponential**   The solution to the individual modal variables $r_i(t)$ was presented in Eq. (5.61). These equations can also be assembled into a revealing matrix form. To do this, define the matrix exponential of $\Lambda$ by

$$e^{\Lambda t} = \text{diagonal } \{e^{\lambda_1 t}, \cdots, e^{\lambda_L t}\} = \begin{bmatrix} e^{\lambda_1 t} & 0 & \cdots & 0 \\ 0 & e^{\lambda_2 t} & \cdots & 0 \\ \vdots & \vdots & \ddots & \vdots \\ 0 & 0 & \cdots & e^{\lambda_L t} \end{bmatrix}$$

$$= \mathbf{I} + \Lambda t + \Lambda^2 \frac{t^2}{2!} + \Lambda^3 \frac{t^3}{3!} + \cdots , \tag{5.90}$$

where the second equality above follows from invoking the Taylor series expansion for a scalar exponential, namely

$$e^{\lambda t} = 1 + \lambda t + \frac{(\lambda t)^2}{2!} + \frac{(\lambda t)^3}{3!} + \cdots . \tag{5.91}$$

This allows us to combine the $L$ equations in Eq. (5.61) into the following single matrix equation:

$$\mathbf{r}(t) = e^{\Lambda t}\mathbf{r}(0) + \int_0^t e^{\Lambda(t-\tau)}\boldsymbol{\beta} x(\tau)\, d\tau, \ \ t \geq 0 , \tag{5.92}$$

where the integral of a vector is interpreted as the component-wise integral, and $\boldsymbol{\beta}$ is as defined earlier in Eq. (5.89). This vector expression for $\mathbf{r}(t)$ is thus the solution to the $L$-dimensional state evolution equation in Eq. (5.87), and has the same form as the scalar solution in Eq. (5.61) for the scalar state evolution equation in Eq. (5.87).

Combining Eq. (5.92) with the expression in Eq. (5.53) that relates $\mathbf{r}(t)$ to $\mathbf{q}(t)$ results in

$$\mathbf{q}(t) = \left(\mathbf{V}e^{\mathbf{\Lambda} t}\mathbf{V}^{-1}\right)\mathbf{q}(0) + \int_0^t \left(\mathbf{V}e^{\mathbf{\Lambda}(t-\tau)}\mathbf{V}^{-1}\right)\mathbf{b}x(\tau)\,d\tau \qquad (5.93)$$

$$= e^{\mathbf{A} t}\mathbf{q}(0) + \int_0^t e^{\mathbf{A}(t-\tau)}\mathbf{b}x(\tau)\,d\tau \ , \quad t \geq 0 \ , \qquad (5.94)$$

where, by analogy with Eq. (5.90), we have defined the matrix exponential

$$e^{\mathbf{A} t} = \mathbf{V}e^{\mathbf{\Lambda} t}\mathbf{V}^{-1}$$

$$= \mathbf{V}\left(\mathbf{I} + \mathbf{\Lambda} t + \mathbf{\Lambda}^2\frac{t^2}{2!} + \mathbf{\Lambda}^3\frac{t^3}{3!} + \cdots\right)\mathbf{V}^{-1}$$

$$= \mathbf{I} + \mathbf{A} t + \mathbf{A}^2\frac{t^2}{2!} + \mathbf{A}^3\frac{t^3}{3!} + \cdots . \qquad (5.95)$$

If the starting time is $t_0$ rather than 0, the solution in Eq. (5.94) gets modified to

$$\mathbf{q}(t) = e^{\mathbf{A}(t-t_0)}\mathbf{q}(t_0) + \int_{t_0}^t e^{\mathbf{A}(t-\tau)}\mathbf{b}x(\tau)\,d\tau \ , \quad t \geq t_0 \ . \qquad (5.96)$$

This equation gives us, in compact matrix notation, the general solution of the CT LTI state evolution description in Eq. (5.1). The correctness of the solution can also be verified by directly substituting it in the state evolution equation, and using the following product and differentiation properties of the matrix exponential:

$$e^{\mathbf{A} t_1}e^{\mathbf{A} t_2} = e^{\mathbf{A}(t_1+t_2)} \ , \quad \text{and} \quad \frac{d}{dt}e^{\mathbf{A} t} = \mathbf{A}e^{\mathbf{A} t} = e^{\mathbf{A} t}\mathbf{A} \ . \qquad (5.97)$$

The expression in Eq. (5.96) also serves to explicitly verify the linearity and time-invariance properties of the solution set of an LTI state evolution equation. The state trajectory is linear in the initial state and input trajectory, taken together. Superposition of the respective initial conditions and input trajectories from two experiments generates a state trajectory that is the same superposition of the state trajectories from the individual experiments.

The matrix exponential of $\mathbf{A}$ is also defined by the infinite series in Eq. (5.95) for the case where $\mathbf{A}$ has repeated eigenvalues. The properties in Eq. (5.97) also hold in this more general case. It follows that the solution given in Eq. (5.96) is actually the solution of the given LTI system for all cases, not just for the case of distinct eigenvalues.

| Example 5.7 | **Sampled-Data Model in Matrix Form** |

In this example we return to the analysis in Example 5.6 of the sampled-data model that results from having a piecewise-constant input $x(t)$ applied to the CT LTI state-space system in Eq. (5.1):

$$x(t) = x[n] \quad \text{for} \quad nT \leq t < nT + T .$$

The sampled-data model for the modal variables is given in Eq. (5.71), repeated here:

$$r_i[n+1] = e^{\lambda_i T} r_i[n] + e^{\lambda_i T}\left(\int_0^T e^{-\lambda_i \sigma}\,d\sigma\right)\beta_i x[n], \quad i = 1, 2, \ldots, L .$$

With the notation and results now available to us, we are able to obtain the following compact and general representation of the sampled-data model, written in the original coordinates, and valid even for the case of repeated eigenvalues:

$$\mathbf{q}[n+1] = \left(e^{\mathbf{A}T}\right)\mathbf{q}[n] + \left(\int_0^T e^{\mathbf{A}(T-\sigma)}\,\mathbf{b}\,d\sigma\right)x[n] . \tag{5.98}$$

This is a DT LTI state-space model, amenable to analysis using the various methods and tools we have been developing.

Suppose, for instance, that we begin with the CT LTI system

$$\dot{\mathbf{q}}(t) = \underbrace{\begin{bmatrix} 0 & 1 \\ 0 & -\gamma \end{bmatrix}}_{\mathbf{A}}\mathbf{q}(t) + \underbrace{\begin{bmatrix} 0 \\ 1 \end{bmatrix}}_{\mathbf{b}} x(t) . \tag{5.99}$$

This simple model arises in describing the motion of a rigid object with velocity-dependent damping but no force other than the external force input $x(t)$—no restorative spring force, for example. The first state variable $q_1(t)$ in this case is the position of the object, and the second state variable $q_2(t)$ is its velocity. The eigenvalues and corresponding eigenvectors of $\mathbf{A}$ are

$$\lambda_1 = 0, \quad \lambda_2 = -\gamma, \quad \mathbf{v}_1 = \begin{bmatrix} 1 \\ 0 \end{bmatrix}, \quad \mathbf{v}_2 = \begin{bmatrix} 1 \\ -\gamma \end{bmatrix}. \tag{5.100}$$

If the force $x(t)$ is set by a computer-controlled actuator and is piecewise constant, taking the value $x[n]$ in the interval $nT \leq t < nT + T$, then a sampled-data state-space model allows us to track the state from time $nT$ to time $nT + T$. The matrices involved in the sampled-data model are provided in Eq. (5.98) and evaluated below, with their approximate values for the case where $\gamma T \ll 1$ listed immediately after:

$$e^{\mathbf{A}T} = \begin{bmatrix} 1 & 1 \\ 0 & -\gamma \end{bmatrix}\begin{bmatrix} 1 & 0 \\ 0 & e^{-\gamma T} \end{bmatrix}\begin{bmatrix} 1 & 1/\gamma \\ 0 & -1/\gamma \end{bmatrix}$$

$$= \begin{bmatrix} 1 & \Delta \\ 0 & e^{-\gamma T} \end{bmatrix} \approx \begin{bmatrix} 1 & T \\ 0 & 1 \end{bmatrix}$$

$$e^{\mathbf{A}T}\left(\int_0^T e^{-\mathbf{A}\sigma}\,d\sigma\right)\mathbf{b} = \begin{bmatrix} \frac{1}{\gamma}(T - \Delta) \\ \Delta \end{bmatrix} \approx \begin{bmatrix} T^2/2 \\ T \end{bmatrix} \tag{5.101}$$

where

$$\Delta = \frac{1}{\gamma}(1 - e^{-\gamma T}) \approx T - \gamma\frac{T^2}{2} . \tag{5.102}$$

The limiting case of $\gamma T \ll 1$ in the case of rigid object motion corresponds to having very low damping. The object then, by Newton's law, has an acceleration that is proportional to the external input $x(t)$—and actually equal to $x(t)$ for the units chosen in the CT model in Eq. (5.99). Over the course of the $n$th sampling interval, therefore, the constant force $x[n]$ causes the velocity to vary linearly with time, from $q_2[n]$ to

$q_2[n+1] = q_2[n] + Tx[n]$, and the position to vary quadratically with time, from $q_1[n]$ to $q_1[n+1] = q_1[n] + Tq_2[n] + (T^2/2)x[n]$. These results are consistent with the state evolution model obtained by substituting the approximate expressions in Eq. (5.101) into the sampled-data model in Eq. (5.98).

**General Solution for DT Systems**    A development parallel to that leading to the CT state solution in Eqs. (5.93) and (5.94) can be carried out for the DT LTI case. The corresponding expression for the solution of (5.3) is

$$\mathbf{q}[n] = \left(\mathbf{V}\mathbf{\Lambda}^n\mathbf{V}^{-1}\right)\mathbf{q}[0] + \sum_{k=0}^{n-1}\left(\mathbf{V}\mathbf{\Lambda}^{n-k-1}\mathbf{V}^{-1}\right)\mathbf{b}x[k] \qquad (5.103)$$

$$= \mathbf{A}^n\mathbf{q}[0] + \sum_{k=0}^{n-1}\mathbf{A}^{n-k-1}\mathbf{b}x[k] \;,\;\; n \geq 0 \;. \qquad (5.104)$$

Equation (5.104) is exactly the expression one would get by simply iterating Eq. (5.3) forward one step at a time, to get $\mathbf{q}[n]$ from $\mathbf{q}[0]$. However, the expression in the modally decomposed form in Eq. (5.103) provides additional insight because it brings out the role of the eigenvalues of $\mathbf{A}$, that is, the natural frequencies of the DT system, in determining the behavior of the system, and in particular its stability properties.

# 5.4 TRANSFER FUNCTIONS, HIDDEN MODES, REACHABILITY, AND OBSERVABILITY

The focus has thus far been on the internal or state behavior of state-space models. We now turn to examining how the input couples to the state, how the state couples to the output, and how the input couples to the output. We start with the latter, by examining the input-output transfer function of an LTI state-space model. This will then lead us to the other couplings of interest. We treat CT systems first, then DT systems.

## 5.4.1 Input-State-Output Structure of CT Systems

**Input-Output Relations**    To analyze the behavior of a causal system in response to initial conditions specified at time $t = 0$ and inputs specified for $t \geq 0$, the unilateral Laplace transform is particularly useful. For a signal $w(t)$, the unilateral Laplace transform is defined by

$$W(s) = \mathcal{L}_+\{w(t)\} = \int_{0-}^{\infty} w(t)e^{-st}\,dt \;, \qquad (5.105)$$

with a region of convergence that is some right half-plane. The lower limit on the integral is written as $0-$ to indicate that the transform captures any impulses at time 0; so if $w(t) = \delta(t)$, then $W(s) = 1$. With this definition, the unilateral Laplace transform of the derivative of the signal $w(t)$ is

$$\mathcal{L}_+\{\dot{w}(t)\} = \int_{0-}^{\infty} \dot{w}(t)e^{-st}\, dt = sW(s) - w(0-) \, . \tag{5.106}$$

If $w(t)$ is continuous at $t = 0$, that is, if $\dot{w}(t)$ has no impulse at $t = 0$, then it suffices to write $w(0)$ instead of $w(0-)$ in this equation. If the initial condition is $w(0) = 0$, then differentiation in time maps to multiplication by $s$ in the transform domain. In the case of the bilateral Laplace transform described in Chapter 2, where the lower limit on the defining integral is $-\infty$ instead of $0-$, differentiation in time maps to multiplication by $s$, with no adjustment for an initial value.

Taking the unilateral Laplace transform of the decoupled modal equations in Eq. (5.58) and rearranging the result shows that

$$R_i(s) = \frac{1}{s - \lambda_i}\left(r_i(0-) + \beta_i X(s)\right), \quad i = 1, 2, \ldots, L \, . \tag{5.107}$$

If $x(t)$ has no impulse at $t = 0$, then $r_i(t)$ is continuous at $t = 0$, and $r_i(0-)$ in Eq. (5.107) can be replaced by $r_i(0)$. The inverse transform of this equation yields precisely the time-domain solution for $r_i(t)$ that was presented earlier in Eq. (5.61), expressing the solution as the sum of the ZIR and the ZSR. We now have a transform-domain derivation of that solution.

Similarly taking the Laplace transform of the output equation in Eq. (5.62), and substituting in the relations from Eq. (5.107), results in

$$Y(s) = \left(\sum_{i=1}^{L} \xi_i R_i(s)\right) + \mathrm{d}X(s)$$

$$= \left(\sum_{i=1}^{L} \frac{\xi_i r_i(0-)}{s - \lambda_i}\right) + \left(\mathrm{d} + \sum_{1}^{L} \frac{\xi_i \beta_i}{s - \lambda_i}\right)X(s) \, . \tag{5.108}$$

The transfer function $H(s)$ between a given input and output of a CT LTI system can be identified as the ratio of the transform of the output to the transform of the input, when all other inputs or excitations, such as initial conditions, are set to zero. Accordingly, we set $r_i(0-) = 0$ in the preceding equations and solve for the ratio $Y(s)/X(s)$ to arrive at the following expression for the transfer function of the state-space model in Eqs. (5.1) and (5.2):

$$H(s) = \left(\sum_{i=1}^{L} \frac{\xi_i \beta_i}{s - \lambda_i}\right) + \mathrm{d} \, . \tag{5.109}$$

Several useful facts can be deduced from this expression. If $H(s)$ in Eq. (5.109) is written as a ratio of polynomials, the denominator polynomial is

$$a(s) = (s - \lambda_1)(s - \lambda_2)\cdots(s - \lambda_L) \, . \tag{5.110}$$

Since this is a degree-$L$ monic polynomial whose roots are precisely the eigenvalues of the matrix $\mathbf{A}$, it must be the characteristic polynomial defined in Eq. (5.13). Thus

$$H(s) = \frac{\eta(s)}{a(s)} . \qquad (5.111)$$

The expression in Eq. (5.109) shows that the numerator polynomial $\eta(s)$ in the above expression has degree $L$ if and only if the direct feedthrough gain d is nonzero. In this case of equal numerator and denominator degree, the transfer function is called exactly proper. If d $= 0$, then $\eta(s)$ can have any degree between 0 and $L - 1$, and then the transfer function is called strictly proper because the numerator degree is strictly smaller than the denominator degree. In any case, the transfer function of a state-space system is a proper rational, that is, the numerator degree does not exceed the denominator degree.

The poles of $H(s)$ are the values of $s$ at which $H(s)$ has infinite magnitude, and the zeros are where it has a magnitude of 0. The expression for $H(s)$ given by Eqs. (5.111) and (5.110) shows that the poles all have to be natural frequencies of the system. However, the converse is not guaranteed: not all natural frequencies have to be poles because $\eta(s)$ and $a(s)$ might have factors $(s - \lambda_j)$ that cancel. This is commonly referred to as a pole-zero cancellation, though the cancelled factor is neither a pole nor a zero of $H(s)$. Each cancelled natural frequency is termed a hidden mode of the system. We shall shortly say more about what gives rise to these pole-zero cancellations or hidden modes. When there are pole-zero cancellations, $H(s)$ is typically written in reduced form, with all cancellations made and with the roots of the resulting denominator (respectively numerator) being precisely the poles (respectively zeros) of $H(s)$.

The form of the transfer function in Eq. (5.109) also provides a route to obtaining a state-space realization of any specified proper rational transfer function $H(s)$ with $L$ distinct poles (after being written in reduced form, so with all cancellations made). What we seek is a state-space model that has the specified transfer function. We first expand $H(s)$ in a partial fraction expansion of the form

$$H(s) = \left( \sum_{i=1}^{L} \frac{k_i}{s - \lambda_i} \right) + \mathrm{d} , \qquad (5.112)$$

where

$$k_i = H(s)(s - \lambda_i)\Big|_{s=\lambda_i} \quad \text{and} \quad \mathrm{d} = H(\infty) . \qquad (5.113)$$

Then choosing an arbitrary set of nonzero numbers $\beta_i$ for $i = 1, \ldots, L$, and subsequently choosing $\xi_i = k_i/\beta_i$, we can assemble a state-space realization in the decoupled form specified in Eqs. (5.58) and (5.62) and Figure 5.6. This state-space model will have the specified transfer function from input $x(\cdot)$ to output $y(\cdot)$.

Given a proper rational transfer function $H(s)$ with $L$ poles, so that its denominator degree is $L$ after all cancellations, an $L$th-order realization of it is called a minimal realization because no realization of $H(s)$ can have lower order. To see this, note that a realization of lower order $L' < L$ would have a transfer function $H'(s) = H(s)$ with no more than $L'$ poles, since it only has $L'$ natural frequencies, but we know $H(s)$ has $L$ poles (after all cancellations). It can be shown that any two minimal state-space realizations of a transfer function are related to each other by a similarity transformation.

The expression for the transfer function in Eq. (5.109) can be rewritten in matrix notation as

$$H(s) = \boldsymbol{\xi}^T (s\mathbf{I} - \boldsymbol{\Lambda})^{-1} \boldsymbol{\beta} + \mathrm{d} \,. \tag{5.114}$$

This expression follows from the definitions of the vectors $\boldsymbol{\xi}^T$ and $\boldsymbol{\beta}$ in Eq. (5.89), and the fact that $s\mathbf{I} - \boldsymbol{\Lambda}$ is a diagonal matrix, whose inverse is therefore again a diagonal matrix, with diagonal entries that are the reciprocals of the corresponding diagonal entries in the original. Substituting in the defining relationships $\boldsymbol{\xi}^T = \mathbf{c}^T \mathbf{V}$ and $\boldsymbol{\beta} = \mathbf{V}^{-1}\mathbf{b}$, and recalling that $\boldsymbol{\Lambda} = \mathbf{V}^{-1}\mathbf{A}\mathbf{V}$, the above expression for $H(s)$ can be written in terms of the matrices defining the state-space description in the original coordinates rather than modal coordinates:

$$H(s) = \mathbf{c}^T \mathbf{V}(s\mathbf{I} - \mathbf{V}^{-1}\mathbf{A}\mathbf{V})^{-1}\mathbf{V}^{-1}\mathbf{b} + \mathrm{d} \,. \tag{5.115}$$

To simplify Eq. (5.115), we use the fact that for invertible matrices $\mathbf{M}, \mathbf{N}$, and $\mathbf{P}$ the inverse $(\mathbf{MNP})^{-1}$ is the product of the individual inverses in reverse order, that is, $\mathbf{P}^{-1}\mathbf{N}^{-1}\mathbf{M}^{-1}$, because multiplying the latter product by $\mathbf{MNP}$ yields the identity matrix. The expression in Eq. (5.115) can thus be reduced further, as follows:

$$H(s) = \mathbf{c}^T \left( \mathbf{V}(s\mathbf{I} - \mathbf{V}^{-1}\mathbf{A}\mathbf{V})\mathbf{V}^{-1} \right)^{-1} \mathbf{b} + \mathrm{d}$$

$$= \mathbf{c}^T (s\mathbf{I} - \mathbf{A})^{-1}\mathbf{b} + \mathrm{d} \,. \tag{5.116}$$

The expression for $H(s)$ in Eq. (5.116) can also be obtained directly by Laplace transforming the original description in Eqs. (5.1) and (5.2) to get

$$s\mathbf{Q}(s) - \mathbf{q}(0) = \mathbf{A}\mathbf{Q}(s) + \mathbf{b}X(s) \,, \tag{5.117}$$

$$Y(s) = \mathbf{c}^T\mathbf{Q}(s) + \mathrm{d}X(s) \,, \tag{5.118}$$

where $\mathbf{Q}(s)$ denotes the component-wise Laplace transform of $\mathbf{q}(t)$, that is, a vector whose $i$th entry is the Laplace transform of $q_i(t)$. The first equation above can be rearranged and solved for $\mathbf{Q}(s)$:

$$\mathbf{Q}(s) = (s\mathbf{I} - \mathbf{A})^{-1}\mathbf{q}(0) + (s\mathbf{I} - \mathbf{A})^{-1}\mathbf{b}X(s) \,. \tag{5.119}$$

Taking the inverse transform of Eq. (5.119) yields the time domain solution previously obtained in Eq. (5.94), but with no need for the assumption of distinct eigenvalues. Comparing the two expressions also shows another way to obtain the matrix exponential:

$$e^{\mathbf{A}t} = \text{inverse Laplace transform of } (s\mathbf{I} - \mathbf{A})^{-1} \,, \tag{5.120}$$

where the inverse transform is computed component-wise, that is, separately for each entry of the matrix $(s\mathbf{I} - \mathbf{A})^{-1}$.

Assuming $\mathbf{q}(0) = \mathbf{0}$ in Eq. (5.119) and substituting the resulting expression for $\mathbf{Q}(s)$ in the output relation in Eq. (5.118) yields the expression for the transfer function $H(s)$ in Eq. (5.116). The advantage of this derivation is again that no assumption is needed regarding the eigenvalues being distinct; the expression holds generally. Our development below, however, is built on the assumption of distinct eigenvalues for simplicity—and the expression for $H(s)$ in modal coordinates, namely Eqs. (5.109) or (5.114), is then very helpful.

**CT System Reachability, Observability, and Hidden Modes**    Equation (5.109) has demonstrated that $H(s)$ will have $L$ poles in general, and precisely at the natural frequencies of the system. Suppose, however, that $\beta_j = 0$ for some $j$. Our definition of the $\{\beta_i\}$ in Eq. (5.56) shows that this happens exactly when $\mathbf{b}$ can be expressed as a linear combination of the eigenvectors other than $\mathbf{v}_j$. In this case, Eq. (5.109) shows that $\lambda_j$ fails to appear as a pole of the transfer function, even though it is still a natural frequency of the system and appears in the ZIR of the state $\mathbf{q}(t)$ for almost all initial conditions.

The underlying cause for this hidden mode—an internal mode that is hidden from the input-output transfer function—is seen in Eqs. (5.58) or (5.107): with $\beta_j = 0$, the input fails to excite the $j$th mode. The mode associated with $\lambda_j$ is said to be an unreachable mode in this case. In contrast, if $\beta_k \neq 0$, the $k$th mode is termed reachable. (The term *controllable* is also used for reachable; strictly speaking there is a slight difference in the definitions of the two concepts in the DT case, but we shall not be concerned about this.)

If all $L$ modes of the system are reachable, then the system itself is termed reachable, otherwise it is called unreachable. In a reachable system, the input can fully excite the state variables, and in fact can transfer the state vector from any specified initial condition to any desired target state in finite time. In an unreachable system, this is not possible. The notion of reachability arises in several places in systems and control theory.

A dual situation happens when $\xi_j = 0$ for some $j$. In this case too, Eq. (5.109) shows that $\lambda_j$ fails to appear as a pole of the transfer function, even though it is still a natural frequency of the system and appears in the ZIR of the state $\mathbf{q}(t)$ for almost all initial conditions. Once again, we have a hidden mode. This time, the cause is evident in Eqs. (5.62) or (5.108): with $\xi_j = 0$, the $j$th mode fails to appear at the output, even when it is present in the state response. The mode associated with $\lambda_j$ is termed unobservable in this case. In contrast, if $\xi_k \neq 0$, then the $k$th mode is called observable.

If all $L$ modes of the system are observable, the system itself is termed observable, otherwise it is called unobservable. In an observable system, the behavior of the state vector can be unambiguously inferred from measurements of the input and output over some interval of time, whereas this is not possible for an unobservable system. The concept of observability also arises repeatedly in systems and control theory.

Hidden modes can cause difficulty, especially if they are unstable. However, if all we are concerned about is representing a transfer function, or equivalently the input-output relation of an LTI system with zero initial conditions, then hidden modes may be of no significance. We can obtain a reduced-order state-space model that has the same transfer function by simply discarding all the equations in Eq. (5.58) that correspond to unreachable or unobservable modes, and discarding the corresponding terms in Eq. (5.62).

The converse is also true: if a state-space model is reachable and observable, then there is no lower-order state-space system that has the same transfer function. In other words, a state-space model that is reachable and observable is a minimal realization of its transfer function.

**Reachability and Observability of Composite Systems**  Consider the composite system shown in Figure 5.7, resulting from the cascade or series interconnection of two subsystems. The equality $x_2(t) = y_1(t)$ defines the interconnection. The input $x_1(t)$ is set equal to the external input $x(t)$, and the overall system output $y(t)$ is set equal to $y_2(t)$. Each subsystem is labeled by its proper rational transfer function, $H_1(s)$ and $H_2(s)$ respectively. The assumption or convention, since we are only given the transfer functions, is that each subsystem is a minimal LTI state-space realization of the associated transfer function, with an $L_1$-component vector $\mathbf{q}_1(t)$ and $L_2$-component vector $\mathbf{q}_2(t)$ as the respective state vectors. The most natural choice of state variables for the interconnected system is then the union of the state variables in the two subsystems, with corresponding state vector

$$\mathbf{q}(t) = \begin{bmatrix} \mathbf{q}_1(t) \\ \mathbf{q}_2(t) \end{bmatrix}. \tag{5.121}$$

Introducing the notation

$$H_1(s) = \frac{\eta_1(s)}{a_1(s)}, \qquad H_2(s) = \frac{\eta_2(s)}{a_2(s)}, \tag{5.122}$$

where $a_1(s)$, $a_2(s)$ have degrees $L_1$ and $L_2$ respectively, and have no cancellations with their respective numerators, it follows that the overall transfer function of the system from input $x(t)$ to output $y(t)$ is

$$H_{\mathrm{ser}}(s) = H_2(s)H_1(s) = \frac{\eta_2(s)\eta_1(s)}{a_2(s)a_1(s)}. \tag{5.123}$$

It must therefore be the case that the characteristic polynomial of the interconnected system is precisely the polynomial $a_1(s)a_2(s)$ of degree $L_1 + L_2$. The natural frequencies of the series combination are then the union of the natural frequencies of the individual systems.

The only possibilities for cancellation are for $\eta_2(s)$ and $a_1(s)$ to have one or more common factors, or for $\eta_1(s)$ and $a_2(s)$ to have one or more common factors. It turns out that the former possibility corresponds to a loss of observability, and the latter to a loss of reachability. The loss of observability is because a zero (or zeros) of the second system blocks the corresponding natural frequency (or frequencies) of the first subsystem from reaching the output.

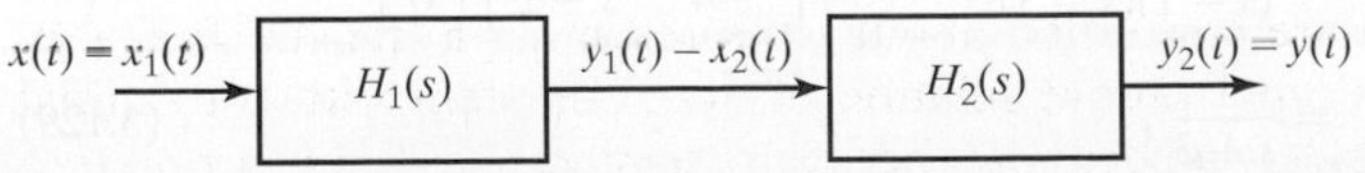

$x(t) = x_1(t)$ → [ $H_1(s)$ ] → $y_1(t) = x_2(t)$ → [ $H_2(s)$ ] → $y_2(t) = y(t)$

**Figure 5.7**  System obtained by cascading two subsystems.

Similarly, the loss of reachability is because a zero (or zeros) of the first subsystem prevents the input from exciting the corresponding natural frequency (or frequencies) of the second subsystem. We illustrate this with a simple example next.

**Example 5.8    Hidden Mode in a Series Combination of Subsystems**

Suppose in Figure 5.7 that

$$H_1(s) = \frac{1}{s-1}, \qquad H_2(s) = \frac{s-1}{s+3} = 1 - \frac{4}{s+3}, \tag{5.124}$$

which yields an overall transfer function of

$$H_{\text{ser}}(s) = \frac{1}{s+3}, \tag{5.125}$$

indicating a hidden mode. To confirm that this is due to a loss of observability and not reachability, we construct first-order state-space models for each of the subsystems, combine them in accordance with the constraints represented in Figure 5.7 to get a second-order state-space model for the combination, and finally check the observability and reachability of the resulting system.

Both the subsystem transfer functions already have the partial fraction expansion form assumed in Eq. (5.112) as the starting point for obtaining a minimal state-space realization of a transfer function. Following the procedure outlined there, the following individual state-space realizations are directly obtained:

$$\dot{q}_1(t) = q_1(t) + x_1(t), \qquad y_1(t) = q_1(t) \tag{5.126}$$

$$\dot{q}_2(t) = -3q_2(t) - 4x_2(t), \qquad y_2(t) = q_2(t) + x_2(t) \tag{5.127}$$

Combining these and taking account of the interconnection constraints generates the following state-space description of the overall system:

$$\begin{bmatrix} \dot{q}_1(t) \\ \dot{q}_2(t) \end{bmatrix} = \underbrace{\begin{bmatrix} 1 & 0 \\ -4 & -3 \end{bmatrix}}_{\mathbf{A}} \begin{bmatrix} q_1(t) \\ q_2(t) \end{bmatrix} + \underbrace{\begin{bmatrix} 1 \\ 0 \end{bmatrix}}_{\mathbf{b}} x(t),$$

$$y(t) = \underbrace{\begin{bmatrix} 1 & 1 \end{bmatrix}}_{\mathbf{c}^T} \begin{bmatrix} q_1(t) \\ q_2(t) \end{bmatrix}. \tag{5.128}$$

The natural frequencies of this system are $\lambda_1 = 1$ and $\lambda_2 = -3$, as expected (and the system is seen to be unstable). Using the formula in Eq. (5.116) to check the overall transfer function, we get

$$H(s) = \begin{bmatrix} 1 & 1 \end{bmatrix} \begin{bmatrix} s-1 & 0 \\ 4 & s+3 \end{bmatrix}^{-1} \begin{bmatrix} 1 \\ 0 \end{bmatrix}$$

$$= \frac{1}{(s-1)(s+3)} \begin{bmatrix} 1 & 1 \end{bmatrix} \begin{bmatrix} s+3 & 0 \\ -4 & s-1 \end{bmatrix} \begin{bmatrix} 1 \\ 0 \end{bmatrix}$$

$$= \frac{1}{s+3}, \tag{5.129}$$

as expected. This computation has used the fact that the inverse of a $2 \times 2$ matrix is given by

$$\begin{bmatrix} a & b \\ c & d \end{bmatrix}^{-1} = \frac{1}{ad - bc} \begin{bmatrix} d & -b \\ -c & a \end{bmatrix}, \tag{5.130}$$

as can be verified on premultiplying both sides by the matrix that is being inverted, yielding the identity matrix on both sides. Note that even though the system is not asymptotically stable internally, it has a bounded output for all bounded inputs, provided it starts from zero initial conditions (or initial conditions that only excite the stable mode). The reason for this BIBO stability is that the pole of $H(s)$ is negative, yielding an absolutely integrable impulse response $h(t)$.

To check observability and reachability, we first determine the eigenvectors, which are

$$\mathbf{v}_1 = \begin{bmatrix} 1 \\ -1 \end{bmatrix} \quad \text{and} \quad \mathbf{v}_2 = \begin{bmatrix} 0 \\ 1 \end{bmatrix}. \tag{5.131}$$

Since $\mathbf{c}^T \mathbf{v}_1 = 0$, the mode associated with $\lambda_1 = 1$ is indeed unobservable, which explains why it is hidden from the transfer function. As for reachability, note that $\mathbf{b}$ cannot be written as a multiple of just $\mathbf{v}_1$ or just $\mathbf{v}_2$—a weighted linear combination of both eigenvectors is required to generate $\mathbf{b}$, which confirms the reachability of both modes. This example also serves to point out two important general facts: an entry of 0 in the input vector $\mathbf{b}$ (in the original, not modal, coordinates) does not signify a loss of reachability; and having all entries nonzero in the output vector $\mathbf{c}^T$ (in the original, not modal, coordinates) does not ensure observability.

If the two subsystems had been connected in the reverse order, the same procedure as above would have led us to the conclusion now that $\lambda_1 = 1$ is observable but unreachable, and therefore hidden again, though for a different reason.

Figure 5.8 shows two other familiar composite systems obtained by interconnecting two subsystems: a closed-loop, or feedback, configuration and a parallel configuration. Both cases can be studied by the same approach used for the series connection above; we simply summarize the results here. Assume as before that $\eta_1(s)$ and $a_1(s)$ have no common factors, and similarly for $\eta_2(s)$ and $a_2(s)$.

For the feedback configuration, the transfer function is

$$H_{\text{fb}}(s) = \frac{\eta_1(s)a_2(s)}{a_1(s)a_2(s) + \eta_1(s)\eta_2(s)}, \tag{5.132}$$

and the denominator $a_1(s)a_2(s) + \eta_1(s)\eta_2(s)$ is the characteristic polynomial of the system. In general, therefore, feedback leads to natural frequencies that are different from those of the subsystems, since the natural frequencies are no longer the roots of $a_1(s)a_2(s)$. Examination of Eq. (5.132) shows that any cancellations between the numerator and denominator are due to common factors between $\eta_1(s)$ and $a_2(s)$. These common factors represent poles or modes of the subsystem in the feedback path that are masked or hidden by the cancelling zeros of the subsystem in the forward path, and that thereby end

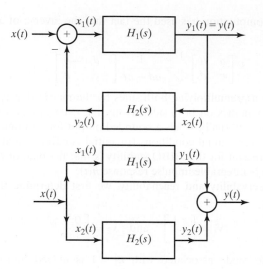

**Figure 5.8** The upper diagram shows a feedback connection of two subsystems; the lower one shows a parallel connection.

up being both unreachable and unobservable. These modes therefore remain as natural frequencies of the closed-loop system, unaffected by the feedback.

For the parallel configuration, the transfer function is

$$H_{\text{par}}(s) = \frac{\eta_1(s)a_2(s) + \eta_2(s)a_1(s)}{a_1(s)a_2(s)}, \qquad (5.133)$$

and the denominator $a_1(s)a_2(s)$ is the characteristic polynomial of the system. The natural frequencies of the parallel connection are thus the union of the subsystem natural frequencies. In this case, cancellations arise from $a_1(s)$ and $a_2(s)$ having common factors. However, such cancellation corresponds to an overall system that has repeated rather than distinct natural frequencies, so the definitions and methods we have developed for studying reachability and observability under the assumption of distinct natural frequencies no longer apply. In the more general framework that includes repeated natural frequencies, it turns out that when $a_1(s)$ and $a_2(s)$ have common factors, the associated hidden modes are both unreachable and unobservable. The reason is that these common modes of the two subsystems are driven in fixed proportion by the input, and seen in fixed linear combination by the output, so they do not get recognized as having multiplicity greater than one. Example 5.13 at the end of this chapter sheds further light on the situation.

## 5.4.2 Input-State-Output Structure of DT Systems

**DT System Transfer Relations**    A very similar development can be carried out for the DT case. To analyze the behavior of DT state-space models in response to initial conditions specified at time $n = 0$ and inputs specified for $n \geq 0$, the most useful transform is the unilateral $z$-transform. For a signal

$w[n]$, the unilateral $z$-transform is defined by

$$W(z) = \sum_{n=0}^{\infty} w[n] z^{-n} , \tag{5.134}$$

with a region of convergence that is the outside of a circle centered at $z = 0$. With this definition, the unilateral $z$-transform of the one-step advanced signal $w[n+1]$ is

$$\sum_{n=0}^{\infty} w[n+1] z^{-n} = zW(z) - zw[0] . \tag{5.135}$$

If the initial condition is $w[0] = 0$, then the one-step advance in time maps to multiplication by $z$ in the transform domain. In the case of the bilateral $z$-transform described in Chapter 1, where the lower limit on the defining sum is $-\infty$ instead of 0, the one-step advance in time maps to just multiplication by $z$, with no adjustment for an initial value.

Taking the unilateral $z$-transform of the decoupled DT modal equations in Eq. (5.67) and rearranging the result shows that

$$R_i(z) = \frac{1}{z - \lambda_i} \Big( z r_i[0] + \beta_i X(z) \Big), \quad i = 1, 2, \ldots, L . \tag{5.136}$$

The inverse transform of this yields precisely the time-domain solution for $r_i[n]$ that was presented earlier in Eq. (5.68), writing the solution as the sum of the ZIR and the ZSR. We now have a transform-domain derivation of that solution.

Similarly taking the $z$-transform of the output equation in Eq. (5.69) and substituting in the above relations produces

$$Y(z) = \Big( \sum_{i=1}^{L} \xi_i R_i(z) \Big) + dX(z)$$

$$= \Big( \sum_{i=1}^{L} \frac{z \xi_i r_i[0]}{z - \lambda_i} \Big) + \Big( d + \sum_{i=1}^{L} \frac{\xi_i \beta_i}{z - \lambda_i} \Big) X(z) . \tag{5.137}$$

The transfer function $H(z)$ is then

$$H(z) = \Big( \sum_{i=1}^{L} \frac{\xi_i \beta_i}{z - \lambda_i} \Big) + d$$

$$= \boldsymbol{\xi}^T (z\mathbf{I} - \boldsymbol{\Lambda})^{-1} \boldsymbol{\beta} + d$$

$$= \mathbf{c}^T (z\mathbf{I} - \mathbf{A})^{-1} \mathbf{b} + d , \tag{5.138}$$

as in the CT case.

The notion of hidden modes, and the definitions of reachability, observability, and minimal realizations, all carry over without change to the DT case. We illustrate these in the following example.

**Example 5.9**  **Evaluating Reachability and Observability of a DT System**

Consider the DT system represented by the state equations

$$\begin{bmatrix} q_1[n+1] \\ q_2[n+1] \end{bmatrix} = \underbrace{\begin{bmatrix} 0 & 1 \\ -1 & \frac{5}{2} \end{bmatrix}}_{\mathbf{A}} \begin{bmatrix} q_1[n] \\ q_2[n] \end{bmatrix} + \underbrace{\begin{bmatrix} 0 \\ 1 \end{bmatrix}}_{\mathbf{b}} x[n] \tag{5.139}$$

$$y[n] = \underbrace{\begin{bmatrix} -1 & \frac{1}{2} \end{bmatrix}}_{\mathbf{c}^T} \begin{bmatrix} q_1[n] \\ q_2[n] \end{bmatrix} + x[n] . \tag{5.140}$$

A delay-adder-gain block diagram representing Eqs. (5.139) and (5.140) is shown in Figure 5.9.

The modes of the system correspond to the roots of the characteristic polynomial, which is

$$\det(\lambda \mathbf{I} - \mathbf{A}) = \lambda^2 - \frac{5}{2}\lambda + 1 . \tag{5.141}$$

The roots are

$$\lambda_1 = 2 , \qquad \lambda_2 = \frac{1}{2} . \tag{5.142}$$

Since $|\lambda_1| > 1$, the system is not asymptotically stable. The corresponding eigenvectors are found by solving

$$(\lambda \mathbf{I} - \mathbf{A})\mathbf{v} = \begin{bmatrix} \lambda & -1 \\ 1 & \lambda - \frac{5}{2} \end{bmatrix} \mathbf{v} = \mathbf{0} \tag{5.143}$$

with $\lambda = \lambda_1 = 2$, and then again with $\lambda = \lambda_2 = \frac{1}{2}$. This yields

$$\mathbf{v}_1 = \begin{bmatrix} 1 \\ 2 \end{bmatrix} , \qquad \mathbf{v}_2 = \begin{bmatrix} 2 \\ 1 \end{bmatrix} . \tag{5.144}$$

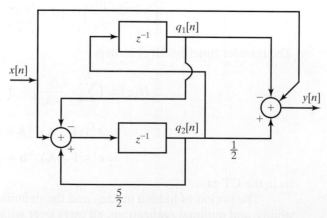

**Figure 5.9**   Delay-adder-gain block diagram for the system in Example 5.9.

The input-output transfer function of the system is

$$H(z) = \mathbf{c}^T (z\mathbf{I} - \mathbf{A})^{-1} \mathbf{b} + \mathrm{d} \tag{5.145}$$

$$(z\mathbf{I} - \mathbf{A})^{-1} = \frac{1}{z^2 - \frac{5}{2}z + 1} \begin{bmatrix} z - \frac{5}{2} & 1 \\ -1 & z \end{bmatrix} \tag{5.146}$$

$$H(z) = \frac{1}{z^2 - \frac{5}{2}z + 1} \left\{ \begin{bmatrix} -1 & \frac{1}{2} \end{bmatrix} \begin{bmatrix} z - \frac{5}{2} & 1 \\ -1 & z \end{bmatrix} \begin{bmatrix} 0 \\ 1 \end{bmatrix} \right\} + 1$$

$$= \frac{1}{2} \frac{z - 2}{z^2 - \frac{5}{2}z + 1} + 1 = \frac{1}{2} \frac{1}{z - \frac{1}{2}} + 1$$

$$= \frac{1}{1 - \frac{1}{2}z^{-1}} . \tag{5.147}$$

Since the pole is inside the unit circle, the system is input-output stable. However, the system has two modes. Consequently one of them must be a hidden mode, that is, it does not appear in the input-output transfer function. Hidden modes are either unreachable from the input or unobservable in the output, or both. To check reachability, note that the input vector $\mathbf{b}$ for this example cannot be written as a scalar multiple of just $\mathbf{v}_1$ or $\mathbf{v}_2$; it takes a linear combination of both eigenvectors to generate $\mathbf{b}$. The system is therefore reachable (despite the fact that $\mathbf{b}$ itself has one of its entries being 0). However, $\mathbf{c}^T \mathbf{v}_1 = 0$, so the first mode is unobservable (despite the fact that both entries of $\mathbf{c}^T$ are nonzero), and this is precisely the mode that is hidden from $H(z)$.

The notion of a minimal realization helps to illuminate Example 4.6, where state-space realizations of different orders were constructed for a second-order difference equation. The following example shows that the third-order realization of this difference equation is nonminimal.

## Example 5.10   A Nonminimal Realization

In Example 4.6 we obtained the third-order realization

$$\mathbf{q}[n+1] = \underbrace{\begin{bmatrix} -a_1 & -a_2 & b_2 \\ 1 & 0 & 0 \\ 0 & 0 & 0 \end{bmatrix}}_{\mathbf{A}} \mathbf{q}[n] + \begin{bmatrix} b_1 \\ 0 \\ 1 \end{bmatrix} x[n] ,$$

$$y[n] = \underbrace{\begin{bmatrix} 1 & 0 & 0 \end{bmatrix}}_{\mathbf{c}^T} \mathbf{q}[n] \tag{5.148}$$

of the second-order difference equation

$$y[n] + a_1 y[n-1] + a_2 y[n-2] = b_1 x[n-1] + b_2 x[n-2] \,. \qquad (5.149)$$

It is easy to see that $\mathbf{A}\mathbf{v}_1 = \mathbf{0}$, where $\mathbf{v}_1 = [0 \ \ b_2 \ \ a_2]^T$. We conclude from this that $\lambda_1 = 0$ is an eigenvalue of $\mathbf{A}$, with associated eigenvector $\mathbf{v}_1$, assuming $b_2$ and $a_2$ are not both 0. Now $\mathbf{c}^T \mathbf{v}_1 = 0$, which shows that this is an unobservable mode of the system (though a very benign one, in that the associated modal contribution vanishes in one time step). We conclude that the third-order system is not minimal, and therefore that the input-output relation in Eq. (5.149) can be realized by a lower-order system.

The concept of reachability—for both CT and DT systems—was introduced in terms of the ability to excite all the modes of the system from the input. The following example expands this view of reachability, by addressing its relevance to the problem of reaching particular target states from the origin of the state space, by appropriate choice of an input signal. We treat the DT case here, but similar results hold for CT.

**Example 5.11   Input Design to Reach a Target State**

Consider an $L$th-order DT LTI state-space model whose state evolves as

$$\mathbf{q}[n+1] = \mathbf{A}\mathbf{q}[n] + \mathbf{b}x[n] \,. \qquad (5.150)$$

Assume $\mathbf{A}$ has distinct eigenvalues $\lambda_1, \ldots, \lambda_L$, and associated independent eigenvectors $\mathbf{v}_1, \ldots, \mathbf{v}_L$. The constants $\beta_1, \ldots, \beta_L$ are defined as in Eq. (5.56) as the weights required to represent $\mathbf{b}$ by a linear combination of the eigenvectors:

$$\mathbf{b} = \mathbf{v}_1 \beta_1 + \cdots + \mathbf{v}_L \beta_L.$$

Suppose the system is reachable, that is, all $\beta_i$ are nonzero, and that we start from the origin at time 0, so $\mathbf{q}[0] = \mathbf{0}$. We want to find a sequence of $L$ inputs, $x[0], x[1], \ldots, x[L-1]$, such that the state at time $L$ takes the value

$$\mathbf{q}[L] = \mathbf{v}_1 \gamma_1 + \cdots + \mathbf{v}_L \gamma_L \qquad (5.151)$$

for specified but arbitrary values of $\gamma_1, \ldots, \gamma_L$. In other words, we want to find a sequence of $L$ inputs that will take us from the origin to any specified target state in $L$ steps.

Equation (5.65) shows that

$$\mathbf{q}[L] = \mathbf{v}_1 r_1[L] + \cdots + \mathbf{v}_L r_L[L] \qquad (5.152)$$

so the sequence of $L$ inputs has to ensure $r_i[L] = \gamma_i$ for $i = 1, \ldots, L$, starting from $r_i[0] = 0$. The explicit solution in Eq. (5.68) shows that

$$\gamma_i = r_i[L] = \sum_{k=0}^{L-1} \lambda_i^{L-1-k} \beta_i x[k] \,. \qquad (5.153)$$

The above equations for all $i$ can be arranged into the following matrix form:

$$
\begin{bmatrix} \gamma_1 \\ \gamma_2 \\ \vdots \\ \gamma_L \end{bmatrix} = \begin{bmatrix} \beta_1 & \lambda_1\beta_1 & \lambda_1^2\beta_1 & \cdots & \lambda_1^{L-1}\beta_1 \\ \beta_2 & \lambda_2\beta_2 & \lambda_2^2\beta_2 & \cdots & \lambda_2^{L-1}\beta_2 \\ \vdots & \vdots & \vdots & \cdots & \vdots \\ \beta_L & \lambda_L\beta_L & \lambda_L^2\beta_L & \cdots & \lambda_L^{L-1}\beta_L \end{bmatrix} \begin{bmatrix} x[L-1] \\ x[L-2] \\ \vdots \\ x[0] \end{bmatrix}
$$

$$
= \begin{bmatrix} \beta_1 & 0 & \cdots & 0 \\ 0 & \beta_2 & \cdots & 0 \\ \vdots & \vdots & \ddots & \vdots \\ 0 & 0 & \cdots & \beta_L \end{bmatrix} \begin{bmatrix} 1 & \lambda_1 & \lambda_1^2 & \cdots & \lambda_1^{L-1} \\ 1 & \lambda_2 & \lambda_2^2 & \cdots & \lambda_2^{L-1} \\ \vdots & \vdots & \vdots & \cdots & \vdots \\ 1 & \lambda_L & \lambda_L^2 & \cdots & \lambda_L^{L-1} \end{bmatrix} \begin{bmatrix} x[L-1] \\ x[L-2] \\ \vdots \\ x[0] \end{bmatrix}
$$

All the $\beta_i$ are nonzero because the system is reachable, so the first matrix on the right of the latter equality is invertible. The second matrix is termed a Vandermonde matrix, which is invertible if (and clearly only if) the $\lambda_i$ are distinct. (If the Vandermonde matrix was not invertible with distinct $\lambda_i$, then some nontrivial linear combination of its columns would add up to the $\mathbf{0}$ vector, but that would allow us to construct a polynomial of degree $L-1$ that had $L$ distinct roots, which is a contradiction.)

It follows that the above set of equations can be solved to find $x[0], \ldots, x[L-1]$:

$$
\begin{bmatrix} x[L-1] \\ x[L-2] \\ \vdots \\ x[0] \end{bmatrix} = \begin{bmatrix} 1 & \lambda_1 & \cdots & \lambda_1^{L-1} \\ 1 & \lambda_2 & \cdots & \lambda_2^{L-1} \\ \vdots & \vdots & \cdots & \vdots \\ 1 & \lambda_L & \cdots & \lambda_L^{L-1} \end{bmatrix}^{-1} \begin{bmatrix} \beta_1 & 0 & \cdots & 0 \\ 0 & \beta_2 & \cdots & 0 \\ \vdots & \vdots & \ddots & \vdots \\ 0 & 0 & \cdots & \beta_L \end{bmatrix}^{-1} \begin{bmatrix} \gamma_1 \\ \gamma_2 \\ \vdots \\ \gamma_L \end{bmatrix}
$$

In the case of a second-order system, for example,

$$
\begin{bmatrix} x[1] \\ x[0] \end{bmatrix} = \begin{bmatrix} 1 & \lambda_1 \\ 1 & \lambda_2 \end{bmatrix}^{-1} \begin{bmatrix} \beta_1 & 0 \\ 0 & \beta_2 \end{bmatrix}^{-1} \begin{bmatrix} \gamma_1 \\ \gamma_2 \end{bmatrix}
$$

$$
= \frac{1}{\lambda_2 - \lambda_1} \begin{bmatrix} \lambda_2 & -\lambda_1 \\ -1 & 1 \end{bmatrix} \begin{bmatrix} \gamma_1/\beta_1 \\ \gamma_2/\beta_2 \end{bmatrix}. \tag{5.154}
$$

This second-order example suggests that the closer $\lambda_1$ and $\lambda_2$ are, the larger the values of the inputs required to attain a given target state. Also, larger values of the ratio $\gamma_i/\beta_i$ will require larger inputs.

Now suppose just the first mode is unreachable, that is, suppose $\beta_1 = 0$, with all other $\beta_i$ nonzero. The preceding formulation shows that the set of states reachable from the origin—from $\mathbf{q}[0] = \mathbf{0}$—by manipulation of the input is confined to the $(L-1)$-dimensional subspace of the state space that is spanned by the eigenvectors associated with the reachable modes. A calculation similar to the above then shows that within this reachable subspace, any target state can be reached from the origin in $L-1$ steps by appropriate choice of the input.

More generally, the part of the state space that is reachable from the origin by manipulation of the input is the subspace spanned by the eigenvectors of the reachable modes. It follows that a system is reachable if and only if one can reach an arbitrary target state from the origin of the state space by appropriate choice of the input signal. (These conclusions hold for CT systems as well, though the precise input history required to reach a target state is not as transparently computed.)

The interplay between the natural dynamics of a system and the motion that an external input attempts to impose on the system is brought out clearly in the following observation. For a reachable DT LTI system of order $\geq 2$, it is impossible to force the state to move from the origin outward strictly along an eigenvector, say its first eigenvector $\mathbf{v}_1$. The reason is that this would require all $r_i[n]$ for $i \neq 1$ to remain at 0 for $n > 0$, while $r_1[n]$ moved from 0 to a succession of nonzero values. This is impossible for a reachable system, since all the $r_i[n]$ are simultaneously excited by the input. Thus, in a reachable $L$th-order DT system, one can get to any point in the state space in $L$ steps, including to a point of the form $\mathbf{q}[L] = \gamma_1 \mathbf{v}_1$ for any $\gamma_1$, but cannot necessarily get to this point in less than $L$ steps.

**Reachability and Observability Matrices**   The question of reaching target states from the origin, treated in Example 5.11, can also be addressed without reference to modal coordinates, instead working directly with the initial system description. For the DT case we start with the LTI state-space model in Eq. (5.3). The set of states reachable in one step from the origin by appropriate choice of $x[0]$ is of the form $\mathbf{q}[1] = \mathbf{b}x[0]$, that is, states on the line spanned by the vector $\mathbf{b}$. The set of states reachable in two steps by appropriate choice of $x[0]$ and $x[1]$ is of the form

$$\mathbf{q}[2] = \mathbf{A}\mathbf{b}x[0] + \mathbf{b}x[1] \,, \tag{5.155}$$

that is, states in the plane spanned by the vectors $\mathbf{b}$ and $\mathbf{A}\mathbf{b}$. Proceeding in this fashion, we conclude—and this is already apparent in Eq. (5.104)—that the set of states reachable from the origin in $L$ steps is the space spanned by the columns of the $L \times L$ matrix

$$\mathcal{R}_L = \begin{bmatrix} \mathbf{A}^{L-1}\mathbf{b} & \mathbf{A}^{L-2}\mathbf{b} & \cdots & \mathbf{A}\mathbf{b} & \mathbf{b} \end{bmatrix} \,. \tag{5.156}$$

This space is referred to as the range of $\mathcal{R}_L$. Note that this reasoning did not assume $\mathbf{A}$ has distinct eigenvalues; the result holds in general. Under our assumption of distinct eigenvalues, the reachable space must be identical with the space spanned by the eigenvectors of the reachable modes of the system. The matrix $\mathcal{R}_L$ is referred to as the $L$-step reachability matrix, or simply the reachability matrix. If only $p < L$ modes of the system are reachable, then the reachable space is already spanned by the columns of $\mathcal{R}_p$, and adding more columns does not increase the set of reachable states.

For a CT system of the form in Eq. (5.1), the development is different. We omit the details, but note that the same conclusion is arrived at as in the DT case: the set of states reachable in finite time, starting at $\mathbf{q}(0) = \mathbf{0}$ and applying a well-chosen input, is precisely the range of the above matrix $\mathcal{R}_L$.

A development similar to Example 5.11 can be used to provide a geometric picture of observability. We define the unobservable subspace to be the subspace spanned by the eigenvectors associated with the unobservable modes. It then turns out that the zero-input response of the system for a given initial condition $\mathbf{q}[0]$ is 0 at all times $n \geq 0$ if and only if this initial condition is in the unobservable subspace. Thus any attempt to determine the initial state of the system from knowledge of the input and output signals for $n \geq 0$ will only be able to determine this initial state modulo the unobservable subspace. In other words, displacing the initial condition by a vector in the unobservable subspace will not change the output, for a given input.

We can also identify the unobservable subspace using a matrix that is "dual" to the $L$-step reachability matrix in Eq. (5.156). Specifically, consideration of Eqs. (5.3) and (5.4) shows that the unobservable subspace is precisely the set of vectors orthogonal to the rows of the following $L \times L$ matrix, referred to as the $L$-step observability matrix:

$$
\mathcal{O}_L =
\begin{bmatrix}
\mathbf{c}^T \\
\mathbf{c}^T \mathbf{A} \\
\vdots \\
\mathbf{c}^T \mathbf{A}^{L-2} \\
\mathbf{c}^T \mathbf{A}^{L-1}
\end{bmatrix}.
\tag{5.157}
$$

This space is referred to as the nullspace of $\mathcal{O}_L$.

While our focus has been on LTI systems so far, some of what we have developed applies to linear periodically varying (LPV) systems. The following example shows how asymptotic stability of an LPV system can be assessed.

**Example 5.12    Asymptotic Stability of an LPV System**

The stability of LPV systems can be analyzed by methods that are close to those used for LTI systems. Suppose, for instance, that

$$
\mathbf{q}[n+1] = \mathbf{A}[n]\mathbf{q}[n] , \quad \mathbf{A}[n] = \mathbf{A}_0 \text{ for even } n, \mathbf{A}[n] = \mathbf{A}_1 \text{ for odd } n.
\tag{5.158}
$$

Then

$$
\mathbf{q}[n+2] = \mathbf{A}_1 \mathbf{A}_0 \mathbf{q}[n]
\tag{5.159}
$$

for even $n$, so the dynamics of the even samples are governed by an LTI model, and the stability of the even samples is accordingly determined by the eigenvalues of the constant matrix $\mathcal{A}_{\text{even}} = \mathbf{A}_1 \mathbf{A}_0$.

The stability of the odd samples is similarly governed by the eigenvalues of the matrix $\mathcal{A}_{\text{odd}} = \mathbf{A}_0 \mathbf{A}_1$. The nonzero eigenvalues of the matrix $\mathcal{A}_{\text{odd}}$ are the same as those of $\mathcal{A}_{\text{even}}$ because if $\lambda$ is a nonzero eigenvalue of $\mathcal{A}_{\text{even}}$, that is, if $(\mathbf{A}_1 \mathbf{A}_0)\mathbf{v} = \lambda \mathbf{v}$ for nonzero $\lambda$ and $\mathbf{v}$, then $\mathbf{A}_0 \mathbf{v}$ must be nonzero and $(\mathbf{A}_0 \mathbf{A}_1)(\mathbf{A}_0 \mathbf{v}) = \lambda(\mathbf{A}_0 \mathbf{v})$, so $\lambda$ is an eigenvalue of $\mathcal{A}_{\text{odd}}$ as well. Thus either matrix can be used for a stability check.

As an example, suppose

$$
\mathbf{A}_0 = \begin{pmatrix} 0 & 1 \\ 0 & 3 \end{pmatrix} , \quad \mathbf{A}_1 = \begin{pmatrix} 0 & 1 \\ 4.25 & -1.25 \end{pmatrix} ,
\tag{5.160}
$$

with respective eigenvalues $(0, 3)$ and $(1.53, -2.78)$, so both matrices have eigenvalues of magnitude greater than 1. Now

$$
\mathcal{A}_{\text{even}} = A_1 A_0 = \begin{pmatrix} 0 & 3 \\ 0 & 0.5 \end{pmatrix} ,
\tag{5.161}
$$

and its eigenvalues are $(0, 0.5)$, which corresponds to an asymptotically stable system.

We conclude this chapter with an example that suggests—but stops well short of fully explaining—how system behavior changes and becomes more intricate when the eigenvalues are not distinct. The example is presented for the DT case, but the situation is similar for CT.

Example 5.13 **Reachability and Observability with Nondistinct Eigenvalues**

In this example, we consider three different third-order DT LTI systems of the form in Eqs. (5.3) and (5.4), with the following respective choices for the associated state evolution matrix $\mathbf{A}$:

$$\mathbf{A}_1 = \begin{bmatrix} \lambda_1 & 1 & 0 \\ 0 & \lambda_1 & 1 \\ 0 & 0 & \lambda_1 \end{bmatrix}, \quad \mathbf{A}_2 = \begin{bmatrix} \lambda_1 & 0 & 0 \\ 0 & \lambda_1 & 1 \\ 0 & 0 & \lambda_1 \end{bmatrix}, \quad \mathbf{A}_3 = \begin{bmatrix} \lambda_1 & 0 & 0 \\ 0 & \lambda_1 & 0 \\ 0 & 0 & \lambda_1 \end{bmatrix}. \quad (5.162)$$

All these matrices have their three eigenvalues at $\lambda_1$, but differ in structure. The structural differences are very apparent in the block diagrams in Figure 5.10, which show the associated systems, but omit the signals feeding in from $x[n]$ or out to $y[n]$, to avoid clutter in the diagrams.

The matrix $\mathbf{A}_1$ has only one independent eigenvector, for example $\mathbf{v}_1 = [1\ 0\ 0]^T$. Any other vector $\mathbf{v}$ that satisfies $(\lambda_1\mathbf{I} - \mathbf{A})\mathbf{v} = \mathbf{0}$ is a linear multiple of $\mathbf{v}_1$.

The matrix $\mathbf{A}_2$ has two independent eigenvectors, for example the same $\mathbf{v}_1$ as before, but also $\mathbf{v}_2 = [0\ 1\ 0]^T$. Any other solution $\mathbf{v}$ of $(\lambda_1\mathbf{I} - \mathbf{A})\mathbf{v} = \mathbf{0}$ is a linear combination of these two.

The matrix $\mathbf{A}_3$ has three independent eigenvectors, for instance the same $\mathbf{v}_1$ and $\mathbf{v}_2$ as before but also $\mathbf{v}_3 = [0\ 0\ 1]^T$. Any other solution $\mathbf{v}$ of $(\lambda_1\mathbf{I} - \mathbf{A})\mathbf{v} = \mathbf{0}$ is a linear combination of these three.

In the case of $\mathbf{A}_3$, the system is already in diagonal form, so some of our analysis in this chapter carries over directly. The ZIR of the associated system state will only involve exponentials of the form $\lambda_1^n$. The transfer function of the system will also only show, at most, a single pole at $\lambda_1$, not repeated poles, suggesting that we have at least two hidden modes in this case. Once our notion of reachability is extended to the repeated-eigenvalue case (which we will not do here), it can be shown that two of the three modes are always unreachable, even if the coupling from $x[n]$ to the three parallel paths in Figure 5.10(c) is nonzero. The reason is that the three branches of the system cannot be excited independently of each other by manipulating the single input signal $x[n]$. Similarly, once our notion of observability is extended to the repeated-eigenvalue case (which we will not do here), we can show that two of the three modes are also unobservable, even if the coupling from the three parallel paths to $y[n]$ in Figure 5.10(c) is nonzero. The reason is that the responses of the three branches cannot be teased apart on the basis of observations of the single output signal $y[n]$. What happens here for $\mathbf{A}_3$ is essentially what would happen for any $3 \times 3$ matrix $\mathbf{A}$ whose eigenvalues are all at $\lambda_1$ and that has three independent eigenvectors associated with $\lambda_1$.

For $\mathbf{A}_2$, the state ZIR will have terms of the form $\lambda_1^n$ as before, but also terms of the form $n\lambda_1^{n-1}$. The transfer function will show, at most, two poles at $\lambda_1$, suggesting that we have at least one hidden mode. In fact, at least one mode is always unreachable, and at least one mode is always unobservable, for this structure. What we see happening here for $\mathbf{A}_2$ is essentially what would happen for any $3 \times 3$ matrix $\mathbf{A}$ whose eigenvalues are all at $\lambda_1$ and that has only two independent eigenvectors associated with $\lambda_1$.

For $\mathbf{A}_1$, the state ZIR will have terms of the form $\lambda_1^n$ as before, but also terms of the form $n\lambda_1^{n-1}$ and $n^2\lambda_1^{n-2}$. The transfer function will show three poles at $\lambda_1$ if and only if there is a nonzero coupling from $x[n]$ to the first of the three stages in

Figure 5.10(a) and a nonzero coupling from the last of the three stages to $y[n]$. Equivalently, we will find three poles in the transfer function if and only if the bottom entry of the input vector $\mathbf{b}$ is nonzero, and the left entry of the output vector $\mathbf{c}^T$ is nonzero. Under these conditions, the system will be reachable and observable. When these conditions fail, we get more limited reachability and/or observability. What we see happening here for $\mathbf{A}_1$ is essentially what would happen for any $3 \times 3$ matrix $\mathbf{A}$ whose eigenvalues are all at $\lambda_1$ and that has only one independent eigenvector associated with $\lambda_1$.

(a) System 1

(b) System 2

(c) System 3

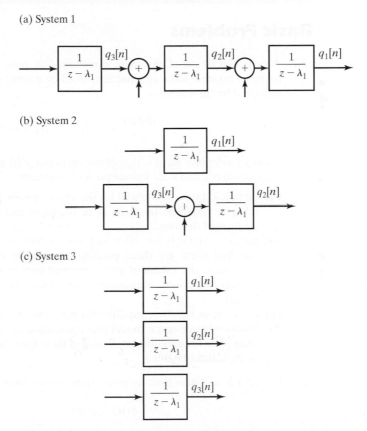

**Figure 5.10**   Block diagrams of the three systems in Example 5.13.

Chapter 6 examines how knowing a system's input and output waveforms, along with a model of the system dynamics, allows one to infer the internal state of the system. This then provides a rational basis for designing a feedback control scheme to regulate the behavior of the system.

## 5.5 FURTHER READING

The further reading suggestions at the end of Chapter 4 are also relevant to this chapter. Foundational material may be found in [Edw]. The dynamics of LTI systems, and usually including the state estimation and state feedback control methods presented in Chapter 6, are treated in numerous books, for instance [Ant], [Ch2], [Ka1], [Lue] and the classic [Zad]. For extensions to nonlinear systems, see [Kha], [Slo], [Vid].

## Problems

## Basic Problems

**5.1.** A particular object of unit mass, constrained to move in a straight line, is acted on by an external force $x(t)$ and restrained by a cubic spring. The system can be described by the equation

$$\frac{d^2p(t)}{dt^2} + kp(t) - \epsilon p^3(t) = x(t) \,,$$

where $p(t)$ denotes the position of the mass and $p^3(t)$ is the cube of the position; the quantities $k$ and $\epsilon$ are known positive constants.

**(a)** Obtain a state-space model for the above system, using physically meaningful state variables. Take $x(t)$ to be the input and let the output $y(t)$ be the position of the mass.

**(b)** Suppose $x(t) = 0$ for all $t$ and the system is in equilibrium. You will find that there are three possible equilibrium conditions of the system. Determine the values of your state variables in each of these three equilibrium conditions, expressing your results in terms of the parameters $k$ and $\epsilon$.

**(c)** For each of the three equilibrium positions you identified in (b), obtain a linearized state-space model that approximately describes small deviations away from the equilibrium. Which of these three linearized models, if any, is asymptotically stable?

**5.2.** Consider a nonlinear time-invariant state-space model described in the form

$$\dot{q}_1(t) = q_2(t)$$

$$\dot{q}_2(t) = -\beta q_1^3(t) + x(t) \,,$$

where $\beta$ is some positive constant.

**(a)** If the input $x(t)$ is fixed at a constant positive value $\bar{x} > 0$, determine the possible equilibrium values $\bar{q}_1$ and $\bar{q}_2$ of $q_1(t)$ and $q_2(t)$ respectively.

**(b)** If the input deviates by a small amount $\tilde{x}(t) = x(t) - \bar{x}$ from its equilibrium value, and if the state variables correspondingly deviate by small amounts $\tilde{q}_1(t) = q_1(t) - \bar{q}_1$ and $\tilde{q}_2(t) = q_2(t) - \bar{q}_2$ respectively from their equilibrium values, find a linearized LTI state-space model that approximately describes how these small deviations are related to each other. In other words, find an LTI state-space model that has $\tilde{q}_1(t)$ and $\tilde{q}_2(t)$ as state variables and $\tilde{x}(t)$ as input.

**(c)** Is your linearized model asymptotically stable?

**5.3.** Consider the linear dynamical system:

$$\dot{\mathbf{q}}(t) = \mathbf{A}\mathbf{q}(t)$$

$$y(t) = \mathbf{c}^T \mathbf{q}(t)$$

where $\mathbf{A}$ is a $2 \times 2$ matrix with eigenvalues $+1$ and $-1$ and corresponding eigenvectors $\begin{bmatrix} 1 \\ 4 \end{bmatrix}$ and $\begin{bmatrix} 1 \\ -4 \end{bmatrix}$, and $\mathbf{c} = \begin{bmatrix} 1 \\ 0 \end{bmatrix}$.

**(a)** Find $\mathbf{q}(t)$ and $y(t)$ for $t \geq 0$, when $\mathbf{q}(0) = \begin{bmatrix} 0 \\ 8 \end{bmatrix}$.

**(b)** Specify all possible initial states $\mathbf{q}(0)$ such that the output for $t \geq 0$ is bounded, i.e., $|y(t)| \leq M$ for some finite $M$ and for all $t \geq 0$.

**5.4.** Consider a system described by a second-order state-space model of the form

$$\dot{\mathbf{q}}(t) = \begin{bmatrix} -5 & 1 \\ 6 & 0 \end{bmatrix} \mathbf{q}(t) + \begin{bmatrix} 0 \\ 1 \end{bmatrix} x(t)$$

$$y(t) = \begin{bmatrix} -1 & 1 \end{bmatrix} \mathbf{q}(t)$$

**(a)** Determine the eigenvalues and corresponding eigenvectors of the system.

**(b)** Determine the response $\mathbf{q}(t)$ for $t \geq 0$ when $x(t) = 0$ for $t \geq 0$ and $\mathbf{q}(0) = \begin{bmatrix} 0 \\ 1 \end{bmatrix}$.

**(c)** Determine a nonzero choice for the initial condition $\mathbf{q}(0)$ such that the ZIR decays to zero as $t \to \infty$.

**5.5.** Consider the electrical circuit shown in Figure P5.5, with the voltage $w(t)$ of the voltage source taken as the input, and the voltage $u_R(t)$ across the resistor taken as the output.

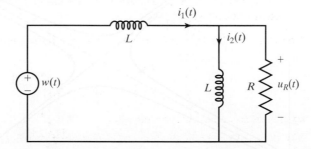

**Figure P5.5**

**(a)** Choosing $i_1(t)$ and $i_2(t)$ as state variables for the circuit, find the corresponding state-space description. In what follows, use $\mathbf{q}(t)$ to denote the state vector in your state-space description. Also assume $R/L = 1$.

**(b)** Find the natural frequencies (i.e., eigenvalues) $\lambda_1$ and $\lambda_2$ of the circuit, and compute the associated eigenvectors, $\mathbf{v}_1$ and $\mathbf{v}_2$ respectively.

**(c)** Specify for what nonzero initial conditions $\mathbf{q}(0)$ the state remains frozen at its initial value, $\mathbf{q}(t) = \mathbf{q}(0)$ for all $t > 0$, when the input is fixed at 0 (i.e., when $w(t) = 0$ for all $t \geq 0$).

**(d)** Using your results from (b), find all initial conditions $\mathbf{q}(0)$ for which each component of $\mathbf{q}(t)$ decays asymptotically to zero as $t \longrightarrow \infty$, when the input is fixed at 0.

**(e)** Using the results from (a) and (b), find all nonzero initial $\mathbf{q}(0)$ for which the output $u_R(t)$ decays asymptotically to zero as $t \longrightarrow \infty$, when the input is fixed at 0.

**5.6.** Consider a second-order state-space model for a particular causal system:

$$\frac{d\mathbf{q}(t)}{dt} = \mathbf{A}\mathbf{q}(t) + \mathbf{b}x(t)$$

$$y(t) = \mathbf{c}^T\mathbf{q}(t) + dx(t)$$

Suppose we know that the transfer function of the system is

$$H(s) = \frac{Y(s)}{X(s)} = \frac{s+1}{(s-3)(s+4)} .$$

Also, if the system is given a zero input, $x(t) = 0$ for all $t \geq 0$, then the state vector $\mathbf{q}(t)$ follows the trajectories (solid lines with arrows) shown in Figure P5.6 for a number of different initial conditions $\mathbf{q}(0) = [q_1(0)\ q_2(0)]^T$, labeled on the graph with circles.

**(a)** Determine the matrix $\mathbf{A}$.

**(b)** Is the system reachable? Is it observable?

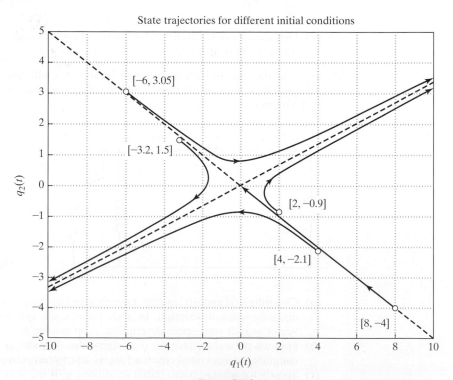

State trajectories for different initial conditions

**Figure P5.6**

**5.7.** You are given the following state-space model for a CT causal system:

$$\dot{\mathbf{q}}(t) = \begin{bmatrix} 0 & 1 \\ 0 & -2 \end{bmatrix} \mathbf{q}(t) + \begin{bmatrix} -2 \\ 4 \end{bmatrix} x(t) \,,$$

$$y(t) = \begin{bmatrix} -1 & 0 \end{bmatrix} \mathbf{q}(t) + x(t) \,.$$

For all the following questions, explain the reasoning behind your answers.

**(a)** Determine the natural frequencies of the system, i.e., the eigenvalues $\lambda_1$ and $\lambda_2$ of the matrix that governs state evolution, and find the associated eigenvectors $\mathbf{v}_1$ and $\mathbf{v}_2$ respectively.

**(b)** Suppose $x(t) = 0$ for $t \geq 0$. Specify the set of all possible nonzero initial conditions $\mathbf{q}(0)$ for which the state is asymptotically zero, i.e., for which the components $q_1(t)$ and $q_2(t)$ of $\mathbf{q}(t)$ go to 0. Is the system asymptotically stable?

**(c)** Still with $x(t) \equiv 0$, suppose the initial condition is not one of those you specified in (b). Specify in as much detail as you can where the state will end up, i.e., what $\mathbf{q}(\infty)$ will be.

**(d)** Show that the system has an unreachable mode, and determine which mode (or eigenvalue) is unreachable. Also, if we start the system out with $\mathbf{q}(0) = \mathbf{0}$ and are allowed to use any $x(t)$ we want, there is some region of the two-dimensional state-space that you can guarantee $\mathbf{q}(t)$ will not be in for any $t$. Specify this unreachable region as fully as you can.

**(e)** Suppose $\mathbf{q}(0) = \mathbf{0}$, and $x(t)$ is known to be bounded for all time, i.e., $x(t) \leq M < \infty$, but is otherwise unknown. Is $q_1(t)$ guaranteed to be bounded? Is $q_2(t)$ guaranteed to be bounded?

**(f)** Show that the system is observable.

**(g)** Using the state-space model above, show that you can express $q_1(t)$ and $q_2(t)$ at any time $t$ in terms of just $x(t)$, $\dot{x}(t)$, $y(t)$, and $\dot{y}(t)$, without knowing $\mathbf{q}(0)$. Explain why this would not be possible if the system were unobservable.

**(h)** Find the (zero-state) transfer function $H(s)$ of the system from its input $x(t)$ to its output $y(t)$. Is the system BIBO stable?

**(i)** Write the state-space description of a first-order system that has the same input and output, and the same transfer function $H(s)$, as the given system. You can use the symbol $q(t)$ to denote the scalar state of this system.

**5.8.** Consider the following state-space model:

$$\dot{\mathbf{q}}(t) = \begin{bmatrix} -3 & 1 \\ 0 & 0 \end{bmatrix} \mathbf{q}(t) + \begin{bmatrix} 1 \\ 0 \end{bmatrix} x(t)$$

$$y(t) = \begin{bmatrix} 1 & 0 \end{bmatrix} \mathbf{q}(t) \,.$$

Answer the following, and explain your reasoning.

**(a)** Is the system asymptotically stable?

**(b)** Is the system observable?

**(c)** If $x(t) = 0$ for $t \geq 0$, is there any nonzero initial condition $\mathbf{q}(0)$ for which the output response $y(t)$ for $t \geq 0$ is identically 0? If so, give an example of such an initial condition.

**(d)** Is the system reachable?

**(e)** Is the system BIBO stable?

**(f)** If $x(t) = e^{-2t}u(t)$ and if the initial state is

$$\mathbf{q}(0) = \begin{bmatrix} 1 \\ 0 \end{bmatrix},$$

what value does the state vector reach asymptotically, i.e., what is $\mathbf{q}(\infty)$?

**(g)** If the input is identically zero, i.e., $x(t) = 0$ for $t \geq 0$, and if the initial state is

$$\mathbf{q}(0) = \begin{bmatrix} 1 \\ 1 \end{bmatrix},$$

what value does the state vector reach asymptotically?

**5.9.** Consider a second-order CT causal LTI system with transfer function

$$H(s) = \frac{s+1}{s+3}$$

and state-space description of the standard form:

$$\frac{d}{dt}\mathbf{q}(t) = \mathbf{A}\mathbf{q}(t) + \mathbf{b}x(t), \qquad y(t) = \mathbf{c}^T\mathbf{q}(t) + dx(t).$$

Suppose we know that

$$\mathbf{b} = \begin{bmatrix} 2 \\ 0 \end{bmatrix}.$$

Also, with some particular initial state $\mathbf{q}(0)$, and with the input $x(t) = 0$ for $t \geq 0$, we get

$$\mathbf{q}(t) = \begin{bmatrix} e^{-3t} \\ e^t - 2e^{-3t} \end{bmatrix}.$$

**(a)** Determine $\mathbf{A}$, $\mathbf{c}^T$, and d.

**(b)** If $\mathbf{q}(0)$ is as in part (a), but now $x(t) = e^{-t}$ for $t \geq 0$, determine $y(t)$.

**(c)** If $\mathbf{q}(0)$ is as in part (a), but now

$$x(t) = q_1(t) - \frac{1}{8}q_2(t)$$

for $t \geq 0$, what is the general form of $y(t)$? You don't need to work out all the constants, but you should at least be able to determine whether $y(t)$ decays, grows, or oscillates, and at what rate, or determine that it stays constant.

**5.10.** The state equations for a causal LTI system are

$$\dot{\mathbf{q}}(t) = \mathbf{A}\mathbf{q}(t) + \mathbf{b}x(t)$$

$$y(t) = \mathbf{c}^T\mathbf{q}(t) + dx(t)$$

where

$$\mathbf{A} = \begin{bmatrix} 0 & 2 \\ 6 & -1 \end{bmatrix}, \quad \mathbf{b} = \begin{bmatrix} 1 \\ 1 \end{bmatrix}, \quad \mathbf{c}^T = \begin{bmatrix} 3 & -2 \end{bmatrix}, \quad d = 0.$$

**(a)** Find the eigenvalues $\lambda_1$ and $\lambda_2$ and the associated eigenvectors $\mathbf{v}_1$ and $\mathbf{v}_2$ of matrix $\mathbf{A}$.

**(b)** Is the system asymptotically stable?

(c) Is the system reachable? Is it observable?

(d) Find the system function $H(s) = Y(s)/X(s)$.

(e) Is the output of the system bounded when the input is bounded and the system is initially in the zero state? Reconcile your answer with your answer in part (b).

**5.11.** Consider the system

$$\frac{d}{dt}\begin{bmatrix} q_1(t) \\ q_2(t) \end{bmatrix} = \begin{bmatrix} -1 & 2 \\ 1 & 0 \end{bmatrix}\begin{bmatrix} q_1(t) \\ q_2(t) \end{bmatrix} + \begin{bmatrix} 1 \\ 0 \end{bmatrix}x(t)$$

$$y(t) = \begin{bmatrix} 1 & -1 \end{bmatrix}\begin{bmatrix} q_1(t) \\ q_2(t) \end{bmatrix}.$$

(a) Is this system asymptotically stable?

(b) Is the system reachable ? Is it observable?

(c) What is the transfer function of the system?

(d) If $x(t) = 0$ and $y(t) = 0$ for $t \geq 0$, can it be concluded that

$$\begin{bmatrix} q_1(t) \\ q_2(t) \end{bmatrix} = \begin{bmatrix} 0 \\ 0 \end{bmatrix}$$

for $t \geq 0$? If so, explain why. If not, find nonzero solutions $q_1(t)$ and $q_2(t)$.

**5.12.** A system with input $x(t)$ and output $y(t)$ is described in terms of the state vector $\mathbf{q}(t)$ as follows:

$$\frac{d\mathbf{q}(t)}{dt} = \begin{bmatrix} 1 & 3 \\ 0 & 1 \end{bmatrix}\mathbf{q}(t) + \begin{bmatrix} 0 \\ -1 \end{bmatrix}x(t)$$

$$y(t) = \begin{bmatrix} 2 & -3 \end{bmatrix}\mathbf{q}(t).$$

(a) Is the system observable? If not, specify the unobservable mode or modes.

(b) Is the system reachable? If not, specify the unreachable mode or modes.

(c) The system is not asymptotically stable. Is the system BIBO stable?

**5.13.** Consider the following state-space description of a CT system:

$$\dot{\mathbf{q}}(t) = \mathbf{A}\mathbf{q}(t) + \mathbf{b}x(t)$$

$$y(t) = \mathbf{c}^T\mathbf{q}(t)$$

where

$$\mathbf{A} = \mathbf{V}\begin{bmatrix} \lambda_1 & 0 \\ 0 & \lambda_2 \end{bmatrix}\mathbf{V}^{-1}.$$

Assume that the system is asymptotically stable. Suppose we would like to simulate this system on a digital computer, using samples of the input, state, and output at times $t = nT$, and approximating $\dot{\mathbf{q}}(t)$ at $t = nT$ by a forward difference (this is the forward-Euler approximation), i.e.,

$$\dot{\mathbf{q}}(nT) \approx \frac{1}{T}[\mathbf{q}(nT + T) - \mathbf{q}(nT)].$$

The resulting DT system can be written in the form

$$\mathbf{q}[n+1] \approx \mathbf{A}_d \mathbf{q}[n] + \mathbf{b}_d x[n]$$

$$y[n] \approx \mathbf{c}_d^T \mathbf{q}[n],$$

where $\mathbf{q}[n] = \mathbf{q}(nT), \quad x[n] = x(nT), \quad y[n] = y(nT).$

**(a)** Determine $\mathbf{A}_d$, $\mathbf{b}_d$, and $\mathbf{c}_d$ in terms of $\mathbf{A}$, $\mathbf{b}$, and $\mathbf{c}$.
**(b)** What are the natural frequencies of the DT system?
**(c)** For what range of values of $T$ is the DT system asymptotically stable?

**5.14.** Consider the second-order difference equation

$$y[n] - 3y[n-1] + 2y[n-2] = x[n-1] + 4x[n-2].$$

**(a)** Specify fourth-order, third-order, and second-order state-space models for the above difference equation.
**(b)** For each of the state-space models in (a), determine all of the modal frequencies, i.e., eigenvalues, and associated mode shapes, i.e., eigenvectors.
**(c)** For each modal frequency of the third-order model from (b), determine:
   (i) if the mode associated with the frequency is reachable; and
   (ii) if it is observable.
   Do your results suggest why there may be a second-order model with the same input-output behavior? Now test your second-order model for reachability and observability. Is there any reason to think that a lower-order model (i.e., a first-order model) could have the same input-output behavior?

**5.15.** Consider the causal DT system represented by the block diagram in Figure P5.15, where D denotes a delay element. The system can be described by the second-order state-space model

$$\mathbf{q}[n+1] = \mathbf{A}\mathbf{q}[n] + \mathbf{b}x[n]$$

$$y[n] = \mathbf{c}^T \mathbf{q}[n] + \mathrm{d}x[n]$$

where

$$\mathbf{q}[n] = \begin{bmatrix} q_1[n] \\ q_2[n] \end{bmatrix}.$$

Suppose the eigenvalues of $\mathbf{A}$ are $\lambda_1 = 2$ and $\lambda_2 = -\frac{1}{2}$. Both eigenvalues are reachable, but suppose $\lambda_1$ is unobservable.

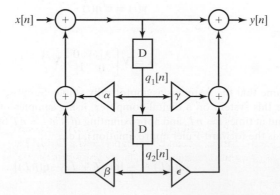

**Figure P5.15**

(a) Is the system asymptotically stable? Is it BIBO stable? You should be able to answer both questions with no computation at all.

(b) Express $\mathbf{A}, \mathbf{b}, \mathbf{c}^T$, and d in terms of the parameters in the block diagram.

(c) Determine $\alpha$ and $\beta$.

(d) If $\mathbf{q}[0] = \mathbf{0}$ and $x[n] = \delta[n]$, what is $\mathbf{q}[1]$? For this case, do $q_1[n]$ and $q_2[n]$ tend to 0 as $n \rightarrow \infty$?

**5.16.** You are given the following state-space model for a causal system:

$$\mathbf{q}[n+1] = \begin{bmatrix} 0 & 1 \\ -\frac{1}{2} & \frac{3}{2} \end{bmatrix} \mathbf{q}[n] + \begin{bmatrix} 1 \\ \frac{1}{2} \end{bmatrix} x[n]$$

$$y[n] = \begin{bmatrix} 0 & 1 \end{bmatrix} \mathbf{q}[n].$$

(a) What are the natural frequencies, or eigenvalues, of the system?

(b) Find a nonzero initial condition $\mathbf{q}[0]$ for which the ZIR of the state vector decays asymptotically to zero. Explain your reasoning.

(c) Find a nonzero initial condition $\mathbf{q}[0]$ from which the ZIR will not move, i.e., $\mathbf{q}[k] = \mathbf{q}[0]$ for $k > 0$.

(d) Is the system reachable?

(e) Will every bounded input produce a bounded state response, i.e., bounded $q_1[n]$ and $q_2[n]$? Explain your answer carefully.

(f) Is the system observable?

(g) What is the transfer function of the system? Is the system BIBO stable?

**5.17.** Consider a DT system with the following state-space description:

$$\mathbf{q}[n+1] = \begin{bmatrix} 3 & 0 \\ -\frac{3}{2} & \frac{1}{2} \end{bmatrix} \mathbf{q}[n] + \begin{bmatrix} 1 \\ 0 \end{bmatrix} x[n]$$

$$y[n] = \begin{bmatrix} \frac{6}{5} & 2 \end{bmatrix} \mathbf{q}[n].$$

(a) Find the natural frequencies of the system and determine whether the system is asymptotically stable.

(b) Determine which natural frequencies of the system are reachable.

(c) Determine which natural frequencies of the system are observable.

(d) Suppose $x[n] = 0$ for $n \geq 0$. Determine all values of of $\mathbf{q}[0]$ that simultaneously satisfy both of the following conditions:
    (i) $y[0] = 5$, and
    (ii) $y[n]$ decays to 0 as $n \rightarrow \infty$.

**5.18.** Consider the following causal DT state-space system

$$\mathbf{q}[n+1] = \mathbf{A}\mathbf{q}[n] + \mathbf{b}x[n]$$

$$y[n] = \mathbf{c}^T\mathbf{q}[n] + \mathrm{d}x[n]$$

where

$$\mathbf{A} = \begin{bmatrix} \frac{1}{2} & \frac{3}{2} \\ 0 & 2 \end{bmatrix}$$

and $\mathbf{b}, \mathbf{c}^T$, and d are unknown.

(a) Determine the eigenvalues $\lambda_1, \lambda_2$ and the eigenvectors $\mathbf{v}_1$ and $\mathbf{v}_2$ of $\mathbf{A}$.

(b) Specify all the possible initial conditions (i.e., $\mathbf{q}[0]$) such that the ZIR will decay asymptotically to zero.

**(c)** Indicate which, if any, of the following could be ZSR system functions for the system:

(i) $H(z) = \dfrac{z^{-1}}{1 - \frac{1}{3}z^{-1}}$ ;

(ii) $H(z) = \dfrac{1 + \frac{1}{2}z^{-1}}{1 - \frac{1}{2}z^{-1}}$ .

**(d)** We want to describe the system in terms of the new set of state variables

$$\mathbf{f}[n] = \begin{bmatrix} f_1[n] \\ f_2[n] \end{bmatrix} ,$$

where $f_1[n] = q_1[n] + q_2[n]$ and $f_2[n] = q_1[n] - q_2[n]$. In other words, we want to describe the system using the following equations:

$$\mathbf{f}[n+1] = \overline{\mathbf{A}}\mathbf{f}[n] + \overline{\mathbf{b}}x[n]$$

$$y[n] = \overline{\mathbf{c}}^T \mathbf{f}[n] + \overline{d}x[n] .$$

(i) Determine $\overline{\mathbf{A}}, \overline{\mathbf{b}}, \overline{\mathbf{c}}^T$, and $\overline{d}$ in terms of $\mathbf{A}, \mathbf{b}, \mathbf{c}^T$, and d.
(ii) We can express the ZIR of the system in the general form

$$\mathbf{f}[n] = \alpha_1 \overline{\lambda}_1^n \overline{\mathbf{v}}_1 + \alpha_2 \overline{\lambda}_2^n \overline{\mathbf{v}}_2 ,$$

where $\alpha_1$ and $\alpha_2$ are constants. Determine $\overline{\lambda}_1, \overline{\lambda}_2, \overline{\mathbf{v}}_1$, and $\overline{\mathbf{v}}_2$.

## Advanced Problems

**5.19.** A particular mechanical system involves a single mass whose position $r(t)$ is governed by the differential equation

$$\frac{d^2}{dt^2}r(t) - 5\left(1 - r^2(t)\right)\frac{d}{dt}r(t) + r(t) = x(t) ,$$

where $x(t)$ denotes some input force acting on the system. Consider

$$y(t) = \int_0^t r(\sigma)\, d\sigma$$

to be the output of interest.

**(a)** Pick appropriate state variables and write a (nonlinear) state-space description of the system for $t \geq 0$, consisting of state evolution equations and an instantaneous output equation. (*Hint:* The model will *not* be second order.) Is your state-space description time-invariant or time-varying?

**(b)** Determine the equilibrium values of the state variables in your model, corresponding to the constant input $x(t) \equiv 0$. Then obtain a linearized model describing the state and output behavior of the system for small deviations of $x(t)$ and of the state variables from their equilibrium values.

**5.20.** Consider a pendulum suspended from a support that allows the pendulum to swing without friction in a vertical plane. We idealize the pendulum as comprising a point mass $m$ at the end of a massless rod of length $R$, and denote the angle the pendulum makes with the downward vertical by $\theta(t)$. Let $\gamma$ denote the acceleration due to gravity. Assume we can exert a torque $x(t)$ on the pendulum. We now arrange for the plane containing the pendulum to be rotated at a constant angular velocity $\omega_0$, as indicated in Figure P5.20.

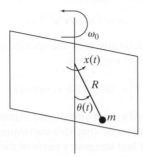

**Figure P5.20**

It can be shown that the motion of this spinning pendulum is then described by the equation

$$\frac{d^2}{dt^2}\theta(t) = \sin\theta(t)\left(\omega_0^2\cos\theta(t) - \frac{\gamma}{R}\right) + \frac{1}{mR^2}x(t) .$$

**(a)** Choose appropriate state variables and write a (nonlinear) state evolution equation and output equation for the system, with $x(t)$ as the input and $\theta(t)$ as the output.

**(b)** Suppose the input torque is identically zero, $x(t) \equiv 0$. If $\omega_0^2 < \gamma/R$, then there are only two equilibrium points, corresponding to the pendulum hanging straight down for all time, i.e., $\theta(t) \equiv 0$, or to its being in the vertical inverted position for all time, $\theta(t) \equiv \pi$. Linearize your state-space model from (a) around the equilibrium corresponding to the pendulum hanging straight down, and compute the natural frequencies of the system (i.e., the eigenvalues of the state evolution matrix). Will the ZIR of the linearized system for an arbitrary initial condition be exponentially decaying, exponentially growing, or oscillatory?

**(c)** If the input torque is identically zero, but $\omega_0^2 > \gamma/R$, there are four equilibrium points. Compute them. One of these equilibrium points again corresponds to the pendulum hanging straight down, and so the linearized model around this equilibrium point still has the same form as that in (b), though a parameter will have changed sign. Compute the new natural frequencies of the system. Will the ZIR of the system for an arbitrary initial condition in this case be exponentially decaying, exponentially growing, or oscillatory? Also determine whether the linearized model is (i) reachable, and (ii) observable.

**5.21.** Consider an undriven CT LTI state-space model of the form $\dot{\mathbf{q}}(t) = \mathbf{A}\mathbf{q}(t)$, and suppose the real matrix $\mathbf{A}$ has a complex eigenvalue $\lambda = \sigma + j\omega$, where $\sigma$ and $\omega$ are both real, with $\omega \neq 0$.

**(a)** Explain why the eigenvector associated with $\lambda$ must be of the form $\mathbf{v} = \mathbf{u} + j\mathbf{w}$ for some real $\mathbf{u}$ and $\mathbf{w}$ that form an independent pair of vectors (which also implies $\mathbf{u} \neq \mathbf{0}$ and $\mathbf{w} \neq \mathbf{0}$). (*Hint:* Show that otherwise one cannot satisfy the defining equation $\mathbf{A}\mathbf{v} - \lambda\mathbf{v}$.)

**(b)** Show that the complex conjugate of $\lambda$, namely $\lambda^* = \sigma - j\omega$, is also an eigenvalue, and that its associated eigenvector is $\mathbf{v}^* = \mathbf{u} - j\mathbf{w}$.

**(c)** In the modal solution to the above undriven system $\dot{\mathbf{q}}(t) = \mathbf{A}\mathbf{q}(t)$ for an arbitrary initial condition $\mathbf{q}(0)$, the complex conjugate eigenvalue pair will contribute a term of the following form:

$$\alpha \mathbf{v} e^{\lambda t} + \alpha^* \mathbf{v}^* e^{\lambda^* t} ,$$

where $\alpha = \gamma + j\xi$, and is determined by the specific value of $\mathbf{q}(0)$. Show that this sum simplifies to the real expression

$$e^{\sigma t}\Big[\mathbf{v}_c \cos(\omega t) + \mathbf{v}_s \sin(\omega t)\Big]$$

for some real vectors $\mathbf{v}_c$ and $\mathbf{v}_s$ that you should express in terms of the given quantities—and you'll then note that this corresponds to a motion in the plane spanned by the real and imaginary parts of the eigenvector $\mathbf{v}$, namely $\mathbf{u}$ and $\mathbf{w}$. Qualitatively describe what this motion looks like for the cases $\sigma < 0$, $\sigma = 0$, and $\sigma > 0$.

**(d)** Suppose that $\mathbf{A}$ is $3 \times 3$, and that in addition to the eigenvalues $\lambda$ and $\lambda^*$ its remaining eigenvalue is at $-10$, for example, with associated real eigenvector $\mathbf{v}_r$. By interpreting the modal solution for the system, qualitatively describe the state trajectory that originates from some arbitrary initial condition $\mathbf{q}(0)$, i.e., describe the trajectory in general geometric terms, relating it to $\mathbf{v}_r$ and the plane spanned by $\mathbf{u}$ and $\mathbf{w}$. Do this for the cases $\sigma < 0$, $\sigma = 0$, and $\sigma > 0$.

**5.22.** Consider the LTI circuit in Figure P5.22.

Use the capacitor voltages $q_1(t)$ and $q_2(t)$ as state variables for this problem.

**(a)** Suppose $x(t) = 0$ for $t \geq 0$ in the circuit. By invoking the symmetry of the resulting circuit, answer the following questions:

(i) Find an initial condition vector $\mathbf{q}(0) = \mathbf{w}_1$ such that the subsequent response, i.e., $\mathbf{q}(t)$ for $t > 0$, involves zero current flowing in either direction through the point $F$. Show that under this condition each entry of $\mathbf{q}(t)$ is just a multiple of a single exponential $e^{\mu_1 t}$, and determine $\mu_1$.

(ii) Find an initial condition vector $\mathbf{q}(0) = \mathbf{w}_2$ such that the subsequent response, i.e., $\mathbf{q}(t)$ for $t > 0$, keeps node $F$ at zero potential with respect to ground. Show that under this condition each entry of $\mathbf{q}(t)$ is just a multiple of a single exponential $e^{\mu_2 t}$, and determine $\mu_2$.

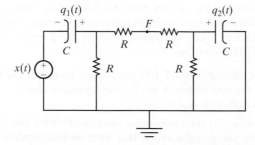

**Figure P5.22**

**(b)** Write down a second-order CT LTI state-space description for the circuit in which the potential at $F$ is specified to be the output of interest $y(t)$. Evaluate $\mathbf{A}$, $\mathbf{b}$, $\mathbf{c}^T$, and d in the resulting model

$$\frac{d\mathbf{q}(t)}{dt} = \mathbf{A}\mathbf{q}(t) + \mathbf{b}x(t)$$

$$y(t) = \mathbf{c}^T\mathbf{q}(t) + dx(t)$$

for the case where $C = 1$ and $R = 4$.

**(c)** Find the eigenvalues $\lambda_1$ and $\lambda_2$ of the matrix $\mathbf{A}$ that you obtained in (b), and also determine their associated eigenvectors, $\mathbf{v}_1$ and $\mathbf{v}_2$ respectively.

**(d)** Explain any relationship you discover between the quantities you computed in (a) and those you computed in (b).

**(e)** Still assuming that $x(t) = 0$ for $t \geq 0$, obtain an expression, in terms of $\lambda_1$, $\lambda_2$, $\mathbf{v}_1$, and $\mathbf{v}_2$, for the general solution of the state evolution equation given in part (b). For the specific case where the initial voltages on the capacitors are given by $q_1(0) = 1$ and $q_2(0) = 5$, what does this expression reduce to?

**(f)** Carry out a similarity transformation for the state-space model in part (b), with the choice

$$\mathbf{M} = [\mathbf{v}_1 \quad \mathbf{v}_2] .$$

The columns of $\mathbf{M}$ are the eigenvectors you calculated in part (c), and

$$\mathbf{q}(t) = \mathbf{M}\mathbf{r}(t) .$$

Determine the resulting state-space model with $\mathbf{r}(t)$ as the new state vector.

**(g)** Using your results from (f), determine the reachability and observability of each of the two modes. Interpret your results physically.

**5.23.** Consider the single-input, single-output $L$th-order CT LTI state-space system

$$\dot{\mathbf{q}}(t) = \mathbf{A}\mathbf{q}(t) + \mathbf{b}x(t) , \quad y(t) = \mathbf{c}^T\mathbf{q}(t) + dx(t) ,$$

whose transfer function is $H(s) = v(s)/a(s)$, where $a(s) = \det(s\mathbf{I} - \mathbf{A})$ is the characteristic polynomial of the system.

**(a)** For $d \neq 0$ the inverse system has a state-space representation involving the same state vector $\mathbf{q}(t)$ but input $y(t)$ and output $x(t)$. Determine this state-space representation, i.e., express the quantities $\mathbf{A}_{in}, \mathbf{b}_{in}, \mathbf{c}_{in}^T,$ and $d_{in}$ in the state-space representation below in terms of $\mathbf{A}, \mathbf{b}, \mathbf{c}^T,$ and d:

$$\dot{\mathbf{q}}(t) = \mathbf{A}_{in}\mathbf{q}(t) + \mathbf{b}_{in}y(t) \quad x(t) = \mathbf{c}_{in}^T\mathbf{q}(t) + d_{in}y(t) .$$

**(b)** Assuming $d \neq 0$, find an expression in terms of the quantities $\mathbf{A}, \mathbf{b}, \mathbf{c}^T,$ and d for the polynomial $v(s)$ defined by the expression for $H(s)$ given above.

**5.24. (a)** Suppose System 1 in Figure P5.24 is described by the first-order state-space model

$$\frac{dq_1(t)}{dt} = \gamma q_1(t) + x_1(t)$$

$$y_1(t) = q_1(t) + x_1(t) ,$$

where $\gamma$ is a parameter.

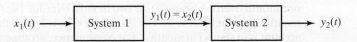

**Figure P5.24**

(i) What is its transfer function $H_1(s)$?

(ii) For what values of $\gamma$, if any, is the state-space model unreachable? Unobservable? Asymptotically stable?

**(b)** Assuming the transfer function of System 2 is $H_2(s) = \dfrac{s+1}{s-2}$, find a first-order state-space model for System 2, using $q_2(t)$ to denote its state variable. Is your model reachable? Is it observable?

**(c)** (i) Combine the state-space models in (a) and (b) to obtain a second-order state-space model for the overall system, using $\begin{bmatrix} q_1(t) \\ q_2(t) \end{bmatrix}$ as the state vector, $x_1(t)$ as the input, and $y_2(t)$ as the output.

(ii) Find the natural frequencies $\lambda_1$ and $\lambda_2$ of the system. For which values of $\gamma$ is the system asymptotically stable? Also explain how to pick the initial conditions $q_1(0)$ and $q_2(0)$, such that the ZIR of the system state vector only contains terms involving $e^{\lambda_1 t}$ and not $e^{\lambda_2 t}$, and vice versa.

(iii) Compute the transfer function $H(s)$ from $x_1(t)$ to $y_2(t)$ using the model in (i), and verify that it equals $H_1(s)H_2(s)$.

(iv) For what values of $\gamma$, if any, is the system unreachable? Which natural frequencies are unreachable? For what values of $\gamma$ is the system unobservable? Which natural frequencies are unobservable? Interpret your results in terms of pole-zero cancellations.

**(d)** Suppose the overall system had System 2 preceding System 1, reversing the order in which the systems are cascaded, so that $x_2(t)$ was the overall input and $y_1(t)$ was the overall output. You could find a state-space model for this interconnection, and assess the reachability and observability of the interconnection in the same manner as above. Without that detailed analysis what is your educated guess as to how the answers to part (iv) of (c) above would change for this case? Present your reasoning.

**5.25.** Figure P5.25 depicts a reachable and observable second-order system placed in series with a reachable and observable third-order system. The overall system is found to have one hidden mode.

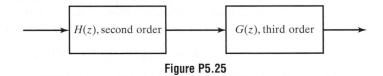

**Figure P5.25**

If the hidden mode corresponds to a pole in the original second-order system, is the hidden mode of the overall system unreachable and/or unobservable? Explain.

# Extension Problems

**5.26.** A bank has developed the following model for the monthly status of the account of a credit-card customer:

The account may be paid up in the current month (status 0), or it may be one month overdue (status 1), or two months overdue (status 2), or three or more months overdue (status 3). Let $q_i[n]$ for $i = 0, 1, 2, 3$ be the probability that the account is in status $i$ in month $n$. For convenience, arrange these status occupancy probabilities in a vector $\mathbf{q}[n]$. Let $p_{ij}$ denote the probability of going to status $i$ at the next time instant $n+1$, from status $j$ at the present time instant $n$, and suppose we have $p_{00} = 0.9$, $p_{01} = 0.5$, $p_{02} = 0.3$, and $p_{03} = 0.2$ for all $n$.

**(a)** Show that $\mathbf{q}[n+1] = \mathbf{A}\mathbf{q}[n]$ for some matrix $\mathbf{A}$ that you should fully specify. Find the eigenvalues and eigenvectors of $\mathbf{A}$. Note that there is one eigenvalue whose value is 1, and whose associated eigenvector $\mathbf{v}_1$ has all its components positive; call this eigenvalue $\lambda_1$. Note also that the other three eigenvalues have magnitudes less than 1. Furthermore, the eigenvectors associated with these three eigenvalues have components that sum to 0; explain this feature of these three eigenvectors, using the definition of eigenvalues, eigenvectors, and the properties of this particular matrix $\mathbf{A}$. (*Hint:* What vector do the columns of $\mathbf{A}$ sum to?)

**(b)** Now explain in detail how the eigenvalues and eigenvectors show you that the status occupancy probabilities will asymptotically approach constant steady-state values for very large $n$, no matter what the initial occupancy probability vector $\mathbf{q}[0]$ is. Find these steady-state probabilities, and denote the corresponding vector by $\mathbf{q}[\infty]$. Next show that a good approximation to $\mathbf{q}[n]$ for large $n$ (but before steady state is reached) takes the form

$$\mathbf{q}[n] \approx \mathbf{q}[\infty] + \mu^n \mathbf{w}$$

for suitably chosen $\mu$ and $\mathbf{w}$. Specify $\mu$. Also, for the case where $q_0[0] = 1$, specify $\mathbf{w}$.

**5.27.** Show that any two minimal LTI state-space realizations of a transfer function are related by a similarity transformation.

**5.28.** The intent of this problem is to give you a feel for how the results in this chapter on modal solutions and reachablity can change when eigenvalues are repeated. The full story is more elaborate, but is based on ideas that are exposed here (also see Example 5.13).

Consider the state-space model

$$\mathbf{q}[n+1] = \mathbf{A}\mathbf{q}[n] + \mathbf{b}x[n], \quad y[n] = \mathbf{c}^T\mathbf{q}[n] + dx[n] ,$$

where

$$\mathbf{A} = \begin{bmatrix} 0 & 0 & 0 \\ 1 & 0 & 0 \\ 0 & 1 & 0 \end{bmatrix} , \quad \mathbf{b} = \begin{bmatrix} 1 \\ 0 \\ 0 \end{bmatrix} , \quad \mathbf{c}^T = \begin{bmatrix} c_1 & c_2 & c_3 \end{bmatrix} ,$$

and d is some scalar constant.

**(a)** What are the eigenvalues of $\mathbf{A}$? How many independent eigenvectors can you find for $\mathbf{A}$? Can $\mathbf{A}$ be transformed to diagonal form, i.e., expressing it as $\mathbf{A} = \mathbf{M}\mathbf{\Lambda}\mathbf{M}^{-1}$ for a diagonal matrix $\mathbf{\Lambda}$ and some invertible matrix $\mathbf{M}$? (*Hint:* If such a representation were possible, then the diagonal elements of $\mathbf{\Lambda}$ would have to be the eigenvalues of $\mathbf{A}$, and the columns of $\mathbf{M}$ would be the associated eigenvectors, with invertibility of $\mathbf{M}$ ensuring that the eigenvectors were independent.)

**(b)** Determine $\mathbf{A}^k$ for all $k > 1$. Now determine, given an arbitrary initial state and zero input, how many steps it takes for the state to go to zero? The system is termed *deadbeat* because the ZIR of its state vector goes to zero in a finite number of steps, rather than decaying asymptotically.

**(c)** Draw a delay-adder-gain block diagram of the system, and using this, or in some other way, determine the unit sample response $h[n]$ and transfer function $H(z)$ from input $x$ to output $y$. Does it appear as though the system has any unreachable or unobservable eigenvalues? We have not defined unreachability or unobservability formally for the case of repeated eigenvalues, so what you are being asked here is to use any informal notion of what these concepts mean in order to guess at a plausible answer, along with an explanation of your reasoning.

**(d)** How do your answers in (c) change if the input vector is changed to

$$\mathbf{b} = \begin{bmatrix} 0 \\ 1 \\ 0 \end{bmatrix} ?$$

**5.29. (a)** Find a state-space description for the circuit in Figure P5.29, in the form

$$\dot{\mathbf{q}}(t) = \mathbf{A}\mathbf{q}(t) + \mathbf{b}i(t)$$

$$v(t) = \mathbf{c}^T\mathbf{q}(t) + \mathrm{d}i(t) \,.$$

Choose as state variables the current in the inductor, $i_L(t)$, and the voltage across the capacitor, $v_C(t)$. For the remainder of this problem, let $L$ and $C$ equal 1.

**(b)** Calculate the eigenvalues $\lambda_1$ and $\lambda_2$ of $\mathbf{A}$ in terms of $R$, and the transfer function $H(s) = V(s)/I(s)$. This transfer function can be computed directly from the circuit by determining its input impedance, or it can be computed from the state-space representation via the expression given within the chapter, namely

$$H(s) = \mathbf{c}^T(s\mathbf{I} - \mathbf{A})^{-1}\mathbf{b} + \mathrm{d} \,.$$

For $R \neq 1$, is the system reachable and observable? You should be able to determine this quite easily from the form of $H(s)$.

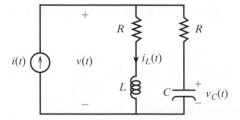

**Figure P5.29**

Given your expression for $H(s)$ in part (b), you might (correctly) think that there are only three possibilities when $R = 1$: (i) both modes are unreachable; (ii) both modes are unobservable; or (iii) one mode is observable but unreachable, while the other one is reachable but unobservable. However, what complicates the discussion of reachability and observability for $R = 1$ is the fact that now $\lambda_1 = \lambda_2$, while our results were developed for the case of distinct eigenvalues. Nevertheless, the basic ideas remain the same: an unreachable mode is one that cannot be excited from the input; an unobservable mode is one that cannot be seen at the output. Proceeding with these notions of what unreachability and unobservability mean in the repeated-eigenvalue case, parts (c) and (d) below are designed to help you rule out possibilities (i) and (ii) above. All three remaining parts of this problem apply to the case $R = 1$.

(c) Find the transfer functions from the input $i(t)$ to each of the two state variables. Can you conclude that at least one of the modes is reachable?

(d) Find the transform-domain expression relating an arbitrary initial state $\mathbf{q}(0)$ to the output $v(t)$, when the input is identically zero, $i(t) \equiv 0$. Can you conclude that at least one of the modes is observable?

(e) Only one of the following equations, for some appropriate choice of the parameters, precisely represents the set of voltage waveforms $v(t)$ that are possible for this circuit, assuming arbitrary initial conditions and an arbitrary input $i(t)$. Determine which one, and specify the coefficients. Explain.

(i) $v(t) = \alpha i(t)$;

(ii) $[dv(t)/dt] + \beta v(t) = \alpha\Big([di(t)/dt] + \beta i(t)\Big)$;

(iii) $[d^2v(t)/dt^2] + \gamma[dv(t)/dt] + \beta v(t) = \alpha\Big([d^2i(t)/dt^2] + \gamma[di(t)/dt] + \beta i(t)\Big)$.

# 6    State Observers and State Feedback

The modal-form representation developed in Chapter 5 for the solution of a linear and time-invariant (LTI) state-space model of a causal system shows explicitly that the state at any given time summarizes everything about the past that is relevant to future behavior of such a model. More specifically, given the value of the state vector at some initial instant, and given the entire input trajectory from the initial instant onward over some interval of time, the entire future state and output trajectories of the model over that interval can be determined. The same general conclusion holds for nonlinear and time-varying state-space models, although they are generally far less tractable analytically. We will continue to focus on LTI state-space models.

It is typically the case that direct measurements of the full state of a system are not available, and therefore the initial state of the system is unknown. Uncertainty about the initial state generates uncertainty about the future state trajectory, even if the model for the system is perfect, and even if we have accurate knowledge of the inputs to the system. The initial state and subsequent trajectory therefore have to be inferred, using the available information, namely the known or measured signals, along with the model of how these signals are interrelated.

Sections 6.1 and 6.2 of this chapter are devoted to addressing the issue of state trajectory estimation, given uncertainty about the initial state of the system. We shall see that the state can actually be determined under appropriate conditions, using what is called a state observer. The observer employs a model of the system along with measurements of both the input and output trajectories of the system, in order to asymptotically infer the state trajectory.

Sections 6.3 and 6.4 of the chapter examine how the input to the system can be controlled in order to yield desirable system behavior. We demonstrate that having knowledge of the present state of the system provides a powerful basis for designing feedback control to stabilize or otherwise improve the behavior of the resulting closed-loop system. When direct measurements of the state are not available, the state estimate provided by an observer turns out to suffice.

## 6.1 PLANT AND MODEL

It is important to make a distinction between the actual, physical, causal system that we are interested in studying, working with, or controlling—this is often termed the plant—and our idealized model for the plant. The plant is usually complex, nonlinear, and time-varying, typically requiring an infinite number, or even a continuum, of state variables and parameters to represent it with ultimate fidelity. The model, on the other hand, is usually an idealized and simplified LTI representation, of relatively low order, that aims to capture the behavior of the plant in some limited regime of its operation, while remaining tractable for analysis, computation, simulation, and design.

The inputs to the model represent the inputs acting on or driving the actual plant, and the outputs of the model represent signals in the plant that are accessible for measurement or are otherwise of interest. In practice not all the inputs to the plant are known. There will generally be unmeasured disturbance inputs that can only be characterized in some general way, perhaps as random processes. Similarly, the measured outputs of the plant will differ from what would be predicted on the basis of our limited model. These differences are partly due to modeling errors and approximations, and partly because of measurement noise.

We focus on the DT case first, but essentially everything carries over in a natural way to the CT case. We shall only treat situations where the underlying plant differs from the model in very limited ways. Specifically, suppose the plant is correctly described by the following $L$th-order LTI state-space equations:

$$q[n+1] = Aq[n] + bx[n] + w[n] , \tag{6.1}$$

$$y[n] = c^T q[n] + dx[n] + \zeta[n] . \tag{6.2}$$

These equations are represented in Figure 6.1. Here $x[n]$ is the known scalar control input, and $w[n]$ denotes a vector of unknown disturbances that drive the plant, not necessarily through the same mechanisms or actuators or channels as the input $x[n]$. For example, perhaps $w[n] = fv[n]$, where $v[n]$ is a scalar disturbance signal and $f$ is a vector describing how this scalar disturbance drives the system, just as $b$ describes how $x[n]$ drives the system. We refer to $w[n]$ as the plant disturbance. The state vector $q[n]$ is also not known. The output $y[n]$ is a known or measured quantity, while $\zeta[n]$ denotes the unknown

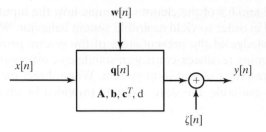

**Figure 6.1** Representation of a DT LTI plant with known control input $x[n]$ and available output measurements $y[n]$, but with unknown disturbance input vector $\mathbf{w}[n]$, state $\mathbf{q}[n]$, and measurement noise $\zeta[n]$.

noise component of this measured output. We refer to $\zeta[n]$ as measurement noise.

With the above equations representing the true plant, we seek a model that might be used to simulate the plant, deduce the plant's internal behavior from available measurements, or support control design. We shall assume that not only $x[n]$ and $y[n]$ are known but also the matrices $\mathbf{A}, \mathbf{b}, \mathbf{c}^T$, and d that govern the plant. We further suppose that nothing is known about the disturbance variables in $\mathbf{w}[n]$ and the measurement noise $\zeta[n]$, or alternatively, that they can be represented as zero-mean random processes. The simplest approach to constructing a model for the plant behavior is to ignore these disturbance and noise variables. The resulting model then takes the following LTI state-space form:

$$\widehat{\mathbf{q}}[n+1] = \mathbf{A}\widehat{\mathbf{q}}[n] + \mathbf{b}x[n] , \tag{6.3}$$

$$\widehat{y}[n] = \mathbf{c}^T\widehat{\mathbf{q}}[n] + \mathrm{d}x[n] . \tag{6.4}$$

The input that drives this model is the same $x[n]$ that is an input to the plant, and is therefore known. However, the state $\widehat{\mathbf{q}}[n]$ and output $\widehat{y}[n]$ of the model will generally differ from the corresponding true state $\mathbf{q}[n]$ and measured output $y[n]$ of the plant, because the true initial state $\mathbf{q}[0]$ of the plant is unknown, and the plant state and output are additionally perturbed by $\mathbf{w}[n]$ and $\zeta[n]$ respectively.

As already noted, several sources of uncertainty are ignored here. At the very least, there will be discrepancies between the actual and assumed parameter values—that is, between the actual entries of $\mathbf{A}, \mathbf{b}, \mathbf{c}^T$, and d in Eqs. (6.1) and (6.2) and the assumed entries of these matrices in Eqs. (6.3) and (6.4) respectively. These discrepancies could have been acknowledged by denoting the matrices in Eqs. (6.3) and (6.4) by $\widehat{\mathbf{A}}, \widehat{\mathbf{b}}, \widehat{\mathbf{c}}^T$, and $\widehat{\mathrm{d}}$ instead of by the same symbols as in Eqs. (6.1) and (6.2), but we shall assume there are no such parameter discrepancies.

More typically, the actual physical system is better represented by a nonlinear, time-varying model of much higher order than the assumed LTI model, and with various other disturbance signals acting on it. The framework of robust control theory is aimed at studying and mitigating the effects of these various additional sources of uncertainty. We limit ourselves here to examining the effects of uncertainty regarding the initial state, plant disturbances, and measurement noise.

## 6.2 STATE ESTIMATION AND OBSERVERS

In this section we focus on using the plant model, along with knowledge of the input and output signals, in order to causally infer the state of the plant. Our primary interest is in permitting the inferred state to be used in real time to generate appropriate control actions. There are applications where causal and real-time operation are not required, but we will not treat those here.

### 6.2.1 Real-Time Simulation

A natural way to obtain an estimate of the current plant state is by running a model forward in real time as a simulator, in synchrony with the operation of the plant. For this, we initialize the model in Eq. (6.3) at some initial time, which can be chosen as $n = 0$ without loss of generality, and pick its initial state $\widehat{\mathbf{q}}[0]$ to be some guess or estimate of the initial state of the plant, for example, $\widehat{\mathbf{q}}[0] = \mathbf{0}$. We then drive the model with the known plant input values $x[n]$ from time $n = 0$ onward, generating an estimated or predicted state trajectory $\widehat{\mathbf{q}}[n]$ for $n > 0$. The corresponding predicted output $\widehat{y}[n]$ can be computed using Eq. (6.4).

In order to examine how well this real-time simulator performs as a state estimator, consider the behavior of the state error vector

$$\widetilde{\mathbf{q}}[n] = \mathbf{q}[n] - \widehat{\mathbf{q}}[n] \,. \tag{6.5}$$

Note that $\widetilde{\mathbf{q}}[n]$ is the difference between the actual and estimated or predicted state trajectories. We will similarly denote the difference between the actual and estimated output trajectories by

$$\widetilde{y}[n] = y[n] - \widehat{y}[n] \,. \tag{6.6}$$

Subtracting Eq. (6.3) from Eq. (6.1) shows that the state estimation error $\widetilde{\mathbf{q}}[n]$ is itself governed by an LTI state-space equation, namely

$$\widetilde{\mathbf{q}}[n + 1] = \mathbf{A}\widetilde{\mathbf{q}}[n] + \mathbf{w}[n] \,, \tag{6.7}$$

with initial condition

$$\widetilde{\mathbf{q}}[0] = \mathbf{q}[0] - \widehat{\mathbf{q}}[0] \,. \tag{6.8}$$

This initial condition represents the uncertainty about the initial state of the plant. The output error can similarly be obtained by subtracting Eq. (6.4) from Eq. (6.2), yielding

$$\widetilde{y}[n] = \mathbf{c}^T \widetilde{\mathbf{q}}[n] + \zeta[n] \,. \tag{6.9}$$

The model in Eq. (6.7) is called the state error model of the real-time simulator. Note that its dynamics are determined by the same matrix $\mathbf{A}$ that governs the plant and model. Consequently, if the plant in Eq. (6.1) is unstable or has otherwise undesirable dynamics, and if either $\widetilde{\mathbf{q}}[0]$ or $\mathbf{w}[n]$ is nonzero, then the error $\widetilde{\mathbf{q}}[n]$ between the actual and estimated state trajectories will grow exponentially, or will have otherwise undesirable behavior. Even if the plant is not unstable, it is apparent from Eq. (6.7) that the state error dynamics are driven by the disturbance process $\mathbf{w}[n]$, and there is no means to shape the effect of this disturbance on the estimation error. The real-time simulator is thus generally an inadequate way of reconstructing the state.

The same development in CT for a plant of the form

$$\dot{\mathbf{q}}(t) = \mathbf{A}\mathbf{q}(t) + \mathbf{b}x(t) + \mathbf{w}(t) , \tag{6.10}$$

$$y(t) = \mathbf{c}^T \mathbf{q}(t) + \mathrm{d}x(t) + \zeta(t) , \tag{6.11}$$

leads to a model and real-time simulator of the form

$$\dot{\widehat{\mathbf{q}}}(t) = \mathbf{A}\widehat{\mathbf{q}}(t) + \mathbf{b}x(t) , \tag{6.12}$$

$$\widehat{y}(t) = \mathbf{c}^T \widehat{\mathbf{q}}(t) + \mathrm{d}x(t) . \tag{6.13}$$

Defining the state estimation error as

$$\widetilde{\mathbf{q}}(t) = \mathbf{q}(t) - \widehat{\mathbf{q}}(t) , \tag{6.14}$$

we obtain the state error equation

$$\dot{\widetilde{\mathbf{q}}}(t) = \mathbf{A}\widetilde{\mathbf{q}}(t) + \mathbf{w}(t) , \tag{6.15}$$

with initial condition

$$\widetilde{\mathbf{q}}(0) = \mathbf{q}(0) - \widehat{\mathbf{q}}(0) . \tag{6.16}$$

The corresponding output error is given by

$$\widetilde{y}(t) = \mathbf{c}^T \widetilde{\mathbf{q}}(t) + \zeta(t) . \tag{6.17}$$

The following example shows how a CT real-time simulator performs at estimating the state of a particular plant.

| **Example 6.1** | **Real-Time Simulation to Estimate the State of a Suspended Pendulum** |

We choose as the plant for this example the linearized representation of pendulum dynamics for small deviations around the hanging or suspended position, as described at the end of Example 5.1. Suppose the specific plant is described by

$$\dot{\mathbf{q}}(t) = \begin{bmatrix} 0 & 1 \\ -8 & -\beta \end{bmatrix} \mathbf{q}(t) + \begin{bmatrix} 0 \\ -1 \end{bmatrix} \big(x(t) + v(t)\big)$$

$$y(t) = \begin{bmatrix} 1 & 0 \end{bmatrix} \mathbf{q}(t) + \zeta(t) , \tag{6.18}$$

where $v(t)$ is a disturbance torque on the plant and $y(t)$ is a noisy measurement of angular position $q_1(t)$, with measurement noise $\zeta(t)$. The parameter $\beta$ determines the damping. For the undamped case where $\beta = 0$, the natural frequencies of the system are $\pm j\sqrt{8}$, which gives rise to a zero-input response (ZIR) that is sinusoidal of period $2\pi/\sqrt{8} \approx 2.22$ sec.

A real-time simulator for this plant takes the form

$$\dot{\widehat{\mathbf{q}}}(t) = \begin{bmatrix} 0 & 1 \\ -8 & -\beta \end{bmatrix} \widehat{\mathbf{q}}(t) + \begin{bmatrix} 0 \\ -1 \end{bmatrix} x(t)$$

$$\widehat{y}(t) = \begin{bmatrix} 1 & 0 \end{bmatrix} \widehat{\mathbf{q}}(t).$$

(6.19)

This can be realized as in Example 4.5 (changing the gain of 8 to –8). The corresponding error dynamics are

$$\dot{\widetilde{\mathbf{q}}}(t) = \begin{bmatrix} 0 & 1 \\ -8 & -\beta \end{bmatrix} \widetilde{\mathbf{q}}(t) + \begin{bmatrix} 0 \\ -1 \end{bmatrix} v(t)$$

$$\widetilde{y}(t) = \begin{bmatrix} 1 & 0 \end{bmatrix} \widetilde{\mathbf{q}}(t) + \zeta(t).$$

(6.20)

Figure 6.2 shows stimulation results obtained for a case with damping parameter $\beta = 0.2$, with a pulsed torque input $x(t)$ that takes the value 1 for the first 5 seconds and is zero thereafter, and assuming no plant disturbance. The plots describe the pendulum angle $q_1(t)$, the estimate $\widehat{q}_1(t)$ of this angle from a real-time simulator, and the associated estimation error $\widetilde{q}_1(t)$. Note that the dynamics of the error appear similar to those of the underlying plant.

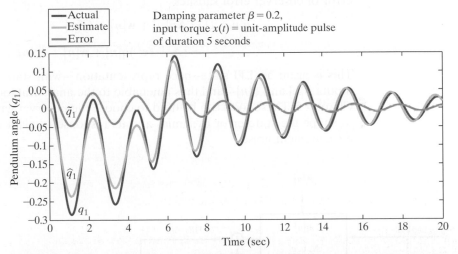

**Figure 6.2**    For the pendulum in Example 6.1 with damping parameter $\beta = 0.2$ and with no plant disturbance, the plots show the trajectories of the pendulum angle $q_1(t)$, the estimate $\widehat{q}_1(t)$ of this angle from the real-time simulator, and the associated estimation error $\widetilde{q}_1(t)$.

## 6.2.2 The State Observer

To do better in the DT case than the real-time simulator in Eq. (6.3), we use not only the input $x[n]$ but also the measured output $y[n]$. The key idea is to use the discrepancy between $y[n]$ and the output $\widehat{y}[n]$ that is predicted by the model or simulator—that is, use the output error $\widetilde{y}[n]$—to generate a correction to the real-time simulator. The same idea applies in CT, but we focus initially on the DT case.

Inserting a scaled version of $\widehat{y}[n]$ as an additive correction to each state evolution equation of the real-time simulator in Eq. (6.3) results in the following system of equations:

$$\widehat{\mathbf{q}}[n+1] = \mathbf{A}\widehat{\mathbf{q}}[n] + \mathbf{b}x[n]$$

$$- \boldsymbol{\ell}\Big(y[n] - \widehat{y}[n]\Big) . \tag{6.21}$$

The resulting system is termed a state observer or state estimator for the plant. The observer equation above has been written in a way that displays its two constituent components: a part that simulates the plant whose state we are trying to estimate, and a part that feeds the correction term $\widetilde{y}[n] = y[n] - \widehat{y}[n]$ into the simulation. This correction term is applied through the $L$-component vector $\boldsymbol{\ell}$, termed the observer gain vector, with $i$th component $\ell_i$. The negative sign in front of $\boldsymbol{\ell}$ is used only to simplify the appearance of some later expressions. Figure 6.3 is a block-diagram representation of the resulting structure.

Subtracting Eq. (6.21) from Eq. (6.1) shows that the state estimation error or observer error satisfies

$$\widetilde{\mathbf{q}}[n+1] = \mathbf{A}\widetilde{\mathbf{q}}[n] + \mathbf{w}[n] + \boldsymbol{\ell}\widetilde{y}[n]$$

$$= (\mathbf{A} + \boldsymbol{\ell}\mathbf{c}^T)\widetilde{\mathbf{q}}[n] + \mathbf{w}[n] + \boldsymbol{\ell}\zeta[n] . \tag{6.22}$$

This is again an LTI state-space representation—with state vector $\widetilde{\mathbf{q}}[n]$ and inputs $\mathbf{w}[n]$ and $\zeta[n]$—and thus amenable to the analytical tools developed in Chapter 5. These tools can be used to choose the observer gain vector $\boldsymbol{\ell}$ so as to shape the state error dynamics and its response to plant disturbance and measurement noise.

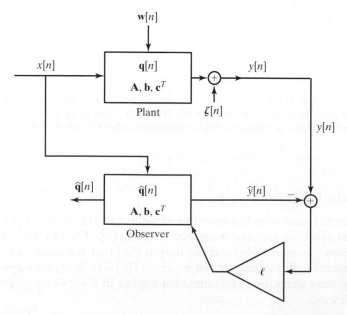

**Figure 6.3** An observer for the plant in the upper part of the diagram comprises a real-time simulation of the plant, driven by the same input, and corrected by a signal derived from the output error.

The development of an observer in the CT case is completely analogous. The real-time simulator in Eq. (6.12) is modified to

$$\dot{\hat{\mathbf{q}}}(t) = \mathbf{A}\hat{\mathbf{q}}(t) + \mathbf{b}x(t)$$

$$- \boldsymbol{\ell}\Big(y(t) - \hat{y}(t)\Big) \tag{6.23}$$

to create the observer. The associated state error equation is then

$$\dot{\tilde{\mathbf{q}}}(t) = \mathbf{A}\tilde{\mathbf{q}}(t) + \mathbf{w}(t) + \boldsymbol{\ell}\tilde{y}(t)$$

$$= (\mathbf{A} + \boldsymbol{\ell}\mathbf{c}^T)\tilde{\mathbf{q}}(t) + \mathbf{w}(t) + \boldsymbol{\ell}\zeta(t) . \tag{6.24}$$

### 6.2.3 Observer Design

Since the structure of the observer is specified, the design of an observer for an LTI plant reduces to choosing the observer gain vector $\boldsymbol{\ell}$. This gain should be chosen to obtain sufficiently rapid decay of the error magnitude, with low sensitivity to plant disturbance, measurement noise, and modeling errors.

**Error Dynamics**    The error dynamics are governed by the natural frequencies of the DT or CT state error equation, Eq. (6.22) or Eq. (6.24) respectively. In both cases these natural frequencies are the $L$ eigenvalues of $\mathbf{A} + \boldsymbol{\ell}\mathbf{c}^T$ or the roots of the characteristic polynomial

$$\kappa(\lambda) = \det\Big(\lambda\mathbf{I} - (\mathbf{A} + \boldsymbol{\ell}\mathbf{c}^T)\Big) \tag{6.25}$$

$$= \lambda^L + \kappa_{L-1}\lambda^{L-1} + \cdots + \kappa_0 . \tag{6.26}$$

For $\boldsymbol{\ell} = \mathbf{0}$, the observer error eigenvalues are simply the eigenvalues $\{\lambda_i\}$ of $\mathbf{A}$, which are the roots of its characteristic polynomial

$$a(\lambda) = \det(\lambda\mathbf{I} - \mathbf{A}) = \prod_{i=1}^{L}(\lambda - \lambda_i) . \tag{6.27}$$

These roots are the natural frequencies of the real-time simulator and of the plant. We next present the key results on how the choice of nonzero observer gains $\boldsymbol{\ell}$ affects the error dynamics, and discuss some of the implications. The analytical demonstrations of these results are presented at the end of this subsection, following Example 6.3.

For nonzero gains $\boldsymbol{\ell}$, the unobservable eigenvalues of the plant remain as eigenvalues of $\mathbf{A} + \boldsymbol{\ell}\mathbf{c}^T$; these eigenvalues are unaffected by the choice of $\boldsymbol{\ell}$. The reason is that information about the unobservable modes does not make its way into the output error signal that is used to correct the real-time simulator. The remaining eigenvalues of the matrix $\mathbf{A} + \boldsymbol{\ell}\mathbf{c}^T$ can be given arbitrary real or complex values by appropriate choice of $\boldsymbol{\ell}$, except that for every complex eigenvalue there has to be a corresponding eigenvalue at the complex conjugate location, since the matrix $\mathbf{A} + \boldsymbol{\ell}\mathbf{c}^T$ is real. Thus, the observable eigenvalues can be moved to any self-conjugate set of locations in the complex plane, while the unobservable eigenvalues remain fixed.

It follows from the preceding statements that the design of an observer whose state error magnitude decays to zero, in the absence of plant disturbance and measurement noise, is possible if and only if all unstable modes of the plant are observable (or equivalently, all unobservable modes are asymptotically stable). This property is termed detectability. For a detectable system, the observer gain $\ell$ can be chosen to produce asymptotically stable error dynamics. From the bounded-input, bounded-state property of asymptotically stable LTI systems, which was proved in Chapter 5, it will then be the case that bounded plant disturbance and bounded measurement noise result in the observer error being bounded.

The preceding results also suggest an alternative way to determine the unobservable eigenvalues of the plant: the roots of $\det[\lambda \mathbf{I} - (\mathbf{A} + \ell \mathbf{c}^T)]$ that cannot be moved, no matter how $\ell$ is chosen, are precisely the unobservable eigenvalues of the plant. This approach to exposing unobservable modes can be easier in many problems than the approach used in Chapter 5, which required first computing the eigenvectors $\mathbf{v}_i$ of the system, and then checking for which $i$ we had $\mathbf{c}^T \mathbf{v}_i = 0$.

In designing observers analytically for low-order systems, we can start by specifying a desired self-conjugate set of natural frequencies $\epsilon_1, \cdots \epsilon_L$ for the observer error dynamics, thus specifying the characteristic polynomial $\kappa(\lambda)$ as

$$\kappa(\lambda) = \prod_{i=1}^{L} (\lambda - \epsilon_i) \,. \tag{6.28}$$

Expanding out this product and equating it to $\det[\lambda \mathbf{I} - (\mathbf{A} + \ell \mathbf{c}^T)]$, as in Eq. (6.25), yields $L$ simultaneous linear equations in the unknown gains $\ell_1, \cdots, \ell_L$. These equations will be consistent and solvable for the observer gains if and only if all the unobservable eigenvalues of the plant are included among the specified observer error eigenvalues $\epsilon_i$. Yet another approach to observer design, involving a transformation to modal coordinates, is described following Example 6.3. For larger systems, specialized computational software would be used.

The above results show that the observer error for an observable LTI plant can be made to decay arbitrarily fast, by choosing the gain vector $\ell$ to place the observer error eigenvalues at appropriate locations. In CT, a very fast decay is obtained by choosing these eigenvalues to have sufficiently negative real parts, while in DT the eigenvalues need to be of sufficiently small magnitude. However, rapid error decay is only part of the story because other factors constrain the choice of $\ell$, as we discuss next.

**Sensitivity to Disturbances, Noise, and Modeling Error**  The observer error representation in Eqs. (6.22) and (6.24) shows that the observer gain $\ell$ enters in two places: it causes the error dynamics to be governed by the state evolution matrix $\mathbf{A} + \ell \mathbf{c}^T$ rather than $\mathbf{A}$, and it serves as the input vector for the measurement noise. This highlights a basic trade-off between error decay and noise immunity. The observer gain can be used to obtain fast error decay, as might be needed in the presence of plant disturbances that continually perturb the system state. However, large entries in $\boldsymbol{\ell}$ may be required to accomplish fast error decay in the CT case, in order to place the eigenvalues of $\mathbf{A} + \boldsymbol{\ell}\mathbf{c}^T$ well into the left half-plane; this is illustrated in Example 6.2 below. Large gains may also be needed in the DT case to obtain fast error decay, if the model is a sampled-data version of some underlying CT system; this is apparent in Example 6.3. These large entries in $\boldsymbol{\ell}$ will have the undesired effect of accentuating the impact of the measurement noise on the state error.

A large observer gain may also lead to large overshoots or oscillations in the observer error, even when the eventual error decay is fast. These transients can cause problems if the state estimates are being used by a feedback controller, for instance. Also, a large observer gain may increase the susceptibility of the observer design to the effects of the various simplifications, approximations, and errors inherent in our using a simple LTI model of the plant. In practice, such considerations would lead us to design conservatively, not attempting to obtain unnecessarily fast error-decay dynamics.

Some aspects of the trade-offs above can be captured in a tractable optimization problem. Modeling the plant disturbance and measurement noise as stationary random processes (which are introduced in Chapter 10), we can pick $\boldsymbol{\ell}$ to minimize some measure of the steady-state variances in the components of the state estimation error. The resulting observer is called a steady-state Kalman filter. We will be in a position to formulate and solve basic problems of this type after we develop the machinery for analyzing stationary random processes. The more general Kalman filter for a state-space system still has the structure of an observer, but with an observer gain that is time-varying, because this filter addresses the more demanding task of optimizing the estimation performance at each instant, not just in the steady state.

---

| **Example 6.2** | **Observer for Undamped Suspended Pendulum** |
|---|---|

We return to the plant considered in Example 6.1, namely the linearized representation of the dynamics of a particular pendulum for small deviations around the normal hanging or suspended position in the undamped case:

$$\dot{\mathbf{q}}(t) = \begin{bmatrix} 0 & 1 \\ -8 & 0 \end{bmatrix} \mathbf{q}(t) + \begin{bmatrix} 0 \\ -1 \end{bmatrix} \big(x(t) + v(t)\big)$$

$$y(t) = \begin{bmatrix} 1 & 0 \end{bmatrix} \mathbf{q}(t) + \zeta(t) . \tag{6.29}$$

As noted earlier, the natural frequencies of this system are $\pm j\sqrt{8}$, and the ZIR is a sinusoid of period $2\pi/\sqrt{8} \approx 2.22$ sec.

An observer for this plant takes the form

$$\dot{\widehat{\mathbf{q}}}(t) = \begin{bmatrix} 0 & 1 \\ -8 & 0 \end{bmatrix} \widehat{\mathbf{q}}(t) + \begin{bmatrix} 0 \\ -1 \end{bmatrix} x(t)$$

$$- \begin{bmatrix} \ell_1 \\ \ell_2 \end{bmatrix} \big(y(t) - \widehat{q}_1(t)\big) . \tag{6.30}$$

The corresponding error dynamics are

$$\dot{\tilde{\mathbf{q}}}(t) = \begin{bmatrix} \ell_1 & 1 \\ -8+\ell_2 & 0 \end{bmatrix} \tilde{\mathbf{q}}(t) + \begin{bmatrix} 0 \\ -1 \end{bmatrix} v(t) + \begin{bmatrix} \ell_1 \\ \ell_2 \end{bmatrix} \zeta(t), \tag{6.31}$$

with characteristic polynomial

$$\kappa(\lambda) = (\lambda - \ell_1)\lambda + (8 - \ell_2) = \lambda^2 - \ell_1\lambda + (8 - \ell_2). \tag{6.32}$$

Note that appropriate choice of $\ell_1$ and $\ell_2$ can convert this into any desired monic polynomial of degree 2, which immediately confirms that the system is observable from the specified output, namely the pendulum angle.

To get an error decay that is a fraction of the oscillatory period of the pendulum, we can pick the natural frequencies of the error dynamics to be $\epsilon_1 = -2$ and $\epsilon_2 = -5$, for example. The corresponding ZIR will be the sum of two exponentials with respective time constants $1/2 = 0.5$ sec and $1/5 = 0.2$ sec, and the transient will essentially vanish in around three time-constants of the dominant mode, hence in around $3 \times 0.5 = 1.5$ seconds. The associated characteristic polynomial is

$$\kappa(\lambda) = (\lambda + 2)(\lambda + 5) = \lambda^2 + 7\lambda + 10. \tag{6.33}$$

Setting this equal to the polynomial in Eq. (6.32) yields $\ell_1 = -7$ and $\ell_2 = -2$.

Figure 6.4 shows plots of plant and observer variables with the preceding choice of observer gains: the pendulum angle $q_1(t)$ under the action of a known pulsed torque $x(t)$; the estimate $\hat{q}_1(t)$ of this angle from the observer; and the associated estimation error $\tilde{q}_1(t)$. The particular results shown here are for the case of no plant disturbance and no measurement noise. As expected, the estimation error essentially vanishes in around 1.5 seconds in this ideal case.

Figure 6.5 illustrates the behavior of the observer in the presence of measurement noise. The input torque $x(t)$ and disturbance torque $v(t)$ are both set to zero for this case, while the measurement noise $\zeta(t)$ takes and holds a random value in

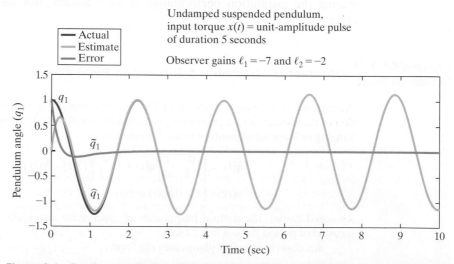

**Figure 6.4** For the undamped pendulum in Example 6.2 with a pulsed input torque $x(t)$ and no plant disturbance or measurement noise, the plots show trajectories of the pendulum angle $q_1(t)$, the estimate $\hat{q}_1(t)$ of this angle from the observer, and the associated estimation error $\tilde{q}_1(t)$, with observer gains chosen to yield error decay with a dominant time constant of 0.5 seconds.

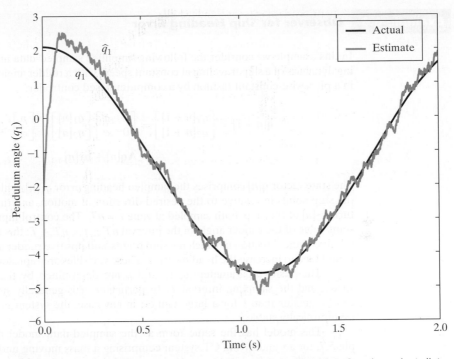

**Figure 6.5**   For the undamped pendulum in Example 6.2 with $x(t) = 0$ and no plant disturbance, but with measurement noise $\zeta(t)$ that takes and holds some random value in the interval
$[-1, 1]$ every millisecond, the plots show the trajectory of the pendulum angle $q_1(t)$ and the
estimate $\widehat{q}_1(t)$ of this angle generated by the observer. The observer gains were chosen to yield
error decay with a dominant time constant of 0.1 seconds.

the interval $[-1, 1]$ every millisecond. The observer gains for this illustration are set at
$\ell_1 = -30$ and $\ell_2 = -192$, which according to the expression in Eq. (6.32) results in the
characteristic polynomial

$$\kappa(\lambda) = \lambda^2 + 30\lambda + 200 = (\lambda + 10)(\lambda + 20) \tag{6.34}$$

for the observer error equation. The associated exponentials have time constants
$1/10 = 0.1$ and $1/20 = 0.05$, so the dominant time constant is 0.1 seconds, and the transients are expected to settle in around $3 \times 0.1 = 0.3$ sec. Note that the noise in the
estimate $\widehat{q}_1(t)$ is considerably attenuated relative to the noise in the original measurement $y(t)$ of the position $q_1(t)$—the measurement noise took values in the interval
$[-1, 1]$ that varied randomly every millisecond, whereas the noise in the estimate
occupies roughly half that range, with much slower variations. The observer has thus
provided some filtering of this measurement noise. Furthermore, the estimate of the
velocity $q_2(t)$ provided by the observer (though not shown here) is of comparable
quality, whereas attempting to estimate the velocity $q_2(t)$ by directly approximating
the derivative of the noisy position measurement $y(t)$ will result in an estimate that is
completely obscured by the noise component.

The next example illustrates observer design for a DT system obtained
as a sampled-data model for an underlying CT system.

**Example 6.3**    **Observer for Ship Heading Error**

In this example we consider the following simplified sampled-data model for the steering dynamics of a ship traveling at constant speed, with a rudder angle that is controlled in a piecewise-constant fashion by a computer-based controller:

$$\mathbf{q}[n+1] = \begin{bmatrix} q_1[n+1] \\ q_2[n+1] \end{bmatrix} = \begin{bmatrix} 1 & \sigma \\ 0 & \alpha \end{bmatrix} \begin{bmatrix} q_1[n] \\ q_2[n] \end{bmatrix} + \begin{bmatrix} \rho \\ \sigma \end{bmatrix} x[n]$$

$$= \mathbf{A}\mathbf{q}[n] + \mathbf{b}x[n] . \tag{6.35}$$

The state vector $\mathbf{q}[n]$ comprises the sampled heading error $q_1[n]$, which is the direction the ship points in relative to the desired direction of motion, and the sampled rate of turn $q_2[n]$ of the ship, both sampled at time $t = nT$. The control input $x[n]$ is the constant value of the rudder angle in the interval $nT \le t < nT + T$; this angle is measured relative to the direction in which the ship points, and positive rudder angle is that which would tend to increase the heading error. These variables are represented in Figure 6.6.

The positive parameters $\alpha$, $\sigma$, and $\rho$ are determined by the type of ship, its speed, and the sampling interval $T$. In particular, $\alpha$ is generally smaller than 1, but can be greater than 1 for a large tanker; in any case, the system in Eq. (6.35) is not asymptotically stable.

This model has the same form as the sampled-data model derived in Example 5.7, for an underlying CT system comprising a mass moving under damping, with an external input force acting on it. A sampled-data model of this form also describes, for instance, the motion of a DC motor whose input is a voltage that is held constant over intervals of length $T$ by a computer-based controller; in this case we have $\alpha = 1$, $\sigma = T$, and (for $x[n]$ in appropriate units) $\rho = T^2/2$ .

Suppose we had noisy measurements of the rate of turn, so in Eq. (6.2)

$$\mathbf{c}^T = \begin{bmatrix} 0 & 1 \end{bmatrix} \tag{6.36}$$

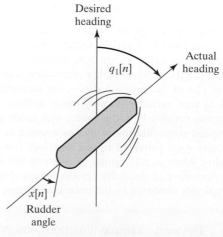

**Figure 6.6**    Heading error and rudder angle for the ship steering system in Example 6.3.

and d = 0. Using this measurement, the error dynamics of an observer for the system would be governed by the matrix

$$\mathbf{A} + \boldsymbol{\ell}\mathbf{c}^T = \begin{bmatrix} 1 & \sigma + \ell_1 \\ 0 & \alpha + \ell_2 \end{bmatrix} . \tag{6.37}$$

The triangular form of this matrix shows that one natural frequency of the error equation is fixed at 1, independent of $\boldsymbol{\ell}$. This natural frequency corresponds to a mode of the original system that is unobservable from rate-of-turn measurements. Moreover, it is not an asymptotically stable mode, so the corresponding observer error will not decay. Physically, the problem is that the rate of turn contains no input from or information about the heading error itself.

Suppose instead that we have noisy measurements of the heading error, so

$$\mathbf{c}^T = \begin{bmatrix} 1 & 0 \end{bmatrix} . \tag{6.38}$$

With this measurement, the associated observer error dynamics would be governed by the matrix

$$\mathbf{A} + \boldsymbol{\ell}\mathbf{c}^T = \begin{bmatrix} 1 + \ell_1 & \sigma \\ \ell_2 & \alpha \end{bmatrix} . \tag{6.39}$$

The characteristic polynomial of this matrix is

$$\kappa(\lambda) = \lambda^2 - \lambda(1 + \ell_1 + \alpha) + \alpha(1 + \ell_1) - \ell_2\sigma , \tag{6.40}$$

which can be made an arbitrary monic polynomial of degree 2 by choice of the gains $\ell_1$ and $\ell_2$. This fact also establishes the observability of the plant model with this output measurement.

One interesting choice of observer gains in this case is $\ell_1 = -1 - \alpha$ along with $\ell_2 = -\alpha^2/\sigma$; for typical parameter values, this results in the value of $\ell_2$ being large. With this choice, the characteristic polynomial of the matrix $\mathbf{A} + \boldsymbol{\ell}\mathbf{c}^T$ is $\kappa(\lambda) = \lambda^2$, so the natural frequencies of the observer error equation are both at 0. We have not treated the repeated-eigenvalue case so far (apart from a brief illustration of possibilities in Example 5.13). However, for this particular case the behavior of the system can easily be deduced from the fact that, with the specified choice of $\boldsymbol{\ell}$,

$$(\mathbf{A} + \boldsymbol{\ell}\mathbf{c}^T)^2 = \begin{bmatrix} -\alpha & \sigma \\ -\alpha^2/\sigma & \alpha \end{bmatrix}^2 = \begin{bmatrix} 0 & 0 \\ 0 & 0 \end{bmatrix} . \tag{6.41}$$

Thus, the ZIR of the observer error dynamics in Eq. (6.22) decays to **0** in two time steps at most, which is the fastest decay possible for this second-order DT system.

We know that the smaller the magnitude of a natural frequency in a DT LTI system, the more rapidly its associated mode decays. It is therefore not surprising that when all natural frequencies are at 0, the system settles quickly. Any $L \times L$ matrix $\mathbf{Z}$ with all its eigenvalues at 0 turns out to satisfy $\mathbf{Z}^L = \mathbf{0}$, the zero matrix. It is also possible, depending on the more detailed structure of such a matrix, that a lower power of the matrix is already **0**. A DT LTI system with all natural frequencies at 0 is sometimes referred to as a deadbeat system, precisely because its ZIR settles to zero in finite time.

In the presence of measurement noise, one may want to choose a slower error decay, so as to keep the observer gain vector $\boldsymbol{\ell}$ smaller than in the deadbeat case, and thereby not accentuate the effects of measurement noise on the estimation error.

**Proofs** Earlier we presented, without proofs, the key results on how the choice of the observer gain $\ell$ affects observer error dynamics. We turn now to establishing those various earlier claims.

To show that each unobservable eigenvalue of the plant remains an eigenvalue of the observer error dynamics, recall from Chapter 5 that for an unobservable eigenvalue $\lambda_j$ of the plant

$$\mathbf{c}^T \mathbf{v}_j = 0 \text{ for some eigenvector } \mathbf{v}_j \neq \mathbf{0}, \text{ with } \mathbf{A}\mathbf{v}_j = \lambda_j \mathbf{v}_j . \tag{6.42}$$

It follows that

$$(\mathbf{A} + \ell\mathbf{c}^T)\mathbf{v}_j = \mathbf{A}\mathbf{v}_j = \lambda_j \mathbf{v}_j \text{ with } \mathbf{v}_j \neq \mathbf{0} , \tag{6.43}$$

no matter how $\ell$ is chosen. The observer error dynamics thus has $\lambda_j$ as an eigenvalue, with associated eigenvector $\mathbf{v}_j$, which proves the desired result.

To see how to place the remaining eigenvalues of $\mathbf{A} + \ell\mathbf{c}^T$ at an arbitrary self-conjugate set of points, it is simplest to work in modal coordinates. Recall from Chapter 5 that $\mathbf{A} = \mathbf{V}\mathbf{\Lambda}\mathbf{V}^{-1}$, where $\mathbf{V}$ is the modal matrix, whose $i$th column is the $i$th eigenvector $\mathbf{v}_i$. We can therefore write

$$\mathbf{A} + \ell\mathbf{c}^T = \mathbf{V}(\mathbf{\Lambda} + \boldsymbol{\psi}\boldsymbol{\xi}^T)\mathbf{V}^{-1} \text{ where } \boldsymbol{\psi} = \mathbf{V}^{-1}\ell, \ \boldsymbol{\xi}^T = \mathbf{c}^T\mathbf{V} . \tag{6.44}$$

Thus $\boldsymbol{\psi}$ is the observer gain vector expressed in modal coordinates, with $i$th entry $\psi_i$. Also $\boldsymbol{\xi}^T$ is $\mathbf{c}^T$ expressed in modal coordinates, which is notation we introduced in Chapter 5. The $j$th mode is observable precisely when the $j$th entry of $\boldsymbol{\xi}$, namely $\xi_j$, is nonzero.

Equation (6.44) establishes that $\mathbf{A} + \ell\mathbf{c}^T$ is related by a similarity transformation to $\mathbf{\Lambda} + \boldsymbol{\psi}\boldsymbol{\xi}^T$. The two matrices therefore have the same characteristic polynomial $\kappa(\lambda)$ and the same eigenvalues. If we can determine how to choose $\boldsymbol{\psi}$ to obtain a desired characteristic polynomial for $\mathbf{\Lambda} + \boldsymbol{\psi}\boldsymbol{\xi}^T$, then we will have determined how to choose $\ell(= \mathbf{V}\boldsymbol{\psi})$ to obtain the same characteristic polynomial for $\mathbf{A} + \ell\mathbf{c}^T$. The first step is to note that

$$\begin{aligned}
\kappa(\lambda) &= \det\left(\lambda\mathbf{I} - (\mathbf{\Lambda} + \boldsymbol{\psi}\boldsymbol{\xi}^T)\right) \\
&= \det\left((\lambda\mathbf{I} - \mathbf{\Lambda}) - \boldsymbol{\psi}\boldsymbol{\xi}^T\right) \\
&= \det\left((\lambda\mathbf{I} - \mathbf{\Lambda})\left[\mathbf{I} - (\lambda\mathbf{I} - \mathbf{\Lambda})^{-1}\boldsymbol{\psi}\boldsymbol{\xi}^T\right]\right) .
\end{aligned} \tag{6.45}$$

Two determinant identities that simplify the preceding expression are stated without proof here. If $\mathbf{M}, \mathbf{P}$ are square matrices of the same dimension,

$$\det(\mathbf{M}\mathbf{P}) = \det(\mathbf{M})\det(\mathbf{P}) . \tag{6.46}$$

Also, for rectangular matrices $\mathbf{R}, \mathbf{S}$ of the same respective dimensions,

$$\det(\mathbf{I} - \mathbf{R}\mathbf{S}^T) = \det(\mathbf{I} - \mathbf{S}^T\mathbf{R}) . \tag{6.47}$$

The latter result is known as Sylvester's identity. The identity matrices $\mathbf{I}$ on the left and right sides of this equation may have different dimension, equal respectively to the dimensions of $\mathbf{R}\mathbf{S}^T$ and $\mathbf{S}^T\mathbf{R}$. If $\mathbf{R}$ and $\mathbf{S}$ are column vectors, for instance, then $\mathbf{S}^T\mathbf{R}$ is a scalar, so the identity matrix on the right of the preceding equation is scalar, hence just the number 1.

The identities in Eqs. (6.46) and (6.47) now justify continuing the chain of equalities in Eq. (6.45) as follows:

$$\kappa(\lambda) = \det(\lambda\mathbf{I} - \mathbf{\Lambda})\det\left(\mathbf{I} - (\lambda\mathbf{I} - \mathbf{\Lambda})^{-1}\boldsymbol{\psi}\boldsymbol{\xi}^T\right)$$

$$= a(\lambda)\det\left(1 - \boldsymbol{\xi}^T(\lambda\mathbf{I} - \mathbf{\Lambda})^{-1}\boldsymbol{\psi}\right). \tag{6.48}$$

Since $\lambda\mathbf{I} - \mathbf{\Lambda}$ is a diagonal matrix, its inverse is also diagonal, with diagonal entries that are the reciprocals of the corresponding diagonal entries in the original matrix. Using this fact to evaluate the last expression above, and rearranging the result, we conclude that

$$\frac{\kappa(\lambda)}{a(\lambda)} = 1 - \sum_{i=1}^{L} \frac{\xi_i \psi_i}{\lambda - \lambda_i}. \tag{6.49}$$

This expression forms the basis for our observer design, as discussed next.

If the desired monic, degree-$L$ characteristic polynomial $\kappa(\lambda)$ of the observer error equation is specified, a standard partial-fraction expansion of the rational function of $\lambda$ on the left in Eq. (6.49) yields

$$\frac{\kappa(\lambda)}{a(\lambda)} = 1 - \sum_{i=1}^{L} \frac{m_i}{\lambda - \lambda_i} \tag{6.50}$$

where

$$m_i = -\frac{\kappa(\lambda)}{a(\lambda)}(\lambda - \lambda_i)\Big|_{\lambda=\lambda_i}. \tag{6.51}$$

Comparing Eqs. (6.49) and (6.50) then shows that the observer gains needed to attain the desired polynomial $\kappa(\lambda)$ are given by

$$\psi_i = m_i/\xi_i. \tag{6.52}$$

However, this is only possible if $\xi_i \neq 0$ for all $i$, that is, if the plant is observable.

If the $j$th mode of the plant is unobservable, corresponding to $\xi_j = 0$, then the pole at $\lambda = \lambda_j$ will not appear in the expansion in Eq. (6.49). The only way to satisfy the equality in that case is to ensure $\kappa(\lambda)$ has a root at $\lambda = \lambda_j$, so that this factor can cancel out on the left side of the equality. In other words, the natural frequencies of the observer error dynamics will have to include every unobservable natural frequency of the plant.

The remaining part of $\kappa(\lambda)$ is unconstrained, and can have arbitrary self-conjugate roots. The partial fraction expansion of $\kappa(\lambda)/a(\lambda)$ in Eq. (6.50) will then only involve terms that correspond to observable modes of the plant, rather than all $L$ modes, but for these modes the prescription in Eq. (6.52) still works. The $\psi_j$ corresponding to unobservable modes can be chosen arbitrarily.

## 6.3 STATE FEEDBACK CONTROL

For a causal system or plant with inputs that we are able to manipulate, it is natural to ask how the inputs should be chosen in order to cause the system to behave in some desirable fashion. Open-loop control uses only information available at the time that one starts interacting with the system. The trouble with open-loop control is that errors, even if recognized, are not corrected or compensated for. If the plant is poorly behaved or unstable, then uncorrected errors can lead to bad or catastrophic consequences. Feedback control, on the other hand, is based on sensing the system's ongoing behavior, and using the measurements of the sensed variables to generate control signals to apply to the system. Feedback control is often also referred to as closed-loop control.

Feedforward control incorporates measurements of signals that affect the plant but that are not themselves affected by the control. For example, in generating electrical control signals for the positioning motor of a steerable radar antenna, the use of measurements of wind velocity would correspond to feedforward control, whereas the use of measurements of antenna position would correspond to feedback control. In general, controls can have both feedback and feedforward components.

We begin our examination of control ideas with the DT case, but the CT case is very similar and is mentioned later. Suppose the DT plant that we want to control is well modeled by the following $L$th-order LTI state-space description:

$$\mathbf{q}[n+1] = \mathbf{A}\mathbf{q}[n] + \mathbf{b}x[n] \tag{6.53}$$

$$y[n] = \mathbf{c}^T\mathbf{q}[n] + \mathrm{d}x[n] . \tag{6.54}$$

We shall also refer to this as the open-loop system. As before, $x[n]$ denotes the control input and $y[n]$ denotes the measured output, both taken to be scalar functions of time. The effects of plant disturbance and measurement noise will be discussed later. The direct feedthrough gain d plays no essential role in what follows, and complicates the appearance of various algebraic expressions, so we shall generally assume d = 0.

### 6.3.1 Open-Loop Control

The following argument provides an illustration of the potential inadequacy of open-loop control, especially when dealing with an unstable plant. Suppose we pick a control input trajectory $x^*[\cdot]$ that would cause the system in Eq. (6.53) to execute some desired state trajectory $\mathbf{q}^*[\cdot]$, provided the system was started in the initial state $\mathbf{q}^*[0]$. Thus

$$\mathbf{q}^*[n+1] = \mathbf{A}\mathbf{q}^*[n] + \mathbf{b}x^*[n] . \tag{6.55}$$

If this control input is now applied when the actual initial state is $\mathbf{q}[0] \neq \mathbf{q}^*[0]$, then the resulting actual state trajectory $\mathbf{q}[\cdot]$ satisfies

$$\mathbf{q}[n+1] = \mathbf{A}\mathbf{q}[n] + \mathbf{b}x^*[n] . \tag{6.56}$$

Subtracting Eq. (6.56) from Eq. (6.55) shows that the difference or error between the desired and actual state trajectories satisfies

$$(\mathbf{q}^*[n+1] - \mathbf{q}[n+1]) = \mathbf{A}(\mathbf{q}^*[n] - \mathbf{q}[n]) \,. \tag{6.57}$$

This is an LTI state-space model for the evolution of the error between the desired and actual state trajectories, $\mathbf{q}^*[n] - \mathbf{q}[n]$, and is governed by the same state evolution matrix $\mathbf{A}$ as the plant. If the original reason for designing a control input was that the plant dynamics were unsatisfactory, then open-loop control produces state error dynamics that will be similarly unsatisfactory. In particular, if the plant is unstable with some eigenvalue of $\mathbf{A}$ having magnitude greater than 1, then the magnitude of the state error will grow geometrically for almost any initial condition.

In open-loop control, we commit to a particular control input at the time we begin interacting with the system, so there is no opportunity to adjust the nominal or baseline control in response to observations of actual system behavior.

### 6.3.2 Closed-Loop Control via LTI State Feedback

As the state variables of a system completely summarize the relevant past of the system, we should expect that knowledge of the state at every instant provides a powerful basis for designing feedback control signals. In this section we consider the use of state feedback for the system in Eq. (6.53), assuming that the entire state vector at each time instant is accessible and measured. Though this assumption is typically unrealistic in practice, it will allow some preliminary results to be developed as a benchmark. A more realistic situation, in which the state cannot be measured but instead has to be estimated, will be treated later. It will turn out in the LTI case that the estimate provided by a state observer will actually suffice to accomplish much of what can be achieved when the actual state is used for feedback.

The particular case of LTI state feedback is represented in Figure 6.7, in which the feedback part of the input $x[n]$ is a weighted linear function of the state variables at that instant:

$$x[n] = \left( \sum_{i=1}^{L} g_i q_i[n] \right) + p[n] = \mathbf{g}^T \mathbf{q}[n] + p[n] \,. \tag{6.58}$$

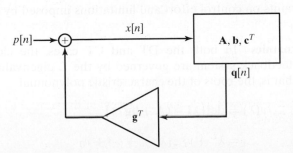

**Figure 6.7**  LTI system with LTI state feedback. Here $\mathbf{g}^T$ is the state feedback gain vector, and $p[n]$ is the new external input signal that augments the feedback signal.

The row vector $\mathbf{g}^T$ is the state feedback gain vector, with $i$th component $g_i$, and $p[n]$ is an external input signal that can be used to augment the feedback signal.

With this choice for $x[n]$, the system in Eqs. (6.53), (6.54) with the assumption $d = 0$ becomes

$$\mathbf{q}[n+1] = \mathbf{A}\mathbf{q}[n] + \mathbf{b}\big(p[n] + \mathbf{g}^T\mathbf{q}[n]\big)$$

$$= \big(\mathbf{A} + \mathbf{b}\mathbf{g}^T\big)\mathbf{q}[n] + \mathbf{b}p[n] \,, \tag{6.59}$$

$$y[n] = \mathbf{c}^T\mathbf{q}[n] \,. \tag{6.60}$$

We refer to this as the closed-loop system. It is again in LTI state-space form, and therefore amenable to analysis by the tools we developed in Chapter 5. Note that since $p[n]$ is the new external input to the closed-loop system, our references below to the ZIR of the closed-loop system will signify the case where $p[n] \equiv 0$ rather than $x[n] \equiv 0$.

The development for CT systems is essentially identical. For an open-loop system that is well modeled by the LTI state-space representation

$$\dot{\mathbf{q}}(t) = \mathbf{A}\mathbf{q}(t) + \mathbf{b}x(t) \,, \tag{6.61}$$

$$y(t) = \mathbf{c}^T\mathbf{q}(t) \,, \tag{6.62}$$

the LTI state feedback control

$$x(t) = \mathbf{g}^T\mathbf{q}(t) + p(t) \tag{6.63}$$

produces the closed-loop LTI state-space system

$$\dot{\mathbf{q}}(t) = (\mathbf{A} + \mathbf{b}\mathbf{g}^T)\mathbf{q}(t) + \mathbf{b}p(t) \,, \tag{6.64}$$

$$y(t) = \mathbf{c}^T\mathbf{q}(t) \,. \tag{6.65}$$

### 6.3.3 LTI State Feedback Design

Since the structure of the closed-loop system under LTI state feedback is specified, the design task is to choose the feedback gain vector $\mathbf{g}^T$. This gain should be picked to obtain desirable closed-loop dynamics, with sufficiently rapid settling of the transient or ZIR behavior of the closed-loop system, but taking account of constraints on control effort and limitations imposed by modeling errors.

**Closed-Loop Dynamics**    In both the DT and CT cases, the closed-loop dynamics of the feedback system are governed by the $L$ eigenvalues of the matrix $\mathbf{A} + \mathbf{b}\mathbf{g}^T$, that is, the roots of the characteristic polynomial

$$\nu(\lambda) = \det\big(\lambda\mathbf{I} - (\mathbf{A} + \mathbf{b}\mathbf{g}^T)\big)$$

$$= \lambda^L + \nu_{L-1}\lambda^{L-1} + \cdots + \nu_0 \,. \tag{6.66}$$

It is reasonable to expect that appropriate choice of the state feedback gain $\mathbf{g}^T$ can result in the natural frequencies of the closed-loop system differing from those of the open-loop system. We describe the possibilities in more detail below, but defer the analytical justification until after Example 6.4.

Note first that the structure of the matrix $\mathbf{A} + \mathbf{b}\mathbf{g}^T$ is analogous to that of the matrix $\mathbf{A} + \boldsymbol{\ell}\mathbf{c}^T$, which governed observer error dynamics. The two structures are said to be dual, meaning that matrix transposition will map the structure of one problem into that of the other. Thus the analysis of how the choice of $\mathbf{g}^T$ affects the eigenvalue placement possibilities for $\mathbf{A} + \mathbf{b}\mathbf{g}^T$ proceeds in close analogy to the earlier arguments for $\mathbf{A} + \boldsymbol{\ell}\mathbf{c}^T$, except that reachability now plays the role that observability did earlier.

When $\mathbf{g}^T = \mathbf{0}$, the eigenvalues of $\mathbf{A} + \mathbf{b}\mathbf{g}^T$ are just those of the matrix $\mathbf{A}$ that governs the open-loop system, that is, the roots of its characteristic polynomial $a(\lambda)$ defined in Eq. (6.27). For nonzero state feedback gains $\mathbf{g}^T$, the unreachable eigenvalues of the plant remain as eigenvalues of $\mathbf{A} + \mathbf{b}\mathbf{g}^T$; these eigenvalues are unaffected by the choice of $\mathbf{g}^T$. The reason is that feedback through the control input has no effect on the unreachable modes. The remaining eigenvalues of $\mathbf{A} + \mathbf{b}\mathbf{g}^T$ can be given arbitrary self-conjugate values, by appropriate choice of $\mathbf{g}^T$.

It follows from the preceding statements that for an LTI system with LTI state feedback, the design of a closed-loop system whose ZIR magnitude decays to 0 is possible if and only if all unstable modes of the plant are reachable (or equivalently, all unreachable modes are asymptotically stable). This property is termed stabilizability. For a stabilizable system, the state feedback gain $\mathbf{g}^T$ can be chosen to produce asymptotically stable closed-loop dynamics. With this, a bounded external signal $p[n]$ will lead to a bounded state trajectory in the closed-loop system; we established this property of asymptotically stable LTI systems in Chapter 5.

The preceding results also suggest an alternative way to determine the unreachable eigenvalues of the system: the roots of $\det[\lambda\mathbf{I} - (\mathbf{A} + \mathbf{b}\mathbf{g}^T)]$ that cannot be modified, no matter how $\mathbf{g}^T$ is chosen, are precisely the unreachable eigenvalues. This route to exposing unreachable modes can be easier in many problems than the approach used in Chapter 5, which required first computing the eigenvectors $\mathbf{v}_i$ of the system and then checking which of these eigenvectors were not needed in writing $\mathbf{b}$ as a linear combination of the eigenvectors.

In designing LTI state feedback analytically for low-order systems, one way to proceed is by specifying a desired self-conjugate set of closed-loop natural frequencies $\mu_1, \cdots \mu_L$, thus specifying the characteristic polynomial $\nu(\lambda)$ as

$$\nu(\lambda) = \prod_{i=1}^{L}(\lambda - \mu_i). \qquad (6.67)$$

Expanding out the product on the right and equating it to $\det[\lambda\mathbf{I} - (\mathbf{A} + \mathbf{b}\mathbf{g}^T)]$, as in Eq. (6.66), yields $L$ simultaneous linear equations in the unknown gains

$g_1, \cdots, g_L$. These equations will be consistent and solvable for the state feed-back gains if and only if all the unreachable eigenvalues of the open-loop system are included among the specified closed-loop eigenvalues $\mu_i$. Another approach to state feedback gain design, involving a transformation to modal coordinates, is outlined after Example 6.4. As in the case of observer design, for state feedback design in larger systems, specialized computational software would be used.

As noted earlier, an unreachable mode of the open-loop system remains fixed at the same frequency in the closed-loop system because the feedback control has no way to access this mode from the input in order to modify it. For the same reason, this unreachable mode of the open-loop system remains unreachable in the closed-loop system. State feedback also cannot pro-duce additional unreachable modes in the closed-loop system beyond those already present in the open-loop system. This is because—again by the same reasoning—these unreachable modes of the closed-loop system would have to then be unreachable modes of the open-loop system as well, since state feedback around the closed-loop system can cancel the original feedback and recover the open-loop system.

The results in this subsection show that the closed-loop ZIR for a reach-able LTI system can be made to decay arbitrarily fast, by choosing the gain $\mathbf{g}^T$ to place the closed-loop eigenvalues at appropriate locations. In CT, a rapid decay is obtained by choosing these eigenvalues to have sufficiently nega-tive real parts, while in DT the eigenvalues need to be of sufficiently small magnitude. However, rapid settling of the closed-loop ZIR is only one con-sideration, because other factors constrain the choice of $\mathbf{g}^T$, as we discuss next.

**Control Effort**    The state feedback gain $\mathbf{g}^T$ affects the closed-loop system in two key ways: first by causing the dynamics to be governed by the eigenvalues of $\mathbf{A} + \mathbf{b}\mathbf{g}^T$ rather than those of $\mathbf{A}$, and second by determining the level of con-trol effort $x[n]$ expended for a given state excursion $\mathbf{q}[n]$, via the relationship in Eq. (6.58). This highlights a basic trade-off between the response rate and the control effort. The state feedback gain can be used to obtain a fast response, bringing the system state from its initially disturbed value rapidly back to the origin. However, large entries in $\mathbf{g}^T$ may be needed to do this—certainly in the CT case, but also in DT if the model is a sampled-data version of some under-lying CT system. These large entries in $\mathbf{g}^T$ result in large control effort being expended. Furthermore, the effects of any errors in measuring or estimating the state vector, or of modeling errors and other discrepancies, are likely to be accentuated with large feedback gains. In practice, these considerations would lead us to design somewhat conservatively, not attempting to obtain excessively fast closed-loop dynamics. Aspects of the trade-offs involved can be captured in tractable optimization problems. Some of these problems are dual to the Kalman filter design problem that we mentioned in connection with optimal observer design.

The following example illustrates the use of state feedback to shape the closed-loop dynamics of a system.

| Example 6.4 | **State Feedback for a Pendulum with Torque Control** |
| --- | --- |

The linearized model of a particular undamped pendulum around either the inverted or suspended positions takes the form

$$\dot{\mathbf{q}}(t) = \begin{bmatrix} 0 & 1 \\ -K & 0 \end{bmatrix} \mathbf{q}(t) + \begin{bmatrix} 0 \\ -1 \end{bmatrix} x(t) , \tag{6.68}$$

where $K > 0$ for the normal suspended position, and $K < 0$ for the inverted position. Both cases will be treated together here.

We could now compute the system eigenvalues and eigenvectors to determine whether the system is reachable and therefore responsive to control. However, this step is actually not necessary. Instead, directly considering the effect of the state feedback $x(t) = \mathbf{g}^T \mathbf{q}(t) + p(t)$ on the system yields the closed-loop description

$$\dot{\mathbf{q}}(t) = \begin{bmatrix} 0 & 1 \\ -K & 0 \end{bmatrix} \mathbf{q}(t) + \begin{bmatrix} 0 \\ -1 \end{bmatrix} [g_1 \quad g_2] \mathbf{q}(t) + \begin{bmatrix} 0 \\ -1 \end{bmatrix} p(t)$$

$$= \begin{bmatrix} 0 & 1 \\ -K - g_1 & -g_2 \end{bmatrix} \mathbf{q}(t) + \begin{bmatrix} 0 \\ -1 \end{bmatrix} p(t) . \tag{6.69}$$

The corresponding characteristic polynomial is

$$\nu(\lambda) = \lambda(\lambda + g_2) + (K + g_1) = \lambda^2 + g_2 \lambda + (K + g_1) . \tag{6.70}$$

Inspection of this expression shows that by appropriate choice of the gains $g_1$ and $g_2$ this polynomial can equal any desired monic second-degree polynomial. In other words, we can obtain any self-conjugate set of closed-loop eigenvalues. This also establishes that the original system is reachable.

Suppose we want the closed-loop eigenvalues at particular values $\mu_1, \mu_2$. This is equivalent to specifying the closed-loop characteristic polynomial to be

$$\nu(\lambda) = (\lambda - \mu_1)(\lambda - \mu_2) = \lambda^2 - \lambda(\mu_1 + \mu_2) + \mu_1 \mu_2 . \tag{6.71}$$

Equating this polynomial to the one in Eq. (6.70) shows that

$$g_1 = \mu_1 \mu_2 - K \quad \text{and} \quad g_2 = -\mu_1 - \mu_2 . \tag{6.72}$$

For the inverted pendulum, both gains are positive when $\mu_1$ and $\mu_2$ form a self-conjugate set in the open left half-plane. For the normal suspended pendulum, $g_1$ may be positive, negative, or zero, while $g_2$ will still be positive.

**Proofs**  We had earlier presented, without proofs, the key results on how the choice of state feedback gain $\mathbf{g}^T$ affects the closed-loop dynamics. We turn now to establishing the various claims. Just as with the dual case of observer error dynamics, it is simplest to work in modal coordinates, noting that

$$\mathbf{A} + \mathbf{b}\mathbf{g}^T = \mathbf{V}(\mathbf{\Lambda} + \boldsymbol{\beta}\boldsymbol{\gamma}^T)\mathbf{V}^{-1} \text{ where } \boldsymbol{\beta} = \mathbf{V}^{-1}\mathbf{b}, \ \boldsymbol{\gamma}^T = \mathbf{g}^T\mathbf{V} . \tag{6.73}$$

Thus $\boldsymbol{\gamma}^T$ is the state feedback gain vector expressed in modal coordinates, with $i$th entry $\gamma_i$. Also $\boldsymbol{\beta}$ denotes the input vector $\mathbf{b}$ expressed in modal coordinates, which is notation introduced in Chapter 5. The $j$th mode is reachable precisely when the $j$th entry of $\boldsymbol{\beta}$, namely $\beta_j$, is nonzero. The above equation shows that the matrix $\mathbf{\Lambda} + \boldsymbol{\beta}\boldsymbol{\gamma}^T$ is related to $\mathbf{A} + \mathbf{b}\mathbf{g}^T$ by a similarity transformation, and therefore the two have the same characteristic polynomial and eigenvalues. If we can determine how to choose $\boldsymbol{\gamma}$ to obtain a desired characteristic polynomial for the matrix $\boldsymbol{\Lambda} + \boldsymbol{\beta}\boldsymbol{\gamma}^T$, then we will have determined how to choose $\mathbf{g}^T \ (= \boldsymbol{\gamma}^T \mathbf{V}^{-1})$ to obtain this same characteristic polynomial for the matrix $\mathbf{A} + \mathbf{b}\mathbf{g}^T$.

Proceeding in a similar vein to the earlier analysis of observer error dynamics produces a helpful expression for the closed-loop characteristic polynomial:

$$v(\lambda) = \det\left(\lambda\mathbf{I} - (\boldsymbol{\Lambda} + \boldsymbol{\beta}\boldsymbol{\gamma}^T)\right)$$

$$= \det(\lambda\mathbf{I} - \boldsymbol{\Lambda}) \det\left(\mathbf{I} - (\lambda\mathbf{I} - \boldsymbol{\Lambda})^{-1}\boldsymbol{\beta}\boldsymbol{\gamma}^T\right)$$

$$= a(\lambda)\left(1 - \sum_{i=1}^{L} \frac{\beta_i\gamma_i}{\lambda - \lambda_i}\right). \tag{6.74}$$

A rearrangement of this result yields

$$\frac{v(\lambda)}{a(\lambda)} = 1 - \sum_{i=1}^{L} \frac{\beta_i\gamma_i}{\lambda - \lambda_i} . \tag{6.75}$$

If the desired monic, degree-$L$ characteristic polynomial $v(\lambda)$ of the closed-loop system is specified, a standard partial fraction expansion of the rational function of $\lambda$ on the left in Eq. (6.75) will take the form

$$\frac{v(\lambda)}{a(\lambda)} = 1 - \sum_{i=1}^{L} \frac{n_i}{\lambda - \lambda_i} , \tag{6.76}$$

where

$$n_i = -\frac{v(\lambda)}{a(\lambda)}(\lambda - \lambda_i)\Big|_{\lambda=\lambda_i} . \tag{6.77}$$

Comparing Eqs. (6.75) and (6.76) then shows that the feedback gains needed to attain the desired polynomial $v(\lambda)$ are given by

$$\gamma_i = n_i/\beta_i . \tag{6.78}$$

However, this is only possible if $\beta_i \neq 0$ for all $i$, that is, if the system is reachable.

If the $j$th mode of the system is unreachable, corresponding to $\beta_j = 0$, then the pole at $\lambda = \lambda_j$ will not appear in the expansion in Eq. (6.75). The only way to satisfy the equality in that case is to ensure $v(\lambda)$ has a root at $\lambda = \lambda_j$, so that this factor can cancel out on the left side of the equality. In other words, the natural frequencies of the closed-loop dynamics will have to include every unreachable natural frequency of the plant.

The remaining part of $v(\lambda)$ is unconstrained, and can have arbitrary self-conjugate roots. The partial fraction expansion of $v(\lambda)/a(\lambda)$ in Eq. (6.76) will then only involve terms that correspond to reachable modes of the system, rather than all $L$ modes, but for these modes the prescription in Eq. (6.78) still works. The $\gamma_j$ corresponding to unreachable modes can be chosen arbitrarily.

**Closed-Loop Transfer Function**    We now examine the input-output characteristics of the closed-loop system by determining its transfer function, focusing on the DT case; the CT version is exactly parallel. The transfer function of the open-loop system in Eqs. (6.1) and (6.2) for the case $d = 0$ is given by

$$H(z) = \mathbf{c}^T(z\mathbf{I} - \mathbf{A})^{-1}\mathbf{b} \qquad (6.79)$$

$$= \frac{\eta(z)}{a(z)}, \qquad (6.80)$$

where $a(z)$ is the characteristic polynomial of the system, defined earlier in Eq. (6.27). Note that there may be pole-zero cancellations involving common roots of $a(z)$ and $\eta(z)$ in Eq. (6.80), corresponding to the presence of system modes that are unreachable, unobservable, or both. Only the uncancelled roots of $a(z)$ remain as poles of $H(z)$, and similarly only the uncancelled roots of $\eta(z)$ remain as zeros of the transfer function.

The closed-loop transfer function from the external input $p[n]$ to the output $y[n]$ is the transfer function of the system in Eqs. (6.59) and (6.60), namely

$$G(z) = \mathbf{c}^T\Big(z\mathbf{I} - (\mathbf{A} + \mathbf{bg}^T)\Big)^{-1}\mathbf{b}. \qquad (6.81)$$

The denominator polynomial of this transfer function, prior to any cancellations with its numerator, is $v(z) = \det(z\mathbf{I} - \mathbf{A} - \mathbf{bg}^T)$. To determine the numerator polynomial of $G(z)$, we follow an indirect route. Note first that the closed-loop transfer function from $p[n]$ to the plant input $x[n]$ is the ratio $X(z)/P(z)$ when the initial condition is zero, $\mathbf{q}[0] = \mathbf{0}$. It follows from Eq. (6.58) and identities we have already established that

$$X(z) = \mathbf{g}^T(z\mathbf{I} - \mathbf{A})^{-1}\mathbf{b}X(z) + P(z)$$

$$= \boldsymbol{\gamma}^T(z\mathbf{I} - \boldsymbol{\Lambda})^{-1}\boldsymbol{\beta}X(z) + P(z)$$

$$= \left(\sum_{i=1}^{L}\frac{\beta_i\gamma_i}{z - \lambda_i}\right)X(z) + P(z). \qquad (6.82)$$

Rearranging this equation to solve for $X(z)$ and then invoking Eq. (6.74) establishes that

$$\frac{X(z)}{P(z)} = \frac{a(z)}{v(z)}, \qquad (6.83)$$

where $v(z)$ is the closed loop characteristic polynomial defined in Eq. (6.66). The transfer function from the input $p[n]$ to the output $y[n]$ of the closed-loop system is therefore

$$G(z) = \frac{Y(z)}{P(z)} = \frac{Y(z)}{X(z)} \frac{X(z)}{P(z)} \tag{6.84}$$

$$= \frac{\eta(z)}{a(z)} \frac{a(z)}{v(z)} \tag{6.85}$$

$$= \frac{\eta(z)}{v(z)} . \tag{6.86}$$

This expression confirms that state feedback has changed the denominator of the input-output transfer function from $a(z)$ in the open-loop case to $v(z)$ in the closed-loop case, and has accordingly modified the characteristic polynomial and poles. However, state feedback has left unchanged the numerator polynomial $\eta(z)$ from which the zeros are selected. The actual zeros of the closed-loop system are those roots of $\eta(z)$ that are not cancelled by roots of the new closed-loop characteristic polynomial $v(z)$, and may therefore differ from the zeros of the open-loop system.

A zero of the open-loop system will disappear from the closed-loop system if state feedback places a closed-loop natural frequency at this location, producing a new cancellation between $\eta(z)$ and $v(z)$. Since state feedback does not affect the reachability or unreachability of modes, this disappearance or cancellation must correspond to the mode having been made unobservable by state feedback. Conversely, a root of $\eta(z)$ that was not a zero of the open-loop system will appear in the closed-loop system if state feedback moves the cancelling open-loop natural frequency to a new location, so the original cancellation between $\eta(z)$ and $a(z)$ does not occur between $\eta(z)$ and $v(z)$. This corresponds to an unobservable but reachable mode of the open-loop system now becoming observable—and remaining reachable—in the closed-loop system.

An exactly parallel development holds for the CT case, with changes that at this point should be very familiar, so we omit the details.

| Example 6.5 | **Ship Steering by State Feedback** |
|---|---|

Consider again the sampled-data DT state-space model in Example 6.3, representing the steering dynamics of a ship traveling at constant speed:

$$\mathbf{q}[n+1] = \begin{bmatrix} q_1[n+1] \\ q_2[n+1] \end{bmatrix} = \begin{bmatrix} 1 & \sigma \\ 0 & \alpha \end{bmatrix} \begin{bmatrix} q_1[n] \\ q_2[n] \end{bmatrix} + \begin{bmatrix} \rho \\ \sigma \end{bmatrix} x[n]$$

$$= \mathbf{A}\mathbf{q}[n] + \mathbf{b}x[n] . \tag{6.87}$$

Recall that the state vector $\mathbf{q}[n]$ comprises the sampled heading error $q_1[n]$ and the sampled rate of turn $q_2[n]$ of the ship, both sampled at time $t = nT$, while the control input $x[n]$ is the constant value of the rudder angle in the interval $nT \le t < nT + T$. The same kind of model could describe the motion of a DC motor under a piecewise-constant applied voltage.

For the purposes of this example, take $\alpha = 1$, $\sigma = T$, and $\rho = T^2/2$, so

$$\mathbf{A} = \begin{bmatrix} 1 & T \\ 0 & 1 \end{bmatrix}, \quad \mathbf{b} = \begin{bmatrix} T^2/2 \\ T \end{bmatrix} . \tag{6.88}$$

Setting

$$x[n] = g_1 q_1[n] + g_2 q_2[n] \qquad (6.89)$$

results in the closed-loop state evolution matrix

$$\mathbf{A} + \mathbf{b}\mathbf{g}^T = \begin{bmatrix} 1 + g_1(T^2/2) & T + g_2(T^2/2) \\ g_1 T & 1 + g_2 T \end{bmatrix}. \qquad (6.90)$$

The characteristic polynomial of this matrix is

$$\nu(\lambda) = \left(\lambda - 1 - g_1(T^2/2)\right)\left(\lambda - 1 - g_2 T\right) - g_1 T\left(T + g_2(T^2/2)\right)$$
$$= \lambda^2 - \lambda\left(2 + g_1(T^2/2) + g_2 T\right) + \left(1 - g_1(T^2/2) + g_2 T\right). \qquad (6.91)$$

For desired closed-loop roots at $\mu_1$ and $\mu_2$, this polynomial will have to be

$$\nu(\lambda) = (\lambda - \mu_1)(\lambda - \mu_2) = \lambda^2 - \lambda(\mu_1 + \mu_2) + \mu_1 \mu_2. \qquad (6.92)$$

Comparing this equation with Eq. (6.91), we arrive at the following pair of simultaneous linear equations that have to be solved for the required feedback gains:

$$\begin{bmatrix} T/2 & 1 \\ -T/2 & 1 \end{bmatrix}\begin{bmatrix} g_1 \\ g_2 \end{bmatrix} = \frac{1}{T}\begin{bmatrix} \mu_1 + \mu_2 - 2 \\ \mu_1 \mu_2 - 1 \end{bmatrix}, \qquad (6.93)$$

so

$$\begin{bmatrix} g_1 \\ g_2 \end{bmatrix} = \frac{1}{T^2}\begin{bmatrix} 1 & -1 \\ T/2 & T/2 \end{bmatrix}\begin{bmatrix} \mu_1 + \mu_2 - 2 \\ \mu_1 \mu_2 - 1 \end{bmatrix}. \qquad (6.94)$$

For illustration, suppose we wish to obtain the fastest possible closed-loop response in this DT model, that is, the deadbeat behavior described earlier in Example 6.3 for the case of the observer error model for this system. This results from placing both closed-loop natural frequencies at 0, so $\mu_1 = 0 = \mu_2$ and $\nu(\lambda) = \lambda^2$. Then Eq. (6.94) shows that

$$g_1 = -1/T^2, \qquad g_2 = -1.5/T. \qquad (6.95)$$

As noted in connection with the deadbeat observer, we have not shown how to analyze system behavior when there are repeated eigenvalues, but in the particular instance of repeated eigenvalues at 0, it is easy to show that the state will die to $\mathbf{0}$ in a finite number of steps—at most two steps, for this second-order system. To establish this here, note that with the above choice of $\mathbf{g}$ we get

$$\mathbf{A} + \mathbf{b}\mathbf{g}^T = \begin{bmatrix} 1/2 & T/4 \\ -1/T & -1/2 \end{bmatrix}, \qquad (6.96)$$

so

$$\left[\mathbf{A} + \mathbf{b}\mathbf{g}^T\right]^2 = 0, \qquad (6.97)$$

which shows that the effects of any nonzero initial condition will vanish in two steps at most.

In practice, such deadbeat behavior may not be attainable, as unduly large control effort—rudder angles, in the case of the ship—would be needed if $T$ was small. Values of $\mu_1$ and $\mu_2$ that had larger magnitudes, though still less than 1, would correspond to slower settling, and would thus require smaller control effort. On the other hand, to the extent we have confidence in how well the underlying CT model predicts behavior over longer intervals, we could in principle use the deadbeat control law with a large enough $T$ to satisfy the control constraints, and then allow the ship to settle with just two motions of the rudder:

$$x[0] = -(1/T^2)q_1[0] - (1.5/T)q_2[0] \,,$$

$$x[1] = -(1/T^2)q_1[1] - (1.5/T)q_2[1] \,,$$

$$x[n] = 0 \quad \text{for} \quad n \geq 2 \,. \tag{6.98}$$

Typically, we do not have direct measurements of the state variables of the plant, but rather knowledge of only the control input, along with noisy measurements of the system output. However, the state can be reconstructed using an observer that produces asymptotically convergent estimates of the state variables, provided the system is well described by an observable (or at least detectable) model of the form in Eqs. (6.53) and (6.54). The next section studies the closed-loop behavior that results from using the state estimates produced by the observer in place of direct state measurements.

## 6.4 OBSERVER-BASED FEEDBACK CONTROL

An obstacle to state feedback is the general unavailability of direct measurements of the state. We focus on the DT case, but the CT case proceeds in a closely parallel way. All we typically have are knowledge of the control signal $x[n]$ that we are applying to the plant, along with possibly noise-corrupted measurements $y[n]$ of the plant output, and a nominal model of the system. We have already seen how to use this information to estimate the state variables, using an observer or state estimator. Let us therefore consider what happens when we substitute the state estimate provided by the observer for the unavailable actual state, in the feedback control law in Eq. (6.58).

With this substitution, the control law is modified to

$$x[n] = \mathbf{g}^T \widehat{\mathbf{q}}[n] + p[n]$$

$$= \mathbf{g}^T (\mathbf{q}[n] - \widetilde{\mathbf{q}}[n]) + p[n] \,. \tag{6.99}$$

The overall closed-loop system is then as shown in Figure 6.8.

A state-space model for this composite closed-loop system is obtained by combining the representations of the subsystems that comprise it, namely the plant in Eq. (6.1) and the observer in Eq. (6.21), factoring in their closed-loop coupling via $x[n]$ and $y[n]$. This would suggest using the vector

$$\begin{bmatrix} \mathbf{q}[n] \\ \widehat{\mathbf{q}}[n] \end{bmatrix} \tag{6.100}$$

as the state vector for the combined system. This is a $2L$-component vector that comprises the entries of $\mathbf{q}[n]$ followed by those of $\widehat{\mathbf{q}}[n]$. However, recalling

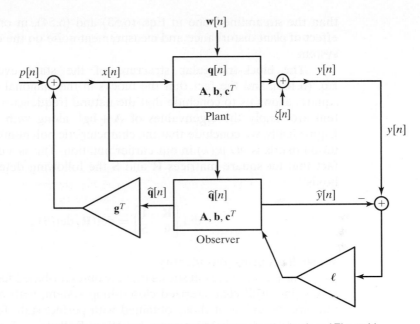

**Figure 6.8**  Observer-based compensator, feeding back an LTI combination of the estimated state variables.

that the observer error state is described by $\tilde{\mathbf{q}}[n] = \mathbf{q}[n] - \hat{\mathbf{q}}[n]$, an equivalent choice of state vector for the combined system is provided by

$$\begin{bmatrix} \mathbf{q}[n] \\ \tilde{\mathbf{q}}[n] \end{bmatrix} , \tag{6.101}$$

because $\hat{\mathbf{q}}[n]$ can be determined from $\mathbf{q}[n]$ and $\hat{\mathbf{q}}[n]$, and conversely $\hat{\mathbf{q}}[n]$ can be determined from $\mathbf{q}[n]$ and $\tilde{\mathbf{q}}[n]$. The choice in Eq. (6.101) of a state vector for the combined system leads to a more easily interpreted system description, namely the following LTI state-space model:

$$\begin{bmatrix} \mathbf{q}[n+1] \\ \tilde{\mathbf{q}}[n+1] \end{bmatrix} = \begin{bmatrix} \mathbf{A} + \mathbf{b}\mathbf{g}^T & -\mathbf{b}\mathbf{g}^T \\ \mathbf{0} & \mathbf{A} + \boldsymbol{\ell}\mathbf{c}^T \end{bmatrix} \begin{bmatrix} \mathbf{q}[n] \\ \tilde{\mathbf{q}}[n] \end{bmatrix} + \begin{bmatrix} \mathbf{b} \\ \mathbf{0} \end{bmatrix} p[n]$$

$$+ \begin{bmatrix} \mathbf{I} \\ \mathbf{I} \end{bmatrix} \mathbf{w}[n] + \begin{bmatrix} \mathbf{0} \\ \boldsymbol{\ell} \end{bmatrix} \zeta[n] . \tag{6.102}$$

The first matrix on the right is a $2 \times 2$ block matrix or partitioned matrix, having $\mathbf{A} + \mathbf{b}\mathbf{g}^T$ as its leading $L \times L$ block, and the $L \times L$ matrix $-\mathbf{b}\mathbf{g}^T$ adjacent to it in the first $L$ rows; the latter matrix in turn has the $L \times L$ matrix $\mathbf{A} + \boldsymbol{\ell}\mathbf{c}^T$ below it in the last $L$ columns. The remaining positions of the matrix are filled with 0s, denoted by the $L \times L$ matrix $\mathbf{0}$. The other block matrices are interpreted similarly. Equation (6.102) combines the plant description in Eqs. (6.1) and (6.2), the observer description in Eq. (6.21), and the feedback control law in Eq. (6.99). Note that we have reverted here to the more elaborate plant representation in Eqs. (6.1) and (6.2) rather than the streamlined one in Eqs. (6.53) and (6.54), in order to display the effect of plant disturbance and measurement noise on the overall closed-loop system.

The block-triangular structure of the state evolution matrix in Eq. (6.102), and the fact that the blocks in the diagonal positions are both square, allows us to conclude that the natural frequencies of the overall system are simply the eigenvalues of $\mathbf{A} + \mathbf{b}\mathbf{g}^T$ along with those of $\mathbf{A} + \boldsymbol{\ell}\mathbf{c}^T$. Equivalently, we conclude that the characteristic polynomial of the state evolution matrix is $\nu(\lambda)\kappa(\lambda)$ in our earlier notation. This is a consequence of the fact that for square matrices $\mathbf{R}$ and $\mathbf{S}$ the following determinantal identity holds:

$$\det \begin{bmatrix} \mathbf{R} & \mathbf{T} \\ \mathbf{0} & \mathbf{S} \end{bmatrix} = \det(\mathbf{R})\det(\mathbf{S}) . \tag{6.103}$$

We shall not prove this identity.

The preceding result shows that the observer-based feedback control law results in a well characterized closed-loop system, with natural frequencies that are the union of those obtained with perfect state feedback and those obtained for the observer error equation. Both sets of natural frequencies can be arbitrarily selected, provided the open-loop system is reachable and observable. One would normally pick the modes that govern observer error decay to be faster than those associated with state feedback, in order to have reasonably accurate estimates available to the feedback control law before the plant state can wander too far away from what is desired.

Another interesting fact is that the transfer function from $p[n]$ to $y[n]$ in the new closed-loop system is exactly what would be obtained with perfect state feedback, namely the transfer function in Eqs. (6.81) and (6.86). The reason is that the condition under which the transfer function is computed — as the input-output response when starting from the zero state, and with other external inputs set to zero — ensures that the observer starts up from the same initial condition, in this case $\widehat{\mathbf{q}}[0] = \mathbf{0}$, as the plant. This in turn ensures that there is no estimation error in the absence of plant disturbance and measurement noise, so the estimated state equals the true state. Another way to reach the same conclusion regarding the closed-loop transfer function is to note that the observer error modes are undriven by the external input $p[n]$ or by anything that $p[n]$ can excite, so these modes are unreachable from $p[n]$; they are therefore hidden from the transfer function.

The preceding observer-based compensator is the starting point for a very general and powerful approach to control design, one that carries over to the multi-input, multi-output case. With the appropriate embellishments around this basic structure, one can obtain every possible stabilizing LTI feedback controller for the system in Eqs. (6.53) and (6.54). Within this class of controllers, we can search for those that have robust properties, in the sense that they are relatively immune to the uncertainties in our models. Further exploration of this is left to more advanced courses.

| **Example 6.6** | **Observer-Based Controller for an Inverted Pendulum** |

Consider as our plant the following linearized model of a particular undamped pendulum around the inverted position:

$$\dot{\mathbf{q}}(t) = \begin{bmatrix} 0 & 1 \\ 8 & 0 \end{bmatrix} \mathbf{q}(t) + \begin{bmatrix} 0 \\ -1 \end{bmatrix} x(t) . \tag{6.104}$$

Its natural frequencies are at $\pm\sqrt{8}$, indicating an unstable system. We saw in Example 6.4 that state feedback with gains

$$\mathbf{g}^T = \begin{bmatrix} g_1 & g_2 \end{bmatrix} = \begin{bmatrix} \mu_1\mu_2 + 8 & -\mu_1 - \mu_2 \end{bmatrix} \tag{6.105}$$

will place the closed-loop natural frequencies at $\mu_1$ and $\mu_2$. Suppose, for example, we choose $\mu_1 = -2$, $\mu_2 = -3$; then $g_1 = 14$, $g_2 = 5$. The associated time constants are $1/2 = 0.5$ and $1/3 = 0.33$, so under perfect state feedback we expect the system to settle in around $3 \times 0.5 = 1.5$ seconds.

Using a measurement of the angular position $q_1(t)$ of the system, an observer for this system takes the form

$$\dot{\hat{\mathbf{q}}}(t) = \begin{bmatrix} 0 & 1 \\ 8 & 0 \end{bmatrix} \hat{\mathbf{q}}(t) + \begin{bmatrix} 0 \\ -1 \end{bmatrix} x(t)$$

$$- \begin{bmatrix} \ell_1 \\ \ell_2 \end{bmatrix} \left( y(t) - \hat{q}_1(t) \right) . \tag{6.106}$$

The corresponding error dynamics are

$$\dot{\tilde{\mathbf{q}}}(t) = \begin{bmatrix} \ell_1 & 1 \\ 8 + \ell_2 & 0 \end{bmatrix} \tilde{\mathbf{q}}(t) , \tag{6.107}$$

with characteristic polynomial

$$\kappa(\lambda) = (\lambda - \ell_1)\lambda - (8 + \ell_2) = \lambda^2 - \ell_1\lambda - (8 + \ell_2) . \tag{6.108}$$

Choosing $\ell_1 = -7$ and $\ell_2 = -18$, for example, this polynomial becomes

$$\kappa(\lambda) = \lambda^2 + 7\lambda + 10 = (\lambda + 2)(\lambda + 5) \tag{6.109}$$

so the error decay time constants are $1/2 = 0.5$ and $1/5 = 0.2$. We thus expect the observer error to settle in around $3 \times 0.5 = 1.5$ seconds.

Now suppose we feed back the estimated state instead of the unavailable actual state, so $x(t) = \mathbf{g}^T\hat{\mathbf{q}}(t) + p(t)$. The response of the resulting closed-loop system is shown by the pair of plots in the upper part of Figure 6.9, with the pendulum angle $q_1(t)$ obtained under this feedback control, and also the corresponding estimate $\hat{q}_1(t)$ of the pendulum angle provided by the observer, starting from its zero state, that is, with $\hat{q}_1(0) = 0$, $\hat{q}_2(0) = 0$. The control $x(t)$ is shown as well.

The pair of plots in the lower part of the figure show, as a benchmark, $q_1(t)$ under the exact state feedback control designed previously, along with the required control effort $x(t)$ under exact state feedback. Note that for a comparable control effort, exact state feedback is able to deal with an intial offset $q_1(0)$ that is 10 times larger than in the observer-based case.

The observer-based controller is associated with a control input that is more oscillatory than under perfect state feedback until the observer estimate is close to

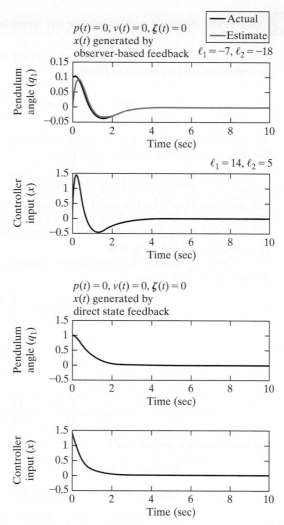

**Figure 6.9**  The upper pair of plots show the response of the inverted pendulum position $q_1(t)$ under feedback control, using an observer-based controller whose estimate $\widehat{q}_1(t)$ of the position is shown. The input $x(t)$ is also shown. For comparison, the response obtained under exact state feedback is displayed in the bottom pair of plots.

the underlying plant state. Once observer convergence is obtained at approximately 1.5 seconds, the feedback is essentially as good as exact state feedback, and the system settles in an additional 1.5 seconds.

As the controller here was designed on the basis of a linearized model, a natural next step would be to simulate the effect of the controller on a more realistic nonlinear model of the pendulum, to ascertain for what range of initial conditions and perturbations the controller performs satisfactorily.

## 6.5 FURTHER READING

Some of the texts suggested for further reading in Chapters 4 and 5 include material on state estimation and state feedback control in LTI systems, see for example [Ant], [Ch2], [Ka1], and [Lue]. A detailed treatment, illustrated by aerospace and other applications, is given in [Frd]. The text [Ast] is an accessible modern introduction to feedback systems, including state-space models, and incorporates key notions in robust control. The control of nonlinear systems is addressed in [Kha], [Slo], [Vid].

# Problems

## Basic Problems

**6.1.** The dynamics of a synchronous electric generator are governed by a model of the form

$$\frac{d^2\theta(t)}{dt^2} + \beta \frac{d\theta(t)}{dt} + \alpha \sin\theta(t) = T(t) ,$$

where $\theta(t)$ is the (relative) angular position of the generator and $T(t)$ is the (normalized) external torque acting on it; the parameter $\alpha$ is positive, but $\beta$ can be positive or negative. Write a state-space model of this system. Then, assuming $T(t)$ is constant at a positive value $T(t) = \overline{T}$, determine for what values of $\overline{T}$ the system will have:

(i) no equilibrium solutions;
(ii) one equilibrium solution with $\theta(t) = \overline{\theta}$ in the range $[0, 2\pi]$;
(iii) two equilibrium solutions with $\theta(t) = \overline{\theta}$ in the range $[0, 2\pi]$.

   Write down the two linearized models computed at the two equilibrium solutions you found in part (iii) above, expressing them in the standard form

$$\dot{\mathbf{q}}(t) = \mathbf{A}\mathbf{q}(t) + \mathbf{b}x(t) .$$

Determine the reachability of each of them by checking whether you can obtain an arbitrary closed-loop characteristic polynomial by LTI state feedback.

**6.2.** Consider the linear dynamic system

$$\dot{\mathbf{q}}(t) = \mathbf{A}\mathbf{q}(t) + \mathbf{b}x(t)$$

$$y(t) = \mathbf{c}^T\mathbf{q}(t)$$

where $\mathbf{A} = \begin{bmatrix} 0 & 1 \\ 2 & -1 \end{bmatrix}$, $\mathbf{b} = \begin{bmatrix} 1 \\ 0 \end{bmatrix}$, and $\mathbf{c} = \begin{bmatrix} 1 \\ 0 \end{bmatrix}$.

**(a)** Is the system reachable? Justify your answer.
**(b)** Assuming that the states of the system are available, we wish to use state feedback to control the modes of the system. The resulting closed-loop system is shown in Figure P6.2.

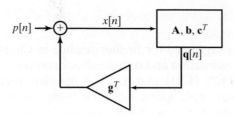

The corresponding state-space equations are

$$\dot{\mathbf{q}}(t) = \mathbf{A}\mathbf{q}(t) + \mathbf{b}\left(p(t) + \mathbf{g}^T\mathbf{q}(t)\right)$$

$$y(t) = \mathbf{c}^T\mathbf{q}(t) .$$

(i) Is it possible to place the natural frequencies or eigenvalues of this system arbitrarily through the choice of an appropriate gain $\mathbf{g}$?

(ii) Find a gain $\mathbf{g}$ such that the natural frequencies or eigenvalues of the closed-loop system are $-2$ and $-1$.

**6.3.** A second-order CT causal LTI system is described by a state-space model of the form:

$$\frac{d\mathbf{q}(t)}{dt} = \mathbf{A}\mathbf{q}(t) + \mathbf{b}x(t)$$

$$y(t) = \mathbf{c}^T\mathbf{q}(t)$$

For $\mathbf{q}(0) = \begin{bmatrix} q_1(0) \\ q_2(0) \end{bmatrix} = \begin{bmatrix} 1 \\ 0 \end{bmatrix}$ and $x(t) = 0$ for $t \geq 0$, we find that

$$\mathbf{q}(t) = \begin{bmatrix} q_1(t) \\ q_2(t) \end{bmatrix} = \begin{bmatrix} 1 \\ 0 \end{bmatrix}e^{2t} .$$

On the other hand, for $\mathbf{q}(0) = \begin{bmatrix} 1 \\ 1 \end{bmatrix}$ and $x(t) = 0$ for $t \geq 0$, we find that

$$\mathbf{q}(t) = \begin{bmatrix} 1 \\ 1 \end{bmatrix}e^{-3t} .$$

**(a)** What are the two eigenvalues of the system and their corresponding eigenvectors?

**(b)** Determine the matrix $\mathbf{A}$.

Now suppose in the state-space description specified above, we are told that

$$\mathbf{b} = \begin{bmatrix} 0 \\ 1 \end{bmatrix}, \quad \mathbf{c}^T = [0 \; 1] .$$

**(c)** Determine whether the system is reachable.

**(d)** Determine whether the system is observable.

**(e)** With $\mathbf{b}$ and $\mathbf{c}^T$ as above, and with $\mathbf{q}(0) = \mathbf{0}$, is the output $y(t)$ bounded for every bounded input $x(t)$, i.e., is the system BIBO stable?

**(f)** We now apply state feedback $x(t) = \mathbf{g}^T\mathbf{q}(t)$, with $\mathbf{g}^T = [g_1 \; g_2]$.

(i) Is it possible to select (real) constants $g_1$ and $g_2$ such that the resulting system is asymptotically stable? You need not actually compute such a $g_1$ and $g_2$ if they do exist.

(ii) If we restrict $g_2$ to zero, is it possible to select a constant $g_1$ such that the resulting system is asymptotically stable? Again, you need not actually compute such a $g_1$ if it exists.

(iii) If we restrict $g_1$ to zero, is it possible to select a constant $g_2$ such that the resulting system is asymptotically stable? You need not actually compute such a $g_2$ if it exists.

**6.4.** Consider a causal CT system described by a state-space model of the form

$$\frac{d\mathbf{q}(t)}{dt} = \mathbf{A}\mathbf{q}(t) + \mathbf{b}x(t), \qquad y(t) = \mathbf{c}^T\mathbf{q}(t) + dx(t) ,$$

with

$$\mathbf{A} = \begin{bmatrix} -1 & 0 \\ -1 & 0 \end{bmatrix}, \quad \mathbf{b} = \begin{bmatrix} \gamma \\ -1 \end{bmatrix}, \quad \mathbf{c}^T = [1 \; 1], \quad d = 1 ,$$

where $\gamma$ is a real parameter.

**(a)** Determine the eigenvalues of $\mathbf{A}$, and the associated eigenvectors.

**(b)** For the case when the input is identically 0, i.e., when $x(t) = 0$ for $t \geq 0$, determine:

(i) all nonzero initial conditions $\mathbf{q}(0)$ for which the state vector does not move from its initial value, i.e., $\mathbf{q}(t) = \mathbf{q}(0)$ for $t > 0$;

(ii) all nonzero initial conditions $\mathbf{q}(0)$ for which the state vector $\mathbf{q}(t)$ decays exponentially to the zero vector $\mathbf{0}$ as $t \to \infty$.

Note from your answer to (b)(i) that the system is not asymptotically stable.

**(c)** Find all values of $\gamma$ for which the state-space model is not reachable, and specify the associated unreachable eigenvalue or eigenvalues for each such $\gamma$.

**(d)** For each $\gamma$ in (c) for which you found the system to be unreachable, determine whether state feedback of the form $x(t) = \mathbf{g}^T\mathbf{q}(t)$ could have made the system asymptotically stable, for some choice of feedback gain vector $\mathbf{g}^T$. Be sure to explain your reasoning.

**6.5.** The state evolution of a DT system is described by the equation

$$\mathbf{q}[n + 1] = \mathbf{A}\mathbf{q}[n] + \mathbf{b}x[n]$$

with

$$\mathbf{A} = \begin{bmatrix} 1 & \beta \\ -\alpha & 1 - \alpha\beta \end{bmatrix}, \quad \mathbf{b} = \begin{bmatrix} 1 \\ 1 \end{bmatrix},$$

where $\alpha$ and $\beta$ are real-valued constants. The matrix $\mathbf{A}$ has one eigenvalue at $\lambda_1 = 0.5$, with associated eigenvector $\mathbf{v_1} = \begin{bmatrix} 2 \\ -1 \end{bmatrix}$.

**(a)** Determine the parameters $\alpha$ and $\beta$, the second eigenvalue $\lambda_2$, and its associated eigenvector $\mathbf{v_2}$.

**(b)** Is the system asymptotically stable? That is, with zero input, will the state vector asymptotically decay to zero for every choice of initial conditions?

The state of the system is measured at each time $n$ and used for state feedback to the input according to the relation

$$x[n] = \mathbf{g}^T \mathbf{q}[n] + p[n] \, ,$$

where $p[n]$ is the input to the closed-loop system.

(c) If possible, determine a $\mathbf{g}^T$ so that the eigenvalues of the closed-loop system are at 0.5 and 0. If it is not possible, clearly explain why not.

**6.6.** Consider the causal DT LTI system

$$\mathbf{q}[n+1] = \begin{bmatrix} 0 & 1 \\ -6 & -5 \end{bmatrix} \mathbf{q}[n] + \begin{bmatrix} 0 \\ 1 \end{bmatrix} x[n] + \begin{bmatrix} 1 \\ 0 \end{bmatrix} w[n] \, ,$$

where $x[n]$ is a control input and $w[n]$ is a disturbance input.

(a) What are the natural frequencies of the system (i.e., the eigenvalues of the state evolution matrix)? Is the system asymptotically stable?

(b) Suppose you use the LTI state feedback

$$x[n] = g_1 q_1[n] + g_2 q_2[n] \, .$$

What choice of the gains $g_1$ and $g_2$ will yield the closed-loop characteristic polynomial $z(z + 0.5)$? For this choice, write down the eigenvalues of the matrix that describes the state evolution of the closed-loop system, and compute the associated eigenvectors.

(c) Suppose the system output is $y[n] = q_1[n]$. With $x[n]$ chosen as in (b), is the closed-loop system observable?

(d) With $x[n]$ as in (b) and $y[n]$ as in (c), what is the transfer function from $w[n]$ to $y[n]$ for the closed-loop system?

(e) Determine in two distinct ways, using the results in (b), (c), and (d), whether the closed-loop system is reachable from the disturbance input $w[n]$.

**6.7.** Consider the DT LTI state-space model

$$\mathbf{q}[n+1] = \begin{bmatrix} -1 & 1 \\ 0 & -2 \end{bmatrix} \mathbf{q}[n] + \begin{bmatrix} 0 \\ 1 \end{bmatrix} x[n]$$

$$y[n] = \begin{bmatrix} -0.5 & 1 \end{bmatrix} \mathbf{q}[n].$$

(a) Explicitly compute its transfer function $H(z)$ from input $x[n]$ to output $y[n]$. Are the poles where you expect them to be? Explain. You should be able to conclude directly from the computed $H(z)$ that the system is reachable and observable—explain your reasoning. Is the system asymptotically stable?

(b) Now suppose we implement LTI state feedback of the form

$$x[n] = \mathbf{g}^T \mathbf{q}[n] + p[n] \, ,$$

where $\mathbf{g}$ is a gain vector and $p[n]$ is a new external input. Determine what choice of $\mathbf{g}$ will result in the closed-loop eigenvalues (or natural frequencies) of the system being at $\pm 0.5$. What are the eigenvectors respectively associated with these eigenvalues? Is the closed-loop system still reachable from $p[n]$? Is it still observable from $y[n]$?

(c) With $\mathbf{g}$ chosen as in (b), what is the transfer function $F(z)$ of the closed-loop system from input $p[n]$ to output $y[n]$? Taking note of your answers in

(a) and (b), explain (in terms of poles, zeros, cancellations, etc.) what has happened to result in this transfer function.

**6.8.** A particular second-order CT causal LTI system has natural frequencies $\lambda_1 = -3$ and $\lambda_2 = -4$; these are the eigenvalues of the matrix that governs state evolution. Their associated eigenvectors are denoted by $\mathbf{v}_1$ and $\mathbf{v}_2$ respectively. The system's input-output transfer function is

$$H(s) = \frac{s+1}{(s+3)(s+4)} .$$

**(a)** Is the system reachable? Is it observable? Explain.

**(b)** Suppose the system is initially at rest, i.e., its initial state is zero. Is it now possible to choose the input in such a way that the state moves out along the eigenvector $\mathbf{v}_1$, with no component along $\mathbf{v}_2$ during the entire motion? Explain your answer carefully.

**(c)** Suppose the output of the above system is applied to the input of another causal second-order LTI system with transfer function

$$G(s) = \frac{s+3}{s(s+5)} .$$

The input to the combined system is then just the original input to the first system, while the output of the combined system is the output of the second system, as shown in Figure P6.8.

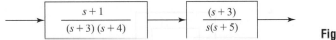

**Figure P6.8**

(i) How many state variables are there in the state-space description of the combined system, and what are the natural frequencies of this combined system?

(ii) Is the combined system asymptotically stable? Explain.

(iii) Is the combined system reachable from the input of the first system? Is it observable from the output of the second system? Explain.

(iv) If you were to build an observer for the combined system using measurements of the input to the first system and the output of the second system, could you get the estimation error of the observer to decay? If not, why not; and if so, could you get the error to decay arbitrarily fast?

**6.9.** Suppose the equations of a DT plant and its observer are as follows.
*Plant:*

$$\mathbf{q}[n+1] = \mathbf{A}\mathbf{q}[n] + \mathbf{b}x[n]$$

$$y[n] = \mathbf{c}^T\mathbf{q}[n] + dx[n] .$$

*Observer:*

$$\hat{\mathbf{q}}[n+1] = \mathbf{A}\hat{\mathbf{q}}[n] + \mathbf{b}x[n] - \ell(y[n] - \hat{y}[n])$$

$$\hat{y}[n] = \mathbf{c}^T\hat{\mathbf{q}}[n] + dx[n] .$$

**(a)** Assume that the initial state of the plant is unknown, that $\ell = 0$ (i.e., no error feedback in the observer), and the initial state of the observer is zero.

Describe the time behavior for $n \geq 0$ of the output error $y[n] - \widehat{y}[n]$ and the state error $\mathbf{q}[n] - \widehat{\mathbf{q}}[n]$ if

$$\mathbf{A} = \begin{bmatrix} 0 & 1 \\ -\frac{1}{2} & \frac{3}{2} \end{bmatrix}, \quad \mathbf{b} = \begin{bmatrix} 1 \\ \frac{1}{2} \end{bmatrix}, \quad \mathbf{c}^T = \begin{bmatrix} 0 & 1 \end{bmatrix}, \quad d = 0.$$

**(b)** Determine whether the system (i.e., the plant) is observable if $\mathbf{A}$, $\mathbf{b}$, $\mathbf{c}$, and d are as given in (a).

**(c)** Determine a choice for the vector $\boldsymbol{\ell}$ so that the the eigenvalues governing state estimation error decay are at $\frac{1}{8}$ and at $\frac{1}{16}$.

**6.10.** Consider the levitated ball in a magnetic suspension system shown in Figure P6.10. The dynamical behavior of the ball is described by the following state equations, where $q_1(t)$ is the vertical position and $q_2(t)$ is the vertical velocity of the ball:

$$\frac{d\mathbf{q}(t)}{dt} = \mathbf{A}\mathbf{q}(t) + \mathbf{b}x(t)$$

$$y(t) = \mathbf{c}^T\mathbf{q(t)}$$

with

$$\mathbf{A} = \begin{bmatrix} 0 & 1 \\ 25 & 0 \end{bmatrix}, \quad \mathbf{b} = \begin{bmatrix} 0 \\ 1 \end{bmatrix}.$$

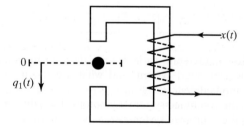

**Figure P6.10**

**(a)** Determine the eigenvalues and eigenvectors of the system.

**(b)** Determine all possible values of $\mathbf{c}$ for which the system is BIBO stable.

**(c)** Determine whether it is possible to place the natural frequencies or eigenvalues of the system at arbitrary self-conjugate locations through the use of state feedback of the form

$$x(t) = \mathbf{g}^T\mathbf{q}(t).$$

**(d)** We assume that the state variables are not directly measurable but both the input $x(t)$ and the output $y(t)$ are with $\mathbf{c}^T = [1\ 0]$. We want to estimate the state of the system based on the following equations:

$$\frac{d\widehat{\mathbf{q}}(t)}{dt} = \mathbf{A}\widehat{\mathbf{q}}(t) + \mathbf{b}x(t) - \boldsymbol{\ell}(y(t) - \widehat{y}(t))$$

$$\widehat{y}(t) = \mathbf{c}^T\widehat{\mathbf{q}}(t)$$

where $\widehat{\mathbf{q}}$ is the estimate for $\mathbf{q}$. Can the gain vector $\boldsymbol{\ell}$ be chosen so that the error in the state estimate decays asymptotically?

**6.11.** A given system is not asymptotically stable but is BIBO stable. Is it possible to implement observer-based feedback to make the system asymptotically stable? Explain.

**6.12.** In this problem we will consider state feedback to stabilize the inverted pendulum shown in Figure P6.12, comprising a mass $m$ at the end of a massless rod of length $L$.

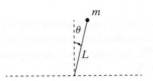

**Figure P6.12**

For small $\theta$, $\ddot{\theta}(t) = \frac{g}{L}\theta(t) + x(t)$, where $g$ is the acceleration due to gravity, and $x(t)$ is the applied torque, in some appropriate units. The natural associated state-space description is

$$\dot{\mathbf{q}}(t) \equiv \begin{bmatrix} \dot{q}_1(t) \\ \dot{q}_2(t) \end{bmatrix} = \begin{bmatrix} 0 & 1 \\ g/L & 0 \end{bmatrix} \begin{bmatrix} q_1(t) \\ q_2(t) \end{bmatrix} + \begin{bmatrix} 0 \\ 1 \end{bmatrix} x(t) = \mathbf{A}\mathbf{q}(t) + \mathbf{b}x(t)$$

where $q_1(t) = \theta(t)$, $q_2(t) = \dot{\theta}(t)$. For this problem pick $g/L = 16$.

**(a)** Determine the eigenvalues and associated eigenvectors of the state evolution matrix $\mathbf{A}$.

**(b)** Determine the input-output (ZSR) transfer function from input $x(t)$ to output $\theta(t)$.

**(c)** Determine a nonzero choice for the initial conditions so that the state vector $\mathbf{q}(t)$ asymptotically decays to zero.

**(d)** Consider the use of state feedback, i.e., choose $x(t) = \mathbf{g}^T\mathbf{q}(t)$ where $\mathbf{g}$ is the feedback gain vector. Determine the constraints on $\mathbf{g}$ so that the system is stabilized.

**(e)** The use of state feedback in this system requires measuring both $\theta(t)$ and $\dot{\theta}(t)$. Determine whether it is possible to stabilize the system by choosing $x(t)$ to be of the form $x(t) = a\theta(t)$ where $a$ is a scalar constant.

**(f)** Suppose now that we only have an inaccurate measurement of the angle $\theta(t)$ (i.e., $q_1(t)$), so the available quantity is

$$y(t) = \begin{bmatrix} 1 & 0 \end{bmatrix} \begin{bmatrix} q_1(t) \\ q_2(t) \end{bmatrix} + \zeta(t) = \mathbf{c}^T\mathbf{q}(t) + \zeta(t),$$

where $\zeta(t)$ denotes the error in the measurement of $\theta(t)$. We would like to stabilize the system by using an estimate of the state rather than the actual state. To obtain an estimate $\hat{\mathbf{q}}(t)$ of the state $\mathbf{q}(t)$, we use our knowledge of the system dynamics (i.e., of the matrix $\mathbf{A}$ and the vector $\mathbf{b}$) and the (inaccurate) measurement $y(t)$. The state-space model that produces $\hat{\mathbf{q}}(t)$ is the following:

$$\frac{d\hat{\mathbf{q}}(t)}{dt} = \mathbf{A}\hat{\mathbf{q}}(t) + \mathbf{b}x(t) - \begin{bmatrix} \ell_1 \\ \ell_2 \end{bmatrix} \left( y(t) - \mathbf{c}^T\hat{\mathbf{q}}(t) \right).$$

Find the state-space description governing the estimation error, defined as $\tilde{\mathbf{q}}(t) = \mathbf{q}(t) - \hat{\mathbf{q}}(t)$, i.e., find $\mathbf{E}$ and $\mathbf{f}$ such that

$$\frac{d\tilde{\mathbf{q}}(t)}{dt} = \mathbf{E}\tilde{\mathbf{q}}(t) + \mathbf{f}\zeta(t).$$

Can you pick $\ell_1$ and $\ell_2$ so that the natural frequencies of the error equation are stable? If so, choose them so that the natural frequencies are at $-4$ and $-3$.

**(g)** Repeat the preceding part (f) for the case when we have a noisy measurement of the angular velocity, i.e.,

$$y(t) = [0 \quad 1]\begin{bmatrix} q_1(t) \\ q_2(t) \end{bmatrix} + \zeta(t) .$$

**(h)** Using $\ell_1$ and $\ell_2$ as found in part (f), provide a state-space description of the system obtained by combining the inverted pendulum state-space description with the state-space description of $\widetilde{\mathbf{q}}(t)$, and using the estimated state as feedback:

$$x(t) = [\gamma_1 \quad \gamma_2]\begin{bmatrix} \widehat{q}_1(t) \\ \widehat{q}_2(t) \end{bmatrix} .$$

What is the characteristic polynomial of the system? Can you choose $\gamma_1$ and $\gamma_2$ so that the overall system is stable with natural frequencies at $-1, -2, -3,$ and $-4$?

## Advanced Problems

**6.13.** Suppose we are given the state-space model $\dot{\mathbf{q}}(t) = \mathbf{A}\mathbf{q}(t) + \mathbf{b}x(t)$ with output equation $y(t) = \mathbf{c}^T\mathbf{q}(t)$, where

$$\mathbf{A} = \begin{bmatrix} 0 & 1 & 0 \\ 0 & 0 & 1 \\ -12 & -19 & -8 \end{bmatrix}, \qquad \mathbf{b} = \begin{bmatrix} 0 \\ 0 \\ 1 \end{bmatrix},$$

$$\mathbf{c}^T = \begin{bmatrix} 2 & 1 & 0 \end{bmatrix} .$$

It turns out for this choice of $\mathbf{A}$ and $\mathbf{b}$, and no matter what the particular numbers are in the last row of $\mathbf{A}$, that

$$(s\mathbf{I} - \mathbf{A})^{-1}\mathbf{b} = \frac{1}{\det(s\mathbf{I} - \mathbf{A})}\begin{bmatrix} 1 \\ s \\ s^2 \end{bmatrix} .$$

It may also help you in this problem to know that

$$s^3 + 8s^2 + 19s + 12 = (s+1)(s+3)(s+4) .$$

**(a)** Find the characteristic polynomial of the system, the natural or modal frequencies (i.e., eigenvalues of $\mathbf{A}$), the associated mode shapes (i.e., eigenvectors), and the transfer function $H(s)$. Is the system reachable? Is it observable? If you find the system to be unreachable or unobservable (or both), specify which mode or modes are respectively unreachable or unobservable (or both).

**(b)** Suppose we implement the state feedback control $x(t) = \mathbf{g}^T\mathbf{q}(t) + p(t)$. Specify what choice of $\mathbf{g}^T$ will make the characteristic polynomial of the closed-loop system equal to

$$(s+1)(s+3)(s+5) .$$

For this choice of state feedback gain, what are the mode shapes of the closed-loop system, and what is the transfer function from $p(t)$ to $y(t)$? Is the closed-loop system reachable from $p(t)$? Is it observable from $y(t)$? If you find the system to be unreachable or unobservable (or both), specify which mode or modes are respectively unreachable or unobservable (or both).

**(c)** Repeat (b) for the case where the desired closed-loop characteristic polynomial is

$$(s+1)(s+2)(s+4).$$

**6.14.** In the following CT second-order LTI state-space system, $x(t)$ is a known scalar input, $\overline{w}$ is an unknown but constant scalar disturbance that adds to the known input, $y(t)$ is a measured output, and $\mathbf{q}(t)$ denotes a state vector with components $q_1(t)$ and $q_2(t)$:

$$\dot{\mathbf{q}}(t) = \begin{bmatrix} 0 & a_1 \\ 1 & a_2 \end{bmatrix} \mathbf{q}(t) + \begin{bmatrix} b_1 \\ b_2 \end{bmatrix} \left( x(t) + \overline{w} \right),$$

$$y(t) = \begin{bmatrix} 0 & 1 \end{bmatrix} \mathbf{q}(t).$$

Our plan is to incorporate $\overline{w}$ as a third state variable, then estimate it using an observer, and finally use this estimate to cancel the effect of the disturbance. Assume $a_1$ and $a_2$ are such that the system has distinct eigenvalues.

**(a)** First determine precisely under what conditions on the parameters the given second-order system is observable from the output $y(t)$.

**(b)** Now determine precisely under what conditions on the parameters the given second-order system is reachable from the input $x(t)$.

Assume in what follows that whatever conditions you identified in (a) and (b) are satisfied, i.e., take the given system as observable and reachable.

**(c)** Taking $q_3(t) = \overline{w}$ as a third state variable, write down a new third-order LTI state-space model that has just the known $x(t)$ as an input (with no disturbance component) and has the same $y(t)$ as its output.

**(d)** Suppose you were now to construct an observer for the third-order system you obtained in (c), with observer gains $\ell_1, \ell_2, \ell_3$ used to feed the output error $y(t) - \widehat{y}(t)$ into the real-time simulator that forms the core of the observer. Write down the $3 \times 3$ matrix whose eigenvalues govern the time evolution of the observer error, and obtain the associated characteristic polynomial.

**(e)** What condition on the parameters $a_1, a_2, b_1,$ and $b_2$ of the original system is necessary and sufficient to guarantee that the eigenvalues governing the error decay of the observer in (d) can be placed at any self-conjugate set of locations by appropriate choice of the observer gains? Interpret this condition as a condition on the poles and/or zeros of the original second-order system.

**(f)** Assume the condition in (e) is satisfied. Determine the observer gains required to place the eigenvalues governing observer error decay at $-1, -2,$ and $-3$, expressing the gains in terms of the system parameters $a_1, a_2, b_1,$ and $b_2$.

**(g)** Our plan now is to set $x(t) = -\widehat{\overline{w}}(t)$, where $\widehat{\overline{w}}(t)$ is the estimate at time $t$ of the constant disturbance $\overline{w}$. When $\widehat{\overline{w}}(t)$ has converged to $\overline{w}$, this will cancel the effect of the constant disturbance $\overline{w}$ on the system. Determine the characteristic polynomial of the resulting overall sixth-order closed-loop system. (*Hint:* it will be the product of two third-degree polynomials that you should be able to write down quite directly.) You should find that one of the roots of this sixth-degree characteristic polynomial is at 0, corresponding to the fact that the disturbance variable $\overline{w}$ remains fixed at some value. What condition will ensure that the remaining roots are all strictly in the left half-plane?

**6.15. (a)** Suppose the transfer function of System 1 in the block diagram shown in Figure P6.15 is

$$H_1(s) = \frac{s}{s-1} = \frac{1}{s-1} + 1.$$

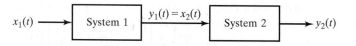

$x_1(t) \longrightarrow$ System 1 $\xrightarrow{y_1(t)=x_2(t)}$ System 2 $\longrightarrow y_2(t)$

**Figure P6.15**

   (i)   Find a first-order state-space model for System 1, using $q_1(t)$ to denote
         its state variable and arranging things such that $y_1(t) = q_1(t) + x_1(t)$.
   (ii)  Is your state-space model for System 1 reachable? Observable?
         Asymptotically stable?
**(b)** Suppose System 2 in Figure P6.15 is described by the first-order state-space
         model

$$\dot{q}_2(t) = \mu q_2(t) + x_2(t)$$

$$y_2(t) = 2q_2(t)$$

where $\mu$ is a parameter, and we are given that $\mu \neq 1$.

   (i)   What is the transfer function $H_2(s)$ of System 2?
   (ii)  For what values of $\mu$, if any, is the state-space model of System 2
         unreachable? Unobservable? Asymptotically stable?
**(c)**  (i)  Combine the state-space models in (a) and (b) to obtain a second-order
         state-space model of the form

$$\dot{\mathbf{q}}(t) = \mathbf{A}\mathbf{q}(t) + \mathbf{b}x_1(t) , \quad y_2(t) = \mathbf{c}^T\mathbf{q}(t) + dx_1(t)$$

   for the overall system in Figure P6.15, using $\begin{bmatrix} q_1(t) \\ q_2(t) \end{bmatrix}$ as the overall state
   vector $\mathbf{q}(t)$, $x_1(t)$ as the overall input, and $y_2(t)$ as the overall output.
   (ii)  Compute the transfer function $H(s)$ from $x_1(t)$ to $y_2(t)$ using the model
         in (c)(i), and verify that it equals $H_1(s)H_2(s)$.
   (iii) What are the eigenvalues of $\mathbf{A}$ in (c)(i)? Check that the eigenvalues you
         obtain are consistent with what you expect from your results in (c)(ii).
         What are the eigenvectors associated with these eigenvalues?
   (iv)  There are values of $\mu$ for which one can find nonzero initial conditions
         $\mathbf{q}(0)$ such that the resulting zero-input solution $\mathbf{q}(t)$ (i.e., the solution
         with $x_1(t) \equiv 0$) decays to $\mathbf{0}$ as $t \to \infty$. Find all such values of $\mu$, and
         for each such $\mu$ specify all initial conditions that lead to such decaying
         zero-input solutions.
   (v)   For what values of $\mu$, if any, is the overall system in (c)(ii):
         (i)  Unreachable? Which natural frequencies are unreachable?
         (ii) Unobservable? Which natural frequencies are unobservable?
         Interpret your results in terms of pole-zero cancellations in the block
         diagram in Figure P6.15.

**(d)** Suppose you can measure both state variables $q_1(t)$ and $q_2(t)$, so that you can choose

$$x_1(t) = g_1 q_1(t) + g_2 q_2(t) .$$

The resulting closed-loop system is still described by a second-order LTI state-space model. What choice of $g_1$ and $g_2$ will result in the closed-loop natural frequencies being at $-1 \pm j1$? You can express your answer in terms of $\mu$. Now determine for what values of $\mu$, if any, your expressions for $g_1$ and/or $g_2$ have infinite magnitude, and reconcile your answer with what you found in (c)(v).

**(e)** Suppose you can only measure the input $x_1(t)$ and the output $y_2(t)$. Fully specify a procedure for estimating the state variables $q_1(t)$ and $q_2(t)$, in such a way that the error between each of the actual and estimated state variables can be expressed as a linear combination of two decaying exponential terms with time constants of 0.5 and 0.25 respectively. Will your estimation scheme work for all values of $\mu$? Again, reconcile your answer with what you found in (c)(v).

## Extension Problems

**6.16.** A model of a rotating machine driven by a piecewise-constant torque takes the state-space form

$$\mathbf{q}[k+1] = \begin{bmatrix} q_1[k+1] \\ q_2[k+1] \end{bmatrix} = \begin{bmatrix} 1 & T \\ 0 & 1 \end{bmatrix} \begin{bmatrix} q_1[k] \\ q_2[k] \end{bmatrix} + \begin{bmatrix} T^2/2 \\ T \end{bmatrix} x[k]$$

$$= \mathbf{A}\mathbf{q}[k] + \mathbf{b}x[k]$$

where the state vector $\mathbf{q}[k]$ comprises the position $q_1[k]$ and velocity $q_2[k]$ of the rotor, sampled at time $t = kT$; $x[k]$ is the constant value of the torque in the interval $kT \le t < kT + T$. Assume for this problem that $T = 0.5$.

**(a)** Is the system asymptotically stable? Note that this model does not have distinct eigenvalues. To answer the question here, you might invoke the stability result that we proved only for the case of distinct eigenvalues, but which we claimed held for the general case as well. Alternatively, for a more direct and more satisfying argument, you could try to find general expressions for the entries of $\mathbf{A}^k$ in this case. Evaluate $\mathbf{A}^{20}$ and $\mathbf{A}^{100}$. Are the entries of $\mathbf{A}^k$ growing linearly with $k$? Quadratically? Exponentially?

**(b)** Suppose we implement a position-feedback control law of the form

$$x[k] = \gamma \, q_1[k] .$$

Write down a state-space model for the closed-loop system, and obtain an expression for its characteristic polynomial. Can $\gamma$ be chosen so as to place the roots of the closed-loop characteristic polynomial at arbitrary self-conjugate locations, i.e., subject to the requirement that complex roots occur in conjugate pairs ?

**(c)** Suppose we implement a state feedback control law of the form

$$x[k] = [\gamma_1 \quad \gamma_2] \begin{bmatrix} q_1[k] \\ q_2[k] \end{bmatrix}.$$

Write down a state-space model of the closed-loop system.

**(d)** What choice of gains $\gamma_1$ and $\gamma_2$ in (c) will result in the natural frequencies of the closed-loop system being at $\pm 0.6$, corresponding to a closed-loop characteristic polynomial of $(z - 0.6)(z + 0.6)$? For this choice of gains, and with $q_1[0] = 4$ and $q_2[0] = 1$, determine and plot $q_1[k]$, $q_2[k]$, and $x[k]$ for $0 \le k \le 20$.

**(e)** The state feedback law in (c) and (d) assumed that we had access to accurate position and velocity measurements. Suppose instead that all we have is a noisy measurement of the position, so the available quantity is

$$y[k] = [1 \quad 0] \begin{bmatrix} q_1[k] \\ q_2[k] \end{bmatrix} + \zeta[k] = \mathbf{c}^T \mathbf{q}[k] + \zeta[k],$$

where $\zeta[k]$ denotes the unknown noise. One way to estimate the actual position and velocity is by using an observer, which has the form

$$\widehat{\mathbf{q}}[k+1] = \mathbf{A}\widehat{\mathbf{q}}[k] + \mathbf{b}x[k] - \begin{bmatrix} \ell_1 \\ \ell_2 \end{bmatrix} \left( y[k] - \mathbf{c}^T \widehat{\mathbf{q}}[k] \right).$$

Here $\widehat{\mathbf{q}}[k]$ is our estimate of $\mathbf{q}[k]$. Let the observer error be denoted by $\widetilde{\mathbf{q}}[k] = \mathbf{q}[k] - \widehat{\mathbf{q}}[k]$. Determine the state-space equation that $\widetilde{\mathbf{q}}[k]$ satisfies.

Show for our rotating machine example that, by proper choice of the observer gain, we can obtain arbitrary self-conjugate natural frequencies for this error equation. What choice of $\ell_1$ and $\ell_2$ will place the natural frequencies of the error equation at 0 and 0.25? For this choice of observer gains, with $q_1[0] = 4$, $q_2[0] = 1$ as in (d), and with $\widehat{q}_1[0] = 0$ and $\widehat{q}_2[0] = 0$, compare $\widehat{q}_1[k]$ and $\widehat{q}_2[k]$ with the plots you obtained for $q_1[k]$ and $q_2[k]$ in (d), still assuming $x[k]$ is generated as specified by (c) and (d), and assuming that the measurement noise is zero, i.e., $\zeta[\cdot] = 0$.

Also explore what happens to the estimation error for zero-mean measurement noise $\zeta[\cdot]$ that is independent and identically distributed at each instant, for example taking the values $+\sigma$ or $-\sigma$ with equal probability at each instant, for some $\sigma > 0$.

**(f)** Obtain a state-space description of the system obtained by combining the observer with the model of the machine, and using the control law

$$x[k] = [\gamma_1 \quad \gamma_2] \begin{bmatrix} \widehat{q}_1[k] \\ \widehat{q}_2[k] \end{bmatrix},$$

rather than the state feedback law of part (c). You may find it most convenient to use the following state vector:

$$\begin{bmatrix} q_1[k] \\ q_2[k] \\ \widetilde{q}_1[k] \\ \widetilde{q}_2[k] \end{bmatrix}.$$

Compute and plot $q_1[k]$, $q_2[k]$, and $x[k]$ for the same initial conditions you used in (d) and (e), and assuming zero measurement noise. Do you notice any consequences of using feedback of the estimated state rather than of the actual state?

Again explore what happens to your plots when there is zero-mean measurement noise.

# 7 Probabilistic Models

In the preceding chapters, we have emphasized deterministic signals. Here and in the remaining chapters, we expand our discussion to include signals based on probabilistic models, referred to as random or stochastic processes or signals. To introduce this important class of signals, we begin in this chapter with a review of the basics of probability and random variables. We assume that you have encountered this foundational material in a previous course, but include a review here for convenient reference and to establish notation. In the following chapters, we apply these concepts to define random signals, explore their properties, and develop methods for signal estimation and detection in this context.

## 7.1 THE BASIC PROBABILITY MODEL

Associated with a basic probability model are the following three components, as indicated in Figure 7.1:

1. The **sample space**, $\Psi$, is the set of all possible outcomes $\psi$ of the probabilistic experiment that the model represents. We require that one and only one outcome be produced in each experiment with the model.

2. An **event algebra** is a collection of subsets of the sample space—referred to as events in the sample space—chosen such that unions of events, intersections of events, and complements of events are themselves events, that is, they are in the collection of subsets. Note that

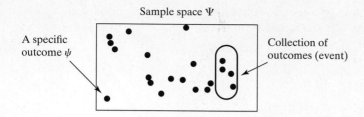

**Figure 7.1**    Sample space and events.

intersections of events can also be expressed in terms of unions and complements. A particular event is said to have occurred if the outcome of the experiment lies in this event subset; thus the set $\Psi$ is the "certain event" because it always occurs, and the empty set $\emptyset$ is the "impossible event" because it never occurs.

3. A **probability measure** associates with each event $A$ a number $P(A)$, termed the probability of $A$, in such a way that
   - $P(A) \geq 0$;
   - $P(\Psi) = 1$; and
   - if $A \cap B = \emptyset$, that is, if events $A$ and $B$ are mutually exclusive, then

$$P(A \cup B) = P(A) + P(B) . \tag{7.1}$$

Note that for any particular case we often have a range of options in specifying what constitutes an outcome, in defining an event algebra, and in assigning a probability measure. It is generally convenient to have as few elements or outcomes as possible in a sample space, but enough of them are needed to enable specification of the events of interest. Typically, the smallest event algebra that contains the events of interest is chosen. An assignment of probabilities to events is also required that is consistent with the above conditions. This assignment may be made on the basis of symmetry arguments or in some other way that is suggested by the particular application.

The joint probability $P(A \cap B)$ is often alternatively written as $P(A, B)$.

## 7.2 CONDITIONAL PROBABILITY, BAYES' RULE, AND INDEPENDENCE

The probability of event $A$, given that event $B$ has occurred, is denoted by $P(A|B)$. Knowledge that $B$ has occurred in effect reduces the sample space to the outcomes in $B$, so a natural definition of the conditional probability is

$$P(A|B) = \frac{P(A, B)}{P(B)} \text{ if } P(B) > 0 . \tag{7.2}$$

It is straightforward to verify that this definition of conditional probability yields a valid probability measure on the sample space $B$. The preceding equation can also be rearranged to the form

$$P(A, B) = P(A|B)P(B) . \tag{7.3}$$

If $P(B) = 0$, then the conditional probability in Eq. (7.2) is undefined.

By symmetry, we can also write

$$P(A, B) = P(B|A)P(A) . \tag{7.4}$$

Combining Eqs. (7.3) and (7.4), we can write

$$P(B|A) = \frac{P(A|B)P(B)}{P(A)} . \tag{7.5}$$

Equation (7.5) plays an essential role in much of the development of methods for signal detection, classification, and estimation.

A more detailed form of Eq. (7.5) can be written for the conditional probability of one of a set of events $\{B_j\}$ that are mutually exclusive and collectively exhaustive, that is, $B_\ell \cap B_m = \emptyset$ if $\ell \neq m$, and $\cup_j B_j = \Psi$. In this case,

$$P(A) = \sum_j P(A, B_j) = \sum_j P(A|B_j)P(B_j) \tag{7.6}$$

so that

$$P(B_\ell|A) = \frac{P(A|B_\ell)P(B_\ell)}{\sum_j P(A|B_j)P(B_j)} . \tag{7.7}$$

The general form of Eq. (7.7) is commonly referred to as Bayes' rule or Bayes' theorem, although that terminology is also often applied to the more specific case of Eq. (7.5).

Events $A$ and $B$ are said to be independent if

$$P(A|B) = P(A) \tag{7.8}$$

or equivalently, from Eq. (7.3), if the joint probability factors as

$$P(A, B) = P(A)P(B) . \tag{7.9}$$

More generally, a collection of events is said to be mutually independent if the probability of the intersection of events from this collection, taken any number at a time, is always the product of the individual probabilities. Note that pairwise independence is not sufficient. Also, two sets of events $A$ and $B$ are said to be independent of each other if the probability of an intersection of events taken from these two sets always factors into the product of the joint probability of those events that are in $A$ and the joint probability of those events that are in $B$.

---

**Example 7.1**    **Transmission Errors in a Communication System**

Consider a communication system that transmits symbols labeled $A$, $B$, and $C$. Because of errors (e.g., noise, dropouts, fading, etc.) introduced by the channel, there is a nonzero probability that for each transmitted symbol, the received symbol differs from the transmitted one. Table 7.1 describes the joint probability for each possible pair of transmitted and received symbols under a certain set of system conditions.

**TABLE 7.1** JOINT PROBABILITY FOR EACH
POSSIBLE PAIR OF TRANSMITTED
AND RECEIVED SYMBOLS

| Symbol Sent | Symbol Received | | |
|:---:|:---:|:---:|:---:|
| | $A$ | $B$ | $C$ |
| $A$ | 0.05 | 0.10 | 0.09 |
| $B$ | 0.13 | 0.08 | 0.21 |
| $C$ | 0.12 | 0.07 | 0.15 |

For notational convenience, we use $A_s$, $B_s$, $C_s$ to denote the events that the symbol $A$, $B$, or $C$, respectively, is sent, and $A_r$, $B_r$, $C_r$ to denote that $A$, $B$, or $C$, respectively, is the symbol received. For example, according to Table 7.1, the probability that symbol $A$ is received and symbol $B$ is sent is $P(A_r, B_s) = 0.13$. Similarly, $P(C_r, C_s) = 0.15$. To determine the marginal probability $P(A_r)$, we sum the probabilities for all the mutually exclusive ways that $A$ is received, so,

$$P(A_r) = P(A_r, A_s) + P(A_r, B_s) + P(A_r, C_s)$$

$$= .05 + .13 + .12 = 0.30 . \tag{7.10}$$

Similarly, we can determine the marginal probability $P(A_s)$ as

$$P(A_s) = P(A_r, A_s) + P(B_r, A_s) + P(C_r, A_s) = 0.24 . \tag{7.11}$$

In a communication context, it may be important to know the probability, for example, that $C$ was sent, given that $B$ was received, that is, $P(C_s|B_r)$. That information is not entered directly in the table but can be calculated from it using Eq. (7.2), which allows the desired conditional probability to be expressed as

$$P(C_s|B_r) = \frac{P(C_s, B_r)}{P(B_r)} . \tag{7.12}$$

The numerator in Eq. (7.12) is given directly in the table as 0.07. The denominator is calculated as $P(B_r) = P(B_r, A_s) + P(B_r, B_s) + P(B_r, C_s) = 0.25$. The result then is that $P(C_s|B_r) = 0.28$.

To determine the probability that symbol $A$ is received, given that symbol $B$ was sent, that is, the conditional probability $P(A_r|B_s)$, we express $P(A_r|B_s)$ as

$$P(A_r|B_s) = \frac{P(A_r, B_s)}{P(B_s)} . \tag{7.13}$$

The numerator is specified in the table as 0.13. The denominator is calculated as $P(B_s) = P(B_s, A_r) + P(B_s, B_r) + P(B_s, C_r) = 0.42$. Consequently, $P(A_r|B_s) = 0.31$.

In communication systems, it is also often of interest to measure or calculate the probability of a transmission error. A transmission error in this example would correspond to any of the following mutually exclusive events happening:

$$(A_s, B_r), (A_s, C_r), (B_s, A_r), (B_s, C_r), (C_s, A_r), (C_s, B_r) . \tag{7.14}$$

The transmission error probability $P_t$ is therefore the sum of the probabilities of these six mutually exclusive events, and all these probabilities can be read directly from the table in the off-diagonal locations, yielding $P_t = 0.72$.

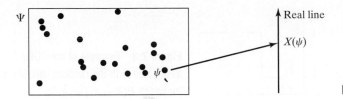

**Figure 7.2**   A random variable.

## 7.3 RANDOM VARIABLES

A real-valued random variable $X(\cdot)$ is a function that maps each outcome $\psi$ of a probabilistic experiment to a real number $X(\psi)$, which is termed the realization of (or value taken by) the random variable in that experiment. This is illustrated in Figure 7.2. An additional technical requirement imposed on this function is that the set of outcomes $\{\psi\}$ which maps to the interval $X \leq x$ must be an event in $\Psi$, for all real numbers $x$. We shall typically just write the random variable as $X$ instead of $X(\cdot)$ or $X(\psi)$.

It is often also convenient to consider mappings of probabilistic outcomes to one of a finite or countable set of categories or labels, say $L_0, L_1, L_2, \ldots$, rather than to a real number. For instance, the random status of a machine may be tracked using the labels Idle, Busy, and Failed. Similarly, the random presence of a target in a radar scan can be tracked using the labels Absent and Present. We can think of these labels as comprising a set of mutually exclusive and collectively exhaustive events, in which each such event comprises all the outcomes that carry that label. Such a mapping associates each outcome $\psi$ of a probabilistic experiment to the category or label $L(\psi)$, chosen from the possible values $L_0, L_1, L_2, \ldots$. We shall typically just write $L$ instead of $L(\psi)$ and refer to this mapping as a categorical random variable or simply as a random variable when the context is clear.

## 7.4 PROBABILITY DISTRIBUTIONS

**Cumulative Distribution Function**   For a (real-valued) random variable $X$, the probability of the event comprising all outcomes $\psi$ for which $X(\psi) \leq x$ is described using the cumulative distribution function (CDF) $F_X(x)$:

$$F_X(x) = P(X \leq x) . \tag{7.15}$$

We can therefore write

$$P(a < X \leq b) = F_X(b) - F_X(a) . \tag{7.16}$$

In particular, if there is a nonzero probability that $X$ takes a specific value $x_1$, that is, if $P(X = x_1) > 0$, then $F_X(x)$ will have a jump at $x_1$ of height $P(X = x_1)$, and $F_X(x_1) - F_X(x_1-) = P(X = x_1)$, as illustrated in Figure 7.3. The CDF is always nondecreasing as a function of $x$; it starts from $F_X(-\infty) = 0$ and rises to $F_X(\infty) = 1$.

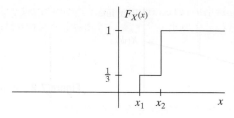

**Figure 7.3**  Example of the CDF associated with the random variable $X$ for which $P(X = x_1) = \frac{1}{3}$, $P(X = x_2) = \frac{2}{3}$.

A related function is the conditional CDF $F_{X|L}(x|L_i)$, used to describe the distribution of $X$ conditioned on some random label $L$ taking the specific value $L_i$, and assuming $P(L = L_i) > 0$:

$$F_{X|L}(x|L_i) = P(X \leq x|L = L_i) = \frac{P(X \leq x, L = L_i)}{P(L = L_i)} . \qquad (7.17)$$

**Probability Density Function**   The probability density function (PDF) $f_X(x)$ of the real random variable $X$ is the derivative of $F_X(x)$:

$$f_X(x) = \frac{dF_X(x)}{dx} . \qquad (7.18)$$

The PDF is always nonnegative because $F_X(x)$ is nondecreasing. At points of discontinuity in $F_X(x)$, corresponding to values of $x$ that have nonzero probability of occurring, there will be (Dirac) impulses in $f_X(x)$, of strength (i.e. area) equal to the height of the discontinuity. We can write

$$P(a < X \leq b) = \int_a^b f_X(x)\,dx . \qquad (7.19)$$

When $a$ and $b$ are minus and plus infinity, respectively, the left side of Eq. (7.19) must be unity and therefore so is the total area under the PDF. Note that because of the structure of the inequalities in the left side of Eq. (7.19), any impulse of $f_X(x)$ at $x = b$ would be included in the integral, while any impulse at $x = a$ would be excluded—that is, the integral actually goes from $a+$ to $b+$.

We can heuristically think of $f_X(x)\,dx$ as corresponding to the probability that $X$ lies in the interval $(x - dx, x]$:

$$P(x - dx < X \leq x) \approx f_X(x)\,dx . \qquad (7.20)$$

Note that at values of $x$ where $f_X(x)$ does not have an impulse, the probability of $X$ having the value $x$ is zero, that is, $P(X = x) = 0$.

A related function is the conditional PDF $f_{X|L}(x|L_i)$, defined as the derivative of $F_{X|L}(x|L_i)$ with respect to $x$.

**Probability Mass Function**   A real-valued discrete random variable $X$ is one that takes only a finite or countable set of real values, $\{x_1, x_2, \cdots\}$. Hence this is actually a categorical random variable—as defined earlier—but specified numerically rather than with labels. The CDF in this case would be a "staircase" function, while the PDF would be zero everywhere, except for impulses at the values $x_j$, with strengths corresponding to the respective probabilities of the $x_j$. These probabilities are conveniently described by the probability mass function (PMF) $p_X(x)$, which gives the probability of the event $X = x_j$:

$$P(X = x_j) = p_X(x_j) . \tag{7.21}$$

## 7.5 JOINTLY DISTRIBUTED RANDOM VARIABLES

Models involving multiple (or compound) random variables are described by joint probabilities. For example, the joint CDF of two random variables $X$ and $Y$ is

$$F_{X,Y}(x,y) = P(X \le x, Y \le y) . \tag{7.22}$$

The corresponding joint PDF is

$$f_{X,Y}(x,y) = \frac{\partial^2 F_{X,Y}(x,y)}{\partial x \, \partial y} , \tag{7.23}$$

which has the heuristic interpretation that

$$P(x - dx < X \le x, y - dy < Y \le y) \approx f_{X,Y}(x,y) \, dx \, dy . \tag{7.24}$$

The marginal PDF $f_X(x)$ is defined as the PDF of the random variable $X$ considered on its own, and is related to the joint density $f_{X,Y}(x,y)$ by

$$f_X(x) = \int_{-\infty}^{+\infty} f_{X,Y}(x,y) \, dy . \tag{7.25}$$

A similar expression holds for the marginal PDF $f_Y(y)$.

We have noted that when the model involves a random variable $X$ and a random label $L$, we may work with the conditional CDF in Eq. (7.17):

$$F_{X|L}(x|L_i) = P(X \le x|L = L_i) = \frac{P(X \le x, L = L_i)}{P(L = L_i)} , \tag{7.26}$$

which is well defined provided $P(L = L_i) > 0$. The derivative of this function with respect to $x$ results in the conditional PDF $f_{X|L}(x|L_i)$. When the model involves two continuous random variables $X$ and $Y$, the corresponding function of interest is the conditional PDF $f_{X|Y}(x|y)$ that describes the distribution of $X$, given $Y = y$. However, for a continuous random variable $Y$, $P(Y = y) = 0$; so even though the following result may seem natural, its justification is more subtle:

$$f_{X|Y}(x|y) = \frac{f_{X,Y}(x,y)}{f_Y(y)} . \tag{7.27}$$

To see the plausibility of Eq. (7.27), note that the conditional PDF $f_{X|Y}(x|y)$ must have the property that

$$f_{X|Y}(x|y) \, dx \approx P(x - dx < X \le x \mid y - dy < Y \le y) , \tag{7.28}$$

but by Bayes' rule, the quantity on the right in the previous equation can be rewritten as

$$P(x - dx < X \leq x \mid y - dy < Y \leq y) \approx \frac{f_{X,Y}(x,y)\, dx\, dy}{f_Y(y)dy} . \tag{7.29}$$

Combining the latter two equations yields Eq. (7.27).

Using similar reasoning, we can obtain relationships such as:

$$P(L = L_i \mid X = x) = \frac{f_{X \mid L}(x \mid L_i) P(L = L_i)}{f_X(x)} . \tag{7.30}$$

Two random variables $X$ and $Y$ are said to be independent or statistically independent if their joint PDF (or equivalently their joint CDF) factors into the product of the individual ones:

$$\begin{aligned} f_{X,Y}(x,y) &= f_X(x)f_Y(y) , \quad \text{or} \\ F_{X,Y}(x,y) &= F_X(x)F_Y(y) . \end{aligned} \tag{7.31}$$

This condition is equivalent to having any collection of events defined in terms of $X$ be independent of any collection of events defined in terms of $Y$.

For a set of more than two random variables to be independent, we require that the joint PDF (or CDF) of random variables from this set factors into the product of the individual PDFs (respectively, CDFs). One can similarly define independence of random variables and random labels.

## Example 7.2   Independence of Events

To illustrate some of the above definitions and concepts, consider two independent random variables $X$ and $Y$ whose individual (i.e., marginal) PDFs are uniform between 0 and 1:

$$f_X(x) = \begin{cases} 1 & 0 \leq x \leq 1 \\ 0 & \text{otherwise} \end{cases} \tag{7.32}$$

$$f_Y(y) = \begin{cases} 1 & 0 \leq y \leq 1 \\ 0 & \text{otherwise.} \end{cases} \tag{7.33}$$

Because $X$ and $Y$ are independent, the joint PDF $f_{X,Y}(x,y)$ is given by

$$f_{X,Y}(x,y) = f_X(x)f_Y(y) . \tag{7.34}$$

We define the events $A, B, C,$ and $D$ as follows:

$$A = \left\{ y > \frac{1}{2} \right\}, \quad B = \left\{ y < \frac{1}{2} \right\}, \quad C = \left\{ x < \frac{1}{2} \right\},$$

$$D = \left\{ x < \frac{1}{2} \text{ and } y < \frac{1}{2} \right\} \cup \left\{ x > \frac{1}{2} \text{ and } y > \frac{1}{2} \right\} . \tag{7.35}$$

These events are illustrated pictorially in Figure 7.4.

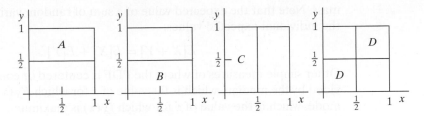

**Figure 7.4**    Illustration of events $A$, $B$, $C$, and $D$ for Example 7.2.

Questions we might ask include whether these events are pairwise independent, for example, whether $A$ and $C$ are independent. To answer such questions, we consider whether the joint probability factors into the product of the individual probabilities. So, for example,

$$P(A, C) = P\left(y > \frac{1}{2}, x < \frac{1}{2}\right) = \frac{1}{4} \tag{7.36}$$

$$P(A) = P(C) = \frac{1}{2}. \tag{7.37}$$

Since $P(A, C) = P(A)P(C)$, events $A$ and $C$ are independent. However,

$$P(A, B) = P\left(y > \frac{1}{2}, y < \frac{1}{2}\right) = 0 \tag{7.38}$$

$$P(A) = P(B) = \frac{1}{2}. \tag{7.39}$$

Since $P(A, B) \neq P(A)P(B)$, events $A$ and $B$ are not independent.

Note that $P(A, C, D) = 0$ since there is no region where all three sets overlap. However, $P(A) = P(C) = P(D) = \frac{1}{2}$, so $P(A, C, D) \neq P(A)P(C)P(D)$ and the events $A$, $C$, and $D$ are not mutually independent, even though they are easily seen to be pairwise independent. For a collection of events to be independent, we require the probability of the intersection of any of the events to equal the product of the probabilities of each individual event. So for the three-event case, pairwise independence is a necessary but not sufficient condition for independence.

# 7.6 EXPECTATIONS, MOMENTS, AND VARIANCE

For many purposes it suffices to have a more aggregated or approximate description than the PDF provides. The expectation—also termed the expected, mean, or average value, or the first moment—of the real-valued random variable $X$ is denoted by $E[X]$ or $\overline{X}$ or $\mu_X$, and defined as

$$E[X] = \overline{X} = \mu_X = \int_{-\infty}^{\infty} x f_X(x)\, dx. \tag{7.40}$$

Considering $f_X(x)$ as a density describing the distribution of a probability "mass" on the real line, the expectation gives the location of the center of mass. Note that the expected value of a sum of random variables is the sum of the individual expected values:

$$E[X + Y] = E[X] + E[Y] . \tag{7.41}$$

Other simple measures of where the PDF is centered or concentrated are provided by the median, which is the value of $x$ for which $F_X(x) = 0.5$, and by the mode, which is the value of $x$ for which $f_X(x)$ is maximum.

The variance or centered second moment of a random variable $X$ is denoted by $\sigma_X^2$ and defined as

$$\sigma_X^2 = E[(X - \mu_X)^2] = \text{expected squared deviation from the mean}$$

$$= \int_{-\infty}^{\infty} (x - \mu_X)^2 f_X(x) \, dx \tag{7.42}$$

$$= E[X^2] - \mu_X^2 ,$$

where the last equation follows on writing $(X - \mu_X)^2 = X^2 - 2\mu_X X + \mu_X^2$ and taking the expectation term by term. We refer to $E[X^2]$ as the second moment of $X$. The square root of the variance, termed the standard deviation, is a widely used measure of the spread of the PDF and is expressed in the same units as the underlying random variable.

The focus of many engineering models that involve random variables is primarily on the means and variances of the random variables. In some cases this is because the detailed PDFs are hard to determine or represent or work with. In other cases, the reason for this focus is that the means and variances completely determine the PDFs, as with the Gaussian (or normal) and uniform PDFs, which are illustrated in the following example.

---

**Example 7.3**    *Gaussian and Uniform Random Variables*

Two common PDFs that we will work with are the Gaussian (or equivalently normal) density and the uniform density:

$$\text{Gaussian:} \quad f_X(x) = \frac{1}{\sigma\sqrt{2\pi}} \exp\left\{-\frac{1}{2}\left(\frac{x-m}{\sigma}\right)^2\right\}$$

$$\tag{7.43}$$

$$\text{Uniform:} \quad f_X(x) = \begin{cases} \frac{1}{b-a} & a < x < b \\ 0 & \text{otherwise.} \end{cases}$$

The two parameters $m$ and $\sigma$ that define the Gaussian PDF can be shown to be its mean and standard deviation respectively. Similarly, though the uniform density can be simply parametrized by its lower and upper limits $a$ and $b$ as above, an equivalent parametrization is via its mean $m = (a + b)/2$ and standard deviation $\sigma = \sqrt{(b-a)^2/12}$.

---

There are useful statements that can be made for general PDFs on the basis of just the mean and variance. Among the most familiar of these is the Chebyshev inequality:

$$P\left(\frac{|X - \mu_X|}{\sigma_X} \geq \alpha\right) \leq \frac{1}{\alpha^2} . \tag{7.44}$$

The inequality in Eq. (7.44) states that for any random variable, the probability that it lies at $\alpha$ or more standard deviations away from the mean (on either side of the mean) is not greater than $1/\alpha^2$. For particular PDFs, much more precise statements can be made, and conclusions derived from the Chebyshev inequality can be very conservative. For instance, choosing $\alpha = 3$ in the case of a Gaussian PDF, the actual probability of being more than three standard deviations away from the mean is only 0.0027, far less than the Chebyshev bound value of $\frac{1}{9}$. Similarly, for a uniform PDF the probability of being more than even two standard deviations away from the mean is precisely 0.

The conditional expectation of the random variable $X$, given that the random variable $Y$ takes the value $y$, is

$$E[X|Y = y] = \int_{-\infty}^{+\infty} x f_{X|Y}(x|y)\, dx = g(y) , \tag{7.45}$$

that is, this conditional expectation takes some value $g(y)$ when $Y = y$. We may also consider the random variable $g(Y)$, namely the function of the random variable $Y$ that, for each $Y = y$, evaluates to the conditional expectation $E[X|Y = y]$. We refer to this random variable $g(Y)$ as the conditional expectation of $X$ "given $Y$" (as opposed to "given $Y = y$"), and denote $g(Y)$ by $E[X|Y]$. Note that the expectation $E[g(Y)]$ of the random variable $g(Y)$—the iterated expectation $E[E[X|Y]]$—is well defined. What we show next is that this iterated expectation works out to something simple, namely $E[X]$. This result will be of particular use in Chapter 8.

Consider first how to compute $E[X]$ when we have the joint PDF $f_{X,Y}(x,y)$. One way is to evaluate the marginal density $f_X(x)$ of $X$, and then use the definition of expectation in Eq. (7.40):

$$E[X] = \int_{-\infty}^{\infty} x \left(\int_{-\infty}^{\infty} f_{X,Y}(x,y)\, dy\right) dx . \tag{7.46}$$

However, it is often simpler to compute the conditional expectation of $X$, given $Y = y$, then average this conditional expectation over the possible values of $Y$, using the marginal density of $Y$. To derive this more precisely, recall that

$$f_{X,Y}(x,y) = f_{X|Y}(x|y)f_Y(y) \tag{7.47}$$

and use this in Eq. (7.46) to deduce that

$$E[X] = \int_{-\infty}^{\infty} f_Y(y) \left(\int_{-\infty}^{\infty} x f_{X|Y}(x|y)\, dx\right) dy = E_Y[E_{X|Y}[X|Y]] . \tag{7.48}$$

We have used subscripts on the preceding expectations in order to make explicit which densities are involved in computing each of them. More simply, one writes

$$E[X] = E[E[X|Y]] . \tag{7.49}$$

The preceding result has an important implication for the computation of the expectation of a function of a random variable. Suppose $X = h(Y)$. then $E[X|Y] = h(Y)$, so

$$E[X] = E[E[X|Y]] = \int_{-\infty}^{\infty} h(y)f_Y(y)\, dy \, . \tag{7.50}$$

This shows that we only need $f_Y(y)$ to calculate the expectation of a function of $Y$; to compute the expectation of $X = h(Y)$, we do not need to determine $f_X(x)$.

Similarly, if $X$ is a function of *two* random variables, $X = h(Y, Z)$, then

$$E[X] = \int_{-\infty}^{\infty} \int_{-\infty}^{\infty} h(y, z)f_{Y,Z}(y, z)\, dy\, dz \, . \tag{7.51}$$

It is easy to show from this that if $Y$ and $Z$ are independent, and if $h(y, z) = g(y)\ell(z)$, then

$$E[g(Y)\ell(Z)] = E[g(Y)]E[\ell(Z)] \, . \tag{7.52}$$

The converse is also true: if Eq. (7.52) holds for all functions $g(\cdot)$ and $\ell(\cdot)$, then $Y$ and $Z$ are independent.

## 7.7 CORRELATION AND COVARIANCE FOR BIVARIATE RANDOM VARIABLES

Consider a pair of jointly distributed random variables $X$ and $Y$. Their marginal PDFs are obtained by projecting the probability mass along the $y$-axis and $x$-axis directions respectively:

$$f_X(x) = \int_{-\infty}^{\infty} f_{X,Y}(x, y)\, dy \, , \qquad f_Y(y) = \int_{-\infty}^{\infty} f_{X,Y}(x, y)\, dx \, . \tag{7.53}$$

In other words, the PDF of $X$ is obtained by integrating the joint PDF over all possible values of the other random variable $Y$—and similarly for the PDF of $Y$.

It is of interest, just as in the single-variable case, to be able to capture the location and spread of the bivariate PDF in some aggregate or approximate way, without having to describe the full PDF. This again suggests focusing on notions of mean and variance. The mean value of the bivariate PDF is specified by giving the mean values of each of its two component random variables: the mean value has an $x$ component that is $E[X]$ and a $y$ component that is $E[Y]$, and these two numbers can be evaluated from the respective marginal densities. The center of mass of the bivariate PDF is thus located at

$$(x, y) = (E[X], E[Y]) \, . \tag{7.54}$$

A measure of the spread of the bivariate PDF in the $x$ direction may be obtained from the standard deviation $\sigma_X$ of $X$, computed from $f_X(x)$; and a measure of the spread in the $y$ direction may be obtained from $\sigma_Y$, computed similarly from $f_Y(y)$. However, these two numbers only offer a partial view. It is also of interest to determine what the spread is in a general direction rather than just along the two coordinate axes. We can consider, for instance, the standard deviation, or equivalently, the variance, of the random variable $Z$ defined as

$$Z = \alpha X + \beta Y \tag{7.55}$$

for arbitrary constants $\alpha$ and $\beta$. Note that by choosing $\alpha$ and $\beta$ appropriately, Eq. (7.55) reduces to $Z = X$ or $Z = Y$, and therefore recovers the special coordinate directions that we have already considered. However, being able to analyze the behavior of $Z$ for arbitrary $\alpha$ and $\beta$ allows assessment of the behavior in all directions.

Before considering the computations involved in determining the variance of $Z$, note that the mean of $Z$ is directly found in terms of quantities already computed, namely $E[X]$ and $E[Y]$:

$$E[Z] = \alpha E[X] + \beta E[Y] . \tag{7.56}$$

As for the variance of $Z$, it is easy to establish from Eqs. (7.55) and (7.56) that

$$\sigma_Z^2 = E\left[(Z - E[Z])^2\right] = \alpha^2 \sigma_X^2 + \beta^2 \sigma_Y^2 + 2\alpha\beta\, \sigma_{X,Y} \tag{7.57}$$

where $\sigma_X^2$ and $\sigma_Y^2$ are the variances along the coordinate directions $x$ and $y$, and $\sigma_{X,Y}$ is the covariance of $X$ and $Y$, also denoted by $\text{cov}(X, Y)$ or $c_{X,Y}$, and defined as

$$\sigma_{X,Y} = \text{cov}(X, Y) = c_{X,Y} = E[(X - E[X])(Y - E[Y])] . \tag{7.58}$$

Equivalently,

$$\sigma_{X,Y} = E[XY] - E[X]E[Y] , \tag{7.59}$$

where Eq. (7.59) follows from multiplying out the terms in parentheses in Eq. (7.58) and then taking term-by-term expectations. Note that when $Y = X$ we recover the familiar expressions for the variance of $X$. The quantity $E[XY]$ that appears in Eq. (7.59), namely, the expectation of the product of the random variables, is referred to as the correlation or second cross-moment of $X$ and $Y$ to distinguish it from the second self-moments $E[X^2]$ and $E[Y^2]$, and will be denoted by $r_{X,Y}$:

$$r_{X,Y} = E[XY] . \tag{7.60}$$

Note also that in Eq. (7.57), the covariance $\sigma_{X,Y}$ is the only new quantity needed when going from mean and spread computations along the coordinate axes to such computations along any axis; we do not need a new quantity for each new direction. In summary, we can express the location of $f_{X,Y}(x, y)$ in an aggregate or approximate way in terms of the first moments, $E[X], E[Y]$; and we can express the spread around this location in an aggregate or approximate way in terms of the (central) second moments, $\sigma_X^2, \sigma_Y^2, \sigma_{X,Y}$.

It is common to work with a normalized form of the covariance, namely the correlation coefficient $\rho_{X,Y}$:

$$\rho_{X,Y} = \frac{\sigma_{X,Y}}{\sigma_X \sigma_Y} \, . \tag{7.61}$$

This normalization ensures that the correlation coefficient is unchanged if $X$ and/or $Y$ is multiplied by any nonzero constant or has any constant added to it. For instance, the centered and normalized random variables

$$V = \frac{X - \mu_X}{\sigma_X} \, , \qquad W = \frac{Y - \mu_Y}{\sigma_Y} \, , \tag{7.62}$$

each of which has mean 0 and variance 1, have the same correlation coefficient as $X$ and $Y$. The correlation coefficient might have been better called the covariance coefficient, since it is defined in terms of the covariance and not the correlation of the two random variables, but this more helpful name is not generally used.

Invoking the fact that $\sigma_Z^2$ in Eq. (7.57) must be nonnegative, and further noting from this equation that $\sigma_Z^2/\beta^2$ is quadratic in $\alpha$, it can be proved by straightforward analysis of the quadratic expression that

$$|\rho_{X,Y}| \leq 1 \, . \tag{7.63}$$

From the various preceding definitions, a positive correlation $r_{X,Y} > 0$ suggests that $X$ and $Y$ tend to take the same sign, on average, whereas a positive covariance $\sigma_{X,Y} > 0$—or equivalently, a positive correlation coefficient $\rho_{X,Y} > 0$—suggests that the deviations of $X$ and $Y$ from their respective means tend to take the same sign, on average. Conversely, a negative correlation suggests that $X$ and $Y$ tend to take opposite signs, on average, while a negative covariance or correlation coefficient suggests that the deviations of $X$ and $Y$ from their means tend to take opposite signs, on average.

Since the correlation coefficient of $X$ and $Y$ captures some features of the relation between their deviations from their respective means, we might expect that the correlation coefficient can play a role in constructing an estimate of $Y$ from measurements of $X$, or vice versa. We will see in Chapter 8, where linear minimum mean square error (LMMSE) estimation is discussed, that this is indeed the case.

The random variables $X$ and $Y$ are said to be uncorrelated (or linearly independent, a less common and potentially misleading term) if

$$E[XY] = E[X]E[Y] \, , \tag{7.64}$$

or equivalently if

$$\sigma_{X,Y} = 0 \quad \text{or} \quad \rho_{X,Y} = 0 \, . \tag{7.65}$$

Thus, uncorrelated does not mean zero correlation unless one of the random variables has an expected value of zero. Rather, uncorrelated means zero covariance. Again, a better term for uncorrelated might have been noncovariant, but this term is not widely used.

Note that if $X$ and $Y$ are independent, then $E[XY] = E[X]E[Y]$ and consequently $X$ and $Y$ are uncorrelated. The converse does not hold in general. For instance, consider the case where the combination $(X, Y)$ takes only the values $(1, 0)$, $(-1, 0)$, $(0, 1)$, and $(0, -1)$, each with equal probability $\frac{1}{4}$. Then $X$ and $Y$ are easily seen to be uncorrelated but not independent.

Additional terminology that we will shortly motivate and find useful occurs in the following definition: two random variables $X$ and $Y$ are referred to as orthogonal if $E[XY] = 0$.

---

**Example 7.4    Correlation and Functional Dependence**

Consider the case in which $Y$ is specified by a deterministic linear function of a random variable $X$, in which case $Y$ is also a random variable:

$$Y = \xi X + \zeta , \tag{7.66}$$

where $\xi$ and $\zeta$ are known constants. Clearly the outcome of the random variable $Y$ is totally determined by the outcome of the random variable $X$, that is, $Y$ is deterministically dependent on $X$. It is straightforward to show that $\rho_{X,Y} = 1$ if $\xi > 0$ and $\rho_{X,Y} = -1$ if $\xi < 0$.

Next, consider the case in which

$$Y = \xi X^2 + \zeta \tag{7.67}$$

and $X$ has a PDF $f_X(x)$ that is even about 0, so $f_X(-x) = f_X(x)$. In this case, $X$ and $Y$ are uncorrelated, even though $Y$ is again completely determined by $X$. As we will see in more detail in Chapter 8, the correlation coefficient is a measure of how well $Y$ is predicted by a linear function of $X$. It is generally not helpful in assessing nonlinear predictability.

---

In Example 7.3, we specified the Gaussian density for a single random variable. In the following example, we describe the bivariate Gaussian density for a pair of random variables.

---

**Example 7.5    Bivariate Gaussian Density**

The random variables $X$ and $Y$ are said to be bivariate Gaussian or bivariate normal if their joint PDF is

$$f_{X,Y}(x, y) = c \, \exp\left\{ -q\left( \frac{x - \mu_X}{\sigma_X}, \frac{y - \mu_Y}{\sigma_Y} \right) \right\} \tag{7.68}$$

where $c$ is a normalizing constant (so that the volume or "mass" under the PDF integrates to 1) and $q(v, w)$ is a quadratic function of its two arguments $v$ and $w$, expressed in terms of the correlation coefficient $\rho$ of $X$ and $Y$:

$$c = \frac{1}{2\pi \sigma_X \sigma_Y \sqrt{1 - \rho^2}} , \tag{7.69}$$

$$q(v, w) = \frac{1}{2(1 - \rho^2)} (v^2 - 2\rho vw + w^2) . \tag{7.70}$$

This density is the natural bivariate generalization of the Gaussian density, and has several nice properties:

- The marginal densities of $X$ and $Y$ are Gaussian.
- The conditional density of $Y$, given $X = x$, is Gaussian with mean

$$\mu_Y + \rho \left( \frac{\sigma_Y}{\sigma_X} \right)(x - \mu_X) \tag{7.71}$$

and variance

$$\sigma_Y^2(1 - \rho^2) \tag{7.72}$$

(which does not depend on the value of $x$); and similarly for the conditional density of $X$, given $Y = y$.

- If $X$ and $Y$ are uncorrelated, that is, if $\rho = 0$, then $X$ and $Y$ are independent, a fact that is not generally true for other bivariate random variables.
- Any two affine (i.e., linear plus constant) combinations of $X$ and $Y$ are themselves bivariate Gaussian (e.g., $Q = X + 3Y + 2$ and $R = 7X + Y - 3$ are bivariate Gaussian).

The bivariate Gaussian PDF and indeed the associated notion of correlation were essentially discovered by the statistician Francis Galton (a first cousin of Charles Darwin) in 1886, while studying the joint distribution of the heights of parents and children. There is a two-dimensional version of the central limit theorem, with the bivariate Gaussian as the limiting density. Consequently, this is a reasonable model for two jointly distributed random variables in many settings. There are also natural generalizations to many variables.

Many of the generalizations of the preceding discussion from two random variables to many random variables are straightforward. In particular, the mean of a joint PDF

$$f_{X_1, X_2, \cdots, X_\ell}(x_1, x_2, \cdots, x_\ell) \tag{7.73}$$

in the $\ell$-dimensional space of possible values has coordinates that are the respective individual means, $E[X_1], \cdots, E[X_\ell]$. The spreads in the coordinate directions are deduced from the individual (marginal) spreads, $\sigma_{X_1}, \cdots, \sigma_{X_\ell}$. To be able to compute the spreads in arbitrary directions, we need all the additional $\ell(\ell - 1)/2$ central second moments, namely $\sigma_{X_i, X_j}$ for all $1 \leq i < j \leq \ell$ (note that $\sigma_{X_j, X_i} = \sigma_{X_i, X_j}$)—but nothing more.

# 7.8 A VECTOR-SPACE INTERPRETATION OF CORRELATION PROPERTIES

A vector-space picture is often a useful aid in recalling the first- and second-moment relationships between two random variables $X$ and $Y$. This picture is not just a mnemonic: there is a very precise sense in which random variables can be thought of as (or are) vectors in a vector space (of infinite dimensions), as long as we are only interested in their first- and second-moment properties. Although we will not develop this correspondence in any depth, it can be very helpful in conjecturing or checking answers in the LMMSE estimation problems that we will encounter in later chapters.

To develop this picture, we represent the random variables $X$ and $Y$ as vectors $\mathbf{X}$ and $\mathbf{Y}$ in some abstract vector space. We define the squared lengths of these vectors as $E[X^2]$ and $E[Y^2]$, respectively the second moments of the associated random variables. Recall that in Euclidean vector space the squared length of a vector is the inner product of the vector with itself. Consistent with this, in our vector-space interpretation we define the inner product $\langle \mathbf{X}, \mathbf{Y} \rangle$ between two general vectors $\mathbf{X}$ and $\mathbf{Y}$ as the correlation (or second cross-moment) of the associated random variables:

$$\langle \mathbf{X}, \mathbf{Y} \rangle = E[XY] = r_{X,Y} \,. \tag{7.74}$$

With this definition, the standard properties required of an inner product in a vector space are satisfied, namely:

- Symmetry: $\langle \mathbf{X}, \mathbf{Y} \rangle = \langle \mathbf{Y}, \mathbf{X} \rangle$;
- Linearity: $\langle \mathbf{X}, a_1 \mathbf{Y_1} + a_2 \mathbf{Y_2} \rangle = a_1 \langle \mathbf{X}, \mathbf{Y_1} \rangle + a_2 \langle \mathbf{X}, \mathbf{Y_2} \rangle$;
- Positivity: $\langle \mathbf{X}, \mathbf{X} \rangle$ is positive for $\mathbf{X} \neq \mathbf{0}$, and 0 otherwise.

This definition of inner product is also consistent with the fact that we refer to two random variables as orthogonal when $E[XY] = 0$.

The centered random variables $X - \mu_X$ and $Y - \mu_Y$ can similarly be represented as vectors $\widetilde{\mathbf{X}}$ and $\widetilde{\mathbf{Y}}$ in this abstract vector space, with squared lengths that are now the variances of the random variables $X$ and $Y$:

$$\sigma_X^2 = E[(X - \mu_X)^2] \,, \qquad \sigma_Y^2 = E[(Y - \mu_Y)^2] \,. \tag{7.75}$$

The lengths of the vectors representing the centered random variables are therefore the standard deviations $\sigma_X$ and $\sigma_Y$ respectively of the associated random variables $X$ and $Y$. The inner product of the vectors $\widetilde{\mathbf{X}}$ and $\widetilde{\mathbf{Y}}$ becomes

$$\langle \widetilde{\mathbf{X}}, \widetilde{\mathbf{Y}} \rangle = E[(X - \mu_X)(Y - \mu_Y)] = \sigma_{X,Y} \,, \tag{7.76}$$

namely the covariance of the random variables.

In Euclidean space, the standard inner product of two vectors is given by the product of the lengths of the individual vectors and the cosine of the angle between them, so

$$\langle \widetilde{\mathbf{X}}, \widetilde{\mathbf{Y}} \rangle = \sigma_{X,Y} = \sigma_X \sigma_Y \cos(\theta) \,. \tag{7.77}$$

Consequently, as depicted in Figure 7.5 the quantity

$$\theta = \cos^{-1}\left(\frac{\sigma_{X,Y}}{\sigma_X \sigma_Y}\right) = \cos^{-1} \rho \tag{7.78}$$

can be thought of as the angle between the vectors, where we can see from Eq. (7.78) that $\rho$ is the correlation coefficient of the two random variables. Correspondingly

$$\rho = \cos(\theta) \,. \tag{7.79}$$

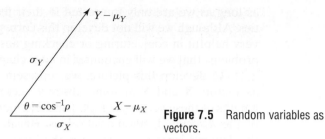

**Figure 7.5**  Random variables as vectors.

Thus, the correlation coefficient is the cosine of the angle between the vectors. It is therefore not at all surprising that

$$-1 \leq \rho \leq 1 . \tag{7.80}$$

When $\rho$ is near 1 the vectors are nearly aligned in the same direction, whereas when $\rho$ is near $-1$ they are close to being oppositely aligned. The correlation coefficient is zero when these vectors $\widetilde{\mathbf{X}}$ and $\widetilde{\mathbf{Y}}$ (which represent the centered random variables) are orthogonal, or equivalently, the corresponding random variables have zero covariance,

$$\sigma_{X,Y} = 0 . \tag{7.81}$$

## 7.9 FURTHER READING

There are numerous books that introduce probability at a level sufficient for this text. The foundations of probabilistic modeling and stochastic processes are presented with notable clarity in [Ber], and illustrated by many worked-out examples that help to develop and anchor the theory. [Gr1] is a more advanced and comprehensive text, accompanied by the extensive collection of solved problems in [Gr2]. The coverage of probability and stochastic processes in texts such as [Coo], [Fin], [He1], [Kay1], [Kri], [Leo], [Mil], [Pa4], [Sha], [Shy], [Th2], [Wll], and [Yat] is slanted towards signals and systems, and includes material treated in later chapters of this text. [Gal] combines intuition and precision in its deeper study of random processes that are important in applications, and also addresses the signal estimation and detection problems considered in our later chapters. We cite it here because it includes (in its Section 10.6) a more detailed description of the vector space picture we have presented in Section 7.8 for the correlation properties of random variables. As noted in the Preface, this text does not venture into the vast domain of information theory opened up by Shannon's seminal 1948 paper [Shn]. Nevertheless, the importance of information theory to a broader understanding of signals, systems and inference is suggested, for example, by the lively treatment in [Mac]. See also [Cov], and [An2] for an introduction in the setting of information transmission.

## Problems

## Basic Problems

**7.1.** Two numbers $x$ and $y$ are selected independently and at random (i.e., with uniform density) between 0 and 1. The events $A, B, C,$ and $D$ depicted in Figure P7.1 are defined as follows:

$$A = \left\{ y > \frac{1}{2} \right\}, \quad B = \left\{ y < \frac{1}{2} \right\}, \quad C = \left\{ x < \frac{1}{2} \right\},$$

$$D = \left\{ x < \frac{1}{2} \text{ and } y < \frac{1}{2} \right\} \cup \left\{ x > \frac{1}{2} \text{ and } y > \frac{1}{2} \right\}.$$

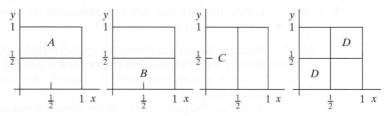

**Figure P7.1**

**(a)** Determine in each of the following three cases whether the indicated pair of events is independent: (i) $A$ and $D$; (ii) $C$ and $D$; and (iii) $A$ and $B$.

**(b)** Are events $B, C,$ and $D$ mutually independent? Remember that for a set of events to be mutually independent, the probabilities of all possible intersections of these events, taken any number at a time, must be given by the products of the individual event probabilities.

**7.2. (a)** A random variable $V$ is uniformly distributed in the interval $[a, b]$. Find its expected value $\mu_V$, its second moment $E[V^2]$, and its variance $\sigma_V^2$.

**(b)** A second random variable $W$ is independent of $V$ but distributed identically to it. Find the mean and variance of the random variable $Y = V + W$. Also determine the covariance of $Y$ and $V$, and their correlation coefficient.

**7.3.** Suppose $X = 2 + V$ and $Y = 2 - V$, where $V$ is a zero-mean Gaussian random variable of variance 4. Determine the correlation of $X$ and $Y$. Are $X$ and $Y$ orthogonal? What is their covariance? What is their correlation coefficient? Are they uncorrelated?

**7.4.** Suppose

$$X = Z + V$$

$$Y = \beta Z + W$$

where the random variables $Z, V,$ and $W$ have respective mean values $\mu_Z, \mu_V,$ and $\mu_W$, and variances $\sigma_Z^2, \sigma_V^2,$ and $\sigma_W^2$, but are mutually uncorrelated (i.e., have zero covariance); the quantity $\beta$ is a scale factor.

**(a)** Determine the covariance $\sigma_{XY}$ of $X$ and $Y$, and the correlation coefficient $\rho_{XY} = \sigma_{XY}/(\sigma_X\sigma_Y)$, in terms of the quantities specified above.

**(b)** Assume $\sigma_V^2 = \sigma_W^2 = \sigma^2$. What does your answer for the correlation coefficient in (a) simplify to? Explicitly check that this simplified answer matches your intuition for what the answer should be when each of the quantities $\sigma^2$ and $\beta$ takes extreme values.

**7.5.** The random variables $X$ and $Y$ have

$$E(X) = -1, \quad E(Y) = 2, \quad E(X^2) = 9, \quad E(XY) = -4, \quad E(Y^2) = 7.$$

**(a)** Calculate the covariance $\sigma_{ZW}$ of the random variables

$$Z = 2X - Y + 5, \quad W = X + \frac{1}{2}Y - 1.$$

**(b)** If $X$ and $Y$ are bivariate Gaussian, what is the joint density of $Z$ and $W$? (Utilize the fact that affine combinations of bivariate Gaussian random variables are bivariate Gaussian.)

**7.6.** A communication system transmits signals labeled 1, 2, and 3. The probability that symbol $j$ is sent and symbol $k$ is received is listed in Table P7.6 for each pair $(j,k)$ of sent and received symbols. For example, the probability is 0.21 that a 3 is sent and 2 is received.

Calculate the probability that the symbol $k$ was sent, given that symbol $k$ is received, for $k = 1, 2, 3$. Also calculate the probability of transmission error incurred in using this system. A transmission error is defined as the reception of any symbol other than the one transmitted.

**TABLE P7.6**

| | k received | | |
|---|---|---|---|
| *j* sent | **1** | **2** | **3** |
| 1 | 0.05 | 0.13 | 0.12 |
| 2 | 0.10 | 0.08 | 0.07 |
| 3 | 0.09 | 0.21 | 0.15 |

# Advanced Problems

**7.7.** Two numbers $x$ and $y$ are selected independently and at random (i.e., with uniform density) between zero and one. Let the events $A$, $B$, $C$, and $D$ depicted in Figure P7.7 be defined as follows:

$$A = \left\{y > \frac{1}{2}\right\}, \quad B = \left\{y < \frac{1}{2}\right\}, \quad C = \left\{x < \frac{1}{2}\right\},$$

$$D = \left\{x < \frac{1}{2} \text{ and } y < \frac{1}{2}\right\} \cup \left\{x > \frac{1}{2} \text{ and } y > \frac{1}{2}\right\}.$$

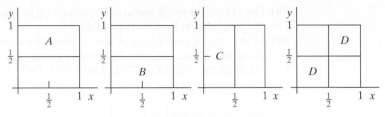

**Figure P7.7**

(a) Using the events defined here, and/or other events defined similarly by geometric regions in the unit square, establish the following facts:

  (i) Two independent events need not remain independent when conditioned on some other event; i.e., define events $E$, $F$, and $G$ such that

$$P(E \cap F) = P(E)P(F) \quad \text{but} \quad P(E \cap F|G) \neq P(E|G)P(F|G) \,.$$

  (ii) Two conditionally independent events need not be independent in the absence of conditioning; i.e., define events $J$, $K$, and $L$ such that

$$P(J \cap K|L) - P(J|L)P(K|L) \quad \text{but} \quad P(J \cap K) \neq P(J)P(K) \,.$$

Sketch the regions corresponding to the events you define.

(b) Determine whether the following statement is true or false for a general choice of events $Q$, $V$, and $W$, explaining your reasoning, and then illustrate your answer with a particular choice of events of the above type (i.e., regions in the unit square, sketched appropriately):

$$P(V \cap W|Q) = P(V|W \cap Q)P(W|Q) \,.$$

**7.8.** Indicate whether each statement below is true or false and give a brief explanation.

(a) If $X$ and $Y$ are uncorrelated random variables, then $X^2$ and $Y^2$ must be uncorrelated.

(b) If $X$ and $Y$ are independent random variables, then

$$E[g(X)h(Y)] = E[g(X)]\,E[h(Y)] \,,$$

where $g(X)$ and $h(Y)$ are arbitrary functions of $X$ and $Y$ respectively.

(c) Consider two random variables $X$ and $Y$ for which the joint density $f_{X,Y}(x,y)$ factors into the product $f_{X,Y}(x,y) = f_{X|Y}(x|y)f_Y(y)$ of the conditional density of $X$ times the marginal density of $Y$. Then $X$ and $Y$ must be independent.

**7.9.** If $\sigma_{XY}$ denotes the covariance of two random variables $X$ and $Y$ whose respective variances are $\sigma_X^2$ and $\sigma_Y^2$, then we know that $\sigma_{XY}^2 \leq \sigma_X^2 \sigma_Y^2$, or equivalently

$$-\sigma_X \sigma_Y \leq \sigma_{XY} \leq \sigma_X \sigma_Y \,. \tag{7.82}$$

(a) Use the above inequality to deduce the inequality $r_{XY}^2 \leq r_X^2 r_Y^2$, where $r_{XY} = E[XY] = \sigma_{XY} + \mu_X \mu_Y$ (the correlation of the two random variables), $r_X^2 = E[X^2]$ (the second moment of $X$), and $r_Y^2 = E[Y^2]$ (the second moment of $Y$). Equivalently, what we want to deduce is

$$-r_X r_Y \leq r_{XY} \leq r_X r_Y \,. \tag{7.83}$$

**(b)** Use the vector-space picture for random variables in Section 7.8 to deduce the inequality in Eq. (7.83) more directly.

**(c)** In the special case where $X$ and $Y$ have the same mean $\mu$ and the same variance $\sigma^2$, one can actually obtain a better (i.e., higher) lower bound on $r_{XY}$ than the one in Eq. (7.83). Utilize Eq. (7.82) to show that

$$-r^2 + 2\mu^2 \leq r_{XY} \leq r^2 \,,$$

where $r^2 = \sigma^2 + \mu^2 = r_X^2 = r_Y^2$.

**7.10.** Figure P7.10 shows the transition probabilities for a communication channel with two possible input symbols, $\{a, b\}$, and two possible output symbols, $\{A, B\}$. We will use this channel to transmit a binary digit, $m$, whose prior probability distribution is $P(m = 0) = 0.6$, $P(m = 1) = 0.4$. When $m = 0$ occurs we transmit $a$, and when $m = 1$ occurs we transmit $b$. Now we must design a decoder. The decoder assigns a decision, $\widehat{m} = 0$ or 1, to each of the symbols $\{A, B\}$.

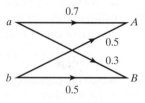

**Figure P7.10**

**(a)** Find the minimum-probability-of-error decoder for the given encoder, i.e., the decision rule that minimizes the error probability, $P_e = P(\widehat{m} \neq m)$.

**(b)** Find the error probability for your decoder from part (a).

**7.11.** Consider the communication system shown in Figure P7.11. The source produces messages whose possible values are from the set $\{-1, 1\}$ according to the following probability distribution:

$$P(X = -1) = g, \ \ P(X = 1) = 1 - g \,.$$

The channel is an additive noise channel, with noise $N$ that is statistically independent of the source messages $X$. The received signal $R$ is

$$R = X + N \,,$$

where

$$N = \begin{cases} +1 & \text{with probability } p \\ -1 & \text{with probability } 1 - p. \end{cases}$$

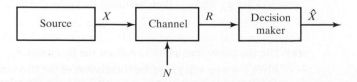

**Figure P7.11**

**(a)** List all the possible values of $R$. For each possible value $r$ of $R$, compute $P(R = r)$.

**(b)** Our estimate, $\widehat{X}$, of the transmitted message $X$ is chosen to be the value of $X$ that is the most probable, given the observation $R = r$. For $\frac{1}{2} < g < 1$ and $\frac{1}{2} < p < 1$, determine the value of $\widehat{X}$ for each possible value of $R$.

**(c)** The probability of being correct for the estimation procedure in (b) is defined as

$$P(\text{correct}) = P(\widehat{X} = -1, X = -1) + P(\widehat{X} = 1, X = 1).$$

Compute $P(\text{correct})$ for the decision rule obtained in (b). Does the procedure in (b) for estimating $X$ maximize the probability of being correct or could a different procedure increase that probability?

**7.12.** In a binary optical communication system, the receiver counts the number of photoelectrons ejected by the light incident on the photocell during an interval $(0, T)$. When no light signal has been transmitted toward the photocell (event $H_0$), the probability that $k$ electrons are counted is

$$P(k \mid H_0) = A_0 v_0{}^k, \qquad k = 0, 1, 2, \dots$$

However, when a signal has been transmitted (event $H_1$),

$$P(k \mid H_1) = A_1 v_1{}^k, \qquad k = 0, 1, 2, \dots$$

with $0 < v_0 < v_1 < 1$. The prior probabilities for the two events are given by $P(H_0)$ and $P(H_1)$, respectively.

**(a)** Determine the two constants $A_0$ and $A_1$.

For parts (b) and (c), assume $P(H_0) = P(H_1) = \frac{1}{2}$.

**(b)** Determine the conditional (i.e., posterior) probability that a signal was sent, given that exactly $m$ photoelectrons were counted.

**(c)** When $k \geq n_0$ the receiver decides that a signal was indeed sent; when $0 \leq k < n_0$, it decides that no signal was sent, where $n_0$ is some positive integer. For this decision rule, calculate the probability $P_e$ of error incurred by the receiver in terms of $n_0$, $v_0$, and $v_1$. For what value of $n_0$ is the probability $P_e$ of error minimum? Is there an alternative decision rule that would lead to a lower $P_e$?

Now assume that $v_0 = 0.3$, $v_1 = 0.7$, and $P(H_1) > 0.7$.

**(d)** How does your choice of $n_0$ (to achieve minimum error) change as $P(H_1)$ increases from 0.7 to 1.0? What is $P_e$ for this range of $P(H_1)$?

**7.13.** A particular communication system has a source that transmits the symbol $X = -1$ and the symbol $X = 1$ with equal probability. The source can use one of two channels, Channel A or Channel B. Figure P7.13 shows the characteristics of each of these channels; the number next to an arrow denotes the probability of receiving the symbol at the right of the arrow, given transmission of the symbol at the left of the arrow. Channel A occasionally loses the symbol completely, while for Channel B the symbol $X = 1$ is at times misinterpreted as a $-1$ at the receiver.

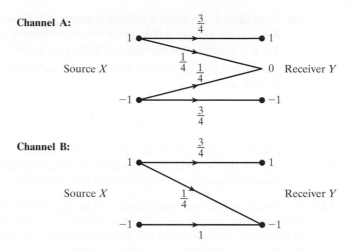

**Figure P7.13**

(a) If the source uses Channel A, what is the probability that $-1$ is received? What are the probabilities for 0 and 1? Repeat your calculations when the source uses Channel B.

(b) Assume now and for the remainder of this problem that the source uses Channel A with probability $\alpha$, and Channel B with probability $1 - \alpha$. Suppose the received symbol is $-1$. What is the probability that Channel A was used, given that $-1$ was received?

(c) Suppose we used the following rule to decide which channel is used when $-1$ is received: if $Y = -1$, we decide in favor of Channel A if

$$P(\text{A used} \mid Y = -1) > P(\text{B used} \mid Y = -1).$$

For what range of $\alpha$ will you decide that Channel A was used when $Y = -1$? For what range of $\alpha$ will you decide that Channel A was used, regardless of $Y$?

(d) Now the same channel is used $N$ times in succession, with each use being independent of the others. We receive $a$ $-1$s, $b$ 0s, and $c$ 1s (so that $a + b + c = N$), and wish to decide which channel was used. Express in terms of $a$, $b$, $c$, and $\alpha$ the decision rule that chooses whichever channel is more probable, given the received sequence.

(e) Assume that $\alpha = 0.8$ with the scheme in (d). Under what conditions will we decide that Channel B was used?

## Extension Problems

**7.14.** Suppose the random variable $Z$ is related to the random variables $Q$ and $V$ by the equation

$$Z = cQ + V,$$

where $c$ is a known constant and

$$E(Q) = 1, \ E(V) = 0 \text{, variance}(Q) = \sigma_Q^2,$$

$$\text{variance}(V) = \sigma_V^2, \ \text{covariance}(Q, V) = \sigma_{Q,V}.$$

(a) Determine $E(Z)$, variance($Z$), covariance($Z, Q$), and covariance($Z, V$) in terms of the above parameters.

(b) We now consider one way of estimating $Q$ from measurements of $Z$. Consider the following linear (actually, linear-plus-constant, or "affine") function of $Z$:

$$\widehat{Q} = a + bZ \,.$$

Find the values of the constants $a$, and $b$ in this expression (expressed in terms of the quantities you computed in (a), and/or in terms of the given parameters of the problem) that will minimize $E[(Q - \widehat{Q})^2]$, and determine this minimum value of $E[(Q - \widehat{Q})^2]$. We refer to $\widehat{Q}$ as the LMMSE estimator of $Q$ in terms of $Z$.

**7.15.** The input to a communication channel at some instant of time is a Gaussian random variable $Q$ with mean value $\mu_Q$ and variance $\sigma_Q^2$. Suppose the corresponding channel output is $X = Q + W$, where the additive disturbance $W$ is a zero-mean Gaussian random variable of variance $\sigma_W^2$, and is independent of $Q$.

(a) Compute the mean $\mu_X$ and variance $\sigma_X^2$ of $X$ in terms of the specified parameters. Will this mean and variance suffice to write down the PDF of $X$? Explain your answer. If your answer is yes, write down the PDF.

(b) Compute the covariance $\sigma_{XQ}$, and then determine the correlation coefficient $\rho_{XQ} = \sigma_{XQ}/(\sigma_X \sigma_Q)$. Over what range does $\rho_{XQ}$ vary as:

   (a) $\mu_Q$ varies from 0 to $\infty$?
   (b) $\frac{\sigma_Q}{\sigma_W}$ varies from 0 to $\infty$?

(c) Determine the joint PDF of $X$ and $Q$, using the fact that $f_{X|Q}(x|q)$ and $f_Q(q)$ are easy to obtain, and combine them appropriately to obtain the desired joint PDF. Confirm that the resulting PDF is a bivariate Gaussian density.

(d) Compute the conditional density $f_{Q|X}(q|x)$ and verify that it is Gaussian. This constitutes a verification of the fact that the conditional densities of bivariate Gaussian variables are Gaussian.

(e) Using the result in (c), compute the MMSE estimate, $\widehat{Q}_{\text{MMSE}}(x)$, of the channel input $Q$, given that $X = x$; i.e., find $\widehat{Q}_{\text{MMSE}}(x)$ that minimizes the conditional MSE

$$E\left\{ \left(Q - \widehat{Q}_{\text{MMSE}}(X)\right)^2 \mid X = x \right\} \,.$$

In the next chapter, you will see that this estimate is given by $E[Q|X = x]$. If you've done this correctly, you'll discover that the estimate is an affine — i.e., linear plus constant — function of $x$. Also determine the corresponding conditional mean square error as given in the above expression.

(f) Suppose $Q$ and $W$ are uncorrelated but no longer independent. Which of your answers in (a) to (d) above would change, if any?

**7.16.** Let $f^+(v, w)$ be the bivariate Gaussian density function

$$f^+(v, w) = (2\pi)^{-1}(1 - \rho^2)^{-1/2} \exp\left[ -\frac{v^2 - 2\rho v w + w^2}{2(1 - \rho^2)} \right],$$

and assume $0 < \rho < 1$. Let $f^-(v, w)$ be the same function, but with correlation coefficient $-\rho$ instead of $+\rho$. The forms of $f^+$ and $f^-$ are suggested by Figure P7.16, which shows contours of equal probability for each of these densities.

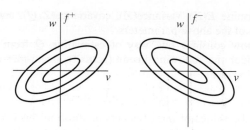

**Figure P7.16**

Now suppose $V$ and $W$ are random variables defined by the joint PDF

$$f_{V,W}(v,w) = \frac{1}{2}\left(f^+(v,w) + f^-(v,w)\right).$$ (7.84)

This corresponds to picking $V$ and $W$ with equal probability from the joint PDF's $f^+$ and $f^-$.

**(a)** Show that $V$ is Gaussian, and that $W$ is Gaussian, but that $V$ and $W$ are not bivariate Gaussian. Present a pictorial argument rather than detailed calculations. Hence the fact that two variables are Gaussian does not necessarily make them bivariate Gaussian. Equivalently, the fact that a bivariate density has Gaussian marginals does not necessarily mean that the density is bivariate Gaussian. However, two independent Gaussian variables are bivariate Gaussian.

**(b)** Show that $V$ and $W$ are not independent.

**(c)** Without integration or any other detailed calculations, argue from the structure of the joint PDF of $V$ and $W$ that $E[VW] = 0$; i.e., these two zero-mean Gaussian random variables are uncorrelated, even though they are not independent. Thus uncorrelated Gaussian variables need not be independent. However, uncorrelated bivariate Gaussian variables are independent.

**7.17.** A communication system transmits signals labeled 1, 2, and 3. The probability that symbol $j$ is sent and symbol $k$ is received is listed in Table P7.17 for each pair $(j,k)$ of sent and received symbols. For example, the probability is 0.21 that a 2 is sent and 3 is received. A receiver decision rule associates one of the transmitted symbols with each possible received symbol. This association specifies, for each possible received symbol, what the receiver's guess, estimate, or decision is regarding the corresponding transmitted symbol. We will consider different receiver decision rules in the remaining parts of this problem.

**TABLE P7.17**

| | \multicolumn{3}{c}{$k$ received} | | |
|---|---|---|---|
| $j$ sent | 1 | 2 | 3 |
| 1 | 0.05 | 0.10 | 0.09 |
| 2 | 0.13 | 0.08 | 0.21 |
| 3 | 0.12 | 0.07 | 0.15 |

**(a)** Suppose the following simple-minded receiver decision rule is used: if $k$ is received, decide $k$ was sent. What is the probability of error (i.e., the probability of making an erroneous decision) with this decision rule?

**(b)** Specify the receiver decision rule that yields the minimum probability of error, and determine the corresponding probability of error.

**(c)** Consider a situation in which we incur a cost $c(j, \ell)$ when $j$ is sent and $\ell$ is the output of the receiver decision rule. Obtain an expression for the expected cost, also called the risk, of the decision rule in (a). One could similarly obtain an expression for the risk associated with any other decision rule, and it is then reasonable to ask what decision rule yields minimum risk. We shall do this in part (d) for a specific choice of the cost function $c(j, \ell)$.

**(d)** With the same setup as in (c), and with $c(j, \ell) = 0$ if $j = \ell$ and $c(j, \ell) = 1$ if $j \neq \ell$, find the decision rule that yields minimum risk. How does minimization of risk for this particular choice of ("all-or-none") cost function relate to minimizing the probability of error as in (b), and how do the resulting decision rules relate to each other?

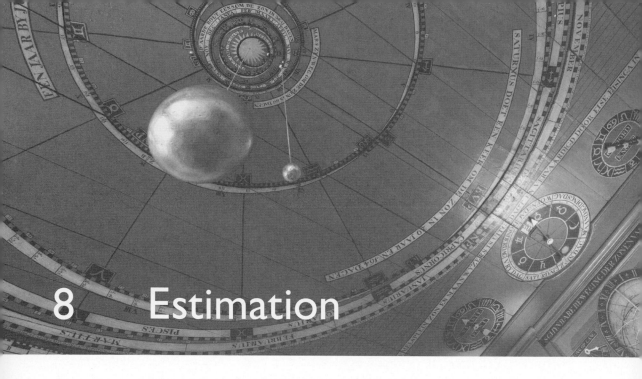

# 8    Estimation

A recurring theme in this text and in much of communication, control, and signal processing is that of making systematic estimates or predictions about some set of quantities, based on information obtained from measurements of other quantities. This process is commonly referred to as inference. Typically, inferring desired information from measurements involves incorporating models that represent prior knowledge or beliefs about how the measurements relate to the quantities of interest.

Inferring the values of a continuous random variable and ultimately those of a random process is the topic of this chapter and several that follow. One key step is the introduction of an error criterion that measures, in a probabilistic sense, the error between the desired quantity and the estimate of it. Throughout the discussion in this and the subsequent related chapters, we focus primarily on choosing the estimate that minimizes the expected or mean value of the square of the error, referred to as a minimum mean square error (MMSE) criterion. In Sections 8.1 and 8.2, we consider the MMSE estimate without imposing any constraint on the form that the estimator takes. In Section 8.3, we restrict the estimator to a linear combination of the measurements, a form of estimation referred to as linear minimum mean square error (LMMSE) estimation.

In Chapter 9, our focus moves from inference problems for continuous random variables to inference problems for discrete random quantities, which may be numerically specified or may be nonnumerical. In the latter case especially, the various possible outcomes are often termed hypotheses, and the inference task in this setting is then referred to as hypothesis testing, that is, the task of deciding which hypothesis applies, given measurements or observations. In such hypothesis testing scenarios the MMSE criterion is often not as meaningful as minimizing the probability of inferring an incorrect hypothesis.

## 8.1 ESTIMATION OF A CONTINUOUS RANDOM VARIABLE

To begin the discussion, assume that $Y$ is a random variable whose value is to be estimated from knowledge of only its probability density function. The discussion will then be broadened to estimation when a measurement or observation of another random variable $X$ is available, together with the joint probability density function of $X$ and $Y$.

Based only on knowledge of the probability density function (PDF) of $Y$, it is desirable to obtain an estimate of $Y$, denoted as $\hat{y}$, that minimizes the mean square error between the actual outcome of the experiment and the estimate $\hat{y}$. Specifically, $\hat{y}$ will be chosen to minimize

$$E[(Y - \hat{y})^2] = \int_{-\infty}^{\infty} (y - \hat{y})^2 f_Y(y)\, dy \,. \tag{8.1}$$

Differentiating Eq. (8.1) with respect to $\hat{y}$ and equating the result to zero results in the equation

$$-2 \int_{-\infty}^{\infty} (y - \hat{y}) f_Y(y)\, dy = 0 \tag{8.2}$$

or

$$\int_{-\infty}^{\infty} \hat{y} f_Y(y)\, dy = \int_{-\infty}^{\infty} y f_Y(y)\, dy \tag{8.3}$$

from which

$$\hat{y} = E[Y] \,. \tag{8.4}$$

The second derivative of $E[(Y - \hat{y})^2]$ with respect to $\hat{y}$ is

$$2 \int_{-\infty}^{\infty} f_Y(y)\, dy = 2 \,, \tag{8.5}$$

which is positive, so Eq. (8.4) does indeed define the minimizing value of $\hat{y}$. Hence the MMSE estimate of $Y$ in this case is simply its mean value, $E[Y]$.

The associated error—the actual MMSE—is found by evaluating the expression in Eq. (8.1) with $\hat{y} = E[Y]$. Thus the MMSE is simply the variance of $Y$, namely $\sigma_Y^2$:

$$\min E[(Y - \hat{y})^2] = E[(Y - E[Y])^2] = \sigma_Y^2 \,. \tag{8.6}$$

In a similar manner, it is possible to show that the median of $Y$, which has half the probability mass of $Y$ below it and the other half above, is the value of $\widehat{y}$ that minimizes the mean absolute deviation, $E[\,|Y - \widehat{y}|\,]$. Also, the mode of $Y$, which is the value of $y$ at which the PDF $f_Y(y)$ is largest, can be shown to minimize the expected value of an all-or-none cost function—a cost that is unity when the error is outside of a vanishingly small tolerance band and is zero within the band. We will not pursue these alternative error metrics further here, but it is important to be aware that the choice of mean square error, while convenient, is only one of many possible error metrics.

The insights from the simple problem leading to Eqs. (8.4) and (8.6) carry over directly to the case in which additional information is available in the form of the measured or observed value $x$ of a random variable $X$ that is related in some way to $Y$. The only change from the previous discussion is that, given the additional measurement, the conditional or *a posteriori* density $f_{Y|X}(y|x)$ is used, rather than the unconditioned density $f_Y(y)$, and now the aim is to minimize

$$E[\{Y - \widehat{y}(x)\}^2 | X = x] = \int_{-\infty}^{\infty} \{y - \widehat{y}(x)\}^2 f_{Y|X}(y|x)\, dy \,. \qquad (8.7)$$

The notation $\widehat{y}(x)$ is introduced to show that, in general, the estimate will depend on the specific value $x$. Exactly the same calculations as in the case of no measurements then show that

$$\widehat{y}(x) = E[Y|X = x] \,, \qquad (8.8)$$

the conditional expectation of $Y$, given $X = x$. The associated MMSE is the variance $\sigma_{Y|x}^2$ of the conditional density $f_{Y|X}(y|x)$, that is, the MMSE is the conditional variance. Thus, the only change from the case of no measurements is that the expectation is now conditioned on the obtained measurement.

Going a step further, if multiple measurements, say $X_1 = x_1$, $X_2 = x_2, \cdots, X_L = x_L$, are available, then one uses the *a posteriori* density $f_{Y|X_1, X_2, \cdots, X_L}(y|x_1, x_2, \cdots, x_L)$. Apart from this modification, there is no change in the structure of the solutions. Thus, without further calculation, it can be stated that the MMSE estimate of $Y$, given $X_1 = x_1, \cdots, X_L = x_L$, is the conditional expectation of $Y$:

$$\widehat{y}(x_1, \cdots, x_L) = E[Y\,|\,X_1 = x_1, \cdots, X_L = x_L] \,. \qquad (8.9)$$

For notational convenience, the measured random variables can be arranged into a column vector $\mathbf{X}$, and the corresponding measurements into the column vector $\mathbf{x}$. The dependence of the MMSE estimate on the measurements can now be indicated by the notation $\widehat{y}(\mathbf{x})$, with

$$\widehat{y}(\mathbf{x}) = \int_{-\infty}^{\infty} y f_{Y|\mathbf{X}}(y\,|\,\mathbf{x})\, dy = E[\,Y\,|\,\mathbf{X} = \mathbf{x}\,] \,. \qquad (8.10)$$

The minimum mean square error for the given value of $\mathbf{X}$ is again the conditional variance, that is, the variance $\sigma_{Y|\mathbf{x}}^2$ of the conditional density $f_{Y|\mathbf{X}}(y\,|\,\mathbf{x})$.

| Example 8.1 | **MMSE Estimate for Discrete Random Variables** |
|---|---|

A discrete-time (DT) discrete-amplitude sequence $s[n]$ is stored on a noisy medium. The retrieved sequence is $r[n]$. The values of $s[n]$ and $r[n]$ at any time instant $n_0$ are random variables, denoted by $S$ and $R$ respectively. It is known that the joint probability mass function (PMF) is as shown in Figure 8.1. In this figure, the small squares denote the outcomes $(-1,-1)$, $(0,0)$, and $(1,1)$, each occurring with probability 0.2, and the small circles denote the outcomes $(0,1)$, $(0,-1)$, $(1,0)$, and $(-1,0)$, each occurring with probability 0.1.

Based on receiving the value $R = 1$, an MMSE estimate $\hat{s}$ of $S$ can be made. From Eq. (8.9), $\hat{s} = E(S|R = 1)$, which can be determined from the conditional PMF $p_{S|R}(s|1)$, which in turn can be expressed as

$$p_{S|R}(s|1) = \frac{p_{S,R}(s,1)}{p_R(1)} \ . \tag{8.11}$$

From Figure 8.1,

$$p_R(1) = 0.3 \tag{8.12}$$

and

$$p_{S,R}(s,1) = \begin{cases} 0 & s = -1 \\ 0.1 & s = 0 \\ 0.2 & s = +1 \ . \end{cases} \tag{8.13}$$

Consequently, using Eqs. (8.12) and (8.13) in Eq. (8.11),

$$p_{S|R}(s|1) = \begin{cases} 1/3 & s = 0 \\ 2/3 & s = +1 \ . \end{cases} \tag{8.14}$$

Using Eq. (8.14), the conditional expectation of $S$ — the MMSE estimate $\hat{s}$ — is

$$\hat{s} = \frac{1}{3} \cdot 0 + \frac{2}{3} \cdot 1 = \frac{2}{3} \ . \tag{8.15}$$

Note that although this estimate minimizes the mean square error, it is not constrained to take account of the fact that $S$ can only have the discrete values of

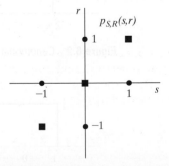

**Figure 8.1**   Joint PMF of $S$ and $R$. The probability associated with the outcome represented by each square is 0.2 and by each circle is 0.1.

+1, 0, or −1. This example will be considered in Chapter 9 from the perspective of hypothesis testing, that is, determining which of the three known possible values will minimize a more suitable error criterion.

In Example 8.1, we considered MMSE estimation of a discrete random variable whose value has been potentially changed, stored, or measured incorrectly. In Example 8.2, we estimate a continuous random variable from a noisy measurement.

## Example 8.2    MMSE Estimate of a Signal in Additive Noise

Consider the noisy measurement $X$ of the angular position of an airport radar antenna. The true position in the absence of noise is a random variable $Y$ and the additive noise is another random variable $W$. Consequently,

$$X = Y + W \,. \tag{8.16}$$

Assume the measurement noise $W$ is independent of the true angular position, that is, $Y$ and $W$ are independent random variables. $Y$ is uniformly distributed in the interval $[-1, 1]$ and $W$ uniformly distributed in the interval $[-2, 2]$. The specific measured value for $X$ is $X = 1$. Consider the MMSE estimate $\widehat{y}$ for the antenna position $Y$, based on this measurement. From Eq. (8.9),

$$\widehat{y} = E(Y|X = 1) \,. \tag{8.17}$$

Equation (8.17) can be evaluated by first obtaining $f_{Y|X}(y|1)$:

$$f_{Y|X}(y|1) = \frac{f_{X|Y}(1|y)f_Y(y)}{f_X(1)} \,. \tag{8.18}$$

The numerator and denominator terms on the right of Eq. (8.18) are next evaluated separately. The PDF $f_{X|Y}(x|y)$ is identical in shape to the PDF of $W$, but with the mean shifted to $y$, as indicated in Figure 8.2. Consequently, $f_{X|Y}(1|y)$ is as shown in Figure 8.3, and $f_{X|Y}(1|y)f_Y(y)$ is as shown in Figure 8.4.

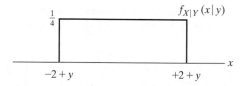

**Figure 8.2**    Conditional PDF of $X$ given $Y$, $f_{X|Y}(x|y)$.

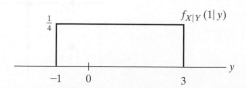

**Figure 8.3**    Plot of $f_{X|Y}(1|y)$.

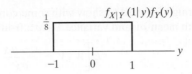

**Figure 8.4**   Plot of $f_{X|Y}(1|y)f_Y(y)$.

To obtain $f_{Y|X}(y|1)$, divide the function in Figure 8.4 by $f_X(1)$, which can easily be obtained by evaluating the convolution of the PDFs of $Y$ and $W$ at the argument 1. More simply, since $f_{Y|X}(y|1)$ must have total area of unity and is the same shape as Figure 8.4 but scaled by $f_X(1)$, it is easily obtained by multiplying Figure 8.4 by 4 to have an area of 1. The resulting value for $\widehat{y}$ is the mean associated with the PDF $f_{Y|X}(y|1)$, which will be

$$\widehat{y} = 0 \,. \tag{8.19}$$

The associated MMSE is the variance of this PDF, namely $\frac{1}{3}$.

---

For the next example, we consider the MMSE estimate of the value of one random variable from the measurement of a second random variable when the two are related through a bivariate Gaussian density.

## Example 8.3    MMSE Estimate for Bivariate Gaussian Random Variables

Consider two random variables $X$ and $Y$ with a bivariate Gaussian joint PDF as defined in Example 7.5, Eq. (7.68). It is convenient to define the centered and normalized bivariate random variables $V$ and $W$ given by

$$V = \frac{X - \mu_X}{\sigma_X}, \qquad W = \frac{Y - \mu_Y}{\sigma_Y} \tag{8.20}$$

with associated PDF

$$f_{V,W}(v,w) = \frac{1}{2\pi\sqrt{1-\rho^2}} \exp\left\{-\frac{(v^2 - 2\rho vw + w^2)}{2(1-\rho^2)}\right\}. \tag{8.21}$$

The number $\rho$ is the correlation coefficient of $X$ and $Y$, and is defined by

$$\rho = \frac{\sigma_{XY}}{\sigma_X \sigma_Y}, \qquad \text{with} \quad \sigma_{XY} = E[XY] - \mu_X \mu_Y \tag{8.22}$$

where $\sigma_{XY}$ is the covariance of $X$ and $Y$.

Now consider $\widehat{y}(x)$, the MMSE estimate of $Y$ given $X = x$, when $X$ and $Y$ are bivariate Gaussian random variables. From Eq. (8.9),

$$\widehat{y}(x) = E[Y \,|\, X = x] \,. \tag{8.23}$$

In terms of $V$ and $W$,

$$\widehat{y}(x) = E\left[(\sigma_Y W + \mu_Y) \,|\, V = \frac{x - \mu_X}{\sigma_X}\right]$$

$$= \sigma_Y E\left[W \,|\, V = \frac{x - \mu_X}{\sigma_X}\right] + \mu_Y \,. \tag{8.24}$$

It is straightforward to show with some computation that $f_{W|V}(w\,|\,v)$ is also Gaussian, but with mean $\rho v$ and variance $1 - \rho^2$, from which it follows that

$$E\left[W \mid V = \frac{x - \mu_X}{\sigma_X}\right] = \rho\left(\frac{x - \mu_X}{\sigma_X}\right). \tag{8.25}$$

Combining Eqs. (8.24) and (8.25),

$$\widehat{y}(x) = E[\,Y\,|\,X = x\,]$$

$$= \mu_Y + \rho\frac{\sigma_Y}{\sigma_X}(x - \mu_X). \tag{8.26}$$

The MMSE estimate in the case of bivariate Gaussian variables has a nice linear (or, more correctly, affine, i.e., linear plus a constant) form.

The MMSE is the variance of the conditional PDF $f_{Y|X}(y|x)$:

$$E[\,(Y - \widehat{y}(x))^2\,|\,X = x\,] = \sigma_Y^2\,(1 - \rho^2). \tag{8.27}$$

Note that $\sigma_Y^2$ is the mean square error in $Y$ in the absence of any additional information. Equation (8.27) shows what the residual mean square error is after a measurement of $X$ is obtained. It is evident and intuitively reasonable that the larger the magnitude of the correlation coefficient between $X$ and $Y$, the smaller the residual mean square error. Also note that in the bivariate Gaussian case, the MMSE of the estimate of $Y$ given $X = x$ does not depend on the specific value $x$.

## 8.2 FROM ESTIMATES TO THE ESTIMATOR

The MMSE estimate of $Y$ in Eq. (8.8) is based on knowing the specific value $x$ that the random variable $X$ takes. While $X$ is a random variable, the specific value $x$ is not, and consequently $\widehat{y}(x)$ is also not a random variable.

It is important in this discussion to draw a distinction between the estimate of a random variable and the procedure by which the estimate is formed for an arbitrary $x$. This is completely analogous to the distinction between the value of a function at a point and the function itself. The procedure or function that produces the estimate is referred to as the estimator.

For instance, in Example 8.1 the MMSE estimate of $S$ for the specific value of $R = 1$ was determined. More generally, an estimate of $S$ for each of the possible values of $R$ could be determined, namely, for $-1$, $0$, and $+1$. A tabulation of these results then allows the MMSE estimate to be looked up when a specific value of $R$ is received. Such a table or, more generally, a function of $R$ would correspond to what is termed the MMSE estimator. The input to the table or estimator would be the specific retrieved value and the output would be the estimate associated with that retrieved value.

The notation $\widehat{y}(x)$ has already been introduced to denote the estimate of $Y$, given $X = x$. The function $\widehat{y}(\cdot)$ determines the corresponding estimator, which is denoted by $\widehat{y}(X)$, or more simply by just $\widehat{Y}$, if it is understood what random variable the estimator is operating on. Note that the estimator $\widehat{Y} = \widehat{y}(X)$ is a random variable. The MMSE estimate $\widehat{y}(x)$ was previously seen to be given by the conditional mean, $E[Y|X = x]$, which suggests yet another natural notation for the MMSE estimator:

$$\widehat{Y} = \widehat{y}(X) = E[Y|X] . \tag{8.28}$$

Note that $E[Y|X]$ denotes a random variable, not a single number.

The preceding discussion applies, essentially unchanged, to the case where several random variables are observed, assembled in the vector $\mathbf{X}$. The MMSE estimator in this case is denoted by

$$\widehat{Y} = \widehat{y}(\mathbf{X}) = E[Y|\mathbf{X}] . \tag{8.29}$$

Perhaps not surprisingly, the MMSE estimator for $Y$ given $\mathbf{X}$ minimizes the mean square error averaged over all $Y$ and $\mathbf{X}$. This is because the MMSE estimator minimizes the mean square error for each particular value $\mathbf{x}$ of $\mathbf{X}$. More formally,

$$E_{Y,\mathbf{X}}\Big( [Y - \widehat{y}(\mathbf{X})]^2 \Big) = E_{\mathbf{X}}\Big( E_{Y|\mathbf{X}}\big( [Y - \widehat{y}(\mathbf{X})]^2 \,|\, \mathbf{X} \big) \Big)$$

$$= \int_{-\infty}^{\infty} \Big( E_{Y|\mathbf{X}}\big( [Y - \widehat{y}(\mathbf{x})]^2 \,|\, \mathbf{X} = \mathbf{x} \big) \Big) f_{\mathbf{X}}(\mathbf{x})\, d\mathbf{x} . \tag{8.30}$$

The subscripts on the expectation operators indicate explicitly which densities are involved in computing the associated expectations; the densities and integration are multivariate when $\mathbf{X}$ is not a scalar. Because the estimate $\widehat{y}(\mathbf{x})$ is chosen to minimize the inner expectation $E_{Y|\mathbf{X}}$ for each value $\mathbf{x}$ of $\mathbf{X}$, it also minimizes the outer expectation $E_{\mathbf{X}}$, since $f_{\mathbf{X}}(\mathbf{x})$ is nonnegative.

---

**Example 8.4    MMSE Estimator for Bivariate Gaussian Random Variables**

In Example 8.3, we constructed the MMSE estimate of one member of a pair of bivariate Gaussian random variables, given a measurement of the other. Using the same notation as in that example, it is evident that the MMSE estimator is simply obtained on replacing $x$ by $X$ in Eq. (8.26):

$$\widehat{Y} = \widehat{y}(X) = \mu_Y + \rho \frac{\sigma_Y}{\sigma_X}(X - \mu_X) . \tag{8.31}$$

The conditional MMSE given $X = x$ was found in the earlier example to be $\sigma_Y^2(1 - \rho^2)$, which did not depend on the value of $x$, so the MMSE of the estimator, averaged over all $X$, is still $\sigma_Y^2(1 - \rho^2)$.

In Example 8.2, we considered the MMSE estimate of the angular position $Y$ of an antenna from a noisy measurement $X$ of the position, that is,

$$X = Y + W \tag{8.32}$$

where $W$ is the random noise. In that example, the MMSE estimate $\widehat{y}$ was determined for a specific measured value of $X$. In the following example, we do not assume a specific measured value for $X$ but instead determine the MMSE estimator.

<br>

**Example 8.5**    **MMSE Estimator for Signal in Additive Noise**

As with Example 8.2, assume that $Y$ and $W$ are independent random variables with $Y$ uniformly distributed in the interval $[-1, 1]$ and $W$ uniformly distributed in the interval $[-2, 2]$. The estimator is

$$\widehat{Y} = \widehat{y}(X) . \tag{8.33}$$

$\widehat{Y}$ is itself a random variable that takes on the specific value $\widehat{y}(x)$ when $X$ takes on the specific value $x$. To develop the estimator, assume that $X = x$ and then determine the MMSE estimate $\widehat{y}(x)$ just as in Example 8.2. As $x$ ranges over the possible values that can occur, the value of $\widehat{y}(x)$ will change. In the discussion below, in addition to determining the MMSE estimator $\widehat{Y} = \widehat{y}(X)$, we can also determine the resulting overall mean square error averaged over all possible values $x$ that the random variable $X$ can take.

Since $\widehat{y}(x)$ is the conditional expectation of $Y$ given $X = x$, $f_{Y|X}(y|x)$ must be determined. For this, first determine the joint density of $Y$ and $W$, and from this the required conditional density.

From the independence of $Y$ and $W$:

$$f_{Y,W}(y,w) = f_Y(y)f_W(w) = \begin{cases} \dfrac{1}{8} & -2 \le w \le 2, -1 \le y \le 1 \\[2mm] 0 & \text{otherwise} \end{cases} \tag{8.34}$$

and is therefore uniform over the rectangle shown in Figure 8.5 and zero outside it.

Conditioned on $Y = y$, $X$ is the same as $y + W$, uniformly distributed over the interval $[y - 2, y + 2]$. Now

$$f_{X,Y}(x,y) = f_{X|Y}(x|y)f_Y(y) = \left(\frac{1}{4}\right)\left(\frac{1}{2}\right) = \frac{1}{8} \tag{8.35}$$

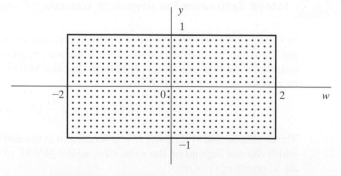

**Figure 8.5**   Joint PDF of $Y$ and $W$ for Example 8.5.

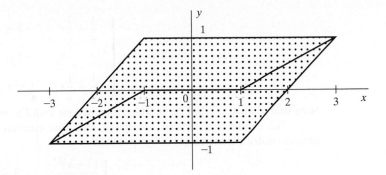

**Figure 8.6**   Joint PDF of X and Y and plot of the MMSE estimator of Y from X for Example 8.5.

for $-1 \leq y \leq 1$, $y - 2 \leq x \leq y + 2$, and zero otherwise. The joint PDF is therefore uniform over the parallelogram shown in Figure 8.6, and zero outside it.

Given $X = x$, the conditional PDF $f_{Y|X}(y|x)$ is uniform on the corresponding vertical section of the parallelogram:

$$f_{Y|X}(y|x) = \begin{cases} \dfrac{1}{3+x} & -3 \leq x \leq -1, -1 \leq y \leq x+2 \\[2mm] \dfrac{1}{2} & -1 \leq x \leq 1, -1 \leq y \leq 1 \\[2mm] \dfrac{1}{3-x} & 1 \leq x \leq 3, x-2 \leq y \leq 1. \end{cases} \tag{8.36}$$

This is shown in Figure 8.7 for various specific values of $x$.

The MMSE estimate $\widehat{y}(x)$ is the conditional mean of $Y$ given $X = x$, which is the midpoint of the corresponding vertical section of the parallelogram. The conditional mean is displayed as the heavy line on the parallelogram in Figure 8.6. In analytical form,

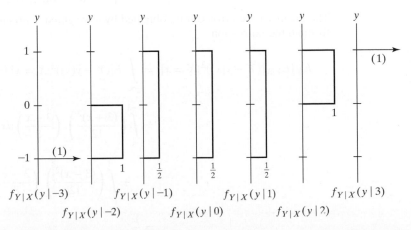

**Figure 8.7**   Conditional PDF $f_{Y|X}(y|x)$ for various realized values of X for Example 8.5.

$$\hat{y}(x) = E[Y|X = x] = \begin{cases} \dfrac{1}{2} + \dfrac{1}{2}x & -3 \leq x < -1 \\[2mm] 0 & -1 \leq x < 1 \\[2mm] -\dfrac{1}{2} + \dfrac{1}{2}x & 1 \leq x \leq 3. \end{cases} \qquad (8.37)$$

Note that when $x = 1$, $\hat{y}(x) = 0$, which is consistent with Example 8.2.

The conditional MMSE associated with this estimate is the variance of the uniform distribution in Eq. (8.36), specifically:

$$E[\{Y - \hat{y}(x)\}^2 | X = x] = \begin{cases} \dfrac{(3+x)^2}{12} & -3 \leq x < -1 \\[2mm] \dfrac{1}{3} & -1 \leq x < 1 \\[2mm] \dfrac{(3-x)^2}{12} & 1 \leq x \leq 3 \end{cases} \qquad (8.38)$$

which again is consistent with Example 8.2, when $X = 1$.

Equation (8.38) specifies the mean square error that results for any specific value $x$ of the measurement of $X$. Since the measurement is a random variable, it is also of interest to know what the mean square error is when averaged over all possible values of the measurement, that is, over the random variable $X$. To determine this, we first determine the marginal PDF of $X$. This can be obtained either by the convolution of $f_Y$ and $f_W$ since $X$ is the sum of the two independent random variables $Y$ and $W$, or through the use of Bayes' rule, with the result that

$$f_X(x) = \frac{f_{X,Y}(x,y)}{f_{Y|X}(y|x)} = \begin{cases} \dfrac{3+x}{8} & -3 \leq x < -1 \\[2mm] \dfrac{1}{4} & -1 \leq x < 1 \\[2mm] \dfrac{3-x}{8} & 1 \leq x \leq 3 \\[2mm] 0 & \text{otherwise}. \end{cases} \qquad (8.39)$$

The mean square error can be obtained by a weighted averaging over all values of $x$ through the expression

$$E_X\{E_{Y|X}[(Y - \hat{y}(x))^2 | X = x]\} = \int_{-\infty}^{\infty} E[(Y - \hat{y}(x))^2 | X = x] f_X(x)\,dx$$

$$= \int_{-3}^{-1} \left(\frac{(3+x)^2}{12}\right) \left(\frac{3+x}{8}\right) dx + \int_{-1}^{1} \left(\frac{1}{3}\right) \left(\frac{1}{4}\right) dx$$

$$+ \int_{1}^{3} \left(\frac{(3-x)^2}{12}\right) \left(\frac{3-x}{8}\right) dx$$

$$= \frac{1}{4}. \qquad (8.40)$$

It is interesting to compare this with the mean square error that would result if $Y$ was estimated by its mean, namely 0. The mean square error would then be the variance $\sigma_Y^2$:

$$\sigma_Y^2 = \frac{[1-(-1)]^2}{12} = \frac{1}{3}, \tag{8.41}$$

so the mean square error is indeed reduced by using knowledge of $X$ and of the probabilistic relation between $Y$ and $X$.

## 8.2.1 Orthogonality

An important property of the MMSE estimator is that the residual error $Y - \widehat{y}(\mathbf{X})$ is orthogonal to any function $h(\mathbf{X})$ of the measured random variable, that is,

$$E_{Y,\mathbf{X}}[\{Y - \widehat{y}(\mathbf{X})\}h(\mathbf{X})] = 0, \tag{8.42}$$

where $\mathbf{X}$ is the vector of measured random variables and the expectation is computed over the joint density of $Y$ and $\mathbf{X}$. This result follows by first expanding the left side of Eq. (8.42) to obtain

$$E_{Y,\mathbf{X}}[\{Y - \widehat{y}(\mathbf{X})\}h(\mathbf{X})] = E_{Y,\mathbf{X}}[Yh(\mathbf{X})] - E_{Y,\mathbf{X}}[\widehat{y}(\mathbf{X})h(\mathbf{X})]. \tag{8.43}$$

Next, apply the following sequence of equalities to the term $E_{Y,\mathbf{X}}[\widehat{y}(\mathbf{X})h(\mathbf{X})]$:

$$E_{Y,\mathbf{X}}[\widehat{y}(\mathbf{X})h(\mathbf{X})] = E_{\mathbf{X}}[E_{Y|\mathbf{X}}[Y|\mathbf{X}]h(\mathbf{X})] \tag{8.44}$$

$$= E_{\mathbf{X}}[E_{Y|\mathbf{X}}[Yh(\mathbf{X})|\mathbf{X}]] \tag{8.45}$$

$$= E_{Y,\mathbf{X}}[Yh(\mathbf{X})]. \tag{8.46}$$

Applying Eq. (8.46) to Eq. (8.43) results in Eq. (8.42).

Equation (8.46) states that the MMSE estimator has the same correlation that $Y$ does with any function of $X$. In particular, choosing $h(\mathbf{X}) = 1$,

$$E_{\mathbf{X}}[\widehat{y}(\mathbf{X})] = E_Y[Y]. \tag{8.47}$$

When the expected value of the estimator $\widehat{y}(\mathbf{X})$ is equal to the expected value of the random variable $Y$, the estimator is referred to as unbiased. Equation (8.47) states that the MMSE estimator is indeed unbiased. This property can be invoked to interpret Eq. (8.42) as stating that the estimation error of the MMSE estimator is uncorrelated with any function of the random variables used to construct the estimator.

# 8.3 LINEAR MINIMUM MEAN SQUARE ERROR ESTIMATION

## 8.3.1 Linear Estimation of One Random Variable from a Single Measurement of Another

In general, the conditional expectation $E(Y|\mathbf{X})$ required for the MMSE estimator developed in the preceding sections is difficult to determine because the conditional density $f_{Y|\mathbf{X}}(y|\mathbf{x})$ is not easily determined. A useful and widely used compromise is to restrict the estimator to be a fixed linear (or more specifically, affine, i.e., linear plus a constant) function of the measured random variables, and to choose the linear relationship so as to minimize the overall mean square error averaged over the values that $Y$ and $\mathbf{X}$ can jointly take. The resulting estimator is called the linear minimum mean square error (LMMSE) estimator. The simplest case is presented first.

Suppose an estimator for the random variable $Y$ is constructed in terms of another random variable $X$, restricting the estimator to the form

$$\widehat{Y}_\ell = \widehat{y}_\ell(X) = aX + b \,, \tag{8.48}$$

where $a$ and $b$ are to be determined so as to minimize the mean square error

$$E_{Y,X}[(Y - \widehat{Y}_\ell)^2] = E_{Y,X}[\{Y - (aX + b)\}^2] \,. \tag{8.49}$$

Note that the expectation is taken over the joint density of $Y$ and $X$, that is, the linear estimator is picked to be optimum when averaged over all possible combinations of $Y$ and $X$ that may occur. The subscripts on the expectation operations in Eq. (8.49) for now make explicit the variables whose joint density the expectation is being computed over; eventually the subscripts will be dropped.

Once the optimum values for the parameters $a$ and $b$ have been chosen in this manner, the estimate of $Y$, given a particular $x$, is simply $\widehat{y}_\ell(x) = ax + b$, computed with the values of $a$ and $b$ already determined. Thus, in the LMMSE case an optimal linear estimator is constructed, and for any particular $x$ this estimator generates an estimate that is not claimed to have any individual optimality property. This is in contrast to the MMSE case considered in the previous sections, where an optimal MMSE estimate for each $x$ was obtained, namely $E[Y|X = x]$, that minimized the mean square error conditioned on $X = x$. The distinction can be summarized as follows: in the unrestricted MMSE case, the optimal estimator is obtained by joining together all the individual optimal estimates, whereas in the LMMSE case the (generally nonoptimal) individual estimates are obtained by simply evaluating the optimal linear estimator.

The expression in Eq. (8.49) is minimized by differentiating it with respect to the parameters $a$ and $b$, and setting each of the derivatives to 0. (Consideration of the second derivatives will show that the values found in this fashion are minimizing, but the demonstration is omitted.) First differentiating Eq. (8.49) with respect to $b$, taking the derivative inside the integral that corresponds to the expectation operation, and then setting the result to 0, the conclusion is

$$E_{Y,X}[Y - (aX + b)] = 0 . \tag{8.50}$$

Equivalently,

$$E[Y] = E[aX + b] = E[\widehat{Y}_\ell] , \tag{8.51}$$

from which

$$b = \mu_Y - a\mu_X \tag{8.52}$$

is deduced, where $\mu_Y = E[Y] = E_{Y,X}[Y]$ and $\mu_X = E[X] = E_{Y,X}[X]$. The optimum value of $b$ specified in Eq. (8.52) in effect serves to make the linear estimator unbiased, that is, the expected value of the estimator is equal to the expected value of the random variable being estimated, as Eq. (8.51) shows.

Using Eq. (8.52) to substitute for $b$ in Eq. (8.48), it follows that

$$\widehat{Y}_\ell = \mu_Y + a(X - \mu_X) . \tag{8.53}$$

In other words, to the expected value $\mu_Y$ of the random variable $Y$ that is being estimated, the optimal linear estimator adds a suitable multiple of the difference $X - \mu_X$ between the measured random variable and its expected value. To find the optimum value of this multiple, $a$, first rewrite the error criterion in Eq. (8.49) as

$$E[\{(Y - \mu_Y) - (\widehat{Y}_\ell - \mu_Y)\}^2] = E[(\widetilde{Y} - a\widetilde{X})^2] , \tag{8.54}$$

where

$$\widetilde{Y} = Y - \mu_Y \quad \text{and} \quad \widetilde{X} = X - \mu_X , \tag{8.55}$$

and where Eq. (8.53) is invoked to obtain the second equality in Eq. (8.54). Taking the derivative of the error criterion in Eq. (8.54) with respect to $a$, and setting it to 0, results in

$$E[(\widetilde{Y} - a\widetilde{X})\widetilde{X}] = 0 . \tag{8.56}$$

Rearranging Eq. (8.56), and recalling that $E[\widetilde{Y}\widetilde{X}] = \sigma_{YX}$, namely the covariance of $Y$ and $X$, and that $E[\widetilde{X}^2] = \sigma_X^2$, we obtain

$$a = \frac{\sigma_{YX}}{\sigma_X^2} = \rho_{YX}\frac{\sigma_Y}{\sigma_X} , \tag{8.57}$$

where $\rho_{YX}$—which will simply be written as $\rho$ when it is clear from context what variables are involved—denotes the correlation coefficient between $Y$ and $X$.

It is also enlightening to interpret Eq. (8.57) in terms of the vector-space picture for random variables developed in Chapter 7. The expression in Eq. (8.54) for the error criterion can be visualized with Figure 8.8. We choose the vector $a\widetilde{X}$, which lies along the vector $\widetilde{X}$, such that the squared length of the error vector $\widetilde{Y} - a\widetilde{X}$ is minimum. The associated vectors are illustrated

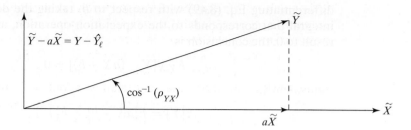

**Figure 8.8**   Expression for *a* from Eq. (8.57) illustrated in vector space.

in Figure 8.8. It follows from familiar geometric reasoning that the optimum choice of $a\widetilde{X}$ must be the orthogonal projection of $\widetilde{Y}$ on $\widetilde{X}$, and that this projection is

$$a\widetilde{X} = \frac{\langle \widetilde{Y}, \widetilde{X} \rangle}{\langle \widetilde{X} \cdot \widetilde{X} \rangle} \widetilde{X} \,. \tag{8.58}$$

Here, as in Chapter 7, $\langle U, V \rangle$ denotes the inner product of the vectors $U$ and $V$, and in the case where the "vectors" are random variables, denotes $E[UV]$. The expressions for $a$ in Eq. (8.57) follow immediately. Recall from Chapter 7 that the correlation coefficient $\rho$ denotes the cosine of the angle between the vectors $\widetilde{Y}$ and $\widetilde{X}$, and that these vectors have respective lengths $\sigma_Y$ and $\sigma_X$.

The preceding projection operation implies that the error $\widetilde{Y} - a\widetilde{X}$, which can also be written as $Y - \widehat{Y}_\ell$, must be orthogonal to $\widetilde{X} = X - \mu_X$. This is precisely what Eq. (8.56) says. In addition, invoking the unbiasedness of $\widehat{Y}_\ell$ shows that $(Y - \widehat{Y}_\ell)$ must be orthogonal to $\mu_X$ (or any other constant), so $(Y - \widehat{Y}_\ell)$ is therefore orthogonal to $X$ itself:

$$E[(Y - \widehat{Y}_\ell)X] = 0 \,. \tag{8.59}$$

In other words, the optimal LMMSE estimator is unbiased and such that the estimation error is orthogonal to the random variable on which the estimator is based. Note that the statement in the case of the MMSE estimator in the previous section was considerably stronger, namely that the error was orthogonal to any function $h(X)$ of the measured random variable, not just to the random variable itself.

The preceding development shows that the properties of (i) unbiasedness of the estimator and (ii) orthogonality of the error to the measured random variable completely characterize the LMMSE estimator. Invoking these properties yields the LMMSE estimator.

Carrying the geometric reasoning further, the Pythagorean theorem applied to the triangle in Figure 8.8 leads to the conclusion that the minimum mean square error (MMSE) obtained through use of the LMMSE estimator is

$$\text{MMSE} = E[(\widetilde{Y} - a\widetilde{X})^2] = E[\widetilde{Y}^2](1 - \rho^2) = \sigma_Y^2(1 - \rho^2) \,. \tag{8.60}$$

This result could also be obtained purely analytically, of course, without recourse to geometric interpretation. The result shows that the mean square error $\sigma_Y^2$ prior to estimation in terms of $X$ is reduced by the factor $1 - \rho^2$ when the observed value of $X$ is used in an LMMSE estimator. The closer that $\rho$ is to $+1$ or $-1$ (corresponding to strong positive or negative correlation respectively), the more that uncertainty about $Y$ is reduced by using an LMMSE estimator to extract information that $X$ carries about $Y$.

Results on the LMMSE estimator can now be summarized in the following expressions for the estimator, with the associated minimum mean square error being given by Eq. (8.60):

$$\widehat{Y}_\ell = \widehat{y}_\ell(X) = \mu_Y + \frac{\sigma_{YX}}{\sigma_X^2}(X - \mu_X) = \mu_Y + \rho \frac{\sigma_Y}{\sigma_X}(X - \mu_X), \qquad (8.61)$$

or the equivalent but perhaps more suggestive form

$$\frac{\widehat{Y}_\ell - \mu_Y}{\sigma_Y} = \rho \frac{X - \mu_X}{\sigma_X}. \qquad (8.62)$$

Equation (8.62) states that the normalized deviation of the estimator from its mean is $\rho$ times the normalized deviation of the observed variable from its mean; the more highly correlated $Y$ and $X$ are, the more closely the two normalized deviations match.

Note that the above expressions for the LMMSE estimator and its mean square error are the same as those obtained in Example 8.4 for the MMSE estimator in the bivariate Gaussian case. The reason is that the MMSE estimator in that case turned out to be linear (actually, affine), as already noted in the example.

---

**Example 8.6    LMMSE Estimator for a Signal in Additive Noise**

In Example 8.5, we determined the MMSE estimator. This example now focuses on the design of an LMMSE estimator. Recall that the random variable $X$ denotes a noisy measurement of the angular position $Y$ of an antenna, so $X = Y + W$, where $W$ denotes the additive noise. The noise is assumed to be independent of the angular position, that is, $Y$ and $W$ are independent random variables, with $Y$ uniformly distributed in the interval $[-1, 1]$ and $W$ uniformly distributed in the interval $[-2, 2]$.

The LMMSE estimator of $Y$ in terms of $X$ requires the respective means and variances, as well as the covariance, of these random variables. It is straightforward to determine that

$$\mu_Y = 0, \quad \mu_W = 0, \quad \mu_X = 0, \quad \sigma_Y^2 = \frac{1}{3}, \quad \sigma_W^2 = \frac{4}{3},$$

$$\sigma_X^2 = \sigma_Y^2 + \sigma_W^2 = \frac{5}{3}, \quad \sigma_{YX} = \sigma_Y^2 = \frac{1}{3}, \quad \rho_{YX} = \frac{1}{\sqrt{5}}. \qquad (8.63)$$

The LMMSE estimator is accordingly

$$\widehat{Y}_\ell = \frac{1}{5}X, \qquad (8.64)$$

and the associated MMSE is

$$\sigma_Y^2(1 - \rho^2) = \frac{4}{15}. \tag{8.65}$$

This MMSE should be compared with the (larger) mean square error of $\frac{1}{3}$ obtained if $\mu_Y = 0$ is used as the estimator for $Y$, and the (smaller) value $\frac{1}{4}$ obtained using the MMSE estimator in Example 8.5.

In the next example we consider a parameterized time function for which the parameters are random variables. The LMMSE estimator is used to estimate the value at one time instant from the observed value at a different time instant.

| **Example 8.7** | **Single-Measurement LMMSE Estimator for Sinusoidal Random Process** |

Consider a sinusoidal signal of the form

$$X(t) = A\cos(\omega_0 t + \Theta) \tag{8.66}$$

where $\omega_0$ is assumed known, while $A$ and $\Theta$ are statistically independent random variables, with the PDF of $\Theta$ being uniform in the interval $[0, 2\pi]$. Thus $X(t)$ is a random signal, or equivalently a set, or "ensemble," of signals corresponding to the various possible outcomes for $A$ and $\Theta$ in the underlying probabilistic experiment. Such signals will be discussed in more detail in Chapter 10, where they will be referred to as random processes. The value that $X(t)$ takes at some particular time $t = t_0$ is simply a random variable, whose specific value will depend on which outcomes for $A$ and $\Theta$ are produced by the underlying probabilistic experiment.

Suppose the LMMSE estimator for $X(t_1)$ is based on a measurement of $X(t_0)$, where $t_0$ and $t_1$ are specified sampling times. In other words, the estimator is of the form

$$\widehat{X}(t_1) = aX(t_0) + b \tag{8.67}$$

with $a$ and $b$ chosen so as to minimize the mean square error between $X(t_1)$ and $\widehat{X}(t_1)$.

It was established that $b$ must be chosen to ensure that the estimator is unbiased:

$$E[\widehat{X}(t_1)] = aE[X(t_0)] + b = E[X(t_1)]. \tag{8.68}$$

Since $A$ and $\Theta$ are independent, and $\Theta$ is uniform in $[0, 2\pi]$,

$$E[X(t_0)] = E[A] \int_0^{2\pi} \frac{1}{2\pi} \cos(\omega_0 t_0 + \theta) \, d\theta = 0 \tag{8.69}$$

and similarly $E[X(t_1)] = 0$, so $b = 0$.

Next, use the fact that the error of the LMMSE estimator is orthogonal to the data:

$$E[(\widehat{X}(t_1) - X(t_1))X(t_0)] = 0 \tag{8.70}$$

and consequently

$$aE[X^2(t_0)] = E[X(t_1)X(t_0)] \tag{8.71}$$

or

$$a = \frac{E[X(t_1)X(t_0)]}{E[X^2(t_0)]} \ . \tag{8.72}$$

The numerator and denominator in Eq. (8.72) are respectively

$$E[X(t_1)X(t_0)] = E[A^2] \int_0^{2\pi} \frac{1}{2\pi} \cos(\omega_0 t_1 + \theta) \cos(\omega_0 t_0 + \theta) \, d\theta$$

$$= \frac{E[A^2]}{2} \cos\{\omega_0(t_1 - t_0)\} \tag{8.73}$$

and $E[X^2(t_0)] = \frac{E[A^2]}{2}$. Thus $a = \cos\{\omega_0(t_1 - t_0)\}$, so the LMMSE estimator is

$$\widehat{X}(t_1) = X(t_0) \cos\{\omega_0(t_1 - t_0)\} \ . \tag{8.74}$$

Note that the distribution of $A$ does not play a role in this equation.

To evaluate the mean square error associated with the LMMSE estimator, compute the correlation coefficient between the samples of the random signal at $t_0$ and $t_1$. It is easily seen that $\rho = a = \cos\{\omega_0(t_1 - t_0)\}$, so the mean square error is

$$\frac{E[A^2]}{2} \left( 1 - \cos^2\{\omega_0(t_1 - t_0)\} \right) = \frac{E[A^2]}{2} \sin^2\{\omega_0(t_1 - t_0)\} \ . \tag{8.75}$$

## 8.3.2 Multiple Measurements

In this section, we extend the LMMSE estimator to the case where the estimation of a random variable $Y$ is based on observations of multiple random variables, say $X_1, \ldots, X_L$, gathered in the vector $\mathbf{X}$. The affine estimator may then be written in the form

$$\widehat{Y}_\ell = \widehat{y}_\ell(\mathbf{X}) = a_0 + \sum_{j=1}^{L} a_j X_j \ . \tag{8.76}$$

The coefficients $a_j$ of this LMMSE estimator can be found by solving a linear system of equations that is completely defined by the first and second moments (i.e., means, variances, and covariances) of the random variables $Y$ and $X_j$. The fact that the model Eq. (8.76) is linear in the parameters $a_j$ is what results in a linear system of equations; the fact that the model is affine in the random variables is why the solution only depends on the first and second moments. Linear equations are easy to solve, and first and second moments are generally easy to determine, hence the popularity of LMMSE estimation.

The development below follows along the same lines as in Section 8.3.1, in which there was a single observed random variable $X$. The opportunity of the extension in this subsection to multiple measurements allows a review of the logic of the development and provides a few additional insights.

The objective is to minimize the mean square error

$$E\left[\left(Y - (a_0 + \sum_{j=1}^{L} a_j X_j)\right)^2\right], \qquad (8.77)$$

where the expectation is computed using the joint density of $Y$ and $\mathbf{X}$. The joint density is used rather than the conditional because the parameters are not going to be chosen to be best for a particular set of measured values $\mathbf{x}$— otherwise the nonlinear estimate would work as well in this case, by setting $a_0 = E[Y \mid \mathbf{X} = \mathbf{x}]$ and setting all the other $a_i$ to zero. Instead, the parameters are chosen to be the best averaged over all possible combinations of $\mathbf{X}$ and $Y$. The linear estimator will in general not be as good as the unconstrained estimator, except in special cases (some of them important, as in the case of multivariate Gaussian random variables), but the linear estimator has the advantage that it is easy to solve for, as is now shown.

The expression in Eq. (8.77) is minimized by differentiation with respect to $a_i$ for $i = 0, 1, \cdots, L$, and setting each of the derivatives to 0. (Again, calculations involving second derivatives establish that indeed minimizing values are obtained, but these calculations are omitted here.) Differentiating with respect to $a_0$ and setting the result to 0 shows that

$$E[Y] = E[a_0 + \sum_{j=1}^{L} a_j X_j] = E[\widehat{Y}_\ell] \qquad (8.78)$$

or

$$a_0 = \mu_Y - \sum_{j=1}^{L} a_j \mu_{X_j}, \qquad (8.79)$$

where $\mu_Y = E[Y]$ and $\mu_{X_j} = E[X_j]$. This optimum value of $a_0$ serves to make the linear estimator unbiased, in the sense that Eq. (8.78) holds, that is, the expected value of the estimator is the expected value of the random variable for which an estimate is desired.

Using Eq. (8.79) to substitute for $a_0$ in Eq. (8.76), it follows that

$$\widehat{Y}_\ell = \mu_Y + \sum_{j=1}^{L} a_j (X_j - \mu_{X_j}). \qquad (8.80)$$

In other words, the estimator adjusts the expected value $\mu_Y$ of the variable being estimated, by a linear combination of the deviations $X_j - \mu_{X_j}$ between the measured random variables and their respective expected values.

Taking account of Eq. (8.80), the mean square error criterion in Eq. (8.77) can be rewritten as

$$E[\{(Y - \mu_Y) - (\widehat{Y}_\ell - \mu_Y)\}^2] = E\left[\left(\widetilde{Y} - \sum_{j=1}^{L} a_j \widetilde{X}_j\right)^2\right], \qquad (8.81)$$

where

$$\tilde{Y} = Y - \mu_Y \quad \text{and} \quad \tilde{X}_j = X_j - \mu_{X_j} . \tag{8.82}$$

Differentiating Eq. (8.81) with respect to each of the remaining coefficients $a_i, i = 1, 2, \ldots L$, and setting the result to zero produces the equations

$$E\left[ \left( \tilde{Y} - \sum_{j=1}^{L} a_j \tilde{X}_j \right) \tilde{X}_i \right] = 0 \quad i = 1, 2, \ldots, L . \tag{8.83}$$

or equivalently, taking into account Eq. (8.80),

$$E[(Y - \widehat{Y}_\ell)\tilde{X}_i] = 0 \quad i = 1, 2, \ldots, L . \tag{8.84}$$

Yet another version follows on noting from Eq. (8.78) that $Y - \widehat{Y}_\ell$ is orthogonal to all constants, in particular to $\mu_{X_i}$, so

$$E[(Y - \widehat{Y}_\ell)X_i] = 0 \quad i = 1, 2, \ldots, L . \tag{8.85}$$

Equations (8.83), (8.84), and (8.85) all express, in slightly different forms, the orthogonality of the estimation error to the random variables used in the estimator. The relationship between these forms follows by invoking the unbiasedness of the estimator. The last of these, Eq. (8.85), is the usual statement of the orthogonality condition that governs the LMMSE estimator. Note once more that the statement in the case of the MMSE estimator in the previous section was considerably stronger, namely that the error was orthogonal to any function $h(\mathbf{X})$ of the measured random variables, not just to the random variables themselves. Rewriting Eq. (8.85) as

$$E[YX_i] = E[\widehat{Y}_\ell X_i] \quad i = 1, 2, \ldots, L \tag{8.86}$$

yields an equivalent statement of the orthogonality condition, namely that the LMMSE estimator $\widehat{Y}_\ell$ has the same correlations as $Y$ with the measured variables $X_i$.

The orthogonality and unbiasedness conditions together determine the LMMSE estimator completely. Also, the preceding development shows that the first moment of $Y$ and the second cross-moments with the $X_i$ are exactly matched by the corresponding first moment of $\widehat{Y}_\ell$ and its second cross-moments with the $X_i$. It follows that $Y$ and $\widehat{Y}_\ell$ cannot be told apart on the basis of only these moments.

Equation (8.83) provides a convenient route to a solution for the coefficients $a_j, j = 1, \ldots, L$. This set of equations can be expressed as

$$\sum_{j=1}^{L} \sigma_{X_i X_j} a_j = \sigma_{X_i Y} , \tag{8.87}$$

where $\sigma_{X_i X_j}$ is the covariance of $X_i$ and $X_j$ (so $\sigma_{X_i X_i}$ is just the variance $\sigma_{X_i}^2$), and $\sigma_{X_i Y}$ is the covariance of $X_i$ and $Y$. Collecting these equations in matrix form results in

$$
\begin{bmatrix}
\sigma_{X_1 X_1} & \sigma_{X_1 X_2} & \cdots & \sigma_{X_1 X_L} \\
\sigma_{X_2 X_1} & \sigma_{X_2 X_2} & \cdots & \sigma_{X_2 X_L} \\
\vdots & \vdots & \ddots & \vdots \\
\sigma_{X_L X_1} & \sigma_{X_L X_2} & \cdots & \sigma_{X_L X_L}
\end{bmatrix}
\begin{bmatrix}
a_1 \\ a_2 \\ \vdots \\ a_L
\end{bmatrix}
=
\begin{bmatrix}
\sigma_{X_1 Y} \\ \sigma_{X_2 Y} \\ \vdots \\ \sigma_{X_L Y}
\end{bmatrix} .
\tag{8.88}
$$

This set of equations is referred to as the normal equations, and can be expressed compactly in matrix notation as

$$
(\mathbf{C_{XX}}) \, \mathbf{a} = \mathbf{c_{XY}}
\tag{8.89}
$$

where the definitions of $\mathbf{C_{XX}}$, $\mathbf{a}$, and $\mathbf{c_{XY}}$ should be evident on comparing Eqs. (8.88) and (8.89). The solution of this set of $L$ equations in $L$ unknowns yields the $\{a_j\}$ for $j = 1, \cdots, L$, and these values may be substituted in Eq. (8.80) to completely specify the estimator. In matrix notation, the solution is

$$
\mathbf{a} = (\mathbf{C_{XX}})^{-1} \mathbf{c_{XY}} .
\tag{8.90}
$$

It can be shown in a straightforward way that the MMSE obtained with the LMMSE estimator is

$$
\sigma_Y^2 - \mathbf{c_{YX}}(\mathbf{C_{XX}})^{-1}\mathbf{c_{XY}} = \sigma_Y^2 - \mathbf{c_{YX}}\mathbf{a} ,
\tag{8.91}
$$

where $\mathbf{c_{YX}}$ is the transpose of $\mathbf{c_{XY}}$. In the case of a single measurement, this reduces to $\sigma_Y^2(1 - \rho^2)$, that is, Eq. (8.60).

---

**Example 8.8**    **Estimation from Two Noisy Measurements**

Assume that a random variable $Y$ is observed through two noisy measurements $X_1$ and $X_2$ so that

$$
X_1 = Y + R_1
$$
$$
X_2 = Y + R_2,
\tag{8.92}
$$

where $Y$ and the two noise variables $R_1$ and $R_2$ are mutually uncorrelated. Assume also that $R_1$ and $R_2$ have zero mean and variance $\sigma_R^2$. The LMMSE estimator for $Y$ can be found, given measurements of $X_1$ and $X_2$. This estimator takes the form $\widehat{Y}_\ell = a_0 + a_1 X_1 + a_2 X_2$. The requirement that $\widehat{Y}_\ell$ be unbiased results in the constraint

$$
a_0 = \mu_Y - a_1 \mu_{X_1} - a_2 \mu_{X_2} = \mu_Y(1 - a_1 - a_2) .
\tag{8.93}
$$

Noting that

$$E[X_i^2] = E[Y^2] + E[R_i^2],$$

$$E[X_1 X_2] = E[Y^2],$$

$$E[X_i Y] = E[Y^2], \tag{8.94}$$

the normal equations for this case become

$$\begin{bmatrix} \sigma_Y^2 + \sigma_R^2 & \sigma_Y^2 \\ \sigma_Y^2 & \sigma_Y^2 + \sigma_R^2 \end{bmatrix} \begin{bmatrix} a_1 \\ a_2 \end{bmatrix} = \begin{bmatrix} \sigma_Y^2 \\ \sigma_Y^2 \end{bmatrix}. \tag{8.95}$$

The solution of Eq. (8.95) results in

$$\begin{bmatrix} a_1 \\ a_2 \end{bmatrix} = \frac{1}{(\sigma_Y^2 + \sigma_R^2)^2 - \sigma_Y^4} \begin{bmatrix} \sigma_Y^2 + \sigma_R^2 & -\sigma_Y^2 \\ -\sigma_Y^2 & \sigma_Y^2 + \sigma_R^2 \end{bmatrix} \begin{bmatrix} \sigma_Y^2 \\ \sigma_Y^2 \end{bmatrix}$$

$$= \frac{\sigma_Y^2}{2\sigma_Y^2 + \sigma_R^2} \begin{bmatrix} 1 \\ 1 \end{bmatrix}. \tag{8.96}$$

Therefore,

$$\widehat{Y}_\ell = \frac{1}{2\sigma_Y^2 + \sigma_R^2} (\sigma_R^2 \mu_Y + \sigma_Y^2 X_1 + \sigma_Y^2 X_2). \tag{8.97}$$

Thus $\widehat{Y}_\ell$ is a weighted linear combination of the prior estimate $\mu_Y$ for $Y$ in the absence of measurements, and the two measurements $X_1$ and $X_2$. The prior estimate $\mu_Y$ is given more weight as the noise in the measurements increases. The measurements are given more weight as the prior uncertainty $\sigma_Y^2$ increases.

Applying Eq. (8.91), the associated MMSE is

$$\frac{\sigma_Y^2 \sigma_R^2}{2\sigma_Y^2 + \sigma_R^2}. \tag{8.98}$$

It is straightforward to check that both the estimator and the associated MMSE take intuitively reasonable values at extreme ranges of the ratio $\sigma_Y^2/\sigma_R^2$.

# 8.4 FURTHER READING

Most of the texts suggested for further reading at the end of Chapter 7 contain material on mean-square-error estimation. See also [Kay2] and [Moo]. We have assumed in our study of MMSE estimation that the requisite probability density functions are known, and similarly in LMMSE estimation that the required first and second moments are known. The subject of statistics is concerned with situations in which such quantities have to be estimated or learned from data. [DeG], [Dek], [Rce] and [Wal] provide good introductions to the statistical approach, while [Cox], [Was], and [Wil] range more widely; none of these has a signals and systems perspective. [St1] and [St2] provide absorbing historical accounts of the development of probability-based statistical methods for describing and quantifying uncertainty, accuracy, and variability in fields ranging from astronomy and geodesy to psychology, biology, and the social sciences.

# Problems

## Basic Problems

**8.1.** For each of the following parts, state whether the given statement is true or false. For a true statement, give a brief but convincing explanation; for a false statement, give a counterexample or convincing explanation.

(a) If $\widehat{Y}_\ell$ is the LMMSE estimator of $Y$ in terms of some other random variable $X$, then the corresponding MMSE $E[(Y - \widehat{Y}_\ell)^2]$ can be expressed as

$$E[(Y - \widehat{Y}_\ell)^2] = E[Y^2] - E[\widehat{Y}_\ell Y].$$

(b) Suppose $X$ and $Y$ are random variables with mean 0 and the same variance $\sigma^2$, and suppose it is known that $E[Y|X = x] = \frac{1}{3}x$ for all values $x$ that the random variable $X$ can take. The correlation coefficient of $X$ and $Y$ must then be $\frac{1}{3}$.

**8.2.** For each of the following parts, state whether the given statement is true or false. For a true statement, give a brief but convincing explanation; for a false statement, give a counterexample or convincing explanation.

(a) Suppose the LMMSE estimator $\widehat{Y}_\ell$ of the random variable $Y$ in terms of $X$ is simply the mean of $Y$, i.e., $\widehat{Y}_\ell = \mu_Y$. Then $X$ and $Y$ must be independent.

(b) Suppose the random variable $X$ is uniformly distributed in the interval $[-1, 1]$, and let $Y = X^2$ (so $Y$ is completely determined by $X$). The LMMSE estimator $\widehat{Y}_\ell$ of $Y$ in terms of $X$ is 0.

(c) Suppose $X_1$ and $X_2$ are uncorrelated random variables. Then the LMMSE estimator for $Y$ in terms of $X_1$ and $X_2$ is given by $\widehat{Y}_\ell = \widehat{Y}_{\ell 1} + \widehat{Y}_{\ell 2}$, where $\widehat{Y}_{\ell 1}$ is the LMMSE estimator of $Y$ in terms of just $X_1$, and similarly $\widehat{Y}_{\ell 2}$ is the LMMSE estimator of $Y$ in terms of just $X_2$.

**8.3.** $X$ and $Y$ are two random variables with unknown PDFs. $X$ is zero-mean. The MMSE estimator $\widehat{Y}$ of $Y$ given $X$ is $\widehat{Y} = 5$. From the information given, specify whether $X$ and $Y$ are definitely statistically independent, definitely not statistically independent, or if it can't be determined from the information given. Explain.

**8.4.** Consider the pair of bivariate Gaussian random variables, $X$ and $Y$, where $\mu_X = 0$. The MMSE estimator for $Y$ in terms of $X$ is $\widehat{Y}_{\text{MMSE}}(X) = 2$.

(a) What is $E[Y]$?

(b) Specify whether $X$ and $Y$ are correlated, uncorrelated, or there isn't enough information to make this determination. Explain.

(c) Specify whether $X$ and $Y$ are independent, dependent, or there isn't enough information to make this determination. Explain.

(d) What is the MMSE estimator of $X$ in terms of $Y$, $\widehat{X}_{\text{MMSE}}(Y)$?

**8.5.** Suppose $X$ and $Y$ are zero-mean unit-variance random variables. If the LMMSE estimator $\widehat{y}_\ell(X)$ of $Y$ in terms of $X$ is given by

$$\widehat{y}_\ell(X) = \frac{3}{4} X \, ,$$

what is its mean square error? Also, suppose the random variable $Q$ is defined by $Q = Y + 3$; what is the LMMSE estimator $\widehat{q}_\ell(X)$ of $Q$ in terms of $X$, and what is its mean square error? Finally, what is the LMMSE estimator $\widehat{x}_\ell(Y)$ of $X$ in terms of $Y$, and what is its mean square error?

**8.6.** The random variables $X$ and $Y$ are uniformly distributed in the shaded region shown in Figure P8.6.

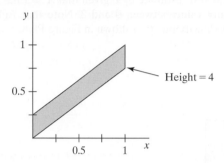

Figure P8.6

**(a)** Determine and sketch the MMSE estimator $\widehat{Y}_{\text{MMSE}}(X)$ of $Y$ given $X$.
**(b)** Determine and sketch the MMSE estimator $\widehat{X}_{\text{MMSE}}(Y)$ of $X$ given $Y$.

**8.7.** Suppose that two random variables $X$ and $Y$ have a joint PDF $f_{X,Y}(x,y)$ that is constant in the shaded region shown in Figure P8.7, and zero elsewhere:

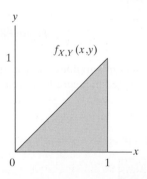

Figure P8.7

**(a)** Make fully labeled sketches of the densities $f_X(x)$ and $f_{Y|X}(y|\frac{1}{3})$.
**(b)** Are $X$ and $Y$ statistically independent? Explain.
**(c)** Determine and make a fully labeled sketch (as a function of $x$) of $\widehat{y}_{\text{MMSE}}(X)$, the MMSE estimator of $Y$ based on observing $X$.
**(d)** To evaluate how well your estimator from (c) will perform on average, determine the mean square error $e^2$ and the bias $b$ associated with the estimator:

$$e^2 = E\left[ \left( \widehat{y}_{\text{MMSE}}(X) - Y \right)^2 \right], \text{ and } b = E\left[ \widehat{y}_{\text{MMSE}}(X) - Y \right],$$

where the expectation is over $X$ and $Y$ jointly.

    **(e)** Determine $\widehat{y}_{\text{LMMSE}}(X)$, the linear MMSE estimator of $Y$, and its associated MMSE.

**8.8.** Two random variables $X$ and $Y$ have a joint PDF $f_{X,Y}(x, y)$ that is equal to a constant $K$ in the shaded region shown in Figure P8.8, and is equal to zero elsewhere.

    **(a)**   (i)   Find $K$.
             (ii)   Make a labeled sketch of the marginal PDF $f_Y(y)$.
            (iii)   Make a labeled sketch of the conditional PDF $f_{Y|X}(y \mid \frac{1}{4})$.
    **(b)** Find the MMSE estimate of $Y$ given that $X = x$ has been observed, i.e., the conditional mean $E(Y \mid X = x)$, where $x$ can be any value between 0 and 2.
    **(c)** Find the LMMSE estimate of $Y$ given that $X = x$ has been observed, where $x$ can be any value between 0 and 2. Note that $E(YX) = \frac{3}{4}$ for the joint probability density function shown in Figure P8.8.

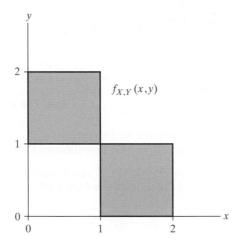

**Figure P8.8**

**8.9.** Suppose that $X$ and $Y$ are random variables with joint PDF $f_{X,Y}(x, y)$ that is constant in the shaded area depicted in Figure P8.9, and zero elsewhere.

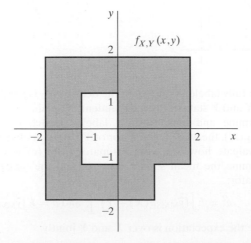

**Figure P8.9**

(a) Find the LMMSE estimator of $Y$ based on measuring $X$.

(b) For $x$ in the range $-2$ to $2$, make a fully labeled sketch of the MMSE estimator of $Y$. How does the corresponding estimator compare with that in (a)?

**8.10.** Consider a sinusoidal signal of the form

$$X(t) = A\cos(\omega_0 t + \Theta)$$

where $\omega_0$ is assumed known, while $A$ and $\Theta$ are statistically independent random variables, with the PDF of $\Theta$ being uniform in the interval $[0, 2\pi]$. Suppose an LMMSE estimator for $X(t_2)$ is to be constructed based on measurements $X(t_0)$ and $X(t_1)$, i.e., an estimator of the form

$$\widehat{X}_\ell(t_2) = a_0 X(t_0) + a_1 X(t_1) + b$$

that minimizes the mean square error

$$E\left[\left(X(t_2) - \widehat{X}_\ell(t_2)\right)^2\right].$$

(a) Determine the optimum value of $b$.

(b) Set up in detail the specific equations you would need to solve in order to obtain the optimum values of $a_0$ and $a_1$, and use these to compute $a_0$ and $a_1$. Check that your answers take reasonable values for the following two cases: (i) $t_2 - t_1$; (ii) $t_2 = t_0$. To handle these computations cleanly, it may help to recall that the inverse of a $2 \times 2$ matrix of the form

$$\begin{bmatrix} p & q \\ r & s \end{bmatrix}$$

is

$$\frac{1}{ps - qr}\begin{bmatrix} s & -q \\ -r & p \end{bmatrix},$$

a claim that you can directly verify by multiplying the two matrices together.

(c) Show that the MMSE associated with this linear estimator is zero.

**8.11.** Consider a digital communication system in which an independent identically distributed (i.i.d.) bit stream $s[n]$ of 1s and 0s is transmitted over a faulty, memoryless channel with 1s and 0s equally probable. The probability of a 1 being received as a 0 is $1/8$ and the probability of a 0 being received as a 1 is $1/4$. This type of channel is referred to as a memoryless binary channel and is depicted in Figure P8.11-1.

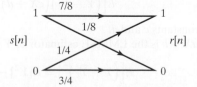

**Figure P8.11-1**

(a) For any time index $n$, determine the joint PMF $P(r,s)$ and the marginal PMF $P(r)$.

**(b)** To obtain an estimate $\widehat{s}[n]$ of $s[n]$ from $r[n]$, the received signal can be pro-cessed through a memoryless, possibly nonlinear system $F$. The memoryless system $F$ in Figure P8.11-2 is to be designed to minimize the mean square error $\varepsilon$ defined as:

$$\varepsilon = E\left\{(s[n] - \widehat{s}[n])^2\right\}.$$

Determine the system $F$.

**(c)** With your system from (b), determine the value $\widehat{s}[n]$ that minimizes

$$E\left\{(s[n] - \widehat{s}[n])^2 \mid r[n] = r\right\}.$$

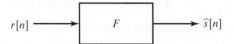

$r[n] \longrightarrow \boxed{\phantom{xx} F \phantom{xx}} \longrightarrow \widehat{s}[n]$

**Figure P8.11-2**

Also determine the probability that at an arbitrary time index $n_0$, the estimate $\widehat{s}[n_0]$ and the true value $s[n_0]$ will be equal.

**8.12.** Consider a communication system in which the random variable $Y$ is transmit-ted through a channel with a random gain $W$, so that the received variable is $X = WY$. Assume that $Y$ and $W$ are independent, and that both of them are uniformly distributed in the range $[1, 2]$.

**(a)** Suppose you are sitting at the receiver and want to estimate the trans-mitted value $Y$ from a measurement of the received value $X$, using the LMMSE estimator $\widehat{Y}_{\ell} = d_1 X + d_2$. Determine what $d_1$ and $d_2$ should be, and compute the associated MMSE.

**(b)** Suppose instead that you are sitting at the transmitter and want to estimate what the received value $X$ will be from a measurement of the transmitted value $Y$. Find the (unconstrained) MMSE estimator $\widehat{X}(Y)$.

## Advanced Problems

**8.13.** Determine if each of the following statements is true or false. For a true state-ment, give a brief but convincing explanation; for a false statement, give a counterexample or convincing explanation.

**(a)** If $\widehat{Y}_{\ell} = aX + b$ is the LMMSE estimator of $Y$ in terms of $X$, then $Y - \widehat{Y}_{\ell}$ is orthogonal to $cX + d$ for any arbitrary constants $c$ and $d$, i.e.,

$$E\left[\left(Y - \widehat{Y}_{\ell}\right)\left(cX + d\right)\right] = 0$$

for any constants $c$ and $d$.

**(b)** If $\widehat{Y}_{\ell} = aX + b$ is the LMMSE estimator of $Y$ in terms of $X$, then

$$E\left[\left(Y - \widehat{Y}_{\ell}\right)^2\right] = E[Y^2] - E\left[\widehat{Y}_{\ell}^2\right].$$

**(c)** Assume $X_1$, $X_2$, and $Y$ are all zero mean. Suppose

- $\widehat{Y}_{1\ell} = a_1 X_1$ is the LMMSE estimator of $Y$ in terms of $X_1$ alone;
- $\widehat{Y}_{2\ell} = a_2 X_2$ is the LMMSE estimator of $Y$ in terms of $X_2$ alone; and

- $\widehat{Y}_\ell = \gamma_1 X_1 + \gamma_2 X_2$, is the LMMSE estimator of $Y$ in terms of $X_1$ and $X_2$.

Then, if $X_1$ and $X_2$ are uncorrelated with each other,

$$\widehat{Y}_\ell = \widehat{Y}_{1\ell} + \widehat{Y}_{2\ell} \,.$$

(In other words, $\gamma_1 = a_1$ and $\gamma_2 = a_2$.)

**(d)** Recall that $\widehat{Y} = E[Y \mid X]$ is the MMSE estimator of $Y$ in terms of $X$. Then $Y - \widehat{Y}$ is orthogonal to $\frac{1}{X^2+1}$.

**(e)** Suppose $X$ and $Y$ have correlation coefficient $\rho = \sigma_{XY}/(\sigma_X \sigma_Y)$. Then

$$\sigma_Y^2(1 - \rho^2) \geq E\left[\left(Y - E[Y \mid X]\right)^2\right].$$

**8.14.** Suppose that $X$ and $Y$ are two random variables for which the LMMSE estimator $\widehat{Y}_\ell = aX + b$ of $Y$ in terms of $X$ is $\widehat{Y}_\ell = 3X$. For each of the following parts, say whether the given statement is true or false. For a true statement, give a brief but convincing explanation; for a false statement, give a counterexample or convincing explanation.

**(a)** It must be the case that $X$ and $Y$ both have mean value $0$.

**(b)** It must be the case that the LMMSE estimator of $X$ in terms of $Y$ is $\widehat{X}_\ell = \frac{1}{3}Y$.

**(c)** It must be the case that $E[Y|X = 2] = 6$.

**(d)** It must be the case that the LMMSE estimator of $Z = 2Y + 7$ in terms of $X$ is $\widehat{Z}_\ell = 6X + 7$.

**8.15.** Suppose $X = aV + bW + c$ and $Y = dV + eW + f$, where $V$ and $W$ are zero mean, unit variance, and uncorrelated with each other, and $a, b, c, d, e,$ and $f$ are some known constants. In this problem, $X$ is a random variable that will be measured and $V$, $W$, and $Y$ are random variables that will be estimated using LMMSE estimators based on measurement of $X$.

**(a)** Find the LMMSE estimator $\widehat{V}_\ell$ of $V$ in terms of $X$, i.e., set $\widehat{V}_\ell = \alpha X + \beta$ and choose the coefficients $\alpha$ and $\beta$ to minimize the mean square error, namely $E[(V - \widehat{V}_\ell)^2]$. Also find this MMSE.

**(b)** Similarly find the LMMSE estimator $\widehat{W}_\ell$ of $W$ in terms of $X$, and find the associated MMSE.

**(c)** You might think that perhaps $X = a\widehat{V}_\ell + b\widehat{W}_\ell + c$, so that the estimates $\widehat{V}_\ell$ and $\widehat{W}_\ell$ are consistent with the fact that $X = aV + bW + c$. Use your results from (a) and (b) to check whether this does indeed turn out to be the case.

**(d)** With similar computations to those you used in (a) and (b), find the LMMSE estimator $\widehat{Y}_\ell$ of $Y$ in terms of $X$, and find the associated mean square error. You might think that $\widehat{Y}_\ell = d\widehat{V}_\ell + e\widehat{W}_\ell + f$, reflecting the fact that $Y = dV + eW + f$. Check whether this is true. *Hint:* To verify that a particular linear function is the LMMSE estimator for a given random variable, you only need to verify that it is unbiased and that the error between the estimator and the given random variable is orthogonal to all the available data.

**8.16.** Let $V$ and $W$ be zero-mean unit-variance random variables that are uncorrelated with each other. Suppose

$$X = V + 4 \,,$$

and

$$Y = -V + 2W - 3 .$$

**(a)** Determine the means, variances, covariance, and correlation coefficient of $X$ and $Y$.

**(b)** Consider the random variables as specified in (a), but with the added constraint that $V$ and $W$ are independent of each other. Suppose it is known after a measurement that $X = 5$. Find the MMSE estimates of $V$, $W$, and $Y$, given this measurement of $X$, and determine the corresponding MMSEs.

**8.17.** Suppose the random variable $Q$ is defined by

$$Q = \begin{cases} 1 & \text{with probability } p \\ 0 & \text{with probability } 1 - p , \end{cases}$$

so the mean of $Q$ is $\mu_Q = p$, and the variance of $Q$ is $\sigma_Q^2 = p(1 - p)$. Now consider random variables $X$ and $Y$ that are two different noise-corrupted observations of $Q$, defined by the following equations:

$$X = Q + V$$

$$Y = Q + W .$$

The noise variables $V$ and $W$ are zero-mean unit-variance bivariate Gaussian random variables with correlation coefficient $r$, i.e.,

$$f_{V,W}(v, w) = \frac{1}{2\pi\sqrt{1 - r^2}} \exp\left\{ -\frac{(v^2 - 2rvw + w^2)}{2(1 - r^2)} \right\}.$$

Take $Q$ to be independent of $V$ and $W$. (Note: All expectations are taken with respect to all of the random variables involved in the expectation.)

**(a)** Compute the following quantities:

(i) $E[VW]$ and $E[QV]$;

(ii) $E[X]$, $E[X^2]$, and $E[XY]$.

**(b)** Suppose $Y$ is to be estimated from $X$ using an estimator $\overline{Y}(X)$ of the form

$$\overline{Y}(X) = \beta X$$

where $\beta$ is a scalar constant. Find an expression for the value of $\beta$ that minimizes the mean square error $E[\{Y - \overline{Y}(X)\}^2]$.

**(c)** Suppose, instead of the estimator in (b), an estimator $\widehat{Y}(X)$ is used whose form is not constrained, but still chosen to minimize the mean square error, i.e., to minimize $E[\{Y - \widehat{Y}(X)\}^2]$. Determine what $\widehat{Y}(X)$ is in the following special cases, without elaborate calculations:

(i) $p = 0$.

(ii) $p = 1$.

**8.18.** Consider a communication system for which the channel gain $g[n]$ is time-varying, as shown in Figure P8.18-1.

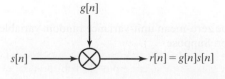

**Figure P8.18-1**

- The transmitted signal $s[n]$ is a random bit stream, modeled as an i.i.d. Bernoulli random process with probability $p$ of being +1 and probability $1 - p$ of being –1 at each time instant.
- The channel gain $g[n]$ is an i.i.d. process, uniformly distributed between 0 and 1, at each instant i.e.,

$$f_G(g[n]) = \begin{cases} 1 & 0 \le g[n] \le 1 \\ 0 & \text{otherwise.} \end{cases} \quad \text{for each } n$$

- $s[n]$ and $g[n]$ are statistically independent.

**(a)** Determine and make a fully labeled sketch of the PDF $f_R(r[n])$.

**(b)** Determine $E\{r[n]\}$ and $E\{r^2[n]\}$.

For parts (c) and (d) only, consider processing $r[n]$ with a memoryless and possibly nonlinear system $F$ to obtain an estimate $\widehat{s}[n]$ of $s[n]$, as indicated in Figure P8.18-2.

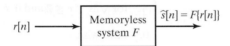

**Figure P8.18-2**

The mean square error $\mathcal{E}$ of the estimator is defined as

$$\mathcal{E} - E\{(s[n] - \widehat{s}[n])^2\} \, .$$

**(c)** For this part only, let $p = \frac{1}{2}$. If $F\{r[n]\}$ is restricted to be of the form

$$F\{r[n]\} = \widehat{s}[n] = a_0 + a_1 r[n]$$

with $a_0$ and $a_1$ being some constants, determine $a_0$ and $a_1$ to minimize the mean square error $\mathcal{E}$ and determine the corresponding $\mathcal{E}$.

**(d)** If there is no restriction on the form of the system $F$ other than it being memoryless, determine $F$ to minimize the mean square error $\mathcal{E}$ and determine the corresponding value of $\mathcal{E}$.

**8.19.** Suppose a random variable $X$ is related to the random variables $G$, $Y$, and $W$ as follows:

$$X = GY + W \, .$$

Here $G$ has mean $\mu_G$ and variance $\sigma_G^2$, and is independent of $Y$ and $W$. Both $Y$ and $W$ have zero mean, and are uncorrelated with each other; their respective variances are $\sigma_Y^2$ and $\sigma_W^2$.

**(a)** Compute the following, in terms of the given quantities:

$$E[G^2], \quad E[X], \quad E[X^2], \quad \sigma_X^2, \quad E[YX], \quad \sigma_{YX} \, .$$

**(b)** Compute the LMMSE estimator of $Y$ given $X$, i.e., determine the constants $a$ and $b$ in $\widehat{y}_\ell(X) = aX + b$ so as to minimize

$$E[\{\widehat{y}_\ell(X) - Y)\}^2] \, .$$

Write down what your expression for $\widehat{y}_\ell(X)$ reduces to in the particular case where $\sigma_W^2 = 0$, and for this particular case find an expression for the corresponding MMSE, $E[\{\widehat{y}_\ell(X) - Y\}^2]$. Do the expressions for the estimator and the MMSE in this particular case further reduce in a reasonable way when you assume (in addition to $\sigma_W^2 = 0$) that $\sigma_G^2 = 0$? Explain.

**8.20.** A signal $s[n]$ to be retrieved from storage is subject to errors due to faulty electronics. The retrieved signal $r[n]$ can be written as

$$r[n] = s[n] + e[n] \,,$$

where $e[n]$ represents the error. Both $s[n]$ and $e[n]$ are i.i.d. random processes. The joint PDF of $r[n]$ and $s[n]$ is shown in Figure P8.20-1.

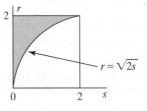

**Figure P8.20-1**

$$f_{R,S}(r,s) = \begin{cases} \frac{3}{4}, & \text{for } \sqrt{2s} \le r \le 2 \text{ and } 0 \le s \le 2 \\ 0, & \text{otherwise.} \end{cases}$$

For the remainder of this problem, you may find some, none, or all of the following useful:

$$f_S(s) = \tfrac{3}{4}(2 - \sqrt{2s}) \quad \text{for } 0 \le s \le 2.$$

$$E\{S \mid R = r\} = r^2/4, \quad E\{S\} = \tfrac{3}{5}, \quad E\{R\} = \tfrac{3}{2},$$

$$E\{RS\} = 1, \quad E\{R^2\} = \tfrac{12}{5}, \quad E\{S^2\} = \tfrac{4}{7}.$$

**(a)** Show that $f_R(r) = \tfrac{3}{8}r^2$ for $0 \le r \le 2$, and 0 elsewhere.

**(b)** Determine $E\{R \mid S = s\}$.

An estimate $\widehat{s}[n]$ of $s[n]$ can be obtained from $r[n]$ as indicated in Figure P8.20-2.

**Figure P8.20-2**

The mean square error $\mathcal{E}$ of the estimator is defined as

$$\mathcal{E} = E\{(s[n] - \widehat{s}[n])^2\} \,.$$

**(c)** Determine the memoryless system $A$ that minimizes the mean square error $\mathcal{E}$.

**(d)** In this part, the output of system $A$ is restricted to be of the form

$$\widehat{s}[n] = a_0 + a_1 r[n] \,,$$

where $a_0$ and $a_1$ are constants. Determine values $a_0$ and $a_1$ that minimize the mean square error $\mathcal{E}$.

**(e)** In this part, the output of system $A$ is of the form

$$\widehat{s}[n] = c r[n - 1]$$

where $c$ is a constant. Determine the value of $c$ that minimizes the mean square error $\mathcal{E}$.

**8.21.** Let $X$ be a noisy measurement of the angular position $Y$ of an antenna: $X = Y + W$, where $W$ denotes the additive noise. Treat $Y$ and $W$ as independent random variables. Suppose $Y$ is uniformly distributed in the interval $[-1, 1]$, and $W$ has the triangular PDF $f_W(w) = [1 - (|w|/2)]/2$ for $|w| \leq 2$ (and $f_W(w) = 0$ elsewhere). Given that $X = 1$, what is the MMSE estimate of $Y$, namely $\widehat{Y}_{\text{MMSE}}(1)$, and what is the corresponding MMSE?

**8.22.** Suppose $X$ and $Y$ are two random variables. The estimator $\widehat{Y}$ of $Y$ from observing $X$ is specified to be of the form $\widehat{Y} = aX^2$, where $a$ is a constant chosen to minimize $E_{X,Y}\{(Y - \widehat{Y})^2\}$. Specify whether with this choice of $a$ it is (necessarily) true or (can be) false that

$$E_{X,Y}\{(Y - \widehat{Y}) \cdot X\} = 0.$$

Explain your reasoning.

**8.23.** Two random variables $X$ and $Y$ have a joint PDF $f_{X,Y}(x, y)$. Assume that their joint moments of all orders are available. The random variable $Y$ can be estimated from the random variable $X$ by the following quadratic estimator:

$$\widehat{y}_q(X) = aX^2.$$

(a) Determine the parameter $a$ so that the quadratic estimator $\widehat{y}_q(X)$ minimizes the mean square error

$$\varepsilon = E_{X,Y}\left\{\left[Y - aX^2\right]^2\right\}.$$

(b) The error of the optimal quadratic estimator is

$$e = Y - \widehat{y}_q(X).$$

Is the error $e$ orthogonal to the measurement $X$?

(c) Is the optimal quadratic estimator $\widehat{y}_q(X)$ unbiased? If yes, justify. If no, determine the bias, i.e., $E_{X,Y}\{e\}$.

**8.24.** A certain coin has a probability $R$ of coming up heads (H), and a probability $1 - R$ of coming up tails (T). However, suppose this probability is completely unknown, so $R$ is modeled as a uniformly distributed random variable in the interval $[0, 1]$, i.e., $f_R(r) = 1$ in this interval. The goal now is to estimate $R$ in some reasonable way by observing the relative number of H's in a series of independent tosses of the coin.

Suppose 11 independent tosses of the coin are made, with $R$ staying fixed from toss to toss, and let $S$ denote the resulting sequence of 11 H's and T's. In the experiment that the calculations will be based on here, the particular sequence turned out to be $s = \text{HHTHTTTHHTH}$ (which happens to have 6 H's and 5 T's).

(a) Determine the conditional probability $P(S = s \mid R = r)$ of observing the particular sequence $s$ above, given $R = r$, i.e., given that the probability of getting H on a toss is $r$.

(b) Determine $P(S = s)$, the probability of getting the particular sequence $s$ above, no longer conditioned on any particular value of $R$. It may help you to know the following for nonnegative integers $m$ and $k$:

$$\int_0^1 p^m (1 - p)^k \, dp = \frac{m! \, k!}{(m + k + 1)!}.$$

(c) Determine the conditional PDF of $R$, given the particular sequence $s$ above, i.e., find the function $f_{R|S}(r\,|\,s)$.

(d) Given the sequence $s$ above, find the unconstrained estimate $\widehat{R}$ of $R$ that minimizes the conditional mean square error: $E[(\widehat{R} - R)^2\,|\,S = s]$.

## Extension Problems

**8.25.** (a) Consider a pair of bivariate Gaussian random variables, $X$ and $Y$. The MMSE estimator for $Y$ in terms of $X$ is $\widehat{y}(X) = 5$. Are $X$ and $Y$ correlated? Explain.

(b) Consider a pair of random variables, $X$ and $Y$, both with unit variance and zero mean. The LMMSE estimator $\widehat{X}_\ell$ of $X$ given $Y$ is $\widehat{X}_\ell = \frac{2}{3}Y$. Can the LMMSE estimator of $Y$ given $X$ be $\widehat{Y}_\ell = \frac{3}{2}X$? Clearly explain why or why not.

(c) If $X$ is a random variable and $Y = X^2$, then can $X$ and $Y$ be uncorrelated random variables?

(d) Let $Y$ be uniform on the interval $[-1, 1]$, let $U$ equal 1 or $-1$ with probability 0.5 each, independent of $Y$, and let $X = UY$. Can the LMMSE estimator of $Y$ based on observation of $X$ be $\widehat{y}_\ell(X) = X$?

(e) If $X$ and $Y$ are uncorrelated random variables, is the LMMSE estimator of $Y$ in terms of $X$ always $\widehat{Y}_\ell = \mu_Y$, the expected value of $Y$?

(f) If $X$ and $Y$ are uncorrelated random variables, is the MMSE estimate of $Y$, given $X = x$, always $\widehat{y}(x) = \mu_Y$, the expected value of $Y$?

**8.26.** The input to a communication channel at some instant of time is a uniformly distributed random variable $Q$ with mean value $\mu_Q$ and variance $\sigma_Q^2$. Suppose the corresponding channel output is $X = Q + W$, where the additive disturbance $W$ is a zero-mean uniformly distributed random variable of variance $\sigma_W^2$, and is independent of $Q$.

(a) We wish to construct a linear estimator of the form $\widehat{q}_\ell(X) = aX + b$ for the random variable $Q$. What choice of $a$ and $b$ minimizes the mean square error of the estimator, $E\left\{(Q - \widehat{q}_\ell(X))^2\right\}$? For this choice of $a$ and $b$, what is the mean square error? Show that the estimator is unbiased, i.e., that $E\left(Q - \widehat{q}_\ell(X)\right) = 0$, and that the estimation error is orthogonal to $X$ (which is an illustration of the orthogonality principle that characterizes linear estimators).

(b) For this part, you are asked to verify experimentally the value for the mean square error obtained in part (a) above, using an appropriate computational package available to you. Assume $\mu_Q = 5$, $\sigma_Q^2 = \frac{64}{3}$, and $\sigma_W^2 = \frac{4}{3}$.

   (i) First, determine the coefficients $a$ and $b$, given the information above.

   (ii) Now, generate 100 sample values of $Q$ and $W$.

   (iii) Then calculate the corresponding values for $X$.

   (iv) Using the coefficients of the estimator, $a$ and $b$ as determined above, calculate the LMMSE estimate of $Q$ for each measured $X$.

   (v) Now compute the empirical mean square error. (The law of large numbers is invoked here to approximate an ensemble mean or probabilistic mean by a sample mean or empirical mean.)

(vi) Repeat this procedure a few times (by generating new sets of values for $Q$ and $W$) and verify that the values for the empirical MSE are close to the mean square error you found in (a), evaluated for these values of $a$ and $b$.

**8.27.** Suppose the random variables $X$ and $Y$ are distributed according to a joint density—known as Morgenstern's bivariate density—that is given by the following expression over the unit square $0 \le x, y \le 1$ of the $(x, y)$-plane, and is 0 everywhere else:

$$f_{X,Y}(x,y) = 1 + \alpha(1 - 2x)(1 - 2y) . \tag{8.99}$$

Plots of this bivariate density are shown in Figure P8.27 for $\alpha = 1, 0, -1$.

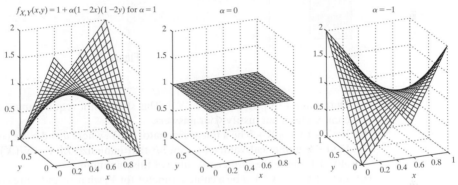

**Figure P8.27**

(a) Determine the marginal density of $X$, namely $f_X(x)$. The marginal density of $Y$ follows simply by symmetry. For what choices of $\alpha$ are $X$ and $Y$ independent?

(b) Determine $\mu_X, \mu_Y, \sigma_X^2$, and $\sigma_Y^2$.

(c) Write down the integral that defines $E[XY]$ for this problem. Evaluating this integral will show that

$$E[XY] = \frac{1}{4} + \frac{\alpha}{36} .$$

Compute $\sigma_{XY}$ and $\rho_{XY}$.

(d) Determine the LMMSE estimator of $Y$ given $X$, and the associated mean square error.

(e) Determine the MMSE estimator, and state whether it performs any better than the LMMSE estimator in this case.

**8.28.** The random variables $X$ and $Y$ have a joint PDF $f_{X,Y}(x,y)$ that is uniformly distributed over the shaded region shown in Figure P8.28, and zero elsewhere. Certain moments of this distribution are as listed here:

$$E(X^2) = 20/3, \quad E(X^6) = 5440/7,$$

$$E(Y^2) = 7, \quad E(XY) = 6, \quad E(X^3Y) = 60.$$

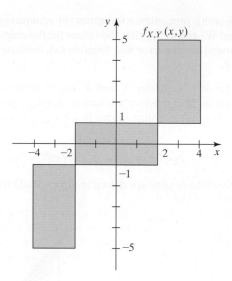

**Figure P8.28**

(a) Determine $E[X]$, $E[Y]$, and $E[X^3]$. This should not require any significant computation, but justify your answers.

(b) Let $\widehat{y}_\ell(X) = aX + b$ be a linear estimator for $Y$, where $a$ and $b$ are constants. Find the numerical values of $a$ and $b$ that minimize $E[(Y - \widehat{y}_\ell(X))^2]$, and also determine the numerical value of this minimum.

(c) Let $\widehat{y}_c(X) = cX^3 + d$ be a cubic estimator for $Y$, where $c$ and $d$ are constants. Find the numerical values of $c$ and $d$ that minimize $E[(Y - \widehat{y}_c(X))^2]$, and write down an expression for this minimum value in terms of the appropriate moments of $X$ and $Y$. Is the cubic estimator better or worse than the linear estimator from (b) in this particular instance?

**8.29.** Consider two random variables $X$ and $Y$ whose joint PDF is constant in the shaded region shown in Figure P8.29, and is zero outside the shaded region.

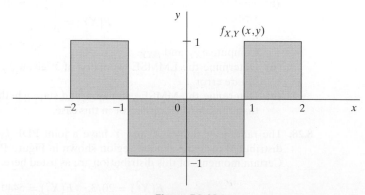

**Figure P8.29**

The following statistics computed from the density function are provided:

$$E[X] = 0 \qquad E[Y] = 0$$
$$E[X^2] = \tfrac{4}{3} \qquad E[Y^2] = \tfrac{1}{3} \qquad E[XY] = 0$$
$$E[X^3] = 0 \qquad E[Y^3] = 0 \qquad E[X^2Y] = \tfrac{1}{2} \qquad E[XY^2] = 0$$
$$E[X^4] = \tfrac{16}{5} \qquad E[Y^4] = \tfrac{1}{5}$$

(a) Are $X$ and $Y$ uncorrelated? Are $X$ and $Y$ independent? Explain.

(b) Give the expression for the LMMSE estimator $\hat{y}_\ell(X)$, and calculate the mean square error for your result, $E[(\hat{y}_\ell(X) - Y)^2]$.

(c) In this part, examine the quadratic estimator $\hat{y}_q(X) = a + bX^2$. Determine the values of $a$ and $b$ that minimize the mean square error, $E[(\hat{y}_q(X) - Y)^2]$. For the values of $a$ and $b$ you just determined, calculate the mean square error.

(d) Sketch and label graphs of the linear and quadratic estimators of parts (b) and (c) respectively, each drawn as a function of the value $x$ of $X$. Also sketch and label the MMSE estimator $\hat{y}(X)$ on the same graph. Explain how you arrive at the MMSE estimator, and calculate the mean square error for this estimator. How does the mean square error of the MMSE estimator compare to that of the linear and quadratic estimators of parts (b) and (c) above?

**8.30.** Let $y[\cdot]$ denote the output of a system driven for all time by a random input signal $w[\cdot]$, and suppose the output at time $n$ is given by

$$y[n] = w[n] + w[n-1]$$

for all $n$. The input $w[k]$ at any time $k$ has a mean value $\mu_w = 0$ and variance $\sigma_w^2$, neither of which varies with $k$. Also, the input at time $k$ is uncorrelated with the input values at all other times. We know nothing else about $w[\cdot]$, and in particular we do not have any measurements of this input signal.

(a) Determine the mean value $\mu_y$ of $y[n]$ and its variance $\sigma_y^2$ (which will not vary with $n$), expressing the latter in terms of $\sigma_w^2$. Also determine the covariance between $y[n+1]$ and $y[n]$, and the covariance between $y[n+1]$ and $y[n-1]$, again expressing your answers in terms of $\sigma_w^2$.

(b) Find the LMMSE estimator of $y[n+1]$ that uses measurements of $y[n]$, written as

$$\hat{y}_1[n+1] = ay[n] + b$$

for optimally chosen $a$ and $b$. Determine $a$ and $b$, and also the resulting mean squared error $E[(y[n+1] - \hat{y}_1[n+1])^2]$.

(c) Find the LMMSE estimator of $y[n+1]$ that uses measurements of $y[n-1]$, written as

$$\hat{y}_2[n+1] = cy[n-1] + d$$

for optimally chosen $c$ and $d$. Determine $c$ and $d$, and also the resulting mean square error.

(d) Determine whether $y[n+1] - \hat{y}_1[n+1]$ is orthogonal to $y[n-1]$, that is, determine if the residual error of the one-step predictor is orthogonal to the value two steps back.

(e) Draw a careful sketch that depicts the mutual relationships that you have exposed in this problem among the three random variables $y[n+1]$, $y[n]$, $y[n-1]$, regarded as vectors (see Section 7.8). Verify that your sketch is consistent with your answers in (a)–(d). Also use it to explain why the LMMSE estimator that uses measurements of both $y[n]$ and $y[n-1]$ will do better than estimators using measurements of $y[n]$ alone or $y[n-1]$ alone.

(f) Find the LMMSE estimator of $y[n+1]$ using measurements of both $y[n]$ and $y[n-1]$, written as

$$\widehat{y}_{12}[n+1] = ey[n] + fy[n-1] + g$$

for optimally chosen $e, f$, and $g$. Determine $e, f$, and $g$, and also the resulting mean square error. Verify that the mean square error is lower than the values you obtained in (b) and (c).

**8.31.** This problem represents the task of estimating what one might call the service rate $R$ of a randomly chosen server at the College Food Court, from measurements of the durations $X_1, X_2, \cdots, X_n$ of his or her service to $n$ customers (who may or may not be successive customers—it doesn't matter for this problem). Thus $X_k$ is measured from the time the $k$th customer places an order with the server to the time that the order is delivered by the server.

Suppose it is believed (on the basis of extensive prior observations of large numbers of servers) that $R$ is well modeled as an exponentially distributed random variable with PDF given by

$$f_R(r) = be^{-br}\, u(r)$$

where the parameter $b$ is assumed known. Here $u(\cdot)$ denotes the usual unit step function (taking the value 0 for negative arguments and the value 1 otherwise).

Assume the random variables $X_1, X_2, \cdots, X_n$, when conditioned on $R = r$ (i.e., assuming the selected server has service rate $r$), are independent and identically distributed, each being independently chosen from the following exponential PDF with parameter $r$:

$$f_{X_i|R}(x\,|\,r) = r e^{-rx}\, u(x), \quad i = 1, 2, \ldots, n\,.$$

You will find the following integral (with appropriate choices of $\alpha$, $y$, and $k$) to be of repeated help in solving this problem:

$$\int_0^\infty \alpha^k e^{-\alpha y}\, d\alpha = \frac{k!}{y^{k+1}}, \quad y > 0\,.$$

(a) What is $E[R]$?

(b) What is $E[X_i|R = r]$, the conditional expectation of $X_i$ given $R = r$?

(c) What is the joint PDF of $R, X_i$? And what is the joint PDF of $R, X_1, X_2, \cdots, X_n$?

(d) What is the PDF of $X_i$, and what is the joint PDF of $X_1, X_2, \cdots, X_n$? Are all the $X_i$ mutually independent (when they are not conditioned on $R = r$)?

(e) What is the conditional density of $R$, given $X_1 = x_1, X_2 = x_2, \cdots, X_n = x_n$?

(f) What is the MMSE estimate of $R$, given $X_1 = x_1, X_2 = x_2, \cdots, X_n = x_n$? Verify that for $n = 0$ your answer simplifies to the expression required by your answer to (a). Also show that your answer simplifies for large $n$ to an expression that appears reasonable in view of your answer to (b).

# 9 Hypothesis Testing

The topic of hypothesis testing arises in many contexts in signal processing and communications, as well as in medicine, statistics, and other settings in which a choice among multiple explanations or hypotheses is made on the basis of limited and noisy data. For example, from tests on such data, we may need to determine whether a person has a particular disease; whether a particular radar return indicates the presence of an aircraft; which of four values was transmitted at a given time in a pulse-amplitude modulation (PAM) system; and so on. Hypothesis testing provides a framework for selecting among $M$ possible explanations for the available data in some principled or optimal way.

## 9.1 BINARY PULSE-AMPLITUDE MODULATION IN NOISE

As a prelude to the broader discussion in later sections of this chapter, we first outline an example of how hypothesis testing naturally arises in the context of pulse-amplitude detection in noise. Chapter 3 introduced the basic principles of PAM, and considered the effects of pulse rate, pulse shape, and channel and receiver filtering in PAM systems. We also developed and discussed the condition for no intersymbol interference (the no-ISI condition). Under the assumption of no ISI, we want to now examine the effects of noise in the channel. Toward this end, we consider the overall PAM model in Figure 9.1, with the channel noise $v(t)$ represented as an additive term.

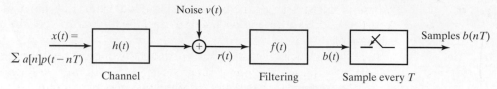

**Figure 9.1**   Overall model of a PAM system.

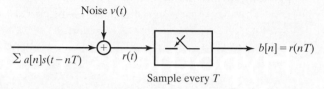

**Figure 9.2**   Simplified representation of a PAM system.

For the present we will assume no postfiltering at the receiver, so $f(t) = \delta(t)$. In Chapter 13 we will see how performance is improved with the use of filtering in the receiver. The pulse $p(t)$ going through the channel with impulse response $h(t)$ produces the signal $s(t) = p(t) * h(t)$ at the channel output. Figure 9.1 thus reduces to the overall system shown in Figure 9.2.

Since we are assuming no ISI, the discussion can focus on a single pulse index $n$, chosen as $n = 0$ for convenience. From Figure 9.2,

$$b[0] = r(0) = a[0]s(0) + v(0) \,. \tag{9.1}$$

Denoting $r(0)$, $a[0]$, and $v(0)$ simply as $r$, $a$, and $v$ respectively, and setting $s(0) = 1$ without loss of generality, Eq. (9.1) becomes

$$r = a + v \,. \tag{9.2}$$

Our broad objective is to determine the value of $a$ as accurately as possible, given the measured value $r$.

There are several variations of this problem, depending on the nature of the transmitted sequence $a[n]$ and the characteristics of the noise. The amplitude $a[n]$ may span a continuous range or it may be discrete, as would be the case, for example, with data represented by a binary code. The amplitude may correspondingly be modeled as a random variable $A$ with a known probability density function (PDF) or probability mass function (PMF); then $a$ is the specific value that $A$ takes in a particular outcome or instance of the probabilistic model. The contribution of the noise also is typically represented as a random variable $V$, usually continuous, with $v$ being the specific value that it takes. We may thus model the quantity $r$ at the receiver as the observation of a random variable $R$, with

$$R = A + V \,. \tag{9.3}$$

We want to then estimate the value of the random variable $A$, given that $R = r$. A further processing step thus needs to be added to our receiver to obtain an estimate of $A$.

In the case of binary signaling, for which the pulse amplitude can be only one of two values, finding an estimate of $A$ reduces to deciding, on the basis of the observed value $r$ of $R$, which of the two possible amplitudes was transmitted. Two common forms of binary signaling in PAM systems are on-off signaling and antipodal signaling. Letting $a_1$ and $a_0$ denote the two possible amplitudes (representing for example a binary "one" or "zero"), in on-off signaling we have $a_0 = 0, a_1 \neq 0$, whereas in antipodal signaling $a_0 = -a_1 \neq 0$.

Thus, in binary signaling, the required post-processing corresponds to deciding between two alternative hypotheses, where the available information may include some prior information along with a measurement $r$ of the single continuous random variable $R$. The hypotheses $H$ are

- $H_0$: the transmitted amplitude $A$ takes the value $a_0$, so $R = a_0 + V$.
- $H_1$: the transmitted amplitude $A$ takes the value $a_1$, so $R = a_1 + V$.

The task is then to decide, given the measurement $R = r$, whether $H_0$ or $H_1$ is responsible for the measurement. Section 9.2 develops a framework for this sort of hypothesis testing or classification task. The extension from two hypotheses to multiple hypotheses and multiple measurements will be straightforward, once the two-hypothesis case is understood.

## 9.2 HYPOTHESIS TESTING WITH MINIMUM ERROR PROBABILITY

We begin with choosing optimally between two hypotheses, and then extend to the case of more than two hypotheses. The general binary hypothesis testing task is to decide, on the basis of a measurement $r$ of a random variable $R$, which of two hypotheses, $H = H_0$ or $H = H_1$, is responsible for the measurement. We shall indicate these decisions by '$H_0$' and '$H_1$' respectively, where the single quotation marks are intended to suggest the announcement of a decision. An alternative common notation is $\widehat{H} = H_0$ and $\widehat{H} = H_1$ respectively, where $\widehat{H}$ denotes the inferred value of the hypothesis $H$. The measurement may be a continuous or discrete random variable.

Suppose $H$ is modeled as a random quantity, and assume we know the *a priori* or prior probabilities

$$P(H_0 \text{ is true}) = P(H = H_0) = P(H_0) = p_0 \qquad (9.4)$$

and

$$P(H_1 \text{ is true}) = P(H = H_1) = P(H_1) = p_1 \,, \qquad (9.5)$$

where the last two equalities in each case simply define streamlined notation that we will be using. When the measured quantity is a continuous random variable, we shall require the conditional densities $f_{R|H}(r|H_0)$ and $f_{R|H}(r|H_1)$ that tell us how the measured variable is distributed under the two respective hypotheses. These conditional densities in effect constitute the relevant model or specification of how the measured data relates to the two hypotheses. For example, in the PAM setting, with $R$ defined as in Eq. (9.3) and assuming $V$ is independent of $A$ under each hypothesis, these conditional densities are simply

$$f_{R|H}(r|H_0) = f_V(r - a_0) \quad \text{and} \quad f_{R|H}(r|H_1) = f_V(r - a_1) \,. \tag{9.6}$$

If $R$ is a discrete random variable, then conditional PMFs are used instead of conditional densities.

It is natural in many settings, as in the case of digital communication using PAM, to want to minimize the probability of picking the wrong hypothesis, that is, to decide with minimum probability of error between the hypotheses, given the measurement $R = r$. Our initial discussion of hypothesis testing focuses on this criterion of minimum probability of error.

### 9.2.1 Deciding with Minimum Conditional Probability of Error

Consider first how one would decide between $H_0$ and $H_1$ with minimum probability of error in the absence of any measurement of $R$. With the choice '$H_0$', we make an error precisely when $H_0$ does not hold, so the probability of error with this choice is $1 - P(H_0) = 1 - p_0$. Similarly, with the choice '$H_1$', the probability of error is $1 - P(H_1) = 1 - p_1$. Thus, for minimum probability of error, we should decide in favor of whichever hypothesis has maximum probability—an intuitively reasonable conclusion. The preceding reasoning extends in the same way to choosing one from among many hypotheses, and leads to the same conclusion.

The same reasoning also applies when the objective is to decide between $H_0$ and $H_1$ with minimum probability of error, knowing that $R = r$. However, in this case all probabilities now need to be conditioned on the measurement $R = r$. The conclusion is that to minimize the conditional probability of error, $P(\text{error}|R = r)$, we need to decide in favor of whichever hypothesis has maximum conditional probability, conditioned on the measurement $R = r$. Thus if $P(H_1|R = r) > P(H_0|R = r)$, we decide '$H_1$', and if $P(H_1|R = r) < P(H_0|R = r)$, we decide '$H_0$'. This choice may be compactly written as

$$P(H_1|R = r) \underset{\substack{< \\ \text{'}H_0\text{'}}}{\overset{\substack{\text{'}H_1\text{'} \\ >}}{}} P(H_0|R = r) \,. \tag{9.7}$$

The corresponding conditional probability of error is

$$P(\text{error}|R = r) = \min\{1 - P(H_0|R = r), 1 - P(H_1|R = r)\} \,. \tag{9.8}$$

If the two conditional probabilities happen to be equal, the same conditional probability of error is obtained whether we choose '$H_0$' or '$H_1$', so the choice is arbitrary. If there were several random variables for which we had measurements, rather than just the single random variable $R$, we would simply condition on all the available measurements in the preceding expressions.

The conditional probabilities $P(H_0|R = r)$ and $P(H_1|R = r)$ that appear in Eq. (9.7) are referred to as the *a posteriori*, or posterior, probabilities of the hypotheses, to distinguish them from the *a priori*, or prior, probabilities, $P(H_0)$ and $P(H_1)$. The decision generated by Eq. (9.7) is accordingly referred to as the maximum *a posteriori* probability decision, usually abbreviated as the MAP decision. This MAP decision minimizes the conditional probability of error, given the measurement.

To evaluate the posterior probabilities in Eq. (9.7), we use Bayes' rule to rewrite them in terms of known quantities, so the optimal decision is determined from the comparison

$$\frac{p_1 f_{R|H}(r|H_1)}{f_R(r)} \underset{\substack{< \\ `H_0'}}{\overset{\substack{`H_1' \\ >}}{}} \frac{p_0 f_{R|H}(r|H_0)}{f_R(r)} , \qquad (9.9)$$

under the reasonable assumption that $f_R(r) > 0$, that is, that the PDF of $R$ is positive at the value $r$ that was actually measured. Since the denominator is the same and positive on both sides of the above expression, the comparison can be further simplified to

$$p_1 f_{R|H}(r|H_1) \underset{\substack{< \\ `H_0'}}{\overset{\substack{`H_1' \\ >}}{}} p_0 f_{R|H}(r|H_0) . \qquad (9.10)$$

### 9.2.2 MAP Decision Rule for Minimum Overall Probability of Error

A decision rule specifies a decision for each possible measurement $r$ that might be obtained. The probability of error $P_e$ associated with such a rule is obtained by averaging the conditional probability of error over all possible measurements $r$, so

$$P_e = \int_{-\infty}^{\infty} P(\text{error}|R = r) f_R(r) \, dr . \qquad (9.11)$$

The fact that $f_R(r) \geq 0$ ensures that $P_e$ is minimized by minimizing $P(\text{error}|R = r)$ for each $r$. The decision rule for minimum $P_e$ is therefore simply given by Eqs. (9.7) or (9.10), applied for each $r$. The expression in Eq. (9.11) also shows that $P_e$ is unaffected by the choice of hypothesis in the case of $f_R(r) = 0$.

The form of the MAP decision rule in Eq. (9.10) is easily visualized and implemented. The prior probabilities $p_i = P(H_i)$ are used to scale the PDFs $f_{R|H}(r|H_i)$ that describe how the measured quantity $R$ is distributed under each of the hypotheses. The optimal decision rule then decides in favor of the hypothesis associated with whichever scaled PDF is largest at the measured value $r$.

The MAP decision rule induces a partitioning of the measurement space, in this case the $r$ axis, into decision regions. The region $\mathcal{D}_1$ comprises those values of $r$ for which the decision rule chooses '$H_1$', and the region $\mathcal{D}_0$ comprises those values of $r$ for which the decision rule chooses '$H_0$'.

The preceding description also applies to choosing with minimum probability of error among multiple hypotheses, rather than just two, and given measurements of several associated random variables, rather than just one. The reasoning is identical.

---

**Example 9.1**    **MAP Rule for On-Off Signaling in Uniform Noise**

Consider a random variable $S$ that can take one of the two values 0 and 1, with corresponding *a priori* probabilities $p_0 = \frac{1}{4}$ and $p_1 = \frac{3}{4}$. This may correspond to the two possible sampled values $a_0$ and $a_1$ at the receiver in an on-off PAM signaling system in the noise-free case. In the presence of additive noise, the noisy received observation is $R = S + N$. Assume $N$ is independent of $S$ and is uniformly distributed in amplitude between $-2$ and $+2$, as indicated in Figure 9.3. Under the two hypotheses $H_0$ and $H_1$, the received value $R$ is

$$H_0: \ R = 0 + N \tag{9.12}$$

$$H_1: \ R = 1 + N . \tag{9.13}$$

For minimum probability of error in deciding between $H_0$ and $H_1$ given an observation of $R$, we use the MAP rule in Eq. (9.7), implemented in the form of Eq. (9.10). The left and right sides of Eq. (9.10) are shown in Figure 9.4. Consequently, the MAP rule specifies the following decisions:

$$\text{decide \ '}H_0\text{' \ for} \quad -2 < r < -1 , \tag{9.14}$$

$$\text{decide \ '}H_1\text{' \ for} \quad -1 < r < 3 . \tag{9.15}$$

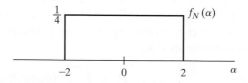

**Figure 9.3**    The PDF of the noise $N$.

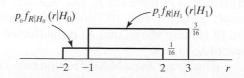

**Figure 9.4**    The scaled conditional probabilities.

Example 9.2 illustrates the use of the MAP rule with more than two hypotheses, and given a measurement of a discrete random variable rather than a continuous one.

| **Example 9.2** | **Use of the MAP Rule with Three Hypotheses** |
|---|---|

In Example 8.1 we considered minimum mean square error (MMSE) estimation of a discrete random variable $S$ from a noisy measurement $R$, where $S$ and $R$ have the joint PMF shown in Figure 8.1 and reproduced in Figure 9.5. Here $(S, R)$ takes the values $(1, 1), (0, 0), (-1, -1)$ with probabilities 0.2 each, and takes the values $(1, 0), (-1, 0), (0, 1), (0, -1)$ with probabilities 0.1 each.

Even though $S$ can only have one of the three discrete values $0$, $+1$, or $-1$, the MMSE criterion in Example 8.1 does not account for this constraint on $S$. The MMSE estimate when $R = 1$ is the conditional mean $\widehat{s} = E[S|R = 1] = \frac{2}{3}$. This estimate minimizes the mean square error but not the conditional probability of error. Given $R = 1$, the conditional probability of error with the choice $\widehat{s} = \frac{2}{3}$ is 1. On the other hand, choosing $\widehat{s} = 0$ leads to a conditional probability of error of $\frac{2}{3}$ because $S = 0$ with probability $\frac{1}{3}$, given that $R = 1$. Choosing $\widehat{s} = 1$ results in the smallest conditional probability of error $\frac{1}{3}$, because $S = 1$ with probability $\frac{2}{3}$, given that $R = 1$.

We now extend this analysis, using the notation of hypothesis testing and invoking the MAP rule to minimize the overall probability of error. The three hypotheses in this case are

$$H_{-1}: S = -1 \tag{9.16}$$

$$H_0: S = 0 \tag{9.17}$$

$$H_1: S = +1 . \tag{9.18}$$

For the MAP rule we determine the three conditional probabilities of these hypotheses, given $R = r$, and choose the hypothesis that corresponds to the maximum conditional (or *a posteriori*) probability. For example, if $P(H_{-1}|R = r)$ is greater than $P(H_0|R = r)$ and $P(H_1|R = r)$, then the optimal decision for $R = r$ is '$H_{-1}$' or equivalently $\widehat{H} = H_{-1}$. This strategy is the extension of Eq. (9.7) for the present ternary hypothesis example.

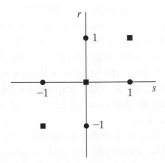

**Figure 9.5**   Joint PMF of $S$ and $R$. The probability associated with the outcome represented by each square is 0.2 and by each circle is 0.1.

We can also implement the preceding MAP rule in the form of Eq. (9.8) by utilizing Bayes' rule to express $P(H_i|R = r)$ as

$$P(H_i|R=r) = \frac{P(S=i)P(R=r|S=i)}{P(R=r)} \qquad i = 0, +1, -1 . \tag{9.19}$$

In comparing the *a posteriori* probabilities for a given $r$, the probability $P(R=r)$ appears as a common scale factor in each expression, so it suffices to compare the three terms $P(S=i)P(R=r|S=i)$ for $i = -1, 0, +1$ and choose the hypothesis $H_i$ corresponding to the maximum of these three terms. This is the extension of Eq. (9.10) for this ternary example.

For this particular case, since Figure 9.5 specifies the joint probabilities, it is easiest to recognize that

$$P(S=i)P(R=r|S=i) = P(R=r, S=i) \qquad i = 0, +1, -1 . \tag{9.20}$$

The probability of error is thus minimized by the following decision rule:

$$\text{If: } r = +1 \quad \text{decide} \quad S = 1, \tag{9.21}$$

$$r = 0 \quad \text{decide} \quad S = 0, \tag{9.22}$$

$$r = -1 \quad \text{decide} \quad S = -1 . \tag{9.23}$$

### 9.2.3 Hypothesis Testing in Coded Digital Communication

The discussion of PAM earlier in this chapter considered binary hypothesis testing on a single received pulse whose two amplitudes represented a 0 or 1 respectively. In modern digital communication systems, the message to be sent uses an alphabet of symbols, with each symbol encoded into a binary sequence of 0s and 1s. Consequently, in addition to making a decision on each received pulse to determine whether it represents a transmitted 0 or 1, we need to further decode a string of such bits to make our best judgement of the transmitted symbol, and perhaps yet further processing to decide on the sequence of symbols that constitutes the entire transmitted message. It would, in principle, be better to take all the raw measurements and then make optimal decisions about the entire sequence of symbols that was transmitted, but this would be a much more complex task, involving many more hypotheses and measurements. In practice, therefore, the task is commonly broken down into stages, with locally optimal decisions made at the single-pulse level to decode sequences of 0s and 1s, then further decisions made to decode at the symbol level, and still further decisions made at the symbol sequence level.

In the following example, we illustrate the second of these decoding stages. The example involves deciding at the receiver which of four possible symbols, each represented by a code word comprising 0s and 1s, was sent from the transmitter over the channel. The example derives the minimum-error-probability decision rule in a slightly different way than in the preceding development, but results again in the MAP rule, embodied in extensions of Eqs. (9.7) and (9.10) that are appropriate to the example.

## Example 9.3    Minimum-Error-Probability Symbol Detection

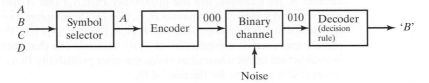

**Figure 9.6**   Communication over a binary channel.

Consider the system in Figure 9.6. Suppose the transmitter randomly selects for transmission one of four possible symbols: $A$, $B$, $C$, or $D$. The probabilities with which these are selected will be denoted by $P(A)$, $P(B)$, $P(C)$, and $P(D)$ respectively. Whatever symbol the transmitter selects is now coded appropriately for transmission over the binary channel. The coding adds some redundancy to provide a basis for error detection or correction at the receiver, in order to combat errors introduced by channel noise that may corrupt the individual bits. The resulting signal is then sent to the receiver. After the receiver decodes the received pulses, attempting to correct for channel noise in the process, it has to arrive at a decision as to which symbol was transmitted.

We model the channel as a binary channel, which accepts a sequence of 0s and 1s from the transmitter, and delivers a sequence of 0s and 1s to the receiver. Suppose that because of the noise in the channel there is a probability $p > 0$ that a transmitted 1 is received as a 0, and that a transmitted 0 is received as a 1. Because the probability is the same for both types of errors, this binary channel is called symmetric. We could treat the nonsymmetric case as easily, apart from some increased notational burden. Implicit in our definition of this channel is the assumption that it is memoryless, that is, its characteristics during any particular transmission slot are independent of what has been transmitted in other time slots. The channel is also assumed time-invariant.

Given such a channel, the transmitter needs to code the selected symbol into binary form. Suppose the transmitter uses three binary digits or bits to code each symbol, as follows:

$$A : 000 \, , \quad B : 011 \, , \quad C : 101 \, , \quad D : 110 \, . \tag{9.24}$$

Because of the nonzero probability of bit errors introduced by the channel, the received sequence for any of these transmissions could be any three-bit binary number:

$$R_0 = 000 \, , \quad R_1 = 001 \, , \quad R_2 = 010 \, , \quad R_3 = 011 \, ,$$
$$R_4 = 100 \, , \quad R_5 = 101 \, , \quad R_6 = 110 \, , \quad R_7 = 111 \, . \tag{9.25}$$

The redundancy introduced by using three bits—rather than the two bits that would suffice to communicate the set of four symbols—is intended to provide some protection against channel noise. Notice that with this particular three-bits/symbol code, a single bit error would be recognized at the receiver as an error because it would result in an invalid code word. It takes two bit errors, which are less likely than single bit errors (for the typical case of $p < 0.5$), to convert any valid code word into another valid one, and thereby elude recognition of the error by the receiver.

The sample space governing the communication of a single symbol across the channel is described as suggested in Table 9.1, listing the probability of every possible combination of transmitted symbol and received sequence. The $(j+1)$th row of column $A$, for example, has the probability $P(A, R_j)$ that $A$ was transmitted and $R_j$ received, and similarly for columns $B$, $C$, and $D$. The simplest way to actually compute this probability is by recognizing that $P(A, R_j) = P(R_j|A)P(A)$; the characterization of the channel permits computation of $P(R_j|A)$, while the characterization of the information source at the transmitter yields the prior probability $P(A)$. The calculations are illustrated in the table for the case of $R_0$.

A decision rule at the receiver selects, for each possible received sequence $R_j$, one of the four possible symbols or hypotheses $A$, $B$, $C$, or $D$. Any such rule can thus be represented in Table 9.1 by selecting one and only one entry in each row. For instance, a particular decision rule may declare $D$ to be the transmitted signal whenever it receives $R_4$; this is indicated in the table by the box around the entry in row $R_4$, column $D$. Each possible decision rule is therefore associated with a table of this form, with precisely one entry boxed in each row. For a given decision rule, the probability of being correct is the sum of the probabilities in all the boxed entries because this sum gives the total probability that the decision rule declares in favor of the same symbol that was transmitted. The probability of error, $P_e$, is therefore 1 minus the probability of being correct.

It follows that to specify the decision rule for minimum probability of error or maximum probability of being correct, we must pick in each row the box that has the maximum entry. If more than one entry has the maximum value, we are free to pick arbitrarily between these; $P_e$ is not affected by which of these we pick. For row $R_j$ in Table 9.1, for the optimum decision rule, we should choose the symbol for which

$$P(\text{symbol}, R_j) = P(R_j|\text{symbol})P(\text{symbol}) \tag{9.26a}$$

$$= P(\text{symbol}|R_j)P(R_j) \tag{9.26b}$$

is maximum. Table 9.2 displays some examples of the required computation, with parameter values as specified in the table caption. The computation in this example is carried out according to the prescription on the right side of Eq. (9.26a).

**TABLE 9.1**   JOINT PROBABILITY OF TRANSMITTED SYMBOL AND RECEIVED BINARY SEQUENCE

|  | $A$: **000** | $B$: **011** | $C$: **101** | $D$: **110** |
|---|---|---|---|---|
| $R_0 = 000$ | $P(A, R_0)$ $= P(R_0\|A)P(A)$ $= (1-p)^3 P(A)$ | $P(B, R_0)$ $= P(R_0\|B)P(B)$ $= p^2(1-p)P(B)$ | $P(C, R_0)$ $= P(R_0\|C)P(C)$ $= p^2(1-p)P(C)$ | $P(D, R_0)$ $= P(R_0\|D)P(D)$ $= p^2(1-p)P(D)$ |
| $R_1 = 001$ | | | | |
| $R_2 = 010$ | | | | |
| $R_3 = 011$ | | | | |
| $R_4 = 100$ | $P(A, R_4)$ | $P(B, R_4)$ | $P(C, R_4)$ | $\boxed{P(D, R_4)}$ |
| $R_5 = 101$ | | | | |
| $R_6 = 110$ | | | | |
| $R_7 = 111$ | | | | |

**TABLE 9.2**   ILLUSTRATION OF THE OPTIMAL (MAP) DECISION RULE WITH $P(A) = \frac{1}{2}$, $P(B) = \frac{1}{4}$, $P(C) = \frac{1}{8}$, $P(D) = \frac{1}{8}$, AND $p = \frac{1}{4}$. *The table entries are calculated as the right side of Eq. (9.26a)*

| | 000<br>$A$ | 011<br>$B$ | 101<br>$C$ | 110<br>$D$ | Decision |
|---|---|---|---|---|---|
| $R_0$<br>000 | | | | | |
| $R_1$<br>001 | | | | | |
| $R_2$<br>010 | $\left(\frac{3}{4}\right)^2 \frac{1}{4}\frac{1}{2}$ | $\left(\frac{3}{4}\right)^2 \frac{1}{4}\frac{1}{4}$ | $\left(\frac{1}{4}\right)^3 \frac{1}{8}$ | $\left(\frac{3}{4}\right)^2 \frac{1}{4}\frac{1}{8}$ | '$A$' |
| $R_3$<br>011 | | | | | |
| $R_4$<br>100 | | | | | |
| $R_5$<br>101 | | | | | |
| $R_6$<br>110 | $\left(\frac{1}{4}\right)^2 \frac{3}{4}\frac{1}{2}$ | $\left(\frac{1}{4}\right)^2 \frac{3}{4}\frac{1}{4}$ | $\left(\frac{1}{4}\right)^2 \frac{3}{4}\frac{1}{8}$ | $\left(\frac{3}{4}\right)^3 \frac{1}{8}$ | '$D$' |
| $R_7$<br>111 | | | | | |

The right side of Eq. (9.26b) permits an intuitive interpretation of what the optimum decision rule does. Since our comparison is being done across the row to find the maximum entry, for a given $R_j$ the term $P(R_j)$ in Eq. (9.26b) stays the same across the row, so all that we need to compare are the *a posteriori* probabilities, $P(\text{symbol}\,|R_j)$, that is, the probabilities of the various symbols, given the data. This is again the MAP decision rule that we derived previously in a slightly different way.

## 9.3 BINARY HYPOTHESIS TESTING

This section focuses on binary hypothesis testing, elaborating on various associated concepts that help provide a more detailed picture. The basic task is to use a measurement $r$ of a random variable $R$ to decide between hypotheses $H_0$ and $H_1$, whose respective prior probabilities are $p_0$ and $p_1 = 1 - p_0$.

### 9.3.1 False Alarm, Miss, and Detection

The sample space that is relevant to evaluating a decision rule for binary hypothesis testing consists of the following four mutually exclusive and collectively exhaustive possibilities: $H_i$ is true and '$H_j$' is declared, $i, j = 1, 2$. Of the four possible outcomes, the two that represent errors are $(H_0, `H_1`)$ and $(H_1, `H_0`)$. Therefore, the probability of error $P_e$—averaged over all possible values of the measured random variable—is

$$P_e = P(H_0, `H_1`) + P(H_1, `H_0`)$$

$$= p_0 P(`H_1`|H_0) + p_1 P(`H_0`|H_1) . \qquad (9.27)$$

The conditional probability $P(`H_1`|H_0)$ is referred to as the conditional probability of a false alarm, and is denoted by $P_{FA}$. The conditional probability $P(`H_0`|H_1)$ is referred to as the conditional probability of a miss, and is denoted by $P_M$. The word *conditional* is usually omitted from these terms in normal use, but it is important to keep in mind that the probability of a false alarm and the probability of a miss are defined as conditional probabilities, and are furthermore conditioned on different events.

The preceding terminology is historically motivated by the radar context, in which $H_1$ represents the presence of a target and $H_0$ the absence of a target. A false alarm then occurs if a target is declared present when it actually is absent, and a miss occurs when a target is declared absent when it actually is present. We will also make reference to the conditional probability of detection,

$$P_D = P(`H_1`|H_1) . \qquad (9.28)$$

In the radar context, this is the probability of declaring a target is present when it actually is present. As with $P_{FA}$ and $P_M$, the word *conditional* is usually omitted in normal use, but it is important to keep in mind that the probability of detection is a conditional probability.

Expressing the probability of error in terms of $P_{FA}$ and $P_M$, Eq. (9.27) becomes

$$P_e = p_0 P_{FA} + p_1 P_M . \qquad (9.29)$$

Also note that

$$P(`H_0`|H_1) + P(`H_1`|H_1) = 1 \qquad (9.30)$$

or

$$P_M = 1 - P_D . \qquad (9.31)$$

To explicitly relate $P_{FA}$ and $P_M$ to whatever the corresponding decision rule is, we recall the notion of a decision region in the measurement space. In the case of a decision rule based on measurement of a single continuous random variable $R$, specifying the decision rule corresponds to choosing a set $\mathcal{D}_1$ of points on the real line such that, when the measured value $r$ of $R$ falls in $\mathcal{D}_1$, we declare '$H_1$', and when $r$ falls outside $\mathcal{D}_1$—in the region that is denoted by $\mathcal{D}_0$—we declare '$H_0$'. This is illustrated in Figure 9.7, for some arbitrary

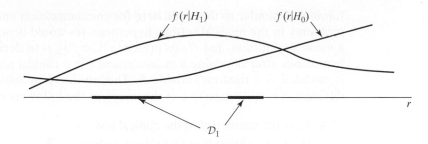

**Figure 9.7** Decision regions. The choice of $\mathcal{D}_1$ marked here is arbitrary, not the optimal choice for minimum probability of error.

choice of $\mathcal{D}_1$. Typically, each of these decision regions comprises a collection of intervals on the real line. There is a direct generalization of this notion to the case where multiple random variables are measured.

With the preceding definitions, we can write

$$P_{FA} = \int_{\mathcal{D}_1} f_{R|H}(r|H_0)\, dr \tag{9.32}$$

and

$$P_M = \int_{\mathcal{D}_0} f_{R|H}(r|H_1)\, dr \,. \tag{9.33}$$

---

**Example 9.4　　False Alarm, Miss, Detection, and Error Probabilities**

We return to Example 9.1 to calculate $P_{FA}, P_M, P_D$, and $P_e$ for the MAP decision rule specified there. With the definitions

$$P_{FA} = P(\text{decide } H_1, \text{given } H_0 \text{ is true}) \tag{9.34}$$

$$P_M = P(\text{decide } H_0, \text{given } H_1 \text{ is true}) \tag{9.35}$$

$$P_D = P(\text{decide } H_1, \text{given } H_1 \text{ is true}) \tag{9.36}$$

we obtain

$$P_{FA} = \frac{3}{4}, \qquad P_M = 0, \qquad P_D = 1,$$

so

$$P_e = p_0 P_{FA} + p_1 P_M = \frac{1}{4} \cdot \frac{3}{4} = \frac{3}{16}.$$

---

While the terminology introduced earlier in this section (e.g., probability of false alarm, miss, and detection) derives from the setting of radar detection, other contexts in which binary hypothesis testing arises have their own terms for these concepts. In the medical literature, for example, the interpretation of clinical results is often understood and described in a hypothesis testing framework similar to that used here for communication and signal detection problems. In the medical setting, hypothesis $H_0$ would denote the absence of a medical condition, and $H_1$ its presence. The task is to decide between these hypotheses after obtaining a measurement $r$ in a clinical test whose outcome is modeled as a random variable $R$. Though the terminology is now slightly different, it is still suggestive of the intent, as the following examples show:

- $P_D$ is the sensitivity of the clinical test;
- $P_{FA}$ is the probability of a false positive;
- $1 - P_{FA}$ is the specificity of the test;
- $P_M$ is the probability of a false negative;
- $P(H_1)$ is the prevalence of the condition that the test is aimed at;
- $P(H_1 | `H_1`)$ is the positive predictive value of the test and $P(H_0 | `H_0`)$ is the negative predictive value.

Some easy exploration using Bayes' rule and the above terminology will show how small the positive predictive value of a test can be if the prevalence of the targeted medical condition is low, even if the test is highly sensitive and specific.

### 9.3.2 The Likelihood Ratio Test

An alternative way of writing the minimum-$P_e$ decision rule in Eq. (9.10) is often useful. Rearranging that equation results in the following equivalent decision rule:

$$\frac{f_{R|H}(r|H_1)}{f_{R|H}(r|H_0)} \underset{`H_0`}{\overset{`H_1`}{\underset{<}{>}}} \frac{p_0}{p_1} \tag{9.37}$$

or

$$\Lambda(r) \underset{`H_0`}{\overset{`H_1`}{\underset{<}{>}}} \eta . \tag{9.38}$$

With $f_{R|H}(r|H_i)$ interpreted as measuring the *likelihood* of the obtained measurement $r$ under the hypothesis $H_i$, the ratio $\Lambda(r)$ is called the likelihood ratio. The above test compares the likelihood ratio with a threshold $\eta$. The higher the prior probability $p_0$ of $H_0$, the greater the threshold $\eta$ is, and the greater the likelihood ratio has to be in order for the test to declare '$H_1$'.

The rewriting of the decision rule as a threshold test on the likelihood ratio is of interest because other formulations of the binary hypothesis testing problem—with criteria other than minimization of $P_e$—also often lead to a decision rule that can be expressed as a likelihood ratio test. The only difference is that the (nonnegative) threshold $\eta$ is picked differently in these other formulations. We describe one of these alternate formulations, the Neyman–Pearson approach, in the next subsection. The last section of the chapter then presents another formulation, which has the objective of minimizing expected cost, and again leads to a likelihood ratio test.

### 9.3.3 Neyman–Pearson Decision Rule and Receiver Operating Characteristic

A difficulty with using the minimization of $P_e$ as the decision criterion in many contexts is that it relies on knowing the *a priori* probabilities $p_0$ and $p_1$, which may be hard to determine. A useful alternative, attributable to Neyman and Pearson, is to maximize the conditional probability of detection $P_D$, while keeping the conditional probability of false alarm $P_{FA}$ below some specified tolerable level. The conditional probabilities are determined by the measurement models under the different hypotheses, and by the decision rule, but not by the *a priori* probabilities governing the selection of hypotheses. This Neyman–Pearson formulation of the hypothesis testing problem in terms of $P_D$ and $P_{FA}$ again leads to a decision rule that involves comparing the likelihood ratio with a threshold, but with the threshold picked differently than for minimum error probability. The approach is developed in more detail next, with the assumption throughout that the measured quantity $R$ is a continuous random variable.

We begin by relating $P_D$ and $P_{FA}$ to the decision regions that define the decision rule. Equation (9.32), repeated here, exhibits this relation for $P_{FA}$:

$$P_{FA} = P(`H_1'|H_0) = \int_{\mathcal{D}_1} f_{R|H}(r|H_0)\, dr\,. \tag{9.39}$$

The analogous integral defining $P_D$ is

$$P_D = P(`H_1'|H_1) = \int_{\mathcal{D}_1} f_{R|H}(r|H_1)\, dr\,. \tag{9.40}$$

Both integrals are over the decision region $\mathcal{D}_1$, and have nonnegative integrands. Thus increasing $P_D$ requires augmenting the set $\mathcal{D}_1$ by adding more of the real axis to it (and correspondingly reducing $\mathcal{D}_0$). However, augmenting the set $\mathcal{D}_1$ only allows $P_{FA}$ to stay unchanged or increase, not to decrease. Our objective is therefore to include in $\mathcal{D}_1$ values of $r$ that contribute as much as possible to the integral that defines $P_D$, but as little as possible to the integral that defines $P_{FA}$. As will be shown shortly, this objective can be achieved by choosing the decision region $\mathcal{D}_1$ to comprise those values of $r$ for which the likelihood ratio $\Lambda(r)$ exceeds a certain threshold $\eta$, so

$$\Lambda(r) = \frac{f_{R|H}(r|H_1)}{f_{R|H}(r|H_0)} \underset{`H_0'}{\overset{\underset{\displaystyle `H_1'}{>}}{\underset{<}{}}} \eta\,. \tag{9.41}$$

Before deriving the above rule, it helps to consider how it operates for values of $\eta$ starting at $\infty$ and dropping steadily from there. With $\eta = \infty$, the decision is always '$H_0$', so the region $\mathcal{D}_1$ is empty and correspondingly $P_D = 0, P_{FA} = 0$. Once $\eta$ drops below the largest value that $\Lambda(r)$ can take, '$H_1$' is declared for some values of $r$, so the decision region $\mathcal{D}_1$ is then nonzero. If an infinitesimal drop in $\eta$ in the vicinity of some value $\eta_0$ causes $P_D$ to increase by $\Delta P_D$ and $P_{FA}$ to increase by $\Delta P_{FA}$, it follows from Eqs. (9.32), (9.40), and (9.41) that

$$\Delta P_D = \eta_0 \, \Delta P_{FA} . \tag{9.42}$$

Hence, at large values of $\eta_0$, $P_D$ increases much faster than $P_{FA}$ for infinitesimal drops in the threshold, while at lower values of $\eta_0$ one obtains a correspondingly smaller increase in $P_D$ relative to the increase in $P_{FA}$.

If $P_{FA}$ increases continuously with decreasing $\eta$, then $\eta$ eventually drops to a value $\bar{\eta}$ at which one of the following happens:

- the specified bound on $P_{FA}$ is attained while $P_D < 1$; or
- $P_D$ reaches 1 before $P_{FA}$ has attained its bound; or
- the bound on $P_{FA}$ is attained simultaneously with $P_D$ reaching the value 1.

This value $\bar{\eta}$ is the threshold to use in the Neyman–Pearson test.

It is possible that as $\eta$ is lowered through some value $\bar{\eta}$, the probability $P_{FA}$ jumps discontinuously from a value below its specified bound, and at which $P_D < 1$, to a value above this bound. In this case, it turns out that a randomized decision rule at the threshold $\bar{\eta}$ allows $P_{FA}$ to attain its specified bound and thereby maximize $P_D$. This randomized rule chooses '$H_1$' with some probability $\alpha$ when $\Lambda(r) > \bar{\eta}$, and otherwise chooses '$H_0$'. The probability $\alpha$ is chosen to obtain a value for $P_{FA}$ that equals the specified bound. Problem 9.20 involves such a randomized decision rule.

The following argument, illustrated in Figure 9.8, suggests in a little more detail why the Neyman–Pearson criterion yields a likelihood ratio test. If the decision region $\mathcal{D}_1$ is optimal for the Neyman–Pearson criterion, then any change in $\mathcal{D}_1$ that keeps $P_{FA}$ the same cannot lead to an improvement in $P_D$. So suppose an infinitesimal segment of width $dr$ at a point $r$ in the optimal $\mathcal{D}_1$ region is converted to be part of $\mathcal{D}_0$. To keep $P_{FA}$ unchanged, an infinitesimal

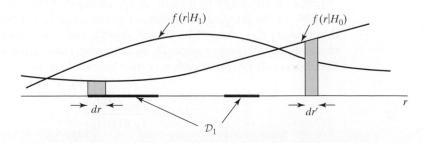

**Figure 9.8**   Illustrating the construction used in deriving the likelihood ratio test for the Neyman–Pearson criterion.

segment of width $dr'$ at an arbitrary point $r'$ in the optimal $\mathcal{D}_0$ region must be converted to be a part of $\mathcal{D}_1$.

The requirement that $P_{FA}$ be unchanged then imposes the condition

$$f_{R|H}(r'|H_0)\, dr' = f_{R|H}(r|H_0)\, dr , \qquad (9.43)$$

while the requirement that the new $P_D$ not be larger than the old implies that

$$f_{R|H}(r'|H_1)\, dr' \leq f_{R|H}(r|H_1)\, dr . \qquad (9.44)$$

Combining Eqs. (9.43) and (9.44), we find

$$\Lambda(r') \leq \Lambda(r) . \qquad (9.45)$$

Equation (9.45) shows that the likelihood ratio cannot be less inside $\mathcal{D}_1$ than it is in $\mathcal{D}_0$. We can therefore conclude that the optimum solution to the Neyman–Pearson formulation is in fact based on a threshold test on the likelihood ratio, where the threshold $\eta$ is picked to obtain the largest possible $P_D$ while ensuring that $P_{FA}$ is not larger than the pre-specified bound.

### Receiver Operating Characteristic

In considering which numerical value to choose as a bound on $P_{FA}$ in the Neyman–Pearson test, it is often useful to look at a plot of $P_D$ as a function $P_{FA}$, as the parameter $\eta$ is varied between 0 and $\infty$. This is referred to as the receiver operating characteristic (ROC). More broadly, the term is used for a plot of $P_D$ versus $P_{FA}$ as some parameter of the decision rule is varied. The ROC can be used to identify whether, for instance, modifying the variable parameter in a given test to permit a slightly higher $P_{FA}$ results in a significantly higher $P_D$. The ROC can also be used to compare different tests.

| Example 9.5 | **Detection and ROC for Signal in Gaussian Noise** |
|---|---|

Consider a scenario in which a radar pulse is emitted from a ground station. If an aircraft is located in the propagation path, a reflected pulse will travel back toward the radar station. We assume that the received signal will then consist of noise alone if no aircraft is present, and noise plus the reflected pulse if an aircraft is present. The processing of the received signal results in a number that we model as the realization of a random variable $R$. If an aircraft is not present, then $R = W$, where $W$ is a random variable denoting the result of processing just the noise. If an aircraft is present, then $R = s + W$, where the constant $s$ is due to processing of the reflected pulse, and is assumed here to be a known value. We thus have the following two hypotheses:

$$H_0: \quad R = W, \qquad (9.46)$$

$$H_1: \quad R = s + W . \qquad (9.47)$$

Assume that the additive noise term $W$ is Gaussian with zero mean and unit variance:

$$f_W(w) = \frac{1}{\sqrt{2\pi}} e^{-w^2/2}. \qquad (9.48)$$

Consequently,

$$f_{R|H}(r|H_0) = \frac{1}{\sqrt{2\pi}} e^{-r^2/2} \tag{9.49}$$

$$f_{R|H}(r|H_1) = \frac{1}{\sqrt{2\pi}} e^{-(r-s)^2/2} . \tag{9.50}$$

The likelihood ratio defined in Eq. (9.37) is then

$$\Lambda(r) = \exp\left[-\frac{(r-s)^2}{2} + \frac{r^2}{2}\right]$$

$$= \exp\left[sr - \frac{s^2}{2}\right] . \tag{9.51}$$

For detection with minimum probability of error, the decision rule compares this likelihood ratio against the threshold $p_0/p_1$, as specified in Eq. (9.37):

$$\exp\left[sr - \frac{s^2}{2}\right] \underset{\text{`}H_0\text{'}}{\overset{\text{`}H_1\text{'}}{\underset{<}{>}}} \eta = \frac{p_0}{p_1} \tag{9.52}$$

It is interesting and important to note that, for this case, the threshold test on the likelihood ratio can be rewritten as a threshold test on the received value $r$. Specifically, Eq. (9.52) can equivalently be expressed as

$$\left[sr - \frac{s^2}{2}\right] \underset{\text{`}H_0\text{'}}{\overset{\text{`}H_1\text{'}}{\underset{<}{>}}} \ln \eta , \tag{9.53}$$

or, if $s > 0$,

$$r \underset{\text{`}H_0\text{'}}{\overset{\text{`}H_1\text{'}}{\underset{<}{>}}} \frac{1}{s}\left[\frac{s^2}{2} + \ln \eta\right] = \gamma , \tag{9.54}$$

where $\gamma$ denotes the threshold on $r$. (If $s < 0$, the inequalities in Eq. (9.54) are reversed.) For example, if both hypotheses are equally likely *a priori*, so that $p_0 = p_1$, then $\ln \eta = 0$ and the decision rule for minimum probability of error when $s > 0$ is

$$r \underset{\text{`}H_0\text{'}}{\overset{\text{`}H_1\text{'}}{\underset{<}{>}}} \frac{s}{2} = \gamma . \tag{9.55}$$

The situation is represented in Figure 9.9.

The receiver operating characteristic displays $P_D$ versus $P_{FA}$ as $\eta$ varies between 0 and $\infty$, or equivalently as $\gamma$ varies between $-\infty$ and $\infty$. For a specified $\gamma$ in this problem, and with $s > 0$,

$$P_{FA} = \frac{1}{\sqrt{2\pi}} \int_{\gamma}^{\infty} e^{-r^2/2} dr \tag{9.56}$$

and

$$P_D = \frac{1}{\sqrt{2\pi}} \int_{\gamma}^{\infty} e^{-(r-s)^2/2} dr . \tag{9.57}$$

Varying $\gamma$ from $-\infty$ to $\infty$ generates the ROC shown in Figure 9.10. The point where $P_{FA} = 0 = P_D$ corresponds to $\gamma = \infty$ and the point where $P_{FA} = 1 = P_D$ corresponds to $\gamma = -\infty$.

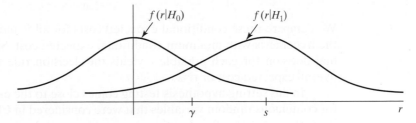

**Figure 9.9**    Threshold $\gamma$ on measured value $r$.

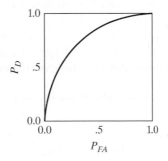

**Figure 9.10**    Receiver operating characteristic.

In a setting more general than the Gaussian case in Example 9.5, a threshold test on the likelihood ratio would not simply translate to a threshold test on the measurement $r$. Nevertheless, we could still decide to use a simple threshold test on $r$ as our decision rule, and then generate and evaluate the associated receiver operating characteristic.

## 9.4 MINIMUM RISK DECISIONS

This section briefly describes a decision criterion, called minimum risk, that includes minimum probability of error as a special case, and that in the binary case again leads to a likelihood ratio test. We describe it for the general case of $M$ hypotheses.

Let the available measurement be the value $r$ of the random variable $R$ (the same development holds if we have measurements of several random variables). Suppose there is a cost $c_{ij}$ associated with each combination of the correct hypothesis $H_j$ and the decision '$H_i$' for $0 \le i, j \le M - 1$, reflecting the costs of actions and consequences that follow from this combination of model and decision. Our objective now is to pick whichever decision has minimum expected cost, or minimum "risk," given the measurement.

The expected cost of deciding '$H_i$', conditioned on $R = r$, is given by

$$E[\text{Cost of '}H_i\text{'}|R = r] = \sum_{j=0}^{M-1} c_{ij}P(H_j|R = r) \,. \qquad (9.58)$$

We compare these conditional expected costs for all $i$, and decide in favor of the hypothesis with minimum conditional expected cost. Specifying this optimal decision for each possible $r$ yields the decision rule that minimizes the overall expected cost or risk.

In this setting hypothesis testing comes close to the estimation problems for continuous random variables that were considered in Chapter 8. We noted there that a variety of such estimation problems can be formulated in terms of minimizing an expected cost function. Establishing an estimate for a continuous random variable is like carrying out a hypothesis test for a continuum of numerically specified hypotheses, rather than just $M$ general hypotheses, with a cost function that reflects some numerical measure of the distance between the actual hypothesis and the one we decide on.

Note that if $c_{ii} = 0$ for all $i$ and if $c_{ij} = 1$ for $j \neq i$, so all errors are penalized equally, then the conditional expected cost in Eq. (9.58) becomes

$$E[\text{Cost of '}H_i\text{'}|R = r] = \sum_{j \neq i} P(H_j|r) = 1 - P(H_i|r) \,. \qquad (9.59)$$

This conditional expected cost is thus precisely the conditional probability of error associated with deciding '$H_i$', conditioned on $R = r$. The right side of the equation then shows that to minimize this conditional probability of error we should decide in favor of the hypothesis with largest conditional probability. In other words, with this choice of costs, the risk (when the expectation is taken over all possible values of $r$) is exactly the probability of error $P_e$, and the optimum decision rule for minimizing this criterion is again seen to be the MAP rule.

Using Bayes' rule to rewrite $P(H_j|R = r)$ in Eq. (9.58), and noting that $f_R(r)$—assumed positive—is common to all the quantities involved in our comparison, we see that an equivalent but more directly implementable procedure is to pick the hypothesis for which

$$\sum_{j=0}^{M-1} c_{ij}f_{R|H}(r|H_j)P(H_j) \qquad (9.60)$$

is minimum. In the case of two hypotheses, and assuming $c_{01} > c_{11}$, it is easy to see that the decision rule based on Eq. (9.60) can be rewritten as

$$\Lambda(r) = \frac{f_{R|H}(r|H_1)}{f_{R|H}(r|H_0)} \underset{\text{`}H_0\text{'}}{\overset{\text{`}H_1\text{'}}{\underset{<}{>}}} \frac{P(H_0)(c_{10} - c_{00})}{P(H_1)(c_{01} - c_{11})} = \eta \,, \qquad (9.61)$$

where $\Lambda(r)$ denotes the likelihood ratio and $\eta$ is the threshold. We have therefore again arrived at a decision rule that involves comparing a likelihood ratio with a threshold. If $c_{ii} = 0$ for $i = 0, 1$ and if $c_{ij} = 1$ for $j \neq i$, then we obtain the threshold associated with the MAP decision rule for minimum $P_e$, as expected.

An issue with the above minimum risk approach to classification, and with the minimum-error-probability formulation that we have examined a few times already, is the requirement that the prior probabilities $P(H_i)$ be known. It is often unrealistic to assume that prior probabilities are known, so we are led to consider alternative criteria. Most important among these alternatives is the Neyman–Pearson approach treated earlier, where the decision is based on the conditional probabilities $P_D$ and $P_{FA}$, thereby avoiding the need for prior probabilities on the hypotheses.

## 9.5 FURTHER READING

Most of the texts suggested for further reading at the end of Chapter 7 contain material on hypothesis testing, but see also [Kay3], [He2], [He3]. Good introductions to the topic from the viewpoint of statistics are in [DeG], [Dek], [Rce], and [Wal]. The challenges involved in meaningful application to an area such as medical statistics are made apparent in [Bl1], [Bl2]. Typically the statistics texts do not address the signals and systems aspects. With the growth of Artificial Intelligence and computer or "machine" vision, hypothesis testing took center place in tasks of classification and pattern recognition from data. These tasks underlie much of machine learning, see for example [Abu], [Alp], [Bis], [Kul], and [Mur].

## Problems

## Basic Problems

**9.1.** A student is taking an exam that she is equally likely to have not studied for (hypothesis $H_0$) or to have studied for (hypothesis $H_1$).

The exam consists of two problems, $a$ and $b$. If the student answers correctly on problem $a$ (respectively problem $b$) we shall say that event $A$ (respectively $B$) has occurred, and otherwise we shall say that event $\overline{A}$ (respectively $\overline{B}$) has occurred. Assume that the student's performance on problem $a$ is independent of performance on problem $b$ if the student has not studied, and also if the student has studied.

Suppose    $P(A|H_1) = 0.8$,    $P(B|H_1) = 0.6$,    $P(A|H_0) = 0.5$,    and $P(B|H_0) = 0.2$.

**(a)** For each possible outcome of the exam, namely each possible combination of $A$ or $\overline{A}$ with $B$ or $\overline{B}$, find the minimum-probability-of-error decision.

**(b)** For the decision rule you developed in (a), find the conditional probability of declaring that the student has not studied ('$H_0$'), given that she actually has ($H_1$).

**9.2.** In choosing with minimum probability of error between the hypothesis $H_0$ that a given measurement $x$ comes from a normal (i.e., Gaussian) distribution with mean 0 and variance 4, and the hypothesis $H_1$ that this measurement comes from a normal distribution with mean 1 and variance 4, we know that the optimal test declares '$H_1$' if $x$ exceeds some threshold $\gamma$. Determine $\gamma$ in each of the following cases: (i) the conditional probability of false alarm is $P_{FA} = 0.5$; and (ii) the conditional probability of a miss is $P_M = 0.5$.

**9.3.** Consider a binary hypothesis testing problem in which we observe a random variable $X$ with the following conditional PDFs specified, and shown in Figure P9.3:

$$f_{X|H}(x \mid H_0) = \frac{1}{\pi(x^2 + 1)} \quad \text{and} \quad f_{X|H}(x \mid H_1) = \frac{2}{\pi(x^2 + 4)} \, .$$

Suppose that the two hypotheses $H_0$ and $H_1$ have prior probabilities $P(H_0) = 0.4$ and $P(H_1) = 0.6$, respectively. We are interested in designing a decision rule for declaring '$H_0$' or '$H_1$', and analyzing its performance.

**(a)** Find the minimum-probability-of-error decision rule, i.e., the decision rule that minimizes $P(H_0, `H_1`) + P(H_1, `H_0`)$. Simplify your answer as much as possible.

**(b)** Indicate, by shading the appropriate regions of the conditional probability density plots in the figure, how you would calculate:

    (i) the (conditional) probability of false alarm, $P_{FA}$; and

    (ii) the (conditional) probability of miss, $P_M$.

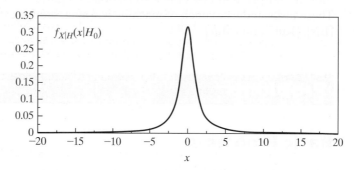

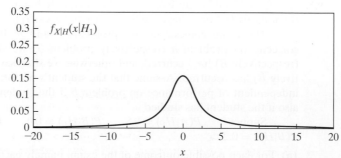

**Figure P9.3**

**9.4.** A random variable $R$ is observed and it is known that with probability $p_0 = \frac{1}{3}$ its PDF is $f_0(r)$, and with probability $p_1 = (1 - p_0) = \frac{2}{3}$ its PDF is $f_1(r)$, as specified below:

$$f_0(r) = \begin{cases} \frac{1}{2}, & -1 \le r \le 1, \\ 0, & \text{otherwise,} \end{cases}$$

$$f_1(r) = \frac{1}{2}e^{-|r|}.$$

We observe the value of $R$ and from this decide that either $f_0(r)$ is the underlying PDF or that $f_1(r)$ is.

For Parts (a) and (b) only, assume that the decision box is specified as follows:

$$\text{if } |r| > \gamma \text{ decide } f_0(r),$$

$$\text{if } |r| \le \gamma \text{ decide } f_1(r).$$

**(a)** For $\gamma = \frac{1}{2}$, determine the probability of error. Clearly show your reasoning.

**(b)** Make a carefully labeled sketch of the ROC for this decision box, as $\gamma$ ranges from 0 to $+\infty$. Clearly show your reasoning.

**(c)** Determine the design for the decision box that minimizes the probability of error. Clearly show your reasoning.

**(d)** For this part of the problem, $p_0$ and $p_1$ are no longer constrained to the values $\frac{1}{3}, \frac{2}{3}$, but $p_0$ cannot be either 0 or 1. For which value(s) of $p_0, 0 < p_0 < 1$, does the decision rule that minimizes the probability of error always decide the same hypothesis, regardless of the value of $R$ observed? Explain.

**9.5.** Consider the following hypothesis testing problem. Under the two hypothesis $H_0$ and $H_1$, the observation $Y$ is

$$H_0: \quad Y = s_0 + N,$$

$$H_1: \quad Y = s_1 + N.$$

Here $s_0$ and $s_1$ are both known constants, and $N$ is a random variable with the PDF $f_N(\alpha)$ shown in Figure P9.5.

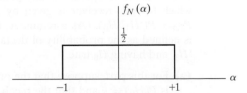

**Figure P9.5**

As a reminder, below are the definitions associated with the decision rule:

(i) $P_0$ and $P_1$ are the prior probabilities for $H_0$ and $H_1$ respectively;

(ii) $P_{FA}$ is $P('H_1'|H_0)$;

(iii) $P_M$ is $P('H_0'|H_1)$;

(iv) $P_D$ is $P('H_1'|H_1)$; and

(v) $P(\text{error}) = P(H_0, 'H_1') + P(H_1, 'H_0')$, i.e., it is the probability that the declared hypothesis is different from the true hypothesis.

**(a)** Must $P_{FA}$ and $P_D$ sum to 1 for every decision rule? Briefly justify your answer.

**(b)** Suppose $P_0 = \frac{1}{4}$ and for a particular decision rule we find that $P_{FA} = \frac{1}{4}$ and $P_D = \frac{3}{4}$. Determine the probability $P(`H_1`)$ that the detector decides $H_1$.

**(c)** Assume the following values:

$$P_0 = \frac{1}{4}, \quad s_0 = 0, \quad s_1 = 1 . \tag{9.62}$$

Determine the range or ranges of values for the observation $y$ for which you would decide '$H_1$' so that the probability of error is minimized.

**(d)** Assume the following values:

$$s_0 = -\frac{1}{2}, \quad s_1 = \frac{1}{2} . \tag{9.63}$$

The decision rule is

$$y \underset{`H_0`}{\overset{`H_1`}{\gtrless}} \gamma .$$

Draw the ROC representing $P_D$ versus $P_{FA}$ as $\gamma$ ranges from $-\infty$ to $+\infty$.

**9.6.** On any particular day, inbound subway trains arrive at the train station according to one of three equally likely schedules: $H_1$, $H_2$, and $H_3$. When schedule $H_i$ is in effect ($i = 1, 2, 3$), the first-order interarrival time $Y$, i.e., the time between a randomly selected pair of consecutive train arrivals, is uniformly distributed in the interval $[0, i]$.

Suppose we make only one observation, i.e., we measure the interarrival time between a randomly selected pair of consecutive trains; let this measured time be $Y = y$. We wish to decide which schedule is in effect.

**(a)** Determine the minimum-error-probability decision rule based on this observation.

**(b)** Find the probability of error for the decision rule you found in (a).

**9.7.** Consider a binary hypothesis testing problem in which a receiver observes a random variable $R$. Based on this observation the receiver decides which one of two hypotheses—denoted by $H_0$ and $H_1$—to declare as true. The receiver can be tuned to operate at any point on the receiver operating characteristic, which for this receiver is given by $P_D = \sqrt{P_{FA}}$, where $P_D = P(`H_1`|H_1)$ and $P_{FA} = P(`H_1`|H_0)$. (As a reminder, the probability of error $P_e$ of the receiver is defined as the probability of declaring '$H_0$' and having $H_1$ true, or declaring '$H_1$' and having $H_0$ true.)

**(a)** For this part, suppose that the prior probability of hypothesis $H_0$ being true is $P(H_0) = \frac{3}{4}$ and that the receiver is tuned to operate at the point $P_D = \frac{1}{2}$ on the ROC curve. Determine $P_{FA}$ and the probability of error $P_e$ at that operating point.

**(b)** For the prior probability of $H_0$ given in (a) (i.e., $P(H_0) = \frac{3}{4}$), there is an operating point on the ROC curve that minimizes the overall probability of error $P_e$. Determine $P_D$ if the receiver operates at that point.

**(c)** Now let $P(H_0) = \frac{1}{4}$. Determine $P_D$ and $P_{FA}$ on the ROC curve and the corresponding $P_e$ such that $P_e$ is minimized.

**9.8.** Consider a digital communication system in which an independent identically distributed (i.i.d.) bit stream $s[n]$ of 1s and 0s is transmitted over a faulty, memoryless channel. $P_0$ denotes the probability that a 0 is sent and $P_1$ denotes the probability that a 1 is sent, with $P_1 = 1 - P_0$. The probability of a 1 being received as a 0 is $\frac{1}{4}$ and the probability of a 0 being received as a 1 is $\frac{1}{4}$. We then process the received signal $r[n]$ through a memoryless, and possibly nonlinear system $H$ to obtain an estimate $\hat{s}[n]$ of $s[n]$ from $r[n]$. The overall system is depicted in Figure P9.8-1.

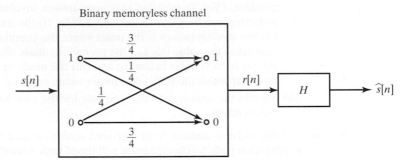

**Figure P9.8-1**

(a) Determine the system $H$ in terms of $P_0$ so that the error probability $P_e$ is minimized, where $P_e$ is defined as the probability that $\hat{s}[n]$ is not equal to $s[n]$ at a given time index $n$.

(b) In this part, assume that the system $H$ has been designed for us and according to the manufacturer it has $P_M = \frac{1}{10}$ and the ROC specified by:

$$\text{ROC:} \quad P_D = (P_{FA})^{\frac{1}{10}}$$

where

$$P_D = \text{Prob (declare that a 1 was sent | a 1 was sent);}$$

$$P_{FA} = \text{Prob (declare that a 1 was sent | a 0 was sent); and}$$

$$P_M = \text{Prob (declare that a 0 was sent | a 1 was sent)}.$$

The overall system in Figure P9.8-1 can then be represented as a new binary memoryless channel as depicted in Figure P9.8-2. Determine the new probabilities $P_a$, $P_b$, $P_c$, and $P_d$.

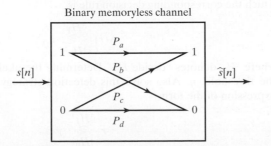

**Figure P9.8-2**

## Advanced Problems

**9.9.** Suppose $X$ and $Y$ are two real random variables. You observe that $X = x$ but don't know the value that $Y$ has taken, and would like to decide with minimum probability of error whether $Y$ is greater than $x$ or less than $x$. (You can assume that the joint, conditional, and marginal probability density functions of $X$ and $Y$ are continuous, i.e., have no jumps or delta functions.)

    **(a)** Specify the appropriate decision rule for the case where $X$ and $Y$ are independent. (You should find that your answer involves one or more of the following numbers associated with a PDF: (i) the mean or expected value; (ii) the median (which is the point where the cumulative distribution function takes the value 0.5, i.e., the probability mass above the median equals the probability mass below it); and (iii) the mode or modes (which are the points at which the PDF takes its maximum value).)

    **(b)** Specify the appropriate decision rule for the case where $X$ and $Y$ are not independent.

**9.10.** A radar pulse is emitted from a ground station. If an aircraft is located in the propagation path, a reflected pulse will travel back toward the radar station. We assume that the received signal consists of noise plus the reflected pulse if an aircraft is present, and noise alone if no aircraft is present. The processing of the received signal produces a number $R$ modeled as a random variable. If an aircraft is present, then $R = s + N$; if an aircraft is not present, then $R = N$. The constant $s > 0$ is due to the reflected pulse and is a known value; the random variable $N$ is due to the noise. We thus have the two hypotheses:

$$H_1: \quad R = s + N, \text{ and}$$

$$H_0: \quad R = N .$$

Assume that the additive noise is Gaussian with zero mean and variance $\sigma^2$, i.e.,

$$f_N(x) = \frac{1}{\sigma\sqrt{2\pi}} e^{-x^2/(2\sigma^2)} .$$

Also assume that the *a priori* probability of an aircraft being in the path of the radar is 0.05, i.e., $P(H_1) = 0.05$.

    **(a)** What is $f_{R|H_1}(r|H_1)$, the PDF for $R$ when an aircraft is present?

    **(b)** What is $f_{R|H_0}(r|H_0)$, the PDF for $R$ when an aircraft is not present?

    **(c)** Suppose we use a minimum-probability-of-error detection strategy for which the corresponding decision rule is

$$\Lambda(r) \underset{'H_0'}{\overset{'H_1'}{\gtrless}} \eta ,$$

where '$H_i$' denotes "decide $H_i$". Determine the likelihood ratio, $\Lambda(r)$, and the threshold, $\eta$. Also solve this detection strategy for $r$ to arrive at an expression of the form

$$r \underset{'H_0'}{\overset{'H_1'}{\gtrless}} \gamma$$

and find $\gamma$.

**(d)** In terms of the standard $Q$-function, defined as

$$Q(x) = \frac{1}{\sqrt{2\pi}} \int_x^\infty e^{-t^2/2}\, dt\,,$$

determine the probability of making an error.

**(e)** Recall that the conditional probability of detection is defined by $P_D = P(\text{'}H_1\text{'} \mid H_1)$, and the probability of false alarm by $P_{FA} = P(\text{'}H_1\text{'} \mid H_0)$.

(i) What is $\gamma$ if $P_{FA} = 1$ and $P_D = 1$?
(ii) What is $\gamma$ if $P_{FA} = 0$ and $P_D = 0$?
(iii) If $P_D = 0.5$, what is $\gamma$?
(iv) If $P_{FA} = 0.5$, what is $\gamma$?

Note that the answers to (i)–(iv) allow you to label, with the corresponding values of $\gamma$, some points on the receiver operating characteristic for this example, as described in the text.

**9.11.** In this problem you will use an appropriate computational platform to get some empirical feel for the performance of a decision rule in a signal detection problem, using Monte Carlo simulation, i.e., by running repeated random trials of the decision rule and computing appropriate statistics.

Suppose, after some processing of a received signal, we obtain a numerical quantity $R$ that, according to some model we have of the signal detection process, equals the sum of a random variable $X$ (which can take the values $\pm A$ with equal probability) and a random variable $N$, i.e.,

$$H_1(x = +A)\ :\ R = +A + N$$

$$H_0(x = -A)\ :\ R = -A + N.$$

The random variable $N$ is Gaussian with zero mean and variance $\sigma^2$:

$$f_N(x) = \frac{1}{\sigma\sqrt{2\pi}} e^{-x^2/(2\sigma^2)}.$$

**(a)** Using $\sigma = 2$ and $A = 2$, generate 10,000 sample values of $N$, and 10,000 sample values of $X$. Then define the received values

$$R = X + N$$

corresponding to the numerical results of 10,000 experiments.

**(b)** Plot the first 50 points of $X$ and $R$ and compare the two plots.

**(c)** Determine the minimum-probability-of-error decision rule that you would use to recover the value of $X$ from $R$. Evaluate the corresponding theoretical probability of error, $P_e$, for this decision rule, using the $Q$-function.

**(d)** Use the rule in (c) to decide (or hypothesize), for each value of $R$ in your data set, what the underlying value of $X$ is (i.e., $\pm A$), and store these decisions in a vector $\widehat{X}$.

**(e)** Compare $X$ and $\widehat{X}$. How many wrong decisions were made? Calculate the empirical probability of error, $\widehat{P}_e$, by dividing the number of wrong decisions your decision rule made by the total number of decisions made.

**(f)** Repeat (a)–(e), but with $\sigma = 2$ and $A = 1$.

**(g)** Repeat (a)–(e), but appropriately modified for the case where $P(X = +A) = 0.75$ rather than 0.5.

**9.12.** Consider the binary hypothesis testing problem in which the conditional PDFs of the data under the two hypotheses are:

$$H_0: \quad f_{R|H}(r \mid H_0) = \frac{1}{\sqrt{2\pi}} e^{-\frac{(r-\mu_0)^2}{2}}, \text{ and}$$

$$H_1: \quad f_{R|H}(r \mid H_1) = \frac{1}{\sqrt{2\pi}} e^{-\frac{(r-\mu_1)^2}{2}}.$$

The *a priori* probabilities are denoted by $p_0 = P(H_0)$ and $p_1 = P(H_1)$, respectively. Assume that $p_0 = p_1 = \frac{1}{2}$. We will also refer throughout this problem to a positive, known value $A > 0$.

Figure P9.12 shows a set of ROC curves for different detection systems. The point $X$ on the solid ROC curve denotes the detection probability or rate ($P_D$) and the false alarm rate ($P_{FA}$) for the minimum-probability-of-error detector that chooses between $H_0$ and $H_1$, under the assumption that $\mu_0 = 0$ and $\mu_1 = +A$.

Answer the following questions and justify your answers.

**(a)** We are now told that $\mu_0 = -A$ (with $\mu_1 = +A$), and the *a priori* probabilities remain equal. From the set $(M, N, O, P)$ of possible points in Figure P9.12, identify all that could correspond to the minimum-error-probability detector under the stated assumptions.

**(b)** We are now told that $\mu_0 = 0$ and $\mu_1 = +A$, but the *a priori* probabilities change to

$$p_0 = 0.38 \quad \text{and} \quad p_1 = 0.62.$$

From the set $(M, N, O, P)$ of possible points in Figure P9.12, identify all that could correspond to the minimum-error-probability detector under the stated assumptions.

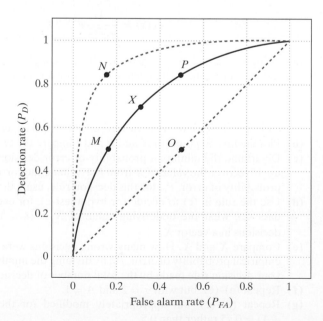

**Figure P9.12**

**9.13.** Consider a system for determining whether a radio channel is being used for a covert communication session. Let $H_1$ denote the hypothesis that such a session is in progress, and let $H_0$ denote the hypothesis that one is not. We will base our decision on a single scalar measurement $R$ of the channel via our antenna. This measurement is a zero-mean Gaussian random variable whose variance is larger when a session is in progress. Specifically,

$$H_0 : \quad f_{R|H}(r \mid H_0) = \frac{1}{\sqrt{2\pi}} e^{-r^2/2}$$

$$H_1 : \quad f_{R|H}(r \mid H_1) = \frac{1}{\sqrt{4\pi}} e^{-r^2/4}.$$

The *a priori* probabilities are denoted by $p_0 = P(H_0)$ and $p_1 = P(H_1)$, respectively.

**(a)** Determine $p_0$ such that the minimum-error-probability decision rule is

$$|r| \underset{\text{'}H_0\text{'}}{\overset{\text{'}H_1\text{'}}{\gtrless}} 1.$$

**(b)** Determine constants $a_1, a_2, b_1, b_2$, and $c$, such that the probability of error $P_e$ for your rule in part (a) can be expressed in the form

$$P_e = a_1 Q(b_1) + a_2 Q(b_2) + c,$$

where the $Q$-function is defined as

$$Q(x) = \frac{1}{\sqrt{2\pi}} \int_x^\infty e^{-t^2/2} dt.$$

**(c)** Is the following statement true or false?
"When $p_0 > 0$, it is possible that the optimal decision rule will always decide $H_1$ irrespective of the observed value of $r$."
If your answer is false, explain. If true, construct an example (i.e., determine a value of $p_0 > 0$ such that the decision rule which minimizes the probability of error always decides $H_1$).

**9.14.** Consider a transmitter that sends either the message $m_0$ (hypothesis $H_0$) or the message $m_1$ (hypothesis $H_1$), where these hypotheses occur with respective probabilities $p_0$ and $p_1$. Assume that a receiver observes $R = r$, where $R$ has the following conditional statistics: when $H_0$ is true, $R$ is Gaussian with mean $s_0$ and variance $\sigma_0^2$, and when $H_1$ is true, $R$ is Gaussian with mean $s_1$ and variance $\sigma_1^2$. Assume that $0 < s_0 < s_1$ and $0 < \sigma_0^2 < \sigma_1^2$.

**(a)** Show that the minimum-probability-of-error decision rule for this problem can be reduced to

$$r^2 + ar \underset{\text{'}H_0\text{'}}{\overset{\text{'}H_1\text{'}}{\gtrless}} \gamma,$$

and provide expressions for the constants $a$ and $\gamma$.

**(b)** Now suppose that we implement a simple threshold decision rule,

$$r \underset{\text{`}H_0\text{'}}{\overset{\text{`}H_1\text{'}}{\underset{<}{\overset{>}{}}}} \xi,$$

instead of the minimum-probability-of-error rule. For this threshold receiver express the (conditional) false-alarm probability $P_{FA}$ and the (conditional) miss probability $P_M$ in terms of

$$Q(x) \equiv \int_x^\infty \frac{e^{-y^2/2}}{\sqrt{2\pi}} \, dy \ .$$

**(c)** Choose the value of $\xi$ for your receiver in part (b) to make $P_{FA} = P_M$. Find the probability of error that results, written in terms of $Q(x)$.

**9.15.** An exponentially distributed random variable $X$ has different variances under hypotheses $H_0$ and $H_1$, as given by the following densities:

$$f_{X|H}(x \mid H_0) = \frac{1}{2}e^{-|x|} \qquad P(H_0) = \frac{3}{5}$$

$$f_{X|H}(x \mid H_1) = \frac{1}{8}e^{-|x|/4} \qquad P(H_1) = \frac{2}{5}$$

**(a)** The decision rule for choosing between $H_0$ and $H_1$ with minimum probability of error, given the measurement $X = x$, involves comparing the absolute value $|x|$ with a threshold $\gamma$. State the precise form of the decision rule and determine $\gamma$.

**(b)** Draw plots of the two conditional densities using the same set of axes (i.e., superimpose them on the same diagram), then indicate and label the areas corresponding to the (conditional) probability of a false alarm, $P_{FA}$, and the (conditional) probability of a miss, $P_M$, for the decision rule you derived in part (a).

**(c)** If $P_{FA} = \alpha$ and $P_M = \beta$, write expressions for the probability $P_e$ of making an error in your decision, and for $P(H_1 \mid \text{`}H_1\text{'})$, the probability that $H_1$ actually holds, given that your decision rule has decided $H_1$ holds.

**(d)** Now suppose the prior probabilities are $P(H_0) = p_0$ and $P(H_1) = 1 - p_0$, instead of the values you considered for the preceding parts. Is there any range of values of $p_0$ for which you would always declare '$H_1$', no matter what the value of $x$? And is there any range of values of $p_0$ for which you would always declare '$H_0$', no matter what $x$ is?

**9.16.** A signal $X[n]$ that we will be measuring for $n = 1, 2$ is known to be generated according to one of the following two hypotheses:

$$H_{\text{neg}}: \quad X[1] = -1 + W[1] \quad \text{and} \quad X[2] = -s + W[2] \quad \text{and}$$

$$H_{\text{pos}}: \quad X[1] = +1 + W[1] \quad \text{and} \quad X[2] = +s + W[2],$$

where $s$ is some known positive number and, under each hypothesis, $W[1]$ and $W[2]$ are i.i.d. random variables uniformly distributed in the interval $[-2,2]$. Given measurements $x[1]$ and $x[2]$ of $X[1]$ and $X[2]$ respectively, we would like to decide between the hypotheses $H_{\text{neg}}$ and $H_{\text{pos}}$.

(a) One *ad hoc* strategy for processing the measurements is to base the decision on the sum of the measurements:

$$r = x[1] + x[2] \, .$$

To analyze decision schemes that are based on consideration of the sum $r$, we first examine the random variable

$$R = X[1] + X[2] \, .$$

For the case where $s = 2$, draw fully labeled sketches of the conditional densities of $R$ under each of the hypotheses, i.e., sketches of $f_{R|H}(r|H_{neg})$ and $f_{R|H}(r|H_{pos})$.

(b) With things still set up as in (a), suppose that the two hypotheses have equal prior probabilities, so $P(H_{neg}) = P(H_{pos}) = \frac{1}{2}$. Specify a decision rule that, on the basis of knowledge that $R = r$, decides between $H_{neg}$ and $H_{pos}$ with minimum probability of error. Also compute the probability of error associated with this decision rule.

(c) If your decision rule in (b) announces in favor of $H_{neg}$, what is the probability that $H_{neg}$ actually holds? In other words, what is $P(H_{neg} | \,'H_{neg}')$?

(d) Now forget about working with the sum of the measurements, and instead take advantage of the fact that you actually have two measurements. Accordingly, first sketch or fully describe the conditional densities

$$f_{X[1],X[2]|H}\big(x[1],x[2] \mid H_{neg}\big) \quad \text{and} \quad f_{X[1],X[2]|H}\big(x[1],x[2] \mid H_{pos}\big) \, ,$$

still for the case where $s = 2$, and use this to specify a decision rule that can actually pick perfectly (i.e., with zero probability of error) between the two hypotheses, no matter what the prior probabilities of the two hypotheses are.

(e) Suppose now that $s = 1$. Again sketch or fully describe the two conditional densities listed in (d). Then, for the case where $P(H_{neg}) = \frac{1}{3}$, so $P(H_{pos}) = \frac{2}{3}$, specify the decision rule that will pick between the two hypotheses with minimum probability of error, on the basis of knowledge that $X[1] = x[1]$ and $X[2] = x[2]$. Also determine the conditional probability of declaring that $H_{pos}$ holds, given that $H_{neg}$ actually holds. Finally, determine the probability of error associated with this decision rule.

## Extension Problems

**9.17.** In a hospital ward for lung disease, a surgeon has discovered a suspicious shadow in the lung X-ray of her patient. The hypothesis $H$ takes the value $H_0$ or $H_1$. Under hypothesis $H_1$ the shadow represents a cancer, and under hypothesis $H_0$ the shadow represents harmless scar tissue. Moreover, assume that the results of the X-ray test are summarized by a random variable $X$, with

$$f_{X|H}(x \mid H_1) = xe^{-x}u(x) \, ,$$

and

$$f_{X|H}(x \mid H_0) = e^{-x}u(x) \, ,$$

where $u(x)$ is the unit step function. The doctor must decide, based on the value of $X$ obtained from the X-ray results, whether or not to operate. Since *a priori* probabilities would be difficult to apply here, we will consider an alternative to the MAP rule for our decision criterion. Specifically, we'll define $P_D$, the conditional probability of correct detection, as

$$P_D = P(\text{decide } H_1 \mid H_1 \text{ true}),$$

and $P_{FA}$, the conditional probability of a false alarm, as

$$P_{FA} = P(\text{decide } H_1 \mid H_0 \text{ true}).$$

Our hypothesis testing rule will aim to maximize $P_D$ with the constraint that $P_{FA} = \alpha = 0.1$. This framework corresponds to the Neyman–Pearson hypothesis test.

**(a)** Show that this maximization problem is equivalent to minimizing

$$\eta(1 - \alpha) + \int_{D_0} \left[ f_{x|H}(x \mid H_1) - \eta f_{x|H}(x \mid H_0) \right] dx,$$

where $D_0$ denotes the range of $X$ values for which we decide in favor of $H_0$ (the complementary range, where we decide in favor of $H_1$, is denoted by $D_1$).

**(b)** It should be clear from (a) that the optimal $D_0$ contains precisely those values of $x$ for which the term in brackets inside the integral above is negative. Show that this is equivalent to

$$\Lambda(x) \underset{`H_0\text'}{\overset{`H_1\text'}{\gtrless}} \eta$$

where $\eta$ is chosen so that $P_{FA} = \alpha$.

**(c)** Using the result in part (b), find the optimal $D_1$ and $D_0$ for deciding whether to operate on the patient. What is the resulting value of $P_D$?

**9.18.** Assume we have to decide between hypotheses $H_0$ and $H_1$ based on a measured random variable $X$. The conditional densities for $X$ given $H_0$ and $H_1$ are shown in Figure P9.18.

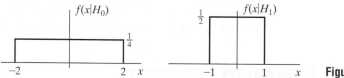

**Figure P9.18**

**(a)** You observe $X = x$. What is the minimum-error-probability decision rule for choosing between '$H_0$' and '$H_1$' when $P(H_0) = P(H_1)$? Find the corresponding (conditional) probability of detection $P_D$, (conditional) probability of false alarm $P_{FA}$, and probability of error $P_e$.

**(b)** For what range of values of $P(H_0)$ does the minimum-error-probability decision rule always choose '$H_0$'?

**(c)** Now consider what happens when you have measurements $x_1$ and $x_2$, respectively, of two random variables $X_1$ and $X_2$. Assume that under each hypothesis, $H_0$ and $H_1$, these random variables are independent, and that each is distributed as shown in Figure P9.18. With $P(H_0) = P(H_1)$, what is the minimum-probability-of-error decision rule? What is the probability of error? Does the value of the probability of error seem reasonable, given the value of $P_e$ obtained in part (a)?

**(d)** Suppose the situation is exactly as in (d) with $P(H_0) = P(H_1)$ again, except that now you are not given $x_1$ and $x_2$ separately, but instead only have the sum $s = x_1 + x_2$, which you can consider as a measurement of the random variable $S = X_1 + X_2$. Determine and make fully labeled sketches of the density of $S$ under each of the hypotheses, and use this information to determine the minimum-probability-of-error decision rule and the probability of error $P_e$.

**9.19.** A continuous random variable $X$ can take values in the range $[0,2]$ and is governed by one of two possible PDFs, corresponding to the following two hypotheses:

$$H_0: \quad f_{X|H}(x|H_0) = \frac{1}{2} \quad \text{for} \quad 0 \le x \le 2$$

$$H_1: \quad f_{X|H}(x|H_1) = \frac{x}{2} \quad \text{for} \quad 0 \le x \le 2.$$

In other words, $X$ is uniformly distributed under $H_0$, and has a "triangular" distribution under $H_1$.

**(a)** Draw fully labeled sketches of the conditional densities

$$f_{X|H}(x|H_0) \quad \text{and} \quad f_{X|H}(x|H_1)$$

on the same figure (i.e., use a common set of horizontal and vertical axes). In a separate figure below the first figure, draw a fully labeled plot of the likelihood ratio

$$\Lambda(x) = \frac{f_{X|H}(x|H_1)}{f_{X|H}(x|H_0)}.$$

**(b)** We have seen that a variety of decision rules take the form

$$\begin{array}{c} \text{`}H_1\text{'} \\ > \\ \Lambda(x) \quad \eta \\ < \\ \text{`}H_0\text{'} \end{array}$$

for some appropriately chosen threshold $\eta$. For what range of $\eta$ will $P_D = P(\text{`}H_1\text{'}|H_1)$ take the value 0, and for what range of $\eta$ will it take the value 1? Also determine the corresponding values of $P_{FA} = P(\text{`}H_1\text{'}|H_0)$ in these two instances.

**(c)** For $0 \le \eta \le 2$, write $P_D$ and $P_{FA}$ in terms of $\eta$, and check that your expressions yield the answers you obtained in (b) for the corresponding values of $\eta$ there.

**(d)** Use your expressions in (c) to compute $P_D$ as a function of $P_{FA}$. If you've done this correctly, you should discover that $dP_D/dP_{FA} = \eta$, which is true much more generally, and is a useful check. Sketch this function in the form of a receiver operating characteristic (ROC) curve. On the same figure, plot $P_M = P(\text{`}H_0\text{'}|H_1)$ as a function of $P_{FA}$.

**(e)** The probability of error for this decision rule depends in general on the prior probabilities $P(H_0) = p_0$ and $P(H_1) = 1 - p_0$. However, if $P_{FA}$ and $P_M$ are related in a particular way, the probability of error will not depend on $p_0$. How should $P_{FA}$ and $P_M$ be related for this to happen? And is there any choice of $\eta$ that will allow $P_{FA}$ and $P_M$ to be related in this way? Your plot of $P_M$ as a function of $P_{FA}$ in part (d) should allow you to decide.

**9.20.** A signal $X[n]$ that we will be measuring for $n = 1, 2$ is known to be generated according to one of the following two hypotheses:

$$H_0: \quad X[n] = s_0[n] + W[n] \quad \text{and}$$

$$H_1: \quad X[n] = s_1[n] + W[n],$$

where $s_0[n]$ and $s_1[n]$ are specified signals and $W[n]$ are i.i.d. Gaussian random variables with mean 0 and variance $\sigma^2$. We would like to decide between the hypotheses $H_0$ and $H_1$ with minimum probability of error. In this Gaussian case the optimum decision rule involves comparing a weighted combination of the measurements with a threshold. The weighted combination takes the form

$$R = v[1]X[1] + v[2]X[2],$$

for appropriately chosen weights $v[1]$ and $v[2]$, and we denote the threshold by $\gamma$. When the measurements are obtained, the random variable $R$ takes a specific value $r$. We decide '$H_1$' if $r > \gamma$, and decide '$H_0$' otherwise.

**(a)** Suppose the hypotheses are equally likely *a priori*, and that $s_0[n] = 0$ for $n = 1, 2$, while $s_1[n] = \delta[n-1] - \delta[n-2]$. What $v[1]$, $v[2]$, and $\gamma$ should be chosen for the optimum decision rule? Also write expressions for the corresponding probability of error $P_e$ and for $P(H_0 \mid {}^{\prime}H_0{}^{\prime})$; your results should be stated in terms of the standard function

$$Q(\alpha) = \frac{1}{\sqrt{2\pi}} \int_\alpha^\infty e^{-t^2/2} \, dt.$$

It may help you to first sketch a figure showing how $R$ is distributed under $H_0$ and $H_1$ respectively.

**(b)** Suppose in (a) that the hypotheses were not equally likely, and that the optimum decision rule resulted in $P({}^{\prime}H_1{}^{\prime} \mid H_0) = 0.5$. What would $v[1]$, $v[2]$ and $\gamma$ be for this case?

**(c)** Suppose in (a) that the hypotheses were not equally likely, but that both had nonzero probability. Are there values of $P(H_0) \neq 0$ and $P(H_1) = 1 - P(H_0) \neq 0$ for which the optimum decision rule ends up always deciding in favor of $H_0$? Explain.

**(d)** Suppose that $s_1[n]$ is as in (a), but $s_0[n] = -s_1[n]$. Assuming that the hypotheses are equally likely, determine $v[1]$, $v[2]$, and $\gamma$. Will the probability of error in this case be greater than, equal to, or smaller than in (a)? Explain your answer.

**9.21.** A signal $X[n]$ that we will be measuring for $n = 1, 2$ is known to be generated according to one of the following two hypotheses:

$$H_{no}: \quad X[n] = W[n] \quad \text{and}$$

$$H_{yes}: \quad X[n] = s[n] + W[n],$$

where $s[1]$ and $s[2]$ are specified (deterministic) numbers, with $0 < s[i] \leq 1$ for $i = 1, 2$, and where $W[1]$ and $W[2]$ are i.i.d. random variables uniformly distributed in the interval $[-1, 1]$ (and hence with mean 0 and variance $\frac{1}{3}$). Given measurements $x[1]$ and $x[2]$ of $X[1]$ and $X[2]$ respectively, we would like to decide between the hypotheses $H_{no}$ and $H_{yes}$.

**(a)** One strategy for processing the measurements is to only look at a linear combination of the measurements, of the form

$$r = g[1]x[1] + g[2]x[2].$$

To analyze decision schemes that are based on consideration of the number $r$, consider the random variable

$$R = g[1]X[1] + g[2]X[2] .$$

Determine the mean and variance of $R$ under each of the hypotheses, and note that the variance does not depend on which hypothesis applies. *Hint:* you do not need to find the densities of $R$ under the two hypotheses in order to find these conditional means and variances.

Now choose $g[1]$ and $g[2]$ to maximize the relative distance between these means, where "relative" signifies that the distance is to be measured relative to the standard deviation of $R$ under hypothesis $H_{no}$ (or equivalently under $H_{yes}$). Equivalently, maximize the following signal-to-noise ratio (SNR):

$$\frac{\left(E[R|H_{yes}] - E[R|H_{no}]\right)^2}{\text{variance}(R|H_{no})} .$$

**(b)** In the particular case where $s[1] = s[2] = 1$, which we shall focus on for the rest of this problem, it turns out that the choice $g[1] = g[2] = c$ will serve, for any nonzero constant $c$, to maximize the SNR in (a). Taking $c = 3$, draw fully labeled sketches of the conditional densities of $R$ under each of the hypotheses, i.e., sketches of $f_{R|H}(r|H_{no})$ and $f_{R|H}(r|H_{yes})$. Suppose now that the prior probabilities on the two hypotheses are $p(H_{no}) = \frac{2}{3}$ and hence $p(H_{yes}) = \frac{1}{3}$. Specify a decision rule that, on the basis of knowledge that $R = r$, decides between $H_{no}$ and $H_{yes}$ with minimum probability of error. Also compute the probability of error associated with this decision rule. (It will probably help you to shade on the appropriate sketch the regions corresponding to the conditional probability of a false yes and to the conditional probability of a false no.)

**(c)** If we did not hastily commit ourselves to working with a scalar measurement obtained by taking a linear combination of the measurements $x[1]$ and $x[2]$, we might perhaps have done better. Accordingly, first sketch or fully describe the conditional densities

$$f_{X[1],X[2]|H}\left(x[1], x[2] \mid H_{no}\right) \quad \text{and} \quad f_{X[1],X[2]|H}\left(x[1], x[2] \mid H_{yes}\right)$$

for the case where $s[1] = s[2] = 1$. Then specify a decision rule that will pick between the two hypotheses with minimum probability of error, on the basis of knowledge that $X[1] = x[1]$ and $X[2] = x[2]$, and still with the prior probabilities specified in (b), namely $p(H_{no}) = \frac{2}{3}$ and hence $p(H_{yes}) = \frac{1}{3}$. Determine the probability of error associated with this decision rule, and compare with your result in (b).

**9.22.** We observe a random variable $Y$ whose statistics are as follows:

$$Y = \begin{cases} W & \text{when } H = H_0 \\ 1 + W & \text{when } H = H_1 , \end{cases}$$

where the hypotheses $H = H_0$ and $H = H_1$ occur with *a priori* probabilities $P_0$ and $P_1 = 1 - P_0$, respectively, and $W$ is a continuous random variable with probability density $f_W(w)$ under both $H_0$ and $H_1$ (i.e., $W$ is independent of $H$).

**(a)** Express the conditional probability densities $f_{Y|H}(y \mid H_0)$ and $f_{Y|H}(y \mid H_1)$ in terms of $f_W(\cdot)$.

**(b)** Suppose we use the threshold test

$$y \quad \begin{array}{c} \text{'}H_1\text{'} \\ > \\ < \\ \text{'}H_0\text{'} \end{array} \quad \gamma,$$

to make a decision based on the observation $Y = y$. The corresponding conditional probability of a miss, $P_M$, is shown in Figure P9.22 as a function of the test's threshold, $\gamma$. Use this $P_M$ versus $\gamma$ behavior and your answers from (a) to determine and sketch the conditional probability densities $f_{Y|H}(y \mid H_0)$ and $f_{Y|H}(y \mid H_1)$.

**(c)** Use your results from (b) to determine the conditional probability of false alarm, $P_{FA}$, as a function of the threshold, $\gamma$, for the threshold test given above. Plot your answer.

**(d)** Suppose that $P_0 = P_1 = \frac{1}{2}$. Find the minimum error probability, $P_{\min}$, that can be realized by optimizing the $\gamma$ value in the threshold test given above.

**(e)** Suppose that $P_0 = P_1 = \frac{1}{2}$. Find the rule for deciding between '$H_0$' and '$H_1$' with minimum probability of error. Your answer should be an explicit pair of decision regions: $D_0 = \{y : \text{'}H_0\text{'}\}$, the set of measurements $y$ for which you will announce '$H_0$'; and $D_1 = \{y : \text{'}H_1\text{'}\}$, the set of measurements for which you will announce '$H_1$'. Note that you are not restricted to using the given threshold test. Evaluate the probability of error, $P_e$, for your optimum decision rule.

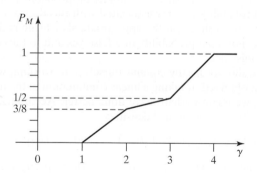

**Figure P9.22**

**9.23.** Suppose that under hypothesis $H_0$ a random variable $X$ is distributed uniformly in the interval $[-2, 2]$, while under hypothesis $H_1$ it is distributed uniformly in the interval $[-1, 1]$. We will be getting a measurement $x$ of $X$, and would like to design a decision rule that will maximize the conditional probability of detection, $P_D = P(\text{'}H_1\text{'}|H_1)$, subject to the conditional probability of false alarm not exceeding a specified level $\beta$, i.e., $P_{FA} = P(\text{'}H_1\text{'}|H_0) \leq \beta$. The Neyman–Pearson result tells us we can do this by choosing a decision rule that announces '$H_1$' if the likelihood ratio

$$\Lambda(x) = \frac{f_{X|H}(x|H_1)}{f_{X|H}(x|H_0)}$$

exceeds a properly selected threshold $\eta$, i.e., if $\Lambda(x) > \eta$; and we announce '$H_0$' if the likelihood ratio falls below the threshold, i.e., if $\Lambda(x) < \eta$. The value we pick for $\eta$ will determine what values of $P_D$ and $P_{FA}$ we get. This problem will explore how we might extend the simple Neyman–Pearson decision rule above if there is a nonzero probability that $\Lambda(X)$ can exactly equal $\eta$.

**(a)** Sketch $\Lambda(x)$ as a function of $x$ for $-2 < x < 2$. Note that we need not spend time wondering what $\Lambda(x)$ is at the edges of the PDFs or for $|x| > 2$, since the probability that $X$ will take any of these specific values is 0. Also note, for use in interpreting (c) below, that $\Lambda(X)$, with the random variable $X$ replacing the argument $x$, is a function of a random variable, hence a random variable itself.

**(b)** For $\eta$ fixed at some value in each of the following ranges, specify $P_D$ and $P_{FA}$:

(1) $\eta$ at some value strictly below 0;
(2) $\eta$ at some value strictly between 0 and 2; and
(3) $\eta$ at some value strictly above 2.

It is clear from the results in (b) that with $\eta$ restricted to the ranges there, we will only get three possible values of $P_{FA}$, with the three values of $P_D$ that go along with these. In other words, the ROC that plots $P_D$ as a function of $P_{FA}$ will only have three points on it. The next part shows how we can get other values of $P_{FA}$ and correspondingly other values of $P_D$ to enable us to meet the specification $P_{FA} \leq \beta$ more closely, thereby getting a higher $P_D$ than if we had to make do with just the three-point ROC above.

**(c)** Suppose we choose $\eta = 2$. What is the probability that we get $\Lambda(X) = 2$ if $H_0$ holds? And what is the probability we get $\Lambda(X) = 2$ if $H_1$ holds? With $\eta = 2$, you should see from the above computations that we will never get $\Lambda(x) > \eta$, but we might well get $\Lambda(x) = \eta$ or $\Lambda(x) < \eta$. Suppose we still announce '$H_0$' when $\Lambda(x) < \eta$; however, when $\Lambda(x) = \eta$ we shall announce '$H_0$' with probability $\alpha$, and otherwise announce '$H_1$'. What are $P_{FA}$ and $P_D$ with this randomized decision rule? Explain carefully. Draw the ROC that you get as $\alpha$ varies from 0 to 1, and also include the three points on the ROC that you computed in (b).

Using a similar randomized decision rule for the threshold $\eta = 0$, we can get additional points on the ROC, but we omit this here.

# 10 Random Processes

The earlier chapters in this text focused on the effect of linear and time-invariant (LTI) systems on deterministic signals, developing tools for analyzing this class of signals and systems, and using these to understand applications in communication (e.g., pulse amplitude modulation), control (e.g., stability of feedback systems), and signal processing (e.g., filtering). It is important to develop a comparable understanding and associated tools for treating the effect of LTI systems on signals modeled as the outcome of probabilistic experiments, that is, the class of signals referred to as random signals, alternatively referred to as random processes or stochastic processes. Such signals play a central role in signal and system analysis and design. In this chapter, we define random processes through the associated ensemble of signals, and explore their time-domain properties. Chapter 11 examines their characteristics in the frequency domain. The subsequent chapters use random processes as models for random or uncertain signals that arise in communication, control and signal processing applications, and study a variety of related inference problems involving estimation and hypothesis testing.

## 10.1 DEFINITION AND EXAMPLES OF A RANDOM PROCESS

In Section 7.3, we defined a random variable $X$ as a function that maps each outcome of a probabilistic experiment to a real number. In a similar manner, a real-valued continuous-time (CT) or discrete-time (DT) random process—$X(t)$ or $X[n]$, respectively—is a function that maps each outcome of a probabilistic experiment to a real CT or DT signal, termed the realization of the random process in that experiment. For any fixed time instant $t = t_0$ or $n = n_0$, the quantities $X(t_0)$ and $X[n_0]$ are simply random variables. The collection of signals that can be produced by the random process is referred to as the ensemble of signals in the random process.

## Example 10.1    Random Oscillators

As an example of a random process, consider a warehouse containing $N$ harmonic oscillators, each producing a sinusoidal waveform of some specific amplitude, frequency, and phase. The three parameters may in general differ between oscillators. This collection constitutes the ensemble of signals. The probabilistic experiment that yields a particular signal realization consists of selecting an oscillator according to some probability mass function (PMF) that assigns a probability to each of the numbers from 1 to $N$, so that the $i$th oscillator is picked with probability $p_i$. Associated with each outcome of this experiment is a specific sinusoidal waveform. Before an oscillator is chosen, there is uncertainty about what the amplitude, frequency, and phase of the outcome of the experiment will be, that is, the amplitude $A$, frequency $\Phi$, and phase $\Theta$ are all random variables. Consequently, for this example, we might express the random process as

$$X(t; A, \Phi, \Theta) = A \sin(\Phi t + \Theta) \qquad (10.1)$$

where, as in Figure 10.1, we have listed after the semi-colon the parameters that are random variables. As the discussion proceeds, we will typically simplify the notation to refer to $X(t)$ when it is clear which parameters are random variables; so, for example, Eq. (10.1) will alternatively be written as

$$X(t) = A \sin(\Phi t + \Theta) . \qquad (10.2)$$

The value $X(t_1)$ at some specific time $t_1$ is also a random variable. In the context of this experiment, knowing the PMF associated with the selection of the numbers 1 to $N$ involved in choosing an oscillator, as well as the specific amplitude, frequency, and phase of each oscillator, we could determine the probability distributions of any of the underlying random variables $A$, $\Phi$, $\Theta$, or $X(t_1)$ mentioned above.

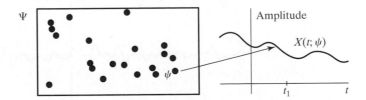

**Figure 10.1**    A random process.

Throughout this and later chapters, we will consider many examples of random processes. What is important at this point, however, is to develop a good mental picture of what a random process is. A random process is not just one signal but rather an ensemble of signals. This is illustrated schematically in Figure 10.2, for which the outcome of the probabilistic experiment could be any of the four waveforms indicated. Each waveform is deterministic, but the process is probabilistic or random because it is not known *a priori* which waveform will be generated by the probabilistic experiment. Consequently, prior to obtaining the outcome of the probabilistic experiment, many aspects of the signal are unpredictable, since there is uncertainty associated with which signal will be produced. After the experiment, or *a posteriori*, the outcome is totally determined.

If we focus on the values that a CT random process $X(t)$ can take at a particular instant of time, say $t_1$—that is, if we look down the entire ensemble at a fixed time—what we have is a random variable, namely $X(t_1)$. If we focus on the ensemble of values taken at an arbitrary collection of $\ell$ fixed time instants $t_1 < t_2 < \cdots < t_\ell$ for some arbitrary positive integer $\ell$, we have a set of $\ell$ jointly distributed random variables $X(t_1), X(t_2), \cdots, X(t_\ell)$, all determined together by the outcome of the underlying probabilistic experiment. From this point of view, a random process can be thought of as a family of jointly distributed random variables indexed by $t$. A full probabilistic characterization of this collection of random variables would require the joint probability density functions (PDFs) of multiple samples of the signal, taken at arbitrary times:

$$f_{X(t_1),X(t_2),\ \cdots,\ X(t_\ell)}(x_1, x_2, \cdots, x_\ell) \tag{10.3}$$

for all $\ell$ and all $t_1, t_2, \cdots, t_\ell$.

Correspondingly, a DT random process consists of a collection of random variables $X[n]$ for all integer values of $n$, with a full probabilistic characterization consisting of the joint PDF

$$f_{X[n_1],X[n_2],\ \cdots,\ X[n_\ell]}(x_1, x_2, \cdots, x_\ell) \tag{10.4}$$

for all $\ell$ and all integers $n_1, \cdots, n_\ell$.

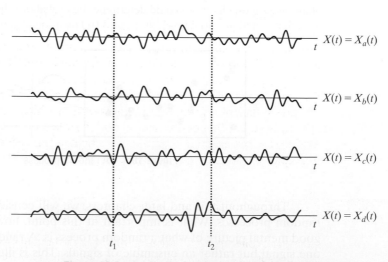

**Figure 10.2**   Realizations of the random process $X(t)$.

In a general context, it would be impractical to have a full characterization of a random process through Eqs. (10.3) or (10.4). As we will see in Example 10.2 and in other examples in this chapter, in many useful cases the full characterization can be inferred from a simpler probabilistic characterization. Furthermore, for much of what we deal with in this text, a characterization of a random process through first and second moments, as discussed in Section 10.2, is useful and sufficient.

**Example 10.2  An Independent Identically Distributed (I.I.D.) Process**

Consider a DT random process whose values $X[n]$ may be regarded as independently chosen at each time $n$ from a fixed PDF $f_X(x)$, so the values are independent and identically distributed, thereby yielding what is called an independent identically distributed (i.i.d.) process. Such processes are widely used in modeling and simulation. For example, suppose a particular DT communication channel corrupts a transmitted signal with added noise. If the noise takes on independent values at each time instant, but with characteristics that seem unchanging over the time window of interest, then the noise may be well modeled as an i.i.d. process. It is also easy to generate an i.i.d. process in a simulation environment, provided a random number generator can be arranged to produce samples from a specified PDF. Processes with more complicated dependence across time samples can then be obtained by filtering or other operations on the i.i.d. process, as we will see in this chapter as well as the next.

For an i.i.d. process, we can write the joint PDF as a product of the marginal densities, that is,

$$f_{X[n_1],X[n_2],\cdots X[n_\ell]}(x_1,x_2,\cdots,x_\ell) = f_X(x_1)f_X(x_2)\cdots f_X(x_\ell) \qquad (10.5)$$

for any choice of $\ell$ and $n_1,\cdots,n_\ell$.

An important set of questions that arises as we work with random processes in later chapters of this text is whether, by observing just part of the outcome of a random process, we can determine the complete outcome. The answer will depend on the details of the random process. For the process in Example 10.1, the answer is yes, but in general the answer is no. For some random processes, having observed the outcome in a given time interval might provide sufficient information to know exactly which ensemble member it corresponds to. In other cases this will not be sufficient. Some of these aspects are explored in more detail later, but we conclude this section with two additional examples that further emphasize these points.

**Example 10.3  Ensemble of Batteries**

Consider a collection of $N$ batteries, with $N_i$ of the batteries having voltage $v_i$, where $v_i$ is an integer between 1 and 10. The plot in Figure 10.3 indicates the number of batteries with each value $v_i$. The probabilistic experiment is to choose one of the batteries, with

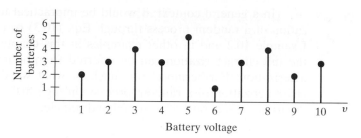

**Figure 10.3**    Plot of battery voltage distribution for Example 10.3.

the probability of picking any specific one being $\frac{1}{N}$, that is, any one battery is equally likely to be picked. Thus, scaling Figure 10.3 by $\frac{1}{N}$ represents the PMF for the battery voltage obtained as the outcome of the probabilistic experiment. Since the battery voltage is a signal (which in this case happens to be constant with time), this probabilistic experiment generates a random process. In fact, this example is similar to the oscillator example discussed earlier, but with frequency and phase both zero so that only the amplitude is random, and restricted to be an integer.

For this example, observation of $X(t)$ at any one time is sufficient information to determine the outcome for all time.

Example 10.3 is a very simple random process that, together with Example 10.4, helps to visualize some important general concepts of stationarity and ergodicity associated with random processes.

## Example 10.4  Ensemble of Coin Tossers

In this example, consider a collection of $N$ people, each independently having written down a long arbitrary string of 1s and 0s, with each entry chosen independently of any other entry in their string (similar to a sequence of independent coin tosses), and with an identical probability of a 1 at each entry. The random process now comprises this ensemble of the strings of 1s and 0s. A realization of the process is obtained by randomly selecting a person (and therefore one of the $N$ strings of 1s and 0s). After selection, the specific ensemble member of the random process is totally determined.

Next, suppose that you are shown only the 10th entry in the selected string. Because of the manner in which the string was generated, it is clearly not possible from that information to determine the 11th entry. Similarly, if the entire past history up to the 10th entry was revealed, it would not be possible to determine the remaining sequence beyond the tenth.

While the entire sequence has been determined in advance by the nature of the experiment, partial observation of a given ensemble member is in general not sufficient to fully specify that member.

Rather than looking at the $n$th entry of a single ensemble member, we can consider the random variable corresponding to the values from the entire ensemble at the $n$th entry. Looking down the ensemble at $n = 10$, for example, we would see 1s and 0s in a ratio consistent with the probability of a 1 or 0 being chosen by each individual at $n = 10$.

## 10.2 FIRST- AND SECOND-MOMENT CHARACTERIZATION OF RANDOM PROCESSES

In the above discussion, we noted that a random process can be thought of as a family of jointly distributed random variables indexed by $t$ or $n$. However it would in general be extremely difficult or impossible to analytically represent a random process in this way. Fortunately, the most widely used random process models have special structure that permits computation of such a statistical specification. Also, particularly when we are processing our signals with linear systems, we often design the processing or analyze the results by considering only the first and second moments of the process.

The first moment or mean function of a CT random process $X(t)$, which we typically denote as $\mu_X(t)$, is the expected value of the random variable $X(t)$ at each time $t$, that is,

$$\mu_X(t) = E[X(t)] . \tag{10.6}$$

The autocorrelation function and the autocovariance function represent second moments. The autocorrelation function $R_{XX}(t_1, t_2)$ is

$$R_{XX}(t_1, t_2) = E[X(t_1)X(t_2)] \tag{10.7}$$

and the autocovariance function $C_{XX}(t_1, t_2)$ is

$$C_{XX}(t_1, t_2) = E[(X(t_1) - \mu_X(t_1))(X(t_2) - \mu_X(t_2))]$$
$$= R_{XX}(t_1, t_2) - \mu_X(t_1)\mu_X(t_2) , \tag{10.8}$$

where $t_1$ and $t_2$ are two arbitrary time instants. The word *auto* (which is sometimes dropped to simplify the terminology) refers to the fact that both samples in the correlation function or the covariance function come from the same process.

One case in which the first and second moments actually suffice to completely specify the process is a Gaussian process, defined as a process whose samples are always jointly Gaussian, represented by the generalization of the bivariate Gaussian to many variables.

We can also consider multiple random processes, for example, two processes, $X(\cdot)$ and $Y(\cdot)$. A full stochastic characterization of this requires the PDFs of all possible combinations of samples from $X(\cdot)$ and $Y(\cdot)$. We say that $X(\cdot)$ and $Y(\cdot)$ are independent if every set of samples from $X(\cdot)$ is independent of every set of samples from $Y(\cdot)$, so that the joint PDF factors as follows:

$$f_{X(t_1), \cdots X(t_k), Y(t'_1), \cdots Y(t'_\ell)}(x_1, \cdots, x_k, y_1, \cdots, y_\ell)$$
$$= f_{X(t_1), \cdots X(t_k)}(x_1, \cdots, x_k) . f_{Y(t'_1), \cdots Y(t'_\ell)}(y_1, \cdots, y_\ell) \tag{10.9}$$

for all $k$, $\ell$, and all choices of sample times.

If only first and second moments are of interest, then in addition to the individual first and second moments of $X(\cdot)$ and $Y(\cdot)$, we need to consider the cross-moment functions. Specifically, the cross-correlation function $R_{XY}(t_1, t_2)$ and the cross-covariance function $C_{XY}(t_1, t_2)$ are defined respectively as

$$R_{XY}(t_1, t_2) = E[X(t_1)Y(t_2)], \text{ and} \tag{10.10}$$

$$C_{XY}(t_1, t_2) = E[(X(t_1) - \mu_X(t_1))(Y(t_2) - \mu_Y(t_2))]$$

$$= R_{XY}(t_1, t_2) - \mu_X(t_1)\mu_Y(t_2) \tag{10.11}$$

for arbitrary time $t_1, t_2$. If $C_{XY}(t_1, t_2) = 0$ for all $t_1, t_2$, we say that the processes $X(\cdot)$ and $Y(\cdot)$ are uncorrelated. Note again that the term *uncorrelated* in its common usage means that the processes have zero covariance rather than zero correlation.

The above discussion carries over to the case of DT random processes, with the exception that now the sampling instants are restricted to integer times. In accordance with our convention of using square brackets $[\cdot]$ around the time argument for DT signals, we will write $\mu_X[n]$ for the mean function of a random process $X[\cdot]$ at time $n$. Similarly, we will write $R_{XX}[n_1, n_2]$ and $C_{XX}[n_1, n_2]$ for the correlation and covariance functions involving samples at times $n_1$ and $n_2$, and $R_{XY}[n_1, n_2]$ and $C_{XY}[n_1, n_2]$ for the cross-moment functions of two random variables $X[\cdot]$ and $Y[\cdot]$ sampled at times $n_1$ and $n_2$ respectively.

## 10.3 STATIONARITY

### 10.3.1 Strict-Sense Stationarity

In general, we would expect that the joint PDFs associated with the random variables obtained by sampling a random process at an arbitrary number $\ell$ of arbitrary times will be time-dependent, that is, the joint PDF

$$f_{X(t_1), \cdots, X(t_\ell)}(x_1, \cdots, x_\ell) \tag{10.12}$$

will depend on the specific values of $t_1, \cdots, t_\ell$. If all the joint PDFs remain the same under arbitrary time shifts, so that if

$$f_{X(t_1), \cdots, X(t_\ell)}(x_1, \cdots, x_\ell) = f_{X(t_1+\alpha), \cdots, X(t_\ell+\alpha)}(x_1, \cdots, x_\ell) \tag{10.13}$$

for arbitrary $\alpha$, then the random process is said to be strict-sense stationary (SSS). Said another way, for an SSS process, the statistics depend only on the relative times at which the samples are taken, not on the absolute times. The processes in Examples 10.2 and 10.3 are SSS. More generally, any i.i.d. process is strict-sense stationary.

### 10.3.2 Wide-Sense Stationarity

Of particular use is a less restricted type of stationarity. Specifically, if the mean value $\mu_X(t)$ is invariant with time and the autocorrelation $R_{XX}(t_1, t_2)$ or, equivalently, the autocovariance $C_{XX}(t_1, t_2)$ is a function of only the time difference $(t_1 - t_2)$, then the process is referred to as wide-sense stationary

(WSS). A process that is SSS is always WSS, but the reverse is not necessarily true. For a WSS random process $X(t)$, we have

$$\mu_X(t) = \mu_X \qquad (10.14)$$

$$R_{XX}(t_1, t_2) = R_{XX}(t_1 + \alpha, t_2 + \alpha) \text{ for every } \alpha$$

$$= R_{XX}(t_1 - t_2, 0)$$

$$= R_{XX}(t_1 - t_2), \qquad (10.15)$$

where the last equality defines a more compact notation since a single argument for the time difference $(t_1 - t_2)$ suffices for a WSS process. Similarly, $C_{XX}(t_1, t_2)$ will be written as $C_{XX}(t_1 - t_2)$ for a WSS process. The time difference $(t_1 - t_2)$ will typically be denoted as $\tau$ and referred to as the lag variable for the autocorrelation and autocovariance functions.

For a Gaussian process, that is, a process whose samples are always jointly Gaussian, WSS implies SSS because jointly Gaussian variables are entirely determined by their joint first and second moments.

Two random processes $X(\cdot)$ and $Y(\cdot)$ are referred to as jointly WSS if their first and second moments, including the cross-covariance, are stationary. In this case, we use the notation $R_{XY}(\tau)$ to denote $E[X(t + \tau)Y(t)]$. It is worth noting that an alternative convention sometimes used elsewhere is to define $R_{XY}(\tau)$ as $E[X(t)Y(t + \tau)]$. In our notation, this expectation would be denoted by $R_{XY}(-\tau)$. It is important to take account of what notational convention is being followed when referencing other sources, and you should also be clear about the notational convention used in this text.

---

**Example 10.5    *Random Oscillators Revisited***

Consider again the harmonic oscillators introduced in Example 10.1:

$$X(t; A, \Theta) = A\cos(\phi_0 t + \Theta) \qquad (10.16)$$

where $A$ and $\Theta$ are independent random variables, and now the frequency is fixed at some known value denoted by $\phi_0$.

If $\Theta$ is also fixed at a constant value $\theta_0$, then every outcome is of the form $x(t) = A\cos(\phi_0 t + \theta_0)$, and it is straightforward to see that this process is not WSS (and consequently also not SSS). For instance, if $A$ has a nonzero mean value, $\mu_A \neq 0$, then the expected value of the process, namely $\mu_A \cos(\phi_0 t + \theta_0)$, is time varying. To show that the process is not WSS even when $\mu_A = 0$, we can examine the autocorrelation function. Note that $x(t)$ is fixed at 0 for all values of $t$ for which $\phi_0 t + \theta_0$ is an odd multiple of $\pi/2$, and takes the values $\pm A$ halfway between such points; the correlation between such samples taken $\pi/\phi_0$ apart in time can correspondingly be 0 (in the former case) or $-E[A^2]$ (in the latter). The process is thus not WSS, even when $\mu_A = 0$.

However, if $\Theta$ is distributed uniformly in $[-\pi, \pi]$, then

$$\mu_X(t) = \mu_A \int_{-\pi}^{\pi} \frac{1}{2\pi} \cos(\phi_0 t + \theta)\, d\theta = 0, \qquad (10.17)$$

$$C_{XX}(t_1, t_2) = R_{XX}(t_1, t_2)$$

$$= E[A^2]E[\cos(\phi_0 t_1 + \Theta)\cos(\phi_0 t_2 + \Theta)]. \qquad (10.18)$$

Equation (10.18) can be evaluated as

$$C_{XX}(t_1, t_2) = \frac{E[A^2]}{2} \int_{-\pi}^{\pi} \frac{1}{2\pi} [\cos(\phi_0(t_2 - t_1)) + \cos(\phi_0(t_2 + t_1) + 2\theta)] \, d\theta \qquad (10.19)$$

to obtain

$$C_{XX}(t_1, t_2) = \frac{E[A^2]}{2} \cos(\phi_0(t_2 - t_1)) . \qquad (10.20)$$

For this restricted case, then, the process is WSS. It can also be shown to be SSS, although this is not totally straightforward to show formally.

For the most part, the random processes that we treat will be WSS. As noted earlier, to simplify notation for a WSS process, we write the correlation function as $R_{XX}(t_1 - t_2)$; the argument $(t_1 - t_2)$ is often denoted by the lag variable $\tau$ at which the correlation is computed. When considering only first and second moments and not the entire PDF or cumulative distribution function (CDF), it will be less important to distinguish between the random process $X(t)$ and a specific realization $x(t)$ of it—so a further notational simplification is introduced by using lowercase letters to denote the random process itself. We shall thus refer to the random process $x(t)$, and—in the case of a WSS process—denote its mean by $\mu_x$ and its correlation function $E[x(t + \tau)x(t)]$ by $R_{xx}(\tau)$. Correspondingly, for DT we refer to the random process $x[n]$ and, in the WSS case, denote its mean by $\mu_x$ and its correlation function $E[x[n + m]x[n]]$ by $R_{xx}[m]$.

### 10.3.3 Some Properties of WSS Correlation and Covariance Functions

For real-valued WSS processes $x(t)$ and $y(t)$, the correlation and covariance functions have the following symmetry properties:

$$R_{xx}(\tau) = R_{xx}(-\tau) , \qquad\qquad C_{xx}(\tau) = C_{xx}(-\tau) , \qquad (10.21)$$

$$R_{xy}(\tau) = R_{yx}(-\tau) , \qquad\qquad C_{xy}(\tau) = C_{yx}(-\tau) . \qquad (10.22)$$

For example, the symmetry in Eq. (10.22) of the cross-correlation function $R_{xy}(\tau)$ follows directly from interchanging the arguments inside the defining expectations:

$$R_{xy}(\tau) = E[x(t)y(t - \tau)] \qquad (10.23a)$$

$$= E[y(t - \tau)x(t)] \qquad (10.23b)$$

$$= R_{yx}(-\tau) . \qquad (10.23c)$$

The other properties in Eqs. (10.21) and (10.22) follow in a similar manner.

Equation (10.21) indicates that the autocorrelation and autocovariance functions have even symmetry. Equation (10.22) indicates that for cross-correlation and cross-covariance functions, interchanging the random variables is equivalent to reflecting the function about the $\tau$ axis. And of course,

Eq. (10.21) is a special case of Eq. (10.22) with $y(t) = x(t)$. Similar properties hold for DT WSS processes.

Another important property of correlation and covariance functions follows from noting that, as discussed in Section 7.7, Eq. (7.63), the correlation coefficient of two random variables has magnitude not exceeding 1. Specifically since the correlation coefficient between $x(t)$ and $x(t + \tau)$ is given by $C_{xx}(\tau)/C_{xx}(0)$, then

$$-1 \leq \frac{C_{xx}(\tau)}{C_{xx}(0)} \leq 1, \tag{10.24}$$

or equivalently,

$$-C_{xx}(0) \leq C_{xx}(\tau) \leq C_{xx}(0). \tag{10.25}$$

Adding $\mu_x^2$ to each term above, we can conclude that

$$-R_{xx}(0) + 2\mu_x^2 \leq R_{xx}(\tau) \leq R_{xx}(0). \tag{10.26}$$

In Chapter 11, we will demonstrate that correlation and covariance functions are characterized by the property that their Fourier transforms are real and nonnegative at all frequencies, because these transforms describe the frequency distribution of the expected power in the random process. The above symmetry constraints and bounds will then follow as natural consequences, but they are worth highlighting here.

We conclude this section with two additional examples. The first, the Bernoulli process, is the more formal name for repeated independent flips of a possibly biased coin. The second example, referred to as the random telegraph wave, is often used as a simplified representation of a random square wave or switch in electronics or communication systems.

## Example 10.6    The Bernoulli Process

The Bernoulli process is an example of an i.i.d. DT process with

$$P(x[n] = 1) = p \tag{10.27}$$

$$P(x[n] = -1) = (1 - p) \tag{10.28}$$

and with the value at each time instant $n$ independent of the values at all other time instants. The mean, autocorrelation, and covariance functions are:

$$E\{x[n]\} = 2p - 1 = \mu_x \tag{10.29}$$

$$E\{x[n + m]x[n]\} = \begin{cases} 1 & m = 0 \\ (2p - 1)^2 & m \neq 0 \end{cases} \tag{10.30}$$

$$C_{xx}[m] = E\{(x[n + m] - \mu_x)(x[n] - \mu_x)\} \tag{10.31}$$

$$= \{1 - (2p - 1)^2\}\delta[m] = 4p(1 - p)\delta[m]. \tag{10.32}$$

## Example 10.7   Random Telegraph Wave

An example of a CT random process that we will make occasional reference to is the random telegraph wave. A representative sample function of a random telegraph wave process is shown in Figure 10.4 and can be defined through the following two properties:

1. $x(0) = \pm 1$ with equal probability 0.5. This property together with the obvious fact that in any time interval the number of sign changes will be either even or odd also implies that $x(t) = \pm 1$ with equal probability 0.5 at any time $t$.

2. $x(t)$ changes polarity at Poisson times, that is, the probability of $k$ sign changes in a time interval of length $T$ is

$$P(k \text{ sign changes in an interval of length } T) = \frac{(\lambda T)^k e^{-\lambda T}}{k!} , \qquad (10.33)$$

where the constant $\lambda$ represents the rate of the transitions.

Property 2 implies that the probability of a (nonnegative), even number of sign changes in an interval of length $T$ is

$$P(\text{even \# of sign changes}) = \sum_{\substack{k=0 \\ k \text{ even}}}^{\infty} \frac{(\lambda T)^k e^{-\lambda T}}{k!} = e^{-\lambda T} \sum_{k=0}^{\infty} \frac{1 + (-1)^k}{2} \frac{(\lambda T)^k}{k!} . \quad (10.34)$$

Using the identity

$$e^{\lambda T} = \sum_{k=0}^{\infty} \frac{(\lambda T)^k}{k!} , \qquad (10.35)$$

Eq. (10.34) becomes

$$P(\text{even \# of sign changes}) = e^{-\lambda T} \frac{(e^{\lambda T} + e^{-\lambda T})}{2}$$

$$= \frac{1}{2}(1 + e^{-2\lambda T}) . \qquad (10.36)$$

Similarly, the probability of an odd number of sign changes in an interval of length $T$ is

$$P(\text{odd \# of sign changes}) = \frac{1}{2}(1 - e^{-2\lambda T}) . \qquad (10.37)$$

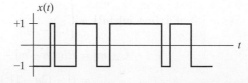

**Figure 10.4**   One realization of a random telegraph wave.

From Eqs. (10.36) and (10.37), we conclude that

$$\mu_X(t) = 0, \text{ and} \tag{10.38}$$

$$
\begin{aligned}
R_{xx}(t_1, t_2) &= E[x(t_1)x(t_2)] \\
&= 1 \times P\left(x(t_1) = x(t_2)\right) + (-1) \times P\left(x(t_1) \neq x(t_2)\right) \\
&= e^{-2\lambda|t_1 - t_2|}.
\end{aligned}
\tag{10.39}
$$

In other words, the process is exponentially correlated and WSS.

## 10.4 ERGODICITY

The formal concept of ergodicity is sophisticated and subtle and beyond the scope of this book, but the essential idea is described here. We typically observe a particular realization of a random process (e.g., we record a noise waveform) and want to characterize the statistics of the random process by measurements on one ensemble member. For instance, we could consider the time average of the waveform to represent the mean value of the process, assuming the mean value is constant for all time. If for (almost) every realization or with probability 1, the time average equals the ensemble mean, the process is referred to as ergodic in mean value.

We could also construct a histogram that represents the fraction of time — rather than the fraction of the ensemble at any given time — for which the waveform lies in different amplitude bins, and then examine whether this temporal amplitude histogram reflects the probability density across the ensemble of the value obtained at a particular sampling time. If the random process is such that the amplitude distribution of (almost) any particular realization over time is representative of the probability distribution down the ensemble, then the process is called ergodic in distribution. More generally, a process is simply termed ergodic if ensemble statistics can be replaced by temporal statistics on (almost) every particular realization. A simple example of a process that is not ergodic is Example 10.3, although the process is SSS. For this example, the behavior of any particular realization is not representative of the behavior down the ensemble.

In our discussion of random processes, we will primarily be concerned with first- and second-order moments. While it is difficult to determine in general whether a random process is ergodic, there are criteria, specified in terms of the moments of the process, that establish ergodicity in the mean and in autocorrelation. Of course, a process with time-varying mean cannot be ergodic in the mean. It can be shown, however, that a WSS process with finite variance at each instant and with an autocovariance function that approaches zero as the lag goes to infinity is ergodic in the mean. Criteria for ergodicity of the mean are explored in Example 11.3 (see also Problem 10.43).

Frequently, ergodicity is simply assumed for convenience, in the absence of evidence that the assumption is not reasonable. Under this assumption, the mean and autocorrelation can be obtained from time averaging on (almost) any single ensemble member or realization, through the following equalities, stated here for the CT case:

$$E[x(t)] = \lim_{T \to \infty} \frac{1}{2T} \int_{-T}^{T} x'(t)\, dt \qquad (10.40)$$

and

$$E[x(t)x(t - \tau)] = \lim_{T \to \infty} \frac{1}{2T} \int_{-T}^{T} x'(t)x'(t - \tau)\, dt , \qquad (10.41)$$

where $x'(t)$ here denotes a particular realization, not the ensemble.

A random process for which Eqs. (10.40) and (10.41) are true for (almost) any realization is referred as second-order ergodic. The DT version of Eq. (10.40), for a process that is ergodic in mean value, states that

$$E\{x[n]\} = \lim_{K \to \infty} \frac{1}{2K + 1} \sum_{-K}^{K} x'[k] , \qquad (10.42)$$

with an analogous version of Eq. (10.41) for ergodicity in correlation.

## 10.5 LINEAR ESTIMATION OF RANDOM PROCESSES

A common class of problems in a variety of aspects of communication, control, and signal processing involves the estimation of one random process from observations of another, or estimating (predicting) future values of a process from observation of its past values. For example, it is common in communication systems that the signal at the receiver is a corrupted version of the transmitted signal, and we would like to estimate the transmitted signal from the received signal. Other examples are predicting weather or financial data from past observations. We will be treating this general topic in more detail in Chapter 12, but a first look at it here can be beneficial in understanding random processes.

We first consider a simple example of linear prediction of a random process, and then a more elaborate example of linear finite-impulse-response (FIR) filtering of a noise-corrupted process to estimate the underlying random signal.

### 10.5.1 Linear Prediction

As a simple illustration of linear prediction, consider a WSS DT process $x[n]$. Knowing the value at time $n_0$, we may wish to predict what the value will be $m$ samples into the future, that is, at time $n_0 + m$. We limit the prediction strategy to an affine one: with $\widehat{x}[n_0 + m]$ denoting the predicted value, we restrict $\widehat{x}[n_0 + m]$ to be of the form

$$\widehat{x}[n_0 + m] = ax[n_0] + b \qquad (10.43)$$

and choose the predictor parameters $a$ and $b$ to minimize the expected value of the square of the error, that is, we choose $a$ and $b$ to minimize

$$\epsilon = E\{(x[n_0 + m] - \widehat{x}[n_0 + m])^2\}, \tag{10.44}$$

or equivalently,

$$\epsilon = E\{(x[n_0 + m] - ax[n_0] - b)^2\}. \tag{10.45}$$

This is identical to the linear minimum mean square error (LMMSE) estimation problem in Section 8.2, with $\widehat{Y}_\ell$ in Eqs. (8.38) and (8.43) now corresponding to $\widehat{x}[n_0 + m]$, and $X$ in those equations corresponding to $x[n_0]$. The development in Section 8.2 concluded that the error $x[n_0 + m] - \widehat{x}[n_0 + m]$ associated with the optimal estimate is orthogonal to the available data $x[n_0]$ and that the estimate is unbiased, that is, the expected value of the error is zero. The corresponding equations are

$$E\{(x[n_0 + m] - ax[n_0] - b)x[n_0]\} = E\{(x[n_0 + m] - \widehat{x}[n_0 + m])x[n_0]\} = 0 \tag{10.46a}$$

$$E\{x[n_0 + m] - ax[n_0] - b\} = E\{x[n_0 + m] - \widehat{x}[n_0 + m]\} = 0. \tag{10.46b}$$

Carrying out the multiplications and expectations in Eqs. (10.46) results in the following, which can be solved for the desired constants:

$$R_{xx}[n_0 + m, n_0] - aR_{xx}[n_0, n_0] - b\mu_x[n_0] = 0 \tag{10.47a}$$

$$\mu_x[n_0 + m] - a\mu_x[n_0] - b = 0. \tag{10.47b}$$

Equivalently, from Eq. (8.57), we obtain

$$a = C_{xx}[n_0 + m, n_0]/C_{xx}[n_0] \tag{10.48}$$

and from Eq. (8.52) we obtain

$$b = \mu_x[n_0 + m] - a\mu_x[n_0]. \tag{10.49}$$

Since the process is assumed to be WSS, $R_{xx}[n_0 + m, n_0] = R_{xx}[m]$. Assume also that it is zero mean, so $\mu_x = 0$. Equations (10.47), (10.48), and (10.49) then reduce to

$$a = R_{xx}[m]/R_{xx}[0] = C_{xx}[m]/C_{xx}[0] \tag{10.50}$$

$$b = 0 \tag{10.51}$$

so that

$$\widehat{x}[n_0 + m] = \frac{C_{xx}[m]}{C_{xx}[0]}x[n_0]. \tag{10.52}$$

When the process is WSS but not necessarily zero mean, the LMMSE predictor is given by

$$\widehat{x}[n_0 + m] = \mu_x + \frac{C_{xx}[m]}{C_{xx}[0]}(x[n_0] - \mu_x). \tag{10.53}$$

An extension of this discussion would consider how to do LMMSE prediction when measurements of several past values are available.

In the next section, we consider another context in which linear estimation of random processes naturally arises, specifically estimation of a random process from noisy measurements.

### 10.5.2 Linear FIR Filtering

As another example, which we will treat in more generality in Chapter 12 on signal estimation, consider a DT signal $s[n]$ that has been corrupted by additive noise $d[n]$. For example, $s[n]$ might be a signal transmitted over a channel and $d[n]$ the noise introduced by the channel. The received signal $r[n]$ is then

$$r[n] = s[n] + d[n] . \qquad (10.54)$$

Assume that $s[\cdot]$ and $d[\cdot]$ are zero-mean jointly WSS random processes and are uncorrelated. At the receiver we would like to process $r[\cdot]$ with a causal FIR filter to estimate the transmitted signal $s[n]$, as indicated in Figure 10.5.

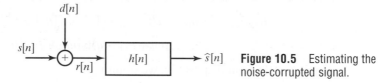

**Figure 10.5**    Estimating the noise-corrupted signal.

If $h[\cdot]$ is a causal FIR filter of length $L$, then

$$\widehat{s}[n] = \sum_{k=0}^{L-1} h[k]r[n-k] . \qquad (10.55)$$

We would like to determine the filter coefficients $h[k]$ to minimize the mean square error between $\widehat{s}[n]$ and $s[n]$, that is, minimize $\epsilon$ given by

$$\epsilon = E\{(s[n] - \widehat{s}[n])^2\}$$

$$= E\{(s[n] - \sum_{k=0}^{L-1} h[k]r[n-k])^2\} . \qquad (10.56)$$

We again apply the results from Section 8.2. Specifically, the error $\{s[n] - \widehat{s}[n]\}$ associated with the optimal estimate is orthogonal to the available data, $r[n-m]$ for $m = 0, \cdots, L-1$. This corresponds to the condition

$$E\{(s[n] - \sum_{k} h[k]r[n-k])r[n-m]\} = 0, \quad m = 0, 1, \cdots, L-1 . \qquad (10.57)$$

Carrying out the multiplications in Eq. (10.57) and taking expectations results in

$$\sum_{k=0}^{L-1} h[k]R_{rr}[m-k] = R_{sr}[m], \quad m = 0, 1, \cdots, L-1 . \tag{10.58}$$

These are the normal equations, Eq. (8.87) and Eq. (8.88), specialized to the context of FIR filtering. They are written here in terms of correlation rather than covariance functions as the two are identical for the zero-mean case that we are assuming. Equation (10.58) constitutes $L$ equations that can be solved for the $L$ parameters $h[k]$. With $r[n] = s[n] + d[n]$, it is straightforward to show that $R_{sr}[m] = R_{ss}[m] + R_{sd}[m]$ and since we assumed that $s[\cdot]$ and $d[\cdot]$ are uncorrelated, then $R_{sd}[m] = 0$ and $R_{sr}[m] = R_{ss}[m]$. Similarly, $R_{rr}[m] = R_{ss}[m] + R_{dd}[m]$.

As will be done in Chapter 12, these results are also easily modified for the case where the processes no longer have zero mean.

## 10.6 LTI FILTERING OF WSS PROCESSES

We will see in later chapters how the correlation properties of a WSS random process, and the effects of LTI systems on these properties, play an important role in understanding and designing systems for such tasks as filtering, signal estimation, signal detection, and system identification. We focus in this section on understanding in the time domain how LTI systems shape the correlation properties of a WSS random process. In Chapter 11, we will develop a parallel picture in the frequency domain, after establishing that the frequency distribution of the expected power in a WSS random signal is described by the Fourier transform of the autocorrelation function.

Consider an LTI system whose input is a sample function of a WSS random process $x(t)$, that is, a signal chosen by a probabilistic experiment from the ensemble that constitutes the random process $x(t)$. More simply, the input is the random process $x(t)$.

Among other considerations, it is of interest to know when the output process $y(t)$—the ensemble of signals obtained as responses to the signals in the input ensemble—will itself be WSS, and to determine its mean and autocovariance or autocorrelation functions, as well as its cross-correlation with the input process. For an LTI system with impulse response $h(t)$, the output $y(t)$ is given by the convolution

$$y(t) = \int_{-\infty}^{+\infty} h(v)x(t-v)\,dv = \int_{-\infty}^{+\infty} x(v)h(t-v)\,dv \tag{10.59}$$

for any specific input signal $x(t)$ for which the convolution is well defined. The convolution is well defined if, for instance, the input $x(t)$ is bounded and the system is bounded-input, bounded-output (BIBO) stable, that is, the system impulse response is absolutely integrable. Figure 10.6 illustrates generically what the two components of the integrand in the convolution integral might look like.

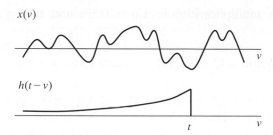

**Figure 10.6**   Illustration of the two terms in the integrand of Eq. (10.59).

Rather than requiring that every sample function of our input process be bounded, it will suffice for the convolution computations below to assume that $E[x^2(t)] = R_{xx}(0)$ is finite. This assumption, assuming also that the system is BIBO stable, ensures that $y(t)$ is a well-defined random process, and that the formal manipulations carried out below—for instance, interchanging expectation and convolution—can all be justified more rigorously by methods that are beyond our scope here. In fact, the results obtained can also be applied, when properly interpreted, to cases where the input process does not have a bounded second moment, for example, when $x(t)$ is so-called CT white noise, for which $R_{xx}(\tau) = \delta(\tau)$. The results can also be applied to a system that is not BIBO stable, as long as it has a well-defined frequency response $H(j\omega)$, as in the case of an ideal low-pass filter, for example.

The convolution relationship Eq. (10.59) can be used to deduce the first- and second-order properties of $y(t)$. What will be established is that $y(t)$ is itself WSS, and that $x(\cdot)$ and $y(\cdot)$ are in fact jointly WSS. Expressions for the autocorrelation of the output and the cross-correlation between input and output will also be developed.

First, consider the mean value of the output. Taking the expected value of both sides of Eq. (10.59),

$$
\begin{aligned}
E[y(t)] &= E\left[\int_{-\infty}^{+\infty} h(v)x(t-v)\,dv\right] \\
&= \int_{-\infty}^{+\infty} h(v)E[x(t-v)]\,dv \\
&= \int_{-\infty}^{+\infty} h(v)\mu_x\,dv \\
&= \mu_x \int_{-\infty}^{+\infty} h(v)\,dv \\
&= H(j0)\,\mu_x = \mu_y\,. \tag{10.60}
\end{aligned}
$$

In other words, the mean of the output process is constant, and equals the mean of the input scaled by the "DC" or zero-frequency gain of the system. This is also what the constant response of the system would be if its input were held constant at the value $\mu_x$.

The preceding result and the linearity of the system also allow us to conclude that applying the zero-mean WSS process $x(t) - \mu_x$ to the input of the stable LTI system will result in the zero-mean process $y(t) - \mu_y$ at the output. This fact will be useful below in converting results that are derived for correlation functions into results that also apply to covariance functions.

Next, consider the cross-correlation between output and input:

$$E[y(t + \tau)x(t)] = E\left[\left\{ \int_{-\infty}^{+\infty} h(v)x(t + \tau - v)\, dv \right\} x(t)\right]$$

$$= \int_{-\infty}^{+\infty} h(v)E[x(t + \tau - v)x(t)]\, dv \,. \tag{10.61}$$

Since $x(t)$ is WSS, $E[x(t + \tau - v)x(t)] = R_{xx}(\tau - v)$, so

$$E[y(t + \tau)x(t)] = \int_{-\infty}^{+\infty} h(v)R_{xx}(\tau - v)\, dv$$

$$= R_{yx}(\tau) \,. \tag{10.62}$$

Note that the cross-correlation depends only on the lag $\tau$ between the output and input processes, not on both $\tau$ and the absolute time location $t$.

We recognize the integral in Eq. (10.62) as the convolution of the system impulse response and the autocorrelation function of the system input. This convolution operation in Eq. (10.62) is a deterministic relation, that is, the cross-correlation between the output and input is deterministically related to the autocorrelation of the input, and can be viewed as the signal that would result if the system input were the autocorrelation function $R_{xx}(\tau)$, as indicated in Figure 10.7. Correspondingly, $R_{xy}(\tau)$ and equivalently $R_{yx}(-\tau)$ would be the output resulting from an input of $R_{xx}(-\tau) = R_{xx}(\tau)$ to a system with an impulse response which is reversed in time, that is, $h(-\tau)$, which we denote as $\overleftarrow{h}(\tau)$. With this notation,

$$R_{xy}(\tau) = \int_{-\infty}^{+\infty} \overleftarrow{h}(v)R_{xx}(\tau - v)\, dv \,. \tag{10.63}$$

The above relations can also be expressed in terms of covariance functions, rather than in terms of correlation functions. For this, simply consider the case where the input to the system is the zero-mean WSS process $x(t) - \mu_x$, with corresponding zero-mean output $y(t) - \mu_y$. Since the correlation function for $x(t) - \mu_x$ is the same as the covariance function for $x(t)$, that is, since

$$R_{x-\mu_x, x-\mu_x}(\tau) = C_{xx}(\tau) \,, \tag{10.64}$$

the results above hold unchanged when every correlation function is replaced by the corresponding covariance function. It therefore follows, for instance, that $C_{yx}(\tau)$ is the convolution of $h(\tau)$ and $C_{xx}(\tau)$, or

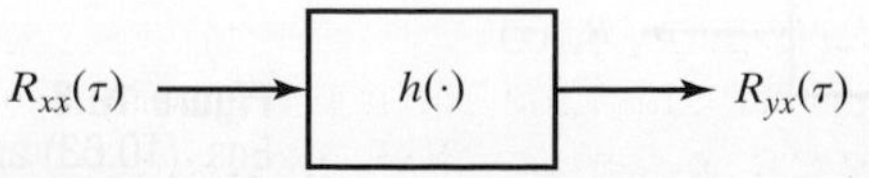

**Figure 10.7**    Representation of Eq. (10.62).

$$C_{yx}(\tau) = \int_{-\infty}^{+\infty} h(v)C_{xx}(\tau - v)\,dv\,. \tag{10.65}$$

Next, we consider the autocorrelation of the output $y(t)$:

$$E[y(t + \tau)y(t)] = E\left[\left\{\int_{-\infty}^{+\infty} h(v)x(t + \tau - v)\,dv\right\}y(t)\right]$$

$$= \int_{-\infty}^{+\infty} h(v)\,\underbrace{E[x(t + \tau - v)y(t)]}_{R_{xy}(\tau - v)}\,dv$$

$$= \int_{-\infty}^{+\infty} h(v)R_{xy}(\tau - v)\,dv\,,$$

$$= R_{yy}(\tau)\,. \tag{10.66}$$

Equation (10.66) states that $R_{yy}(\tau)$ is the convolution of the system impulse response and the cross-correlation $R_{xy}(\tau)$. Note that the autocorrelation of the output depends only on $\tau$, and not on both $\tau$ and $t$. Combining this with the earlier results, we conclude that $x(\cdot)$ and $y(\cdot)$ are jointly WSS, as claimed. The corresponding result for covariances is

$$C_{yy}(\tau) = \int_{-\infty}^{+\infty} h(v)C_{xy}(\tau - v)\,dv\,. \tag{10.67}$$

The combination of Eqs. (10.63) and (10.66) can be represented pictorially as shown in Figure 10.8, and corresponds to the relation

$$R_{yy}(\tau) = \overline{R}_{hh}(\tau) * R_{xx}(\tau). \tag{10.68}$$

The composite impulse response $\overline{R}_{hh}(\tau)$ is the deterministic autocorrelation function of $h(t)$ introduced in Chapter 1, and is given by

$$\overline{R}_{hh}(\tau) = \overleftarrow{h}(\tau) * h(\tau) = \int_{-\infty}^{+\infty} h(t + \tau)h(t)\,dt\,. \tag{10.69}$$

For the covariance function version of Eq. (10.68), we have

$$C_{yy}(\tau) = \overline{R}_{hh}(\tau) * C_{xx}(\tau)\,. \tag{10.70}$$

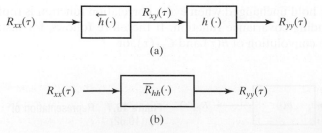

(a)

(b)

**Figure 10.8**   Combination of Eqs. (10.63) and (10.66).

Note that the deterministic correlation function of $h(t)$ is still what we use, even when relating the covariances of the input and output. Only the means of the input and output processes are adjusted in arriving at the present result; the impulse response is untouched.

The correlation relations in Eqs. (10.62), (10.63), (10.66), and (10.68), as well as their covariance counterparts, are very powerful, and we will make considerable use of them. Of equal importance are their statements in the Fourier and Laplace transform domains. Denoting the Fourier and Laplace transforms of the correlation function $R_{xx}(\tau)$ by $S_{xx}(j\omega)$ and $S_{xx}(s)$ respectively, and similarly for the other correlation functions of interest, we have

$$S_{yx}(j\omega) = H(j\omega)S_{xx}(j\omega) , \qquad S_{yy}(j\omega) = |H(j\omega)|^2 S_{xx}(j\omega) ,$$

$$S_{yx}(s) = H(s)S_{xx}(s) , \qquad S_{yy}(s) = H(s)H(-s)S_{xx}(s) . \qquad (10.71)$$

We can denote the Fourier and Laplace transforms of the covariance function $C_{xx}(\tau)$ by $D_{xx}(j\omega)$ and $D_{xx}(s)$, respectively, and similarly for the other covariance functions of interest, and then write the same sorts of relationships as above.

Exactly parallel results hold in the DT case. Consider a stable DT LTI system whose impulse response is $h[n]$ and whose input is the WSS random process $x[n]$. Then, as in the CT case, we can conclude that the output process $y[\cdot]$ is jointly WSS with the input process $x[\cdot]$, and

$$\mu_y = \mu_x \sum_{n=-\infty}^{\infty} h[n] \qquad (10.72)$$

$$R_{yx}[m] = h[m] * R_{xx}[m] \qquad (10.73)$$

$$R_{yy}[m] = \overline{R}_{hh}[m] * R_{xx}[m] , \qquad (10.74)$$

where $\overline{R}_{hh}[m]$ is the deterministic autocorrelation function of $h[m]$, defined as

$$\overline{R}_{hh}[m] = \sum_{n=-\infty}^{+\infty} h[n+m]h[n] . \qquad (10.75)$$

The corresponding Fourier and $z$-transform statements of these relationships are

$$\mu_y = H(e^{j0})\mu_x , \quad S_{yx}(e^{j\Omega}) = H(e^{j\Omega})S_{xx}(e^{j\Omega}) , \quad S_{yy}(e^{j\Omega}) = |H(e^{j\Omega})|^2 S_{xx}(e^{j\Omega}) ,$$

$$\mu_y = H(1)\mu_x , \qquad S_{yx}(z) = H(z)S_{xx}(z) , \qquad S_{yy}(z) = H(z)H(1/z)S_{xx}(z) .$$
$$(10.76)$$

All of these expressions can also be rewritten for covariances and their transforms.

In Chapter 11, we will use these relationships to show that the Fourier transform of the autocorrelation function describes how the expected power of a WSS process is distributed in frequency. For this reason, the Fourier transform of the autocorrelation function is termed the power spectral density (PSD) of the process.

The relationships developed in this chapter are also very important in using random processes to measure or identify the impulse response of an LTI system. For example, from Eq. (10.73), if the input $x[n]$ to a DT LTI system is a WSS random process with autocorrelation function $R_{xx}[m] = \delta[m]$, then by measuring the cross-correlation between the input and output, we obtain a measurement of the system impulse response. It is easy to construct a DT input process with autocorrelation function $\delta[m]$, for example, an i.i.d. process that is equally likely to take the values $+1$ and $-1$ at each time instant, such as that in Example 10.6.

As another example, suppose the input $x(t)$ to a CT LTI system with impulse response $h(t)$ is a WSS random telegraph wave, as in Example 10.7. The process $x(t)$ has zero mean and autocorrelation function $R_{xx}(\tau) = e^{-2\lambda|\tau|}$. If we determine the cross-correlation $R_{yx}(\tau)$ with the output $y(t)$ and then use the relation

$$R_{yx}(\tau) = R_{xx}(\tau) * h(\tau) \,, \tag{10.77}$$

we can obtain the system function $H(s)$ and the system impulse response $h(\tau)$. Specifically, if $S_{yx}(s), S_{xx}(s)$, and $H(s)$ denote the associated Laplace transforms, then

$$H(s) = \frac{S_{yx}(s)}{S_{xx}(s)} \,. \tag{10.78}$$

Note that $S_{xx}(s)$ is a well-behaved function of the complex variable $s$ in this case, whereas any particular sample function of the process $x(t)$ would not have such a well-behaved transform. The same comment applies to $S_{yx}(s)$.

As a third example, suppose that we know the autocorrelation function $R_{xx}[m]$ of the input $x[n]$ to a DT LTI system, but do not have access to $x[n]$ and therefore cannot determine the cross-correlation $R_{yx}[m]$ with the output $y[n]$, but can determine the output autocorrelation $R_{yy}[m]$. For example, if

$$R_{xx}[m] = \delta[m] \tag{10.79}$$

and we determine $R_{yy}[m]$ to be $R_{yy}[m] = \left(\frac{1}{2}\right)^{|m|}$, then

$$R_{yy}[m] = \left(\frac{1}{2}\right)^{|m|} = \overline{R}_{hh}[m] = \overleftarrow{h}[m] * h[m] \,. \tag{10.80}$$

Equivalently, $H(z)H(z^{-1})$ is the $z$-transform $S_{yy}(z)$ of $R_{yy}[m]$. Additional assumptions or constraints, for instance on the stability and causality of the system and its inverse, may allow one to recover $H(z)$ from knowledge of $H(z)H(z^{-1})$.

## 10.7 FURTHER READING

Several references suggested for further reading at the end of Chapter 7 contain detailed discussions of specific classes of random processes—Bernoulli, Poisson, renewal, Markov, Gaussian and martingale processes. WSS processes are much more simply characterized, and well suited to analysis as random signals that can be the inputs and outputs of LTI systems. (Since Gaussian processes are characterized by their first and second moments, they are amenable to similar analysis.) Thus references in Chapter 7 that include material on WSS processes generally describe LTI filtering of WSS processes as in this chapter, examine the frequency domain or spectral characterization of WSS processes as we do in Chapter 11, and treat some of the signal estimation and detection problems we consider in Chapters 12 and 13. Additional references for this and later chapters include classics such as [Dav] and [Van], as well as more recent texts such as [Gar], [Had], [Hay], [Jan], [Pur], and [Shi]. Treatments of random processes, their spectral characteristics, and signal estimation—material related to Chapters 10, 11, and 12 here—are also found in books devoted to time series and forecasting, such as [Blo], [Bro], [Cht], [Dur], and [Woo]. These are typically more grounded in statistics, with applications as varied as econometrics, climatology and process control. [Kle] gives an absorbing history of the development of time series, spanning a period that includes the era of statistics described in [St1] and [St2].

## Problems

## Basic Problems

**10.1.** For the random telegraph wave introduced in Example 10.7, evaluate (as a function of $T$ for $T > 0$) the conditional probability that $X(t_0 + T) = +1$, given that $X(t_0) = +1$. For what range of $T > 0$ is this conditional probability higher than the conditional probability that $X(t_0 + T) = -1$? If, for a given $T > 0$, you were to predict that $X(t_0 + T) = +1$, given that $X(t_0) = +1$, what would be the probability that your prediction is wrong? How does this probability vary with $T$, and does this seem reasonable?

**10.2.** As shown in Figure P10.2, a particular random process $X(t)$ is represented by a sample space with three possible time functions as outcomes. The probabilities for the three outcomes $x_1(t)$, $x_2(t)$, and $x_3(t)$ are

$$P\{x_1(t)\} = \frac{1}{3}, \quad P\{x_2(t)\} = \frac{1}{4}, \quad \text{and } P\{x_3(t)\} = \frac{5}{12}.$$

**(a)** Determine the PMF for the random variable $X(t_1)$.
**(b)** Determine the joint PMF $p_{X(t_1),X(t_2)}(x_1, x_2)$ for the two random variables $X(t_1)$ and $X(t_2)$.
**(c)** Determine the autocorrelation $R_{XX}(t_1, t_2) = E[x(t_1)x(t_2)]$.

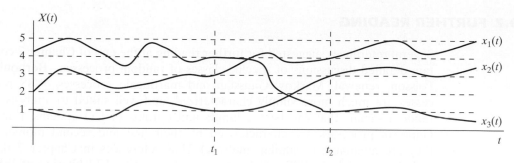

**Figure P10.2**

**10.3.** A random process $W(t)$ can have four different time functions as outcomes, as shown in Figure P10.3. The probabilities for the four outcomes $w_1(t), w_2(t), w_3(t)$, and $w_4(t)$ are

$$P\{w_1(t)\} = \frac{1}{3}, \quad P\{w_2(t)\} = \frac{1}{4}, \quad P\{w_3(t)\} = \frac{1}{4}, \quad \text{and } P\{w_4(t)\} = \frac{1}{6}.$$

Given that $W(t_1) = 6$ and $W(t_2) = 4$, find the minimum mean square error (MMSE) estimate for $W(t_3)$.

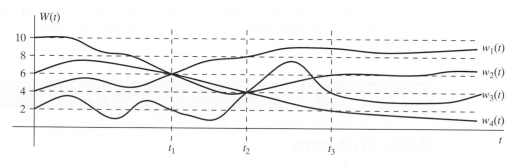

**Figure P10.3**

**10.4.** If $x(t)$ and $y(t)$ are two zero-mean WSS random processes with $R_{xy}(\tau) = 0$ for all $\tau$, is it always true that

$$E\{x^2(t+\tau)y^2(t)\} = R_{xx}(0)R_{yy}(0)?$$

Explain.

**10.5.** Is an SSS random process $x(t)$ necessarily mean-ergodic? Explain.

**10.6.** Let $\{\theta_k\}$ be a set of i.i.d. random variables, uniformly distributed on the interval $[0, 2\pi]$. Let the process $x[n]$ be formed by

$$x[2n] = \cos\theta_n ,$$

$$x[2n+1] = \sin\theta_n ,$$

so that, for example, $x[-2] = \cos\theta_{-1}, x[-1] = \sin\theta_{-1}, x[0] = \cos\theta_0, x[1] = \sin\theta_0,$ $x[2] = \cos\theta_1, x[3] = \sin\theta_1$, and so on.

(a) Is the process $x[n]$ WSS? Explain.

(b) Is the process $x[n]$ i.i.d.? Explain.

**10.7.** (a) If $x[n]$ is a DT WSS random process with autocorrelation function $R_{xx}[m]$, then either

$$E\left[(x[n] - x[k])^2\right] = 2(R_{xx}[n] - R_{xx}[k])$$

or

$$E\left[(x[n] - x[k])^2\right] = 2(R_{xx}[0] - R_{xx}[n-k]) \,.$$

Choose the correct equality, and explain.

(b) True or false: If $x(t)$ is a zero-mean WSS process with autocovariance function $C_{xx}(\tau) = e^{-|\tau|}$, and $y(t) = V + x(t)$, where $V$ is a zero-mean random variable uncorrelated with $x(\cdot)$, then for almost all sample functions $y(t)$

$$\lim_{T \to \infty} \frac{1}{2T} \int_{-T}^{T} y(t)\, dt = 0 \,.$$

**10.8.** (a) Consider a random process $X(t)$ that is defined by $X(t) = A\cos(\omega_0 t)$, where $\omega_0$ is a constant.

  (i) Suppose that $A$ is a random variable uniformly distributed over $[0,1]$. Determine the autocorrelation $R_{XX}(t_1,t_2)$ and autocovariance $C_{XX}(t_1,t_2)$ of $X(t)$.

  (ii) Repeat part (i) for the case when $A$ is a Gaussian random variable with mean $\mu_A = 0.5$ and variance $\sigma_A^2 = \frac{1}{12}$.

  (iii) Is $X(t)$ WSS?

(b) Now let $X(t) = A\cos(\omega_0 t + \theta_0) + B\cos(\omega_1 t + \theta_1)$, where $\omega_0 \neq \omega_1$ are constants, $A$, $B$, $\theta_0$, and $\theta_1$ are all random variables that are mutually independent, and both $\theta_0$ and $\theta_1$ are uniformly distributed over the interval $0 \leq \theta < 2\pi$. What are the first- and second-order moments of the process $X(t)$? In other words, find $E[X(t)]$ and $E[X(t_1)X(t_2)]$. Is the process WSS?

**10.9.** Consider a CT random process $x(t)$ defined as follows:

$$x(t) = \cos(Wt + \Theta), \quad \text{for } -\infty < t < \infty,$$

where $W$ and $\Theta$ are statistically independent random variables, with $W$ taking values $\omega$ uniformly distributed on the interval $[-\omega_o, \omega_o]$ and $\Theta$ taking values $\theta$ uniformly distributed on the interval $[0, 2\pi]$.

The trigonometric identity below might be helpful:

$$\cos(A)\cos(B) = \frac{\cos(A+B) + \cos(A-B)}{2}.$$

(a) Determine the following ensemble-average statistics of the random process $x(t)$:

  (i) the mean function $\mu_x(t) \equiv E[x(t)]$, and

  (ii) the correlation function $R_{xx}(t_1,t_2) \equiv E[x(t_1)x(t_2)]$.

(b) Is the process WSS? Is the process ergodic in the mean?

(c) Determine in terms of $\omega$ and $\theta$ the following time averages of a single realization of the random process (for notational simplicity we use $x(t)$ to denote the particular realization):

$$\langle x(t) \rangle \equiv \lim_{T \to \infty} \frac{1}{T} \int_{-T/2}^{T/2} x(t)\, dt$$

$$\langle x(t+\tau_o)x(t) \rangle \equiv \lim_{T \to \infty} \frac{1}{T} \int_{-T/2}^{T/2} x(t+\tau_o)x(t)\, dt, \quad \text{where } \tau_o \text{ is a positive constant.}$$

(d) Can a really long record of a single realization be used with appropriate time averaging to calculate, at least approximately, $R_{xx}(t_1,t_2)$?

**10.10.** Consider the random process

$$X(t) = \cos(\omega t + \theta)\,,$$

where $\omega$ and $\theta$ are independent random variables, with $\omega$ uniformly distributed over $[-B_0, B_0]$ and $\theta$ uniformly distributed over $[-\pi, \pi]$.

(a) Determine $E[X(t)]$, the expected value of $X(t)$.
(b) Determine the autocorrelation function $R_{XX}(t_1, t_2) = E[X(t_1)X(t_2)]$.
(c) Is the process $X(t)$ WSS? Clearly explain why or why not.

**10.11. (a)** Suppose $x(t)$ is a WSS random process with mean $\mu_x$ and autocovariance function $C_{xx}(\tau) = 2e^{-|\tau|}$. What feature of this characterization guarantees that the process $x(t)$ is ergodic in the mean, i.e., that the time average equals the ensemble mean for almost every sample function $x(t)$:

$$\lim_{T \to \infty} \frac{1}{2T} \int_{-T}^{T} x(t)\, dt = \mu_x \,.$$

**(b)** If now $y(t) = x(t) + Z$, where $Z$ is a zero-mean random variable with variance $\sigma_Z^2$, and $Z$ is uncorrelated with the process $x(t)$, determine the mean $\mu_y$ and autocovariance function $C_{yy}(\tau)$ of the process $y(t)$. Also determine what the time average

$$\lim_{T \to \infty} \frac{1}{2T} \int_{-T}^{T} y(t)\, dt$$

would be for a general sample function of the process $y(t)$. Using this result or otherwise, determine if the process $y(t)$ is ergodic in the mean.

**10.12.** A DT zero-mean WSS process $e[n]$ has autocorrelation function

$$R_{ee}[m] = \frac{\sin(\pi m/3)}{m}\,,$$

for $m \neq 0$, and $R_{ee}[0] = \pi/3$. The process $x[n]$ is defined by the relation

$$x[n] = (-1)^n\, e[n]\,.$$

for all $n$. Show that $x[n]$ is WSS and sketch its PSD $S_{xx}(e^{j\Omega})$ in the region $|\Omega| \leq \pi$. Is $x[\cdot]$ also jointly WSS with $e[\cdot]$?

**10.13.** Suppose $x[n]$ is a zero mean, WSS random sequence with autocorrelation $R_{xx}[m] = \delta[m]$ and is the input to an LTI system with impulse response

$$h[n] = \begin{cases} 1 & n = 0, 1, 2 \\ 0 & \text{otherwise .} \end{cases}$$

The system output is $y[n]$. Determine $R_{yy}[m]$ and $R_{xy}[m]$ as defined below:

$$R_{yy}[m] \triangleq E(y[n+m]y[n])$$

$$R_{xy}[m] \triangleq E(x[n+m]y[n]).$$

**10.14.** Suppose the DT random process $w[n]$ is WSS with mean $\mu$, and with autocovariance function $C_{ww}[m] = \sigma^2\delta[m]$. Let $w[n]$ be applied to the input of a stable DT LTI system with unit sample response $h[n] = \beta^n u[n]$, where $u[n]$ is the unit step function. Denote the random process at the output of this system by $y[n]$. Now suppose we generate another random process $x[n]$ from $w[n]$ according to the equation $x[n] = b[n]w[n]$, where $b[n]$ is a Bernoulli process whose value at any time is 1 with probability $p$, and otherwise 0. You can think of $x[n]$ as a corrupted version of $w[n]$, in which random samples of $w[n]$ are set to 0. Assume the process $b[\cdot]$ is independent of $w[\cdot]$. Write down expressions for:

**(a)** the autocorrelation function $R_{ww}[m]$ of the process $w[n]$;
**(b)** the mean $\mu_y$ and autocovariance function $C_{yy}[m]$ of the process $y[n]$;
**(c)** the cross-covariance function $C_{yw}[m]$ of the processes $y[\cdot]$ and $w[\cdot]$;
**(d)** the mean $\mu_x$ and autocovariance function $C_{xx}[m]$ of the process $x[n]$;
**(e)** the cross-covariance function $C_{xw}[m]$ of the processes $x[\cdot]$ and $w[\cdot]$; and
**(f)** the cross-covariance function $C_{yx}[m]$ of the processes $y[\cdot]$ and $x[\cdot]$.

**10.15.** As depicted in Figure P10.15, $p(t)$ is the output of a stable LTI system with impulse response $h(\cdot)$ and WSS input $x(\cdot)$, so

$$p(t) = \int_{-\infty}^{\infty} h(\alpha)x(t-\alpha)\,d\alpha .$$

Suppose $y(t) = p(t) + e(t)$ for some zero-mean WSS process $e(\cdot)$ that is uncorrelated with $x(\cdot)$. Let $e(\cdot)$ and $x(\cdot)$ have autocorrelation functions $R_{ee}(\tau)$ and $R_{xx}(\tau)$, respectively.

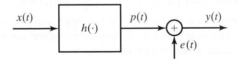

**Figure P10.15**

**(a)** Express $R_{px}(\tau)$ in terms of an appropriate combination of $h(\cdot)$ and $R_{xx}(\cdot)$.
**(b)** Determine $R_{xe}(\tau)$ and explain why $R_{pe}(\tau) = 0$.
**(c)** Write $R_{ye}(\tau)$, $R_{yp}(\tau)$, $R_{yx}(\tau)$, and $R_{yy}(\tau)$ in terms of appropriate combinations of $h(\cdot)$, $R_{ee}(\cdot)$, and $R_{xx}(\cdot)$.

**10.16.** For each of the following parts, indicate whether the given statement is true or false. For a true statement, give a brief but convincing explanation; for a false statement, give a counterexample or convincing explanation.

**(a)** Consider a CT LTI system whose (unit) impulse response is $-\delta(t-17)$. If the input to this system is a WSS process $x(t)$ with autocorrelation function $R_{xx}(\tau)$, then the corresponding WSS output process $y(t)$ has autocorrelation function $R_{yy}(\tau) = R_{xx}(\tau)$.

**(b)** Suppose the WSS input $x(t)$ to a stable CT LTI system has autocorrelation function $R_{xx}(\tau) = e^{-|\tau|}$. It is possible for the corresponding WSS output process $y(t)$ to have autocorrelation function $R_{yy}(\tau) = e^{-3|\tau|}$.

**(c)** Suppose $x(t)$ is a CT WSS random process with autocorrelation function $R_{xx}(\tau)$, and let $y(t)$ be defined as

$$y(t) = \frac{dx(t)}{dt}.$$

Then

$$R_{yx}(\tau) = \frac{dR_{xx}(\tau)}{d\tau}.$$

**10.17.** For the system in Figure P10.17, $x(t)$ is WSS with PSD $S_{xx}(j\omega) = N_0$. Determine the expected value of $r(t)$, $E\{r(t)\}$, in terms of $N_0$, $t_0$ and $h(t)$.

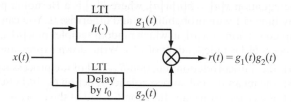

**Figure P10.17**

**10.18.** Consider a stable LTI system with impulse response $h(t)$, such that if a WSS signal $x(t)$ with autocorrelation function $R_{xx}(\tau) = e^{-|\tau|}$ is applied to the input of the system, the resulting output $y(t)$ has autocorrelation function $R_{yy}(\tau) = e^{-|\tau|}$. Can $y(t)$ always be written in the form $y(t) = \alpha x(t - t_0)$ for some constants $\alpha, t_0$? Explain.

## Advanced Problems

**10.19.** Give a simple example of each of the following. If it is not possible to specify such an example, clearly state, in one or two sentences, why not:

**(a)** a random process that is SSS but not strict-sense or wide-sense ergodic;

**(b)** a random process that is wide-sense ergodic but not wide-sense stationary;

**(c)** a nonstationary random process.

**10.20.** Suppose $x[n]$ is a zero-mean WSS random process with autocorrelation function

$$R_{xx}[k] = \rho^{|k|} \qquad 0 < \rho < 1.$$

Find the LMMSE predictor of $x[n]$ given $x[n-1]$ and $x[n-2]$. Also determine its corresponding mean square error.

**10.21.** A particular WSS process $x[n]$ has mean denoted by $\mu_x$ and has autocovariance function $C_{xx}[m]$ whose value at the origin is $C_{xx}[0] = 4$. It is also known that the LMMSE predictor of $x[n+1]$ based on measurements of the current and previous values of the process, namely $x[n]$ and $x[n-1]$, is

$$\widehat{x}_{n,n-1}[n+1] = \frac{1}{3}x[n] + \frac{1}{3}x[n-1] + 2. \tag{10.81}$$

(a) Determine $\mu_x$.

(b) Determine $C_{xx}[1]$ and $C_{xx}[2]$.

(c) Even without computation of the mean square error, we can guarantee that it will not exceed 4—why is that? Now actually compute the mean square error associated with the above estimator.

(d) Suppose instead that you used an LMMSE predictor of the form

$$\widehat{x}_{n-K}[n+1] = \gamma x[n-K] + \zeta \,,$$

for some fixed $K \geq 0$, with $\gamma$ and $\zeta$ chosen to minimize the mean square error. For what choices of $K$ are you guaranteed to not do better than the estimator in Eq. (10.81)? (For all other $K$, it might be possible to do better, depending on what the rest of the autocovariance function is.)

**10.22.** (a) The process $x[n]$ is WSS, with mean 0 and autocovariance function $C_{xx}[m] = \alpha^{|m|}$ for some $\alpha$ of magnitude less than one. If you compute the quantity

$$\frac{1}{2N+1} \sum_{k=-N}^{N} x[k]$$

for a particular realization of the process $x[n]$, will the quantity converge to any specific value as $N$ becomes large, and if so, to what value, and why?

(b) Suppose

$$y[n] = W + x[n] \,,$$

where $x[n]$ is the process defined in (a), and $W$ is a random variable with mean 0 and variance $\sigma_W^2$, and is uncorrelated with $x[k]$ for all $k$. Show that $y[n]$ is WSS, and determine its mean $\mu_y$ and autocovariance function $C_{yy}[m]$.

(c) With $x[n]$ and $y[n]$ defined as in (a) and (b), show that the two processes are jointly WSS, and determine the cross-covariance function $C_{yx}[m]$. Also compute:

    (i) the LMMSE estimator of $y[1]$ in terms of $x[0]$ (in the form $\widehat{y}[1] = ax[0] + b$ for some optimally chosen $a$ and $b$), along with the associated mean square error;

    (ii) the LMMSE estimator of $x[0]$ in terms of $y[1]$, along with the associated mean square error.

(d) For the process defined in (b), if you compute the quantity

$$\frac{1}{2N+1} \sum_{k=-N}^{N} y[k]$$

for a particular realization of the process $y[n]$, will the quantity converge to any specific value as $N$ becomes large, and if so, to what value, and why?

**10.23.** Consider a CT zero-mean WSS process $x(t)$, with correlation function

$$R_{xx}(\tau) = r_0 \, e^{-\beta |\tau|} \,.$$

(a) Show that the LMMSE estimator of $x(t)$, given measurements of $x(t - \tau_k)$ for $k = 1, 2, \cdots, N$ and $0 < \tau_1 < \tau_2 < \cdots < \tau_N$, is $\widehat{x}(t) = ax(t - \tau_1)$ for some appropriate constant $a$. In other words, the estimator depends only on the most recent past measurement. Determine the constant $a$ and the MMSE. One approach is to show that an appropriate choice of $a$ will ensure the conditions that define the LMMSE estimator, namely unbiasedness and the orthogonality condition between the estimation error and the data,

$$E[\{x(t) - ax(t - \tau_1)\}x(t - \tau_k)] = 0,$$

or equivalently

$$E[x(t)x(t - \tau_k)] = aE[x(t - \tau_1)x(t - \tau_k)], \qquad k = 1, 2, \cdots, N.$$

Alternatively, but equivalently, set up the associated normal equations and show that there is a solution with only one component of the estimator's parameter vector being nonzero, namely the one corresponding to the quantity $x(t - \tau_1)$.

**(b)** Show that the LMMSE estimator of $x(t)$, given measurements of both $x(t - \tau_k)$ and $x(t + \tau_k')$ for $k = 1, 2, \cdots, N$ and $0 < \tau_1 < \tau_2 < \cdots < \tau_N$, $0 < \tau_1' < \tau_2' < \cdots < \tau_N'$, is of the form $\widehat{x}(t) = ax(t - \tau_1) + bx(t + \tau_1')$ for some appropriate constants $a$ and $b$. Again, set up the normal equations, and note that there is a solution with only $x(t - \tau_1)$ and $x(t + \tau_1')$ having nonzero weights.

**10.24. (a)** Consider the random process $X(t) = A\cos(\Omega t + \Theta)$, where $A$, $\Omega$, and $\Theta$ are all random variables that are independent of each other, with $\Theta$ uniformly distributed over the interval $0 \leq \theta \leq 2\pi$. Suppose the PDF of $\Omega$, which we denote by $f_\Omega(\omega)$, is an even function of its argument, i.e., $f_\Omega(-\omega) = f_\Omega(\omega)$. (There is actually no loss of generality in assuming this, since $\cos(\omega t)$ is an even function of $\omega$.)

Denote the inverse Fourier transform of this PDF, namely $\frac{1}{2\pi}\int_{-\infty}^{\infty} f_\Omega(\omega)e^{j\omega\tau}\,d\omega$, by $g(\tau)$. Determine the mean function $E[X(t)]$ and autocorrelation function $E[X(t+\tau)X(t)]$, expressing the latter in terms of $E[A^2]$ and $g(\tau)$. Is the process WSS?

Also determine the autocorrelation function of the process in the following cases:

   (i) $f_\Omega(\omega) = \frac{1}{2}[\delta(\omega - \omega_0) + \delta(\omega + \omega_0)]$;
   (ii) $f_\Omega(\omega)$ uniform in the interval $[-\omega_0, \omega_0]$;
   (iii) $f_\Omega(\omega) = \frac{1}{2\pi}\frac{2a}{a^2+\omega^2}$, with $a > 0$.

**(b)** Consider the random process $X(t) = A\cos(\omega_0 t + \Theta_0) + B\cos(\omega_1 t + \Theta_1)$, where $\omega_0 \neq \omega_1$, but both of these numbers are fixed and known. Suppose $\Theta_0$ and $\Theta_1$ are random variables that are independent of the random variables $A$ and $B$; also assume $A$ and $B$ are orthogonal to each other, i.e., $E[AB] = 0$. If both $\Theta_0$ and $\Theta_1$ are uniformly distributed over the interval $0 \leq \theta < 2\pi$, find $E[X(t)]$ and $E[X(t+\tau)X(t)]$. Is the process WSS?

**10.25.** The signal $x(t)$ is a zero-mean WSS random process with autocorrelation function $R_{xx}(\tau)$. Consider the random process $y(t)$ defined in terms of $x(t)$ as:

$$y(t) = x(t) \cdot \cos(2\pi t + \phi).$$

Specify for the following cases whether $y(t)$ is WSS. Clearly justify your answers in a few lines. You may find the following trigonometric identity useful:

$$\cos(\alpha \pm \beta) = \cos\alpha\cos\beta \mp \sin\alpha\sin\beta.$$

**(a)** If $\phi = 0$.

**(b)** If $\phi$ is uniformly distributed in the interval $[0, 2\pi]$ and is independent of $x(t)$.

**10.26.** Suppose $x(t) = y(t)\cos(\omega_o t + \Theta)$, where $y(t)$ is a WSS process with mean $\mu_y$ and autocovariance function $C_{yy}(\tau)$; $\omega_o$ is a known constant; and $\Theta$ is a random variable that is independent of $y(\cdot)$ and is uniformly distributed in the interval $[0, 2\pi]$. You might find it helpful in one or more parts of the problem to recall that

$$\cos(A)\cos(B) = \frac{1}{2}[\cos(A + B) + \cos(A - B)].$$

**(a)** Find the mean $\mu_x(t)$ and autocorrelation function $E[x(t + \tau)x(t)]$ of the process $x(t)$. Also find the cross-correlation function $E[y(t + \tau)x(t)]$. Explain precisely what features of your answers tell you that (i) $x(\cdot)$ is a WSS process; and (ii) $x(\cdot)$ and $y(\cdot)$ are jointly WSS.

**(b)** Suppose $C_{yy}(\tau) = e^{-|\tau|}$ and $\mu_y \neq 0$. Obtain an expression for the PSD $S_{yy}(j\omega)$ in this case, and draw a fully labeled sketch of it. Also obtain an expression for the PSD $S_{xx}(j\omega)$, and draw a fully labeled sketch of it.

**(c)** With the properties of $y(t)$ specified as in (b), is $y(t)$ ergodic in mean value? Be sure to give a reason for your answer. A somewhat harder question: is $x(t)$ ergodic in mean value? Again, describe your reasoning. If you are able to evaluate either of the following integrals on the basis of your answers here, please do so:

$$\lim_{T \to \infty} \frac{1}{2T} \int_{-T}^{T} y(t)\, dt, \qquad \lim_{T \to \infty} \frac{1}{2T} \int_{-T}^{T} x(t)\, dt,$$

where $y(t)$ and $x(t)$ here should be interpreted as the specific realizations taken by these quantities in a particular experiment.

**10.27.** Suppose $x_c(t)$ is a zero-mean WSS random process with autocorrelation function $R_{x_c x_c}(\tau)$. The C/D converter in Figure P10.27-1 is ideal, i.e., $x_d[n] = x_c(nT)$.

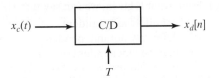

**Figure P10.27-1**

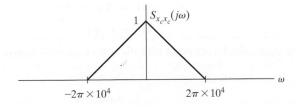

**Figure P10.27-2**

**(a)** Determine the mean $E\{x_d[n]\}$, and also express the autocorrelation function $E\{x_d[n + m]x_d[n]\}$ of the DT random process $x_d[n]$ in terms of $R_{x_c x_c}(\tau)$. Is $x_d[n]$ necessarily WSS if $x_c(t)$ is WSS?

**(b)** Assume that $S_{x_c x_c}(j\omega)$, the CT Fourier transform of $R_{x_c x_c}(\tau)$, is as shown in Figure P10.27-2. Determine $R_{x_c x_c}(\tau)$ and $E\{x_c^2(t)\}$.

**(c)** Determine $E\{x_d^2[n]\}$ for $\frac{1}{T} = 40$ kHz and for $\frac{1}{T} = 15$ kHz.

**10.28.** We wish to generate a WSS random process $y[n]$ by processing another random process $x[n]$ with an LTI system shown in Figure P10.28.

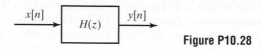

$$x[n] \quad \boxed{H(z)} \quad y[n]$$

**Figure P10.28**

The autocorrelation of the input random process $x[n]$ is

$$R_{xx}[m] = \sigma_x^2 \delta[m].$$

We choose the system function $H(z)$ so that the cross-correlation between the processes $x[n]$ and $y[n]$ is

$$R_{xy}[m] = (0.5)^{-m} u[-m].$$

The corresponding cross-PSD is

$$S_{xy}(e^{j\Omega}) = \frac{1}{1 - 0.5 e^{j\Omega}}.$$

In each of the following parts, choose the correct statement: (i), (ii), (iii). If more than one is correct, indicate all that are correct. Explain your reasoning succinctly.

**(a)** To obtain the desired $y[n]$, $H(z)$ must represent
   (i) a stable minimum phase system.
   (ii) a system that is stable, but is not minimum phase.
   (iii) a system that is not uniquely specified by the given information.

**(b)** From what is given, we can say that $x[n]$ and $x[n+k]$, for $k \neq 0$, are
   (i) definitely independent.
   (ii) definitely not independent.
   (iii) may be independent or not.

**(c)** From what is given, we can say that $y[n]$ and $y[n+1]$ are
   (i) definitely independent.
   (ii) definitely not independent.
   (iii) may be independent or not.

**10.29.** Consider the second-order state-space model for a particular linear, CT system:

$$\dot{\mathbf{q}}(t) = \mathbf{A}\mathbf{q}(t) + \mathbf{b}x(t)$$

$$y(t) = \mathbf{c}^\mathsf{T}\mathbf{q(t)}.$$

The eigenvalues associated with the state transition matrix $\mathbf{A}$ are

$$\lambda_1 = -1$$

$$\lambda_2 = -2.$$

The mode $\lambda_1 = -1$ is unreachable from the input.
The mode $\lambda_2 = -2$ is unobservable from the output.
   Let $r_1(t), r_2(t)$ be the state variables in modal form associated with $\lambda_1 = -1$ and $\lambda_2 = -2$, respectively.
   Assume the system started a long time ago, effectively at $t = -\infty$, and is already in steady state. The input $x(t)$ is zero-mean CT noise with constant power spectral density $S_{xx}(j\omega)$.

Determine the following to within a scale factor:

**(a)** $S_{r_2 r_2}(j\omega)$, the PSD of $r_2(t)$;

**(b)** $S_{yy}(j\omega)$, the PSD of the output $y(t)$.

**10.30.** For each of the following parts, state whether the given statement is true or false. For a true statement, give a brief but convincing explanation; for a false statement, give a counterexample or convincing explanation.

**(a)** It is possible to generate two jointly WSS DT processes $x[\cdot]$ and $y[\cdot]$ whose cross-correlation function $R_{yx}[m]$ is any specified function $g[m]$, provided that

$$\sum_{m=-\infty}^{\infty} |g[m]| < \infty .$$

**(b)** If a CT WSS process $y(t)$ has expected value $\mu_y$ and autocovariance function $C_{yy}(\tau) = 3^{-|\tau|} + 2$, then

$$\lim_{T\to\infty} \frac{1}{2T} \int_{-T}^{T} y(t)\,dt = \mu_y$$

(in the sense that the variance of the difference between the left and right sides goes to 0).

**10.31.** Suppose $x(t)$ is a real-valued, zero-mean WSS process with autocorrelation function given by $R_{xx}(\tau) = e^{-|\tau|}$. Let $x(t)$ be processed by a stable, LTI system with real-valued impulse response $h(t)$ as shown in Figure P10.31.

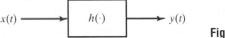

$x(t) \longrightarrow \boxed{h(\cdot)} \longrightarrow y(t)$

**Figure P10.31**

**(a)** We are told that $h(t)$ is causal and stable and that the output $y(t)$ has autocorrelation

$$R_{yy}(\tau) = 3e^{-3|\tau|} .$$

(i) Find $|H(j\omega)|$, i.e., the magnitude of the Fourier transform of $h(t)$.

(ii) Suppose the system has a stable and causal inverse. Find a possible $h(t)$ for the LTI system. Is your answer unique to within a scaling factor? If yes, explain why. If no, give an example of another possible $h(t)$.

**(b)** Now suppose that we have no other information about the LTI system $h(t)$ beside its stability, but we know the cross-correlation function

$$R_{yx}(\tau) = e^{-\tau} u(\tau) - 2e^{-2\tau} u(\tau) + e^{-3\tau} u(\tau) .$$

Find a possible impulse response $h(t)$. Is your answer unique? If yes, explain why. If no, specify another possible $h(t)$.

**10.32. (a)** Suppose $x(t)$ is a WSS random process and is the input to a stable and causal LTI system with impulse response $h(t) = e^{-t}u(t)$. If $\mu_x = 3$, determine the mean of the output.

**(b)** Consider two zero-mean and jointly WSS processes $x(t)$ and $w(t)$ with autocorrelation functions $R_{xx}(\tau)$ and $R_{ww}(\tau)$. Their cross-correlation is $R_{wx}(\tau)$. The process $x(t)$ is filtered by a stable and causal LTI system with impulse response $h(t)$ as indicated in Figure P10.32.

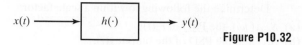

**Figure P10.32**

The output is the WSS random process $y(t)$. Determine the cross-correlation function $R_{wy}(\tau)$ between $w(\cdot)$ and $y(\cdot)$ in terms of $R_{xx}(\tau)$, $R_{ww}(\tau)$, $R_{wx}(\tau)$, and $h(t)$. Choose the correct form from the list below. Explain your reasoning (i.e., derive your answer) succinctly.

(i)  $R_{xx}(\tau)$ convolved with $h(-\tau)$.
(ii)  $R_{xx}(\tau)$ convolved with $h(\tau)$.
(iii)  $R_{wx}(\tau)$ convolved with $h(-\tau)$.
(iv)  $R_{wx}(\tau)$ convolved with $h(\tau)$.
(v)  $R_{ww}(\tau)$ convolved with $h(-\tau)$.
(vi)  $R_{ww}(\tau)$ convolved with $h(\tau)$.

# Extension Problems

**10.33.** Suppose we are given a pair of jointly WSS random processes $x(t)$ and $w(t)$ with known cross-correlation $R_{xw}(\tau)$. If $y(t)$ is the WSS random process obtained by passing $x(t)$ through a stable CT LTI system with known impulse response $h(t)$, derive an expression for $R_{yw}(\tau)$ in terms of the known functions. Also derive an expression for $R_{wy}(\tau)$ if we are given $R_{wx}(\tau)$. As a check, be sure that you recover the results that you expect when: (i) $w(t) = x(t)$; and (ii) $w(t) = y(t)$.

**10.34.** In a wide variety of real situations, two or more uncorrelated sources are received through a channel or system with cross talk. In the receivers or sensors, the signals interfere with each other and it is of interest to process the received signals to separate them. One such scenario is depicted in Figure P10.34-1.

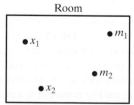

**Figure P10.34-1**

The two sources $x_1(t)$ and $x_2(t)$ might be speech or music; the received signals $m_1(t)$ and $m_2(t)$ could correspond to the outputs of two microphones. Room acoustics can be effectively modeled in terms of LTI systems. The impulse response from source $i$ to microphone $k$ is denoted as $h_{ik}(t)$. A two-input, two-output LTI model for the two received signals is shown in Figure P10.34-2.

If microphone 1 is close to source 1 and microphone 2 is close to source 2, it is reasonable (and simplifies the algebra in this problem) to assume that $h_{11}(t) = h_{22}(t) = \delta(t)$. We also assume that the two sources $x_1(t)$ and $x_2(t)$ are WSS, zero-mean, uncorrelated, and that their autocorrelation functions are known.

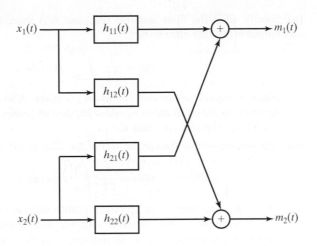

**Figure P10.34-2**

Approaches to recover $x_1(t)$ and $x_2(t)$ from $m_1(t)$ and $m_2(t)$ typically require a way of estimating the impulse responses $h_{12}(t)$ and $h_{21}(t)$. Some information can be obtained by measuring the autocorrelation and cross-correlation functions for the microphone outputs.

**(a)** Determine $R_{m_1 m_1}(\tau)$, $R_{m_2 m_2}(\tau)$, and $R_{m_1 m_2}(\tau)$ in terms of $R_{x_1 x_1}(\tau)$, $R_{x_2 x_2}(\tau)$, $h_{12}(t)$, and $h_{21}(t)$.

**(b)** If it is known that $h_{12}(t) = h_{21}(t)$, and you could measure only one of $R_{m_1 m_1}(\tau)$, $R_{m_2 m_2}(\tau)$, or $R_{m_1 m_2}(\tau)$, which one would be the most useful in determining $h_{12}(t)$?

**10.35.** Consider a stable, causal, DT system with input $w[n]$ and output $y[n]$ that are related by

$$y[n] = -\left(\sum_{k=1}^{N} a_k y[n-k]\right) + w[n] .$$

The output is said to be governed by an $N$th-order auto-regressive model in this case, because the output depends on past values of itself as well as on the present input. Suppose $w[n]$ is an i.i.d. process that at each time (and independently of what happens at other times) takes the values $+M$ and $-M$ with equal probability. Assume $M$ and the coefficients $a_k$ are all known. Determine the LMMSE estimator $\hat{y}[n]$ of $y[n]$, in terms of all past values of $y[\cdot]$, and find the associated mean square error. (*Hint*: Study the above equation to come up with a guess for what $\hat{y}[n]$ might be, then verify it by invoking the orthogonality principle that governs LMMSE estimation; the causality and stability of the system also play a role.)

**10.36.** Suppose the CT WSS random process $x(t)$ has some mean $\mu_x$ and autocovariance function $C_{xx}(\tau)$ that are both unknown. We want to estimate the ensemble mean $\mu_x$ by taking a time average of $x(t)$. To do this, we will pass $x(t)$ through an LTI filter with impulse response $h(t)$ that you will specify in (a) below, and then examine its sampled output $y(0)$ as specified in (b). What will emerge in (c) is then a condition for ergodicity in the mean of the process $x(t)$, i.e., a condition for the time average of $x(t)$, taken over an infinitely long interval, to equal its ensemble average $\mu_x$.

**(a)** Specify what the filter impulse response $h(t)$ should be in order for the output $y(t)$ of the filter to be

$$y(t) = \frac{1}{2T} \int_{t-T}^{t+T} x(\alpha)\, d\alpha \;,$$

where $T$ is given. Show that the mean $\mu_y$ of the WSS output process $y(t)$ is the same as the input mean $\mu_x$. Also express the output covariance function $C_{yy}(\tau)$ in terms of $C_{xx}(\cdot)$ and $T$.

**(b)** Suppose we sample the output $y(t)$ of the filter in (a) at time $t = 0$, so

$$y(0) = \frac{1}{2T} \int_{-T}^{+T} x(\alpha)\, d\alpha \;.$$

This is the time average of $x(t)$ over a finite window of length $2T$, centered at $0$. The mean of this random variable $y(0)$ is evidently still $\mu_y = \mu_x$. Express the variance $\sigma^2_{y(0)}$ of this random variable in terms of $C_{xx}(\tau)$ and $T$. The random variable $y(0)$ will be centered around $\mu_x$ and will have a spread that is indicated by the variance $\sigma^2_{y(0)}$. Each time we run this experiment we will get a different $y(0)$, i.e., a different finite time average of $x(t)$, but staying within a few standard deviations of $\mu_x$ with high probability (invoke the Chebyshev inequality to infer this).

**(c)** In general, the variance you computed in (b) is positive for any finite $T$, so the finite time average of $x(t)$ does not equal the ensemble average. However, if the variance computed in (b) tends to $0$ as $T$ tends to $\infty$, then effectively $y(0)$ will tend to $\mu_x$ as $T$ tends to $\infty$, i.e.,

$$\lim_{T \to \infty} \frac{1}{2T} \int_{-T}^{+T} x(\alpha)\, d\alpha = \mu_x \;,$$

which is what we would like. For which of the following $C_{xx}(\tau)$ does the variance you computed in (b) go to $0$ as $T$ tends to $\infty$?

  (i)  $C_{xx}(\tau) = e^{-|\tau|}$.
  (ii)  $C_{xx}(\tau) = e^{-|\tau|} + A$, where $A$ is a positive constant.
  (iii)  $C_{xx}(\tau) = \frac{1}{1+|\tau|}$.
  (iv)  $C_{xx}(\tau) = \cos|\tau|$.

Whether the variance $\sigma^2_{y(0)}$ tends to $0$ as $T$ tends to $\infty$ can often be decided with only approximate information on the covariance $C_{xx}(\tau)$. We started off this problem saying that the covariance was unknown; in practice, you would probably have approximate information (e.g., bounds on rate of decay with $|\tau|$) that would suffice to check the above condition.

This problem could have been phrased for the case of a DT WSS process, and you would then perhaps have recognized its conclusion as a generalization of the (weak) law of large numbers to the case of a correlated process $x[n]$, whereas your previous encounter with this "law" was for the case of an i.i.d. process.

**10.37.** For each of the following random process examples described in this chapter, determine whether the process is second-order ergodic.

**(a)** Random phase sinusoid $x(t) = A\cos(\omega_o t + \theta)$, where $A, \omega_o$ are nonzero constants, and $\theta$ is a random variable distributed uniformly over the interval $0 \le \theta < 2\pi$.

(b) Random amplitude and phase sinusoid $x(t) = A\cos(\omega_0 t + \theta)$, where $\omega_0$ is a nonzero constant, $A$ is a random variable uniformly distributed over $-2 \le A < 2$, and $\theta$ is a random variable uniformly distributed over the interval $0 \le \theta < 2\pi$.

(c) The Bernoulli process $x[n]$, where at each time instant $n$,

$$P(x[n] = 1) = p, \qquad P(x[n] = -1) = 1 - p;$$

the value at each time instant $n$ is independent of the values at all other time instants.

(d) The random telegraph wave.

**10.38.** Suppose the random process $y[n]$ is given by

$$y[n] = w[n] + \beta w[n-1]$$

where $w[n]$ is an i.i.d. process that takes the values $\pm 1$ with equal probability at each time.

(a) Show that $y[n]$ is WSS with mean value 0. Also determine and plot the autocorrelation function $R_{yy}[m]$ of $y[n]$ for $|m| \le 4$, in each of the following four cases: $\beta = 3, 1, -1,$ and $-3$.

(b) Use an appropriate computational package to generate a sample realization of the process $w[n]$ for $n = 0$ to $100$. Then for each of the four values of $\beta$ given in (a), use the $w[n]$ you have generated in order to obtain a corresponding realization of the process $y[n]$ for $n = 1$ to $100$. Show plots of your realization of $w[n]$ and these four realizations of $y[n]$. Are there qualitative differences among the plots of $y[n]$ for the different $\beta$? If so, describe them.

**10.39.** Suppose the input $x[n]$ and output $y[n]$ of a causal system are related by

$$y[n] = x[n] + \alpha x[n-D],$$

where $x[n]$ is i.i.d. and uniformly distributed in the interval $[-1, 1]$ at each time step, and $\alpha, D$ are parameters denoting a scale factor and (integer) delay respectively. Note that this system is stable for all $\alpha$.

(a) Compute the means $\mu_x$ and $\mu_y$, as well as the autocovariance functions $C_{xx}[m]$ and $C_{yy}[m]$, in term of $\alpha$ and $D$.

(b) Using an appropriate computational package, generate a segment of one realization of the signal $x[n]$ for $n = 1, \ldots, 500$. With computations on this single realization, and assuming ergodicity of the mean and of the covariance (both assumptions actually hold in this i.i.d. case), numerically estimate $\mu_x$ and $C_{xx}[m]$ for values of $m$ in the range $|m| \le 10$. For instance, a plausible estimate of the autocorrelation function at lag 2 would be

$$R_{xx}[2] \approx \frac{1}{498} \sum_{k=1}^{498} x[k+2]x[k].$$

Do you come close to what you expect to get?

(c) With your realization of $x[n]$ from (b), and choosing $\alpha = 1$ and $D = 5$, compute the corresponding realization of $y[n]$ using the input-output relation above for $n = 6, \cdots, 500$. Now suppose you were given only this $y[n]$ and the form of the model above that relates $x[n]$ to $y[n]$, but did not know $\alpha$, $D$, or $x[n]$ (except for the fact that it is a realization of an i.i.d. process). Plot a segment of $y[n]$, say for $n = 6, \cdots, 60$, and see if you are able to deduce anything about $\alpha$ and $D$ by inspection of this.

(d) Assuming ergodicity of $y[n]$ for mean and covariance computations, estimate $\mu_y$ and $C_{yy}[m]$ for $|m| \leq 10$. Plot the resulting autocovariance, and see if you can deduce from the plot what $\alpha$ and $D$ might be.

**10.40.** Suppose $w[n]$ is a WSS random process, with mean $\mu_w$ and autocovariance function $C_{ww}[m] = \sigma_w^2 \delta[m]$.

(a) The random signal $w[n]$ is the input at time $n$ to a causal system whose output $y[n]$ at that time satisfies

$$y[n] = w[n] + \alpha w[n-1].$$

What is the unit sample response, $h_1[n]$, of this system? Explain how you know that the system is BIBO stable for all $\alpha$.

(b) For the system in (a), determine $\mu_y$, $C_{yw}[m]$, and $C_{yy}[m]$ in terms of $\mu_w$, $\sigma_w^2$, and $\alpha$.

(c) Using your results from (b), compute the correlation coefficient $\rho$ between $y[n+1]$ and $y[n]$ in terms of the parameters of the problem. Then determine for what respective values of $\alpha$ this correlation coefficient $\rho$ takes its maximum and minimum values, and determine what these extreme values of $\rho$ are.

(d) Suppose instead that $w[n]$ is the input at time $n$ to a causal and stable system whose output $y[n]$ at that time satisfies

$$y[n] = \alpha y[n-1] + w[n-1].$$

Find the impulse response $h_2[n]$ of this system, and specify what constraint on $\alpha$, if any, is needed for stability of this causal system.

(e) Repeat (b), but this time for the system in (d) rather than the one in (a), and assuming it started operation infinitely far back in the past (so it is in steady state at any finite time).

(f) Repeat (c), but this time for the system in (d), and using your results from (e). For this system, though, what you will actually be computing for the extreme values of $\rho$ will be the least upper bound and greatest lower bound, rather than the maximum and minimum, because though you will be able to approach these extreme values arbitrarily closely, you cannot attain them exactly—explain why not.

**10.41.** Suppose $x[n]$ is an i.i.d. random process whose PDF at each value of $n$ is a Gaussian density of mean 0 and variance $\sigma^2$.

(a) Determine the mean $\mu_x[n]$ and autocovariance function of the process $x[n]$:

$$E\{(x[n+m] - \mu_x[n+m])(x[n] - \mu_x[n])\}.$$

(b) A random process $y[n]$ is obtained from $x[n]$ as

$$y[n] = x[n] + \alpha x[n-1],$$

where $\alpha$ is a known constant. Determine the mean $\mu_y[n]$ of the output process $y[n]$, its autocovariance function

$$E\{(y[n+m] - \mu_y[n+m])(y[n] - \mu_y[n])\},$$

the correlation coefficient $\rho_{n+m,n}$ between $y[n+m]$ and $y[n]$, and the cross-covariance between the output and input, namely

$$E\{(y[n+m] - \mu_y[n+m])(x[n] - \mu_x[n])\} .$$

Which of your answers, if any, depends on $n$? Also check that your various answers take reasonable values (i.e., values that you can verify directly by some explicit reasoning) in the following three limiting cases:

$$\alpha \to -\infty, \qquad \alpha = 0, \qquad \alpha \to +\infty.$$

(c) Suppose we want to estimate $y[n+1]$ given a measurement of $y[n]$, i.e., we want to do a one-step prediction using a linear estimator of the form

$$\hat{y}[n+1] = ay[n] + b .$$

Find $a$ and $b$ to minimize the mean square error

$$E\{(\hat{y}[n+1] - y[n+1])^2\} ,$$

and also determine the value of this MMSE. Again, check that your answers are reasonable in the three limiting cases mentioned in (b).

(d) Repeat (c) for the case of a two-step prediction, with

$$\hat{y}[n+2] = ay[n] + b$$

chosen to minimize

$$E[(\hat{y}[n+2] - y[n+2])^2] .$$

(e) It can be shown that $y[n+m]$ and $y[n]$ are bivariate Gaussian. Use this to write down, or fully describe in some other convincing fashion, the joint density

$$f_{y[n+m],y[n]}(y_1, y_2)$$

for (i) $m = 1$, and (ii) $m \geq 2$. What does your result in part (ii) here tell you about whether you could have done a better job of two-step prediction in part (d) by using some fancier estimator?

(f) Repeat the problem where $x[n]$ is an i.i.d. random process with the PDF at each value of $n$ a Gaussian density of mean $\mu$ and variance $\sigma^2$.

**10.42.** Suppose a CT signal $x(t)$ is broadcast by a transmitter, and let

$$y(t) = x(t) + \alpha\, x(t - \Delta)$$

denote the signal picked up at the antenna of a particular receiver. You can think of the received signal $y(t)$ as comprising the transmitted signal $x(t)$ arriving on a direct path and without noticeable delay or attenuation, followed by an echo arriving with delay $\Delta$ and attenuation factor $\alpha$. Suppose $x(t)$ is modeled as a WSS random process, with mean $\mu_x$ and autocovariance function $C_{xx}(\tau)$.

(a) Determine (i) the mean of $y(t)$, (ii) the cross-covariance function between $y(\cdot)$ and $x(\cdot)$, (iii) the autocovariance function of $y(t)$, all expressed in terms of $C_{xx}(\tau)$ and the constants $\mu_x$, $\alpha$, $\Delta$. Explain on what basis you conclude that $x(\cdot)$ and $y(\cdot)$ are jointly WSS.

(b) Suppose we know that $\mu_x = 0$, but we don't know any of the quantities $C_{xx}(\tau)$, $\alpha$, and $\Delta$. However, assume we are able to use extensive measurements of $y(t)$ to obtain its autocovariance $C_{yy}(\tau)$, as shown in Figure P10.42.

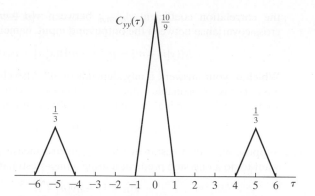

**Figure P10.42**

Using your results in (a), draw a fully labeled sketch of a choice for $C_{xx}(\tau)$ and specify numerical values for $\alpha$ and $\Delta$ that are all consistent with the $C_{yy}(\tau)$ in the figure. Also, explain what tests you would make on the $C_{xx}(\tau)$ you have come up with, to confirm that it is a valid autocovariance function; then verify that your choice of $C_{xx}(\tau)$ does indeed satisfy those tests.

**10.43.** Underwater acoustic communication systems have many difficulties imposed by fluctuations in the environment as well as multipath phenomena. In particular, when a signal is emitted from a source, there are multiple paths over which the sound can propagate as it travels to the receiver. For example, two common paths of propagation include the direct, or line-of-sight, path and a bottom-reflected path. At the receiver, it is often difficult to separate the multiple paths if they overlap in time. However, if we can accurately estimate the time difference of the arrivals, we may be able to correctly recover the transmitted signal. Consider the model shown in Figure P10.43-1 for a signal transmitted through a multipath environment. Assume that $\alpha_1$ is the attenuation and $\Delta_1$ is the delay associated with the direct path of propagation, corresponding to a channel system function of $\alpha_1 e^{-s\Delta_1}$. Similarly, $\alpha_2$ is the attenuation and $\Delta_2$ is the delay associated with the bottom-reflected path corresponding to a system function of $\alpha_2 e^{-s\Delta_2}$. Also assume that the direct arrival is stronger than the bottom-reflected signal, i.e., $\alpha_1 > \alpha_2$.

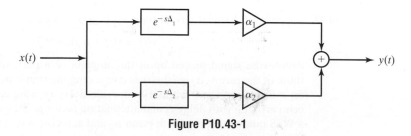

**Figure P10.43-1**

Although we may not have explicit measurements of the transmitted signal, $x(t)$, we can often model the signal as a random process of known autocorrelation. Based on our receiver measurements, we may also obtain an estimate of the autocorrelation of the received signal, $y(t)$. The autocorrelations of both $x(t)$ and $y(t)$, $R_{xx}(\tau)$, and $R_{yy}(\tau)$, are shown in Figure P10.43-2. The units of $\tau$ are seconds.

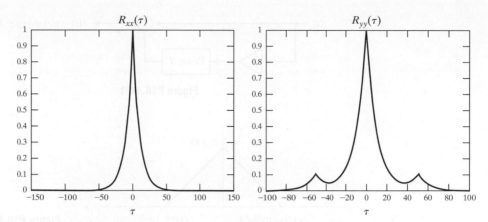

Figure P10.43-2

(a) Find expressions for $R_{yx}(\tau)$ and $R_{yy}(\tau)$ in terms of $R_{xx}(\tau)$ and the problem parameters $\alpha_1, \alpha_2, \Delta_1,$ and $\Delta_2$.

(b) If $R_{yx}(\tau)$ is as shown in Figure P10.43-3, estimate the values of $\alpha_1, \alpha_2, \Delta_1,$ and $\Delta_2$.

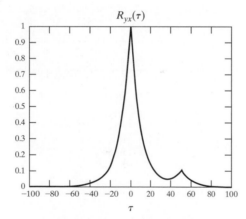

Figure P10.43-3

10.44. A signal $s(t)$ is transmitted over a multipath channel as depicted in Figure P10.44-1, where $T = 10^{-3} = 1$ msec. This $s(t)$ is a zero-mean WSS random process with autocorrelation function shown in Figure P10.44-2. A receiver is to be designed to compensate for the multipath propagation. Assume that the receiver has the structure shown in Figure P10.44-3. Determine the gain constant $\alpha$ in the receiver so that the mean square error between $\widehat{s}(t)$ and $s(t)$ is minimized, i.e., determine $\alpha$ to minimize $\varepsilon$ defined as:

$$\varepsilon = E\left\{\left(s(t) - \widehat{s}(t)\right)^2\right\} .$$

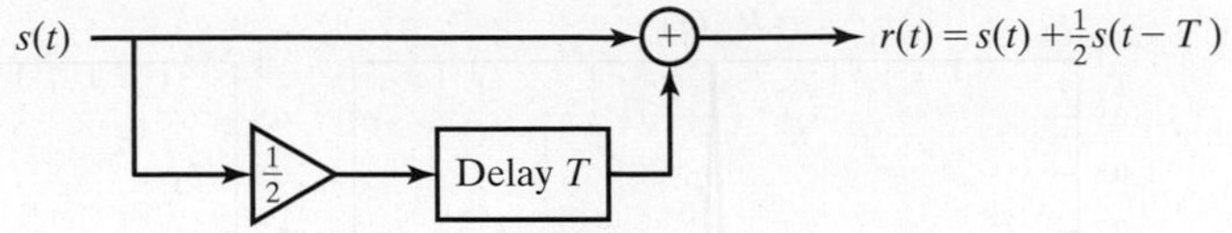

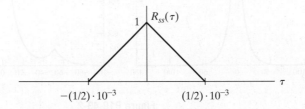

Figure P10.44-1

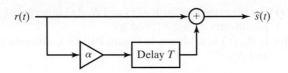

Figure P10.44-2

Figure P10.44-3

# 11 Power Spectral Density

Understanding a signal in the frequency domain is central to the design of any linear and time-invariant (LTI) system intended to extract, enhance, or suppress the signal. We know this well in the case of deterministic signals, and it is just as true in the case of random signals. For instance, to extract an audio signal from background disturbance or noise, one might want to build an LTI filter that enhances the audio component of the received signal and suppresses the noise. The design of the filter characteristic requires understanding the spectral or frequency distribution of the audio and noise components, both of which are often modeled as random processes, since the specific signal and noise waveforms are not known.

There are challenges in trying to find an appropriate frequency-domain description for a wide-sense stationary (WSS) random process. The individual sample functions of such a process typically extend over the entire time axis; they generally have nonzero instantaneous power at all times and nonzero time-averaged power over all intervals, and thus infinite energy. They are therefore unlikely to have Fourier transforms that are ordinary, well-behaved functions of frequency. Instead, the transforms of individual sample functions are generalized functions of frequency that have to be interpreted in terms of frequency-domain integrals. This is in contrast to the finite-energy signals that we treated in Chapter 1, which have zero time-averaged power and well-behaved Fourier transforms. In addition, since a particular sample function of a WSS process is specified as the outcome of a probabilistic experiment, its features will have a random component, so it is necessary to search for transform features that are representative of the whole class of sample functions, that is, of the random process itself.

A particularly useful approach is to focus on the frequency distribution of the expected value of the signal power. This measure of signal strength is well suited to the second-moment characterizations of WSS processes. Section 11.1 examines the spectral distribution of expected instantaneous power, while Section 11.2 explores the spectral distribution of expected time-averaged power. The Einstein–Wiener–Khinchin theorem outlined in Section 11.2 shows the two notions of power spectral density (PSD) are identical. Section 11.3 then presents some examples of how PSDs can be applied. The PSD also plays an important role in our treatment of signal estimation and signal detection in Chapters 12 and 13 respectively.

## 11.1 SPECTRAL DISTRIBUTION OF EXPECTED INSTANTANEOUS POWER

In this section, we develop the continuous-time (CT) case in some detail; the discrete-time (DT) version is very similar. Motivated by situations in which $x(t)$ is the voltage across or current through a unit resistor, the quantity $x^2(t)$ is typically referred to as the instantaneous power in the signal $x(t)$. Suppose $x(t)$ is a WSS process with finite expected instantaneous power, so $E[x^2(t)] = R_{xx}(0) < \infty$. We can then write that

$$E[x^2(t)] = R_{xx}(0) = \frac{1}{2\pi} \int_{-\infty}^{\infty} S_{xx}(j\omega)\, d\omega \ , \qquad (11.1)$$

where $S_{xx}(j\omega)$ is the CT Fourier transform (CTFT) of the autocorrelation function $R_{xx}(\tau)$. We shall assume throughout that the autocorrelation function behaves well enough to possess a Fourier transform, though we will allow the transform to have components that are impulses. Since the autocorrelation is a real and even function of the lag $\tau$, that is, since $R_{xx}(\tau) = R_{xx}(-\tau)$, the transform $S_{xx}(j\omega)$ is real and even in $\omega$. Although this fact could be used to simplify our notation, we shall stay with the notation $S_{xx}(j\omega)$ to avoid a proliferation of notational conventions, and to keep apparent the fact that this quantity is the Fourier transform of $R_{xx}(\tau)$.

### 11.1.1 Power Spectral Density

A physical interpretation of the function $S_{xx}(j\omega)$ can be obtained by considering the result of passing the signal $x(t)$ through an ideal bandpass filter, as indicated in Figure 11.1. The frequency response $H(j\omega)$ of this filter has the value 1 in the passband, and the value 0 outside.

Because of the way in which $y(t)$ is obtained from $x(t)$, the expected value of the instantaneous power in the output $y(t)$ can be interpreted as the expected instantaneous power that $x(t)$ has in the selected passband. Using the fact that

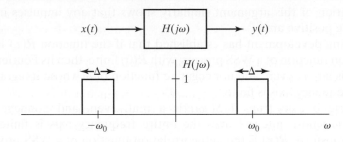

**Figure 11.1**   Ideal bandpass filter to extract a band of frequencies from a WSS input process.

$$S_{yy}(j\omega) = |H(j\omega)|^2 S_{xx}(j\omega) , \qquad (11.2)$$

which was established in Eq. (10.72) in Chapter 10, this expected power can be computed as

$$E[y^2(t)] = R_{yy}(0) = \frac{1}{2\pi} \int_{-\infty}^{+\infty} S_{yy}(j\omega)\, d\omega = \frac{1}{2\pi} \int_{\text{passband}} S_{xx}(j\omega)\, d\omega . \qquad (11.3)$$

The first two equalities are the result of applying the relations in Eq. (11.1) to the signal $y(t)$ instead of $x(t)$, while the last equality follows from invoking Eq. (11.2) and the characteristics of the ideal bandpass filter. Thus the integral

$$\frac{1}{2\pi} \int_{\text{passband}} S_{xx}(j\omega)\, d\omega \qquad (11.4)$$

is the expected value of the instantaneous power of $x(t)$ in the passband, no matter how narrow the passband is, or where it is on the frequency axis. It is therefore reasonable to think of $S_{xx}(j\omega)$ as describing how the expected instantaneous power of the WSS process $x(t)$ is distributed over frequency. Thus $S_{xx}(j\omega)$, which is the transform of the autocorrelation function of the WSS process $x(t)$, is referred to as the power spectral density or PSD of $x(t)$. This is analogous to the definition in Chapter 1 of the energy spectral density (ESD) for finite-energy signals.

Note that the instantaneous power of $y(t)$, and hence the expected instantaneous power $E[y^2(t)]$, is always nonnegative, no matter how narrow the passband. For a sufficiently narrow passband comprising sections of width $\Delta$ centered on some arbitrary nonzero frequency pair $\pm\omega_0$ at which $S_{xx}(j\omega)$ is continuous, we can write

$$0 \le E[y^2(t)] \approx \frac{1}{\pi} S_{xx}(j\omega_0)\Delta . \qquad (11.5)$$

Similarly, if $S_{xx}(j\omega)$ is continuous at $\omega = 0$, then for a narrow passband of total width $\Delta$ centered at $\omega_0 = 0$ we can write

$$0 \le E[y^2(t)] \approx \frac{1}{2\pi} S_{xx}(j0)\Delta . \qquad (11.6)$$

It follows that, in addition to being real and even in $\omega$, the PSD is nonnegative at all frequencies:

$$S_{xx}(j\omega) \ge 0 \quad \text{for all } \omega . \qquad (11.7)$$

A small modification of this argument similarly shows that any impulses in $S_{xx}(j\omega)$ must have positive area.

The preceding development has established that if the function $R(\tau)$ is the autocorrelation function of a WSS process, with $R(0)$ finite, then its Fourier transform $S(j\omega)$ is a real, even, and nonnegative function of $\omega$ whose integral over the entire frequency axis is finite.

The converse is also true: if $S(j\omega)$ is a real, even, and nonnegative function of $\omega$ whose integral over the entire frequency axis is finite, then its inverse transform $R(\tau)$ is the autocorrelation function of a WSS process, with $R(0)$ finite. The idea behind this converse result—though for the DT version—is described in Section 11.3.2, where we discuss modeling filters, which allow us to construct a process with a specified PSD or autocorrelation function. The fact that $R(\tau)$ is the autocorrelation function of a WSS process if and only if its transform $S(j\omega)$ is real, even, and nonnegative is known as Bochner's theorem, and the corresponding result in the DT case is Herglotz's theorem. This result immediately shows, for example, that the "rectangular" function defined by

$$F(\tau) = K > 0 \text{ for } |\tau| < \tau_o \quad \text{and} \quad = 0 \text{ for } |\tau| \geq \tau_o \qquad (11.8)$$

cannot be the autocorrelation function of a WSS process, because its transform $G(j\omega)$ is a sinc function and therefore is negative for certain ranges of $\omega$. On the other hand, the "triangular" function

$$F(\tau) = 1 - \frac{|\tau|}{\tau_o} \text{ for } |\tau| < \tau_o \quad \text{and} \quad = 0 \text{ for } |\tau| \geq \tau_o \qquad (11.9)$$

is a valid autocorrelation function because its transform is the square of a sinc function, hence real, even, and nonnegative.

While the PSD $S_{xx}(j\omega)$ is the Fourier transform of the autocorrelation function, it is useful to have a name for the Laplace transform of the autocorrelation function; we shall refer to $S_{xx}(s)$ as the complex PSD.

## Example 11.1   PSD of a Sinusoidal Random Process

The sinusoidal random process

$$X(t) = A\sin(\omega_0 t + \Theta), \qquad (11.10)$$

with a specified frequency $\omega_0$, but with amplitude $A$ and phase angle $\Theta$ that are independent random variables, and with $\Theta$ uniformly distributed in $[-\pi, \pi]$, was shown in Example 10.5 to be WSS, with mean value $\mu_X = 0$ and autocovariance function

$$C_{XX}(\tau) = \frac{E[A^2]}{2}\cos(\omega_0\tau) = R_{XX}(\tau). \qquad (11.11)$$

The definition of $X(t)$ in Example 10.5 was actually written in terms of a cosine rather than a sine, but the derivation used there carries over to this case with minor modifications and yields the same result. The corresponding PSD is the CTFT of the autocorrelation function $R_{XX}(\tau)$ in Eq. (11.11):

$$S_{XX}(j\omega) = \frac{\pi E[A^2]}{2}\Big(\delta(\omega - \omega_0) + \delta(\omega + \omega_0)\Big) . \qquad (11.12)$$

Thus, as might have been anticipated, the expected instantaneous power is concentrated at the frequency of the sinusoid, that is, at $\pm\omega_0$, and the strength of the associated impulses is linear in the second moment of the amplitude, $E(A^2)$.

This example can be generalized in various directions. Consider, for instance, the sum of sinusoids

$$X(t) = \sum_i A_i \sin(\omega_i t + \Theta_i) , \qquad (11.13)$$

with specified frequencies $\omega_i$, but with amplitude $A_j$ and phase angle $\Theta_j$ that are independent random variables for each $j$, with each $\Theta_j$ uniformly distributed in $[-\pi, \pi]$, and with either the set of amplitudes $\{A_i\}$ being pairwise orthogonal—that is, $E[A_i A_j] = 0$ for $i \neq j$—or the set of $\{\Theta_i\}$ being pairwise independent. Computations similar to those in Example 10.5 then show that

$$R_{XX}(\tau) = \sum_i \frac{E[A_i^2]}{2} \cos(\omega_i \tau) . \qquad (11.14)$$

The CTFT of $R_{XX}(\tau)$ is the corresponding PSD:

$$S_{XX}(j\omega) = \frac{\pi}{2} \sum_i E[A_i^2]\Big(\delta(\omega - \omega_i) + \delta(\omega + \omega_i)\Big) . \qquad (11.15)$$

Example 11.1 exhibits a discrete or "line" spectrum, reflecting the fact that the random process is constructed from a discrete set of sinusoids. In contrast, the next example has a continuous spectrum, corresponding to frequency content in all frequency bands.

**Example 11.2  PSD of Exponentially Correlated Process**

Consider a CT WSS process $x(t)$ with autocorrelation function

$$R_{xx}(\tau) = e^{-\alpha|\tau|} , \quad \alpha > 0 . \qquad (11.16)$$

The autocorrelation of the random telegraph wave in Example 10.7 is of this form, with $\alpha = 2\lambda$, where $\lambda$ is the rate of the Poisson switching between the levels $\pm1$. In that case, the expected duration of a cycle was $2/\lambda$, which suggests that the frequency

$$\omega_o = \frac{2\pi}{2/\lambda} = \frac{\pi\alpha}{2} \qquad (11.17)$$

roughly demarcates the range of frequencies of interest.

The CTFT of $R_{xx}(\tau)$ is the PSD of the process, and is given by

$$S_{xx}(j\omega) = \frac{2\alpha}{\omega^2 + \alpha^2} . \qquad (11.18)$$

This function $S_{xx}(j\omega)$ is plotted in Figure 11.2 for the case $\alpha = 6$. The points corresponding to $\omega_o = \pi\alpha/2 = 3\pi$ and $-\omega_0$ are marked.

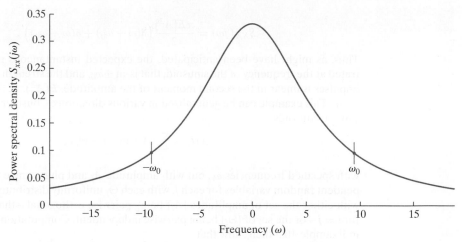

**Figure 11.2**   PSD of a WSS process with exponential autocorrelation $R_{xx}(\tau) = e^{-\alpha|\tau|}$, for the case $\alpha = 6$.

## 11.1.2 Fluctuation Spectral Density

The autocovariance function $C_{xx}(\tau)$ of a WSS process $x(t)$ is also the autocorrelation function of the WSS process $x(t) - \mu_x$, which represents deviations or fluctuations of the process from its mean value $\mu_x$. It follows that the CTFT of $C_{xx}(\tau)$, denoted by $D_{xx}(j\omega)$ in Chapter 10, is also a power spectral density, but represents the frequency distribution of the expected instantaneous power in the fluctuations. We shall refer to $D_{xx}(j\omega)$ as the fluctuation spectral density, or FSD, of the process $x(t)$. The same reasoning as was used in the case of $S_{xx}(j\omega)$ leads to the constraint

$$D_{xx}(j\omega) \geq 0 . \tag{11.19}$$

An immediate implication of this is an inequality we deduced in Chapter 10, namely

$$|C_{xx}(\tau)| \leq C_{xx}(0) , \tag{11.20}$$

which follows from the following set of relations:

$$|C_{xx}(\tau)| = \frac{1}{2\pi}\left|\int_{-\infty}^{\infty} D_{xx}(j\omega)e^{j\omega\tau}\,d\omega\right|$$

$$\leq \frac{1}{2\pi}\int_{-\infty}^{\infty} |D_{xx}(j\omega)e^{j\omega\tau}|\,d\omega$$

$$= \frac{1}{2\pi}\int_{-\infty}^{\infty} D_{xx}(j\omega)\,d\omega = C_{xx}(0) . \tag{11.21}$$

**Example 11.3** **Conditions for Ergodicity of the Mean**

This example develops time- and spectral-domain conditions for ergodicity of the mean of a WSS process $x(t)$, that is, for convergence of the time average of a sample function to the ensemble average $\mu_x$. Specifically, consider the stable LTI system whose impulse response is

$$h(t) = \frac{1}{2T} \quad \text{for} \quad |t| < T \quad \text{and} \quad = 0 \quad \text{elsewhere}. \tag{11.22}$$

If $x(t)$ is the input to this system, then the results in Chapter 10 show that its output $y(t)$ is also WSS, with mean

$$\mu_y = \mu_x \int_{-\infty}^{\infty} h(t)\, dt = \mu_x \tag{11.23}$$

and autocovariance function

$$C_{yy}(\tau) = \overline{R}_{hh}(\tau) * C_{xx}(\tau), \tag{11.24}$$

where $\overline{R}_{hh}(\tau)$ is the deterministic autocorrelation of the impulse response in Eq. (11.22). This deterministic autocorrelation is a triangular pulse of height $1/(2T)$ at the origin that drops linearly on both sides to the value 0 at $|\tau| = 2T$.

The output $y(t)$ at time $t = 0$ is given by

$$y(0) = \frac{1}{2T} \int_{-T}^{T} x(v)\, dv, \tag{11.25}$$

and is therefore the time average computed over a window of length $2T$. The expected value of this time average is $E[y(0)] = \mu_y = \mu_x$, and is therefore the ensemble average of interest, $\mu_x$. However, for any given sample function, the time average will vary randomly around this value $\mu_x$. The associated variance of this time average is $Var\{y(0)\} = C_{yy}(0)$. With $\Lambda(\tau)$ denoting a triangular pulse of height 1 at the origin that drops linearly to 0 at $|\tau| = 2T$, we can use Eq. (11.24) to write

$$Var\{y(0)\} = C_{yy}(0) = \frac{1}{2T} \int_{-2T}^{2T} \Lambda(\tau) C_{xx}(\tau)\, d\tau. \tag{11.26}$$

The time average of $x(t)$ will tend to its expected value, namely $\mu_x$, as $T \to \infty$ if and only if the variance in Eq. (11.26) tends to 0 as $T \to \infty$. (Convergence to the expected value that results from the variance going to 0 is referred to as mean square convergence to the expected value.) This condition can be tested explicitly for any specified $C_{xx}(\tau)$ by substituting it in the right side of Eq. (11.26) and checking whether $C_{yy}(0) \to 0$ as $T \to \infty$. A sufficient condition for the variance in Eq. (11.26) to go to 0 as $T \to \infty$ is that $C_{xx}(\tau) \to 0$ as $\tau \to \infty$, meaning that samples taken increasingly far apart are asymptotically uncorrelated.

To deduce a spectral condition for ergodicity of the mean, note that $C_{yy}(0)$ in Eq. (11.26) can be written as

$$C_{yy}(0) = \frac{1}{2\pi} \int_{-\infty}^{\infty} D_{yy}(j\omega)\, d\omega, \tag{11.27}$$

and the CTFT of Eq. (11.24) yields

$$D_{yy}(j\omega) = |H(j\omega)|^2 D_{xx}(j\omega). \tag{11.28}$$

The quantity $|H(j\omega)|^2$ in this equation is a sinc-squared function in the frequency domain, with height $|H(j0)|^2 = 1$ at the origin, and its first null at $|\omega| = \pi/T$. Thus, as $T \to \infty$, this function is more narrowly concentrated around the origin, but its height at the origin remains fixed at 1. It follows from this and Eqs. (11.27) and (11.28) that $C_{yy}(0) \to 0$ as $T \to \infty$ if and only if the FSD $D_{xx}(j\omega)$ has no impulse at $\omega = 0$. Thus a process is ergodic in mean value if and only if its FSD has no impulse at the origin.

The focus so far has been entirely on CT WSS processes. Exactly parallel results apply in the DT case, with the conclusion that $S_{xx}(e^{j\Omega})$—the real, even, and nonnegative function of $\Omega$ obtained as the DT Fourier transform (DTFT) of the autocorrelation function $R_{xx}[m]$—is the power spectral density of $x[n]$. Similarly, $D_{xx}(e^{j\Omega})$ is the PSD of the fluctuations $x[n] - \mu_x$. The examples below illustrate how the DT results can be applied.

---

**Example 11.4**  **PSD of a One-Step Correlated Process**

Consider a WSS DT process $x[n]$ with mean $\mu_x$ and autocovariance function

$$C_{xx}[m] = \sigma_x^2(\rho\delta[m-1] + \delta[m] + \rho\delta[m+1]) . \tag{11.29}$$

Adjacent time samples of the process are therefore correlated, with correlation coefficient $\rho$, while samples that are two or more time instants apart are uncorrelated. The corresponding FSD is the DTFT of this autocovariance function, hence

$$D_{xx}(e^{j\Omega}) = \sigma_x^2\left(1 + 2\rho\cos(\Omega)\right) , \tag{11.30}$$

which is nonnegative for all $\Omega$ if and only if $|\rho| \leq 0.5$. This is therefore also the condition for the function in Eq. (11.29) to be a valid autocovariance function.

Note that if $\rho$ is near $+0.5$ then the FSD peaks around the frequency $\Omega = 0$, while if $\rho$ is near $-0.5$ then it peaks around $\Omega = \pm\pi$. This reflects the fact that a positive covariance between adjacent samples is suggestive of an ensemble of signals that are generally slowly varying, while a negative covariance between adjacent samples suggests an ensemble of signals whose signs tend to alternate from one time step to the next.

The special case of $\rho = 0$ in the above example is worth treating separately, which we do in the next example.

---

**Example 11.5**  **PSD of an I.I.D. Process**

Consider an independent identically distributed (i.i.d.) process $x[n]$, where the value at each time is drawn independently of all others, using a distribution $f_X(x)$ with mean $\mu_x$ and variance $\sigma_x^2$. In the special case of the Bernoulli process of Example 10.6, for instance, where $x[n] = 1$ with probability $p > 0$ and $x[n] = -1$ with probability $1 - p$ at each time instant, independently of all other times, we have $\mu_x = 2p - 1$ and

$$\sigma_x^2 = E[X^2] - \mu_x^2 = 1 - (2p-1)^2 = 4p(1-p) . \tag{11.31}$$

The autocovariance function of such an i.i.d. process is

$$C_{xx}[m] = E[(x[n+m] - \mu_x)(x[n] - \mu_x)] = \sigma_x^2 \delta[m] , \qquad (11.32)$$

since for $m \neq 0$ the two deviations from the mean in the preceding expression are independent, while for $m = 0$ the above expression is simply the defining equation for $\sigma_x^2$. The FSD of the process is accordingly

$$D_{xx}(e^{j\Omega}) = \sigma_x^2 , \qquad (11.33)$$

which is constant or flat over the entire frequency range for $\Omega$, namely $[-\pi, \pi]$.

The autocorrelation function of the i.i.d. process is

$$R_{xx}[m] = C_{xx}[m] + \mu_x^2 , \qquad (11.34)$$

so the PSD of the process is

$$S_{xx}(e^{j\Omega}) = \sigma_x^2 + 2\pi \mu_x^2 \delta(\Omega) . \qquad (11.35)$$

This PSD is constant over frequency, apart from the impulse at $\Omega = 0$ that reflects the power contributed by the mean value of the process. If the mean value is $\mu_x = 0$, then the PSD is constant over the entire frequency range $[-\pi, \pi]$.

Note that because the conclusions in this example were built on first and second moments, we did not actually require the process to be i.i.d. It would have sufficed for the process values at distinct times just to be uncorrelated, not necessarily independent.

**DT White Process** The preceding example shows that a zero-mean DT WSS process $x[n]$ with values that are uncorrelated across time — that is, with $C_{xx}[m] = K\delta[m]$ for some $K > 0$—has a spectral density that is constant or flat at the value $K$ over all frequencies. A WSS process with this characteristic is said to be a white process of intensity $K$ (by analogy with the notion that white light is an equal-intensity mixture of all colors). A nonwhite WSS process is called colored.

The converse is also true: a WSS process $x[n]$ with a flat spectrum has zero mean and is uncorrelated across time. To see this, suppose

$$S_{xx}(e^{j\Omega}) = K > 0 \qquad (11.36)$$

for all $\Omega$, where $K$ is some positive constant. The corresponding autocorrelation function is determined by taking the inverse DTFT of the PSD, and is therefore

$$R_{xx}[m] = K\delta[m] . \qquad (11.37)$$

Suppose the mean value of the process is $\mu_x$. Then

$$C_{xx}[m] = R_{xx}[m] - \mu_x^2 = K\delta[m] - \mu_x^2 , \qquad (11.38)$$

with corresponding DTFT

$$D_{xx}(e^{j\Omega}) = K - 2\pi \mu_x^2 \delta(\Omega) . \qquad (11.39)$$

Since this FSD has to be nonnegative at all $\Omega$, including at $\Omega = 0$, it must be the case that $\mu_x = 0$ to avoid a negative impulse at $\Omega = 0$. Thus the white process has zero mean, and its autocovariance function is $C_{xx}[m] = K\delta[m]$, that is, the process is uncorrelated across time.

The same argument used above to establish that a white process necessarily has zero mean can be used to demonstrate that any WSS process whose PSD has no impulse at $\Omega = 0$ must correspond to a zero-mean process. The converse is not true, however: a zero-mean process could have a PSD with an impulse at $\Omega = 0$, as the following example demonstrates. An impulse in the PSD at $\Omega = 0$ indicates a nonzero mean value if and only if an impulse of the same strength is not present in the FSD at $\Omega = 0$.

---

**Example 11.6**   **Zero-Mean Process Whose PSD has an Impulse at $\Omega = 0$**

Suppose $x[n]$ is a white process, therefore with a mean value of 0 and autocorrelation function $R_{xx}[m] = K\delta[m]$, with $K > 0$. Let

$$y[n] = A + x[n] , \tag{11.40}$$

where $A$ is a zero-mean random variable with variance $\sigma_A^2 > 0$ and uncorrelated with $x[\cdot]$. The mean of the process $y[n]$ is therefore also 0, and its autocorrelation function is

$$R_{yy}[m] = \sigma_A^2 + R_{xx}[m] = \sigma_A^2 + K\delta[m] = C_{yy}[m] . \tag{11.41}$$

The corresponding PSD is determined by computing the DTFT of the preceding equation:

$$S_{yy}(e^{j\Omega}) = 2\pi\sigma_A^2\delta(\Omega) + K = D_{yy}(e^{j\Omega}) . \tag{11.42}$$

Thus, although $y[n]$ has zero mean, its PSD has an impulse at $\Omega = 0$.

---

The result in the above example relates to the fact that $y[n]$ is not ergodic in mean value, though $x[n]$ is. The time average of any realization of $y[\cdot]$ is the particular value of $A$ that corresponds to that realization, and this will be nonzero for almost all realizations. The expected value of the instantaneous power accordingly has a nonzero "DC" or zero-frequency component. The lack of ergodicity of the mean, and its association with the impulse in the FSD $D_{yy}(e^{j\Omega})$ at $\Omega = 0$, also follow from the DT versions of the results of Example 11.3.

**CT White Process**   As in the DT case, a CT WSS process $x(t)$ is termed white, with intensity $K$, if its PSD is flat over all frequencies: $S_{xx}(j\omega) = K > 0$ for all $\omega$. The corresponding autocorrelation function is $R_{xx}(\tau) = K\delta(\tau)$. However, Eq. (11.1)—with the integral being taken over the entire $\omega$ axis—shows that the expected instantaneous power of such a process must be infinite. Thus a white CT process is unrealizable, and does not satisfy the condition $E[x^2(t)] < \infty$ that we have been assuming. It is nevertheless a useful idealization, for the same reasons that an impulse, though unrealizable, is useful as an idealization in many areas of analysis.

A physically important CT WSS process, whose PSD is indeed essentially constant out to the terahertz range at room temperature, is generated by the thermal fluctuations of the electrons in a resistor held at constant temperature. This process manifests itself as a fluctuating voltage across the terminals of an open-circuit resistor, and was experimentally discovered by J. B. Johnson. A theoretical derivation on the basis of classical physics was subsequently given by H. Nyquist, who was Johnson's colleague at Bell Labs. The PSD of this Johnson–Nyquist noise for a resistor of value $R$ held at temperature $T$ degrees Kelvin, and with $k$ denoting Boltzmann's constant, is flat at the value

$$S_{xx}(j\omega) = 2kTR \tag{11.43}$$

out to very high frequencies. The PSD begins to decay in magnitude when quantum effects set in, at frequencies of the order of $kT/h$, where $h$ is Planck's constant.

A related process in the CT case is what is referred to as a bandlimited white process. Its spectrum is nonzero and flat in some finite region of the frequency axis, and zero outside this: $S_{xx}(j\omega) = K > 0$ for $|\omega| < \omega_m$, and $S_{xx}(j\omega) = 0$ for $|\omega| \geq \omega_m > 0$. The corresponding autocorrelation function is

$$R_{xx}(\tau) = K\frac{\sin(\omega_m\tau)}{\pi\tau}. \tag{11.44}$$

### 11.1.3 Cross-Spectral Density

In the context of two jointly WSS processes $x(\cdot)$ and $y(\cdot)$, Chapter 10 introduced the cross-spectral densities $S_{yx}(j\omega)$ and $D_{yx}(j\omega)$ as the respective Fourier transforms of the cross-correlation functions $R_{yx}(\tau) = R_{xy}(-\tau)$ and $C_{yx}(\tau) = C_{xy}(-\tau)$. These arise, for example, in determining the frequency response of an LTI system whose input is $x(t)$ and output is $y(t)$, as illustrated at the end of Chapter 10. They are also crucial in linear minimum mean square error (LMMSE) estimation of a process $y(t)$ from measurements of a process $x(\cdot)$, as the next chapter shows in the setting of Wiener filtering. The following example illustrates how the cross-spectral density arises when computing the spectral density of the sum of two jointly WSS random signals.

**Example 11.7    FSD of a Sum of Two Jointly WSS Signals**

Suppose a sensor produces an output signal $z(t)$ that is related to two jointly WSS input signals $x(t)$ and $y(t)$ as follows:

$$z(t) = x(t) + \alpha y(t - \Delta). \tag{11.45}$$

The signal $x(t)$ is from a source whose transmissions are received at the sensor without attenuation or delay, while the signal $y(t)$ is from a source whose transmissions are received with attenuation $\alpha$ and delay $\Delta$.

Taking the expected value of Eq. (11.45) produces the relationship

$$\mu_z = \mu_x + \alpha\mu_y, \tag{11.46}$$

and subtracting this relation from Eq. (11.45) then yields the following relation among the deviations from the respective means:

$$\tilde{z}(t) = \tilde{x}(t) + \alpha\tilde{y}(t - \Delta). \tag{11.47}$$

We use this equation to compute the autocovariance of $z(t)$ as follows:

$$E[\tilde{z}(t+\tau)\tilde{z}(t)] = C_{xx}(\tau) + \alpha[C_{xy}(\tau+\Delta) + C_{yx}(\tau-\Delta)] + \alpha^2 C_{yy}(\tau) = C_{zz}(\tau) .$$
$$(11.48)$$

The last equality in Eq. (11.48) introduces streamlined notation, taking account of the observation that the autocovariance function of $z(t)$ does not depend on $t$.

Taking the CTFT of Eq. (11.48) now shows that

$$D_{zz}(j\omega) = D_{xx}(j\omega) + \alpha \underbrace{[e^{j\omega\Delta}D_{xy}(j\omega) + e^{-j\omega\Delta}D_{yx}(j\omega)]}_{2\text{Re}\{e^{-j\omega\Delta}D_{yx}(j\omega)\}} + \alpha^2 D_{yy}(j\omega) \geq 0 , \quad (11.49)$$

where the last inequality invokes the nonnegativity of any FSD, Eq. (11.19).

**A Fundamental Inequality**  The preceding example allows us to derive an important inequality that constrains the cross-FSD of two jointly WSS processes. The inequality in Eq. (11.49) holds for all values of $\alpha$ and $\Delta$. For this quadratic expression in $\alpha$ to always be nonnegative, the following condition must be satisfied:

$$\left(\text{Re}\{e^{-j\omega\Delta}D_{yx}(j\omega)\}\right)^2 \leq D_{xx}(j\omega)D_{yy}(j\omega) . \quad (11.50)$$

The largest value the left-hand side of this inequality can take is $|D_{yx}(j\omega)|^2$, so we arrive at the fundamental inequality

$$|D_{yx}(j\omega)|^2 \leq D_{xx}(j\omega)D_{yy}(j\omega) . \quad (11.51)$$

This inequality is the extension of a very similar—and familiar—inequality for two random variables $X$ and $Y$, namely

$$\sigma_{YX}^2 \leq \sigma_{XX}\sigma_{YY} , \quad (11.52)$$

where $\sigma_{XX} = \sigma_X^2$ and $\sigma_{YY} = \sigma_Y^2$. The similarity is not accidental: by going to the frequency domain, the analysis of jointly WSS processes is made as simple, at each frequency, as the analysis of two random variables. This theme will appear again in the next chapter, in connection with Wiener filtering.

The preceding development shows that the bound in Eq. (11.51) is a necessary condition for the function $D_{yx}(j\omega)$ to be the cross-fluctuation density of two jointly WSS processes $x(\cdot)$ and $y(\cdot)$. The bound is also sufficient, in the sense that processes $x(\cdot)$ and $y(\cdot)$ with specified FSDs $D_{xx}(j\omega)$ and $D_{yy}(j\omega)$ can be constructed to be jointly WSS with a prescribed cross-fluctuation density $D_{yx}(j\omega)$ if the bound in Eq. (11.51) is satisfied. The proof of this fact is outlined in Problem 11.20, and again in the context of the estimation results in Chapter 12, specifically in Problem 12.26.

## 11.2 EXPECTED TIME-AVERAGED POWER SPECTRUM AND THE EINSTEIN–WIENER–KHINCHIN THEOREM

Section 11.1 showed that the PSD of a WSS process, defined as the transform of its autocorrelation function, describes the spectral or frequency distribution of the expected instantaneous power in the process. In the case of a CT process $x(t)$, this is the spectral distribution of $E[x^2(t)]$. If $x(t)$ is ergodic in correlation, so that time averages and ensemble averages are almost always equal when correlations are evaluated, then $E[x^2(t)]$ is also the time-averaged power in almost any ensemble member. This suggests that an alternate route to the PSD for a WSS process might be based on analyzing the spectral distribution of the time-averaged power. Such an approach is described here for the CT case, as this is notationally simpler, but the development for DT WSS processes is very similar.

As noted at the beginning of this chapter, individual sample functions of a WSS process are unlikely to have well-behaved Fourier transforms. However, truncating a sample function to a finite-duration window typically leads to a finite-energy signal, which then does have a well-defined Fourier transform. Let $x_T(t)$ be the signal obtained by rectangular windowing of $x(t)$ to the interval $(-T, T)$. Thus $x_T(t) = x(t)$ in the interval $(-T, T)$ but is 0 outside this interval, so

$$x_T(t) = w_T(t) x(t) , \qquad (11.53)$$

where the window function $w_T(t)$ is defined to be 1 for $|t| < T$ and 0 otherwise. Denoting the Fourier transform of $x_T(t)$ by $X_T(j\omega)$, the results in Section 1.3.2 of Chapter 1 show that the energy spectral density (ESD) of $x_T(t)$ is given by

$$\overline{S}_{xx}(j\omega) = |X_T(j\omega)|^2 . \qquad (11.54)$$

Recall that the ESD of a signal describes how the energy of the signal is distributed over frequency. The results in Chapter 1 were for the case of deterministic signals, so the preceding equation should be read as being applied to individual sample functions of the WSS process.

The ESD in Eq. (11.54) is the Fourier transform of the deterministic autocorrelation of $x_T(t)$, as was established for the DT case in Chapter 1, but the CT case is derived the same way. The following CTFT pair captures this relation:

$$\int_{-\infty}^{\infty} w_T(\alpha)w_T(\alpha - \tau)x(\alpha)x(\alpha - \tau)\, d\alpha \iff |X_T(j\omega)|^2 , \qquad (11.55)$$

where the double arrow denotes a Fourier transform pair. Dividing both sides of this by $2T$ (which is valid because scaling a signal by a constant scales its Fourier transform by the same factor) produces the transform pair

$$\frac{1}{2T}\int_{-\infty}^{\infty} w_T(\alpha)w_T(\alpha - \tau)x(\alpha)x(\alpha - \tau)\, d\alpha \iff \frac{1}{2T}|X_T(j\omega)|^2 . \qquad (11.56)$$

The quantity on the right—the ESD normalized by the window length—is called the periodogram of the finite-duration signal $x_T(t)$. The units associated with the ESD are "energy/Hz," so the units of the periodogram are "power/Hz." The periodogram accordingly describes how the time-averaged power—the total energy divided by the total time—is distributed over frequency for the particular sample function.

To move beyond a relationship that holds for sample functions and arrive at one that characterizes the process as a whole, we have to average over the ensemble, that is, take expectations. Because of the linearity of the Fourier transform, the expected value of the Fourier transform of a signal is the Fourier transform of the expected value, so we can take expectations of both sides in Eq. (11.56) to obtain a new Fourier transform pair. Moving the expectation inside the integral on the left side produces the transform pair

$$\frac{1}{2T} \int_{-\infty}^{\infty} w_T(\alpha) w_T(\alpha - \tau) E[x(\alpha) x(\alpha - \tau)] \, d\alpha \iff \frac{1}{2T} E[|X_T(j\omega)|^2] \,.$$
(11.57)

Now invoking the fact that $E[x(\alpha) x(\alpha - \tau)] = R_{xx}(\tau)$, which does not depend on $\alpha$, we can move the autocorrelation out of the integral, so the left side of Eq. (11.57) becomes

$$\frac{1}{2T} R_{xx}(\tau) \int_{-\infty}^{\infty} w_T(\alpha) w_T(\alpha - \tau) \, d\alpha = R_{xx}(\tau) \Lambda(\tau) \,,$$
(11.58)

where $\Lambda(\tau)$ is a triangular pulse of height 1 at the origin that decays linearly to 0 at $|\tau| = 2T$, as in Example 11.3, and is the result of carrying out the deterministic autocorrelation of $w_T(\tau)$ and dividing by $2T$. The preceding steps allow the transform pair in Eq. (11.57) to be written as

$$R_{xx}(\tau) \Lambda(\tau) \iff \frac{1}{2T} E[|X_T(j\omega)|^2] \,.$$
(11.59)

Now taking the limit as $T$ goes to $\infty$ leads to the desired result:

$$R_{xx}(\tau) \iff S_{xx}(j\omega) = \lim_{T \to \infty} \frac{1}{2T} E[|X_T(j\omega)|^2] \,.$$
(11.60)

The interpretation of this limit is most direct when $S_{xx}(j\omega)$ is a continuous function of $\omega$, as is the case when $R_{xx}(\tau)$ is absolutely integrable. The result in Eq. (11.60) is the Einstein–Wiener–Khinchin (EWK) theorem, proved by Wiener, and independently by Khinchin, in the early 1930s, but stated by Einstein in 1914. It shows that the PSD can indeed also be interpreted as the spectral distribution of the expected time-averaged power in the process.

Minor changes to the above derivation produce a similar result for the cross-spectral density between jointly WSS processes $x(\cdot)$ and $y(\cdot)$. Instead of Eq. (11.56), the relevant equation is

$$\frac{1}{2T} \int_{-\infty}^{\infty} w_T(\alpha) w_T(\alpha - \tau) x(\alpha) y(\alpha - \tau) \, d\alpha \iff \frac{1}{2T} X_T(j\omega) Y_T(-j\omega) \,,$$
(11.61)

where $Y_T(j\omega)$ is the CTFT of the windowed signal $y_T(t) = w_T(t) y(t)$. Taking the expected value of both sides and then the limit as $T \to \infty$ results in

$$R_{xy}(\tau) \iff S_{xy}(j\omega) = \lim_{T \to \infty} \frac{1}{2T} E[X_T(j\omega)Y_T(-j\omega)] . \qquad (11.62)$$

**Spectral Estimation**     The task of estimating $S_{xx}(j\omega)$ or $S_{xy}(j\omega)$ from experimental or simulated data is referred to as spectral estimation. The topic is vast, but the EWK result underlies one important practical approach to the task that is worth sketching out here. In the case of $S_{xx}(j\omega)$, the basic idea is—for a well-chosen and fixed value of $T$—to compute the periodogram on each of $M$ nonoverlapping windows. Averaging these $M$ results produces an averaged periodogram estimate of the PSD. This estimate will approximate the expected value of the periodogram, namely

$$\frac{1}{2T} E[|X_T(j\omega)|^2] , \qquad (11.63)$$

with an error whose variance can decrease as $1/M$ if the windows are sufficiently uncorrelated. The question then is how well the expected periodogram approximates the PSD of interest.

To answer this, we return to Eq. (11.59). The transform of the left side of that equation is the frequency-domain convolution of $S_{xx}(j\omega)$ with the squared-sinc transform of the triangular pulse, so

$$\frac{1}{2T} E[|X_T(j\omega)|^2] = \frac{1}{2\pi} S_{xx}(j\omega) * \frac{2\sin^2(\omega T)}{\omega^2 T} . \qquad (11.64)$$

Thus the expected value of the periodogram is a smoothed or blurred version of the PSD, where the smoothing is the result of convolving the true PSD with the squared sinc function. The expected value of the periodogram therefore does not exactly equal the PSD.

The main lobe of the squared sinc in Eq. (11.64) extends $\pm\pi/T$ on either side of the maximum, so this gives some indication of the frequency resolution attained by a periodogram-based estimate. Two sharp peaks in $S_{xx}(j\omega)$ that are separated by less than $\pi/T$ will probably not be recognized as two separate peaks in the estimate. This limit on resolution also follows from the observation that the lowest-frequency spectral component that can be detected using a window of length $2T$ will need to contribute around a period's worth of data to the window—and this period corresponds to a frequency of $\omega_{\min} = \pi/T$.

For a fixed $T$, the situation can be improved by carefully choosing windows other than the rectangular window $w_T(t)$. Windows that fall off less sharply in the time domain will yield sharper cutoffs in the frequency domain. When such tapered windows are employed, it can also be useful to overlap the $M$ windows somewhat, to allow more effective use of the available data and thereby reduce the error variance. A widely used approach with tapered overlapping windows is Welch's method (developed for the DT case). Beyond these measures, improved resolution requires a longer window, that is, larger $T$.

It is interesting to note that in the limit of $T \to \infty$, the squared sinc function in Eq. (11.64) behaves as an impulse; more precisely,

$$\lim_{T \to \infty} \frac{1}{2\pi} \frac{2\sin^2(\omega T)}{\omega^2 T} \to \delta(\omega) . \tag{11.65}$$

The EWK theorem is then recovered from Eq. (11.64)—through a slightly different argument than the one we invoked with Eq. (11.59)—by taking $T \to \infty$, since $S_{xx}(j\omega) * \delta(\omega) = S_{xx}(j\omega)$.

A similar process yields an estimate of the cross-spectral density $S_{xy}(j\omega)$, but now based on approximating the expectation in Eq. (11.62) by averaging $\frac{1}{2T} X_T(j\omega) Y_T(-j\omega)$.

The DT versions of all the above results can be obtained similarly, so the details are not presented here, but the following example illustrates how the DT results are applied.

## Example 11.8  Periodogram Averaging for a ±1 Bernoulli Process

Example 11.5 established that a Bernoulli process $x[n]$ taking the values $\pm 1$ with equal probability at each instant, independently of the values at other instants, has mean $\mu_x = 0$ and variance $\sigma_x^2 = 1$ at each instant, and has a flat PSD, in this case

$$S_{xx}(e^{j\Omega}) = \sigma_x^2 = 1 . \tag{11.66}$$

Although the analysis is straightforward in this simple case, that may not be so in more complicated cases. Instead, we may only have measurements of the process, or access to a simulator that can generate sample functions of the process. In these cases, an estimate of the PSD could be obtained by some version of periodogram averaging. For our Bernoulli i.i.d. example, the simulation of the process is very easy.

Each plot in the first row of Figure 11.3 shows a single periodogram without averaging (so $M = 1$) for a window length of 50, but computed using four distinct segments of a single simulation run. The plots illustrate how different the results can be from one case to the next because of the random variations. It would be a mistake to attempt an interpretation of any specific peak in any of these plots as indicative of anything fundamental about the process. The vertical spread of values obtained in each case reflects the standard deviation (i.e., square root of the variance) in the periodogram estimate. The variation in the periodogram waveform as a function of $\Omega$ is commensurate with the smoothing expected for a window of size 50, namely variations on a scale of $\pi/50$.

The second row of plots in Figure 11.3 shows the periodograms for window lengths of 50 and 200, but now averaged over $M = 4$ windows, which reduces the variance by a factor of 4 and the standard deviation by a factor of 2 in both cases. The spread of values is similar for the two window lengths, as they have the same value of $M$; and this spread is around half of that seen in the top row, as anticipated. What differs between the two window lengths is the rate of variation with $\Omega$: variations on a scale of $\pi/50$ for the shorter window, and $\pi/200$ for the longer one, reflecting the potentially better resolution obtainable with the longer window.

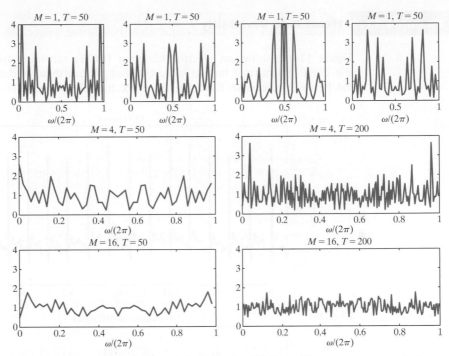

**Figure 11.3** Spectral estimation of the PSD of a $\pm 1$ Bernoulli process, using periodogram averaging. Each plot in the first row shows a single periodogram for a window length of 50, computed using distinct segments of a single simulation run. The second row of plots shows the periodograms for window lengths of 50 and 200, but now averaged over $M = 4$ windows. The bottom row repeats the experiment of the second row, but now averaging the periodograms over $M = 16$ realizations. All plots are over the frequency range $[0, 2\pi]$.

The bottom row repeats the experiment of the second row, but now averaging the periodograms over $M = 16$ realizations, which reduces the standard deviation by 4 over the original single-periodogram case.

The behavior seen in these various plots is consistent with the above discussion of spectral estimation based on the EWK result.

# 11.3 APPLICATIONS

In this section we give a few additional examples of settings in which PSDs provide useful analytical or computational tools and insights.

## 11.3.1 Revealing Cyclic Components

An important application of PSD computations is to reveal the presence of cyclic behavior, which is associated with strong peaks in the PSD. The following example from biomedical signal processing is representative of this kind of investigation.

Example 11.9   **Heart Rate Variability**

Figure 11.4(a) shows an electrocardiogram (ECG) recording from electrodes on the chest of a normal human subject at rest. A heartbeat is defined as the interval between successive R waves; these are the tall spikes in the recording, which

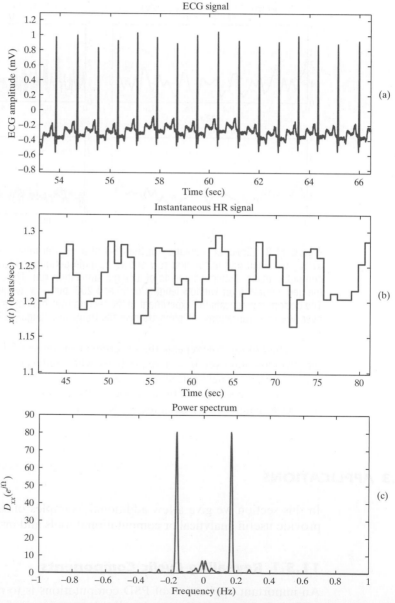

**Figure 11.4**   (a) Electrocardiogram trace; (b) CT instantaneous heart rate signal $x(t)$; and (c) heart rate variability spectrum, showing strong peak at the respiratory frequency, due to systematic variation of heart rate over the course of a respiratory cycle.

result from the electrical wave that causes the heart ventricles to contract on each beat. Computing the reciprocal of the duration of a heartbeat yields the instantaneous heart rate (HR) in beats per second, which we associate with all time points within that heartbeat. This computation produces the piecewise constant HR signal $x(t)$ shown in Figure 11.4(b).

For the PSD estimation, we generate a DT signal from $x(t)$ by sampling it at a rate at least equal to the fastest heart rate in the data, so that the DT signal $x[n] = x(nT)$ is defined at least once per heartbeat. Here $T$ is the smallest heartbeat duration in the data record, or equivalently, $1/T$ is the fastest heart rate in the data record. A reasonable choice for this data is to sample $x(t)$ at 2 Hz. To avoid aliasing effects, $x(t)$ should actually be bandlimited to frequencies below $1/2T$ by a preliminary filtering operation, before sampling. Alternatively, we could first sample $x(t)$ at a considerably higher rate for which the aliasing is negligible, say $K/T$ for some integer $K \gg 1$, then band-limit the DT signal to $1/(2TK)$ by appropriate DT filtering, before finally downsampling by the factor $K$ to recover the original sampling rate of $1/T$.

To avoid having the "DC" or zero-frequency component dominate the PSD, we first—in each window for which the periodogram is computed—subtract out the time average of $x[n]$ in that window, to obtain the zero-average signal $\tilde{x}[n]$. In effect, we end up estimating the FSD, $D_{xx}(e^{j\Omega})$. The PSD of the resulting signal $\tilde{x}[n]$ is now estimated by periodogram averaging.

Figure 11.4(c) shows the PSD estimate obtained by averaging four windows ($M = 4$) of 50 points each, using Welch's method, with a tapered rather than a rectangular window, and with some overlap of the windows. The frequency axis is marked from $-1/(2T)$ to $1/(2T)$ Hz, rather than from $-\pi$ to $\pi$ radians, to enable interpretation in terms of the original CT HR signal.

A particularly interesting feature of this PSD is the prominent peak at the frequency 0.166 Hz, which corresponds to the average respiratory rate of the subject in this experiment. The peak indicates that HR varies cyclically over the course of a respiratory period. The physiological explanation for this coupling from respiration to HR involves both the mechanical effect of respiration on filling and emptying of the heart, and the action of control reflexes that regulate blood pressure in the body, even over the time scale of a single respiratory cycle, by modulating heart rate.

### 11.3.2 Modeling Filters

Generating sample functions of a unit-intensity DT white process $w[\cdot]$ is straightforward, for instance as a $\pm 1$ Bernoulli process of the sort described in Example 11.5. This can be useful in simulating stochastic systems and in other applications. In many situations, however, what one wants are sample functions of a colored DT WSS process $x[\cdot]$ with specified mean $\mu_x$ and specified autocovariance function $C_{xx}[m] = R_{xx}[m] - \mu_x^2$ or FSD $D_{xx}(e^{j\Omega}) = S_{xx}(e^{j\Omega}) - 2\pi \mu_x^2 \delta(\Omega)$.

A natural way to generate such a colored process is to pass a white process through an appropriately chosen stable filter with unit sample response $h[n]$ or frequency response $H(e^{j\Omega})$. Applying the above unit-intensity white process $w[\cdot]$ to the input of such a filter produces at the output a zero-mean WSS process $\tilde{x}[\cdot]$ whose PSD is given by

$$S_{\tilde{x}\tilde{x}}(e^{j\Omega}) = D_{xx}(e^{j\Omega}) = H(e^{j\Omega})H(e^{-j\Omega}) = |H(e^{j\Omega})|^2 . \tag{11.67}$$

The second equality in Eq. (11.67) shows that the desired $H(e^{j\Omega})$ is a spectral factor of the FSD $D_{xx}(e^{j\Omega})$, in the sense described in Chapter 2. The spectral factor constitutes a generalized square root of the FSD.

If an appropriate frequency response $H(e^{j\Omega})$ can be chosen to satisfy Eq. (11.67) for the specified $D_{xx}(e^{j\Omega})$, then sample functions of the process $\tilde{x}[\cdot]$ are produced at the output of the filter when the filter is driven by unit-intensity white noise. Sample functions of $x[\cdot]$ can be obtained by adding the constant $\mu_x$ to this output, thus obtaining the correct mean value. The filter $H(e^{j\Omega})$ is therefore referred to as a modeling or shaping filter because it models the given colored process or shapes the constant-intensity white process into the given colored process.

A spectral factor can sometimes be determined by inspection, or with elementary computations, as in Example 11.10. More generally, since $D_{xx}(e^{j\Omega})$ is real, even, and nonnegative, and assuming that it contains no impulses, it has a square root $\sqrt{D_{xx}(e^{j\Omega})}$ that is also real, even, and nonnegative. The corresponding noncausal filter is guaranteed to be bounded-input, bounded-output (BIBO) stable under an appropriate continuity condition on $\sqrt{D_{xx}(e^{j\Omega})}$ that limits how fast it can change with $\Omega$. As the discussion of spectral factorization and all-pass filters in Chapter 2 shows, all possible choices for $H(e^{j\Omega})$ are then given by

$$H(e^{j\Omega}) = A(e^{j\Omega})\sqrt{D_{xx}(e^{j\Omega})} \,, \tag{11.68}$$

where $A(e^{j\Omega})$ is the frequency response of an all-pass filter, that is, a stable filter satisfying $|A(e^{j\Omega})| = 1$. The all-pass factor is chosen to obtain a frequency response $H(e^{j\Omega})$ that has desirable characteristics for the implementation or application.

If the original $D_{xx}(e^{j\Omega})$ had impulses, at $\Omega = 0$ or at pairs of nonzero frequencies $\Omega = \pm\Omega_i$, then the preceding modeling filter would be used first to generate the process corresponding to the nonimpulsive portion of $D_{xx}(e^{j\Omega})$. To this process we could then add a zero-mean random constant to generate a suitable impulse at $\Omega = 0$ in $D_{xx}(e^{j\Omega})$, following the pattern in Example 11.6. Similarly, we could add a random-phase cosine at frequency $\Omega = \Omega_i$ to generate a pair of impulses in the original $D_{xx}(e^{j\Omega})$ at the frequencies $\pm\Omega_i$, following the DT version of the pattern in Example 11.1.

**Example 11.10   Modeling a One-Step Correlated Process**

Suppose we want to generate sample functions of the zero-mean DT WSS process $x[n]$ described in Example 11.4. Its autocovariance function was specified as

$$C_{xx}[m] = \sigma_x^2(\rho\delta[m-1] + \delta[m] + \rho\delta[m+1]) \,, \tag{11.69}$$

so the process at any time is correlated across adjacent time steps, but uncorrelated beyond that. This autocovariance is simple enough that a time-domain approach works well, so we describe that first, and then the spectral version that parallels the time-domain approach very closely.

The process $x[n]$ is to be generated as the output of a stable modeling filter with unit sample response $h[\cdot]$ and with input $w[n]$ that is a unit-intensity white process, so $C_{ww}[m] = \delta[m]$. It follows from the relations developed in Chapter 10 that

$$C_{xx}[m] = \overline{R}_{hh}[m] , \qquad (11.70)$$

where the right side of this equation is the deterministic autocorrelation of $h[n]$. With $C_{xx}[m]$ being nonzero at only three values of $m$, the preceding equation suggests that we can choose $h[n]$ to be nonzero at only two instants of time, for instance

$$h[n] = a\delta[n] + b\delta[n-1] . \qquad (11.71)$$

The corresponding $\overline{R}_{hh}[m]$ is then

$$\overline{R}_{hh}[m] = ab\,\delta[m-1] + (a^2 + b^2)\,\delta[m] + ab\,\delta[m+1] . \qquad (11.72)$$

Taking account of Eqs. (11.69) and (11.70), the coefficients $a$ and $b$ have to satisfy

$$a^2 + b^2 = \sigma_x^2 , \qquad ab = \sigma_x^2 \rho . \qquad (11.73)$$

Since $a^2 + b^2 \pm 2ab = (a \pm b)^2 \geq 0$, it follows that $1 \pm 2\rho \geq 0$, or $|\rho| \leq 0.5$. This constraint on $\rho$ that is necessary for realizability of a modeling filter is also—and not coincidentally—the same constraint discovered in Example 11.4 for nonnegativity of the associated FSD. Furthermore, the constraint suffices to guarantee that a real $a$ and $b$ can be found to simultaneously solve the two relations in Eq. (11.73). Combining the relations leads to

$$a^4 - \sigma_x^2 a^2 + (\sigma_x^2 \rho)^2 = 0 , \qquad (11.74)$$

from which

$$a^2 = \frac{\sigma_x^2}{2}\left(1 \pm \sqrt{1 - 4\rho^2}\right) \quad \text{and} \quad b = \sigma_x^2 \rho / a . \qquad (11.75)$$

In general, four possible real values for $a$, and the corresponding real values for $b$, can be extracted from Eq. (11.75). The existence of multiple solutions shows that different choices of $h[n]$ can give rise to the same $\overline{R}_{hh}[m]$. As the spectral-domain analysis below makes evident, convolving any given solution $h[n]$ with the unit sample response of an all-pass filter yields another choice for $h[n]$ that will provide the same modeling or shaping effect on the PSD.

The spectral-domain argument for this example closely parallels the above time-domain development. Note that the FSD of the process $x[n]$ is

$$D_{xx}(e^{j\Omega}) = \sigma_x^2\left(1 + 2\rho\cos(\Omega)\right) , \qquad (11.76)$$

with the restriction that $|\rho| \leq 0.5$. One way to determine a convenient spectral factor is by working with the complex form of the FSD, obtained by setting $e^{j\Omega} = z$. The result for this example is

$$D_{xx}(z) = \sigma_x^2\left(1 + \rho(z + z^{-1})\right)$$

$$= (a + bz)(a + bz^{-1}) = (a^2 + b^2) + ab(z + z^{-1}) , \qquad (11.77)$$

so $a$ and $b$ again need to satisfy Eq. (11.73). The modeling filter transfer function can thus be written as

$$H(z) = A(z)(a + bz^{-1}),  \tag{11.78}$$

where $A(z)$ is the transfer function of a (stable) all-pass system, so $A(z)A(z^{-1}) = 1$. When $A(z) = 1$, the filter unit sample response is exactly that in Eq. (11.71). Other choices of all-pass $A(z)$ for a particular $a$ and $b$ that satisfy Eq. (11.73) will produce the filters corresponding to the other solutions of Eq. (11.73). Choices of the form $A(z) = z^{-N}$ will produce delayed (for $N > 0$, otherwise advanced) versions of the unit sample response in Eq. (11.71). An endless set of further choices is possible, though all others will have unit sample response that are no longer of finite duration.

Figure 11.5(a) shows two sample functions of this process obtained by passing two independently generated Bernoulli $\pm 1$ sequences through the preceding modeling filter, for the extreme case of $\rho = 0.5$. Figure 11.5(b) repeats this for $\rho = -0.5$. The sample functions for the two cases bear out the expectations mentioned at the end of Example 11.4.

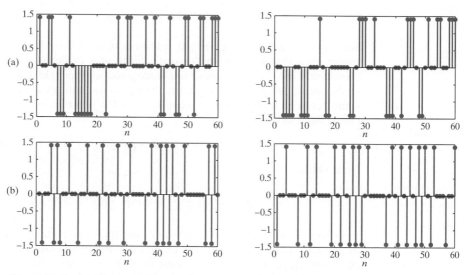

**Figure 11.5**  (a) The upper figures show two sample functions of a one-step correlated process obtained by passing two independently generated Bernoulli $\pm 1$ sequences through the modeling filter in Example 11.10, for the case $\rho = 0.5$. (b) The lower figures repeat this for $\rho = -0.5$.

The PSD of a DT WSS process—that is, the transform of its autocorrelation function—is necessarily real, even, and nonnegative. The above development of the modeling filter suggests why the following converse, referred to earlier as Herglotz's theorem, holds: if $S(e^{j\Omega})$ is a real, even, and nonnegative function of $\Omega$, with period $2\pi$, and if its integral over $[-\pi, \pi]$ is finite, then its inverse transform $R[m]$ is the autocorrelation function of a WSS process, with $R[0] < \infty$. We had earlier stated the CT version of this result, Bochner's theorem. More careful statements and proofs of these results are beyond the scope of this text.

### 11.3.3 Whitening Filters

Some of the estimation and detection problems that we examine in the following chapters are relatively easy to formulate, analyze, and solve when a random process involved in the problem—for instance, the set of measurements—is white, that is, has a flat spectral density. When the process is colored rather than white, the results from the white case can still often be invoked in some appropriate way under one of the following two circumstances.

- The colored process is modeled as the result of passing a white process through some LTI modeling or shaping filter, as described in the preceding subsection. The modeling filter shapes the white process at the input into one that has the spectral characteristics of the given colored process at the output.

- The colored process is transformable into a white process by passing it through an appropriately chosen LTI filter. This filter flattens the spectral characteristic of the colored process presented at the input, resulting in the flat spectral characteristic of a white process at the output. Such a filter is referred to as a whitening filter.

To obtain the equation governing the design of a whitening filter, suppose that the filter input $x[n]$ is a WSS process with a PSD $S_{xx}(e^{j\Omega})$ that has no impulses. We would like the output $w[n]$ to be a white process with variance $\sigma_w^2$. Since

$$S_{ww}(e^{j\Omega}) - |H(e^{j\Omega})|^2 S_{xx}(e^{j\Omega}) , \qquad (11.79)$$

it follows that

$$|H(e^{j\Omega})|^2 = \frac{\sigma_w^2}{S_{xx}(e^{j\Omega})}. \qquad (11.80)$$

This tells us what the squared magnitude of the frequency response of the LTI whitening filter must be to obtain a white output with variance $\sigma_w^2$. If the complex PSD $S_{xx}(z)$ is available as a rational function of $z$, then we can obtain $H(z)$ by appropriate factorization of $H(z)H(z^{-1})$, as discussed in Chapter 2.

---

**Example 11.11    Whitening Filter**

To illustrate the determination of a whitening filter, suppose that

$$S_{xx}(e^{j\Omega}) = \frac{5}{4} - \cos(\Omega) = (1 - \frac{1}{2}e^{j\Omega})(1 - \frac{1}{2}e^{-j\Omega}) . \qquad (11.81)$$

To whiten $x[n]$, we require a stable LTI filter for which

$$|H(e^{j\Omega})|^2 = \frac{1}{(1 - \frac{1}{2}e^{j\Omega})(1 - \frac{1}{2}e^{-j\Omega})}, \qquad (11.82)$$

or equivalently,

$$H(z)H(z^{-1}) = \frac{1}{(1 - \frac{1}{2}z)(1 - \frac{1}{2}z^{-1})} . \qquad (11.83)$$

The filter is constrained to be stable in order to produce a WSS output. One choice of $H(z)$ that results in a causal filter is

$$H(z) = \frac{1}{1 - \frac{1}{2}z^{-1}}, \tag{11.84}$$

with region of convergence given by $|z| > \frac{1}{2}$. This system function could be multiplied by the system function $A(z)$ of any all-pass system, that is, a system function satisfying $A(z)A(z^{-1}) = 1$, and still produce the same whitening action because $|A(e^{j\Omega})|^2 = 1$.

### 11.3.4 Sampling Bandlimited Random Processes

A WSS random process is termed bandlimited if its PSD is bandlimited, that is, is zero for frequencies outside some finite band. For deterministic signals that are bandlimited, we can sample at or above the Nyquist rate and recover the signal exactly. When properly interpreted, a similar result holds for bandlimited random processes, as described below.

Assume the CT WSS random process $x_c(t)$ has autocorrelation function $R_{x_c x_c}(\tau)$ and corresponding PSD $S_{x_c x_c}(j\omega)$ that is bandlimited to $|\omega| < \frac{\pi}{T}$:

$$S_{x_c x_c}(j\omega) = 0 \quad \text{for} \quad |\omega| \geq \frac{\pi}{T}. \tag{11.85}$$

The sampled DT process is defined by $x[n] = x_c(nT)$, and is easily seen to be WSS, with autocorrelation function $R_{xx}[m] = R_{x_c x_c}(mT)$.

The CT signal reconstructed from the DT samples is defined—as in the deterministic case—by the expression

$$y_c(t) = \sum_n x[n] \frac{\sin(\pi(t - nT)/T)}{\pi(t - nT)/T}. \tag{11.86}$$

The sampling theorem in the stochastic case then asserts that the mean square value of the error between $x_c(t)$ and $y_c(t)$ is zero:

$$E\{[x_c(t) - y_c(t)]^2\} = 0. \tag{11.87}$$

In other words, there is zero expected power in the error between the original process and the signal reconstructed from its samples. One proof of this result is outlined in Problem 11.21.

## 11.4 FURTHER READING

Several of the references suggested for further reading at the end of Chapters 7 and 10 address WSS processes in detail, including the frequency domain characteristics captured by the power spectral density. The beginnings of such treatments in the engineering literature can be seen in [Dav] and [Lee]. Texts such as [Jen], [Kay4], [Koo], [Ma2], [Mar], [Per], [Por], and

[Sto] are primarily devoted to the study and application of spectral analysis for WSS processes. The classical periodogram-averaging approach to estimating PSDs that we have outlined in this chapter is a prime example of a nonparametric approach to the problem. Parametric methods instead typically aim to directly identify the parameters of a modeling filter for the WSS process of interest, using measured sample functions of the process. The modeling filter is assumed to come from some particular family of parameterized models (for instance, autoregressive models), and its parameters are chosen to minimize a cost function that reflects how well the modeling filter accounts for the observed data while using only a modest number of parameters. Both the parametric and non-parametric viewpoints are represented in the above references.

# Problems

## Basic Problems

**11.1.** **(a)** Suppose $x(\cdot)$ and $y(\cdot)$ are independent random processes, and each is WSS. Show that $z(t) = x(t)y(t)$ is also WSS, and write its PSD in terms of the PSDs $S_{xx}(j\omega)$ and $S_{yy}(j\omega)$.

  **(b)** Suppose $x(t)$ is a WSS process and $y(t) = x(t - \tau_1)$. Is $C_{yx}(\tau_1) \geq C_{yx}(\tau)$ for all $\tau$? Express $S_{yx}(j\omega)$ in terms of $S_{xx}(j\omega)$.

**11.2.** **(a)** Figure P11.2 shows three candidates (labeled [A], [B], and [C]) for the autocorrelation function $R_{xx}(\tau)$ of a WSS CT random process $x(t)$. For each candidate $R_{xx}(\tau)$, state whether it is a possible autocorrelation function for a WSS random process $x(t)$. Give brief justifications for your answers.

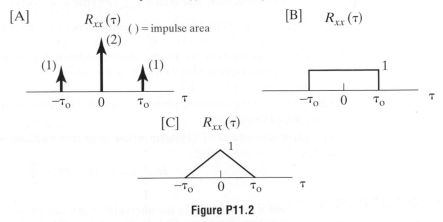

**Figure P11.2**

  **(b)** For each of the following functions $R[m]$, state whether it can be the autocorrelation function of a DT WSS random process, where $m$ denotes the lag. If it cannot be, explain why not. If it can be, explain in detail how you would obtain such a process by appropriately filtering a Bernoulli process that takes values at each time instant of $+1$ or $-1$, with equal probability.

   (i) $R[m] = 1$ for $m = 0$; 0.7 for $|m| = 1$; and 0 elsewhere.
   (ii) $R[m] = 2$ for $m = 0$; $-1$ for $|m| = 1$; and 0 elsewhere.

**11.3.** Figure P11.3-1 shows a sampling system whose input, $x_c(t)$, is a zero-mean WSS random process with PSD as shown in Figure P11.3-2. Assume that the continuous-to-discrete (C/D) box in Figure P11.3-1 is an ideal sampler for which the output is $x_d[n] = x_c(nT)$.

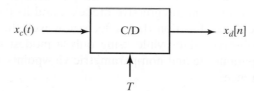

**Figure P11.3-1**

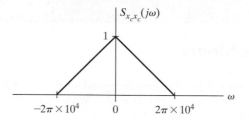

**Figure P11.3-2**

**(a)** Determine $E[x_c^2(t)]$, the mean-squared value of the input process $x_c(t)$.
**(b)** Show that $R_{x_d x_d}[m] = R_{x_c x_c}(\tau)|_{\tau=mT}$. State whether there are any restrictions on the value of $T$ for this to be true, and if so, what they are.
**(c)**  (i) Determine and sketch $S_{x_d x_d}(e^{j\Omega})$ for $\frac{1}{T} = 40\,\text{kHz}$.
       (ii) Determine and sketch $S_{x_d x_d}(e^{j\Omega})$ for $\frac{1}{T} = 15\,\text{kHz}$.

**11.4.** We are given a DT LTI system whose frequency response $H(e^{j\Omega})$ over the interval $[-\pi, \pi]$ is 1 for $(\pi/4) < |\Omega| \le \pi$, and is 0 for $|\Omega| \le (\pi/4)$; in other words, this system functions as an ideal high-pass filter. The input to the system is a white-noise process $w[n]$ with $E\{w^2[n]\} = 10$. If $v[n]$ denotes the output of the system, what is $E\{v^2[n]\}$?

**11.5.** For each of the following parts, state whether the claim is true or false, and give a brief explanation.

**(a)** Consider a DT LTI system whose frequency response is

$$H(e^{j\Omega}) = 2 \quad \text{for} \quad |\Omega| < \frac{\pi}{2},$$

and is 0 elsewhere in the interval $[-\pi, \pi]$, i.e., for $\frac{\pi}{2} < |\Omega| \le \pi$. If the system is driven by an i.i.d. input signal $x[n]$ that takes the values $\pm 1$ with equal probability, then the output $y[n]$ of the system has unit variance, i.e., $\sigma_{y[n]}^2 = 1$.

**(b)** If the autocorrelation function of a WSS random process $x[n]$ is given by

$$R_{xx}[m] = \delta[m] - 0.3\big(\delta[m-1] + \delta[m+1]\big),$$

then the frequency distribution of the expected instantaneous power of the process is more concentrated at low frequencies than high frequencies.

**11.6.** Suppose the WSS random processes $g[\cdot]$ and $v[\cdot]$ are zero mean and uncorrelated. Let $x[n] = g[n] + v[n]$. We are told that the complex PSD of this sum is given by

$$S_{xx}(z) = \frac{(1 - \frac{1}{3}z)(1 - \frac{1}{3}z^{-1})}{(1 - \frac{1}{2}z)(1 - \frac{1}{2}z^{-1})},$$

and that the autocorrelation of $v[n]$ is

$$R_{vv}[m] = \frac{2}{3}\delta[m].$$

Determine $S_{gg}(z)$ and $S_{gx}(z)$.

**11.7.** Suppose $w[n]$ is a zero-mean WSS random process, with $C_{ww}[m] = \sigma^2 \delta[m]$. If $w[n]$ is the input to a causal system whose output $y[n]$ satisfies

$$y[n] = w[n] + w[n-1] + w[n-2],$$

determine the unit sample response $h[\cdot]$ of the system, and also the covariance functions $C_{yw}[m]$ and $C_{yy}[m]$ in terms of $\sigma^2$. Then compute and sketch the PSD $S_{yy}(e^{j\Omega})$ of the output, for $|\Omega| \leq \pi$, and taking $\sigma^2 = 1$.

**11.8.** Suppose $x(\cdot)$ and $y(\cdot)$ are two real-valued jointly WSS random processes. The autocorrelation function of $x(t)$ is $R_{xx}(\tau) = e^{-|\tau|}$. State whether it is possible to specify a choice for $y(t)$ so that the cross spectral density $S_{xy}(j\omega)$ is as shown in Figure P11.8. Note that the amplitude at $\omega = 1$ is $j = \sqrt{-1}$. If your answer is no, explain why not. If your answer is yes, explain how you might specify or construct $y(t)$.

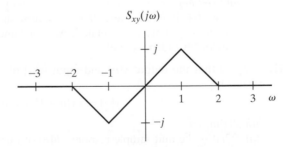

Figure P11.8

**11.9.** Figure P11.9 depicts a stable LTI system with input $x[n]$ and output $y[n]$, which are real-valued jointly WSS random processes with PSDs $S_{xx}(e^{j\Omega})$ and $S_{yy}(e^{j\Omega})$ and cross-spectral density $S_{xy}(e^{j\Omega})$.

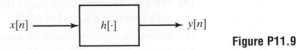

Figure P11.9

For each of the statements below, specify whether it is true or false. Clearly show your reasoning.

**(a)** At any value of $\Omega$ for which $S_{yy}(e^{j\Omega})$ is not zero, $S_{xy}(e^{j\Omega})$ is necessarily not zero.

**(b)** At any value of $\Omega$ for which $S_{xy}(e^{j\Omega})$ is not zero, $S_{xx}(e^{j\Omega})$ is necessarily not zero.

**(c)** At any value of $\Omega$ for which $S_{yy}(e^{j\Omega})$ is zero, $S_{xy}(e^{j\Omega})$ is necessarily zero.

**(d)** The real part of the cross spectral density $S_{xy}(e^{j\Omega})$ must always be nonnegative.

**11.10.** Suppose $q_1(t)$ is obtained from $x_1(\cdot)$ by filtering through a stable system with frequency response $\frac{1-j\omega}{1+j\omega}$, and $q_2(t)$ is obtained from $x_2(\cdot)$ by filtering through another stable system with the same frequency response $\frac{1-j\omega}{1+j\omega}$. Express the cross-spectral density $S_{q_1q_2}(j\omega)$ in terms of $S_{x_1x_2}(j\omega)$. Assume $x_1(\cdot)$ and $x_2(\cdot)$ are jointly WSS.

**11.11.** Let $x(t)$ be a real-valued, zero-mean WSS process with autocorrelation $R_{xx}(\tau)$; the Fourier transform of this autocorrelation is the PSD $S_{xx}(j\omega)$. Suppose $x(t)$ is processed by a pair of stable LTI systems as shown in Figure P11.11. The impulse responses are known to be real.

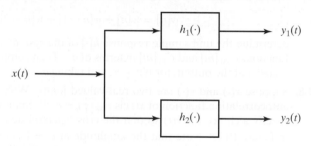

**Figure P11.11**

**(a)** Find $R_{y_1y_2}(\tau)$ and $S_{y_1y_2}(j\omega)$ in terms of $R_{xx}(\tau), h_1(t), h_2(t), S_{xx}(j\omega), H_1(j\omega)$, and $H_2(j\omega)$.

**(b)** Show that if $H_1(j\omega)$ and $H_2(j\omega)$ occupy disjoint frequency bands then $y_1(\cdot)$ and $y_2(\cdot)$ are uncorrelated. Are $y_1(\cdot)$ and $y_2(\cdot)$ also guaranteed to be statistically independent?

**11.12.** Suppose that the output $y[n]$ and input $w[n]$ of a causal LTI DT system are related by

$$y[n] = \beta y[n-1] + w[n]$$

for all times $n$.

**(a)** What is the unit sample response $h[n]$ of this system, and what condition on $\beta$ will ensure the system is BIBO stable?

Assume for the rest of this problem that the stability condition you identified in (a) is satisfied. Also, suppose the input $w[n]$ is actually a WSS process whose PSD $S_{ww}(e^{j\Omega})$ is constant at some value $M > 0$ for all frequencies $\Omega$.

**(b)** What is the mean value $\mu_w$ of the input $w[n]$? And what is the autocovariance function $C_{ww}[m]$ of $w[n]$?

All remaining answers should be expressed in terms of $\beta$ and $M$.

**(c)** Determine the PSD $S_{yy}(e^{j\Omega})$ of the output. Assuming $\beta > 0$, determine at what frequencies in the range $|\Omega| \leq \pi$ this output PSD takes its maximum and minimum values, and find these maximum and minimum values.

**(d)** Using any method you choose, determine $C_{yy}[0]$ and $C_{yy}[1]$, where $C_{yy}[m]$ denotes the autocovariance function of the output.

**(e)** Determine the linear minimum mean square error (LMMSE) estimator

$$\hat{y}[4] = c\,y[3] + d$$

of $y[4]$ in terms of $y[3]$, i.e., find the constants $c$ and $d$ for which the mean square error

$$E[(y[4] - \hat{y}[4])^2]$$

is minimized. Also determine the associated mean square error.

**(f)** Determine the LMMSE estimator $\hat{y}[3]$ of $y[3]$ in terms of $y[4]$, and the associated mean square error.

**(g)** Determine the LMMSE estimator of $y[4]$ in terms of *all* past values $y[k]$, $k \le 3$, and also determine the associated mean square error. (*Hint:* Use what you know of the relation between $y[n]$ and $w[n]$ to conjecture a form for this estimator, then verify that the requisite orthogonality conditions are satisfied.)

**11.13.** We wish to produce a WSS stochastic process $y[n]$ with a specified autocorrelation function $R_{yy}[m]$. The approach is to apply an LTI filter to a white random process $x[n]$ as indicated in Figure P11.13.

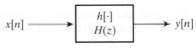

**Figure P11.13**

The process $x[n]$ is zero mean and has autocorrelation function:

$$R_{xx}[m] = \delta[m] = \begin{cases} 1 & m = 0 \\ 0 & m \neq 0. \end{cases}$$

We will choose the filter transfer function $H(z)$ so that $R_{yy}[m] = 0.5^{|m|}$, with corresponding PSD

$$S_{yy}(e^{j\Omega}) = \frac{1}{1 - \frac{1}{2}e^{-j\Omega}} + \frac{\frac{1}{2}e^{j\Omega}}{1 - \frac{1}{2}e^{j\Omega}} = \frac{3}{5 - 4\cos\Omega}.$$

**(a)** Choose the correct statement and explain your reasoning. To obtain the desired $y[n]$, $H(z)$ must represent a

   (i) stable, minimum-phase system.

   (ii) system that is stable, but doesn't need to be minimum phase.

   (iii) system that doesn't need to be stable or minimum phase.

**(b)** Choose the correct statement and explain your reasoning. From what is given, we can say that $x[n]$ and $x[n+k]$, $k \neq 0$, are:

   (i) definitely independent;

   (ii) definitely not independent;

   (iii) may be independent.

**(c)** Choose the correct statement and explain your reasoning. From what is given, we can say that $y[n]$ and $y[n+k]$, $k \neq 0$, are:

   (i) definitely independent;

   (ii) definitely not independent;

   (iii) may be independent.

**(d)** Determine one choice for $H(z)$ (including its region of convergence) that produces a process $y[n]$ with the desired $R_{yy}[m]$.

**11.14.** The input to a particular stable and causal first-order system with transfer function $H(z)$ is a unit-intensity white-noise process $w[n]$, i.e., a process with PSD $S_{ww}(e^{j\Omega}) = 1$. The corresponding output $y[n]$ is a WSS process with PSD

$$S_{yy}(e^{j\Omega}) = 16\frac{(1 - 3z^{-1})(1 - 3z)}{(1 - 4z^{-1})(1 - 4z)}\bigg|_{z=e^{j\Omega}}.$$

**(a)** Plot this PSD as a function of $\Omega$ for $|\Omega| \leq \pi$.
**(b)** Suppose we know that the system has a stable and causal inverse, also of first order. Find a choice of $H(z)$ that is consistent with this information.

**11.15.** A measured PSD for a CT random process is modeled as

$$S(j\omega) = \frac{\omega^2 + 1}{\omega^2 + 100}.$$

We would like to represent the process as the output of an LTI filter with the transfer function $H(s)$ excited by a white process noise $w(t)$, where

$$S_{ww}(j\omega) = 1.$$

**(a)** Assume that $H(s)$ is minimum phase. Determine a choice for $H(s)$.
**(b)** Assume that $H(s)$ is only constrained to be causal and stable, rather than minimum phase. Suppose it is also known that $h(t)$ asymptotically decays as $e^{-t}$ as $t \to \infty$, i.e., as $t \to \infty$, $h(t)$ is approximately proportional to $e^{-t}$. Determine a choice for $H(s)$.

**11.16.** Consider the LTI system shown in Figure P11.16-1.

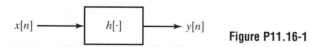

**Figure P11.16-1**

Here $x[n]$ is an i.i.d. process with mean $\mu_x = 1$ and variance $\sigma_x^2 = \frac{1}{4}$. The impulse response $h[n]$ of the system is given in Figure P11.16-2.

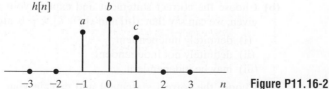

**Figure P11.16-2**

The process $y[n]$ at the output has zero mean and variance $\sigma_y^2 = \frac{3}{2}$. The cross-spectrum $S_{yx}(e^{j\Omega})$ between the input and the output of the system is a real-valued function of $\Omega$.

(a) Determine the values of $a, b$, and $c$ consistent with the information given.

(b) Determine and sketch the spectrum $S_{yy}(e^{j\Omega})$ of the process $y[n]$ for $\Omega \in [-\pi, \pi]$. What kind of filter is $h[n]$ (low-pass, high-pass, band-pass, band-stop, all-pass)?

(c) If possible, find the impulse response $g[n]$ of a causal, stable LTI system whose output $w[\cdot]$ is a white process when its input is the process $y[\cdot]$. If this is not possible, explain why not.

**11.17.** Consider a WSS random process $x[n]$ for which the complex PSD is

$$S_{xx}(z) = \frac{(1 - \frac{1}{3}z)(1 - \frac{1}{3}z^{-1})}{(1 - \frac{1}{2}z)(1 - \frac{1}{2}z^{-1})} \ .$$

Find a whitening filter $H_w(z)$ for the process $x[n]$, choosing it to be stable and causal, and to have a stable and causal inverse. Is your answer unique to within a constant scale factor? If yes, explain why. If not, construct a second whitening filter.

## Advanced Problems

**11.18.** Denote the PSD of a DT WSS random process $x[n]$ by $S_{xx}(e^{j\Omega})$, and assume the mean value of the process is $\mu_x$.

(a) Let $q[n] = x[n] - \mu_x$. Express $S_{qq}(e^{j\Omega})$ in terms of $S_{xx}(e^{j\Omega}), \mu_x$.

(b) Suppose

$$S_{xx}(e^{j\Omega}) = 10 + Ke^{j\Omega} + Ke^{-j\Omega}$$

for some constant $K$. Determine the following quantities:

  (i) The autocorrelation function $R_{xx}[m]$.

  (ii) $E\{x^2[n]\}$.

  (iii) The mean $\mu_x$ (your answer to (a) may help you here, invoking $S_{qq}(e^{j\Omega}) \geq 0$).

  (iv) The most positive and most negative values that $K$ can take.

  (v) The time average

$$\lim_{N \to \infty} \frac{1}{2N+1} \sum_{n=-N}^{N} x[n] \ ,$$

where $x[n]$ now denotes a particular sample function of the random process.

(c) Suppose $K = -2$ in (b). Write an expression for the LMMSE estimator of $x[n+1]$ in terms of $x[n]$, and for its mean square error. Also write an expression for the LMMSE estimator of $x[n+2]$ in terms of $x[n]$, and for its mean square error.

(d) Suppose the process $x[n]$ above for some $K > 0$ is applied to a filter with unit sample response

$$h[n] = \delta[n - 100] + \delta[n - 101] \ ,$$

resulting in the output process $y[n]$. Determine the autocorrelation function $R_{yy}[m]$, and provide a fully labeled sketch of it.

**11.19.** A DT WSS stationary random process $v[n]$ has PSD given by

$$S_{vv}(e^{j\Omega}) = K + e^{j2\Omega} + e^{-j2\Omega},$$

where $K$ is some constant.

**(a)** What is the smallest value that $K$ can take? Sketch $S_{vv}(e^{j\Omega})$ in the range $|\Omega| \leq \pi$ for the case where $K$ takes this smallest possible value, and state in what frequency range or ranges—low, middle, and/or high—the expected instantaneous power of $v[n]$ is concentrated for this $K$.

For the rest of this problem, use the value $K = 4$.

**(b)** Determine the mean value $\mu_v$ of $v[n]$ (explaining your reasoning carefully), and also determine and sketch the autocorrelation function $R_{vv}[m]$.

**(c)** Obtain the following four LMMSE estimators of $v[n+1]$, and their associated mean square errors:

   (i)  the estimator of $v[n+1]$ using measurement of $v[n]$;
   (ii)  the estimator of $v[n+1]$ using measurement of $v[n-1]$;
   (iii)  the estimator of $v[n+1]$ using measurement of $v[n]$ and $v[n-1]$; and
   (iv)  the estimator of $v[n+1]$ using measurement of $v[n-1]$ and $v[n-3]$.

**(d)** Define the random process $x[n]$ by

$$x[n] = v[n] + A,$$

where $v[n]$ is as specified above, and $A$ is a zero-mean random variable, uncorrelated with $v[n]$ for all $n$, and of variance $\sigma_A^2 > 0$ (so a random choice of $A$ is made for each realization of the random process, and this $A$ does not vary with $n$). Determine:

   (i)  the mean $\mu_x$ and autocorrelation function $R_{xx}[m]$;
   (ii)  the LMMSE estimator of $x[n+1]$ using measurement of $x[n]$;
   (iii)  the power spectral density $S_{xx}(e^{j\Omega})$, specified for $|\Omega| \leq \pi$; and
   (iv)  whether the process $x[n]$ is ergodic in mean value.

**11.20.** Suppose $x[n]$ is a zero-mean WSS process with $R_{xx}[m] = N_0\delta[m]$, so that the Fourier transform of the autocorrelation, i.e., the PSD, is flat at all frequencies: $S_{xx}(e^{j\Omega}) = N_0$. In other words, $x[n]$ is a white process. Assume this process $x[n]$ is applied to the input of a stable system with transfer function

$$H(z) = \frac{z+5}{z-\frac{1}{2}}.$$

**(a)** What is the PSD $S_{yy}(e^{j\Omega})$ of the output $y[n]$?

The filter $H(z)$ is referred to as a shaping filter for the process $y[n]$ because it shapes the flat spectrum of the input into that of the output, impressing on the output its own frequency characteristics. It is also called a modeling filter for the process $y[n]$, since it models $y[n]$ by relating it to the simpler white process.

**(b)** Specify the transfer function of a stable and causal first-order (i.e., single pole) filter $G(z)$ that will produce a WSS white-noise process at its output when its input is the process $y[n]$ defined in (a). Such a filter is termed a causal whitening filter for $y[n]$.

**(c)** Are there other stable, causal first-order filters that could serve as whitening filters for $y[n]$ in (a)? If so, how are they related to $G(z)$ in (b)?

**(d)** Are there higher-order causal filters that could serve as whitening filters for $y[n]$ in (a)? If so, again specify how they are related to $G(z)$.

**(e)** Are there first-order noncausal filters that could serve as whitening filters for $y[n]$? If so, give an example of one.

**11.21.** Consider a DT WSS process $x[n]$ with PSD given by

$$S_{xx}(e^{j\Omega}) = \frac{9}{(\frac{1}{2} + e^{j\Omega})(\frac{1}{2} + e^{-j\Omega})}.$$

Find the unit impulse response $h[n]$ of one possible whitening filter satisfying $h[n] = 0$ for $n \leq 1$, so that the PSD $S_{yy}(e^{j\Omega})$ of $y[n] = x[n] * h[n]$ is a (nonzero) constant for all $\Omega$.

**11.22.** Assume that $x[n]$ in Figure P11.22 is zero-mean WSS, with correlation function $R_{xx}[m]$ and PSD $S_{xx}(e^{j\Omega})$. Suppose that the process $w[\cdot]$ is independent of the process $x[\cdot]$, and at any time instant takes the value 1 with probability $p$ or the value 0 with probability $1 - p$; also assume that the values of $w[\cdot]$ at different time instants are independent. Thus the signal $y[n] = x[n]w[n]$ is obtained by setting random components of $x[\cdot]$ to zero.

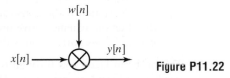

Figure P11.22

Explicitly check all your answers to the questions below, to be sure that they take reasonable values for the extreme cases of $p = 1$ and $p = 0$. Make sure to explain why you think these extreme values are reasonable.

**(a)** Find the mean $\mu_w$ of the WSS process $w[n]$. Show that its autocorrelation function has the form $R_{ww}[m] = \alpha\delta[m] + \beta$, where $\alpha$ and $\beta$ are constants that you should determine, and $\delta[m]$ is the unit sample function. Also find an expression for the PSD $S_{ww}(e^{j\Omega})$ of the process $w[n]$.
**(b)** Verify that $y[\cdot]$ and $x[\cdot]$ are jointly WSS, and compute $R_{yx}[m]$, $R_{yy}[m]$ and the PSD $S_{yy}(e^{j\Omega})$.

**11.23.** Let $x[n]$ be a WSS DT process, with PSD given by $S_{xx}(e^{j\Omega})$. Also assume this PSD contains no impulses.

**(a)** We would like to filter $x[n]$ through the LTI system in Figure P11.23-1, which has frequency response $H(e^{j\Omega})$, producing an output $z[n]$ whose cross-spectral density $S_{zx}(e^{j\Omega})$ is some specified function $T(e^{j\Omega})$. Determine the $H(e^{j\Omega})$ that will accomplish this, expressing your answer in terms of $S_{xx}(e^{j\Omega})$ and $T(e^{j\Omega})$.

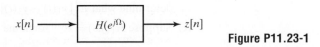

Figure P11.23-1

**(b)** Let $e[n]$ be a zero-mean WSS process that is uncorrelated with the $x[n]$ defined above, so $R_{ex}[m] = 0$, and denote its PSD by $S_{ee}(e^{j\Omega})$. With $z[n]$ defined as in (a), let $y[n] = z[n] + e[n]$. The relation of the various signals is indicated in Figure P11.23-2. Show that $S_{yx}(e^{j\Omega})$ can be expressed in terms of just $S_{zx}(e^{j\Omega})$ alone. Also evaluate $S_{yy}(e^{j\Omega})$ in terms of $S_{xx}(e^{j\Omega})$, $H(e^{j\Omega})$, and $S_{ee}(e^{j\Omega})$.

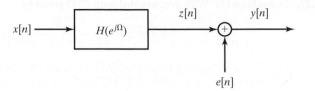

**Figure P11.23-2**

**(c)** Suppose that in addition to picking $T(e^{j\Omega})$—and thus specifying both $S_{zx}(e^{j\Omega})$ and $S_{yx}(e^{j\Omega})$—we also wanted to make $S_{yy}(e^{j\Omega})$ equal to some specified real, nonnegative, even function of frequency, $U(e^{j\Omega})$. Use your result from (b) to show what choice of $S_{ee}(e^{j\Omega})$ is needed to accomplish this, expressing $S_{ee}(e^{j\Omega})$ in terms of $S_{xx}(e^{j\Omega})$, $T(e^{j\Omega})$, and $U(e^{j\Omega})$. Using the fact that $S_{ee}(e^{j\Omega})$ is a PSD will show you that we cannot actually have $U(e^{j\Omega})$ be an arbitrary real, nonnegative, and even function of frequency; write down a constraint involving $S_{xx}(e^{j\Omega})$, $T(e^{j\Omega})$, and $U(e^{j\Omega})$ that must be satisfied.

   Note that if the constraint you identified at the end of (c) is satisfied, then the scheme above provides a way of generating a process $y[n]$ of specified cross-correlation with $x[n]$, and specified autocorrelation.

**11.24.** This problem leads you through one derivation of the sampling theorem for bandlimited random processes. The common input to the two systems in Figure P11.24 is a WSS random process $x(t)$ with power spectral density $S_{xx}(j\omega)$.

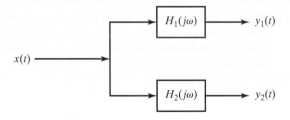

**Figure P11.24**

**(a)** Express $E\left\{[y_1(t) - y_2(t)]^2\right\}$ in terms of $H_1(j\omega)$, $H_2(j\omega)$, and $S_{xx}(j\omega)$.

Suppose for the remainder of this problem that $S_{xx}(j\omega)$ is nonzero only in the frequency range $|\omega| < \omega_m$, for some $\omega_m$. The process $x(t)$ is then referred to as a bandlimited random process.

**(b)** Suppose $H_1(j\omega) = H_2(j\omega)$ for $|\omega| < \omega_m$, but that these two frequency responses possibly differ outside this band. Using your result in (a), determine what $E\left\{[y_1(t) - y_2(t)]^2\right\}$ is for this case.

**(c)** Suppose $H_1(j\omega) = e^{j\omega\tau}$ for some $\tau$, and

$$H_2(j\omega) = \sum_{n=-\infty}^{\infty} s[n]e^{j\omega nT}, \quad T = \pi/\omega_m ,$$

where $s[n]$ is a sequence of real numbers. Express $y_1(t)$, $y_2(t)$ in terms of $x(\cdot)$.

(d) Suppose now that the sequence $s[n]$ in (c) is given by

$$s[n] = \frac{\sin\left(\omega_m(\tau - nT)\right)}{\omega_m(\tau - nT)} .$$

Carefully explain why in this case $H_1(j\omega) = H_2(j\omega)$ for $|\omega| < \omega_m$. (*Hint:* $H_2(j\omega)$ in (c) is periodic in $\omega$.)

(e) Put the preceding calculations together to deduce the sampling theorem for bandlimited WSS processes, expressing $x(\tau)$ in terms of its samples.

## Extension Problems

**11.25.** An analog-to-digital (A/D) converter can be represented as a C/D converter followed by a quantizer. A useful model for quantization error in a linear quantizer is to represent the error as a zero-mean i.i.d. process with variance $\sigma_e^2$, and uncorrelated with the DT quantizer input $x_d[\cdot]$. This leads to the model for an A/D converter shown in Figure P11.25-1, where $q[n]$ represents the quantized signal and $e[n]$ represents the error introduced by quantization.

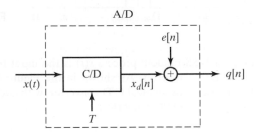

**Figure P11.25-1**

Assume:

  (i) $x(t)$ is a zero-mean WSS random process;

 (ii) $e[n]$ is i.i.d., zero mean, with variance $\sigma_e^2 = \frac{1}{5} \times 10^{-3}$;

(iii) $x_d[\cdot]$ and $e[\cdot]$ are uncorrelated random processes;

(iv) the PSD, $S_{xx}(j\omega)$, of $x(t)$ is as shown in Figure P11.25-2; and

 (v) the sampling period is $T = \frac{1}{4} \times 10^{-5}$, i.e., the sampling frequency is $\omega_s = \frac{2\pi}{T} = 8\pi \times 10^5$.

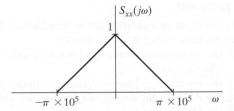

**Figure P11.25-2**

(a) Determine and make a labeled sketch of $S_{qq}(e^{j\Omega})$, the PSD of $q[n]$, in the range $|\Omega| \le 4\pi$.

**(b)** The signal-to-noise ratio, $\mathrm{SNR}_q$, of the quantized signal $q[n]$ is defined as:

$$\mathrm{SNR}_q \triangleq \frac{E\{x_d^2[n]\}}{E\{e^2[n]\}}.$$

Determine $\mathrm{SNR}_q$.

**(c)** Suppose $q[n]$ is now processed as shown in Figure P11.25-3, where $H(e^{j\Omega})$ is the ideal low-pass filter shown in Figure P11.25-4, $x_r[n]$ is due only to $x_d[n]$, and $e_r[n]$ is due only to $e[n]$. What value of $\Omega_{co}$ would you choose so that $R_{x_r x_r}[n] = R_{x_d x_d}[n]$ and $E\{x_r^2[n]\}/E\{e_r^2[n]\}$ (i.e., the SNR after filtering) is maximized? State the maximized SNR value.

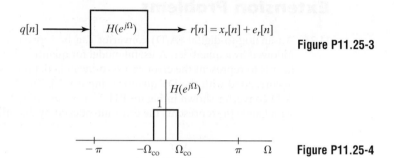

**Figure P11.25-3**

**Figure P11.25-4**

**11.26.** A zero-mean WSS random process $s(t)$ is the input to an A/D converter followed by a D/A converter. As shown in the Figure P11.26-1, the A/D–D/A cascade is modeled by an ideal C/D converter, with additive quantization noise, followed by an ideal D/C converter.

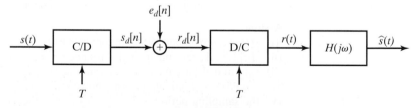

**Figure P11.26-1**

In particular:

(i) the input process $s(t)$ has the bandlimited PSD shown in Figure P11.26-2;
(ii) the C/D output is $s_d[n] = s(nT)$;
(iii) the quantization noise $e_d[n]$ is a zero-mean, WSS, white-noise process with PSD $S_{e_d e_d}(e^{j\Omega}) = \sigma_e^2$;
(iv) the process $e_d[n]$ is uncorrelated with the process $s_d[\cdot]$;
(v) the A/D output is $r_d[n] = s_d[n] + e_d[n]$; and
(vi) the D/A output is

$$r(t) = \sum_{n=-\infty}^{\infty} r_d[n] \frac{\sin[\pi(t - nT)/T]}{\pi(t - nT)/T}.$$

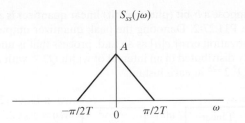

Figure P11.26-2

**(a)** Determine and make a labeled sketch of $S_{rr}(j\omega)$, the PSD of $r(t)$.
**(b)** Determine and make a labeled sketch of $S_{sr}(j\omega)$, the cross-spectral density of $s(t)$ and $r(t)$.

To reduce the effect of the quantization noise, we want to pass $r(t)$ through an LTI filter with frequency response $H(j\omega)$ to obtain an estimate $\widehat{s}(t)$ of $s(t)$. The error measure that we want to minimize is the mean square error

$$\mathcal{E} = E\{[s(t) - \widehat{s}(t)]^2\}.$$

**(c)** In this part we restrict the filter $H(j\omega)$ to be an ideal low-pass filter with unity gain and cutoff frequency $\omega_c$, as shown in Figure P11.26 3. For $\Lambda > \sigma_e^2 T$, determine the $\omega_c$ value, in terms of $A$, $T$, and $\sigma_e^2$, that minimizes the mean square error $\mathcal{E}$ defined above.

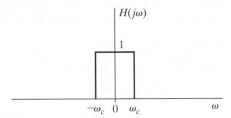

Figure P11.26-3

**11.27.** The audio on a compact disc is stored at a sampling rate of 44.1 kHz (on the CD it's stored in a coded format that allows for significant error detection and correction). As part of the digital-to-analog (D/A) conversion, the data is upsampled to a much higher sampling rate. As we explore in this problem, this allows for a much simpler, and therefore less expensive D/A converter.

The basic strategy is suggested in the block diagram shown in Figure P11.27-1. Here $x[n]$ represents the audio on the compact disc, which is an appropriately bandlimited $x_c(t)$ sampled at 44.1 kHz, and is so finely quantized that for the purposes of this problem we can ignore the quantization of $x[n]$. The upsampler converts $x[n]$ to samples that correspond to having sampled $x_c(t)$ at a rate of $L \cdot 44.1$ kHz, i.e., if $x[n] = x_c(nT)$ then $g[n] = x_c(nT/L)$.

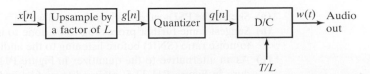

Figure P11.27-1

Suppose a $b$-bit (plus sign bit) linear quantizer is applied to $g[n]$ as shown in Figure P11.27-2. Denoting the peak quantizer output by $Q$, we shall model the quantization error $e[n]$ as an i.i.d. process that is uncorrelated with $g[\cdot]$ and uniformly distributed in an interval of width $Q2^{-b}$ with zero mean and variance $\sigma_e^2 = \frac{1}{12} Q^2 \, 2^{-2b}$ at each instant.

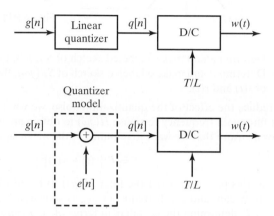

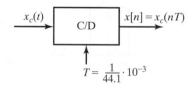

**Figure P11.27-2**

As shown in Figure P11.27-3, assume that $x[n]$ is a random process obtained by sampling a CT signal $x_c(t)$ at a sampling frequency of 44.1 kHz, and that the PSD of $x_c(t)$ is as shown in Figure P11.27-4.

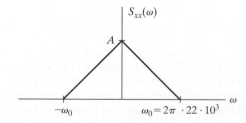

**Figure P11.27-3**

**Figure P11.27-4**

**(a)** Sketch the PSDs of $x[n]$, $g[n]$, $e[n]$, and $w(t)$.

**(b)** Suggest some further processing to be done to $w(t)$ to improve the signal-to-noise ratio (SNR) before listening to the audio.

**(c)** As an alternative to the quantizer in Figure P11.27-2, consider the procedure in Figure P11.27-5, referred to as first-order noise shaping. For this quantizer sketch the PSD of $q[n]$ and $w(t)$.

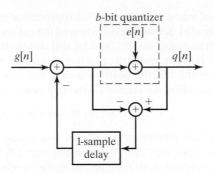

**Figure P11.27-5**

**(d)** In a high-quality CD player, $L = 256$ and $b = 1$ bit. Which of the two quantizers do you think would be preferable?

**11.28.** Figure P11.28 depicts a DT $\Sigma - \Delta$ quantizer where the random process $e[n]$ denotes the error introduced by a linear quantizer.

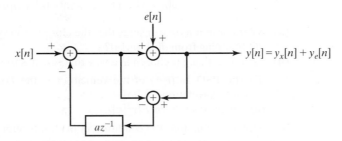

**Figure P11.28**

Suppose the signal $x[n]$ is a WSS process with a triangular PSD that has the value $M$ at $\Omega = 0$, and decays linearly to the value 0 for $\Omega = \pm\pi/2$. The quantization noise $e[n]$ is modeled as a (zero-mean) white-noise random process that is uncorrelated with $x[\cdot]$ and has autocorrelation function $R_{ee}[m] = \sigma_e^2 \delta[m]$.

The output process $y[n]$ is represented as the sum

$$y[n] = y_x[n] + y_e[n],$$

where

$$y_x[n] \text{ denotes the output due to } x[n],$$

$$y_e[n] \text{ denotes the output due to } e[n].$$

In the block diagram in Figure P11.28, $a$ is an adjustable parameter.

**(a)** Determine the transfer functions $H_x(z)$ from $x[n]$ to $y_x[n]$ and $H_e(z)$ from $e[n]$ to $y_e[n]$.

**(b)** Determine the PSDs of $y_x[n]$ and $y_e[n]$ in terms of $M$, $\sigma_e^2$, and $a$. Will the sum of these PSDs yield $S_{yy}(e^{j\Omega})$, the PSD of $y[\cdot]$?

**(c)** Determine $E\{y_x^2[n]\}$ and $E\{y_e^2[n]\}$, expressing them in terms of appropriate integrals in the frequency domain.

**(d)** Determine the value of $a$ that maximizes the in-band output SNR, defined as the ratio of $E\{y_x^2[n]\}$ to the expected squared value of the noise component $y_e[n]$ that is contained in the signal band $|\Omega| \leq \pi/2$.

**11.29.** You are asked to design the microprocessor-based cruise control system for a car. Let $v[n]$ denote the deviation of the car's velocity from the desired value at the $n$th sampling instant, and let $x[n]$ denote the deviation of the throttle position at this time from the position nominally required to maintain the desired velocity; the throttle position is only varied at the sampling instants. Assume these variables are related via the first-order state-space description

$$v[n+1] = (1-\alpha)v[n] + \beta x[n] + d[n],$$

where $d[n]$ represents the effects of external disturbances acting on the car. Assume also that $d[n]$ is zero-mean white noise, i.e., its power spectral density is constant at all frequencies, $S_{dd}(e^{j\Omega}) = \sigma^2$, and its correlation function is correspondingly $R_{dd}[m] = \sigma^2 \delta[m]$. Take $\alpha$ (which represents the effect of frictional damping) to be positive but less than 1, and $\beta$ to be positive; assume both these parameters are known. Now suppose that you pick the control to be in state-feedback form, i.e., $x[n] = gv[n]$, where $g$ is a constant gain, so that the closed-loop system is actually

$$v[n+1] = (1-\alpha+\beta g)v[n] + d[n].$$

**(a)** What condition on $g$ ensures that the closed-loop system is a BIBO stable LTI mapping from $d[n]$ to $v[n]$?

**(b)** Assuming that $g$ is chosen in a way that satisfies the condition in (a), find:

   (i) the PSD $S_{vv}(e^{j\Omega})$ of the velocity, i.e., the Fourier transform of the autocorrelation $R_{vv}[m]$;
   (ii) an expression for $E\{v^2[n]\}$.

Now, still assuming that the condition in (a) holds, suppose our control design task is formulated as that of choosing $g$ to minimize $E\{v^2[n] + rx^2[n]\}$. This criterion reflects our desire to keep both the velocity variations and the throttle variations close to 0. The positive parameter $r$ allows us to reflect how undesirable throttle variations are relative to velocity variations; a large $r$ would be used if we did not want excessive throttle variations (e.g., for reasons of fuel economy or emission control).

**(c)** Use your expression in (b)(ii) above to find an equation that could be solved to determine the optimum $g$.

**11.30.** A particular DT WSS random process $\kappa[n]$ has autocorrelation function

$$R_{\kappa\kappa}[m] = 10\delta[m] + 3\gamma\Big(\delta[m-1] + \delta[m+1]\Big).$$

**(a)** What are the most positive and most negative values that $\gamma$ can take in this instance? Determine the mean and variance of $\kappa[n]$, and also the correlation coefficient between $\kappa[n]$ and $\kappa[n-1]$.

For the rest of this problem, assume $\gamma = 1$.

**(b)** Show that $\kappa[n]$ can be generated as the output of an appropriate stable first-order state-space system driven from time $-\infty$ by a (zero-mean) white process $w[n]$ of unit intensity, so $w[n]$ has variance 1. (*Hint:* First consider what unit sample response or transfer function you would want this system to have.) Explicitly write down this state-space system in the following form:

$$q[n+1] = \alpha q[n] + \beta w[n], \quad \kappa[n] = \xi q[n] + dw[n],$$

with appropriately chosen values of the coefficients $\alpha, \beta, \xi$, and d. Explain your reasoning.

You might find more than one first-order state-space model that will accomplish the objective, but any one of them will suffice as an answer for our purposes.

(c) Suppose we have another first-order state-space system, driven by the colored process $\kappa[n]$ that was produced by the system in (b). Let $p[n]$ denote the state variable of this system, and assume the output $y[n]$ of this system can be measured. The system thus takes the form

$$p[n+1] = ap[n] + b\kappa[n], \quad y[n] = p[n] + v[n] .$$

Here $a$ and $b$ are some fixed nonzero scalar parameters whose precise values don't matter to us, and $v[n]$ is a (zero-mean) white measurement-noise process with variance $\sigma^2$ and is uncorrelated with $w[\cdot]$. Combine this system description with your result from (b) to write down a second-order state-space model with state variables $q[n]$ and $p[n]$, white input $w[n]$, and measured output $y[n]$. Also determine the eigenvalues and eigenvectors associated with the system, i.e., the eigenvalues and eigenvectors of the one-step state transition matrix of this system. As a check, one of the eigenvalues of your model should turn out to be 0.

(d) Determine what conditions, if any, have to be satisfied by the various coefficients in this problem for the combined system in (c) to be:

  (i) reachable from the input signal $w[n]$?
  (ii) observable in the output signal $y[n]$?

For each of the above cases, also specify which modes become unreachable or unobservable when the respective conditions are not satisfied.

(e) Suppose $w[n]$, $\kappa[n]$, and $v[n]$ cannot be measured, although their properties specified above are known. However, as mentioned before, $y[n]$ is measured. Write down in detail the equations of a second-order observer to propagate estimates $\widehat{q}[n]$ and $\widehat{p}[n]$ of $q[n]$ and $p[n]$ respectively for all $n \geq 0$. Also write down a second-order state-space model describing the evolution of the errors $\widetilde{q}[n] = q[n] - \widehat{q}[n]$ and $\widetilde{p}[n] = p[n] - \widehat{p}[n]$. Pick the observer gains to put both eigenvalues of the error model at 0.

(f) For the observer you designed in (e), obtain an expression for the steady-state variance of $\widetilde{p}[n]$, expressed in terms of $a$, $b$, and $\sigma$.

**11.31. (a)**  (i) Numerically generate a 100-point segment of a random signal $v[n]$, $n = 1, \cdots, 100$, constructed as a realization of an i.i.d. process whose values at any time are chosen uniformly in the interval $(-1, 1)$. Also determine analytically the mean $\mu_v$ and autocorrelation function $R_{vv}[m]$ of this i.i.d. process, as well as its PSD $S_{vv}(e^{j\Omega})$.

  (ii) Plot the 100-point signal $v[n]$ for two different realizations. Compute the time average for each of these realizations, and compare with the theoretical (ensemble) mean $\mu_v$. Also compute the deterministic autocorrelation for each of your realizations, for $m = 0, 1, 2, 3, 4$ (and additional values of $m$, if you like):

$$\overline{R}_{vv}[m] = \frac{1}{N_m} \sum_{n} v[n+m]v[n] ,$$

where the sum ranges over the $N_m$ terms for which both $v[n+m]$ and $v[n]$ are defined (this is needed since you have only defined

$v[n]$ for 100 values of $n$). Compare your results with the theoretical (ensemble) autocorrelation function $R_{vv}[m]$ computed earlier. Does the evidence suggest the process $v[n]$ is ergodic, as far as means and autocorrelations are concerned?

(iii) For each realization in (ii), compute and plot the periodogram, defined as the squared magnitude of the DTFT of the 100-point signal, divided by 100. Comment on any significant similarities and differences between the periodograms for the two realizations. Also compute and plot the empirical (i.e., experimental) average periodogram over 200 realizations. What, if anything, changes with the preceding results if you use a signal segment of duration 400 instead of 100 time instants?

**(b)** (i) Consider the random signal defined by $x[n] = 3 + (-1)^n + v[n]$, where $v[n]$ is constructed as in (a). Analytically determine the mean $\mu_x[n]$, autocorrelation $R_{xx}[n+m,n]$, and autocovariance $C_{xx}[n+m,n]$ of this random signal. Determine which of these quantities, if any, does not depend on $n$.

(ii) Generate and plot a 150-point realization of $x[n]$, then apply it to the input of a causal filter whose transfer function $H(z)$ has zeros at $z = 1$ and $z = -1$, and poles at $z = 0.8 \pm j0.2$. Plot the output $y[n]$ for $n = 1, \cdots, 150$. Comment on any features of $y[n]$ that seem to reflect characteristics imposed by the filter. Compute and plot the periodogram of $y[n]$, omitting any initial filter transient.

(iii) Repeat this experiment for 200 realizations, then compute and plot the averaged periodogram of $y[n]$. Compare this plot with the theoretical expression for the PSD of the output, namely $S_{yy}(e^{j\Omega}) = \sigma_v^2|H(e^{j\Omega})|^2$, where $\sigma_v^2 = R_{vv}[0]$ is the variance of $v[n]$ at any time.

**11.32.** Consider a zero-mean WSS process $y[n]$ with $E\{y^2[n]\} = \sigma^2$. Suppose signal values at adjacent instants have a correlation coefficient of $\rho$, but that signal values more than one instant apart are uncorrelated.

**(a)** Use an appropriate modeling filter driven by a white process to numerically generate and plot 100 consecutive values of $y[n]$, for $\rho = \pm0.1$ and $\rho = \pm0.4$, with $\sigma^2 = 9$. For each of these four cases, also evaluate

$$\frac{1}{99}\sum_k y[k]y[k+1] \quad \text{and} \quad \frac{1}{98}\sum_k y[k]y[k+2],$$

where $k$ in the first sum ranges over the 99 values in your data set for which both $y[k]$ and $y[k+1]$ are defined, and in the second sum ranges over the 98 values for which both $y[k]$ and $y[k+2]$ are defined. Are the values you get close to what you would expect to find?

**(b)** (i) Design an LMMSE estimator $\hat{y}[k+1]$ for $y[k+1]$, based on measurement of $y[k]$, and test it out on your data set for the case where $\rho = 0.4$. In particular, compute

$$\frac{1}{99}\sum_k (\hat{y}[k+1] - y[k+1])^2,$$

where $k$ ranges over the first 99 values of your data set. Is the value you get close to what you would expect to find?

(ii) Design an LMMSE estimator $\hat{y}[k+1]$ for $y[k+1]$ based on measurement of both $y[k]$ and $y[k-1]$, and test it out on your data set. In particular, compute

$$\frac{1}{98} \sum_k (\hat{y}[k+1] - y[k+1])^2 \,,$$

where $k$ ranges over the middle 98 values of your data set. Do you seem to be doing better than in (b)(i), or are these numbers about the same? Would you expect, in the limit of an infinitely long experiment rather than one involving just 100 points, that this number will be smaller than the corresponding quantity you computed in (b)(i), or do you think the two numbers should be the same?

# 12 Signal Estimation

In the preceding two chapters we developed concepts and tools for dealing with random processes in both the time and frequency domains. We are now equipped to address some prototypical problems of inference involving such processes. This chapter deals with linear minimum mean square error (LMMSE) estimation of a wide-sense stationary (WSS) process from measurements of another process that is jointly WSS with it. Such estimation is generally referred to as Wiener filtering, after Norbert Wiener, who in the early 1940s solved the challenging causal version of this problem. In causal Wiener filtering, past and present measurements are processed by an optimally chosen causal linear and time-invariant (LTI) filter to produce LMMSE estimates of the process of interest. The derivation of the causal Wiener filter makes use of a minimum-phase LTI modeling filter whose input is some white process and whose output is the WSS measured process.

Rudolph Kalman in the 1960s extended such LMMSE estimation to nonstationary processes, with the measured process now modeled as the noisy output of a time-varying linear state-space model driven by some white disturbance process. The objective in Kalman filtering is to use the measurements to obtain LMMSE estimates of the state variables. In the case of an LTI system, the resulting filter takes the form of the state-space observer studied in Chapter 6, but with a possibly time-varying observer gain vector. If the disturbance and noise processes are stationary, and if the system is reachable from the disturbance input and is observable from the measured output, then the observer gain or Kalman gain converges to some constant vector, exactly yielding an observer of the sort studied in Chapter 6.

The current chapter focuses on Wiener filtering, though a simple example of a Kalman filter, viewed as an optimal observer, is also presented at the end.

## 12.1 LMMSE ESTIMATION FOR RANDOM VARIABLES

As LMMSE estimation for jointly WSS processes is closely related to LMMSE estimation for random variables, it is helpful to first briefly summarize the key results from Chapter 8. That chapter showed how to construct an LMMSE estimator of a random variable $Y$ in terms of some collection of $L$ measured random variables $\{X_i\}_{i=1}^{L}$. This estimator $\widehat{Y}$ is a linear combination of the measured random variables (actually a linear combination plus a constant, also referred to as an affine combination):

$$\widehat{Y} = \mu_Y + \sum_{i=1}^{L} a_i(X_i - \mu_{X_i}) = \mu_Y + \mathbf{a}^T(\mathbf{X} - \mu_\mathbf{X}), \qquad (12.1)$$

as seen in Eq. (8.79). We had previously written $\widehat{Y}_\ell$ for the LMMSE estimator, but since this chapter deals only with linear estimators, we drop the subscript $\ell$ throughout for notational simplicity. The quantities $\mu_Y$ and $\mu_{X_i}$ in Eq. (12.1) denote the expected or mean values of the indicated random variables, and we also use the following vector notation:

$$\mathbf{a} = \begin{bmatrix} a_1 \\ a_2 \\ \vdots \\ a_L \end{bmatrix}, \qquad \mathbf{X} = \begin{bmatrix} X_1 \\ X_2 \\ \vdots \\ X_L \end{bmatrix}, \qquad \mu_\mathbf{X} = \begin{bmatrix} \mu_{X_1} \\ \mu_{X_2} \\ \vdots \\ \mu_{X_L} \end{bmatrix}. \qquad (12.2)$$

The form of the estimator in Eq. (12.1) shows that it is unbiased, that is, $E[\widehat{Y}] = \mu_Y = E[Y]$.

The optimal vector $\mathbf{a}$ is obtained by solving the normal equations given in Eq. (8.87), which are the result of invoking the orthogonality of the estimation error $Y - \widehat{Y}$ to each measured random variable $X_i$. In matrix form, the normal equations are written as

$$(\mathbf{C}_\mathbf{XX})\mathbf{a} = \mathbf{c}_\mathbf{XY}. \qquad (12.3)$$

Here $\mathbf{C}_\mathbf{XX}$ is the symmetric $L \times L$ covariance matrix of the random vector $\mathbf{X}$. Its $i,j$th element is the covariance $\sigma_{X_iX_j}$ between the measured random variables $X_i$ and $X_j$, and equals its $j,i$th element, namely $\sigma_{X_jX_i}$. Similarly, $\mathbf{c}_\mathbf{XY}$ is an $L \times 1$ vector whose $i$th component is the covariance $\sigma_{X_iY}$ between the measured random variable $X_i$ and the random variable $Y$ that we are estimating. Equivalently, this matrix and vector of covariances can respectively be defined by the expressions

$$\mathbf{C}_\mathbf{XX} = E[(\mathbf{X} - \mu_\mathbf{X})(\mathbf{X} - \mu_\mathbf{X})^T], \qquad \mathbf{c}_\mathbf{XY} = E[(\mathbf{X} - \mu_\mathbf{X})(Y - \mu_Y)] = \mathbf{c}_{Y\mathbf{X}}^T,$$
$$(12.4)$$

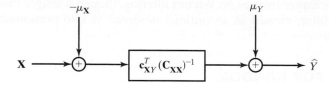

**Figure 12.1**  Schematic representation of Eq. (12.5), showing the LMMSE estimator of a random variable $Y$ in terms of a measured random vector $\mathbf{X}$.

where the expectation of the matrix or vector is taken entry-by-entry. The LMMSE estimator is thus completely specified in terms of the first and second moments of the constituent random variables.

The preceding equations allow us to write

$$\widehat{Y} = \mu_Y + \mathbf{c}_{\mathbf{X}Y}^T (\mathbf{C}_{\mathbf{XX}})^{-1}(\mathbf{X} - \mu_{\mathbf{X}}) \,. \tag{12.5}$$

This relationship is shown schematically by the block diagram in Figure 12.1, which will form a useful point of comparison for our later results.

The associated minimum mean square error (MMSE) can be computed directly, and written in at least the following ways:

$$
\begin{aligned}
\text{MMSE} &= E[(Y - \widehat{Y})^2] \\
&= E[(Y - \widehat{Y})Y] \\
&= E[Y^2] - E[\widehat{Y}Y] \\
&= \sigma_Y^2 - \mathbf{a}^T \mathbf{c}_{\mathbf{X}Y} \\
&= \sigma_Y^2 - \mathbf{c}_{\mathbf{X}Y}^T (\mathbf{C}_{\mathbf{XX}})^{-1} \mathbf{c}_{\mathbf{X}Y} \\
&= \sigma_Y^2 \left( 1 - \frac{1}{\sigma_Y^2} \mathbf{c}_{\mathbf{X}Y}^T (\mathbf{C}_{\mathbf{XX}})^{-1} \mathbf{c}_{\mathbf{X}Y} \right) .
\end{aligned}
\tag{12.6}
$$

The second equality in the chain above follows from noting that since the zero-mean error $Y - \widehat{Y}$ is orthogonal to each of the measured random variables, it is also orthogonal to $\widehat{Y}$, that is, $E[(Y - \widehat{Y})\widehat{Y}] = 0$. The fourth equality is the result of subtracting and adding $\mu_Y^2$ to the expression that precedes it, then grouping terms to form variances and covariances. The nonnegative quantity

$$\rho_{\mathbf{X}Y}^2 = \frac{1}{\sigma_Y^2} \mathbf{c}_{\mathbf{X}Y}^T (\mathbf{C}_{\mathbf{XX}})^{-1} \mathbf{c}_{\mathbf{X}Y} \tag{12.7}$$

that appears in the last line of Eq. (12.6) must be no greater than 1 because the MMSE must be nonnegative. It plays a role analogous to $\rho_{XY}^2$, the square of the correlation coefficient between random variables $X$ and $Y$ when performing LMMSE estimation of $Y$ from a single random variable $X$. In the vector space language for LMMSE estimation introduced in Chapter 8, the quantity $\rho_{\mathbf{X}Y}^2$ is the squared cosine of the angle between the vector $Y$ and the subspace spanned by the vectors $\{X_i\}$, or equivalently between $Y$ and $\widehat{Y}$.

Our goal in this chapter is to extend LMMSE estimation to the case where the measurements come from a WSS process, and the random variable we are estimating is the value at some instant of another WSS process that is jointly WSS with the measured process. As expected with LMMSE estimation problems, the solution will only require first and second moments of the involved random processes, specifically only their means, autocovariance functions, and cross-covariance function. We treat only the discrete-time (DT) case in detail, as the development in DT is more transparent. However, most of the DT results have close continuous-time (CT) parallels that are briefly stated at the end of the chapter.

## 12.2 FIR WIENER FILTERS

Let $x[\cdot]$ and $y[\cdot]$ be two jointly WSS random processes, with respective mean values $\mu_x$, $\mu_y$, autocovariance functions $C_{xx}[m]$ and $C_{yy}[m]$, and cross-covariance function $C_{xy}[m]$. Suppose initially that we want to construct an LMMSE estimator of $y[n]$ for some specific $n$, using measurements of the $L$ samples $x[n], x[n-1], \ldots, x[n-L+1]$ from the process $x[\cdot]$. This estimator then takes the form

$$\widehat{y}[n] = \mu_y + \sum_{j=0}^{L-1} h[j]\Big(x[n-j] - \mu_x\Big) \tag{12.8}$$

for some optimally chosen set of coefficients $h[0], h[1], \ldots, h[L-1]$. This is essentially Eq. (12.1), apart from notational changes. Furthermore, since only relative positions on the time axis matter for jointly WSS processes, the same choice of $h[j]$ will be optimum for any value of $n$ because changing $n$ does not change the relative times of the samples involved in the problem.

The optimum values of the $h[j]$ are obtained by solving the associated normal equations, which will take the form in Eq. (12.3), with the appropriate substitutions. In particular, the $i, j$th and $j, i$th entries of the covariance matrix $\mathbf{C_{XX}}$ in Eq. (12.3) will be $C_{xx}[i-j]$, that is, the autocovariance of the process $x[\cdot]$ at lag $i-j$. Similarly, the $i$th entry of $\mathbf{c_{XY}}$ in Eq. (12.3) will be $C_{xy}[1-i]$, that is, the cross-covariance of the processes $x[\cdot]$ and $y[\cdot]$ at lag $i-1$. The following example, which is similar to the one in Section 10.5.2, illustrates the procedure.

## Example 12.1    FIR Estimation of a Signal Corrupted by Additive Noise

Suppose we have noise-corrupted measurements $x[n]$ of a scaled version of a WSS process $y[n]$:

$$x[n] = 2y[n] + v[n] , \tag{12.9}$$

where $v[n]$ is a white-noise process (and thus necessarily zero mean) with autocovariance function $C_{vv}[m] = \sigma^2 \delta[m]$, and is uncorrelated with $y[\cdot]$. Assume that $y[n]$ has mean $\mu_y = 1$ and autocovariance function $C_{yy}[m] = (0.3)^{|m|}$. Then $x[n]$ is WSS, with $\mu_x = 2$, and

$$C_{xx}[m] = 4C_{yy}[m] + C_{vv}[m] = 4(0.3)^{|m|} + \sigma^2 \delta[m] . \tag{12.10}$$

Also, $x[\cdot]$ is jointly WSS with $y[\cdot]$, with cross-covariance

$$C_{xy}[m] = 2C_{yy}[m] = 2(0.3)^{|m|} . \qquad (12.11)$$

The LMMSE estimator for $y[n]$ that is based on $x[n], x[n-1], x[n-2]$ takes the form

$$\widehat{y}[n] = 1 + h[0]\Big(x[n] - 2\Big) + h[1]\Big(x[n-1] - 2\Big) + h[2]\Big(x[n-2] - 2\Big), \qquad (12.12)$$

and the values of $h[j]$ are determined by solving the associated normal equations, namely

$$4 \begin{bmatrix} 1 + (\sigma^2/4) & 0.3 & (0.3)^2 \\ 0.3 & 1 + (\sigma^2/4) & 0.3 \\ (0.3)^2 & 0.3 & 1 + (\sigma^2/4) \end{bmatrix} \begin{bmatrix} h[0] \\ h[1] \\ h[2] \end{bmatrix} = 2 \begin{bmatrix} 1 \\ 0.3 \\ (0.3)^2 \end{bmatrix} . \qquad (12.13)$$

As a check on these equations, note that if there is no noise, that is, if $\sigma^2 = 0$, then the unique solution to this system of equations is $h[0] = 1/2, h[1] = 0, h[2] = 0$, so

$$\widehat{y}[n] = 1 + (1/2)\Big(x[n] - 2\Big) = x[n]/2 ,$$

which, in the noise-free case, precisely equals $y[n]$, and must therefore be the optimal estimator for that case. At the other extreme, if the noise intensity is high, that is, $\sigma^2 \gg 1$, then all the $h[j]$ are small, so $\widehat{y}[n] \approx 1 = \mu_y$. In other words, the best estimator of $y[n]$ in the high-noise case approaches its mean value $\mu_y$.

The summation in Eq. (12.8) is easily recognized as a convolution sum, which suggests the LTI filter implementation shown in Figure 12.2. This is the Wiener filter for the problem—an LMMSE estimator that works for all $n$ because the processes involved are jointly WSS. The estimator first removes the mean of the WSS measured process to obtain the zero-mean WSS fluctuation process $\widetilde{x}[n] = x[n] - \mu_x$, then feeds this to the input of an LTI system whose unit sample response $h[\cdot]$ is the set of optimal $h[j]$ values computed for $j = 0, 1, \ldots, L-1$ and is 0 for all other $j$. This filter thus has a unit sample response of finite duration—it is a finite impulse response or FIR filter—and is causal. The output of the LTI system is the zero-mean WSS fluctuation process $\widehat{y}[n] - \mu_y$, so adding $\mu_y$ produces the LMMSE estimate $\widehat{y}[n]$.

The LTI system associated with Eq. (12.8), which filters the measurement fluctuations to produce the LMMSE estimate fluctuations, is a simple example of a Wiener filter, namely a causal FIR Wiener filter. There are many variations on this that are amenable to the same sort of solution. For instance, we could have specified an estimator of the form

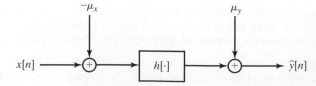

**Figure 12.2**   Structure of the Wiener filter. Optimal choice of the unit sample response $h[\cdot]$ generates the LMMSE estimate of a WSS process $y[\cdot]$ from measurements of a related process $x[\cdot]$.

$$\widehat{y}[n] = \mu_y + \sum_{j=-L}^{L} h[j]\big(x[n-j] - \mu_x\big),  \tag{12.14}$$

which uses measurements $x[\cdot]$ over the time window $[n-L, n+L]$ to estimate the value $y[n]$ of the process $y[\cdot]$ at the center of the window. Determination of the optimal values of the $h[j]$ again simply involves setting up the corresponding $2L+1$ normal equations and solving them. The result is once more an FIR Wiener filter with the structure shown in Figure 12.2, except that the filter is no longer causal. The following example illustrates yet another variation, namely FIR Wiener prediction of a future value of a process from measurements of the present and past values of the process over a finite window.

**Example 12.2   FIR Prediction**

For a WSS process $x[\cdot]$, consider obtaining the LMMSE estimator of $x[n+1]$ using measurements of $x[n], x[n-1], \cdots, x[n-L+1]$. This corresponds to one-step LMMSE prediction of a WSS process, using measurements of the $L$ most recent values. If we define the WSS process $y[n]$ by the relation $y[n] = x[n+1]$ for all $n$, then this reduces exactly to the problem addressed at the beginning of this section. The estimator takes the form of Eq. (12.8), but with $\widehat{y}[n] = \widehat{x}[n+1]$. We can thus write

$$\widehat{x}[n+1] = \mu_x + \sum_{j=0}^{L-1} h[j]\big(x[n-j] - \mu_x\big)  \tag{12.15}$$

and determine the optimal values of the $h[j]$ using the normal equations. Since $C_{xy}[m] = C_{xx}[m-1]$, these take the form

$$
\begin{bmatrix}
C_{xx}[0] & C_{xx}[1] & C_{xx}[2] & \cdots & C_{xx}[L-1] \\
C_{xx}[-1] & C_{xx}[0] & C_{xx}[1] & \cdots & C_{xx}[L-2] \\
\vdots & \cdots & \cdots & \cdots & \vdots \\
C_{xx}[1-L] & C_{xx}[2-L] & \cdots & \cdots & C_{xx}[0]
\end{bmatrix}
\begin{bmatrix}
h[0] \\
h[1] \\
\vdots \\
h[L-1]
\end{bmatrix}
=
\begin{bmatrix}
C_{xx}[-1] \\
C_{xx}[-2] \\
\vdots \\
C_{xx}[-L]
\end{bmatrix}.
$$
$$\tag{12.16}$$

As $C_{xx}[-m] = C_{xx}[m]$, the square matrix on the left is symmetric. The relations gathered in Eq. (12.16) are referred to as the Yule–Walker equations.

It is interesting to consider two special cases: a geometrically (i.e., exponentially) correlated process and a finitely correlated process.

**Geometrically Correlated Process**   For the first special case, suppose $C_{xx}[m] = C_0(\rho)^{|m|}$ for some $C_0 > 0$ and $\rho$ of magnitude less than 1. This corresponds to a geometrically correlated process—one that is exponentially correlated but in discrete time. The normal or Yule–Walker equations in Eq. (12.16) then become

$$
\begin{bmatrix}
1 & \rho & \rho^2 & \cdots & \rho^{L-1} \\
\rho & 1 & \rho & \cdots & \rho^{L-2} \\
\vdots & \cdots & \cdots & \cdots & \vdots \\
\rho^{L-1} & \rho^{L-2} & \cdots & \cdots & 1
\end{bmatrix}
\begin{bmatrix}
h[0] \\
h[1] \\
\vdots \\
h[L-1]
\end{bmatrix}
=
\begin{bmatrix}
\rho \\
\rho^2 \\
\vdots \\
\rho^L
\end{bmatrix}.  \tag{12.17}
$$

For $|\rho| < 1$, the matrix on the left is invertible, and therefore these equations have a unique solution. We see by inspection that the choice $h[0] = \rho$ and $h[j] = 0$

for $1 \leq j \leq L - 1$ satisfies Eq. (12.17). Thus, for an exponentially correlated process, the LMMSE one-step predictor is

$$\widehat{x}[n + 1] = \mu_x + \rho(x[n] - \mu_x) = \rho x[n] + (1 - \rho)\mu_x \tag{12.18}$$

and therefore involves only the most recent measured value $x[n]$. All previous measurements are ignored, even though $x[n + 1]$ is correlated with each previous measurement. Knowledge of $x[n]$ carries all that is relevant to estimating $x[n + 1]$.

The corresponding MMSE is most directly determined from the fourth equality in Eq. (12.6), and results in

$$\text{MMSE} = C_{xx}[0] - h[0]C_{xx}[-1] = C_0(1 - \rho^2) . \tag{12.19}$$

The additional terms of the form $h[j]C_{xx}[-j - 1]$ that would have appeared in the preceding equation are absent because $h[j] = 0$ for $j > 0$. The expression in Eq. (12.19) is consistent with what we know from Chapters 7 and 8 for LMMSE estimation of a random variable from measurement of a single other random variable.

**Proximally Correlated Process**   As the second special case, suppose

$$C_{xx}[m] = C_0\Big(\rho\delta[m + 1] + \delta[m] + \rho\delta[m - 1]\Big) , \tag{12.20}$$

so that the value of the process at any time is only correlated with the values one step before and after, and is uncorrelated for lags greater than 1. (More generally, one could consider a process whose nonzero correlations are confined to some finite set of lags.) As we saw in Example 11.4, the constraint $|\rho| \leq 0.5$ is required for this to be a valid autocovariance function, with a DT Fourier transform (DTFT) that is nonnegative, as a spectral density is required to be. For this case, and restricting ourselves to $L = 3$ in order to keep the calculations simple, the normal or Yule–Walker equations in Eq. (12.16) become

$$\begin{bmatrix} 1 & \rho & 0 \\ \rho & 1 & \rho \\ 0 & \rho & 1 \end{bmatrix} \begin{bmatrix} h[0] \\ h[1] \\ h[2] \end{bmatrix} = \begin{bmatrix} \rho \\ 0 \\ 0 \end{bmatrix} . \tag{12.21}$$

The form of the matrix on the left motivates an alternative name for a proximally correlated process, namely a banded process. Solving this system of three equations in three unknowns shows that

$$h[0] = \frac{\rho(1 - \rho^2)}{1 - 2\rho^2}; \quad h[1] = -\frac{\rho^2}{1 - 2\rho^2}; \quad h[2] = \frac{\rho^3}{1 - 2\rho^2} . \tag{12.22}$$

Note that both $h[1]$ and $h[2]$ are nonzero: both $x[n - 1]$ and $x[n - 2]$ are used along with $x[n]$ in the estimator for $x[n + 1]$, even though $x[n + 1]$ is uncorrelated with both $x[n - 1]$ and $x[n - 2]$.

The corresponding MMSE is again easiest to determine from the fourth equality in Eq. (12.6), which yields

$$\text{MMSE} = C_{xx}[0] - h[0]C_{xx}[-1] = C_0\Big(1 - \frac{\rho^2(1 - \rho^2)}{1 - 2\rho^2}\Big) . \tag{12.23}$$

The additional terms of the form $h[j]C_{xx}[-j - 1]$ that would have appeared in the preceding equation are again absent, but now because $C_{xx}[-j - 1] = 0$ for $j > 0$.

The MMSE in Eq. (12.23) is smaller than the MMSE of $C_0(1 - \rho^2)$ that would have been obtained if the prediction of $x[n + 1]$ had used a measurement of $x[n]$ alone. For example, with $|\rho| = 0.5$, the MMSE in Eq. (12.23) evaluates to $0.625C_0$, whereas with measurement of $x[n]$ alone the MMSE is $0.75C_0$. This corresponds to a 16.7% reduction in the variance (and 8.7% reduction in the standard deviation) of the error, by going from measurement of $x[n]$ alone to measurement of $x[n]$, $x[n-1]$, $x[n-2]$. Such a reduction could be significant in a particular application. For smaller values of $|\rho|$, the reduction in MMSE will be less.

A particular realization of a zero-mean process $x[n]$ with the autocovariance function in Eq. (12.20) is shown in Figure 12.3(a), for the case of $\rho = 0.5$, with $C_0 = 1$. This signal is generated as described in Example 11.10. The middle panel (b) in the figure shows the one-step predictions produced for this process by the LMMSE estimator specified via Eqs. (12.15) and (12.22), which give $h[0] = 0.75$, $h[1] = -0.5$, and $h[2] = 0.25$. The corresponding estimates are labeled $\widehat{x}_3[n+1]$ in the figure. The MMSE associated with the estimator is $0.625$, while the empirical time-averaged square error across all the time points shown for the particular realization of the estimator in panel (b) is $0.614$.

For comparison, Figure 12.3(c) shows the one-step predictions produced by an LMMSE estimator that only uses a measurement of $x[n]$ to predict $x[n+1]$. The corresponding estimates are labeled $\widehat{x}_1[n+1]$ in the figure, and are simply given by $\widehat{x}_1[n+1] = \rho x[n] = 0.5x[n]$. The associated MMSE for this estimator is $0.75$, while the empirical time-averaged square error for the particular realization of the estimator in panel (c) is $0.746$.

It is visually apparent from the plots in Figure 12.3 that the estimates in panels (b) and (c) tend to have smaller amplitudes than the true signal $x[n+1]$. The reason is that the LMMSE estimator in Eq. (12.15) factors in both the available measurements

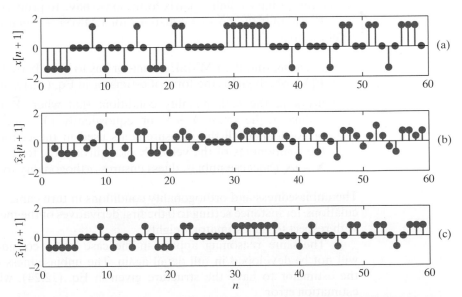

**Figure 12.3**    Panel (a) shows a realization of a zero-mean process $x[\cdot]$ with autocovariance function $C_{xx}[m] = 0.5\delta[m+1] + \delta[m] + 0.5\delta[m-1]$, generated as in Example 11.10. Panel (b) shows the one-step predictions of $x[n+1]$ using the LMMSE estimator designed for this finitely correlated case in Example 12.2, with measurements of $x[n], x[n-1]$, and $x[n-2]$: these estimates are denoted by $\widehat{x}_3[n+1]$. Panel (c) shows the one-step predictions of $x[n+1]$ using an LMMSE estimator that just measures $x[n]$; these estimates are denoted by $\widehat{x}_1[n+1]$.

and the mean $\mu_x$, which is the MMSE estimate in the absence of measurements. In this case $\mu_x = 0$, so the estimates have some tendency to be pulled toward 0. As more measurements are folded into the estimator, that is, as $L$ in Eq. (12.15) increases, the importance of the mean $\mu_x$ decreases, because the estimates are then more determined by the measurements.

## 12.3 THE UNCONSTRAINED DT WIENER FILTER

With jointly WSS processes $x[\cdot]$ and $y[\cdot]$ as in the preceding section, suppose we again want to construct an LMMSE estimator of $y[n]$ for some specific $n$, but now using measurements of all values of the process $x[\cdot]$, so

$$\widehat{y}[n] = \mu_y + \sum_{j=-\infty}^{\infty} h[j]\Big(x[n-j] - \mu_x\Big) \tag{12.24}$$

for some optimally chosen set of coefficients $h[j]$. The difference from Eq. (12.8) is that the summation now ranges from $-\infty$ to $\infty$. As before, the same set of coefficients will work for all $n$ because the processes involved are jointly WSS. The relation in the preceding equation still involves a convolution, and has the structure shown in Figure 12.2. However, now the filter has a unit sample response of potentially infinite duration. The normal equations can no longer have a finite matrix form, so we have to proceed differently.

Recall that the normal equations for LMMSE estimation of $Y$ from $\mathbf{X}$ resulted from

- deducing that the LMMSE estimator has to be unbiased, that is, $E[\widehat{Y}] = E[Y]$, which yields the form of estimator in Eq. (12.1); and then
- invoking the orthogonality condition: that when $\widehat{Y}$ is the LMMSE estimator, the error $Y - \widehat{Y}$ or equivalently $(Y - \mu_Y) - (\widehat{Y} - \mu_Y)$ is orthogonal to all the measurements used in the estimator, that is, to the components of the vector $\mathbf{X}$ or, equivalently, to the components of $\mathbf{X} - \mu_{\mathbf{X}}$ (because unbiasedness ensures orthogonality to any constant).

The unbiasedness and orthogonality conditions in turn came from simple calculations, for instance setting to 0 the first derivatives of the mean square error with respect to the estimator weights.

The same reasoning applies in the case being considered here, and will not be developed in full detail again. The unbiasedness condition leads the estimator to have the structure given in Eq. (12.24), which causes the estimation error

$$e[n] = y[n] - \widehat{y}[n] = (y[n] - \mu_y) - (\widehat{y}[n] - \mu_y) \tag{12.25}$$

to have a mean value of 0. The orthogonality condition requires

$$E\Big[e[n]x[n-m]\Big] = 0 = E\Big[e[n]\Big(x[n-m] - \mu_x\Big)\Big] \tag{12.26}$$

for all $m$. The range $m \geq 0$ ensures orthogonality of the error $e[n]$ to present and past values of $x[\cdot]$, while the range $m < 0$ ensures orthogonality to future values of $x[\cdot]$.

Under the assumption that the filter with unit sample response $h[\cdot]$ in Figure 12.2 is stable, or at least has a well-defined frequency response, it follows that its output process $\widehat{y}[n] - \mu_y$ is jointly WSS with the measured input process. The above orthogonality condition can therefore be written more simply as

$$C_{ex}[m] = C_{yx}[m] - C_{\widehat{y}x}[m] = 0 , \tag{12.27}$$

or

$$C_{\widehat{y}x}[m] = C_{yx}[m] \tag{12.28}$$

for all $m$. The latter equation provides an alternative way of stating the orthogonality condition for the optimal system, namely that the cross-covariance between the estimate and the measurements (i.e., the left side of the preceding equation) equals the cross-covariance between the process being estimated and the measurements (the right side of the preceding equation).

To compute the unit sample response $h[\cdot]$, we observe that since $\widehat{y}[n]$ is obtained by filtering $x[n]$ through an LTI system with unit sample response $h[\cdot]$, the following relationship applies, obtained by arguments similar to those used to establish Eq. (10.65):

$$C_{\widehat{y}x}[m] = \sum_{j=-\infty}^{\infty} h[j]C_{xx}[m-j] = h[m] * C_{xx}[m] . \tag{12.29}$$

This result may be derived directly from Eq. (12.24) by rewriting it as

$$\widehat{y}[n] - \mu_y = \sum_{j=-\infty}^{\infty} h[j]\Big(x[n-j] - \mu_x\Big) , \tag{12.30}$$

then multiplying both sides by $(x[n-m] - \mu_x)$, moving this term inside the summation on the right side, and taking the expected value.

Combining the relation in Eq. (12.29) with the equality in Eq. (12.28) then yields

$$h[m] * C_{xx}[m] = C_{yx}[m] \tag{12.31}$$

for all $m$, where $C_{xx}[m]$ and $C_{yx}[m]$ are known, and $h[m]$ is to be determined. Transforming this time-domain convolution relationship to the frequency domain results in the simple multiplicative relation

$$H(e^{j\Omega})D_{xx}(e^{j\Omega}) = D_{yx}(e^{j\Omega}) . \tag{12.32}$$

Rearranging this equation, we get the desired expression for the frequency response of the unconstrained LMMSE or Wiener filter at all $\Omega$ for which $D_{xx}(e^{j\Omega}) \neq 0$:

$$H(e^{j\Omega}) = \frac{D_{yx}(e^{j\Omega})}{D_{xx}(e^{j\Omega})} . \tag{12.33}$$

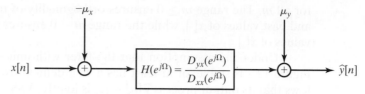

**Figure 12.4** Representation of the unconstrained Wiener filter in the frequency domain.

At frequencies $\Omega$ for which $D_{xx}(e^{j\Omega}) = 0$ we are guaranteed by Eq. (11.51) that $D_{yx}(e^{j\Omega}) = 0$ also, so $H(e^{j\Omega})$ can be chosen arbitrarily at these frequencies; the choice does not affect the mean square error because the measurement process $x[n]$ has no spectral content at these frequencies. The resulting expression for $H(e^{j\Omega})$ is guaranteed to correspond to a stable filter under a mild continuity condition.

The overall structure is shown in Figure 12.4. Note the close similarity to the LMMSE estimator structure and gain shown in Figure 12.1. The terms $D_{xx}(e^{j\Omega})$ and $D_{yx}(e^{j\Omega})$ play roles analogous to $\mathbf{C_{XX}}$ and $\mathbf{c}_{XY}^T = \mathbf{c}_{YX}$ respectively. By expressing the result in the frequency domain, we have in effect made the answer to the LMMSE estimation problem for two jointly WSS processes as simple as the answer to the LMMSE estimation problem for two random variables. Specifically, in the frequency domain, the action of the LTI filter at each frequency $\Omega$ is decoupled from the action at other frequencies. This familiar pattern will become apparent again in the following paragraphs.

The mean square error corresponding to the optimum filter, that is, the MMSE, is

$$\text{MMSE} = E\left[e^2[n]\right] = C_{ee}[0] \,, \tag{12.34}$$

so we begin by determining an expression for $C_{ee}[m]$, similar to the fourth equality in Eq. (12.6):

$$C_{ee}[m] = C_{yy}[m] - C_{\hat{y}y}[m] = C_{yy}[m] - (h[m] * C_{xy}[m]) \,, \tag{12.35}$$

where $h[m]$ is the unit sample response of the optimum filter. The terms on the right are all known, so $C_{ee}[m]$ and thus $C_{ee}[0]$ are now computable.

A frequency-domain expression for the MMSE is also useful. Taking the DTFT of Eq. (12.35), we get the error fluctuation spectral density (FSD)

$$D_{ee}(e^{j\Omega}) = D_{yy}(e^{j\Omega}) - H(e^{j\Omega})D_{xy}(e^{j\Omega})$$

$$= D_{yy}(e^{j\Omega})\left(1 - \rho_{yx}(e^{j\Omega})\rho_{yx}^*(e^{j\Omega})\right) \,, \tag{12.36}$$

where

$$\rho_{yx}(e^{j\Omega}) = \frac{D_{yx}(e^{j\Omega})}{\sqrt{D_{yy}(e^{j\Omega})D_{xx}(e^{j\Omega})}} \,. \tag{12.37}$$

The function $\rho_{yx}(e^{j\Omega})$ plays the role of a complex frequency-domain correlation coefficient and is referred to as the coherence function of the two processes. The fact that $D_{ee}(e^{j\Omega})$ in Eq. (12.36) is nonnegative at all frequencies ensures $|\rho_{yx}(e^{j\Omega})| \leq 1$ at all $\Omega$; this observation provides an alternative route to deriving the bound in Eq. (11.51). We can now write

$$\text{MMSE} = C_{ee}[0] = \frac{1}{2\pi} \int_{-\pi}^{\pi} D_{ee}(e^{j\Omega}) \, d\Omega$$

$$= \frac{1}{2\pi} \int_{-\pi}^{\pi} D_{yy}(e^{j\Omega})\left(1 - |\rho_{yx}(e^{j\Omega})|^2\right) d\Omega . \tag{12.38}$$

The closer the magnitude of the coherence function is to 1 in some frequency range, the better $y[\cdot]$ is approximated by the LMMSE estimate $\widehat{y}[\cdot]$ in that frequency range, that is, by appropriate LTI filtering of the components of $x[\cdot]$ in that frequency range.

Again, note the similarity of the integrand $D_{yy}(e^{j\Omega})(1 - |\rho_{yx}(e^{j\Omega})|^2)$ in the last line of Eq. (12.38) to the expression $\sigma_Y^2(1 - \rho_{YX}^2)$ that gives the MMSE after LMMSE estimation of a random variable $Y$ using measurements of a random vector $\mathbf{X}$. The difference in the context of WSS processes is that the expression is integrated over all frequencies to get the MMSE.

The following examples show how the preceding development is applied in various settings. The first example serves mainly as a simple check on the Wiener filter solution.

## Example 12.3 Noncausal Prediction

Suppose the measured process $x[n]$ is WSS, and we wish to predict it $n_0 > 0$ steps ahead, so

$$y[n] = x[n + n_0] . \tag{12.39}$$

This is straightforward for an unconstrained Wiener filter, as it suffices to choose the filter's unit sample response to be $h[n] = \delta[n + n_0]$. Alternatively, we can use the expression in Eq. (12.33). Note that

$$C_{yx}[m] = C_{xx}[m + n_0] \tag{12.40}$$

so

$$D_{yx}(e^{j\Omega}) = e^{j\Omega n_0} D_{xx}(e^{j\Omega}) . \tag{12.41}$$

The optimum filter therefore has frequency response

$$H(e^{j\Omega}) = D_{yx}(e^{j\Omega})/D_{xx}(e^{j\Omega}) = e^{j\Omega n_0} . \tag{12.42}$$

The corresponding unit sample response is thus $\delta[n + n_0]$, as expected; this filter shifts its input forward in time by $n_0$ steps. The corresponding MMSE is easily determined to be 0, using the expressions in Eq. (12.38).

Causal prediction is more challenging and interesting, and we will examine it later in this chapter.

The canonical Wiener filtering problem is examined in the following example, where the aim is to estimate a signal from a set of measurements of the signal that are corrupted by additive noise.

**Example 12.4** **Unconstrained Estimation of a Signal in Additive Noise**

Suppose, as in Example 12.1, that we have noise-corrupted measurements $x[n]$ of a scaled version of an underlying WSS random signal $y[n]$ of interest to us, with

$$x[n] = 2y[n] + v[n] , \qquad (12.43)$$

where $v[n]$ is a WSS noise process. Assume that $v[\cdot]$ is uncorrelated with $y[\cdot]$, so their covariance is zero, that is, $C_{yv}[m] = 0$ for all $m$. Assume also that we know the means $\mu_y$, $\mu_v$ and autocovariances $C_{yy}[m]$, $C_{vv}[m]$. We would like to construct an LMMSE estimator of $y[n]$ using the entire record of the measured signal $x[\cdot]$. This is provided by a Wiener filter acting on the measurement process $x[\cdot]$, using the structure in Figure 12.4 with $\mu_x = 2\mu_y + \mu_v$, and producing the output $\widehat{y}[n]$.

The quantities required to determine the Wiener filter are very directly computed, invoking the fact that $C_{yv}[m] = 0$ in order to simplify the expressions:

$$C_{xx}[m] = 4C_{yy}[m] + C_{vv}[m] + 2C_{yv}[m] + 2C_{yv}[-m]$$

$$= 4C_{yy}[m] + C_{vv}[m] \qquad (12.44)$$

and

$$C_{yx}[m] = 2C_{yy}[m] + C_{yv}[m] = 2C_{yy}[m] . \qquad (12.45)$$

Taking transforms and substituting into Eq. (12.35), the frequency response of the Wiener filter is found to be

$$H(e^{j\Omega}) = \frac{2D_{yy}(e^{j\Omega})}{4D_{yy}(e^{j\Omega}) + D_{vv}(e^{j\Omega})} . \qquad (12.46)$$

Note that the filter in this case must have a unit sample response that is an even function of time, since its frequency response is a real, and hence even, function of frequency. The filter is thus noncausal.

For $\Omega$ where the power in the signal fluctuations is much greater than the power in the noise fluctuations, $H(e^{j\Omega}) \approx 1/2$, and for $\Omega$ where the fluctuation noise power is much greater than the fluctuation signal power, $H(e^{j\Omega}) \approx 2D_{yy}(e^{j\Omega})/D_{vv}(e^{j\Omega}) \approx 0$. For example, when the noise is white with variance $\sigma^2 = 4$ and $C_{yy}[m]$ has the proximally correlated form in Eq. (12.20), with $C_0 = 2$ and $\rho = 0.5$, we have

$$D_{yy}(e^{j\Omega}) = 2(1 + \cos \Omega) . \qquad (12.47)$$

The optimal filter then has the low-pass frequency response shown in Figure 12.5.

Figure 12.6 shows a simulation of this filter. The plots show a realization of the underlying signal $y[n]$; a realization of the corresponding measured signal $x[n]$; and the corresponding LMMSE estimate of $y[n]$, namely $\widehat{y}[n]$.

The MMSE for this filter is found by making the appropriate substitutions in Eq. (12.38). The result is

$$\text{MMSE} = \frac{1}{2\pi} \int_{-\pi}^{\pi} \frac{D_{yy}(e^{j\Omega})D_{vv}(e^{j\Omega})}{4D_{yy}(e^{j\Omega}) + D_{vv}(e^{j\Omega})} \, d\Omega . \qquad (12.48)$$

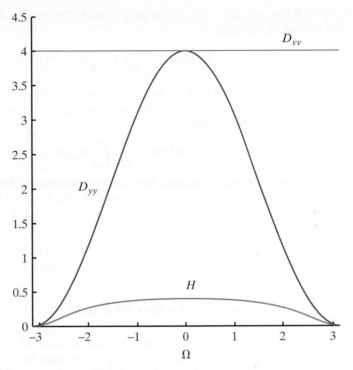

**Figure 12.5**    Plots of the signal FSD $D_{yy}(e^{j\Omega})$, the white noise FSD $D_{vv}(e^{j\Omega})$, and the frequency response $H(e^{j\Omega})$ of the Wiener filter.

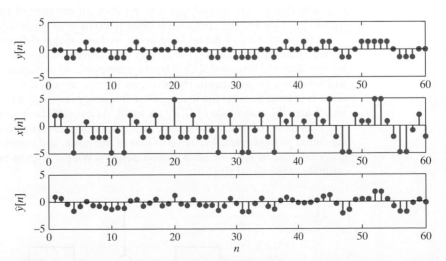

**Figure 12.6**    Estimating a signal corrupted by additive noise. From top to bottom: a realization of the underlying signal $y[n]$; a realization of the corresponding measured signal $x[n]$; and the corresponding LMMSE estimate of $y[n]$, namely $\widehat{y}[n]$.

If the noise intensity is low relative to the FSD of the signal, then the above expression can be simplified to

$$\text{MMSE} \approx \frac{1}{2\pi} \int_{-\pi}^{\pi} \frac{1}{4} D_{vv}(e^{j\Omega}) \, d\Omega = \frac{1}{4} C_{vv}[0] \,. \qquad (12.49)$$

This is consistent with the fact that in the low-noise regime $\widehat{y}[n] \approx \frac{1}{2}x[n]$, so the error is essentially $\frac{1}{2}v[n]$. At the other extreme, when the noise intensity is large relative to the FSD of the signal,

$$\text{MMSE} \approx \frac{1}{2\pi} \int_{-\pi}^{\pi} D_{yy}(e^{j\Omega}) \, d\Omega = C_{yy}[0] \,, \qquad (12.50)$$

which is the signal variance. Again, this is consistent with the fact that in the high-noise regime $\widehat{y}[n] \approx \mu_y$, so the error is essentially the deviation of $y[n]$ from its mean.

The following example goes a step further than the previous one, and addresses a situation that arises in many applications.

## Example 12.5  Deconvolution of a Blurred Signal

In Figure 12.7, $r[n]$ represents a filtered or "blurred" version of the WSS signal of interest, $y[n]$, with $G(z)$ representing the transfer function of the stable LTI system or sensor that is used to measure $y[n]$. Thus $r[\cdot]$ is the result of convolving $y[\cdot]$ with the unit sample response $g[\cdot]$. The process $v[\cdot]$ is WSS additive noise that is uncorrelated with $y[\cdot]$, so $C_{yv}[m] = 0$. The noise is therefore uncorrelated with $r[\cdot]$ as well, because $C_{rv}[m] = g[m] * C_{yv}[m]$. We only have access to the blurred and noise-corrupted measurement process $x[\cdot]$, and wish to design an LTI system with transfer function $H(z)$ that will filter this measured signal to produce an estimate of the underlying input signal $y[n]$. This filtering operation is often referred to as deconvolution or deblurring.

In the absence of the additive noise, we have $x[n] = r[n]$, so a stable inverse filter with transfer function $H(z) = 1/G(z)$ will recover the input $y[\cdot]$ exactly. This stable inverse filter will be causal if $G(z)$ is minimum phase, as explained in Section 2.3.2, but will otherwise be noncausal. However, an inverse filter is not a good solution when noise is present because the inverse filter frequency response $H(e^{j\Omega})$ has large magnitude where the magnitude of $G(e^{j\Omega})$ is small. As a result, the inverse filter accentuates precisely those frequencies where the power in the signal-related component $r[n]$ of the measured process is small relative to that of the noise, and therefore produces a very noisy estimate of $y[n]$. We shall instead design a Wiener filter to produce an LMMSE estimate of the signal $y[n]$.

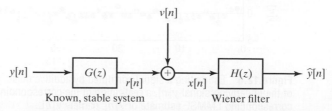

**Figure 12.7**   Wiener filtering of a blurred and noisy signal.

The expression in Eq. (12.33) for the frequency response of the Wiener filter shows that we need to determine $D_{xx}(e^{j\Omega})$ and $D_{yx}(e^{j\Omega})$. Straightforward calculations show that

$$D_{xx}(e^{j\Omega}) = D_{rr}(e^{j\Omega}) + D_{vv}(e^{j\Omega})$$

$$= G(e^{j\Omega})G(e^{-j\Omega})D_{yy}(e^{j\Omega}) + D_{vv}(e^{j\Omega}) \tag{12.51}$$

and

$$D_{yx}(e^{j\Omega}) = D_{yr}(e^{j\Omega}) = G(e^{-j\Omega})D_{yy}(e^{j\Omega}) . \tag{12.52}$$

The frequency response of the Wiener filter is then

$$H(e^{j\Omega}) = \frac{G(e^{-j\Omega})D_{yy}(e^{j\Omega})}{G(e^{j\Omega})G(e^{-j\Omega})D_{yy}(e^{j\Omega}) + D_{vv}(e^{j\Omega})} . \tag{12.53}$$

Rewriting the above expression as

$$H(e^{j\Omega}) = \frac{1}{G(e^{j\Omega})} \frac{D_{rr}(e^{j\Omega})}{D_{rr}(e^{j\Omega}) + D_{vv}(e^{j\Omega})} \tag{12.54}$$

shows that the same Wiener filter is obtained if we first find the LMMSE estimate $\widehat{r}[n]$ from $x[n]$ (as in Example 12.4), and then pass $\widehat{r}[n]$ through the inverse filter $1/G(e^{j\Omega})$.

In the limiting case of low noise, where $D_{vv}(e^{j\Omega}) \to 0$, this becomes just the inverse filter, with frequency response $1/G(e^{j\Omega})$. If the noise is white, with an intensity $D_{vv}(e^{j\Omega}) = \sigma_v^2$ that dominates the other term in the denominator, then

$$H(e^{j\Omega}) \approx \frac{1}{G(e^{j\Omega})} \frac{D_{rr}(e^{j\Omega})}{\sigma_v^2} , \tag{12.55}$$

which corresponds to first filtering the measured process $x[n]$ to preferentially pass those frequencies that are strongly present in $r[n]$, and subsequently applying the inverse filter.

The next example deals with Wiener filtering for a situation involving multiplicative rather than additive uncertainty or noise.

## Example 12.6  Demultiplication

A WSS signal $s[n]$ with mean $\mu_s$ and autocovariance function $C_{ss}[m]$ is transmitted over a multiplicative channel that causes the received signal $r[n]$ at time $n$ to be given by

$$r[n] = f[n]s[n] , \tag{12.56}$$

where $f[n]$ is the channel's multiplicative factor at time $n$. A channel with time-varying transmission characteristics, as here, is also referred to as a fading channel. For example, $s[n]$ might be the intensity of an optical source at the $n$th transmission, $f[n]$ the attenuation experienced by transmission through some turbulent medium, and $r[n]$ the received intensity. Assume that $f[n]$ is also a WSS process, with mean $\mu_f$ and autocovariance function $C_{ff}[m]$, and that it is independent of the input signal $s[\cdot]$. We wish to estimate $s[n]$ from $r[\cdot]$ using a Wiener filter.

Note that $r[n]$ is also WSS, with mean value $\mu_r = \mu_f \mu_s$ and autocovariance function

$$C_{rr}[m] = R_{rr}[m] - \mu_r^2$$

$$= R_{ff}[m]R_{ss}[m] - \mu_f^2 \mu_s^2$$

$$= (C_{ff}[m] + \mu_f^2)(C_{ss}[m] + \mu_s^2) - \mu_f^2 \mu_s^2$$

$$= C_{ff}[m]C_{ss}[m] + \mu_f^2 C_{ss}[m] + \mu_s^2 C_{ff}[m] . \tag{12.57}$$

It is also jointly WSS with $s[\cdot]$, with cross-covariance given by

$$C_{sr}[m] = R_{sr}[m] - \mu_s \mu_r$$

$$= R_{ss}[m]\mu_f - \mu_s^2 \mu_f$$

$$= C_{ss}[m]\mu_f . \tag{12.58}$$

The frequency response of the Wiener filter is now given by

$$H(e^{j\Omega}) = \frac{D_{sr}(e^{j\Omega})}{D_{rr}(e^{j\Omega})}$$

$$= \frac{D_{ss}(e^{j\Omega})\mu_f}{\frac{1}{2\pi}[D_{ff}(e^{j\Omega}) \circledast D_{ss}(e^{j\Omega})] + \mu_f^2 D_{ss}(e^{j\Omega}) + \mu_s^2 D_{ff}(e^{j\Omega})} , \tag{12.59}$$

where $\circledast$ denotes periodic convolution:

$$D_{ff}(e^{j\Omega}) \circledast D_{ss}(e^{j\Omega}) = \int_{<2\pi>} D_{ff}(e^{j\nu})D_{ss}(e^{j(\Omega-\nu)}) \, d\nu . \tag{12.60}$$

Evaluating this integral requires invoking the fact that both $D_{ff}(e^{j\Omega})$ and $D_{ss}(e^{j\Omega})$ are periodic, with period $2\pi$.

For a special case that serves as a check on this answer, suppose $f[n]$ is fixed at the value $\mu_f$. Then $D_{ff}(e^{j\Omega}) = 0$, and we simply get $H(e^{j\Omega}) = 1/\mu_f$, which is easily verified to yield $\widehat{s}[n] = s[n]$, so the input signal is exactly recovered.

Another special case occurs when the deviations of $f[n]$ from its mean are white, so $D_{ff}(e^{j\Omega}) = \sigma_f^2$. Then

$$H(e^{j\Omega}) = \frac{D_{ss}(e^{j\Omega})\mu_f}{\sigma_f^2 E\{s^2[n]\} + \mu_f^2 D_{ss}(e^{j\Omega})} . \tag{12.61}$$

## 12.4 CAUSAL DT WIENER FILTERING

The previous two sections treated cases of DT Wiener filtering that had relatively direct solutions. In the FIR case, this involved setting up the appropriate normal equations and solving. In the unconstrained case, the orthogonality condition led to a convolution relationship that held on the entire time axis, and transforming this to the frequency domain then permitted a simple algebraic solution for the optimal filter. In both cases, the filter implementation involved the structure in Figure 12.2, with the LTI filter being FIR in the first case, and unconstrained in the second case.

Other scenarios for Wiener filtering can be more subtle. The most important of these is the case where the LTI filter is constrained to be causal, so $h[j] = 0$ for $j < 0$. The estimator is thus restricted to have the form

$$\hat{y}[n] = \mu_y + \sum_{j=0}^{\infty} h[j]\left(x[n-j] - \mu_x\right),$$  (12.62)

with the lower limit on the summation now being 0. The estimate of $y[n]$ in this case is determined by present and past values of $x[\cdot]$, but future values are excluded. The form of this filter already ensures that the estimator is unbiased. The requirement that the error $e[n] = y[n] - \hat{y}[n]$ be orthogonal to present and past values of $x[\cdot]$ is written as

$$E\Big[e[n]x[n-m]\Big] = 0 = E\Big[e[n]\big(x[n-m] - \mu_x\big)\Big], \ m \geq 0,$$  (12.63)

or equivalently as

$$C_{\hat{y}x}[m] = C_{yx}[m], \ m \geq 0.$$  (12.64)

Note that the identities in Eqs. (12.63) and (12.64) now hold only for $m \geq 0$. Similarly, the relation that we derived from this equation in the unconstrained case now becomes

$$h[m] * C_{xx}[m] = C_{yx}[m], \ m \geq 0.$$  (12.65)

The restriction of the equalities in Eqs. (12.63)–(12.65) to nonnegative $m$ means that the Fourier transforms of the two sides are in general not equal, so our previous solution for the frequency response of the Wiener filter no longer holds. This is not an impediment when $C_{xx}[m] = \sigma_x^2 \delta[m]$, that is, when the measured process is white. The reason is that the values for $m < 0$ are irrelevant in this case, since now $h[m] * C_{xx}[m] = \sigma_x^2 h[m]$. Using this in Eq. (12.65) shows that

$$h[m] = \frac{1}{\sigma_x^2} C_{yx}[m]$$  (12.66)

for $m \geq 0$, and $h[m] = 0$ for $m < 0$. One approach to solving the general case is to first whiten the measurements in a causal and causally invertible way, then apply the preceding solution, and finally work back to what filtering is implied for the original measured process. We shall instead proceed more directly, essentially following Wiener's original solution.

Returning to the general relation in Eq. (12.65), the challenge now is to isolate the part of the left side of Eq. (12.65) that relates to $m \geq 0$, while still preserving enough information to solve for $h[\cdot]$ or its transform. Wiener's solution of this problem involved recognizing that under appropriate assumptions the autocovariance function $C_{xx}[m]$ can be written as

$$C_{xx}[m] = f[m] * \overleftarrow{f}[m],$$  (12.67)

where $\overleftarrow{f}[k] = f[-k]$ is the time-reversed version of $f[k]$, and where $f[\cdot]$ is the unit sample response of a stable, causal system with a stable, causal inverse. In the transform domain, this translates to having

$$D_{xx}(e^{j\Omega}) = F(e^{j\Omega})F^*(e^{j\Omega}) = F(e^{j\Omega})F(e^{-j\Omega}) , \qquad (12.68)$$

where the superscript * denotes the complex conjugate. Using $F(z)$ to denote the $z$-transform of $f[n]$, we can write

$$D_{xx}(z) = F(z)F(z^{-1}) . \qquad (12.69)$$

The stability and causality conditions translate to requiring that $F(z)$ have all its poles and zeros inside the unit circle. Thus $F(z)$ is a minimum-phase function, as discussed in Section 2.3.2. The decomposition of $D_{xx}(e^{j\Omega})$ or $D_{xx}(z)$ into a product of the form in Eq. (12.68), respectively Eq. (12.69), is referred to as minimum-phase spectral factorization, and was discussed in Section 2.4. The existence of a minimum-phase spectral factor is guaranteed under a condition due to Paley and Wiener on the FSD $D_{xx}(e^{j\Omega})$, requiring that $|\log D_{xx}(e^{j\Omega})|$ have a finite integral over an interval of length $2\pi$. The condition rules out, for instance, a $D_{xx}(e^{j\Omega})$ that is identically zero over some finite interval in $[-\pi, \pi]$. This is reasonable because the inverse of a spectral factor of such an FSD will not have a well-defined frequency response, and therefore could not be minimum phase.

Given the properties above, the function $\overleftarrow{f}[\cdot]$ in Eq. (12.67) corresponds to the unit sample response of a stable anticausal system with a stable anticausal inverse. Denote the unit sample response of this inverse by $\overleftarrow{g}[\cdot]$. Thus $\overleftarrow{f}[j] = 0 = \overleftarrow{g}[j]$ for $j > 0$, and both $\overleftarrow{f}[\cdot]$ and $\overleftarrow{g}[\cdot]$ are absolutely summable, with $\overleftarrow{f}[k] * \overleftarrow{g}[k] = \delta[k]$. The absolute summability of $\overleftarrow{g}[\cdot]$ is what guarantees that the convolutions carried out below are well behaved.

We are now in a position to determine $h[\cdot]$ from Eq. (12.65). Convolving the left side of that equation with $\overleftarrow{g}[\cdot]$ produces the function $h[m] * f[m]$. The results of this convolution for $m \geq 0$ only depend on values of $h[m] * C_{xx}[m]$ for $m \geq 0$, as $\overleftarrow{g}[\cdot]$ is anticausal. Observe also that since $h[m] * f[m]$ is the convolution of two causal functions, it is itself causal, that is, 0 for $m < 0$.

Convolving the right side of Eq. (12.65) with $\overleftarrow{g}[\cdot]$ produces the function $\overleftarrow{g}[m] * C_{yx}[m]$, which is in general nonzero for all $m$. Again, however, its values for $m \geq 0$ depend only on values of $C_{yx}[m]$ for $m \geq 0$ because $\overleftarrow{g}[\cdot]$ is anticausal. As the equality in Eq. (12.65) holds for $m \geq 0$, we can now write

$$h[m] * f[m] = \left( \overleftarrow{g}[m] * C_{yx}[m] \right)u[n] , \qquad (12.70)$$

where $u[n]$ is the unit step function. Now taking the DTFT on both sides and rearranging the result, we have the desired expression for the frequency response of the causal Wiener filter:

$$H(e^{j\Omega}) = \frac{1}{F(e^{j\Omega})}\left[\frac{D_{yx}(e^{j\Omega})}{F(e^{-j\Omega})}\right]_+ , \qquad (12.71)$$

where the notation $[P(e^{j\Omega})]_+$ denotes the transform of the causal part of the signal $p[n]$, that is, the transform of $p[n]u[n]$. In terms of $z$-transforms, we can write the transfer function of the causal Wiener filter as

$$H(z) = \frac{1}{F(z)} \left[ \frac{D_{yx}(z)}{F(z^{-1})} \right]_+ . \tag{12.72}$$

The corresponding MMSE is given by

$$\text{MMSE} = C_{ee}[0] = \frac{1}{2\pi} \int_{-\pi}^{\pi} D_{ee}(e^{j\Omega}) \, d\Omega$$

$$= \frac{1}{2\pi} \int_{-\pi}^{\pi} \Big( D_{yy}(e^{j\Omega}) - H(e^{j\Omega})D_{xy}(e^{j\Omega})$$

$$- H(e^{-j\Omega})D_{yx}(e^{j\Omega}) + |H(e^{j\Omega})|^2 D_{xx}(e^{j\Omega}) \Big) \, d\Omega . \tag{12.73}$$

This involves more terms than the corresponding expression for the unconstrained case, namely Eq. (12.38), because the relation in Eq. (12.65) only holds for $m \geq 0$ in the causal case and therefore cannot be used to condense the above expression. It can be shown fairly directly, though we omit the derivation, that this MMSE exceeds the MMSE of the unconstrained Wiener filter by the amount

$$\Delta \text{MMSE} = \frac{1}{2\pi} \int_{-\pi}^{\pi} \left| \left[ \frac{D_{yx}(e^{j\Omega})}{F(e^{-j\Omega})} \right]_- \right|^2 d\Omega , \tag{12.74}$$

where the notation $[P(e^{j\Omega})]_-$ denotes the transform of the strictly anticausal part of the signal $p[n]$, that is, the transform of $p[n](1 - u[n])$.

---

**Example 12.7    The Causal DT Wiener Predictor**

Consider a measured process $x[n]$ that is the result of passing (zero-mean) white noise of unit variance through a modeling or shaping filter with transfer function

$$F(z) = \alpha_0 + \alpha_1 z^{-1} , \tag{12.75}$$

where both $\alpha_0$ and $\alpha_1$ are assumed nonzero. The filter is stable (and causal), so the process $x[n]$ is WSS, has zero mean, and has PSD

$$D_{xx}(e^{j\Omega}) = F(e^{j\Omega})F^*(e^{j\Omega}) = F(e^{j\Omega})F(e^{-j\Omega}) . \tag{12.76}$$

If $|\alpha_1| < |\alpha_0|$ then the inverse of the filter $F(z)$ is also stable and causal. Under this inequality condition, $F(z)$ constitutes a minimum-phase spectral factor of $D_{xx}(z)$. Assume for now that this inequality holds (we shall see later in this example what to do if it does not hold).

Suppose we want to perform causal one-step prediction for this process $x[n]$: obtaining the LMMSE estimator $\widehat{y}[n]$ for $y[n] = x[n+1]$ in terms of $x[k]$ for $k \leq n$. Then $C_{yx}[m] = C_{xx}[m+1]$, so

$$D_{yx}(z) = zD_{xx}(z) = zF(z)F(z^{-1}) . \tag{12.77}$$

Therefore

$$\left[ \frac{D_{yx}(z)}{F(z^{-1})} \right]_+ = \left[ zF(z) \right]_+ = \alpha_1 . \tag{12.78}$$

Now using the expression in Eq. (12.72), the transfer function of the causal Wiener predictor is

$$H(z) = \frac{1}{F(z)}\left[zF(z)\right]_+ = \frac{\alpha_1}{\alpha_0 + \alpha_1 z^{-1}} . \tag{12.79}$$

The MMSE of the unconstrained Wiener predictor is 0, as was pointed out in Example 12.3, so the MMSE of the causal filter is given by the expression in Eq. (12.74), which evaluates to $\alpha_0^2$. For comparison, estimating $x[n+1]$ by its mean value of 0 results in a mean square error of $C_{xx}[0] = \alpha_0^2 + \alpha_1^2$. Also, the LMMSE estimator of $x[n+1]$ in terms of a measurement of just $x[n]$ would have produced a mean square error of

$$\alpha_0^2 + \alpha_1^2 - \frac{\alpha_0^2 \alpha_1^2}{\alpha_0^2 + \alpha_1^2} . \tag{12.80}$$

For $|\alpha_1| \approx |\alpha_0| = \alpha$, the mean square error for the causal Wiener predictor is $\alpha^2$, while predicting $x[n+1]$ by its mean value results in a mean square error of $2\alpha^2$, and prediction in terms of just $x[n]$ results in a mean square error of $1.5\alpha^2$. Thus going from a predictor based on the most recent measurement to the full causal Wiener predictor results in a one-third reduction in mean square error in this case, from $1.5\alpha^2$ to $\alpha^2$.

Figure 12.8 revisits Example 12.2, to compare the results obtained there using an FIR prediction filter with results obtained using the full causal Wiener filter in Eq. (12.79). We again choose $\rho = 0.5$ and $C_{xx}[0] = 1$, obtained by having $\alpha_0 = \alpha_1 = \alpha = 1/\sqrt{2}$. This does not quite satisfy the condition $|\alpha_1| < |\alpha_0|$ required for stability of the causal inverse, $1/F(z)$, of the modeling filter. As a result, the transfer function of the causal Wiener filter that results from formal substitution in Eq. (12.79) is

$$H(z) = \frac{1}{1 + z^{-1}} , \tag{12.81}$$

which has its single pole on the unit circle, and is thus only marginally stable. The situation can be remedied by choosing $|\alpha_1|$ to be slightly less than $|\alpha_0|$ throughout (i.e., restricting ourselves to the case of $|\rho| < 1$), but we proceed pragmatically and approximate the transfer function of the predictor by

$$H(z) \approx \frac{1}{1 + 0.99z^{-1}} . \tag{12.82}$$

We will see below that this filter provides performance consistent with our earlier calculations.

Figure 12.8(a) shows a realization of the finitely correlated process in Example 12.2, though a different realization from the one in Figure 12.3(a). Figure 12.8(b) shows the predictions $\widehat{x}_\infty[n+1]$ produced by the approximation in Eq. (12.82) to the full causal Wiener filter in Eq. (12.79). The theoretical MMSE for this case is $\alpha^2 = 0.5$, while the empirical time-averaged square error for the specific realization here is 0.497. Panels (c) and (d) in Figure 12.8 show the prediction results for this specific realization using respectively the FIR predictors $\widehat{x}_3[n+1]$ and $\widehat{x}_1[n+1]$ developed in Example 12.2. The respective empirical time-averaged square errors for these predictors are 0.650 (compared to the theoretical MMSE of 0.625) and 0.754 (compared to the theoretical MMSE of 0.75).

The derivation above assumed that $|\alpha_1| < |\alpha_0|$, which guarantees that the modeling filter $F(z)$ is minimum phase. If $|\alpha_1| > |\alpha_0|$, then $F(z) = \alpha_0 + \alpha_1 z^{-1}$ is still stable and causal, with a pole at $z = 0$ and a zero at $-\alpha_1/\alpha_0$, but this zero is now

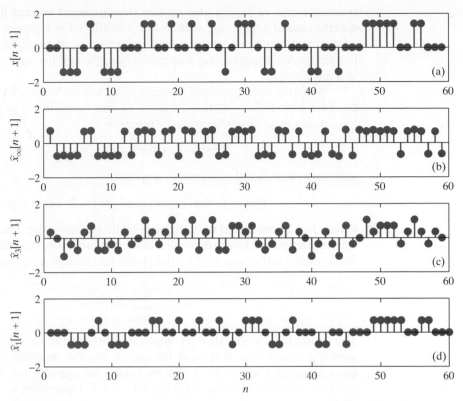

**Figure 12.8**  Panel (a) shows a realization of a zero-mean process $x[n]$ with autocovariance function $C_{xx}[m] = 0.5\delta[m+1] + \delta[m] + 0.5\delta[m-1]$, generated as in Example 11.10. Panel (b) shows the predictions $\widehat{x}_\infty[n+1]$ of $x[n+1]$ using the full causal Wiener filter. Panel (c) shows the predictions $\widehat{x}_3[n+1]$ obtained using the LMMSE estimator designed in Example 12.2, with measurements of $x[n], x[n-1]$, and $x[n-2]$. Panel (d) shows the predictions $\widehat{x}_1[n+1]$ using the LMSSE estimator with a measurement of just $x[n]$.

outside the unit circle. The causal inverse of $F(z)$ is therefore not stable because it has a pole outside the unit circle. The required minimum-phase factor $\widetilde{F}(z)$ is obtained on multiplying $F(z)$ by an appropriately chosen all-pass factor $A(z)$ of the form given in Eq. (2.27), which replaces the zero at $-\alpha_1/\alpha_0$ by a zero at its reciprocal location:

$$\widetilde{F}(z) = F(z)A(z) \tag{12.83}$$

$$= \alpha_0\Big(1 + (\alpha_1/\alpha_0)z^{-1}\Big)\Big(\frac{z^{-1} + (\alpha_1/\alpha_0)}{1 + (\alpha_1/\alpha_0)z^{-1}}\Big) \tag{12.84}$$

$$= \alpha_1 + \alpha_0 z^{-1}. \tag{12.85}$$

It is easily verified that $\widetilde{F}(z)\widetilde{F}(z^{-1}) = F(z)F(z^{-1}) = D_{xx}(z)$.

**The Role of the Modeling Filter in Prediction**    The minimum-phase modeling filter for the measured process $x[n]$, or equivalently the minimum-phase spectral factor of $D_{xx}(z)$, plays a key role in causal Wiener filtering. For the specific case of prediction, such as that considered in Example 12.7, there is a more direct way to understand the significance of this modeling filter, as outlined next. We focus on one-step prediction, though the approach generalizes in a natural way.

Suppose the measured process $x[k]$, whose values for $k \leq n$ will be used for LMMSE estimation of $x[n+1]$, can be modeled as the output of a minimum-phase filter with transfer function

$$F(z) = f_0 + f_1 z^{-1} + f_2 z^{-2} + \cdots \tag{12.86}$$

and associated causal unit sample response

$$f[n] = f_n \text{ for } n \geq 0 \quad \text{and } f[n] = 0 \text{ otherwise,} \tag{12.87}$$

driven by a (zero-mean) unit-variance white process $w[n]$. Then $x[n]$ for any $n$ can be determined from knowledge of $w[j]$ for all $j \leq n$, by taking the weighted linear combination specified by the above unit sample response, that is,

$$x[n] = \sum_{j=-\infty}^{n} f_{n-j} w[j] . \tag{12.88}$$

The causal invertibility of $F(z)$, which is equivalent to having $F(\infty) = f_0 \neq 0$, similarly guarantees that $w[n]$ for any $n$ can be uniquely recovered from knowledge of $x[k]$ for all $k \leq n$ by taking an appropriate weighted linear combination. Thus, having $w[j]$ for all $j \leq n$ is equivalent to having $x[k]$ for all $k \leq n$.

Note now that an expression similar to Eq. (12.88) holds for $x[n+1]$ as well:

$$x[n+1] = f_0 w[n+1] + \sum_{j=-\infty}^{n} f_{n-j+1} w[j] . \tag{12.89}$$

Knowledge of $x[k]$ for all $k \leq n$, along with the causal invertibility of $F(z)$, ensures that all $w[j]$ for $j \leq n$ are known, but $w[n+1]$ is not, so the summation term in the preceding equation is known but the first term is not. The fact that $w[\cdot]$ is white means that the known $w[j]$ are uncorrelated with $w[n+1]$, and therefore yield no information about it to an LMMSE estimator. It follows that our LMMSE estimate of $x[n+1]$ is

$$\widehat{x}[n+1] = \sum_{j=-\infty}^{n} f_{n-j+1} w[j] . \tag{12.90}$$

This can be formally verified by checking that $x[n+1] - \widehat{x}[n+1]$ is orthogonal to all $w[j]$ for $j \leq n$, and hence to $x[k]$ for all $k \leq n$. The associated MMSE is simply the variance of the term $f_0 w[n+1]$ that was eliminated in going from Eq. (12.89) to Eq. (12.90):

$$\text{MMSE} = f_0^2 = [F(\infty)]^2 . \tag{12.91}$$

The summation in Eq. (12.90) can be seen as the result at time $n$ of convolving $w[\cdot]$ with a causal filter whose transform is $[zF(z)]_+$, in the notation introduced earlier. To generate $w[\cdot]$ causally from $x[\cdot]$, we pass $x[\cdot]$ through a filter with transfer function $1/F(z)$. Putting all this together, it follows that the causal one-step Wiener prediction filter has transfer function

$$H(z) = \frac{1}{F(z)}\Big[zF(z)\Big]_+ . \tag{12.92}$$

This is precisely the result that was obtained in the preceding example in Eq. (12.79).

# 12.5 OPTIMAL OBSERVERS AND KALMAN FILTERING

This section revisits the canonical problem considered in Example 12.4—estimating a DT signal from measurements corrupted by additive noise. We will now restrict the estimation filter to be causal, however, so its design will require the results developed in Section 12.4. Furthermore, we shall assume the signal of interest is the result of filtering a white DT process through a causal and stable LTI state-space system. This formulation opens the door to an important set of extensions of Wiener filtering, referred to as Kalman filtering.

## 12.5.1 Causal Wiener Filtering of a Signal Corrupted by Additive Noise

As in Example 12.4, suppose we have available the measured process

$$x[n] = y[n] + v[n] , \tag{12.93}$$

where the zero-mean WSS process $y[n]$ that we wish to estimate from our measurements is the result of passing a white process $w[\cdot]$ of intensity $\sigma_w^2 > 0$ through a causal and stable modeling filter, with transfer function

$$G(z) = \frac{\eta(z)}{a(z)} . \tag{12.94}$$

In the next subsection we shall explicitly consider a reachable and observable $L$th-order state-space realization of this modeling filter, but for now it suffices to note that $a(z)$ is the monic degree-$L$ characteristic polynomial of the system, and $\eta(z)$ has degree less than $L$, corresponding to the simplifying assumption we shall make that there is no direct feedthrough from input to output in the state-space model. The reachability and observability of the system ensure that the polynomials $\eta(z)$ and $a(z)$ have no common factors, and the stability of the system means that the roots of $a(z)$ are all within the unit circle. It follows that the underlying signal of interest, $y[n]$, is a zero-mean WSS process with complex FSD given by

$$D_{yy}(z) = \sigma_w^2 \frac{\eta(z)\eta(z^{-1})}{a(z)a(z^{-1})} \ . \tag{12.95}$$

Assume the additive noise process $v[\cdot]$ is also white, with intensity $\sigma_v^2 > 0$, and is uncorrelated with the driving process $w[\cdot]$, and therefore with the signal $y[\cdot]$. Under the preceding assumptions, $x[\cdot]$ is also zero-mean and WSS. With the definition $r = \sigma_w^2/\sigma_v^2 > 0$, the complex PSD of $x[n]$ is given by

$$D_{xx}(z) = \sigma_w^2 \frac{\eta(z)\eta(z^{-1})}{a(z)a(z^{-1})} + \sigma_v^2$$

$$= \sigma_v^2 \frac{r\eta(z)\eta(z^{-1}) + a(z)a(z^{-1})}{a(z)a(z^{-1})}$$

$$= \sigma_v^2 \frac{v(z)v(z^{-1})}{a(z)a(z^{-1})} \ , \tag{12.96}$$

where $v(z)$ is a degree-$L$ monic polynomial with roots inside the unit circle, satisfying

$$v(z)v(z^{-1}) = r\eta(z)\eta(z^{-1}) + a(z)a(z^{-1}) \ . \tag{12.97}$$

This equation embodies the minimum-phase spectral factorization step that is expected in such problems.

The existence of a spectral factor $v(z)$ in Eq. (12.97) that has all its roots within the unit circle follows from the fact that the expression on the right side of the equation has self-reciprocal zeros, meaning that if it has a zero at $z = z_o$, then it also has a zero at $z = z_o^{-1}$. Furthermore, this expression cannot have zeros on the unit circle because a zero at $z = e^{j\Omega_0}$ for $r > 0$ would imply that

$$r|\eta(e^{j\Omega_0})|^2 + |a(e^{j\Omega_0})|^2 = 0 \ ,$$

which means $\eta(z)$ and $a(z)$ would each individually have a zero at $z = e^{j\Omega_0}$, contradicting the fact that $\eta(z)$ and $a(z)$ have no common roots. The polynomial $v(z)$ is then defined by the requirement that it contain precisely the zeros of the right side of Eq. (12.97) that lie inside the unit circle. Note also that $v(z)$ and $a(z)$ will not have any common zeros because Eq. (12.97) shows that any value of $z$ where both these polynomials go to 0 must also be a $z$ at which $\eta(z)$ goes to 0—but that would again contradict the fact that $\eta(z)$ and $a(z)$ have no common factors. Finally, since the degree of $\eta(z)$ is less than $L$, it follows that the degree-$L$ polynomial $v(z)$ must be monic, that is, have the coefficient of its highest-degree term be 1, just as $a(z)$ has.

From Eq. (12.96), the minimum-phase spectral factor for $D_{xx}(z)$ is

$$F(z) = \sigma_v \frac{v(z)}{a(z)} \ . \tag{12.98}$$

The remaining quantity required to compute the causal Wiener filter is $D_{yx}(z)$. Since $y[\cdot]$ and $v[\cdot]$ are uncorrelated, it follows that

$$D_{yx}(z) = D_{yy}(z) \tag{12.99}$$

and is thus given by Eq. (12.95).

Substituting the results of the preceding calculations into Eq. (12.72) and using Eq. (12.97) to make a simplification along the way, we see that the transfer function of the required causal Wiener filter is

$$
\begin{aligned}
H(z) &= \frac{1}{F(z)}\left[\frac{D_{yx}(z)}{F(z^{-1})}\right]_+ \\
&= \frac{a(z)}{\sigma_v v(z)}\left[\sigma_w^2 \frac{\eta(z)\eta(z^{-1})}{a(z)a(z^{-1})}\frac{a(z^{-1})}{\sigma_v v(z^{-1})}\right]_+ \\
&= \frac{a(z)}{v(z)}\left[\frac{r\eta(z)\eta(z^{-1})}{a(z)v(z^{-1})}\right]_+ \\
&= \frac{a(z)}{v(z)}\left[\frac{v(z)}{a(z)} - \frac{a(z^{-1})}{v(z^{-1})}\right]_+ .
\end{aligned}
\tag{12.100}
$$

The term $v(z)/a(z)$ survives the $[\cdot]_+$ operation intact, as its inverse transform is causal. Since the term $a(z^{-1})/v(z^{-1})$ has an anticausal inverse transform, its only contribution to the $[\cdot]_+$ operation is the value of the associated time function at time 0. This value is 1 because both $a(z)$ and $v(z)$ are monic polynomials in $z$. It follows therefore that the optimum causal filter is given by the simple expression

$$
H(z) = 1 - \frac{a(z)}{v(z)} = \frac{v(z) - a(z)}{v(z)} ,
\tag{12.101}
$$

which is completely determined by the result of the spectral factorization in Eq. (12.97).

The Wiener filter we have arrived at in Eq. (12.101) takes the measured process $x[\cdot]$ as input and causally generates the LMMSE estimate $\widehat{y}[n]$ of the signal $y[n]$, thereby extracting the signal from the additive noise (with some residual error, of course). The next subsection shows that the same effect can be obtained by using an appropriately designed observer for the underlying state-space system that generates the signal $y[n]$.

### 12.5.2  Observer Implementation of the Wiener Filter

Observers were introduced in Chapter 6 as a mechanism for state estimation in LTI state-space systems. It is therefore perhaps not surprising that an observer is relevant to estimating a signal obtained as the output of such a system. The notation used for various signals in the present context differs from that introduced in Chapter 6. We therefore set up the framework and notation again here.

The signal of interest to us, $y[n]$, is the output of a causal and stable LTI state-space system driven by an unknown white process $w[\cdot]$ of known variance $\sigma_w^2 > 0$. The measured signal $x[n]$ is the result of an unknown additive white-noise process $v[n]$ of known variance $\sigma_v^2 > 0$, and uncorrelated with $w[\cdot]$, corrupting the signal $y[n]$. The relevant $L$th-order reachable and observable state-space model takes the form

$$\mathbf{q}[n+1] = \mathbf{A}\mathbf{q}[n] + \mathbf{b}w[n] \,, \tag{12.102}$$

$$y[n] = \mathbf{c}^T\mathbf{q}[n] \,, \tag{12.103}$$

$$x[n] = y[n] + v[n] \,. \tag{12.104}$$

Note that our use of $y[n]$ to denote the actual system output and $x[n]$ to represent its noise-corrupted version differs from our use of these symbols in Chapter 6. We also assume that this system has been operating since time $-\infty$, as our analysis is restricted to WSS processes. The system transfer function from $w[n]$ to $y[n]$ is

$$G(z) = \mathbf{c}^T(z\mathbf{I} - \mathbf{A})^{-1}\mathbf{b} = \frac{\eta(z)}{a(z)} \,, \tag{12.105}$$

as in Eq. (12.94), where $a(z) = \det(z\mathbf{I} - \mathbf{A})$ is the characteristic polynomial of the system.

The observer for this system takes the form

$$\widehat{\mathbf{q}}[n+1] = \mathbf{A}\widehat{\mathbf{q}}[n] - \boldsymbol{\ell}\Big(x[n] - \widehat{y}[n]\Big)$$

$$= (\mathbf{A} + \boldsymbol{\ell}\mathbf{c}^T)\widehat{\mathbf{q}}[n] - \boldsymbol{\ell}x[n] \,, \tag{12.106}$$

where $\boldsymbol{\ell}$ is the observer gain vector and

$$\widehat{y}[n] = \mathbf{c}^T\widehat{\mathbf{q}}[n] \,. \tag{12.107}$$

Note that the observer is not driven by the underlying system output $y[n]$ but rather by the measured signal $x[n]$, and that the unknown noise processes $w[n]$ and $v[n]$ do not enter the construction of the observer either. The quantity $\widehat{y}[n]$ denotes the observer's estimate of $y[n]$, but with no implication yet that this is the LMMSE estimate, given $x[k]$ for $k \leq n$. One motivation for constructing an observer in Chapter 6 was to overcome uncertainty about the initial state $\mathbf{q}[0]$, whereas here the uncertainty in the state is a consequence of $w[n]$ and $v[n]$ being unknown. The initial condition plays no role because we have assumed a stable system operating since time $-\infty$.

Equations (12.106) and (12.107) show that the observer in the current context can be viewed as an LTI state-space system that has $x[n]$ as input and $\widehat{y}[n]$ as output. The interesting and important fact now is that the observer gain $\boldsymbol{\ell}$ can be chosen such that this system has precisely the transfer function $H(z)$ of the Wiener filter designed in the Section 12.5.1, as specified in Eq. (12.101). With this choice of $\boldsymbol{\ell}$ the quantity $\widehat{y}[n]$ does indeed become the causal LMMSE estimate of $y[n]$.

The denominator of the observer transfer function is the characteristic polynomial of $\mathbf{A} + \boldsymbol{\ell}\mathbf{c}^T$. The results on observer design in Chapter 6 show that this characteristic polynomial can be made equal to any monic polynomial of degree $L$ by appropriate choice of $\boldsymbol{\ell}$ because the given state-space system is observable. Thus $\boldsymbol{\ell}$ can be chosen to make the denominator of the transfer function equal to $v(z)$, as required by Eq. (12.101), with $v(z)$ defined through the spectral factorization in Eq. (12.97). If this choice of $\ell$ results in the numerator polynomial of the observer transfer function being $v(z) - a(z)$, as also required by Eq. (12.101), then the desired result would be established.

We use an indirect argument to deduce what the numerator polynomial of the observer transfer function must be. Note from Eqs. (12.106) and (12.107) that implementing a unit-gain output feedback around this observer by setting

$$x[n] = \widehat{y}[n] + p[n],\qquad (12.108)$$

where $p[n]$ is now some new external input, results in the state-space system

$$\widehat{\mathbf{q}}[n+1] = \mathbf{A}\widehat{\mathbf{q}}[n] - \boldsymbol{\ell}p[n]\qquad (12.109)$$

with the same output as before, specified in Eq. (12.107). Such unit-gain feedback produces a new denominator polynomial for the transfer function, changing it from $v(z)$ to the difference between $v(z)$ and the original numerator polynomial. However, inspection of the preceding equation shows that its characteristic polynomial, which is also the denominator polynomial of the new system, is $a(z) = \det(z\mathbf{I} - \mathbf{A})$. The conclusion is that the original numerator polynomial of the observer must have been $v(z) - a(z)$, which is exactly the numerator polynomial of the Wiener filter derived in the previous subsection, see Eq. (12.101).

In summary, if the signal $y[n]$ that we wish to estimate from measurements corrupted by additive white noise has been generated by driving a stable LTI state-space system with another white process that is uncorrelated with the measurement noise, then the requisite Wiener filter can be realized as an observer for the state-space system. The observer gain is chosen to make the observer characteristic polynomial equal to the $v(z)$ that results from the spectral factorization in Eq. (12.97). The observer output $\widehat{y}[n]$ is then the LMMSE estimate of $y[n]$.

### 12.5.3 Optimal State Estimates and Kalman Filtering

It is natural now to wonder whether the components $\widehat{q}_i[n]$ of the state estimate $\widehat{\mathbf{q}}[n]$ generated by the observer in Eq. (12.106) are optimal estimates, that is, if they are the causal LMMSE estimates of the underlying state variables $q_i[n]$. It can be shown that this is indeed the case: the optimal observer generates the causal LMMSE estimate $\widehat{\mathbf{q}}[n]$ of the state $\mathbf{q}[n]$ and not just the causal LMMSE estimate $\widehat{y}[n]$ of the output $y[n]$. We will not demonstrate this, but offer the following intuitive explanation. Because the state-space system generating $y[n]$ is reachable and observable, its state vector is fully excited, and fully reflected in the output measurements over time. Hence the only way for the optimal output estimate $\widehat{y}[n]$ to be generated by the state observer is for the underlying observer state $\widehat{\mathbf{q}}[n]$ to also be the optimal estimate.

The optimal observer developed in Section 12.5.2 is the simplest form of Kalman filter. In the usual development of a Kalman filter, one begins with a state-space model such as in Eqs. (12.102), (12.103), and (12.104), then develops the causal LMMSE estimator for the underlying state vector $\mathbf{q}[n]$, discovering in the process that this estimator takes the form of an observer for the initial state-space model, with an optimally chosen observer gain. We have reversed the process in our development, in order to arrive at this via Wiener filtering.

The Kalman filter can take much more general forms, with an underlying causal state-space model that can be time-varying and even unstable, with no requirement of wide-sense stationarity for the associated processes, and with multiple disturbance inputs and measurement outputs allowed. The spectral factorization step that characterizes causal Wiener filtering is replaced by solution of a so-called Riccati equation that, in the case where the disturbance and noise processes are Gaussian, propagates the error covariance matrix of the state estimate.

## 12.6 ESTIMATION OF CT SIGNALS

A very similar development to the DT case can be carried out in CT for several of the prototype signal estimation problems we have considered, though the details of the derivations may differ. Also, in DT the notion of white noise is simple and accessible, whereas in CT white noise is an extreme idealization, with infinite expected power at every instant. We briefly summarize the CT parallels below for some of the preceding results on DT Wiener filtering, but omit all derivations.

For the unconstrained CT case, where measurements $x(\cdot)$ of some WSS process are filtered to construct the LMMSE estimate $\widehat{y}(t)$ for some process $y(t)$ that is jointly WSS with $x(\cdot)$, the frequency response of the Wiener filter is given by

$$H(j\omega) = \frac{D_{yx}(j\omega)}{D_{xx}(j\omega)} , \qquad (12.110)$$

and the associated MMSE is

$$\text{MMSE} = \frac{1}{2\pi} \int_{-\infty}^{\infty} \Big( D_{yy}(j\omega) - H(j\omega)D_{xy}(j\omega) \Big) \, d\omega$$

$$= \frac{1}{2\pi} \int_{-\pi}^{\pi} \Big( D_{yy}(j\omega) - \frac{D_{yx}(j\omega)D_{xy}(j\omega)}{D_{xx}(j\omega)} \Big) \, d\omega . \qquad (12.111)$$

For causal CT Wiener filtering, the frequency response of the optimum filter is given by

$$H(j\omega) = \frac{1}{F(j\omega)} \left[ \frac{D_{yx}(j\omega)}{F(-j\omega)} \right]_+ , \qquad (12.112)$$

where $F(j\omega)$ is the frequency response of a causal and causally invertible spectral factor of $D_{xx}(j\omega)$.

Finally, for prediction of a WSS CT process a time $T > 0$ into the future, the filter transfer function is given by

$$H(s) = \frac{1}{F(s)}\left[e^{sT}F(s)\right]_{+}.\tag{12.113}$$

## 12.7 FURTHER READING

The design of optimal systems for LMMSE estimation of wide-sense stationary signals was sparked by Wiener's work in [Wie]. The topic was reflected in engineering textbooks such as [Dav] and [Lee] a decade later. A readable recent biography of Wiener is [Con]. Several of the texts cited in Chapters 7, 8, 10, and 11 for their inclusion of WSS processes and power spectral density devote particular attention to Wiener filtering. The spectral factorization required for causal Wiener filtering is treated in [Moo], [Op2], [Pa1], [Pa3], [Pa4], and [Th1], for example. Kalman filtering is covered by some of these references, also [Poo], and by others cited in Chapter 6 in the context of optimal observers for feedback control. The relation between Wiener and Kalman filtering is addressed in [And], [Ka2], and [Kam]. Filtering and prediction for finite-state, discrete-time Markov processes (including hidden Markov models or HMMs) is treated in [Frs].

## Problems

## Basic Problems

**12.1.** A certain zero-mean WSS signal $y(t)$ with autocorrelation function $R_{yy}(\tau)$ and corresponding PSD $S_{yy}(j\omega)$ is transmitted through a channel that has a fixed but random gain $G$, whose mean and variance are $\mu_G$ and $\sigma_G^2$ respectively. Due to noise at the receiver, the received signal $x(t)$ takes the form

$$x(t) = Gy(t) + w(t),$$

where $w(t)$ is a zero-mean WSS noise process with autocorrelation function $R_{ww}(\tau)$ and corresponding PSD $S_{ww}(j\omega)$. The transmitted process $y(\cdot)$ and the noise process $w(\cdot)$ are uncorrelated with each other, i.e., $R_{yw}(\tau) = 0$, and are independent of $G$.

**(a)** Determine the following in terms of the given quantities:
  (i) $E[G^2]$;
  (ii) mean value of $x(t)$;
  (iii) autocorrelation function $R_{xx}(\tau)$ of the process $x(t)$; and
  (iv) cross-correlation function $R_{yx}(\tau)$ between $x(\cdot)$ and $y(\cdot)$.
**(b)** Compute the frequency response $H(j\omega)$ of a stable and possibly noncausal LTI Wiener filter that takes as input the received signal $x(\cdot)$ and produces as output the LMMSE estimate $\hat{y}(t)$ of the transmitted signal $y(t)$, i.e., find the filter that minimizes $E[\{\hat{y}(t) - y(t)\}^2]$.

(i) What does $H(j\omega)$ reduce to for those frequencies $\omega$, if any, where the PSD of the noise process $S_{ww}(j\omega)$ is zero, but the PSD of the transmitted process, $S_{yy}(j\omega)$ is nonzero? Is this the answer you would have expected? Explain.

(ii) What does $H(j\omega)$ reduce to for those frequencies $\omega$, if any, where the PSD of the transmitted process is zero, but the PSD of the noise process is nonzero? Is this the answer you would have expected? Explain.

**12.2.** A certain zero-mean CT WSS signal $y(t)$ with autocorrelation function $R_{yy}(\tau)$ and corresponding PSD $S_{yy}(j\omega)$ is transmitted through a channel. The characteristics of the channel and receiver are such that the received signal $x(t)$ is of the form

$$x(t) = by(t) + v(t) \ .$$

The quantity $v(t)$ represents receiver noise, and is a zero-mean WSS noise process with autocorrelation function $R_{vv}(\tau)$ and corresponding PSD $S_{vv}(j\omega)$, and is uncorrelated with $y(\cdot)$, i.e., $R_{yv}(\tau) = 0$. The quantity $b$ is a random variable that is independent of $y(\cdot)$ and $v(\cdot)$, and that takes the value 1 or 0 for all time; it can be thought of as indicating whether the channel works ($b = 1$) or doesn't ($b = 0$). The probability that $b = 1$ is $p$.

**(a)** Compute $S_{yx}(j\omega)$ and $S_{xx}(j\omega)$, then find the frequency response $H(j\omega)$ of a stable and possibly noncausal LTI (Wiener) filter that takes as input the received signal $x(\cdot)$ and produces as output the LMMSE estimate $\widehat{y}(t)$ of the transmitted signal $y(t)$, i.e., find the filter that minimizes $E[\{y(t) - \widehat{y}(t)\}^2]$. Express your answer in terms of quantities specified in the problem statement. Check that your filter specializes to what you expect when $p = 1$ and $p = 0$.

**(b)** Find an expression for the PSD $S_{ee}(j\omega)$ of the error $e(t) = y(t) - \widehat{y}(t)$ associated with the optimum filter you designed in (a), again expressing your answer in terms of quantities specified in the problem statement. Again check that your expression reduces to what you expect when $p = 1$ and $p = 0$.

**12.3.** In your new job as a research scientist at the the Oceanographic Institute, you have access to recorded measurements of a random process $x(t)$. What you are really interested in, however, is the zero-mean WSS random process $y(t)$, which is related to $x(t)$ as follows:

$$x(t) = y(t) + w(t) \ ,$$

where $w(\cdot)$ is a zero-mean WSS noise process that is uncorrelated with the process $y(\cdot)$. You want to design a (possibly noncausal) LTI filter with impulse response $h(t)$ that will filter $x(t)$ and produce the LMMSE estimate of $y(t)$, i.e., you want a Wiener filter to produce $\widehat{y}(t)$. However, suppose you have no measurements of $y(t)$ or $w(t)$ from which to directly compute the correlation information that will allow you to design the optimum filter. What you do have are extensive records of measurements taken by your predecessor, using an old sensor, of the filtered signals

$$v(t) = g(t) * w(t) \qquad \text{and} \qquad m(t) = g(t) * \left( y(t) + w(t) \right),$$

where $g(t)$ is the impulse response of the old sensor, but is unknown to you, and $*$ denotes convolution. You can use these old records to compute good approximations of $R_{vv}(\tau)$ and $R_{mm}(\tau)$. The question now is whether these correlation functions suffice to design the Wiener filter.

(a) Express the PSDs $S_{vv}(j\omega)$ and $S_{mm}(j\omega)$ in terms of $S_{yy}(j\omega)$, $S_{ww}(j\omega)$, and the frequency response $G(j\omega)$ of the old sensor.

(b) Express the frequency response $H(j\omega)$ of the desired Wiener filter in terms of only $S_{vv}(j\omega)$ and $S_{mm}(j\omega)$.

(c) Let $S_{ee}(j\omega)$ denote the PSD of the error signal $e(t) = y(t) - \hat{y}(t)$. Express the ratio $S_{ee}(j\omega)/S_{yy}(j\omega)$ in terms of only $H(j\omega)$ and/or the PSDs $S_{vv}(j\omega)$ and $S_{mm}(j\omega)$. This ratio gives some idea of the quality of the Wiener filter at each frequency because it compares the spectral power of the error after estimation to the spectral power of the error before estimation.

**12.4.** The random process $r[n]$ is a zero-mean, unit-variance, white process. The random process $y[n]$ is obtained by filtering $r[n]$ through a filter with frequency response $G(e^{j\Omega})$, as depicted in Figure P12.4-1. Assume all signals and system impulse responses are real-valued.

(a) What is the PSD of $y[n]$, $S_{yy}(e^{j\Omega})$, expressed in terms of $G(e^{j\Omega})$?

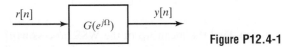

Figure P12.4-1

The process $x[n]$ is obtained from the multiplication of the process $r[n]$ specified above and a process $w[n]$ as shown in Figure P12.4-2. The process $w[\cdot]$ is independent of $r[n]$ and takes the value 1 with probability $p$, and 0 with probability $(1-p)$, independently for each $n$:

$$w[n] = \begin{cases} 1 & \text{with probability } p \\ 0 & \text{with probability } 1-p . \end{cases}$$

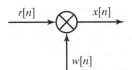

Figure P12.4-2

(b) Calculate the mean and autocovariance functions of $x[n]$. Is $x[n]$ a white process?

(c) Design the LTI filter $H_1(e^{j\Omega})$ in Figure P12.4-3 with the input $x[n]$, so that the output process $q[n]$ has the same PSD as $y[n]$, your result from part (a).

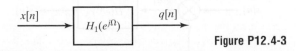

Figure P12.4-3

(d) Design the LTI filter $H_2(e^{j\Omega})$ in Figure P12.4-4 for which the input $x[n]$ will produce an output $\hat{y}[n]$ that at every instant is the LMMSE estimate of $y[n]$.

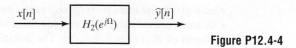

**Figure P12.4-4**

**(e)** For your answer in part (d), calculate the resulting mean square error. What is the mean square error when $p = 0$ and when $p = 1$? Comment on these answers; do they seem reasonable to you?

**12.5.** Assume that $y[n]$ in Figure P12.5 is zero-mean WSS, with correlation function $R_{yy}[m]$ and PSD $S_{yy}(e^{j\Omega})$. Suppose that the process $w[\cdot]$ is independent of the process $y[n]$, and at any time instant takes the value 1 with probability $p$ or the value 0 with probability $1 - p$; also assume that the values of $w[\cdot]$ at different time instants are independent. Thus the signal $x[n] = y[n]w[n]$ is obtained by setting random components of $y[n]$ to zero.

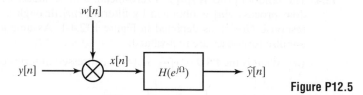

**Figure P12.5**

**(a)** Find the mean $\mu_w$ of the WSS process $w[n]$, and show that the correlation function of $w[n]$ is of the form $R_{ww}[m] = \alpha\delta[m] + \beta$, where $\alpha$ and $\beta$ are constants that you should determine, and $\delta[m]$ is the unit sample function. Also find an expression for the PSD $S_{ww}(e^{j\Omega})$ of the process $w[n]$.

**(b)** Compute $R_{yx}[m]$ and $R_{xx}[m]$.

**(c)** Specify the frequency response $H(e^{j\Omega})$ of a stable LTI filter that will take $x[n]$ as its input and produce an estimate $\widehat{y}[n]$ of $y[n]$ at its output, with $H(e^{j\Omega})$ chosen such that the mean square error, namely $E[(y[n] - \widehat{y}[n])^2]$, is minimized. Your answer can be specified in terms of the PSDs of $y[n]$ and $w[n]$. What would you expect your expression for $H(e^{j\Omega})$ to reduce to when $p = 1$? Does it indeed reduce to what you expect?

**(d)** Find an expression for the mean square error that results from application of the filter in (c). Your answer can be specified in terms of integrals involving the PSDs of $y[n]$ and $w[n]$. What would you expect your expression to reduce to when (i) $p = 0$, and (ii) $p = 1$? Does it indeed reduce to what you expect in these two cases?

**12.6.** Figure P12.6 is a block diagram of a faulty DT memory system in which sample values are randomly set to 0 ("dropped out") when they are retrieved, and a post-retrieval estimation filter that is used to estimate the correct value of the signal stored in memory. It is known that

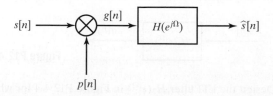

**Figure P12.6**

(i) $s[n]$ is the correct signal that is stored in memory; it is a zero-mean WSS random process with autocorrelation function $R_{ss}[n]$, and with PSD $S_{ss}(e^{j\Omega})$, specified as follows:

$$R_{ss}[n] = \frac{16}{15} \cdot \left(\frac{1}{4}\right)^{|n|}, \quad S_{ss}(e^{j\Omega}) = \frac{16}{|4 - e^{-j\Omega}|^2} \quad \text{for } |\Omega| \le \pi.$$

(ii) $p[n]$ is a random sequence of 0s and 1s that models memory dropouts; $p[n]$ has the following properties:
  - $p[\cdot]$ and $s[\cdot]$ are statistically independent;
  - $p[\cdot]$ is i.i.d., i.e., each time sample of $p[\cdot]$ is statistically independent of all other time samples; and
  - $\text{Prob}(p[n] = 1) = \frac{3}{4}$, $\text{Prob}(p[n] = 0) = \frac{1}{4}$.

(iii) $g[n] = s[n]p[n]$ is the corrupted signal that is retrieved from the memory in the presence of dropouts; $g[n]$ can always be written in the form

$$g[n] = ks[n] + r[n],$$

where $k$ is a constant, and $r[n]$ is the WSS random process $r[n] = s[n](p[n] - k)$.

(iv) $H(e^{j\Omega})$ is the frequency response of an LTI (but not necessarily causal) filter that is used to estimate $s[n]$ from $g[\cdot]$.

Answer the following questions with the information given above:

(a) Determine the constant $k$ for which the process $r[n]$ is zero mean and uncorrelated with the process $s[n]$.

(b) Determine $R_{rr}[n]$, the autocorrelation function of $r[n]$, when $r[n]$ is zero mean and uncorrelated with the process $s[n]$.

(c) Determine the filter frequency response $H(e^{j\Omega})$ that minimizes the mean square estimation error, $E\left[(s[n] - \hat{s}[n])^2\right]$.

**12.7. (a)** The frequency response of a particular DT LTI system is

$$H(e^{j\Omega}) = \frac{e^{j2\Omega}}{1 - \frac{1}{2}e^{-j\Omega}}.$$

Determine its unit sample response $h[n]$. If you do this correctly, you will find that the system is neither causal nor anticausal. Also determine

$$\sum_{k=-\infty}^{\infty} h[k] \quad \text{and} \quad \int_0^\pi |H(e^{j\Omega})|^2 \, d\Omega.$$

Recall:

$$\sum_{i=0}^{\infty} r^i = \frac{1}{1-r}, \quad |r| < 1.$$

**(b)** If $x[n]$ denotes a WSS process with mean value $\mu_x$ and autocovariance function $C_{xx}[m] = \sigma_x^2 \delta[m]$, what is the LMMSE estimate of $x[n+2]$ in terms of $x[n]$? In other words, find $\alpha$ and $\beta$ in $\hat{x}[n+2] = \alpha x[n] + \beta$ such that $E\{(x[n+2] - \hat{x}[n+2])^2\}$ is minimized. Also find the associated MMSE.

**(c)** If the process $x[n]$ in (b) is applied to the input of the system in (a), what is the PSD $S_{yy}(e^{j\Omega})$ of the output process $y[n]$? Also evaluate $E\{y[n]\}$, $E\{y^2[n]\}$, and

$$\lim_{N \to \infty} \frac{1}{2N+1} \sum_{k=-N}^{N} y[k] \ .$$

**(d)** For this part of the question, assume $\mu_x = 0$ for simplicity. With all quantities as previously defined, suppose what you can measure is $q[n] = y[n] + v[n]$ for all $n$, where $v[n]$ is (zero-mean) white noise of intensity $\sigma_v^2$, and is uncorrelated with the process $x[k]$. Compute the frequency response $W(e^{j\Omega})$ of the noncausal Wiener filter that has $q[n]$ as input at time $n$ and produces the LMMSE estimate $\widehat{x}[n+2]$ as output at time $n$. Explicitly check that your answer reduces to something that you expect in the case of $\sigma_v^2 = 0$.

**12.8.** The message signal $y[n]$ in Figure P12.8 is to be encrypted and transmitted across a noisy channel, then decrypted and filtered at the receiver. We model $y[n]$ as a zero-mean WSS random process with autocorrelation function $R_{yy}[m]$ and corresponding PSD $S_{yy}(e^{j\Omega})$. The signal $p[n]$ is used for both the encryption at the transmitter and the decryption at the receiver, and is an i.i.d. process that takes the values $+1$ or $-1$ with equal probability at each time; it is independent of the process $y[\cdot]$. Note that $p^2[n] = 1$ for all $n$. The transmitted signal $q[n]$ is the product $p[n]y[n]$.

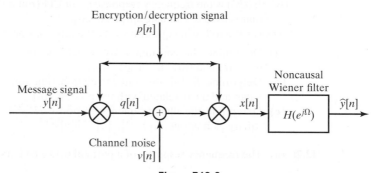

**Figure P12.8**

**(a)** Determine the respective means $\mu_p$ and $\mu_q$ of the processes $p[n]$ and $q[n]$, their respective autocorrelations $R_{pp}[m]$ and $R_{qq}[m]$ (expressed in terms of $R_{yy}[\cdot]$), and also the cross-correlation $R_{yq}[m]$ between the message signal and the transmitted signal. Would an intruder who was able to intercept the transmitted process $q[\cdot]$ have any use for a (possibly noncausal) linear estimator of $y[n]$ based on measurements of $q[\cdot]$? Explain your answer.

The channel adds a noise signal $v[n]$ to the transmitted signal, so that the received signal is

$$q[n] + v[n] = p[n]y[n] + v[n] \ .$$

Assume $v[n]$ is a zero-mean and white WSS process, with $R_{vv}[m] = \sigma_v^2\delta[m]$; suppose it is uncorrelated with $y[\cdot]$, and both processes are independent of $p[\cdot]$. We assume, as indicated in Figure P12.8, that the intended receiver knows the specific encryption signal $p[n]$, i.e., the specific sample function from the ensemble that was used for encryption. If there was no channel noise (i.e., if we had $v[n] = 0$), the decryption would then simply involve multiplying the received signal by $p[n]$ because

$$p[n]q[n] = p[n]\left(p[n]y[n]\right) = p^2[n]y[n] = y[n] \ ,$$

where the last equality is a consequence of having $p^2[n] = 1$. In the presence of noise, we can still attempt to decrypt in the same manner, but will follow it up by a further stage of filtering. The signal to be filtered is thus

$$x[n] = p[n]\left(p[n]y[n] + v[n]\right) = y[n] + p[n]v[n] .$$

**(b)** Determine $\mu_x$, $R_{xx}[m]$, and $R_{yx}[m]$.

**(c)** Suppose the filter at the receiver is to be a (stable) noncausal Wiener filter, constructed so as to produce the LMMSE estimate $\widehat{y}[n]$ of $y[n]$. Determine the frequency response $H(e^{j\Omega})$ of this filter, and explicitly check that it is what you would expect it to be in the two limiting cases of $\sigma_v^2 = 0$ and $\sigma_v^2 \to \infty$. Also write an expression, in terms of $S_{yy}(e^{j\Omega})$ and $\sigma_v^2$, for the mean square error obtained with this filter, and explicitly check that it is what you would expect it to be in the preceding two limiting cases.

**12.9.** Consider a zero-mean WSS process $y[n]$ with $E\{y^2[n]\} = \sigma^2$. Suppose signal values at adjacent instants have a correlation coefficient of $\rho$, but that signal values more than one instant apart are uncorrelated. We already know how to construct a one-step LMMSE predictor for the process using just the present value, i.e., picking $\widehat{y}[n + 1] = ay[n]$ with an optimally chosen $a$. It is also easy to see that prediction based on a single measurement that is strictly in the past is not useful: if $\widehat{y}[n + 1] = by[n - k]$ for some $k > 0$, then the optimal choice is $b = 0$.

Suppose now that we construct a one-step LMMSE predictor using the present and most recent past value, i.e., suppose we choose $\widehat{y}[n + 1] = cy[n] + dy[n - 1]$. You might think, based on what is said in the preceding paragraph, that we would discover $c = a$ and $d = 0$, where $a$ is the optimum value referred to in the previous paragraph. If you did think so, you'd be wrong! Explain intuitively why the optimum $d$ might end up being nonzero, then find the best choices of $c$ and $d$, and determine the associated mean square error.

**12.10.** Consider the system described by the block diagram in Figure P12.10-1.

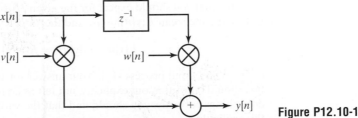

**Figure P12.10-1**

The random sequences $x[\cdot]$, $v[\cdot]$, and $w[\cdot]$ are mutually independent, WSS, with autocorrelations $R_{xx}[m]$, $R_{vv}[m]$, and $R_{ww}[m]$ respectively, and PSDs $S_{xx}(e^{j\Omega})$, $S_{vv}(e^{j\Omega})$, and $S_{ww}(e^{j\Omega})$ respectively. The sequence $x[n]$ is zero mean. The sequences $v[n]$ and $w[n]$ have means $\mu_v$ and $\mu_w$, respectively.

Design the noncausal Wiener filter that estimates $x[n]$ for all $n$ from the measurements $\{y[k], -\infty < k < +\infty\}$, i.e., the noncausal linear filter $H_{WF}(e^{j\Omega})$ in Figure P12.10-2. Express your answer in terms of the available statistics for $x[n]$, $w[n]$, $v[n]$, and any algebraic or trigonometric operations needed, including but not limited to addition, scalar multiplication, multiplication by a complex number, and convolution.

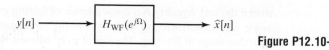

**Figure P12.10-2**

**12.11.** Consider a stable and causal DT system with input $w[n]$ and output $y[n]$ that are related by

$$y[n] = -\sum_{k=1}^{N} a_k y[n-k] + w[n].$$

The output is said to be governed by an $N$th-order autoregressive model in this case, because the output depends on past values of itself, as well as on the present input. Suppose $w[n]$ is known to be a (zero-mean) white process with variance $\sigma_w^2$, but is otherwise unknown, and assume the $a_k$ are known. Determine the LMMSE estimator $\hat{y}[n]$ of the output $y[n]$, in terms of measurements of all past values of $y[\cdot]$. Also find the associated MMSE. Be sure to explain where/how you used the stability and causality of the system in your reasoning. *Hint:* To find the LMMSE estimator, study the governing equation above and come up with a plausible guess for what you think $\hat{y}[n]$ will be, then verify that this guess satisfies the orthogonality conditions associated with this problem.

**12.12.** The input to a particular stable LTI filter with frequency response

$$H(e^{j\Omega}) = \frac{1}{1 - \frac{1}{2}e^{-j\Omega}}$$

is a white DT WSS process $w[n]$ whose PSD is $S_{ww}(e^{j\Omega}) = 9$ for all $\Omega$. Denote the output of the system at time $n$ by $y[n]$.

**(a)** Find a first-order difference equation relating the input and output of the system, and also explicitly determine the unit sample response $h[n]$ of the system. As a check, explicitly compute $\sum h[n]$ and compare the value you get with what you should expect for the given $H(e^{j\Omega})$. Is the system causal?

**(b)** Determine the mean $E\{y[n]\} = \mu_y$ and the autocorrelation function

$$E\{y[n+m]y[n]\} = R_{yy}[m]$$

of the WSS output process $y[\cdot]$. Your answer for the autocorrelation function should be written out explicitly, not left as an integral or sum. If you've done things correctly, you should find that the variance of $y[n]$ is 12; verify this explicitly.

**(c)** Specify completely the LMMSE causal one-step predictor for the process $y[\cdot]$. This predictor forms the LMMSE estimator $\hat{y}[n+1]$ for $y[n+1]$, using all values of $y[k]$ for $k \le n$. One way to do this is using your input–output equation from (a) to conjecture the form of this predictor, and then to verify your conjecture using the orthogonality condition that characterizes LMMSE estimation. Another way is to design an appropriate causal Wiener filter. Use either of the above approaches to find the predictor, showing the main steps of your calculation. Also find the predictor using the other way, to check that you get the same answer either way. Finally, determine the MMSE associated with the predictor. Could the correct answer for the MMSE be larger than 12?

You may find it helpful to recall the following identity for geometric series:

$$1 + \alpha + \cdots + \alpha^{m-1} = \frac{1 - \alpha^m}{1 - \alpha}.$$

**12.13.** Let $y[n]$ be a WSS process with autocorrelation function

$$R_{yy}[m] = 9\Big(\delta[m] - \alpha\delta[m-1] - \alpha\delta[m+1]\Big)$$

where $\alpha > 0$.

**(a)** What is the maximum value $\alpha$ can take? Explain your reasoning. If $\alpha$ is increased toward its maximum value, does the power of the signal shift to lower or higher frequencies?

**(b)** Determine the following (expressed in terms of $\alpha$, if necessary):

(i) $E\{y[n]\}$ and $E\{y^2[n]\}$;

(ii) the correlation coefficient $\rho$ between $y[4]$ and $y[5]$.

**(c)** Suppose we are told that we will be given the measurement $y[4]$, and we want to find the LMMSE estimator of $y[5]$ in terms of $y[4]$. Find the estimator, and determine the associated MMSE.

**(d)** Suppose $x[n] = y[n] + w[n]$, where $w[n]$ is a white process that is uncorrelated with $y[\cdot]$ and has PSD $S_{ww}(e^{j\Omega}) = 9\alpha^2$. Determine the PSD $S_{xx}(e^{j\Omega})$ and show that it can be written in the form

$$S_{xx}(e^{j\Omega}) = K(1 - \beta e^{-j\Omega})(1 - \beta e^{j\Omega})$$

for $K$ and $\beta$ that you should determine, expressed in terms of $\alpha$ if necessary. Also determine the cross spectral density $S_{yx}(e^{j\Omega})$ in terms of $\alpha$.

**(e)** Determine the frequency response $H(e^{j\Omega})$ of the noncausal Wiener filter that produces the LMMSE estimate $\widehat{y}[n]$ of $y[n]$ in terms of measurements of the entire process $x[\cdot]$.

**(f)** Determine the frequency response $G(e^{j\Omega})$ of the causal Wiener filter that at time $n$ uses measurements of $x[k]$ for all present and past times $k \le n$ to produce an LMMSE prediction of the measurement at the next step, i.e., an LMMSE estimate $\widehat{x}[n+1]$ of $x[n+1]$. Also determine the associated mean square error.

**12.14.** Suppose $y[n]$ is a zero-mean WSS random process with PSD given by $S_{yy}(e^{j\Omega}) = 5 + 4\cos\Omega$ and the corresponding autocorrelation function shown in Figure P12.14-1.

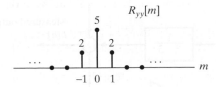

**Figure P12.14-1**

**(a)** Sketch the pole-zero plot corresponding to $S_{yy}(z)$. Be sure to plot all the poles and zeros.

**(b)** Suppose $y[n]$ is generated via the system shown in Figure P12.14-2. where $w[n]$ is a WSS white process with unit PSD, i.e., $S_{ww}(e^{j\Omega}) = 1$. Determine one possible unit sample response $g[\cdot]$.

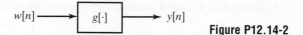

**Figure P12.14-2**

We now want an optimum causal LTI filter with impulse response $h[\cdot]$ for obtaining a one-step prediction of $y[n]$. The desired system is shown in Figure P12.14-3, with $h[\cdot]$ chosen to minimize $E\left[(\widehat{y}[n+1] - y[n+1])^2\right]$.

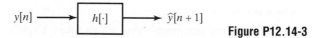

**Figure P12.14-3**

**(c)** If $h[n]$ is restricted to have length two as shown in Figure P12.14-4, determine $h[n]$, i.e., find $a$ and $b$.

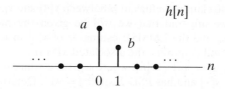

**Figure P12.14-4**

**(d)** Restricting $h[n]$ to be causal but of possibly infinite length, determine $h[n]$, the impulse response of the causal Wiener filter.

**12.15.** The system shown in Figure P12.15-1 comprises a causal plant embedded in a feedback loop. The input signal, $a[\cdot]$, and the noise disturbance, $w[\cdot]$, are zero-mean, uncorrelated, WSS white processes, with respective autocorrelation functions:

$$R_{aa}[m] = \sigma_a^2 \delta[m], \quad R_{ww}[m] = \sigma_w^2 \delta[m] .$$

The transfer function $E(z)$ from $a[n]$ to $b[n]$ and the transfer function $F(z)$ from $w[n]$ to $b[n]$ are easily seen to be

$$E(z) = \frac{1}{z}, \quad F(z) = \frac{z-3}{z} = 1 - 3z^{-1} .$$

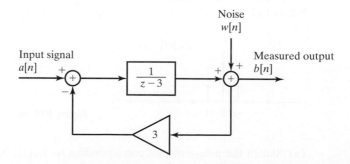

**Figure P12.15-1**

This problem concerns the design of an LTI filter with system function $H(z)$ that will use measurements of the output signal $b[n]$ to generate $\widehat{a}[n]$, the LMMSE estimate of the input $a[n]$, as shown in Figure P12.15-2:

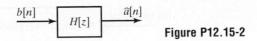

**Figure P12.15-2**

**(a)** Suppose there is no noise $w[n]$, i.e., $\sigma_w^2 = 0$, and that the filter $h[n]$ is allowed to be noncausal. Determine, without much work, what $H(z)$ should be, and find the corresponding mean square error $E[(a[n] - \widehat{a}[n])^2]$.

**(b)** Suppose now that the filter $h[n]$ is still allowed to be noncausal, but that $\sigma_w^2$ is no longer restricted to 0, i.e., $\sigma_w^2 \geq 0$. Determine $H(z)$. Make sure to check that your answer reduces to what you obtained in part (a), when $\sigma_w^2 = 0$.

**(c)** Suppose again that there is no noise $w[n]$, i.e., $\sigma_w^2 = 0$, but the estimation filter $H(z)$ is restricted to be causal. Again, find $H(z)$ and the corresponding mean square error.

**(d)** Find the causal Wiener filter when $\sigma_w^2 > 0$.

**12.16.** Suppose the zero-mean WSS process $x[n]$ is obtained by applying a zero-mean WSS white process $w[n]$ with PSD $S_{ww}(e^{j\Omega}) = \sigma^2$ to the input of a (stable, causal) filter with system function

$$M(z) = 1 - 3z^{-1} .$$

**(a)** If $S_{xx}(e^{j\Omega})$ denotes the PSD of $x[n]$, find $S_{xx}(z)$. Also find the autocovariance function $C_{xx}[m]$ of the process $x[n]$, the variance of the random variable $x[n+1]$, and the correlation coefficient $\rho$ between $x[n]$ and $x[n+1]$.

**(b)** Specify the LMMSE estimator of $x[n+1]$ based on a measurement of $x[n]$, and compute the associated mean square error. Is it less than the variance of $x[n+1]$ that you computed in (a)?

**(c)** Find the system function $F(z)$ of a stable and causal filter whose inverse $1/F(z)$ is also stable and causal, such that $S_{xx}(z) = F(z)F(z^{-1})$.

**(d)** Find the system function of the causal Wiener filter that generates an estimate of $x[n+1]$ based on the present and all past $x[k]$, $k \leq n$, i.e., find the system function of the one-step predictor. Do you expect that the mean square error for this case will be less than, equal to, or greater than what you computed in (b)? Determine the mean square error to confirm whether your expectation is correct.

**12.17.** We have measurements of a WSS random process $x[n]$ that is modeled as the output of a minimum-phase LTI system whose input is a white process $w[n]$, with $E\{w^2[n]\} = 1$. (Recall that a minimum-phase DT system is defined as stable, causal, and with a stable, causal inverse.) The situation is shown in Figure P12.17-1.

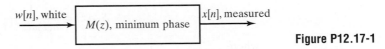

**Figure P12.17-1**

Suppose the transfer function of the system above is

$$M(z) = \frac{\gamma}{z - \lambda} + d ,$$

where $\gamma \neq 0$ and $d \neq 0$. We would like to pass the process $x[n]$ through a stable LTI filter with system function $H(z)$ that is chosen to make this filter the LMMSE estimator of $x[n+1]$, i.e., the LMMSE one-step predictor, as shown in Figure P12.17-2.

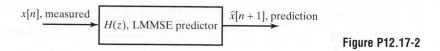

x[n], measured → H(z), LMMSE predictor → x̂[n + 1], prediction

**Figure P12.17-2**

**(a)** Suppose there are no constraints on the LTI filter $H(z)$ beyond stability. Determine the optimum filter and the associated MMSE,

$$E\{(x[n + 1] - \hat{x}[n + 1])^2\} .$$

**(b)** Suppose now that we constrain the filter $H(z)$ to not only be stable but also causal. Again determine the optimum filter and the associated mean square error. Your answers will be expressed in terms of the given parameters, namely $\gamma$, $\lambda$, and d.

**12.18. (a)** Suppose $x[n]$ is a zero-mean WSS random sequence with autocorrelation $R_{xx}[m] = (\frac{1}{3})^{|m|}$. We want to design a causal LTI filter $h[n]$, as shown in Figure P12.18, with only one nonzero value, i.e., with $h[n] = a\delta[n - n_o]$, where $n_o$ is an integer greater than or equal to 0. The output $g[n]$ is to be the best such linear one-step predictor of $x[n]$, i.e., it is chosen to minimize the mean square prediction error, $E[(g[n] - x[n + 1])^2]$. If $n_o$ is fixed, determine the value of $a$, in terms of $n_o$, to minimize this error.

x[n] ⟶ h[·] ⟶ g[n]                     **Figure P12.18**

**(b)** A zero-mean WSS process $x(t)$ is known to have autocorrelation function $R_{xx}(\tau) = 6e^{-3|\tau|}$. Determine the optimum causal LTI Wiener filter to obtain the LMMSE estimate of $x(t + T)$ for a fixed $T > 0$, using measurements of $x(\cdot)$ from the infinite past up to time $t$. Also compute the associated MMSE. State in words what your answer for the Wiener filter tells you about LMMSE prediction for an exponentially correlated process.

## Advanced Problems

**12.19.** Suppose the autocorrelation function $R_{xx}[m]$ of a zero-mean WSS process $x[n]$ has the following z-transform:

$$S_{xx}(z) = \frac{1}{a(z)a(z^{-1})} ,$$

where

$$a(z) = z^L + a_1 z^{L-1} + a_2 z^{L-2} + \cdots + a_L$$

is a polynomial of degree $L$ whose roots are all inside the unit circle. Note that we can also write $S_{xx}(z)$ as

$$S_{xx}(z) = \frac{z^L}{a(z)} \frac{z^{-L}}{a(z^{-1})} = \frac{1}{1 + a_1 z^{-1} + \cdots + a_L z^{-L}} \frac{1}{1 + a_1 z + \cdots + a_L z^L} .$$

**(a)** Find the system function $M(z)$ of a stable and causal filter with a stable and causal inverse such that $M(z)M(z^{-1}) = S_{xx}(z)$.

**(b)** Find the system function $H_1(z)$ and the corresponding unit sample response $h_1[n]$ of a (stable) causal Wiener filter that uses measurements of $x[\cdot]$ up to and including time $n$ in order to produce the LMMSE estimate of $x[n+1]$, so the filter is the one-step Wiener predictor. (*Hint:* Depending on how you tackle the problem, you may or may not find it convenient to use the relation

$$\frac{z^{L+1}}{a(z)} = z - \frac{a_1 z^L + a_2 z^{L-1} + \cdots + a_L z}{a(z)} ,$$

along with the observation that

$$\frac{a_1 z^L + a_2 z^{L-1} + \cdots + a_L z}{a(z)}$$

has an inverse transform that is a causal and stable, or absolutely summable, signal.)

**(c)** Find the system function $H_2(z)$ and unit sample response $h_2[n]$ of the causal two-step Wiener predictor for LMMSE estimation of $x[n+2]$ from measurements of $x[\cdot]$ up to and including time $n$. You may leave your answer in terms of the coefficients $p_1, \cdots, p_L$ defined through the identity below:

$$\frac{z^{L+2}}{a(z)} = z^2 - a_1 z - \frac{p_1 z^L + p_2 z^{L-1} + \cdots + p_L z}{a(z)} .$$

Note these coefficients $p_1, \cdots, p_L$ can easily be written explicitly in terms of $a_1, \cdots, a_L$, but that's not important to do here.

**12.20.** A DT random process $x[n]$ is the sum of a zero-mean DT WSS random process $y[n]$ and a zero-mean DT WSS white-noise process $v[n]$, so $x[n] = y[n] + v[n]$. Assume the following:

(i) The PSD of $y[n]$ is $S_{yy}(e^{j\Omega}) = 4 - 4\cos(\Omega)$.
(ii) The PSD of $v[n]$ is $S_{vv}(e^{j\Omega}) = 1$.
(iii) The random processes $y[\cdot]$ and $v[\cdot]$ are uncorrelated.

We are interested in finding a causal Wiener filter to generate a one-step prediction of $y[n]$, i.e., to obtain the LMMSE estimate of $p[n] = y[n+1]$, based on measurements of $x[k]$ for all $k \le n$. Denote this LMMSE estimate by $\hat{y}[n+1]$. It may help you in this problem to recall that a causal system with transfer function

$$\frac{1}{1 - \alpha z^{-1}}$$

has unit sample response $\alpha^n u[n]$.

**(a)** Find a spectral factor (or "generalized square root") $M_{xx}(e^{j\Omega})$ with all of the following properties:

(i) $S_{xx}(e^{j\Omega}) = |M_{xx}(e^{j\Omega})|^2$;
(ii) $M_{xx}(e^{j\Omega})$ could be the frequency response of an LTI system that is both causal and stable; and
(iii) $1/M_{xx}(e^{j\Omega})$ could be the frequency response of an LTI system that is both causal and stable.

**(b)** Using your answer from (a), find the inverse transform $w_{xx}[n]$ of

$$W_{xx}(e^{j\Omega}) = \frac{1}{M_{xx}(e^{j\Omega})} .$$

**(c)** Determine $R_{px}[m]$ in terms of $R_{yy}[m]$.

**(d)** Let $h[n]$ be the impulse response of the causal Wiener filter whose output is the process $\hat{y}[n+1]$ when its input is the process $x[n]$. Find an explicit expression for $h[n]$ in terms of $w_{xx}[n]$. For the purposes of this part, the most helpful expression for $S_{yy}(e^{j\Omega})$ will probably be

$$S_{yy}(e^{j\Omega}) = 4 - 2e^{j\Omega} - 2e^{-j\Omega} .$$

**(e)** By modifying your calculations in (d)—or in some other fashion—determine the causal Wiener filter for two-step prediction, i.e., for obtaining the LMMSE estimate of $y[n+2]$, based on measurements of $x[k]$ for all $k \le n$, and explain why this result is reasonable.

**12.21.** A particular causal first-order DT LTI system is governed by a model in state-space form:

$$q[n+1] = 3q[n] + x[n] + d[n]$$

where $x[n]$ is a known control input and $d[n]$ is an unknown zero-mean, WSS white-noise disturbance input with $E(d^2[n]) = \sigma_d^2$. We would like to use an observer to construct an estimate $\hat{q}[n]$ of $q[n]$, using the noisy output measurements

$$y[n] = 2q[n] + v[n] ,$$

where the measurement noise $v[n]$ is also an unknown zero-mean, WSS white-noise process with $E(v^2[n]) = \sigma_v^2$. Assume the measurement noise is uncorrelated with the system disturbance: $E(v[n]d[k]) = 0$ for all $n, k$.

**(a)** Specify which of the following equations you would implement as your (causal) observer, explaining your reasoning. In each case, $\ell$ denotes the observer gain.

    (i) $\hat{q}[n+1] = 3\hat{q}[n] + x[n] + d[n] - \ell(y[n] - 2\hat{q}[n] - v[n])$ .
    (ii) $\hat{q}[n+1] = 3\hat{q}[n] + x[n] - \ell(y[n] - 2\hat{q}[n] - v[n])$ .
    (iii) $\hat{q}[n+1] = 3\hat{q}[n] + x[n] - \ell(y[n] - 2\hat{q}[n])$ .
    (iv) $\hat{q}[n+1] = 3\hat{q}[n] - \ell(y[n] - 2\hat{q}[n])$ .
    (v) $\hat{q}[n+1] = 3\hat{q}[n] - \ell(y[n] - 2\hat{q}[n] - v[n])$ .
    (vi) Something other than the above (specify).

**(b)** Obtain a state-space model for the observer error, $\tilde{q}[n] = q[n] - \hat{q}[n]$, writing it in the form

$$\tilde{q}[n+1] = \alpha \tilde{q}[n] + p[n] ,$$

with $\alpha$ and $p[n]$ expressed in terms of the parameters and signals specified in the problem statement (but with $p[n]$ not involving $\tilde{q}[n]$, of course). Check: If you have done things correctly, you should find that $\alpha = 0$ when $\ell = -\frac{3}{2}$.

**(c)** Determine the system function of the error system in (b) and the corresponding impulse response, i.e., find the system function and corresponding impulse response that relate $\tilde{q}[n]$ to the input $p[n]$.

**(d)** Note that the input process $p[n]$ in (b) is WSS and zero-mean. Determine its autocovariance function $C_{pp}[m]$ in terms of parameters specified in the problem statement.

**(e)** For $\ell = -\frac{3}{2}$, determine the mean $E(\tilde{q}[n])$ of the observer error, its second moment $E(\tilde{q}^2[n])$, and its variance.

**(f)** If we no longer fix $\ell$ to have the value specified in (e), what constraints must $\ell$ satisfy if the observer error $\tilde{q}[n]$ is to be a zero-mean WSS process (assuming the observer has been running since the beginning of time, i.e., starting infinitely far in the past)? Verify that the choice of $\ell$ in (e) satisfies the constraints that you specify here.

**(g)** Assume the constraints on $\ell$ that you specified in (f) are satisfied and that the observer has been running since the beginning of time. Find a general expression for the mean $E(\tilde{q}[n])$ of the observer error, its second moment $E(\tilde{q}^2[n])$, and its variance. You might find it helpful to recall that for $|\alpha| < 1$

$$\sum_{k=0}^{\infty} \alpha^{2k} = \frac{1}{1-\alpha^2} \ .$$

**(h)** Evaluate your variance expression in (g) for the case $\sigma_d^2 = 0$ and $\ell = -\frac{4}{3}$, and show that the error variance in this case is smaller that what you get (still for $\sigma_d^2 = 0$) with the earlier choice in (e) of $\ell = -\frac{3}{2}$.

**(i)** Find a quadratic equation satisfied by the value of $\ell$ that minimizes the variance expression you obtained in part (g).

**12.22.** We have measurements of a WSS random process $x[n]$ that is modeled as the output of a causal and BIBO-stable LTI system whose input is a (zero-mean) WSS white process $w[n]$, with $E\{w^2[n]\} = 1$.

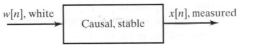

**Figure P12.22**

Suppose the above model has a reachable and observable state-space representation of the form

$$\mathbf{q}[n+1] = \mathbf{A}\mathbf{q}[n] + \mathbf{b}w[n] \tag{12.114}$$

$$x[n] = \mathbf{c}^T\mathbf{q}[n] + dw[n] \ , \tag{12.115}$$

where the first equation is the state evolution equation, and the second equation is the output equation. The corresponding system function is

$$M(z) = \mathbf{c}^T(z\mathbf{I} - \mathbf{A})^{-1}\mathbf{b} + d \ . \tag{12.116}$$

**(a)** Explain why the above BIBO-stable state-space model is guaranteed to be asymptotically stable.

Note that we don't have access to the processes $w[n]$ or $\mathbf{q}[n]$; all we have is the measured output $x[n]$. This problem will make use of the inverse system for the above model. We assume the above model has a causal and BIBO-stable inverse, i.e., the model is minimum-phase. Causality of the inverse requires $d \neq 0$, which we will assume from now on.

The rest of this problem is concerned with using the present and all past values of the measured process, i.e., $x[k]$ for $k \leq n$, to construct the LMMSE estimator $\hat{x}[n+1]$ for the next value of the measured process, namely $x[n+1]$. In other words, we will be constructing an optimal one-step predictor. Our strategy is based on writing Eq. (12.115) at time $n+1$ rather than $n$, from which one can deduce (though we don't ask you to do so here) that

$$\widehat{x}[n+1] = \mathbf{c}^T\widehat{\mathbf{q}}[n+1] + d\widehat{w}[n+1] = \mathbf{c}^T\widehat{\mathbf{q}}[n+1] , \tag{12.117}$$

where the $\widehat{\phantom{x}}$ on the other quantities in the preceding equation again denotes the LMMSE estimator based on $x[k]$ for $k \le n$ in each case, and where the last equality follows from the fact that $\widehat{w}[n+1] = 0$.

**(b)** Explain carefully why $\widehat{w}[n+1] = 0$, i.e., why the LMMSE estimate of $w[n+1]$, based on measurements of $x[k]$ for $k \le n$, turns out to be zero.

You will discover in what follows how one can, using only the available measurements $x[k]$ for $k \le n$, construct an estimate $\widehat{\mathbf{q}}[n+1]$, not necessarily LMMSE, that converges to $\mathbf{q}[n+1]$ exponentially fast. Thus, following a transient interval, having this estimate is essentially as good as having $\mathbf{q}[n+1]$ itself.

**(c)** Show that a state-space model for the inverse system of the model in Eqs. (12.114) and (12.115) can be written in the form

$$\mathbf{q}[n+1] = \mathbf{A}_{inv}\mathbf{q}[n] + \mathbf{b}_{inv}x[n] \tag{12.118}$$

$$w[n] = \mathbf{c}_{inv}^T\mathbf{q}[n] + d_{inv}x[n] , \tag{12.119}$$

for some appropriate $\mathbf{A}_{inv}$, $\mathbf{b}_{inv}$, $\mathbf{c}_{inv}^T$, and $d_{inv}$ that you should write in terms of the quantities $\mathbf{A}$, $\mathbf{b}$, $\mathbf{c}^T$, and $d$ in Eq. (12.114) that describe the original system. Note that the above model has the same state vector as the original system, but its input and output are interchanged from the original.

Our assumption of a BIBO-stable inverse, combined with the reasoning that went into answering part (a), guarantees that the above inverse system is asymptotically stable.

Suppose we now build a real-time simulator for the state evolution of the inverse system, in the form

$$\widehat{\mathbf{q}}[n+1] = \mathbf{A}_{inv}\widehat{\mathbf{q}}[n] + \mathbf{b}_{inv}x[n] . \tag{12.120}$$

If we were to start this with the correct initial condition $\widehat{\mathbf{q}}[0] = \mathbf{q}[0]$ at time 0, and drive it with $x[k]$ for $0 \le k \le n$, we would have $\widehat{\mathbf{q}}[k+1] = \mathbf{q}[k+1]$ for $0 \le k \le n$. However, since we don't have access to $\mathbf{q}[0]$, we use some guessed initial condition $\widehat{\mathbf{q}}[0]$ in our simulator.

**(d)** Find a state evolution equation for the error $\widetilde{\mathbf{q}}[n] = \mathbf{q}[n] - \widehat{\mathbf{q}}[n]$ for $n \ge 0$, and use the equation to explain why this error goes to zero exponentially fast, at a rate determined by the eigenvalues of the inverse system, i.e., by the zeros of the system function $M(z)$ in Eq. (12.116).

In view of the above facts, we can use $\mathbf{c}^T\widehat{\mathbf{q}}[n+1]$ as our approximation of $\widehat{x}[n+1]$. Invoking Eq. (12.120), we have

$$\widehat{x}[n+1] \approx \mathbf{c}^T\mathbf{A}_{inv}\widehat{\mathbf{q}}[n] + \mathbf{c}^T\mathbf{b}_{inv}x[n] . \tag{12.121}$$

This approximation becomes good exponentially fast. The state evolution Eq. (12.120) along with its output Eq. (12.121) is thus a state-space model of the one-step predictor.

To find the MMSE of the estimator, note that after the transient period the error is

$$x[n+1] - \widehat{x}[n+1] = dw[n+1] , \tag{12.122}$$

so the MMSE is simply $d^2E\{w^2[n+1]\} = d^2$.

(e) Specialize to the case of a first-order model, replacing $\mathbf{A}, \mathbf{b}$, and $\mathbf{c}^T$ by scalars $a$, $b$, and $c$ respectively, and check that you obtain the causal Wiener one-step predictor and associated MMSE that you know how to derive by other methods.

**12.23.** Consider three jointly WSS zero-mean CT random processes $y(\cdot)$, $x_1(\cdot)$, and $x_2(\cdot)$. Suppose we wish to construct an LMMSE estimator for $y(t)$, using all past, present, and future values of $x_1(\cdot)$ and $x_2(\cdot)$. We do this by computing the estimate $\widehat{y}(t)$ of $y(t)$ as follows:

$$\widehat{y}(t) = h_1 * x_1(t) + h_2 * x_2(t) , \tag{12.123}$$

then choosing $h_1(\cdot)$ and $h_2(\cdot)$ to minimize $E[\{y(t) - \widehat{y}(t)\}^2]$. We can think of $h_1(\cdot)$ and $h_2(\cdot)$ as the impulse responses of two stable LTI systems that respectively take as inputs the measurements $x_1(\cdot)$ and $x_2(\cdot)$; the outputs of these two systems are then added to form the estimate of interest. Our objective is to determine these two impulse responses, or equivalently their Fourier transforms $H_1(j\omega)$ and $H_2(j\omega)$.

(a) Invoking the orthogonality condition that governs the solution of LMMSE problems (you need not derive this condition), write down the equation that relates the correlation functions $R_{yx_1}(\tau)$ and $R_{\widehat{y}x_1}(\tau)$ when the estimator is LMMSE, and explain your answer.

Similarly write down the equation that relates the correlation functions $R_{yx_2}(\tau)$ and $R_{\widehat{y}x_2}(\tau)$.

(b) Using your results from (a) and the expression in Eq. (12.123), obtain a formula expressing $R_{yx_1}(\tau)$ as appropriate convolutions and/or sums involving the functions $R_{x_1x_1}(\cdot)$, $R_{x_1x_2}(\cdot)$, $h_1(\cdot)$, and $h_2(\cdot)$—or functions closely related to these.

Similarly obtain an expression for $R_{yx_2}(\tau)$ in terms of the given functions.

How do your expressions simplify if the processes $x_1(\cdot)$ and $x_2(\cdot)$ are uncorrelated, i.e., if $R_{x_1x_2}(\tau) = 0$ for all $\tau$?

(c) Taking Fourier transforms of your equations in (b) will give you two linear simultaneous equations in the transform domain for the desired frequency responses $H_1(j\omega)$ and $H_2(j\omega)$. Assemble these equations in the following matrix form, filling in the asterisks appropriately:

$$\begin{bmatrix} * & * \\ * & * \end{bmatrix} \begin{bmatrix} H_1(j\omega) \\ H_2(j\omega) \end{bmatrix} = \begin{bmatrix} * \\ * \end{bmatrix} .$$

(As a check, if you've done things correctly then the determinant of the 2×2 matrix will be a real function of $\omega$.)

There is no need to solve this set of equations in the general case, but write down the solution for the special case you considered in (b), namely where the two measured processes are uncorrelated: $R_{x_1x_2}(\tau) = 0$ for all $\tau$.

**12.24.** Figure P12.24 shows the channel model for a wireless communication system with a direct path and a reflected path.

In this figure:

(i) The channel input $s(t)$ is a zero-mean WSS random process whose PSD is

$$S_{ss}(j\omega) = \frac{2\lambda\sigma_s^2}{\omega^2 + \lambda^2} ,$$

where $\lambda$ and $\sigma_s^2$ are positive constants.

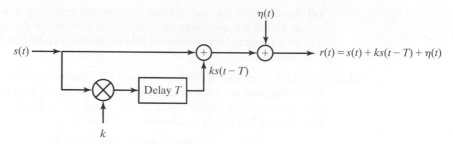

**Figure P12.24**

(ii) The channel output is $r(t) = s(t) + ks(t - T) + \eta(t)$, where $k$ is a positive constant representing the strength of the reflected path, $T$ is a positive constant representing the delay of the reflected path relative to the direct path, and $\eta(t)$ is the channel noise.

(iii) The channel noise $\eta(t)$ is zero-mean WSS white noise that is uncorrelated with the process $s(\cdot)$, and has PSD $S_{\eta\eta}(j\omega) = N$.

Answer the following questions, given the information above:

(a) Determine $S_{rr}(j\omega)$, the power spectral density of $r(t)$.

(b) We want to pass $r(t)$ through an LTI filter with frequency response $H(j\omega)$ to obtain an estimate of the channel input $s(t)$. Determine the frequency response $H(j\omega)$ that minimizes the mean square error of this estimate.

## Extension Problems

**12.25.** Suppose we have access to a WSS process $x[n]$ with mean $\mu_x$ and FSD given by $D_{xx}(e^{j\Omega})$. Describe how you would generate another process $y[\cdot]$ that is jointly WSS with $x[n]$, and that has a specified mean $\mu_y$, FSD $D_{yy}(e^{j\Omega})$, and cross fluctuation density $D_{yx}(e^{j\Omega})$. (*Hint:* Start by generating $\widehat{y}[n]$, the LMMSE estimate of $y[n]$ in terms of $x[\cdot]$.) Also explicitly identify what constraint the specified $D_{yx}(e^{j\Omega})$ has to satisfy, relative to the given $D_{xx}(e^{j\Omega})$ and $D_{yy}(e^{j\Omega})$, in order for your method to work.

**12.26.** (a) Suppose an estimator $\widehat{y}[n]$ for a zero-mean process $y[n]$ is constructed by LTI filtering of a zero-mean measured process $x[n]$ using a system with frequency response $H(e^{j\Omega})$, where $x[\cdot]$ and $y[\cdot]$ are jointly WSS. Derive the expression in Eq. (12.73) for the mean square error of this estimator. (This result applies even if the estimator is not LMMSE, and also to processes whose means are nonzero, after the standard adjustments in constructing the estimator: subtract the mean $\mu_x$ of $x[n]$ at the input to the filter, and add back the mean $\mu_y$ of $y[n]$ at the output of the filter.)

(b) For the case where the filter in (a) is the causal Wiener filter for LMMSE estimation of $y[n]$ using all $x[k]$, $k \leq n$, derive the expression in Eq. (12.74) for the "price of causality," that is, the additional mean square error over that of the unrestricted Wiener filter.

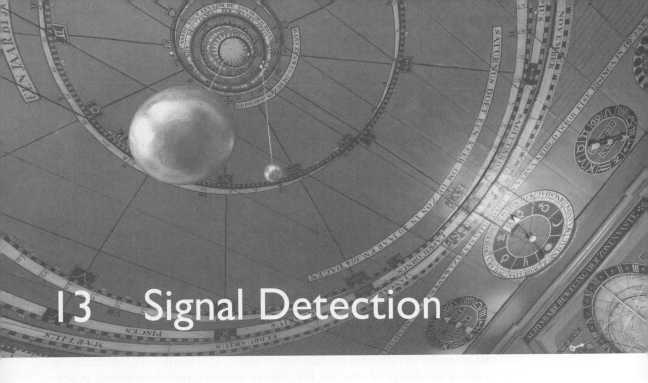

# 13  Signal Detection

In Chapter 9 we treated the problem of deciding optimally among multiple hypotheses $H_i$, given a measurement $r$ of a related random variable $R$. In a digital communication setting, for example, the measurement could be a sample of a processed signal at a receiver, and the hypotheses might then relate to which of several candidate symbols was transmitted by the sender in a particular time slot. It was shown in Chapter 9 that the probability of error in such hypothesis testing problems is minimized by picking whichever hypothesis has the largest posterior probability, that is, the largest probability conditioned on the measurement.

The reasoning that yielded the maximum *a posteriori* probability (or MAP) rule for minimum-error-probability decisions also extends to the case where we have measurements of more than one random variable. The difference with multiple measurements is that the posterior probabilities are conditioned on all the available measurements. This extension was noted in Chapter 9, and is explored in more detail here.

In many applications multiple measurements are obtained as a sequence of samples of one or more received waveforms over an interval of time. The typical problem involves deciding whether a measured waveform is just noise or is a particular signal of interest that is hidden in noise. This signal detection problem arises in many forms in radar, sonar, and communications applications, for example, and is the focus of this chapter.

Though we start the chapter quite generally, the emphasis—just as in the case of LMMSE signal estimation in Chapter 12—will be on problems where the noise is only described by its first and second moments, either because it is Gaussian, or because only first and second moments are known.

In the Gaussian case, we can obtain minimum-error-probability solutions. In the non-Gaussian case, and given only first and second moments, one has to settle for weaker notions of optimality, typically maximizing some measure of signal-to-noise ratio (SNR) involving a decision variable derived from the measurements.

## 13.1 HYPOTHESIS TESTING WITH MULTIPLE MEASUREMENTS

Consider the problem of choosing with minimum probability of error between two possible hypotheses, $H = H_0$ and $H = H_1$, knowing that a set of random variables $R_i$ takes respective values $r_i$, for $i = 0, 1, 2, \ldots, L-1$. More compactly, let $\mathbf{R}$ denote the vector of random variables $R_i$, and $\mathbf{r}$ denote the corresponding vector of values $r_i$ taken by these random variables. Then to minimize the conditional probability of error, $P(\text{error}|\mathbf{R} = \mathbf{r})$, we decide in favor of whichever of the two hypotheses has maximum conditional probability, conditioned on $\mathbf{R} = \mathbf{r}$. The optimum decision rule for binary hypothesis testing thus takes the form

$$P(H_1|\mathbf{R} = \mathbf{r}) \underset{`H_0\text{'}}{\overset{`H_1\text{'}}{\gtrless}} P(H_0|\mathbf{R} = \mathbf{r}) . \tag{13.1}$$

The notation '$H_i$' was introduced in Chapter 9 as a shorthand to indicate that our choice $\widehat{H}$ of hypothesis $H$ is $H_i$, or $\widehat{H} = H_i$. The associated decision regions $\mathcal{D}_i$—defined such that we declare '$H_i$' precisely when $\mathbf{r}$ falls in the region $\mathcal{D}_i$—are now regions in an $L$-dimensional space, rather than segments of the real line. If more than two hypotheses are involved, the minimum-error-probability decision rule chooses whichever hypothesis $H_i$ has the largest posterior probability $P(H_i|\mathbf{R} = \mathbf{r})$ among all the available hypotheses.

As seen in Chapter 9, a rule equivalent to the one above follows from applying Bayes' rule to the posterior probabilities on both sides of Eq. (13.1), then canceling the factor $f_\mathbf{R}(\mathbf{r})$ that appears in both denominators. The result is

$$p(H_1)f_{\mathbf{R}|H}(\mathbf{r}|H_1) \underset{`H_0\text{'}}{\overset{`H_1\text{'}}{\gtrless}} p(H_0)f_{\mathbf{R}|H}(\mathbf{r}|H_0) . \tag{13.2}$$

This latter form of the MAP rule is the most useful for applications, as the quantities that appear in it are typically known or deducible from the problem description. In particular, the two conditional densities constitute the probabilistic models for the measurements under each of the hypotheses. Again, for minimum-error-probability decisions among more than two hypotheses, we would pick whichever hypothesis $H_i$ has the largest value of $p(H_i)f_{\mathbf{R}|H}(\mathbf{r}|H_i)$.

A further rearrangement of Eq. (13.2) yields the likelihood-ratio form of the optimal binary decision rule:

$$\frac{f_{\mathbf{R}|H}(\mathbf{r}|H_1)}{f_{\mathbf{R}|H}(\mathbf{r}|H_0)} \underset{`H_0'}{\overset{`H_1'}{\gtrless}} \eta , \tag{13.3}$$

where the threshold $\eta$ is chosen to be the ratio $P(H_0)/P(H_1)$ for minimum-error-probability decisions, but could be specified differently for criteria such as Neyman–Pearson or minimum-risk detection.

---

**Example 13.1**  **Two Measurements are Better Than One**

Assume that we have to decide between hypotheses $H_0$ and $H_1$ based on a measured random variable $X$, with the conditional densities for $X$, given $H_0$ and $H_1$ respectively, being those shown in Figure 13.1. We can think of $H_1$ as corresponding to the presence of some underlying condition that the hypothesis-testing problem is intended to detect, on the basis of a measurement of $X$; thus $H_0$ corresponds to the absence of this condition. The prior probabilities of the two hypotheses are $P(H_0) = \frac{3}{4}$ and $P(H_1) = \frac{1}{4}$.

The minimum-error-probability decision rule for choosing between the two hypotheses is found by direct application of Eq. (13.2). Since

$$\tfrac{1}{4}f_{X|H}(x|H_1) = \tfrac{1}{8} \quad < \quad \tfrac{3}{4}f_{X|H}(x|H_0) = \tfrac{3}{16} \tag{13.4}$$

for all $x$ in the region $|x| < 2$, we declare '$H_0$' always, no matter what the obtained value of $x$ is. Thus the conditional probability of false alarm is $P_{FA} = P(`H_1'|H_0) = 0$, but the conditional probability of detection is $P_D = P(`H_1'|H_1) = 1 - P(`H_0'|H_1) = 1 - P_M = 0$ as well, where $P_M$ denotes the conditional probability of a miss. As indicated in Eq. (9.29), the associated overall probability of error, averaged over all possible $X$, can be written in terms of $P_{FA}$ and $P_M$ as

$$P_e = P(H_0)P_{FA} + P(H_1)P_M = P(H_1) = \tfrac{1}{4} . \tag{13.5}$$

Now suppose instead that we have measurements $x_1$ and $x_2$, respectively, of two random variables $X_1$ and $X_2$ that are each distributed in the same way as the random variable $X$ above, and are independent of each other under either hypothesis. We arrange these two random variables in the random vector $\mathbf{X}$ for notational convenience. Since the random variables are independent under each hypothesis,

$$f_{\mathbf{X}|H}(x_1, x_2|H_i) = f_{X|H}(x_1|H_i)f_{X|H}(x_2|H_i) \qquad \text{for } i = 0, 1 . \tag{13.6}$$

This equation shows that the conditional joint probability density function (PDF) of the two random variables for $H = H_0$ is nonzero and uniform at the value $\frac{1}{16}$ over the

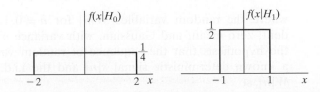

**Figure 13.1**    The conditional densities for $X$ given $H_0$ and $H_1$.

region $|x_1| < 2$, $|x_2| < 2$. For $H = H_1$ the conditional joint PDF is nonzero and uniform at the value $\frac{1}{4}$ over the region $|x_1| < 1$, $|x_2| < 1$. Both conditional densities are 0 outside the specified regions. Again applying Eq. (13.2), we find that in the region $|x_1| < 1$, $|x_2| < 1$,

$$\frac{1}{4} f_{\mathbf{X}|H}(x_1, x_2 | H_1) = \frac{1}{16} \quad > \quad \frac{3}{4} f_{\mathbf{X}|H}(x_1, x_2 | H_0) = \frac{3}{64} , \tag{13.7}$$

so the optimum decision rule declares '$H_1$' in this region. In the remainder of the bigger region $|x_1| < 2$, $|x_2| < 2$,

$$\frac{1}{4} f_{\mathbf{X}|H}(x_1, x_2 | H_1) = 0 \quad < \quad \frac{3}{4} f_{\mathbf{X}|H}(x_1, x_2 | H_0) = \frac{3}{64} , \tag{13.8}$$

so the optimum decision rule declares '$H_0$'. The corresponding $P_{FA} = \frac{4}{16} = \frac{1}{4}$ and $P_D = 1$, so the associated overall probability of error is

$$P(\text{error}) = P(H_0) P_{FA} + P(H_1) P_M = P(H_0) P_{FA} = \frac{3}{16} . \tag{13.9}$$

With two measurements, therefore, the probability of error is reduced, from $\frac{1}{4}$ to $\frac{3}{16}$.

A nonoptimal use of two independent measurements could in principle lead to performance that is no better, and possibly worse, than using a single measurement optimally. For instance, in this particular example it turns out that a decision rule based on using the average of $x_1$ and $x_2$ does best if it always declares '$H_0$', which yields the same performance as the optimal decision rule for a single measurement.

The remainder of this chapter deals with hypothesis tests for which optimal implementations involve more elaborate processing of measurements than in the above example. The formulation and approach, however, are essentially as straightforward as in the preceding example.

## 13.2 DETECTING A KNOWN SIGNAL IN I.I.D. GAUSSIAN NOISE

The prototype detection problem that we study in this section involves a discrete-time (DT) signal $r[n]$ measured over a finite-length time window, say $n = 0, 1, 2, \cdots, L - 1$. We consider these measurements to be the realized values of a set of random variables $R[n]$.

Let $H_0$ denote the hypothesis that the random variables $R[n]$ constitute independent, identically distributed (i.i.d.) zero-mean Gaussian noise, so

$$H_0: \quad R[n] = W[n] , \tag{13.10}$$

where the random variables $W[n]$ for $n = 0, 1, 2, \cdots, L - 1$ are independent, zero mean, and Gaussian, with variance $\sigma^2$. Similarly, let $H_1$ denote the hypothesis that the sequence of random variables $R[n]$ is the sum of a known deterministic signal $s[n]$ and the i.i.d. Gaussian noise sequence $W[n]$, so

$$H_1: \quad R[n] = s[n] + W[n] , \tag{13.11}$$

where the $W[n]$ are again distributed as above. We typically refer to $s[n]$ as the target signal. The signal detection problem here is to decide in favor of $H_0$ or $H_1$ on the basis of the measurements $r[n]$, that is, to decide whether the target signal is absent from or present in the received measurements.

The simplest version of this problem was presented in Example 9.5 (see also Problem 9.9), which dealt with the special case of a single measurement, $L = 1$. The problem there reduced to deciding whether a particular measurement $r$ was the realized value of a Gaussian random variable with mean 0 and specified standard deviation (taken to be 1 in that example), or of another Gaussian random variable with the same standard deviation of 1 but a nonzero mean $s$.

The optimal solution that we develop for $L > 1$ reduces the analysis to calculations as simple as in the case of $L = 1$. The eventual task ends up being to decide whether a particular quantity $g$—derived from the measurements $r[n]$ and the signal $s[n]$—is the realized value of a zero-mean Gaussian random variable with specified standard deviation, or of another Gaussian random variable with the same standard deviation but a nonzero mean. The details follow.

### 13.2.1 The Optimal Solution

For detection with minimum probability of error, the MAP rule in Eq. (13.2) is again used, comparing the values of

$$P(H_i)f(r[0], r[1], \ldots, r[L-1] \mid H_i) \tag{13.12}$$

for $i = 0, 1$, and deciding in favor of whichever hypothesis yields the maximum value of this expression. We have dropped the subscripts **R** and $H$ on the PDF for notational simplicity.

With $W[n]$ being i.i.d. and Gaussian, the conditional densities in Eq. (13.12) are easy to evaluate, and take the form

$$f(r[0], r[1], \ldots, r[L-1] \mid H_0) = \frac{1}{(2\pi\sigma^2)^{(L/2)}} \prod_{n=0}^{L-1} \exp\left\{ -\frac{(r[n])^2}{2\sigma^2} \right\}$$

$$= \frac{1}{(2\pi\sigma^2)^{(L/2)}} \exp\left\{ -\sum_{n=0}^{L-1} \frac{(r[n])^2}{2\sigma^2} \right\} \tag{13.13}$$

and

$$f(r[0], r[1], \ldots, r[L-1] \mid H_1) = \frac{1}{(2\pi\sigma^2)^{(L/2)}} \prod_{n=0}^{L-1} \exp\left\{ -\frac{(r[n] - s[n])^2}{2\sigma^2} \right\}$$

$$= \frac{1}{(2\pi\sigma^2)^{(L/2)}} \exp\left\{ -\sum_{n=0}^{L-1} \frac{(r[n] - s[n])^2}{2\sigma^2} \right\}. \tag{13.14}$$

The test in Eq. (13.2) will still hold if a nonlinear, strictly increasing function is applied equally to both sides. Because of the exponential form of the expressions in Eqs. (13.13) and (13.14) it is particularly convenient to take the natural logarithm of both sides of the test for our particular case. After some subsequent simplification and rearrangement, and also denoting $P(H_i)$ by $p_i$ for notational simplicity, the MAP test takes the equivalent form

$$g = \sum_{n=0}^{L-1} r[n]s[n] \overset{'H_1'}{\underset{'H_0'}{\gtrless}} \sigma^2 \ln(p_0/p_1) + \frac{1}{2}\sum_{n=0}^{L-1} s^2[n] , \tag{13.15}$$

where all computations that involve the measurements $r[n]$ have been gathered on the left side of the inequalities, and the quantity on the right side can be precomputed from the problem specifications. The summation on the right side of Eq. (13.15) is the energy of the deterministic signal $s[n]$, which we denote by $\mathcal{E}$:

$$\mathcal{E} = \sum_{n=0}^{L-1} s^2[n] . \tag{13.16}$$

The test in Eq. (13.15) now becomes

$$g \overset{'H_1'}{\underset{'H_0'}{\gtrless}} \gamma \tag{13.17}$$

where the threshold $\gamma$ is given by

$$\gamma = \sigma^2 \ln(p_0/p_1) + \frac{\mathcal{E}}{2} . \tag{13.18}$$

The case $L = 1$ was treated in Example 9.5 and also Problem 9.9, where $g = r[0]s[0]$ or more simply $g = rs$. Under the assumption that $s > 0$, the threshold test for this case can be rewritten as

$$r \overset{'H_1'}{\underset{'H_0'}{\gtrless}} \gamma/s = \gamma' . \tag{13.19}$$

This is the form in which the threshold test was written in Example 9.5 and Problem 9.9, except that the symbol $\gamma$ was used previously for what we are calling $\gamma'$ now. If $s < 0$, the only change in the above test is that '$H_1$' and '$H_0$' are interchanged.

   If the criterion changes from minimum-error-probability detection to minimum-risk or Neyman–Pearson detection, the optimal decision rule is still of the form in Eq. (13.19), except that the term $p_0/p_1$ in the expression for the threshold in Eq. (13.18) is replaced by some other appropriate constant $\eta$. In the Neyman–Pearson setting, for instance, this $\eta$—or equivalently $\gamma$ itself—would be chosen as low as possible, subject to the specified upper bound on $P_{FA}$.

## 13.2.2 Characterizing Performance

The performance of the optimum solution to the signal detection problem in this chapter can be assessed through its associated probability of error. This requires considering the range of possible values that the quantity $g = \sum r[n]s[n]$ in Eq. (13.15) can take under each hypothesis, and computing the probability that the value lies on the wrong side of the threshold $\gamma$, thus leading to an incorrect decision. We next describe the associated reasoning and computations.

If hypothesis $H_0$ is true, the sequence $r[n]$ on the left-hand side of Eq. (13.15) will consist only of the values realized by i.i.d. Gaussian noise. Correspondingly, $g$ will be the realized value of the random variable

$$G = \sum_{n=0}^{L-1} W[n]s[n] . \tag{13.20}$$

Since $W[n]$ at each instant $n$ is Gaussian and independent of the $W[\cdot]$ at other times, and since a weighted linear combination of independent Gaussian random variables is also Gaussian, the random variable $G$ is Gaussian. From Eq. (13.20), its mean value is zero and its variance is

$$\sigma^2 \sum_{n=0}^{L-1} s^2[n] = \sigma^2 \mathcal{E} , \tag{13.21}$$

so its standard deviation is $\sigma\sqrt{\mathcal{E}}$.

If hypothesis $H_1$ is true, then the signal is in fact present along with the additive zero-mean noise. The sequence $r[n]$ on the left-hand side of Eq. (13.15) in this case will consist of the signal values $s[n]$ perturbed additively by the values realized by Gaussian white noise. Correspondingly, $g$ will be the realized value of the random variable

$$G = \sum_{n=0}^{L-1} \big(s[n] + W[n]\big)s[n] = \mathcal{E} + \sum_{n=0}^{L-1} W[n]s[n] . \tag{13.22}$$

Thus the random variable $G$ is the sum of a known constant $\mathcal{E}$ and a linear combination of independent Gaussian variables, and thus is Gaussian itself. Its mean value is $\mathcal{E}$, but its variance is still as in Eq. (13.21), hence its standard deviation is still $\sigma\sqrt{\mathcal{E}}$.

The optimal test in Eq. (13.15) is therefore described by Figure 13.2. This figure is essentially identical to the one presented in Example 9.5, except that the figure there had the received value $r$ on the horizontal axis, whereas Figure 13.2 has $g$. The hypothesis test for the presence of a known signal in i.i.d. Gaussian noise is equivalent to using the threshold $\gamma$ to decide which of two Gaussian distributions—of respective means

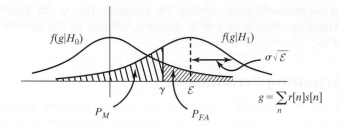

**Figure 13.2**   Evaluating the performance of the threshold test for minimum-error-probability detection of a known signal in i.i.d. Gaussian noise.

0 (for $H_0$) and $\mathcal{E}$ (for $H_1$) but equal standard deviations $\sigma\sqrt{\mathcal{E}}$—has produced the value $g$. The value of $g$ is derived according to the summation on the left of Eq. (13.15); it is a weighted combination of the measured values $r[n]$, with the respective weights being the corresponding signal values $s[n]$.

If the prior probabilities of the hypotheses are equal, the threshold $\gamma = \mathcal{E}/2$, exactly halfway between the two means. If $H_0$ is more likely than $H_1$, so $p_0 > p_1$, then $\gamma$ shifts to a larger value, indicating that $g$ needs to be even larger before the test is willing to declare '$H_1$', which makes sense.

From Figure 13.2 we see that the conditional probability of a false alarm, $P_{FA} = P(\text{'}H_1\text{'}|H_0)$, is the area in the tail to the right of $\gamma$ under a Gaussian PDF of mean 0 and standard deviation $\sigma\sqrt{\mathcal{E}}$. This area cannot be computed in closed form, but can be written in terms of the tabulated tail-probability function $Q(x)$ for a standard (i.e., zero mean, unit variance) Gaussian:

$$Q(x) = \frac{1}{\sqrt{2\pi}} \int_x^\infty e^{-v^2/2}\, dv . \tag{13.23}$$

It is helpful in applications to know the following bounds on $Q(x)$ for $x > 0$:

$$\frac{x}{(1+x^2)} \frac{e^{-x^2/2}}{\sqrt{2\pi}} < Q(x) < \frac{1}{x} \frac{e^{-x^2/2}}{\sqrt{2\pi}} , \quad x > 0 . \tag{13.24}$$

It is also useful to recognize, from the symmetry of the Gaussian PDF, that

$$Q(-x) = 1 - Q(x) . \tag{13.25}$$

A simple change of variables demonstrates that for a Gaussian random variable of mean value $\alpha$ and standard deviation $\beta$, the area under the PDF to the right of some value $\gamma$ can be written in terms of $Q(\cdot)$ as follows:

$$\frac{1}{\beta\sqrt{2\pi}} \int_\gamma^\infty e^{-(w-\alpha)^2/(2\beta^2)}\, dw = Q\!\left(\frac{\gamma - \alpha}{\beta}\right) . \tag{13.26}$$

The argument of $Q(\cdot)$ is the distance of $\gamma$ from the mean $\alpha$, measured in units of the standard deviation $\beta$. With this notation, $P_{FA}$ for our problem is

$$P_{FA} = Q\left(\frac{\gamma}{\sigma\sqrt{\mathcal{E}}}\right)$$

$$= Q\left(\frac{\sigma}{\sqrt{\mathcal{E}}}\ln(p_0/p_1) + \frac{\sqrt{\mathcal{E}}}{2\sigma}\right), \qquad (13.27)$$

where the second equation is the result of substituting in the expression for $\gamma$ from Eq. (13.18).

Similarly, the conditional probability of a miss, $P_M = P(`H_0'|H_1)$, is the area in the tail portion to the left of $\gamma$ under a Gaussian PDF of mean $\mathcal{E}$ and standard deviation $\sigma\sqrt{\mathcal{E}}$. Using Eq. (13.26) and then Eq. (13.25), this area is

$$P_M = 1 - Q\left(\frac{\gamma - \mathcal{E}}{\sigma\sqrt{\mathcal{E}}}\right)$$

$$= 1 - Q\left(\frac{\sigma}{\sqrt{\mathcal{E}}}\ln(p_0/p_1) - \frac{\sqrt{\mathcal{E}}}{2\sigma}\right)$$

$$= Q\left(-\frac{\sigma}{\sqrt{\mathcal{E}}}\ln(p_0/p_1) + \frac{\sqrt{\mathcal{E}}}{2\sigma}\right). \qquad (13.28)$$

Note that both $P_{FA}$ and $P_M$ depend only on the relative prior probabilities of the two hypotheses and on the signal-energy-to-noise-power ratio (SNR) $\mathcal{E}/\sigma^2$—or equivalently the square root of this SNR. In the particular setting of the signal detection problem being considered here, the variation of the known signal $s[n]$ as a function of time does not affect the performance of the optimal decision rule; only the signal energy is relevant. We will later see modifications of the problem in which the signal shape does indeed matter.

The probability of error over all possible outcomes can now be computed as shown in Chapter 9, Eq. (9.29):

$$P_e = p_0 P_{FA} + p_1 P_M . \qquad (13.29)$$

Thus the probability of error depends only on the relative prior probabilities of the two hypotheses and on the SNR.

As a simple illustration, consider the special case where the two hypotheses are equally likely, so $p_0 = p_1$ and $\ln(p_0/p_1) = 0$. Then

$$P_{FA} = P_M = P_e = Q\left(\frac{\sqrt{\mathcal{E}}}{2\sigma}\right) < \frac{1}{\sqrt{2\pi}}\frac{e^{-\mathcal{E}/(8\sigma^2)}}{\sqrt{\mathcal{E}}/(2\sigma)}, \qquad (13.30)$$

where we have invoked the upper bound from Eq. (13.24). Thus $P_{FA}, P_M$, and the overall error probability $P_e$ fall off somewhat faster than exponentially with increasing SNR.

### 13.2.3 Matched Filtering

The decision variable $g$ in Eq. (13.15) is linearly dependent on the measurements $r[n]$. This allows the value of $g$ to be computed by means of a linear and time-invariant (LTI) filter whose input is $r[n]$, and whose output $g[n]$ is sampled at an appropriate time to form the desired value. Specifically, consider an LTI filter with unit sample response $h[\cdot]$. When the input $r[\cdot]$ is applied to this filter, the output $g[n]$ at any arbitrary time $n$ is given by the following convolution:

$$g[n] = \sum_{k=-\infty}^{\infty} r[k]h[n-k] \,. \tag{13.31}$$

If we sample the filter output at time $n = 0$, the result is

$$g[0] = \sum_{k=-\infty}^{\infty} r[k]h[-k] \,. \tag{13.32}$$

Choosing the filter's unit sample response such that

$$h[-k] = s[k] \tag{13.33}$$

for $0 \le k \le L-1$, with $h[\cdot] = 0$ elsewhere, causes the filter output $g[0]$ in Eq. (13.32) to be

$$g[0] = \sum_{k=0}^{L-1} r[k]s[k] = g \,, \tag{13.34}$$

where $g$ is the quantity defined in Eq. (13.15) and required for the threshold test. The unit sample response of this filter, as specified in Eq. (13.33), is thus the time reversal of the target signal. The filter is said to be matched to the target signal, or to be the matched filter for the target signal.

Putting together the above results, an implementation of the optimum detector for a known and finite-duration signal in i.i.d. Gaussian noise can be constructed as in Figure 13.3. The matched filter $h[\cdot]$ specified above in Eq. (13.33) is anticausal. For a purely causal implementation of the optimal detector, all that is needed is for $h[\cdot]$ to be delayed by $L$ steps, and the output correspondingly sampled $L$ steps later. We shall generally work with the anti-causal, i.e., unshifted matched filter, for notational simplicity.

**Scaling the Matched Filter**   If the unit sample response of the matched filter is scaled by some positive number $K > 0$, so it changes from the $h[\cdot]$ defined above to $Kh[\cdot]$, then the output of the filter at time 0 will be $Kg$ rather than simply $g$. Equation (13.17) shows that the only modification required in the detection procedure to account for this is a corresponding change in the threshold, from $\gamma$ to $K\gamma$. The underlying reason is that the distance

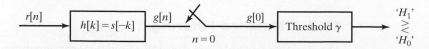

**Figure 13.3**   Optimum detector for a known signal in additive i.i.d. Gaussian noise.

between the mean values of the random variable $G$ under the two hypotheses gets scaled by $K$, as does the common standard deviation. With the threshold scaled by the same factor as the filter, there is no change in detection performance.

**Properties of the Matched Filter**    The matched filter was derived above as a device for generating the quantity $g$ that is compared with a threshold to decide optimally whether or not the target signal is present in the measurements. Further examination of this filter's time-domain and frequency-domain characteristics gives additional insight on why matched filtering is a very reasonable way to process the received measurements, prior to sampling and thresholding in order to decide between the two hypotheses.

Suppose the noise-free signal $s[n]$ is applied to the input of a filter that is matched to it. Denoting the output of the filter in this noise-free case by $\bar{g}[n]$, the output at time $n$ is given by

$$\bar{g}[n] = \sum_{k=-\infty}^{\infty} s[k]h[n-k] = \sum_{k=-\infty}^{\infty} s[k]s[k-n] = \bar{R}_{ss}[n] . \qquad (13.35)$$

(In writing the second summation, we are taking $s[j] = 0$ for $j$ outside the interval $0, 1, 2, \ldots, L-1$.) Thus the output of the matched filter when $s[n]$ is applied to it is the deterministic autocorrelation function $\bar{R}_{ss}[n]$ of $s[\cdot]$. The peak value of $\bar{R}_{ss}[n]$ is the signal energy $\mathcal{E}$, and occurs at zero lag:

$$\bar{R}_{ss}[0] = \mathcal{E} > \bar{R}_{ss}[n] \text{ for } n \neq 0 . \qquad (13.36)$$

The best time at which to sample the output of the matched filter is therefore at $n = 0$, from the viewpoint of obtaining the largest possible contribution from the signal (if the signal is present), relative to the contribution of the noise component of the received measurements. The noise contribution does not depend on when the sampling is done.

We next consider the matched filter in the frequency domain. Taking the transform of Eq. (13.33), the frequency response of the matched filter is

$$H(e^{j\Omega}) = S(e^{-j\Omega}) = |S(e^{j\Omega})| e^{-j\angle S(e^{j\Omega})} , \qquad (13.37)$$

where $S(e^{j\Omega})$ is the DT Fourier transform (DTFT) of the target signal $s[n]$. Hence the magnitude $|H(e^{j\Omega})|$ of the matched filter's frequency response equals the magnitude $|S(e^{j\Omega})|$ of the target signal's spectral distribution. The matched filter therefore accentuates those frequencies where the target signal has strong spectral content, and attenuates those frequencies where the signal has relatively little content. This seems reasonable, given that the i.i.d. noise $W[n]$, when considered for all $n$ rather just for $n$ in $[0, L-1]$, has flat spectral content.

On the other hand, Eq. (13.37) shows that the phase $\angle H(e^{j\Omega})$ of the matched-filter's frequency response is the negative of the phase $\angle S(e^{j\Omega})$ of the target signal's DTFT. In effect, the matched filter adjusts the phase of the frequency components of the target signal, when this signal is present, so that the components all add up constructively for the sampling at the filter output at time 0. The spectrum of the noise component of the received signal is unaffected by the phase characteristic of the filter.

## 13.3 EXTENSIONS OF MATCHED-FILTER DETECTION

The prototype signal detection problem of the preceding section can be developed further in various ways, building on the matched-filter structure. Some of these extensions are considered briefly in the subsections below. Section 13.4 then extends the discussion to discriminating among multiple known signals observed with additive i.i.d. Gaussian noise, where the solution again involves matched filtering.

### 13.3.1  Infinite-Duration, Finite-Energy Signals

The detection problem of Section 13.2 involved a signal $s[n]$ and measurements $r[n]$ that were, for convenience, defined for $n = 0, 1, 2, \ldots, L - 1$. However, any other interval of finite length $L$ could have been chosen in setting up the problem. With $s[n]$ defined to be 0 outside of whatever interval is chosen, the optimum solution would still involve matched filtering using an LTI filter with unit sample response satisfying $h[n] = s[-n]$. The filter's output sampled at time 0 is then compared to the threshold $\gamma$, as before, to arrive at a decision.

We noted briefly in Section 13.2 that an alternate choice for the unit sample response of the matched filter is $h[n] = s[-n + D]$ for some fixed time shift $D$, with the output of this shifted matched filter accordingly sampled at time $D$ rather than 0. For a signal of finite duration, the value of $D$ can always be chosen such that the time-shifted matched filter is causal.

Our derivation of the matched filter solution for minimum-error-probability detection of a signal in additive i.i.d. Gaussian noise applies for an arbitrarily large $L$, and with the interval of interest being arbitrarily positioned in time. What is less obvious, however, is that the result also applies to the case of a signal $s[n]$ of infinite duration, provided the signal has finite energy $\mathcal{E}$, meaning it is an $\ell^2$ signal. We shall simply assume this extension, without attempting a rigorous demonstration. The matched filter's unit sample response $h[\cdot]$ in this case will be of infinite duration, and there will typically be no time shift that makes it causal.

### 13.3.2  Maximizing SNR for Signal Detection in White Noise

In the signal detection problem of Section 13.2, the fact that the noise was i.i.d. and Gaussian allowed the solution of the minimum-error-probability detection problem to be described in detail. It also permitted a simple implementation of this solution: LTI processing of the measurements by a matched filter, followed by sampling, and then by thresholding. In addition, it guaranteed that the sampled output of the matched filter was Gaussian under each hypothesis, and could therefore be characterized entirely by its mean and variance under each hypothesis. With the means and the common variance determined, the computation of the error probability proceeded directly, exposing the dependence of this error probability on the SNR and the prior probabilities of the two hypotheses.

Suppose now that the noise component of the received measurements is only known to come from a (zero-mean) white process of intensity $\sigma^2$, but is otherwise unknown. An i.i.d. Gaussian process of mean 0 and variance $\sigma^2$ is only one of an unlimited number of ways of producing such noise. Without a more detailed description of the noise, it is not possible in general to evaluate and implement the MAP rule for minimum-error-probability detection. However, it is reasonable to consider what can be accomplished using the same appealing detector structure as in the Gaussian case, namely processing by an LTI detection filter with some well-chosen unit sample response $h[\cdot]$, followed by sampling and thresholding, as in Figure 13.4. We assume the target signal $s[n]$ whose presence we are checking for has possibly infinite duration, as allowed in Section 13.3.1.

To pursue this idea, consider the quantity $g[0]$ that appears at the output of the sampler at time 0, namely

$$g[0] = \sum_{-\infty}^{\infty} r[n]h[-n] \,, \tag{13.38}$$

considered as the realized value of some random variable $G$. Though the PDF of $G$ under each hypothesis cannot be determined without more detailed information on the noise PDF, the mean and variance of $G$ under each hypothesis can be computed in terms of known quantities. This computation requires determining what values the sequence $r[n]$ in Eq. (13.38) can take under each of the hypotheses, and noting that $g[0]$ is a linear combination of these values, with weight $h[-n]$ on the value $r[n]$. We conclude that under $H_0$, the mean of $G$ is 0, while under $H_1$ the mean is

$$\mu = \sum_{-\infty}^{\infty} s[n]h[-n] \,. \tag{13.39}$$

To simplify the following discussion, but without loss of generality, let us assume $\mu$ is positive rather than negative.

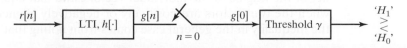

**Figure 13.4**  Structure of detector based on thresholding of sampled output of an LTI filter acting on the received signal.

From Eq. (13.38), the variance of $G$ under either hypothesis is similarly computed to be

$$\sigma_G^2 = \sigma^2 \sum h^2[-n] \,. \tag{13.40}$$

If the separation between the means associated with $H_0$ and $H_1$, namely the distance $\mu - 0 = \mu$, is large compared to the common standard deviation $\sigma_G$, then one might expect to distinguish quite well between a sample $g[0]$ that was obtained under $H_0$ versus one obtained under $H_1$. This is suggested, for instance, by the Chebyshev inequality from elementary probability, which states that the probability of $g[0]$ under $H_0$ being more than $C\sigma_G$ away from its mean value of 0 (in either direction) is less than $1/C^2$ — and this probability could be much less because the Cheybshev inequality is generally quite conservative. Similarly, the probability of $g[0]$ under $H_1$ being more than $C\sigma_G$ away from its mean value $\mu$ is less than $1/C^2$. Thus, it is reasonable to suppose that the larger the ratio $\mu/\sigma_G$, the better our ability to distinguish between $H_0$ and $H_1$ on the basis of a measurement of $g[0]$.

As both $\mu$ and $\sigma_G$ depend on the filter's unit sample response $h[\cdot]$, according to the expressions in Eqs. (13.39) and (13.40) respectively, we can look for the $h[\cdot]$ that maximizes $\mu/\sigma_G$, or equivalently maximizes the square of this, namely the SNR at the output of the filter, defined by

$$\mathrm{SNR}_{\mathrm{out}} = \frac{\mu^2}{\sigma_G^2} = \frac{\left( \sum_{-\infty}^{\infty} s[n] h[-n] \right)^2}{\sigma^2 \sum h^2[-n]} \,. \tag{13.41}$$

The reason for calling this an SNR is that $\mu^2$ reflects the signal energy while $\sigma_G^2$ reflects the noise power or intensity. Note that maximizing the output SNR will not in general — except in the Gaussian case — minimize the probability of error. Maximization of the SNR in Eq. (13.41) is easily carried out by invoking the Cauchy–Schwarz inequality, which we state and prove next.

**Cauchy–Schwarz Inequality**   For real $\ell^2$ functions $x[n]$ and $y[n]$ the Cauchy–Schwarz inequality is

$$\left( \sum_{-\infty}^{\infty} x[n] y[n] \right)^2 \le \left( \sum_{-\infty}^{\infty} x^2[n] \right) \left( \sum_{-\infty}^{\infty} y^2[n] \right) , \tag{13.42}$$

with equality if and only if $y[n] = Kx[n]$ for all $n$ and some constant $K$. In the case of functions that are nonzero only over some finite interval, this inequality reduces to the statement that the inner product or dot product of two vectors in real Euclidean space is bounded in magnitude by the product of the lengths of the individual vectors, and that equality is attained only when the two vectors are positively or negatively aligned. A direct proof of the inequality in the general case comes from noting that

$$\sum_{-\infty}^{\infty} (\alpha x[n] - y[n])^2 \ge 0 \tag{13.43}$$

for all real $\alpha$. Expanding this out, we get

$$\underbrace{\left(\sum_{-\infty}^{\infty} x^2[n]\right)}_{a} \alpha^2 - \underbrace{\left(2\sum_{-\infty}^{\infty} x[n]y[n]\right)}_{b} \alpha + \underbrace{\left(\sum_{-\infty}^{\infty} y^2[n]\right)}_{c} \geq 0 . \qquad (13.44)$$

This quadratic function of $\alpha$, namely $a\alpha^2 - b\alpha + c$, is nonnegative for large $|\alpha|$ because $a > 0$. It remains nonnegative for all $\alpha$ precisely under the condition $b^2 \leq 4ac$ that guarantees the quadratic has at most one real root. Using the definitions in Eq. (13.44) to substitute for $a$, $b$, and $c$ in the inequality $b^2 \leq 4ac$ produces the Cauchy–Schwarz inequality in Eq. (13.42). Also, $b^2 = 4ac$ precisely when the quadratic attains the value 0 for some $\alpha = K$, and Eq. (13.43) shows that this happens precisely when $y[n] = Kx[n]$.

Returning now to the problem of choosing the unit sample response $h[\cdot]$ in Eq. (13.41) so as to maximize the output SNR, we apply the Cauchy–Schwarz inequality:

$$\mathrm{SNR}_{\mathrm{out}} = \frac{\left(\sum_{-\infty}^{\infty} s[n]h[-n]\right)^2}{\sigma^2 \sum h^2[-n]} \leq \frac{\sum_{-\infty}^{\infty} s^2[n]}{\sigma^2} = \frac{\mathcal{E}}{\sigma^2} = \mathrm{SNR}_{\mathrm{in}} , \qquad (13.45)$$

where $\mathcal{E}$ denotes the energy of the target signal, as before. Equality is attained precisely when $h[-n] = Ks[n]$ for some $K$, which we can take without loss of generality to be 1. Thus the optimum filter is again the matched filter. With matched filtering, the sample at the output of the filter attains its highest possible SNR, which equals the SNR of the input signal, namely $\mathcal{E}/\sigma^2$.

Without further information about the PDF of the noise, there is little to guide the choice of a detection threshold $\gamma$ for the sampled output of the matched filter. However, if the matched filter has been effective in separating the distribution of $g[0]$ under $H_0$ from its distribution under $H_1$, then a threshold at some location between the two conditional means may be appropriate, for instance at $\mu/2$.

The matched filter can thus be seen as simply a device for maximizing the SNR of the sampled output of an LTI detection filter, even in the case where the noise is only known to be white, but possibly non-Gaussian. We will assume this interpretation for the rest of this chapter whenever matched filtering is discussed without an explicit assumption that the noise is Gaussian. In the Gaussian case, the matched filter (with the correct threshold on its sampled output) also minimizes the probability of error.

### 13.3.3 Detection in Colored Noise

We now consider the case where the noise in our detection problem, rather than being a white process $w[n]$, is a zero-mean, wide-sense stationary (WSS) process $v[n]$ with fluctuation spectral density (FSD) given by $D_{vv}(e^{j\Omega})$. If we think of this colored noise as obtained by passing a white noise process $w[n]$ of intensity $\sigma^2$ through a stable modeling or shaping filter with frequency response $M(e^{j\Omega})$, then

$$D_{vv}(e^{j\Omega}) = \sigma^2 M(e^{j\Omega})M(e^{-j\Omega}) \ . \tag{13.46}$$

Thus $\sigma M(e^{j\Omega})$ is a spectral factor of $D_{vv}(e^{j\Omega})$. We assume that $M(e^{j\Omega})$ and therefore $D_{vv}(e^{j\Omega})$ are nonzero at all frequencies. If this was not the case, there would be noise-free frequency components in the received measurements, which leads to degeneracies in the solution. Note also that we could have assumed $w[n]$ had unit intensity, as the shaping filter can incorporate any scaling that is required—but the assumption of intensity $\sigma^2$ will allow more transparent comparison with expressions written down earlier for the case of white noise with intensity $\sigma^2$.

If the underlying white-noise process $w[n]$ is actually zero-mean i.i.d. and Gaussian, then the process $v[n]$ obtained by filtering $w[n]$ through the shaping filter is colored Gaussian noise. It has the property, for example, that the values at any two distinct times are bivariate Gaussian; more generally, the values at an arbitrary set of times are multivariate Gaussian. For this case, the matched-filter solution we develop below will result in the minimum-error-probability decision, after appropriate sampling and thresholding of the matched-filter output. If it is only known that the process $w[n]$ is white, then the matched-filter solution below will only guarantee maximization of the SNR of an output sample, as in Section 13.3.2. The validity of these claims follows from the way the colored-noise problem is converted below—in a reversible way—to the white-noise problems treated earlier.

We again wish to decide between the hypothesis $H_0$ that the received measurements $r[n]$ constitute only the colored noise $v[n]$, and the hypothesis $H_1$ that the measurements contain a known signal $s[n]$ corrupted additively by this noise. The setting is illustrated by the diagram in Figure 13.5(a). The diagram also shows the filtering of the received signal $r[n]$ by an LTI filter of frequency response $H(e^{j\Omega})$, followed by sampling, which yields the quantity $g[0]$ that is compared against a threshold in the optimum decision rule.

The optimum solution to this problem is obtained by transforming it to the problem solved in Section 13.2. The first step is to transform the portion of the diagram in Figure 13.5(a) that generates $r[n]$ into the equivalent diagram in Figure 13.5(b). This diagram makes apparent the presence of an underlying signal $q[n]$ that is either white noise of intensity $\sigma^2$, or is a signal $p[n]$ additively corrupted by this noise. This signal $p[n]$ is the result of passing $s[n]$ through a filter with frequency response $1/M(e^{j\Omega})$.

The optimum solution to this transformed problem has already been determined in Section 13.2.3: we process $q[n]$ through a filter matched to the target signal $p[n]$, then sample at time 0 and compare with an appropriate threshold. In the system shown in Figure 13.5, the filtering of $q[n]$ prior to the sampling is performed by the series combination of $M(e^{j\Omega})$ and $H(e^{j\Omega})$. It follows that

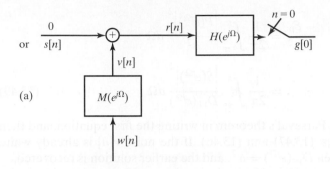

(a)

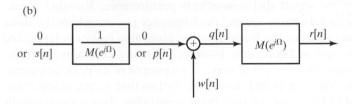

(b)

**Figure 13.5** (a) Detecting a signal $s[n]$ in additive colored noise $v[n]$, using measurements $r[n]$. (b) The equivalent problem of detecting a signal $p[n]$ in additive white noise $w[n]$, where $p[n]$ is the result of filtering $s[n]$ through a filter with frequency response $1/M(e^{j\Omega})$.

$$H(e^{j\Omega})M(e^{j\Omega}) = P(e^{-j\Omega}) = \frac{S(e^{-j\Omega})}{M(e^{-j\Omega})}, \tag{13.47}$$

where $S(e^{j\Omega})$ denotes the transform of the original signal $s[n]$. Hence the frequency response of the desired optimum filter is given by

$$H(e^{j\Omega}) = \frac{S(e^{-j\Omega})}{M(e^{j\Omega})M(e^{-j\Omega})} = \frac{S(e^{-j\Omega})}{D_{vv}(e^{j\Omega})/\sigma^2}. \tag{13.48}$$

This is the matched-filter frequency response for colored noise; the earlier result for white noise is recovered on setting $D_{vv}(e^{j\Omega}) = \sigma^2$ in the last expression in Eq. (13.48).

The generalized matched-filter frequency response in Eq. (13.48) has its magnitude determined by the signal spectrum magnitude $|S(e^{j\Omega})|$ in the numerator and also the FSD of the noise (measured relative to the FSD of white noise of intensity $\sigma^2$) in the denominator. Just as in the white-noise case, therefore, the detection filter has its highest magnitude response at those frequencies where the input SNR is large, and lowest response at those frequencies where the input SNR is small. The phase characteristic of the generalized matched filter is still the negative of the target signal's phase characteristic. The noise plays no role because the phase of the FSD $D_{vv}(e^{j\Omega})$ is 0 at all frequencies.

Invoking our earlier results shows that the performance of the optimum solution developed here is determined by the ratio of the energy of the prefiltered signal $p[n]$ to the variance $\sigma^2$ of the white noise $w[n]$. Denoting the energy of $p[n]$ as $\mathcal{E}_p$, the performance of the system—probability of error in the case $w[n]$ is Gaussian white noise, or output sample SNR otherwise—is thus determined by the ratio

$$\frac{\mathcal{E}_p}{\sigma^2} = \frac{1}{2\pi} \int_{-\pi}^{\pi} \frac{\left|P(e^{j\Omega})\right|^2}{\sigma^2}\, d\Omega$$

$$= \frac{1}{2\pi} \int_{-\pi}^{\pi} \frac{\left|S(e^{j\Omega})\right|^2}{D_{vv}(e^{j\Omega})}\, d\Omega\,, \tag{13.49}$$

where we have used Parseval's theorem in writing the first equation, and then substituted from Eqs. (13.47) and (13.46). If the noise $v[n]$ is already white with intensity $\sigma^2$, then $D_{vv}(e^{j\Omega}) = \sigma^2$, and the earlier solution is recovered.

The expression in Eq. (13.49) shows that, unlike in the white-noise case, the shape of the signal $s[n]$ now affects performance. Roughly speaking, concentrating signal energy around the frequency regions where the noise intensity is minimum makes the integral in the preceding equation large, and thereby results in improved detection. However, in many situations the transmitted signal is constrained in other ways, for example in its peak amplitude and time duration. The task then is to choose $s[n]$ so that its transform maximizes the integral in Eq. (13.49) under these constraints. There are generally no closed-form solutions to this optimization problem.

### 13.3.4 Continuous-Time Matched Filters

We have so far focused on DT rather than continuous-time (CT) signal detection, largely because CT white noise is an idealization that is much harder to visualize than DT white noise or even the CT unit impulse. For example, the expected power in CT white noise at each instant of time is infinite. However, the colored-noise case developed for DT in Section 13.3.3 can be directly carried over to CT, with straightforward changes.

Consider, for example, a zero-mean CT WSS noise process $v(t)$ with FSD given by $D_{vv}(j\omega)$. A known finite-energy signal $s(t)$ is either added to this (hypothesis $H_1$) or not (hypothesis $H_0$), and the resulting measured signal $r(t)$ has to be processed in order to decide which hypothesis applies. Suppose we commit to processing $r(t)$ through some detection filter with impulse response $h(\cdot)$ and frequency response $H(j\omega)$, then sampling the output $g(t)$ of this at time 0, and finally comparing the sample value with a threshold to arrive at a decision, as in Figure 13.6.

In this setting, the quantity

$$g(0) = \int_{-\infty}^{\infty} r(\tau)h(-\tau)\, d\tau \tag{13.50}$$

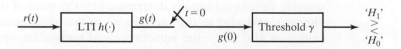

**Figure 13.6**   CT signal detection.

is the realized value of a random variable $G$ that has expected value 0 under $H_0$, and expected value

$$\int_{-\infty}^{\infty} s(\tau)h(-\tau)\,d\tau = \frac{1}{2\pi}\int_{-\infty}^{\infty} S(j\omega)H(j\omega)\,d\omega \qquad (13.51)$$

under $H_1$. The variance of $G$ under either hypothesis is

$$C_{gg}(0) = \frac{1}{2\pi}\int_{-\infty}^{\infty} |H(j\omega)|^2 D_{vv}(j\omega)\,d\omega . \qquad (13.52)$$

The output SNR can now be defined in a natural way as the squared distance between the means under the two hypotheses, divided by the common variance:

$$\frac{1}{2\pi}\frac{\left(\int_{-\infty}^{\infty} S(j\omega)H(j\omega)\,d\omega\right)^2}{\int_{-\infty}^{\infty} |H(j\omega)|^2 D_{vv}(j\omega)\,d\omega} . \qquad (13.53)$$

The optimization task is to pick $H(j\omega)$ to maximize this output SNR.

The above problem formulation is quite similar to what was treated in Section 13.3.2, although now in the frequency domain, and involving integrals rather than summations. The solution can be carried out using the appropriate form of the Cauchy–Schwarz inequality. However, the DT version of this problem was already solved in Section 13.3.3 using a different (noise-whitening) argument, so we shall simply state the CT solution by analogy with Eq. (13.48):

$$H(j\omega) = \frac{S(-j\omega)}{D_{vv}(j\omega)/\sigma^2} , \qquad (13.54)$$

where $\sigma^2$ is interpreted here as an arbitrary positive parameter whose specific value needs to be taken account of in setting the threshold $\gamma$. Substituting this in Eq. (13.53) shows that the optimum output sample SNR is given by

$$\frac{1}{2\pi}\int_{-\infty}^{\infty} \frac{\left|S(j\omega)\right|^2}{D_{vv}(j\omega)}\,d\omega , \qquad (13.55)$$

which is the CT analog of Eq. (13.49).

### 13.3.5 Matched Filtering and Nyquist Pulse Design

In Chapter 3 we assumed a noise-free setting in describing the design of a Nyquist pulse for zero intersymbol interference (ISI) in pulse amplitude modulation (PAM). Section 3.2 concluded with the statement that in the presence of noise it is desirable to perform part of the pulse shaping at the receiver. We are now in a position to better understand this statement.

Consider an on-off signaling scheme in which, for each bit, a pulse is either transmitted or not. At the receiver we wish to detect the presence or absence of a transmitted pulse. If a pulse with transform $P(j\omega)$ passes through a baseband channel with frequency response $H(j\omega)$, the pulse that the receiver is aiming to detect will have transform $P(j\omega)H(j\omega)$. Suppose the channel introduces additive white noise. The optimum receiver filter, matched to this pulse, should then have frequency response $H(-j\omega)P(-j\omega)$, in order to produce an output sample whose SNR is maximized when a pulse is present.

Let us assume that a matched filter is implemented at the receiver. The design of the transmitted pulse transform $P(j\omega)$ for zero ISI at the receiver now requires that the pulse transform after filtering by the channel and processing at the receiver, namely

$$P(j\omega)H(j\omega)H(-j\omega)P(-j\omega) = |P(j\omega)|^2|H(j\omega)|^2 \,, \tag{13.56}$$

satisfies the Nyquist zero-ISI condition. If $|H(j\omega)|$ is constant in its passband, then the quantity $|P(j\omega)|^2$ has to satisfy the Nyquist zero-ISI condition, that is, its periodic replications at integer multiples of the (angular) signaling frequency $2\pi/T$ should add up to a constant. Once such a $P(j\omega)$ is designed, it determines both the transmitted pulse and the receiver filter. The task of pulse shaping is thus equally shared between the transmitter and receiver.

### 13.3.6 Unknown Arrival Time and Pulse Compression

In an application such as radar or sonar, a known signal pulse—electromagnetic or acoustic respectively—is propagated out from the transmitter. If a reflecting object is present in the propagation path, an attenuated and noise-corrupted version of the transmitted signal is returned to the receiver. If no object is present, the device simply measures noise during this interval. The delay from initial transmission to the arrival of any reflected pulse determines the round-trip distance to the object, on multiplication by the speed of propagation of the signal. The propagating signals in these applications are CT, but the analysis below is phrased in terms of the DT signals that correspond to samples of the CT waveforms at some regular sampling rate.

The signal processing task at the radar or sonar device is to determine whether the received measurements during some window of time constitute only noise or the reflected signal plus noise. Assuming the noise is i.i.d. Gaussian, and that the received signal shape differs from the transmitted shape only by amplitude scaling through an attenuation factor $\alpha$, we are faced with essentially the signal detection problem studied earlier. The optimal detector will use an amplitude-scaled version of a filter matched to the transmitted signal, and an appropriately chosen threshold.

In practice, an upper limit on the transmitted signal amplitude $A$ is determined by the peak transmitter power, so the only way to increase the signal energy $\mathcal{E}$ for better detection is by increasing $L$, that is, by sending a longer pulse. A longer pulse, however, implies that the next interrogating pulse will have to be correspondingly delayed. More importantly, if a second reflecting object is a small further distance away in the path of the original pulse, its reflection may overlap with that of the first object, and the ability to resolve the two objects—to recognize them as two objects rather than one—will be compromised. We shall see below that clever signal design can greatly improve the resolution.

There are at least two key differences from the signal detection problem considered in Section 13.2. First, the location of the reflected signal in time—if there is a reflected signal—is unknown because the position of the reflecting object is unknown. As a result, the time at which the output of the matched-filter should be sampled is unknown. Second, the attenuation factor $\alpha$ is unknown, so the required scaling of the detection filter (or equivalently, the scaling of the threshold) is unknown.

A strategy for dealing with these two issues comes from returning to the discussion of matched filter properties in Section 13.2.3. A calculation similar to what generated Eqs. (13.35) and (13.36) shows that if the matched filter input is the delayed noise-free signal $s[n - D]$, where $D$ is the delay, then the filter output is given by

$$\overline{g}[k] = \sum_{n=-\infty}^{\infty} s[n - D]h[k - n] = \sum_{n=-\infty}^{\infty} s[n - D]s[n - k] = \overline{R}_{ss}[k - D] .$$

$$(13.57)$$

The output is therefore the delayed deterministic autocorrelation of the received signal, and its peak value $\mathcal{E}$ occurs when its argument is 0, hence at $k = D$. As a consequence, in the case where the noise-free signal is received, one can deduce the value of the delay $D$ simply by noting at what instant the matched filter output attains its maximum.

If the reflected signal is received in the presence of additive noise, the matched filter output at each time will be perturbed by a noise component. If the noise is not excessive, then the matched filter output is not greatly perturbed from the noise-free case. The maximum output value can then be sampled and passed on for comparison to the selected threshold. In the absence of a reflected signal, the sample value will reflect only the noise, and this will typically fall below the detection threshold.

If the signal is received with a higher noise intensity, the maximum output of the matched filter may occur at a different value of $k$ than in the noise-free case, because the maximum of the component due to the reflection has been masked by the component due to the noise. This will lead to an incorrect choice of sampling time and sample value. It may also happen in this high-noise case that the noise alone, with no reflected signal present, causes a prominent peak in the matched detector output, which then gets misinterpreted as indicating the presence of the reflected signal.

The preceding considerations support performing what is known as pulse compression: shaping the length-$L$ signal $s[n]$ so that $\overline{R}_{ss}[k] \ll \mathcal{E}$ for $k \neq 0$, or for $k$ beyond some small region ($\ll L$) around 0, so that the matched-filter output is concentrated at—or narrowly around—the peak value. The location of the maximum in $\overline{R}_{ss}[k - D]$ for this case will be well preserved even when the signal is corrupted by noise. Furthermore, having a maximum value that is much larger than the noise-perturbed values adjacent to it may suffice for one to decide that an underlying signal is present, despite uncertainty about the attenuation factor and therefore about the appropriate threshold.

An example of a signal that performs well in this respect and is widely used is the sequence

$$+A, +A, +A, +A, +A, -A, -A, +A, +A, -A, +A, -A, +A \,, \qquad (13.58)$$

(known as the Barker-13 sequence) extended on both sides by zeros. The deterministic autocorrelation function of this signal has the value $13A^2$ at zero lag, while the value at all other lags has magnitude $A$ or 0. Another popular way to achieve pulse compression is by using a so-called chirp signal: a quasi-sinusoidal signal whose frequency is swept linearly—or in some other monotonic fashion—over time. Because different subintervals of the chirp waveform are composed of essentially sinusoidal segments of different frequencies, the deterministic autocorrelation is small at lags greater than the length of such a segment.

## 13.4 SIGNAL DISCRIMINATION IN I.I.D. GAUSSIAN NOISE

The two-hypothesis signal detection problems considered in the preceding sections can be easily extended to the case of several hypotheses. Suppose, as earlier, that the values of the DT signal $r[n]$ measured over $n = 0, 1, 2, \cdots, L - 1$ are the realized values of a set of random variables $R[n]$. Now, however, let $H_i$, $i = 0, 1, \cdots, M - 1$, denote the hypothesis that the variables $R[n]$ are the result of additive i.i.d. Gaussian noise corrupting the $i$th signal, $s_i[n]$, out of a set of $M$ known deterministic signals, so

$$H_i: \quad R[n] = s_i[n] + W[n] \,. \qquad (13.59)$$

Here again, the quantities $W[n]$ under each hypothesis denote independent, zero-mean, Gaussian random variables with variance $\sigma^2$. We assume the prior probability of hypothesis $H_i$ is $P(H_i)$. The task is to decide, on the basis of the measured $r[n]$ and with minimum probability of error, which hypothesis holds. This may be regarded as a problem of discriminating among $M$ possible choices for the underlying signal, given noise-corrupted measurements of one of these signals.

The above situation arises, for example, in digital communication. Assume one of $M$ symbols is selected for transmission during an allocated time slot. Each symbol is mapped to a distinct signal that is suited to the transmission characteristics of the particular communication channel. Let $s_i[n]$ for $n = 0, 1, 2, \cdots, L - 1$ denote the sequence of measurements that would be received in the allocated time slot in the noise-free case, if the signal associated with the symbol $i$ is transmitted. Assume the transmission channel corrupts the signal at each instant with additive i.i.d. Gaussian noise. The task at the receiver is then to decide which of the $M$ signals was actually transmitted, given the received measurements $r[n]$ for $n = 0, 1, 2, \cdots, L - 1$ in a particular time slot. This is exactly the signal discrimination task formulated here as a hypothesis testing problem.

For minimum error probability, we again use the MAP rule in Eq. (13.2), in this case comparing the values of

$$P(H_i)f(r[0], r[1], \ldots, r[L-1] \,|\, H_i) \tag{13.60}$$

for $i = 0, 1, 2, \cdots, M-1$, and deciding in favor of whichever hypothesis yields the maximum value of this expression. Equivalently, given that the natural logarithm is a monotonically increasing function of its argument, the logs of the above expressions can be compared:

$$\ln\{P(H_i)\} + \ln\{f(r[0], r[1], \ldots, r[L-1] \,|\, H_i)\}, \qquad i = 0, 1, 2, \cdots M-1. \tag{13.61}$$

Making the appropriate substitutions, and discarding terms that are common to all of the $M$ expressions in Eq. (13.61), it is straightforward to conclude that the optimal test requires comparison of the quantities

$$\left( \sum_{n=0}^{L-1} r[n]s_i[n] \right) + \sigma^2 \ln\{P(H_i)\} - \frac{\mathcal{E}_i}{2}, \qquad i = 0, 1, 2, \cdots M-1, \tag{13.62}$$

where $\mathcal{E}_i$ denotes the energy of the $i$th signal:

$$\mathcal{E}_i = \sum_{n=0}^{L-1} s_i^2[n]. \tag{13.63}$$

The largest of the expressions in Eq. (13.62), for $i = 0, 1, \cdots, M-1$, determines which hypothesis is selected by the decision rule.

If the signals have equal energies and equal prior probabilities, then the above comparison reduces to deciding in favor of the signal with the highest value of

$$g_i = \sum_{n=0}^{L-1} r[n]s_i[n]. \tag{13.64}$$

The computations in Eq. (13.64) can again be carried out using matched filters whose outputs are sampled at the appropriate time.

As done earlier with the case of signal detection, we can generalize the signal discrimination problem here in several ways. In particular, the signals can be allowed to exist for all time rather than just the interval $[0, L-1]$, as long as they have finite energy. We shall assume this more general setting in what follows, and omit writing explicit limits on the summations in our expressions.

## Example 13.2  Binary Signal Discrimination in I.I.D. Gaussian Noise

We consider the case in which there are only two candidate signals that need to be distinguished from each other in the presence of additive i.i.d. Gaussian noise. This is a special case of the preceding results, with $M = 2$. The optimal test becomes

$$\left(\sum_n r[n]s_1[n]\right) + \sigma^2 \ln p_1 - \frac{\mathcal{E}_1}{2} \underset{\substack{< \\ `H_0`}}{\overset{\substack{`H_1` \\ >}}{\gtrless}} \left(\sum_n r[n]s_0[n]\right) + \sigma^2 \ln p_0 - \frac{\mathcal{E}_0}{2}, \qquad (13.65)$$

where $p_0 = P(H_0)$ and $p_1 = P(H_1)$, as before. This can be rewritten as

$$g = \sum_n r[n]\left(s_1[n] - s_0[n]\right) \underset{\substack{< \\ `H_0`}}{\overset{\substack{`H_1` \\ >}}{\gtrless}} \sigma^2 \ln(p_0/p_1) + \frac{\mathcal{E}_1 - \mathcal{E}_0}{2} = \gamma . \qquad (13.66)$$

The quantity on the left is what would be computed for detection of the difference signal $s_1[n] - s_0[n]$, and can be obtained using a filter that is matched to this difference signal, that is, with unit sample response $h[n] = s_1[-n] - s_0[-n]$.

The performance analysis of the above decision rule can be carried out exactly as described in Section 13.2.2, by examining the distribution of the random variable

$$G = \sum_n R[n]\left(s_1[n] - s_0[n]\right) \qquad (13.67)$$

under the two hypotheses. Before doing so, we introduce the following notation for a quantity that arises in the analysis:

$$\mathcal{X} = \sum_n s_0[n]s_1[n] . \qquad (13.68)$$

From the Cauchy–Schwarz inequality in Eq. (13.42) we can write

$$-\sqrt{\mathcal{E}_0 \mathcal{E}_1} \le \mathcal{X} \le \sqrt{\mathcal{E}_0 \mathcal{E}_1} , \qquad (13.69)$$

where we attain the upper bound if $s_0[n] = Ks_1[n]$ for some positive $K$ (which must equal $\sqrt{\mathcal{E}_0/\mathcal{E}_1}$), and attain the lower bound if $s_0[n] = -Ks_1[n]$ for this same positive $K$.

Under $H_0$ the random variable $G$ in Eq. (13.67) is Gaussian of mean value $-\mathcal{E}_0 + \mathcal{X}$ and variance $\sigma^2 \mathcal{E}_d$, where $\mathcal{E}_d$ is the energy of the difference signal:

$$\mathcal{E}_d = \sum_n \left(s_1[n] - s_0[n]\right)^2 = \mathcal{E}_1 + \mathcal{E}_0 - 2\mathcal{X} . \qquad (13.70)$$

Under $H_1$, the random variable $G$ is again Gaussian, with mean $\mathcal{E}_1 - \mathcal{X}$, and still with variance $\sigma^2 \mathcal{E}_d$. This leads us back to the now familiar task of assessing the performance of a hypothesis test for separating two Gaussians of equal variance but different means, using a specified threshold $\gamma$, given in this case by the right side of Eq. (13.66).

Consider two special cases. If $s_0[n]$ is the zero signal and $s_1[n] = s[n]$ for all $n$, we recover the problem considered in Section 13.2, namely detecting a known signal in i.i.d. Gaussian noise. In the context of digital communication, this would correspond to on-off signaling, where the presence of the signal $s[n]$ indicates binary digit 1, and its absence indicates binary digit 0. The expressions here reduce appropriately to those obtained earlier. If the two hypotheses are equally likely, so $p_0 = p_1$, then the probability of error is again given by the expression in Eq. (13.30), namely

$$P_{\text{on-off}} = Q\left(\frac{\sqrt{\mathcal{E}}}{2\sigma}\right) . \qquad (13.71)$$

For a second special case, suppose the two signals have equal energy, so $\mathcal{E}_1 = \mathcal{E}_2 = \mathcal{E}$, fixed at some specified value. Also assume the two hypotheses have equal prior probability, so $p_0 = p_1$. Under these conditions, the threshold $\gamma$ is 0. Now $P_{FA} = P_M = P_e$, so the probability of error is

$$P_e = Q\left(\frac{\mathcal{E} - \mathcal{X}}{\sigma\sqrt{\mathcal{E}_d}}\right) = Q\left(\frac{\sqrt{\mathcal{E}} - \mathcal{X}}{\sigma\sqrt{2}}\right), \tag{13.72}$$

where the second equality follows from using Eq. (13.70).

Suppose we are free to choose $s_0[n]$ and $s_1[n]$, subject only to the constraint that their respective energies equal some given common value $\mathcal{E}$. The smallest probability of error then results from choosing these signals to give the largest possible argument in the $Q(\cdot)$ function above. Noting the bounds in Eq. (13.69) and the signal choices that cause the bounds to be attained, the probability of error is minimized by choosing $\mathcal{X} = -\sqrt{\mathcal{E}_0 \mathcal{E}_1} = -\mathcal{E}$, obtained by setting $s_1[n] = -s_0[n]$ for all $n$. This yields what is known as bipolar or antipodal signaling. Note that antipodal signaling involves only the same peak power at the transmitter as in the case of on-off signaling, though the average power and the average energy per bit are doubled (at least when $p_0 = p_1 = 0.5$).

With antipodal signaling, therefore, the probability of error is

$$P_{\text{antip}} = Q\left(\frac{\sqrt{\mathcal{E}}}{\sigma}\right). \tag{13.73}$$

Compared with the on-off signaling case, where $s_0[\cdot] = 0$ and where the probability of error $P_{\text{on-off}}$ is in Eq. (13.71), the effect of antipodal signaling is to double the distance between the means of the two Gaussians that have to be separated, for a fixed noise intensity. The argument of the $Q(\cdot)$ function is thereby doubled, and the effective SNR is increased by a factor of four. Thus antipodal signaling yields substantially better error performance than on-off signaling.

The following example demonstrates the above results more concretely in the setting of PAM communication, which was discussed in Chapter 3. Though the problem is stated in CT, it is DT in the numerical simulation.

## Example 13.3  On-Off and Antipodal PAM, with Matched-Filter Detector

We illustrate some of the ideas in this chapter through numerical simulation of a PAM strategy for CT communication of a binary DT signal $a[n]$. The basic signal or pulse $p(t)$ being communicated in a given time slot of duration $T = 1$ time unit is rectangular in this example: $p(t) = A$ for $|t| < T/4$, which is the middle half of the time slot, and $p(t) = 0$ elsewhere, for some $A > 0$. The full PAM signal is constructed from a concatenation of such pulses, with the $n$th pulse $p(t - nT)$ being amplitude-scaled by the value $a[n]$ that is to be transmitted in the $n$th time slot. The PAM signal is therefore

$$x(t) = \sum_n a[n]p(t - nT). \tag{13.74}$$

For on-off signaling, $a[n]$ is either 0 or 1, whereas for antipodal signaling $a[n]$ is $+1$ or $-1$; we assume the two levels are equally likely in each case. Antipodal signaling entails the same peak power as on-off signaling, but is expected to provide better noise immunity, though at the cost of higher average power. This example ignores distortion on the communication channel and focuses on dealing with the effects of noise. In each time slot, the receiver has to decide which of the two possible signal values was sent, based on the received noisy waveform.

The top two panels in Figure 13.7 show schematically how the noise-free PAM signal can be generated by sending an appropriate impulse train into a filter with impulse response $p(t)$. The PAM waveform here corresponds to on-off signaling of the sequence of 13 bits shown at the top of the figure, with the choice $A = 1$. The time axis indicates the sample number, with 100 samples corresponding to one unit of time in CT.

We assume additive wideband Gaussian noise in CT, producing i.i.d. Gaussian noise of intensity $\sigma^2$ at the DT sampling times in the simulation. The third panel in Figure 13.7 has the same underlying PAM signal as in the second panel, but with the added noise shown for $\sigma = 3.5$. The noise is seen to swamp out the signal, giving little clue as to what the PAM signal is. If we did not do any filtering of this noise, but instead just took a single sample every $T$ units of time and compared it with a threshold of $A/2 = 0.5$, we would not expect to do much better than if we decided randomly. Our analysis in this chapter shows that the probability of error for this situation is $Q(A/(2\sigma)) = Q(1/7) = 0.443$.

Recognizing that the decisions in distinct bit slots are independent in our setting, the decisions constitute a Bernoulli process, and the number of errors is governed by a binomial distribution. The expected number of errors in 13 bits is therefore given by $0.443 \times 13 = 5.76$ bits, and the standard deviation in this number is computed as $\sqrt{13 \times 0.433 \times (1 - 0.433)} = 1.79$ bits.

The result of using a matched filter on the received signal is shown in the bottom panel of Figure 13.7. In CT, the matched-filter impulse response is $h(t) = p(-t)$ in each time slot. Since $p(t)$ is a rectangular pulse that is nonzero only over an interval of length $T/2$ around 0, the action of the corresponding matched filter is simply a windowed integration, sometimes referred to as "boxcar" integration, or a sliding average of the received signal over a window of length $T/2$, with its output sampled every $T$ units of

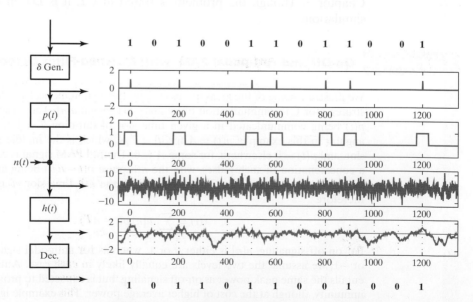

**Figure 13.7**    Binary detection with on-off signaling.

time. The output of the matched filter at these sampling times has therefore accumulated the effect of the underlying PAM signal in the corresponding bit slot, while simultaneously averaging out the noise, thereby improving the SNR over direct sampling of the received signal. Our DT simulation uses a scaled version of the matched filter, with a correspondingly scaled threshold, shown in the bottom panel by the horizontal line at the value 0.5.

The decoded binary sequence in the simulation has 2 bit errors out of 13. Taking account of the fact that we have a pulse that is 50 samples long, the expression in Eq. (13.71) gives the probability of error for our on-off signaling as $Q(\sqrt{50}/(2 \times 3.5)) = 0.156$, so the expected number of errors in 13 bits is $0.156 \times 13 = 2.03$ bits and the corresponding standard deviation is 1.31 bits, which matches what is observed.

In Figure 13.8, we show the corresponding results when antipodal rather than on-off signaling is used. The top two panels depict the generation of the transmitted PAM waveform with the same binary sequence as was used in Figure 13.7, and the third panel shows the received signal, including the additive noise. Despite the two underlying signal levels being twice as far apart as in on-off signaling, the noise still obscures the underlying signal. We again do not expect to do well by directly comparing one sample of this noisy signal in each bit slot against a threshold of 0. The probability of error for this direct approach would be $Q(A/\sigma) = Q(1/3.5) = 0.388$, with the expected number of errors in 13 bits then being $0.388 \times 13 = 5.03$ bits, and the associated standard deviation being 1.76.

The matched filter for this antipodal case has impulse response $h(t)$ equal to the difference between the two target signals, namely $p(t)$ and $-p(t)$. This again causes the matched-filter to perform windowed integration of the received signal over an interval of $T/2$. The resulting matched-filter output is shown in the bottom panel of Figure 13.8, after scaling. The decoded binary sequence in this case happens to have no bit errors. The expression in Eq. (13.73) gives the probability of error for our antipodal signaling as $Q(\sqrt{50}/3.5) = 0.022$, so the expected number of errors in 13 bits is $0.022 \times 13 = 0.28$ bits, with standard deviation 0.53 bits.

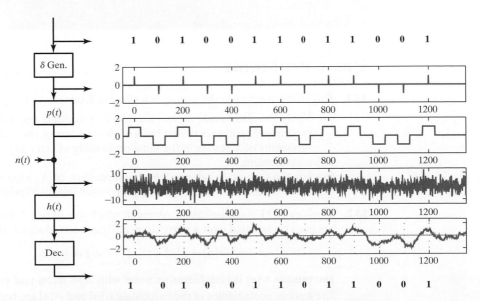

**Figure 13.8**   Binary detection with antipodal signaling.

In Table 13.1 we summarize the results for this example, showing the advantage of using matched filtering, and of antipodal over on-off signaling.

**TABLE 13.1** COMPUTED BIT-ERROR RATES FOR THE PAM SIGNALS IN EXAMPLE 13.3.

|  | No matched filter | With matched filter |
|---|---|---|
| On-off signaling | 0.443 | 0.156 |
| Antipodal signaling | 0.388 | 0.022 |

## 13.5 FURTHER READING

Much of the development of the theory of signal detection has been motivated by application contexts such as radar, treated in the early classic [Wod] and more recently in texts such as [Rch] and [Sko], also sonar and communications systems. However, the formulation, approach and terminology established in these domains have extended broadly into other applications. Several of the texts cited in Chapters 7–12 for their inclusion of WSS processes, power spectral density and signal estimation also address signal detection. Books that focus primarily on signal detection include [He2], [He3], [Kay3], [Lev], and [McD]. For a fascinating exploration of the Cauchy–Schwarz inequality and its applications, see [Ste].

# Problems

## Basic Problems

**13.1.** This problem refers to the setting described in Example 13.1.

   **(a)** Verify the claim made in the example, that deciding optimally between the two given hypotheses on the basis of the average of two independent measurements of $X$ results in the same probability of error as just using a single measurement.

   **(b)** For the case of $K$ independent measurements of $X$, what is the optimum decision rule, and what is the associated probability of error?

**13.2.** Consider a DT communication system in which one of two deterministic signals has been transmitted over a noisy channel. The received signal, $r[n]$, is given by

$$r[n] = s_i[n] + w[n], \quad i = 1 \text{ or } i = 2 .$$

The process $w[n]$ is i.i.d. Gaussian noise with zero mean and variance $\sigma_w^2 = \frac{1}{2}$. The *a priori* probabilities of the two pulses $s_1[n]$ and $s_2[n]$ are both equal to $\frac{1}{2}$.

The pulses have the following properties:

$$\sum_{n=-\infty}^{\infty} s_1^2[n] = \sum_{n=-\infty}^{\infty} s_2^2[n] = 1 \quad \text{and} \quad \sum_{n=-\infty}^{\infty} s_1[n]s_2[n] = \frac{1}{2}.$$

Consider the following two proposed receivers, shown in Figure P13.2:

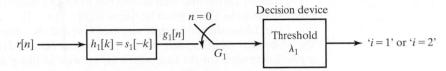

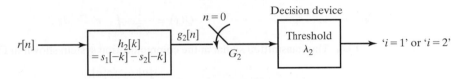

**Figure P13.2**

(a) On the same plot, sketch the PDF of $G_1$ given that $i = 1$ and the PDF of $G_1$ given that $i = 2$ when detection scheme 1 is used. On a second plot, sketch the PDF of $G_2$ given that $i = 1$ and the PDF of $G_2$ given that $i = 2$ when the detection scheme 2 is used.

In detection scheme 1 of Figure P13.2, the value of $G_1$ is tested against the threshold $\lambda_1$, and in detection scheme 2 of Figure P13.2, $G_2$ is tested against the threshold $\lambda_2$, i.e.,

$$G_1 \underset{'i=2'}{\overset{'i=1'}{\gtrless}} \lambda_1, \quad G_2 \underset{'i=2'}{\overset{'i=1'}{\gtrless}} \lambda_2.$$

(b) Determine the value of $\lambda_1$ that minimizes the probability of error and the value of $\lambda_2$ that minimizes the probability of error.
(c) Choose the correct statement below, and clearly and succinctly explain your reasoning.
   (i) Detection scheme 1 achieves a lower probability of error than detection scheme 2.
   (ii) Detection scheme 2 achieves a lower probability of error than detection scheme 1.
   (iii) Detection scheme 1 and detection scheme 2 achieve the same probability of error.

**13.3.** Consider the following signal detection problem, which is based on measurements of a received signal $r[n]$ under two possible hypothesis, $H_0$ and $H_1$:

$$H_0: r[n] = -s[n] + v[n],$$
$$H_1: r[n] = s[n] + v[n],$$

where $s[n]$ is a known pulse, and the noise samples, $v[n]$, are independent, Gaussian, and zero-mean random variables with variance $\sigma^2$. At the receiver we decide to use the strategy shown in Figure P13.3.

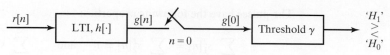

**Figure P13.3**

Here $h[\cdot]$ is the unit sample response of an LTI system and $\gamma$ is a constant. Suppose the pulse $s[n]$ is given by $s[n] = 2\delta[n] - \delta[n-1]$.

Determine the unit sample response $h[n]$ and the value of $\gamma$ that minimize the probability of error, first for the case where $H_0$ and $H_1$ are equally likely *a priori*, and then for the case where $H_1$ is twice as likely as $H_0$. In each case, compute the corresponding probability of error, expressing your answer in terms of the standard $Q(\cdot)$ function,

$$Q(x) = \frac{1}{\sqrt{2\pi}} \int_x^\infty e^{-z^2/2} dz \ .$$

**13.4.** The transmitted signal in the communication system shown in Figure P13.4 is

$$s[n] = Ap[n],$$

where $A = 0$ with probability $\frac{1}{3}$ and $A = 1$ with probability $\frac{2}{3}$. The two hypotheses are thus

$$H_0: \quad A = 0 \ \text{and}$$

$$H_1: \quad A = 1 \ .$$

The pulse $p[n]$ has unit energy, i.e.,

$$\sum_{n=-\infty}^{+\infty} p^2[n] = 1 \ .$$

The noise $w[n]$ introduced by the channel is zero-mean, i.i.d. Gaussian noise with variance $\sigma^2$, and independent of the transmitted signal.

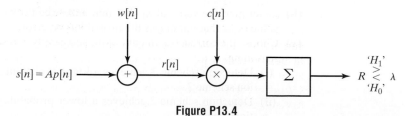

**Figure P13.4**

The received signal $r[n]$ is

$$r[n] = Ap[n] + w[n].$$

The system acting on the received signal in Figure P13.4 is referred to as a correlation receiver and is often used in radar and communications sytems. The pulse $c[n]$ is a finite-energy pulse, the choice for which is considered in (c). The block denoted by $\sum$ in the figure computes the quantity $R$ defined by

$$R = \sum_{n=-\infty}^{+\infty} r[n]c[n] \ .$$

The detector applies a threshold test to the random variable $R$:

$$R \begin{array}{c} \text{`}H_1\text{'} \\ > \\ < \\ \text{`}H_0\text{'} \end{array} \lambda$$

**(a)** In terms of $p[n]$, $c[n]$, and $\sigma^2$, determine the expected value of $R$ under $H_0$ and $H_1$.

**(b)** In terms of $p[n]$, $c[n]$, and $\sigma^2$, determine the variance of $R$ under $H_0$ and $H_1$.

**(c)** Suppose the energy of $c[n]$ is specified to be some value $K$. Determine the choice for $c[n]$ to maximize

$$\left( E[R|H_0] - E[R|H_1] \right)^2 .$$

**(d)** Will your choice for $c[n]$ in (c) with an appropriate choice of $\lambda$ necessarily minimize the probability of error? Recall that

$$\text{Probability of error} = P(\text{`}H_0\text{'}, H_1) + P(\text{`}H_1\text{'}, H_0) .$$

Explain, and if your answer is affirmative, determine $\lambda$ in terms of the specified parameters.

**13.5.** The diagram in Figure P13.5-1 represents a system in which the signal $d[n]$ is transmitted through a noisy communications channel, and $r[n]$ is received:

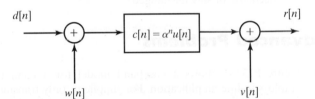

**Figure P13.5-1**

The parameter $\alpha$ that specifies the impulse response of the channel is some known number, with magnitude less than 0.5. The noise process $w[n]$ is such that its value at each instant is a zero-mean Gaussian random variable with known variance $\sigma_w^2$, and the values at different instants are independent of each other; in other words, $w[n]$ is a zero-mean i.i.d. Gaussian process, with autocovariance function $C_{ww}[m] = \sigma_w^2 \delta[m]$. The noise process $v[n]$ is also a zero-mean i.i.d. Gaussian process, independent of the process $w[\cdot]$ and with known variance $\sigma_v^2$ at each instant, i.e., $C_{vv}[m] = \sigma_v^2 \delta[m]$.

The signal $d[n]$ can be either 0 for all time (hypothesis $H_0$), or can be the unit sample function (hypothesis $H_1$):

$$H_1: \quad d[n] = \delta[n] \qquad P(H_1) = p_1 ,$$
$$H_0: \quad d[n] = 0 \qquad P(H_0) = p_0 .$$

In each of the two cases specified in (a) and (b) below, you are to design a receiver that takes $r[n]$ as input and decides between $H_0$ and $H_1$ with minimum probability of error. The optimum receiver in each case involves the following steps, shown in Figure P13.5-2: LTI (but possibly noncausal) filtering of $r[n]$; sampling the output $g[n]$ of the filter at some appropriate time $n_0$; and deciding in favor of $H_0$ or $H_1$, based on where the sample value is relative to a threshold $\gamma$.

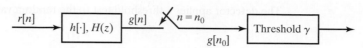

**Figure P13.5-2**

Thus, in order to specify the minimum-error-probability receiver in each of the following two cases, you will need to specify: (i) the filter impulse response $h[\cdot]$ or system function $H(z)$; (ii) the instant $n_0$ at which you sample the output $g[n]$ of the filter; (iii) the threshold $\gamma$ that you compare the sample with; and (iv) what the decisions are for sample values above and below the threshold, respectively.

**(a)** Suppose $\sigma_w^2 = 0$ and $\sigma_v^2 > 0$. Specify the minimum-error-probability receiver. If the channel impulse response were changed such that the magnitude of $\alpha$ was doubled, would the probability of error increase, decrease, or stay unchanged? If you believe it would change, by what factor would the noise variance $\sigma_v^2$ have to be multiplied in order to bring the probability of error back to its original value?

**(b)** Suppose $\sigma_w^2 > 0$ and $\sigma_v^2 = 0$. Specify the minimum-error-probability receiver. Find an expression for the probability of error in the case where $\sigma_w^2 = 1$, writing this probability in terms of the standard $Q$-function defined in Eq. (13.23). If the channel impulse response were changed such that the magnitude of $\alpha$ was doubled, would the probability of error increase, decrease, or stay unchanged?

## Advanced Problems

**13.6.** Figure P13.6-1 shows a baseband model for a communication system that employs in-line amplification. For simplicity, only transmission of a single bit is considered.

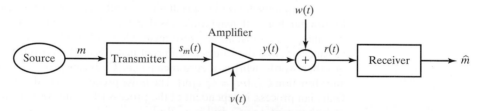

**Figure P13.6-1**

In Figure P13.6-1:

  (i) $m$ is a binary symbol that is equally likely to be 0 or 1.

 (ii) $s_m(t)$ is the transmitter waveform that is used to convey the message $m$. The waveforms $s_0(t)$ and $s_1(t)$ associated with messages $m = 0$ and $m = 1$, respectively, are shown in Figure P13.6-2.

(iii) The amplifier output $y(t)$ is given by

$$y(t) = \sqrt{G} s_m(t) + v(t),$$

where $G > 1$ is the power gain of the amplifier and $v(t)$ is noise that the amplifier injects. $v(t)$ is a zero-mean, white Gaussian noise process that

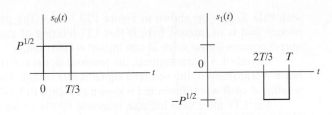

**Figure P13.6-2**

is statistically independent of the message $m$ and has PSD $S_{vv}(j\omega) = N_v$. A Gaussian process has the property (among many others) that filtering it through an LTI system yields a process whose value at any time instant is Gaussian random variable.

(iv) $w(t)$ is a zero-mean, white Gaussian noise process that is generated by the receiver electronics. It is statistically independent of the message $m$ and the amplifier's noise process $v(\cdot)$. The PSD of $w(t)$ is $S_{ww}(j\omega) = N_w$.

(v) The receiver shown in detail in Figure P13.6-3 filters and then samples the received waveform, $r(t) = y(t) + w(t)$, to obtain the random variable $Z = z(T)$. The receiver's output is its decision, $\hat{m} = 0$ or 1, as to which message was sent, based on the value of $Z$.

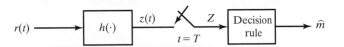

**Figure P13.6-3**

(vi) The receiver filter has impulse response

$$h(t) = \begin{cases} 1/\sqrt{T}, & \text{for } 0 \le t \le T, \\ 0, & \text{otherwise}, \end{cases}$$

as shown in Figure P13.6-4.

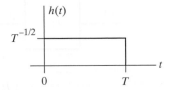

**Figure P13.6-4**

(a) Determine the conditional probability densities for $Z$ given $m = 0$ and $m = 1$, i.e., determine $f_{Z|m}(z\,|0)$ and $f_{Z|m}(z\,|1)$.

(b) Determine the minimum-error-probability decision rule for deciding $\hat{m} = 0$ or $\hat{m} = 1$. Reduce your rule to a threshold test on $Z$.

**13.7.** One of two equally likely symbols $A$ and $B$ is sent over a noisy channel. The symbol $A$ is represented by sending the pulse $p(t)$ whose Fourier transform is $P(j\omega)$ as shown in Figure P13.7-1(a), so

$$P(j\omega) = \begin{cases} (1 - \frac{|\omega|}{2\pi \cdot 10^3}) & |\omega| < 2\pi \cdot 10^3 \\ 0 & |\omega| \ge 2\pi \cdot 10^3. \end{cases}$$

The symbol $B$ is represented by sending no pulse (i.e., zero). The noise $n(t)$ on the channel is an additive zero-mean WSS bandlimited Gaussian noise process with PSD $S_{nn}(j\omega)$ as shown in Figure P13.7-1(b). The property of a Gaussian process that is of interest here is that LTI filtering of such a process yields an output process whose value at any instant is Gaussian.

If symbol $A$ is transmitted, the received signal is $r(t) = p(t) + n(t)$. If symbol $B$ is transmitted, the received signal is $r(t) = n(t)$. The system for deciding whether $A$ or $B$ was transmitted is shown in Figure P13.7-2.

The LTI filter with impulse response $h(\cdot)$ is an ideal low-pass filter with frequency response shown in Figure P13.7-3.

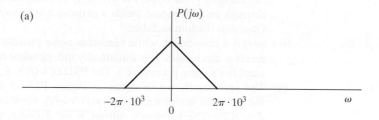

(a)

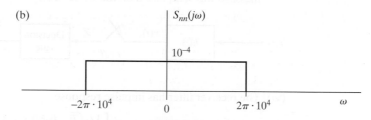

(b)

**Figure P13.7-1**

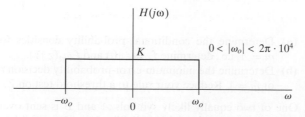

**Figure P13.7-2**

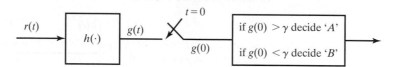

**Figure P13.7-3**

The impulse response $h(t)$ is scaled to have unit energy, i.e.,
$\int_{-\infty}^{+\infty} h^2(t)\, dt = 1$

**(a)** Determine $p(0)$.

**(b)** Determine the variance of the noise $n(t)$.

**(c)** Determine the relationship between $K$ and $\omega_o$ to ensure that
$\int_{-\infty}^{+\infty} h^2(t)\, dt = 1$.

**(d)** Assume that $|\omega_o| < 2\pi \cdot 10^4$. The filter output $g(t)$ can be represented as

$$g(t) = p_h(t) + n_h(t),$$

where $p_h(t) = p(t) * h(t)$ is the signal component of $g(t)$ and $n_h(t) = n(t) * h(t)$ is the noise component. Determine the variance of the noise component $n_h(t)$.

**(e)** Again assume that $|\omega_o| < 2\pi \cdot 10^4$. Determine the values of $\omega_o$ and $\gamma$ that minimize the probability of error.

**13.8.** Two measurements, $R_1$ and $R_2$, are taken at the receiving end of a communication channel at two different times. Under hypothesis $H_0$, "noise alone is present at the receiver," the values of $R_1$ and $R_2$ are given by

$$R_1 = X_1$$
$$R_2 = X_2$$

where the noise samples $X_1$ and $X_2$ have means 0, variances $\sigma^2$, but are not independent, instead being governed by a bivariate Gaussian density, i.e.,

$$f_{X_1,X_2}(x_1,x_2|H_0) = \frac{1}{2\pi\sigma^2\sqrt{1-\rho^2}} \exp\left[-\frac{x_1^2 - 2\rho x_1 x_2 + x_2^2}{2(1-\rho^2)\sigma^2}\right].$$

The quantity $\rho$ turns out to be the correlation coefficient between $X_1$ and $X_2$.

Under hypothesis $H_1$, "both signal and noise are present," the received signal samples are given by

$$R_1 = s_1 + X_1$$
$$R_2 = s_2 + X_2$$

where $s_1$ and $s_2$ are the signal samples (which are known constants).

The prior probabilities of the hypotheses are

$$p_0 = P(H_0), \quad p_1 = P(H_1), \quad p_0 + p_1 = 1.$$

Suppose the receiver implements the MAP decision rule for minimum error probability, deciding that a signal is present—i.e., declaring '$H_1$'—for each pair $(r_1, r_2)$ of samples for which $P(H_1|r_1, r_2) > P(H_0|r_1, r_2)$, and otherwise deciding that no signal is present.

**(a)** Show that the decision rule is of the form

$$k_1 r_1 + k_2 r_2 \underset{'H_0'}{\overset{'H_1'}{\gtrless}} \gamma,$$

and obtain expressions for $\gamma$, $k_1$, and $k_2$. Describe how the decision varies as $p_0$ varies between 0 and 1.

**(b)** When $\rho = 0$ in the bivariate Gaussian case, the associated random variables are not just uncorrelated but are also independent. For this case, your solution in (a) should reduce to the matched-filter solution developed in this chapter for minimum-error-probability detection. Verify that it does.

**13.9.** This problem involves the Laplacian PDF, defined as

$$f_X(x) = \frac{\alpha}{2} e^{-\alpha|x-m|}$$

where $X$ is a random variable, $\alpha$ is a parameter related to the variance $\sigma^2$ of $X$ (actually $\alpha = \sqrt{2}/\sigma$), and $m$ is the mean of $X$. The Laplacian PDF is often used in models for speech and images.

A received signal $R[n]$ is known to be given by one of the following two models, with equal probability:

$$H_0 : R[n] = W[n],$$

$$H_1 : R[n] = s[n] + W[n].$$

Here each $W[n]$ is a Laplacian random variable, with the value at each time $n$ given by the Laplacian PDF with zero mean ($m = 0$); all of the $W[n]$ are independent. The signal $s[n]$ is deterministic and known.

**(a)** Given only a measurement $r[1]$ of $R[1]$, explain how to decide between $H_0$ and $H_1$ with minimum probability of error, and compute this probability.

**(b)** Determine the joint PDF of $R[1], R[2], \ldots, R[L]$ given $H_1$, i.e., find

$$f_{\mathbf{R}|H_1}(r[1], r[2], \ldots, r[L] \mid H_1).$$

**(c)** Given measurements of $R[n]$ for $n = 1, 2, \ldots, L$, obtain the decision rule that will decide between $H_0$ and $H_1$ with minimum probability of error. For the case of $L = 2$, try and interpret your answer geometrically, with a sketch to show the decision regions associated with '$H_0$' and '$H_1$' in the $(r[1], r[2])$ plane.

**13.10.** Consider a (memoryless) communication channel whose input $X_j$ and output $Y_j$ at time $j$ are random variables related for $1 \le j \le 4$ by

$$Y_j = G_j X_j + W_j, \qquad 1 \le j \le 4,$$

where the gains $G_j$ are i.i.d. Gaussian random variables of mean 0 and variance $\sigma_G^2$, so $G_j$ is distributed as $\mathcal{N}(0, \sigma_G^2)$. The random variables $W_j$ denote channel noise, assumed to be i.i.d. Gaussians distributed as $\mathcal{N}(0, \sigma_W^2)$ and independent of the $G_k$. The situation is shown in Figure P13.10.

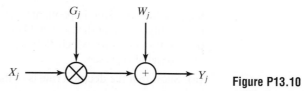

**Figure P13.10**

We wish to transmit a single binary random variable $H = \{0, 1\}$ across this channel, using the following scheme:

- Given that $H = 0$, the inputs are selected to be $(X_1, X_2, X_3, X_4) = (1, 1, 0, 0)$.
- Given that $H = 1$, the inputs are selected to be $(X_1, X_2, X_3, X_4) = (0, 0, 1, 1)$.

Assume that the prior probabilities for the two possible choices of $H$ are $P(H_0) = p_0$ and $P(H_1) = p_1 = 1 - p_0$. Based on the observation of the channel outputs $Y_j$ for $1 \le j \le 4$, we need to make a decision on which binary value $H$ was transmitted.

**(a)** Explain why under each of the hypotheses $H = 0$ and $H = 1$, the random variables $Y_j$ for $1 \le j \le 4$ are individually Gaussian and independent of each other. Determine the corresponding means and variances of these conditionally Gaussian output variables under each hypothesis.

**(b)** Your friend suggests passing the received sequence of channel outputs through an LTI filter whose unit sample response is matched to the difference between the two known input signals, then comparing the output of the matched filter at some appropriate time with a well-chosen threshold. You think this is worth a try, even though the situation here is not quite the one for which your friend's prescription would be optimal (because the channel gain varies randomly in time here). As a result, you try to use a measurement of the random variable

$$D = Y_1 + Y_2 - Y_3 - Y_4$$

to decide between the hypotheses. Determine how this random variable is distributed under $H = 0$ and under $H = 1$. Does this tell you why looking at the measured value of $D$ will not help distinguish between the two possibilities for $H$? Explain. So it's time to go back to basics!

**(c)** Determine the PDFs

$$f_{Y_1,Y_2,Y_3,Y_4|H}(y_1,y_2,y_3,y_4|0)$$

and

$$f_{Y_1,Y_2,Y_3,Y_4|H}(y_1,y_2,y_3,y_4|1) \ .$$

**(d)** Find the likelihood ratio

$$\Lambda(y_1,y_2,y_3,y_4) = \frac{f_{Y_1,Y_2,Y_3,Y_4|H}(y_1,y_2,y_3,y_4|1)}{f_{Y_1,Y_2,Y_3,Y_4|H}(y_1,y_2,y_3,y_4|0)}$$

and also the log-likelihood ratio $\ln \Lambda(y_1,y_2,y_3,y_4)$.

**(e)** Determine the rule that uses the measured values $y_j$ of the random variables $Y_j$ for $1 \leq j \leq 4$ to decide with minimum probability of error what bit $H$ was sent, either 0 or 1 (denote this decision by $\widehat{H} = 0$ and $\widehat{H} = 1$ respectively). Make clear in the way you write the decision rule that it only needs the quantities $q = y_1^2 + y_2^2$ and $r = y_3^2 + y_4^2$ rather than the individual output measurements.

It can be shown that under the condition $H = 0$ the quantity

$$Q = Y_1^2 + Y_2^2$$

is an exponentially distributed random variable with mean $\mu_{Q0} = 2(\sigma_G^2 + \sigma_W^2)$, i.e., its conditional PDF is

$$f_{Q|H}(q|0) = \frac{1}{\mu_{Q0}} \exp\left(-\frac{q}{\mu_{Q0}}\right) \qquad \text{for } q > 0$$

and 0 for $q < 0$. Similarly, still under the condition $H = 0$, the quantity

$$R = Y_3^2 + Y_4^2$$

is exponentially distributed with mean $\mu_{R0} = 2\sigma_W^2$, and is independent of $Q$.

You may also recall from your probability class that for such (conditionally) independent exponential random variables, the probability that $R$ is larger than $Q$, conditioned on $H = 0$, is given by

$$P(R > Q|H = 0) = \frac{\mu_{R0}}{\mu_{Q0} + \mu_{R0}} \ .$$

Under $H = 1$, the roles are reversed, i.e., $Q$ and $R$ are still independent and exponentially distributed, but the conditional mean of $Q$ is $\mu_{Q1} = 2\sigma_W^2$ and the conditional mean of $R$ is $\mu_{R1} = 2(\sigma_G^2 + \sigma_W^2)$, while

$$P(Q > R|H = 1) = \frac{\mu_{Q1}}{\mu_{Q1} + \mu_{R1}} .$$

**(f)** Use these facts in the case where the hypotheses are equally likely, i.e., $p_0 = p_1 = 0.5$, to obtain expressions (stated in terms of the problem parameters) for:

  (i) the conditional probabilities

$$P(\widehat{H} = 1|H = 0)$$

  and

$$P(\widehat{H} = 0|H = 1) ;$$

  (ii) the probability of error $P_e$.

**13.11.** A signal $X[n]$ that we will be measuring for $n = 1, 2, \ldots, L$ is known to be generated according to one of the following two hypotheses:

  $H_0:$    $X[n] = W[n]$   holds with *a priori* probability $P(H_0) = p_0$ ,

  $H_1:$    $X[n] = V[n]$   holds with *a priori* probability $P(H_1) = p_1 = 1 - p_0$ .

Here $W[n]$ is a zero-mean i.i.d. Gaussian process with known constant variance $\sigma_0^2$ at each time instant, i.e., the value at each time instant is governed by the probability density function

$$f_W(w) = \frac{1}{\sigma_0\sqrt{2\pi}} \exp\left\{-\frac{w^2}{2\sigma_0^2}\right\}$$

and the values at different times are independent of each other. Similarly, $V[n]$ is a zero-mean Gaussian process, taking values that are independent at distinct times, but with a variance that changes in a known manner over time, so the variance at time $n$ is known to be $\sigma_n^2$. We will find it notationally helpful in working through this problem to use the definition

$$h[n] = \left(\frac{1}{\sigma_0^2} - \frac{1}{\sigma_n^2}\right).$$

Note that $h[n]$ may be positive for some $n$ but negative or zero for others, corresponding to having $\sigma_0 < \sigma_n$, $\sigma_0 > \sigma_n$ or $\sigma_0 = \sigma_n$ respectively.

**(a)** Suppose we only have a measurement at $n = 1$, with $X[1] = x[1]$. Show that the decision rule for choosing between $H_0$ and $H_1$ with minimum probability of error, given this measurement, takes the form

$$h[1]\left(x[1]\right)^2 \underset{`H_0'}{\overset{`H_1'}{\gtrless}} \gamma$$

for some appropriately chosen threshold $\gamma$. Also specify $\gamma$ in terms of the problem parameters.

**(b)** With your result from (a), but now assuming $h[1] > 0$, sketch and label the two conditional densities—namely $f_{X[1]|H}(x|H_0)$ and $f_{X[1]|H}(x|H_1)$—that govern $X[1]$ under the two respective hypotheses.

Assuming that the two hypotheses are equally likely so $p_0 = p_1$, mark in the points $\pm\sqrt{\gamma/h[1]}$ on the horizontal (i.e., $x$) axis, then shade in the region or regions whose total area yields the conditional probability $P('H_1'|H_0)$, and express this conditional probability in terms of the standard $Q$ function,

$$Q(\alpha) = \frac{1}{\sqrt{2\pi}} \int_\alpha^\infty e^{-v^2/2}\, dv\ .$$

(c) With the same situation as in (b), but with the hypotheses no longer restricted to be equally likely *a priori*, specify the range of values for $p_0$ in which the optimal decision will always be '$H_1$', no matter what the measured value $x[1]$.

(d) Now suppose we have measurements at $n = 1, 2, \ldots, L$, i.e., we know $X[1] = x[1], X[2] = x[2], \ldots, X[L] = x[L]$. Determine the decision rule for minimum probability of error, writing it in a form that generalizes your result from (a).

(e) Suppose that in fact

$$V[n] = S[n] + W[n]$$

where $S[n]$ is a zero-mean i.i.d. Gaussian process that is independent of $W[\cdot]$ and has variance $\alpha_n^2$, so $\sigma_n^2 = \alpha_n^2 + \sigma_0^2$. Show that your decision rule from (d) can be written as a comparison of the quantity

$$\sum_{n=1}^{L} x[n]\, \widehat{s}_n(x[n]) \tag{13.75}$$

with a fixed threshold, where $\widehat{s}_n(X[n])$ denotes the LMMSE estimator of $S[n]$ from a measurement of $X[n]$ under hypothesis $H_1$. This form of the decision rule is similar to what we obtained in the case of a deterministic signal, see Eq. (13.15).

**13.12.** In a particular binary communication system, we are interested in detecting the presence or absence of a known pulse $s[n]$ in a received signal $r[n]$. If the pulse is present, the received signal is $r[n] = s[n] + v[n]$; if the pulse is absent, the received signal is $r[n] = v[n]$. Thus

$$H_0(\text{pulse absent}): r[n] = v[n]$$

$$H_1(\text{pulse present}): r[n] = s[n] + v[n].$$

The following is known about $v[n]$:
  (i) $v[n]$ is a WSS Gaussian random process with mean value $\bar{v}$. Equivalently, $v[n] - \bar{v}$ is the result of passing unit-variance, zero-mean, i.i.d. Gaussian noise through an appropriate LTI shaping filter with unit sample response $b[n]$ and frequency response $B(e^{j\Omega})$.
  (ii) $R_{vv}[m] - \bar{v}^2 = E\{(v[n+m] - \bar{v})(v[n] - \bar{v})\} = C_{vv}[m] = \frac{5}{2}\delta[m] + \delta[m-1] + \delta[m+1]$.

(a) Find a choice of $b[n]$ that results in the above $C_{vv}[m]$.
(b) Assume $\bar{v} = 0$. Determine the unit sample-response of a whitening filter for $v[n]$, i.e., determine $h_w[\cdot]$ in Figure P13.12-1 such that $R_{ww}[m] = \delta[m]$. Is your choice unique?

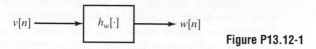

$$v[n] \longrightarrow \boxed{h_w[\cdot]} \longrightarrow w[n]$$

**Figure P13.12-1**

The signal $r[n]$ will be processed as shown in Figure P13.12-2. The following is known about $s[n]$ and $h[n]$:
  (i) $s[n] = \delta[n]$; and
  (ii) $\sum\limits_{n=-\infty}^{\infty} h^2[n] = 1$.

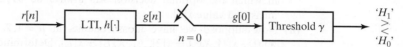

**Figure P13.12-2**

**(c)** For a fixed probability of false alarm, and still with $\bar{v} = 0$, determine a choice for $h[n]$ and a corresponding $\gamma$ so that the probability of detection is maximized. Also, state whether or not your choice is unique.

**(d)** Now suppose that $\bar{v} = 1$. Find the mean value of $g[n]$, the output of the LTI filter, under each of the hypotheses.

**(e)** Repeat part (c) for $\bar{v} = 1$.

**13.13.** Consider the communication system shown in Figure P13.13-1. A binary message $m$, which is equally likely to be 0 or 1, is encoded into DT waveform $s_m[n]$, where

$$s_0[n] = 0, \text{ for all } n$$

$$s_1[0] = \sqrt{\mathcal{E} - a^2}, \quad s_1[1] = a, \quad \text{and} \quad s_1[n] = 0 \text{ otherwise,}$$

with $-\sqrt{\mathcal{E}} \le a \le \sqrt{\mathcal{E}}$, as shown in Figure P13.13-2. Here $\mathcal{E}$ denotes the energy of the signal $s_1[n]$.

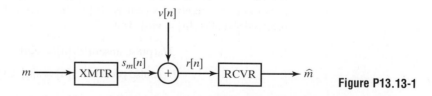

**Figure P13.13-1**

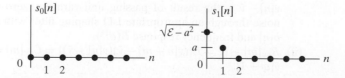

**Figure P13.13-2**

The receiver observes the signal

$$r[n] = s_m[n] + v[n],$$

where $v[n]$ is a zero-mean, WSS colored Gaussian noise process that is statistically independent of the message $m$, and whose correlation function is

$$R_{vv}[k] = E(v[n+k]v[n]) = \begin{cases} \sigma_v^2, & \text{for } k = 0 \\ \sigma_v^2/4 & \text{for } k = \pm 1 \\ 0, & \text{otherwise.} \end{cases}$$

You can think of a colored Gaussian process as being the result of LTI filtering of an i.i.d. Gaussian process. In this problem you will need to invoke the fact that every linear combination of samples of a Gaussian process yields a Gaussian random variable.

Assume that the receiver is constructed as shown in Figure P13.13-3. Here, the LTI filter has impulse response $h[n] = s_1[1-n]$. The output of this filter, $y[n]$, is sampled at $n = 1$ to yield $Y = y[1]$, and used in the decision rule

$$Y \underset{\hat{m}=0}{\overset{\hat{m}=1}{\underset{<}{\overset{>}{\gtrless}}}} \gamma.$$

This would have been the optimum filter structure for minimum probability of error if $v[n]$ had been white Gaussian noise. Even though this structure is not the optimum for the colored noise $v[n]$ that we have here, we shall try and do the best we can with it.

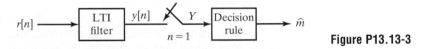

**Figure P13.13-3**

(a) Find $E(Y \mid m = 1)$ and $\text{var}(Y \mid m = 1)$, the mean and variance of $Y$ given that $m = 1$. What is the probability density function of $Y$, conditioned on $m = 1$?

(b) For the same questions as in (a), but now for the case of $m = 0$, state (but there is no need to derive) the corresponding answers.

(c) What value of $\gamma$ in this structure minimizes the probability of error, $P_e = \Pr(\hat{m} \neq m)$? Briefly justify your answer.

(d) Find the error probability of the receiver in Figure P13.13-3 when the optimum $\gamma$ value is employed. Express your result in terms of the $Q$-function,

$$Q(x) = \int_x^\infty \frac{e^{-t^2/2}}{\sqrt{2\pi}} \, dt \,.$$

If you have done this part correctly, you should find that you get the same error probability $P_e$ whether $a = 0$, $\sqrt{\mathcal{E}}$, or $-\sqrt{\mathcal{E}}$. However, there is a choice of $a$ in the allowed range that yields a smaller $P_e$; find this optimum value of $a$.

(e) Instead of the $h[n]$ specified earlier, determine the $h[n]$ and associated threshold $\gamma$ that will actually minimize the probability of error for this colored-noise problem. Again express the resulting probability of error in terms of the $Q$-function.

**13.14.** The DT signal emitted by a particular transmitter is either $x_0[n] = 0$ or $x_1[n] = \delta[n] + \delta[n-1]$, and these two possibilities are equally likely. The signal is sent over an LTI channel with impulse response $c[n] = \delta[n] + 0.2\,\delta[n-1]$. The received signal $y[n]$ is given by

$$y[n] = c[n] * x_i[n] + v[n]$$

with either $i = 0$ or $i = 1$, where $v[n]$ is a zero-mean WSS colored Gaussian noise process with autocorrelation function $R_{vv}[m] = (0.5)^{|m|}$. A colored Gaussian process may be thought of as resulting from LTI filtering of an i.i.d. Gaussian process through a shaping or modeling filter. The received signal $y[n]$ is processed to decide with minimum probability of error whether $x_0[n]$ or $x_1[n]$ was transmitted.

**(a)** Compute and sketch $c[n] * x_1[n]$.

**(b)** What is the PSD, $S_{vv}(e^{j\Omega})$, of the noise process $v[n]$?

**(c)** Suppose we first pass $y[n]$ through an LTI prefilter with impulse response $\phi[n]$ chosen so as to make the "noise component" of the prefilter's output, namely $v_\phi[n] = \phi[n] * v[n]$, into a WSS white process with autocorrelation function $R_{v_\phi v_\phi}[m] = \delta[m]$. Find a causal $\phi[n]$ that accomplishes this.

**(d)** With $\phi[n]$ as in (c), compute and sketch the signal component of the prefilter's output if $x_1[n]$ is sent, i.e., find $x_{1p}[n] = \phi[n] * c[n] * x_1[n]$.

**(e)** What is the smallest interval over which you need to observe the prefiltered signal $y_\phi[n] = \phi[n] * y[n]$ in order to decide whether $x_0[n]$ or $x_1[n]$ was sent, with minimum probability of error?

**(f)** Suppose you are now permitted to process $y_\phi[n]$ through an LTI filter, and to base your choice of hypothesis on the value, relative to some threshold, of the filter output $g[n]$ at some instant, say $n = 0$. What choice of filter will maximize the SNR in $g[0]$? If you omitted the prefiltering, and instead directly filtered $y[n]$ to produce an output $g[n]$, could there be a choice of filter that improved the SNR in $g[0]$? Is there a choice of threshold for which the SNR-maximizing filter also minimizes the probability of error? If so, what threshold value is needed?

## Extension Problems

**13.15.** Consider the following causal DT system. The input, $x[n]$, is Gaussian i.i.d. noise with zero mean and $R_{xx}[m] = \delta[m]$. The input is sent simultaneously through two channels producing $w_0[n]$ and $w_1[n]$, as indicated in Figure P13.15.

Assume the system started a long time ago and is already in steady state. For each $n$, the output $y[n]$ is independently chosen to be either $w_0[n]$ or $w_1[n]$.

**(a)** Compute the impulse response of the filters from $x[n]$ to $w_0[n]$ and from $x[n]$ to $w_1[n]$, i.e., determine $h_0[n]$ and $h_1[n]$.

**(b)** Given a particular measurement $y[n]$, design a test to decide whether the output came from $w_0[n]$ or $w_1[n]$, and determine the corresponding range of values for $y[n]$ for which the test decides $w_1[n]$. Your decision should

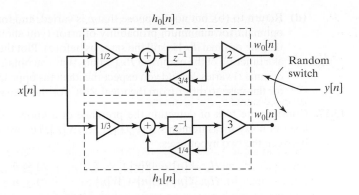

**Figure P13.15**

be based on a minimum-probability-of-error test, assuming the *a priori* probabilities associated with $w_0[n]$ and $w_1[n]$ are each equal to $\frac{1}{2}$.

**(c)** Now assume instead that the switch chooses $w_0[n]$ with probability $P_0$ and $w_1[n]$ with probability $P_1 = 1 - P_0$. The probability of false alarm, $P_{FA}$, and the probability of detection, $P_D$, are defined as follows:

$$P_{FA} = P(\text{declare } w_1[n]|\text{switch chose } w_0[n])$$

$$P_D = P(\text{declare } w_1[n]|\text{switch chose } w_1[n]).$$

**(i)** Find the largest $P_0$ such that $P_{FA} = P_D = 1$.
**(ii)** Find the smallest $P_0$ such that $P_{FA} = P_D = 0$.

**13.16.** Using a suitable computational package, write a program that generates a signal $r[n]$ by first randomly choosing, with respective probabilities $p$ and $1 - p$, one of a given pair of signals $s_1[n]$ and $s_0[n]$, then adding zero-mean i.i.d. Gaussian noise of a specified variance to the selected signal. Your given signals should be of comparable energy, though not necessarily identical energy. Now write a program to perform detection on $r[n]$ by correlating it with a specified signal $d[n]$, i.e., computing $g = \sum r[n]d[n]$, comparing the result with a specified threshold $t$, declaring "$s_1$" when $r > t$, and "$s_0$" otherwise. Arrange to perform the detection a large number of times (e.g., 10,000), so that empirical statistics regarding the probability of detection, probability of false alarm, and probability of error can be determined and compared with their theoretical values.

**(a)** Plot $s_1[n]$ and $s_0[n]$, and also plot a typical $r[n]$ when it is (i) a noise-corrupted version of $s_1[n]$, and (ii) a noise-corrupted version of $s_0[n]$.

**(b)** With $p = 0.5$, run your program repeatedly with $d[n]$ and $t$ chosen appropriately for minimum-error-probability detection. Determine the empirical values of $P_D$, $P_{FA}$, and $P_{\text{error}}$ and compare them with the theoretical values. Repeat this for a few different choices of noise variance.

**(c)** Keep $d[n]$ and $p$ the same as in (b), and fix the noise variance, but now run your program with a range of values of $t$, from very negative to very positive. Plot the empirical values of $P_D$ as a function of the empirical values of $P_{FA}$ that you obtain as you vary $t$; this is the empirical receiver operating characteristic, or ROC. Finally, plot the empirical value of $P_{\text{error}}$ as a function of $t$, and check if your choice of $t$ in part (b) indeed seems plausible as the one that gives minimum error probability (keep in mind that the empirical results will have some variability, and will not exactly line up with theoretical results). What is the effect on your results of changing the noise variance to some other value?

**(d)** Return to (b), but now suppose that $p$ is varied, and for each $p$ you pick the optimal $t$ for minimium probability of error (you should convince yourself that the optimal $d[n]$ is still the same as before). Plot the empirical $P_D$ values as a function of the empirical $P_{FA}$ values that you obtain as $p$ (and with it, the optimal $t$) varies. Would you expect the ideal (as opposed to empirical) ROC for this case to differ from the ideal ROC for the scenario in (c)? Explain.

**13.17.** Consider the task of detecting the presence of a known, finite-length signal in a set of noisy measurements $X[1], X[2], \ldots, X[L]$, i.e., we would like to choose between the two hypotheses:

$$H_0: X[n] = W[n], \qquad\qquad 1 \leq n \leq L$$
$$H_1: X[n] = s[n] + W[n], \qquad 1 \leq n \leq L.$$

Here each $W[n]$ is a zero-mean Gaussian random variable—the noise—with variance $\sigma^2$, and the $W[\cdot]$ at distinct times are independent. The signal $s[n]$ is deterministic and known. Denote the prior probabilities of hypotheses $H_0$ and $H_1$ by $p_0$ and $p_1 = 1 - p_0$, respectively.

**(a)** Using an appropriate computational package, write a program to create a realization of $X[n]$ under one of the two hypotheses above, with $p_0 = p_1 = \frac{1}{2}$, $L = 5000$, $\sigma^2 = 1$, and

$$s[n] = 0.5 \cos(\pi (n - \Delta)/2)$$

where $\Delta$ can be set as 0, 1, 2, or 3, but will not be known to the receiver.

**(b)** Run your program from (a) and plot the resulting measurements. Can you decide which hypothesis (i.e., signal absent or signal present) is appropriate just by visually examining the plot? And if so, can you determine the underlying value of $\Delta$?

**(c)** Implement a minimum-error-probability decision rule in order to select an appropriate hypothesis (signal present or signal absent) for the data generated in (b). There is a complication here in that $s[n]$ is not completely known because of the uncertainty in $\Delta$. However, a reasonable thing to do is evaluate each of the four candidate signals (namely $s[n]$ with $\Delta = 0, 1, 2, 3$) using your decision rule, and decide in favor of the candidate signal (if any) that satisfies by the largest margin whatever condition you developed for deciding '$H_1$'.

**(d)** Write a program to repeat (b) and (c) automatically a large number of times (e.g., 10,000), and determine the empirical probability of correctly deciding the hypothesis (i.e., the fraction of outcomes in which the hypothesis is correctly decided), and of correctly choosing $\Delta$ when '$H_1$' is correctly decided. How does the empirical probability of a correct decision on the hypothesis compare with the theoretical probability for the case of known $\Delta$?

# Bibliography

**Abu:** Y. S. Abu-Mostafa, M. Magdon-Ismail, and H.-T. Lin, *Learning from Data: A Short Course*, AMLBook 2012.

**Alp:** E. Alpaydin, *Introduction to Machine Learning*, 3rd ed., MIT Press 2014.

**And:** B. D. O. Anderson and J. B. Moore, *Optimal Filtering*, Prentice Hall 1979 (reprinted by Dover 2005).

**An1:** J. B. Anderson, *Digital Transmission Engineering*, 2nd ed., Wiley-IEEE Press 2005.

**An2:** J. B. Anderson and R. Johannesson, *Understanding Information Transmission*, IEEE Press 2005.

**Ant:** P. J. Antsaklis and A. M. Michel, *A Linear Systems Primer*, Birkhäuser 2007.

**Ast:** K. J. Åström and R. M. Murray, *Feedback Systems: An Introduction for Scientists and Engineers*, Princeton University Press 2008.

**Bar:** J. R. Barry, E. A. Lee, and D. G. Messerschmitt, *Digital Communication*, 3rd ed., Kluwer Academic Publishers 2004.

**Ber:** D. P. Bertsekas and J. N. Tsitsiklis, *Introduction to Probability*, 2nd ed., Athena Scientific 2008.

**Bis:** C. M. Bishop, *Pattern Recognition and Machine Learning*, Springer 2006.

**Bl1:** M. Bland, *An Introduction to Medical Statistics*, 4th ed., Oxford University Press 2015.

**Bl2:** M. Bland and J. Peacock, *Statistical Questions in Evidence-Based Medicine*, Oxford University Press 2000.

**Blo:** P. Bloomfield, *Fourier Analysis of Time Series: An Introduction*, 2nd ed., Wiley 2000.

**Bra:** R. N. Bracewell, *The Fourier Transform and its Applications*, 3rd ed., McGraw-Hill 1999.

**Bro:** P. J. Brockwell and R. A. Davis, *Introduction to Time Series and Forecasting*, 2nd ed., Springer 2002.

**Buc:** J. R. Buck, M. M. Daniel, and A. C. Singer, *Computer Explorations in Signals and Systems Using MATLAB*, 2nd ed., Prentice Hall 2001.

**Cas:** H. Caswell, *Matrix Population Models: Construction, Analysis, and Interpretation*, 2nd ed., Sinauer 2001.

**Cha:** D. C. Champeney, *A Handbook of Fourier Theorems*, Cambridge University Press 1987.

**Cht:** C. Chatfield, *The Analysis of Time Series: An Introduction*, 6th ed., Chapman & Hall/CRC Press 2004.

**Ch1:** C.-T. Chen, *Signals and Systems*, 3rd ed., Oxford University Press 2004.

**Ch2:** C.-T. Chen, *Linear System Theory and Design*, 4th ed., Oxford University Press 2013.

**Chu:** L. O. Chua and P.-M. Lin, *Computer Aided Analysis of Electronic Circuits: Algorithms and Computational Techniques*, Prentice Hall 1975.

**Clo:** C. M. Close, D. K. Frederick, and J. C. Newell, *Modeling and Analysis of Dynamic Systems*, 3rd ed., Wiley 2002.

**Con:** F. Conway and J. Siegelman, *Dark Hero of the Information Age: In Search of Norbert Wiener, the Father of Cybernetics*, Basic Books 2005.

**Coo:** G. R. Cooper and C. D. McGillem, *Probabilistic Methods of Signal and System Analysis*, 3rd ed., Oxford University Press 1999.

**Cov:** T. M. Cover and J. A. Thomas, *Elements of Information Theory*, 2nd ed., Wiley 2006.

**Cox:** T. F. Cox, *An Introduction to Multivariate Data Analysis*, Hodder Arnold 2005.

**Dal:** D. J. Daley and J. Gani, *Epidemic Modelling: An Introduction*, Cambridge University Press 1999.

**Dav:** W. B. Davenport and W. L. Root, *An Introduction to the Theory of Random Signals and Noise*, McGraw-Hill 1958 (reprinted by IEEE Press 1987).

**DeG:** M. H. DeGroot and M. J. Schervish, *Probability and Statistics*, 4th ed., Pearson 2012.

**Dek:** F. M. Dekking, C. Kraaikamp, H. P. Lopuhaä, and L. E. Meester, *A Modern Introduction to Probability and Statistics: Understanding Why and How*, Springer 2005.

**Dur:** J. Durbin and S. J. Koopman, *Time Series Analysis by State Space Methods*, 2nd ed., Oxford University Press 2012.

**Edw:** C. H. Edwards and D. E. Penney, *Differential Equations and Linear Algebra*, 3rd ed., Pearson 2008.

**Fin:** T. L. Fine, *Probability and Probabilistic Reasoning for Electrical Engineering*, Pearson 2006.

**Frd:** B. Friedland, *Control System Design: An Introduction to State-Space Methods*, McGraw-Hill 1986 (reprinted by Dover 2005).

**Frs:** B. Fristedt, N. Jain, and N. Krylov, *Filtering and Prediction: A Primer*, American Mathematical Society 2007.

**Gal:** R. G. Gallager, *Stochastic Processes: Theory for Applications*, Cambridge University Press 2013.

**Gar:** W. A. Gardner, *Introduction to Random Processes, With Applications to Signals & Systems*, 2nd ed., McGraw-Hill 1990.

**Gib:** J. D. Gibson, *Principles of Digital and Analog Communications*, 2nd ed., Prentice Hall 1993.

**Gr1:** G. R. Grimmett and D. R. Stirzaker, *Probability and Random Processes*, 3rd ed., Oxford University Press 2001.

**Gr2:** G. R. Grimmett and D. R. Stirzaker, *One Thousand Exercises in Probability*, Oxford University Press 2001.

**Gui:** E. A. Guillemin, *Theory of Linear Physical Systems: Theory of Physical Systems from the Viewpoint of Classical Dynamics, Including Fourier Methods*, Wiley 1963 (reprinted by Dover 2013).

**Had:** A. H. Haddad, *Probabilistic Systems and Random Signals*, Prentice Hall 2006.

**Hay:** M. Hayes, *Statistical Digital Signal Processing and Modeling*, Wiley 1996.

**Ha1:** S. Haykin and B. Van Veen, *Signals and Systems*, 2nd ed., Wiley 2002.

**Ha2:** S. Haykin, *Digital Communication Systems*, Wiley 2014.

**He1:** C. W. Helstrom, *Probability and Stochastic Processes for Engineers*, 2nd ed., Macmillan 1991.

**He2:** C. W. Helstrom, *Statistical Theory of Signal Detection*, Pergamon 1960.

**He3:** C. W. Helstrom, *Elements of Signal Detection and Estimation*, Prentice Hall 1995.

**Jan:** J. Jan, *Digital Signal Filtering, Analysis and Restoration*, IET 2000.

**Jen:** G. M. Jenkins and D. G. Watts, *Spectral Analysis and its Applications*, Holden-Day 1968.

**Ka1:** T. Kailath, *Linear Systems*, Prentice Hall 1980.

**Ka2:** T. Kailath, *Lectures on Wiener and Kalman Filtering*, 2nd ed., Springer 1981.

**Kam:** E. W. Kamen and J. K. Su, *Introduction to Optimal Estimation*, Springer 1999.

**Kay1:** S. M. Kay, *Intuitive Probability and Random Processes Using MATLAB*, Springer 2006.

**Kay2:** S. M. Kay, *Fundamentals of Statistical Signal Processing: Estimation Theory*, Prentice Hall 1993.

**Kay3:** S. M. Kay, *Fundamentals of Statistical Signal Processing: Detection Theory*, Prentice Hall 1998.

**Kay4:** S. M. Kay, *Modern Spectral Estimation: Theory and Application*, Prentice Hall 1988.

**Kha:** H. K. Khalil, *Nonlinear Control*, Pearson 2015.

**Kle:** J. L. Klein, *Statistical Visions in Time: A History of Time Series Analysis, 1662–1938*, Cambridge University Press 1997.

**Koo:** L. H. Koopmans, *The Spectral Analysis of Time Series*, Academic Press 1995.

**Kri:** V. Krishnan, *Probability and Random Processes*, Wiley 2006.

**Kul:** S. Kulkarni and G. Harman, *An Elementary Introduction to Statistical Learning Theory*, Wiley 2011.

**Kwa:** H. Kwakernaak and R. Sivan, *Modern Signals and Systems*, Prentice Hall 1991.

**La1:** B. P. Lathi, *Linear Systems and Signals*, 2nd ed., Oxford University Press 2005.

**La2:** B. P. Lathi and Z. Ding, *Modern Digital and Analog Communication Systems*, 4th ed., Oxford University Press 2009.

**Lee:** Y. W. Lee, *Statistical Theory of Communication*, Wiley 1960 (reprinted by Dover 2005).

**Leo:** A. Leon-Garcia, *Probability, Statistics, and Random Processes for Electrical Engineering*, 3rd ed., Prentice Hall 2008.

**Lev:** B. C. Levy, *Principles of Signal Detection and Parameter Estimation*, Springer 2008.

**Lue:** D. G. Luenberger, *Introduction to Dynamic Systems: Theory, Models, and Applications*, Wiley 1979.

**Lyn:** S. Lynch, *Dynamical Systems with Applications using MATLAB*, 2nd ed., Birkhäuser 2014.

**Mac:**   D. J. C. MacKay, *Information Theory, Inference, and Learning Algorithms*, Cambridge University Press 2003.

**Ma1:**   D. G. Manolakis and V. K. Ingle, *Applied Digital Signal Processing: Theory and Practice*, Cambridge University Press 2011.

**Ma2:**   D. G. Manolakis, V. K. Ingle, and S. M. Kogon, *Statistical and Adaptive Signal Processing: Spectral Estimation, Signal Modeling, Adaptive Filtering and Array Processing*, Artech House 2005.

**Mar:**   S. L. Marple, Jr., *Digital Spectral Analysis with Applications*, Prentice Hall 1987.

**McC:**   J. H. McClellan, C. S. Burrus, A. V. Oppenheim, T. W. Parks, R. W. Schafer, and H. W. Schuessler, *Computer-Based Exercises for Signal Processing Using MATLAB 5*, Prentice Hall 1998.

**McD:**   R. N. McDonough and A. D. Whalen, *Detection of Signals in Noise*, 2nd ed., Academic Press 1995.

**Mil:**   S. Miller and D. Childers, *Probability and Random Processes, with Applications to Signal Processing and Communications*, 2nd ed., Academic Press 2012.

**Mit:**   S. K. Mitra, *Digital Signal Processing: A Computer-Based Approach*, 4th ed., McGraw-Hill 2011.

**Moo:**   T. K. Moon and W. C. Stirling, *Mathematical Methods and Algorithms for Signal Processing*, Prentice Hall 2000.

**Mur:**   K. P. Murphy, *Machine Learning: A Probabilistic Perspective*, MIT Press 2012.

**Op1:**   A. V. Oppenheim and A. S. Willsky, with S. H. Nawab, *Signals & Systems*, 2nd ed., Prentice Hall 1997.

**Op2:**   A. V. Oppenheim and R. W. Schafer, *Discrete-Time Signal Processing*, 3rd ed., Pearson 2010.

**Pa1:**   A. Papoulis, *Signal Analysis*, McGraw-Hill 1977.

**Pa2:**   A. Papoulis, *Circuits and Systems: A Modern Approach*, Holt, Rinehart, and Winston 1980.

**Pa3:**   A. Papoulis, *The Fourier Integral and its Applications*, McGraw-Hill 1962.

**Pa4:**   A. Papoulis and S. U. Pillai, *Probability, Random Variables and Stochastic Processes*, 4th ed., McGraw-Hill 2002.

**Per:**   D. B. Percival and A. T. Walden, *Spectral Analysis for Physical Applications: Multitaper and Conventional Univariate Techniques*, Cambridge University Press 1993.

**Phi:**   C. L. Phillips, J. M. Parr, and E. A. Riskin, *Signals, Systems, and Transforms*, 5th ed., Pearson 2014.

**Poo:**   H. V. Poor, *An Introduction to Signal Detection and Estimation*, 2nd ed., Springer 1994.

**Por:**   B. Porat, *Digital Processing of Random Signals: Theory and Methods*, Prentice Hall 1994 (reprinted by Dover 2008).

**Pra:**   P. Prandoni and M. Vetterli, *Signal Processing for Communications*, EPFL Press 2008.

**Pr1:**   J. G. Proakis and D. G. Manolakis, *Digital Signal Processing: Principles, Algorithms, and Applications*, 4th ed., Prentice Hall 2007.

**Pr2:**   J. G. Proakis and M. Salehi, *Digital Communications*, 5th ed., McGraw-Hill 2008.

**Pur:**   M. B. Pursley, *Random Processes in Linear Systems*, Prentice Hall 2002.

**Rce:**   J. A. Rice, *Mathematical Statistics and Data Analysis*, 3rd ed., Brooks/Cole 2007.

**Rch:** M. A. Richards, *Fundamentals of Radar Signal Processing*, 2nd ed., McGraw-Hill 2014.

**Rob:** M. J. Roberts, *Signals and Systems: Analysis Using Transform Methods and MATLAB*, 2nd ed., McGraw-Hill 2012.

**Sch:** L. L. Scharf, *Statistical Signal Processing: Detection, Estimation, and Time Series Analysis*, Addision-Wesley 1991.

**Sha:** K. S. Shanmugan and A. M. Breipohl, *Random Signals: Detection, Estimation and Data Analysis*, Wiley 1988.

**Shn:** C. E. Shannon and W. Weaver, *The Mathematical Theory of Communication*, University of Illinois Press 1949 (reprinted in 1998).

**Shi:** R. Shiavi, *Introduction to Applied Statistical Signal Analysis: Guide to Biomedical and Electrical Engineering Applications*, 3rd ed., Elsevier 2007.

**Shy:** J. J. Shynk, *Probability, Random Variables, and Random Processes: Theory and Signal Processing Applications*, Wiley 2013.

**Sie:** W. McC. Siebert, *Circuits, Signals, and Systems*, McGraw-Hill/MIT Press 1986.

**Sko:** M. I. Skolnik, *Introduction to Radar Systems*, 3rd ed., McGraw-Hill 2001.

**Slo:** J.-J. E. Slotine and W. Li, *Applied Nonlinear Control*, Prentice Hall, 1991.

**Ste:** J. M. Steele, *The Cauchy-Schwarz Master Class: An Introduction to the Art of Mathematical Inequalities*, Cambridge University Press 2004.

**St1:** S. M. Stigler, *The History of Statistics: The Measurement of Uncertainty before 1900*, Harvard University Press 1986.

**St2:** S. M. Stigler, *Statistics on the Table: The History of Statistical Concepts and Methods*, Harvard University Press 1999.

**Sto:** P. Stoica and R. L. Moses, *Spectral Analysis of Signals*, Prentice Hall 2005.

**Str:** S. H. Strogatz, *Nonlinear Dynamics and Chaos, with Applications to Physics, Biology, Chemistry, and Engineering*, 2nd ed., Westview Press 2014.

**Th1:** C. W. Therrien, *Discrete Random Signals and Statistical Signal Processing*, Prentice Hall 1992.

**Th2:** C. W. Therrien and M. Tummala, *Probability and Random Processes for Electrical and Computer Engineers*, 2nd ed., CRC Press 2012.

**Van:** H. L. Van Trees and K. L. Bell, with Z. Tian, *Detection, Estimation, and Modulation Theory, Part I: Detection, Estimation, and Filtering Theory*, 2nd ed., Wiley 2013.

**Vet:** M. Vetterli, J. Kovačević, and V. K. Goyal, *Foundations of Signal Processing*, Cambridge University Press 2014.

**Vid:** M. Vidyasagar, *Nonlinear Systems Analysis*, 2nd ed., Prentice Hall 1993 (reprinted by SIAM 2002).

**Wal:** R. E. Walpole, R. H. Myers, S. L. Myers, and K. E. Ye, *Probability and Statistics for Engineers and Scientists*, 9th ed., Pearson 2011.

**Was:** L. Wasserman, *All of Statistics: A Concise Course in Statistical Inference*, Springer 2004.

**Wie:** N. Wiener, *Extrapolation, Interpolation, and Smoothing of Stationary Time Series, with Engineering Applications*, MIT Press 1949 (reprinted in 1964).

**Wil:** R. R. Wilcox, *Basic Statistics: Understanding Conventional Methods and Modern Insights*, Oxford University Press 2009.

**Wll:** R. H. Williams, *Probability, Statistics, and Random Processes for Engineers*, Brooks/Cole 2003.

**Wod:** P. M. Woodward, *Probability and Information Theory, With Application to Radar*, 2nd ed., Pergamon 1964.

**Woo:**    W. A. Woodward, H. L. Gray, and A. C. Elliott, *Applied Time Series Analysis*, CRC Press 2011.

**Yat:**    R. D. Yates and D. J. Goodman, *Probability and Stochastic Processes: A Friendly Introduction for Electrical and Computer Engineers*, 2nd ed., Wiley 2005.

**Zad:**    L. A. Zadeh and C. A. Desoer, *Linear System Theory: The State Space Approach*, McGraw-Hill 1963 (reprinted by Dover 2008).

**Zie:**    R. E. Ziemer and W. H. Tranter, *Principles of Communications: Systems, Modulation, and Noise*, 7th ed., Wiley 2014.

# INDEX